Springer Optimization and Its Applications

Volume 154

Aims and Scope
Optimization has continued to expand in all directions at an astonishing rate. New algorithmic and theoretical techniques are continually developing and the diffusion into other disciplines is proceeding at a rapid pace, with a spot light on machine learning, artificial intelligence, and quantum computing. Our knowledge of all aspects of the field has grown even more profound. At the same time, one of the most striking trends in optimization is the constantly increasing emphasis on the interdisciplinary nature of the field. Optimization has been a basic tool in areas not limited to applied mathematics, engineering, medicine, economics, computer science, operations research, and other sciences.

The series *Springer Optimization and Its Applications (SOIA)* aims to publish state-of-the-art expository works (monographs, contributed volumes, textbooks, handbooks) that focus on theory, methods, and applications of optimization. Topics covered include, but are not limited to, nonlinear optimization, combinatorial optimization, continuous optimization, stochastic optimization, Bayesian optimization, optimal control, discrete optimization, multi-objective optimization, and more. New to the series portfolio include Works at the intersection of optimization and machine learning, artificial intelligence, and quantum computing.

Volumes from this series are indexed by Web of Science, zbMATH, Mathematical Reviews, and SCOPUS.

More information about this series at http://www.springer.com/series/7393

Themistocles M. Rassias • Panos M. Pardalos
Editors

Mathematical Analysis and Applications

Editors
Themistocles M. Rassias
Department of Mathematics
Zografou Campus
National Technical University of Athens
Athens, Greece

Panos M. Pardalos
Department of Industrial
and Systems Engineering
University of Florida
Gainesville, FL, USA

ISSN 1931-6828 ISSN 1931-6836 (electronic)
Springer Optimization and Its Applications
ISBN 978-3-030-31338-8 ISBN 978-3-030-31339-5 (eBook)
https://doi.org/10.1007/978-3-030-31339-5

Mathematics Subject Classification: 26-XX, 28-XX, 30-XX, 32-XX, 35-XX, 91-XX

This Springer imprint is published by the registered company Springer Nature Switzerland AG.
The registered company address is: Gewerbestrasse 11, 6330 Cham, Switzerland

Preface

Mathematical Analysis and Applications is devoted to the presentation of high-quality research and survey papers belonging to a broad spectrum of areas in which *Analysis* plays a central role. The book in hand provides an insight into the investigation of several problems and theories in real and complex analysis, functional analysis, approximation theory, operator theory, analytic inequalities, Radon transform, nonlinear analysis, and various applications of interdisciplinary research. The contributing papers have been written by eminent scientists from the international mathematical community who are experts in the individual subjects.

In this book, some papers are devoted to certain applications as for example in the three-body problem, finite element analysis in fluid mechanics, algorithms for difference of monotone operators, a vibrational approach to a financial problem, etc.

This publication provides valuable and up-to-date information as well as research results which are hoped to be useful to graduate students and researchers working in Mathematics, Physics, Engineering, and Economics.

It is our pleasure to express our thanks to all the contributors of chapters in this book who participated in this collective effort.

Last but not least, we would like to acknowledge the superb assistance that the staff of Springer has provided for the publication of this work.

Athens, Greece — Themistocles M. Rassias
Gainesville, FL, USA — Panos M. Pardalos

Contents

Exact Solution to Systems of Linear First-Order Integro-Differential Equations with Multipoint and Integral Conditions

M. M. Baiburin and E. Providas

Abstract This paper is devoted to the study of nonhomogeneous systems of linear first-order ordinary integro-differential equations of Fredholm type with multipoint and integral boundary constraints. Sufficient conditions for the solvability and correctness of the problem are established and the unique solution is provided in closed-form. The approach followed is based on the extension theory of operators.

1 Introduction

Mathematical modeling in the theory of automatic control, the theory of oscillation, mathematical physics, biology, applied mathematics, and economics, very often, leads to the study of multipoint boundary value problems for differential, functional-differential, and integro-differential equations. These types of boundary value problems and their solutions have been investigated by many researchers, for example, [1, 2, 5, 8, 9, 21]. Of special interest are the multipoint boundary value problems for a system of differential equations (DEs) and integro-differential equations (IDEs), see, for example, [4, 6, 11, 12, 22]. It should be noted that obtaining exact solutions even to multipoint boundary value problems for a differential, or an integro-differential equation, is a difficult task. Therefore, usually numerical methods are employed as in [3, 7, 15] and elsewhere.

Recently, in [6] the solution to a class of boundary value problems for a system of linear first-order DEs coupled with multipoint and integral conditions has been obtained in closed-form. Here, we continue this study to systems of linear first-order

M. M. Baiburin
Department of Fundamental Mathematics, L.N. Gumilyov Eurasian National University, Astana, Republic of Kazakhstan
e-mail: merkhasyl@mail.ru

E. Providas (✉)
University of Thessaly, Larissa, Greece
e-mail: providas@uth.gr

T. M. Rassias, P. M. Pardalos (eds.), *Mathematical Analysis and Applications*, Springer Optimization and Its Applications 154,
https://doi.org/10.1007/978-3-030-31339-5_1

ordinary IDEs of Fredholm type with multipoint and integral boundary constraints. The method proposed is based on the extension theory of linear operators in a Banach space, as it has been developed in terms of inverse operators [5, 13] and in terms of direct operators [14], and has been used to investigate the correctness properties to some extensions of operators [10, 16, 20] and more recently to solve exactly initial and two-point boundary value problems for integro-differential equations [17–19].

We first examine the solvability conditions and then obtain the exact solution of the following system of IDEs subject to multipoint and integral boundary conditions:

$$y'(x) - Ay(x) - \sum_{i=0}^{m} G_i(x) \int_0^1 H_i(t)y(t)dt = f(x), \quad x \in [0, 1],$$

$$\sum_{i=0}^{m} A_i y(x_i) + \sum_{j=0}^{s} B_j \int_{\xi_j}^{\xi_{j+1}} C_j(t)y(t)dt = \mathbf{0}, \tag{1}$$

where $A,\ A_i,\ B_j$ are $n \times n$ constant matrices, $G_i(x),\ H_i(x),\ C_j(x)$ are variable $n \times n$ matrices, whose elements are continuous functions on $[0, 1]$, $f(x)$ is a vector of n continuous functions on $[0, 1]$, and $y(x)$ is a vector of n sought continuous functions with continuous derivatives on $[0, 1]$; the points x_i, ξ_j satisfy the conditions $0 = x_0 < x_1 < \cdots < x_{m-1} < x_m = 1$, $0 = \xi_0 < \xi_1 < \cdots < \xi_s < \xi_{s+1} = 1$. The problem (1) may be obtained as a perturbation of a corresponding boundary value problem for a system of first-order DEs, specifically

$$y'(x) - Ay(x) = f(x),$$

$$\sum_{i=0}^{m} A_i y(x_i) + \sum_{j=0}^{s} B_j \int_{\xi_j}^{\xi_{j+1}} C_j(t)y(t)dt = \mathbf{0}, \tag{2}$$

whose solvability and the construction of the exact solution were investigated in [6].

The rest of the paper is organized as follows. In Section 2 some necessary definitions are given and preliminary results are derived. In Section 3 the two main theorems for the existence and the construction of the exact solution are presented. Lastly, some conclusions are drawn in Section 4.

2 Definitions and Preliminary Results

Let X, Y be complex Banach spaces. Let $P : X \to Y$ denote a linear operator and $D(P)$ and $R(P)$ its domain and the range, respectively. An operator P is called an *extension* of the operator $P_0 : X \to Y$ if $D(P_0) \subseteq D(P)$ and $Pu = P_0u$, for all $u \in D(P_0)$. An operator $P : X \to Y$ is called *correct* if $R(P) = Y$ and the inverse operator P^{-1} exists and is continuous on Y.

We say that the problem $Pu = f$, $f \in Y$, is correct if the operator P is correct. The problem $Pu = f$ with a linear operator P is uniquely solvable on $R(P)$ if the corresponding homogeneous problem $Pu = 0$ has only a zero solution, i.e. if $\ker P = \{0\}$. The problem $Pu = f$ is said to be everywhere solvable on Y if it admits a solution for any $f \in Y$.

Throughout this paper, we use lowercase letters and brackets to designate vectors and capital letters and square brackets to symbolize matrices. The unit and zero matrices are denoted by $\mathbf{I}$ and $[0]$, respectively, and the zero column vector by $\mathbf{0}$.

The set of all complex numbers is specified by $\mathbf{C}$. If $c_i \in \mathbf{C}$, $i = 1, \dots, n$, then we write $c = (c_1, \dots, c_n) \in \mathbf{C}^n$. By $C_n[0, 1]$, we mean the space of continuous vector functions $f = f(x) = (f_1(x), \dots, f_n(x))$ with norm

$$\|f\|_{C_n} = \|f_1(x)\| + \|f_2(x)\| + \dots + \|f_n(x)\|, \quad \|f(x)\| = \max_{x\in[0,1]} |f(x)|. \tag{3}$$

Let $f = f(x) = col\,(f_1(x), \dots, f_n(x)) \in C_n[0, 1]$. Further, let the operators $L, K, H : C_n[0, 1] \to C_n[0, 1]$ be defined by the matrices

$$L(x) = \begin{bmatrix} l_{11}(x) & \cdots & l_{1n}(x) \\ \vdots & \dots & \vdots \\ l_{n1}(x) & \cdots & l_{nn}(x) \end{bmatrix}, \quad K(x) = \begin{bmatrix} k_{11}(x) & \cdots & k_{1n}(x) \\ \vdots & \dots & \vdots \\ k_{n1}(x) & \cdots & k_{nn}(x) \end{bmatrix},$$

$$H(x) = \begin{bmatrix} h_{11}(x) & \cdots & h_{1n}(x) \\ \vdots & \dots & \vdots \\ h_{n1}(x) & \cdots & h_{nn}(x) \end{bmatrix},$$

where $l_{ij}, k_{ij}, h_{ij} \in C[0, 1]$. Let $l_0 = \max |l_{ij}|$, $k_0 = \max |k_{ij}|$, $h_0 = \max |h_{ij}|$, $i, j = 1, \dots, n$. Finally, consider the points ξ_j, $j = 0, \dots, s+1$ satisfying the conditions $0 = \xi_0 < \xi_1 < \dots < \xi_s < \xi_{s+1} = 1$.

We now prove the next lemma which is used several times in the sequel.

Lemma 1 *The next estimates are true*

$$\|Lf\|_{C_n} \le l_0 n \|f\|_{C_n}, \tag{4}$$

$$\| \int_0^x L(t)f(t)dt \|_{C_n} \le l_0 n \|f\|_{C_n}, \quad x \in [0, 1], \tag{5}$$

$$\|K(x) \int_0^x L(t)f(t)dt \|_{C_n} \le k_0 l_0 n^2 \|f\|_{C_n}, \quad x \in [0, 1], \tag{6}$$

$$\| \int_0^1 K(x) \int_0^x L(t)f(t)dtdx \|_{C_n} \le k_0 l_0 n^2 \|f\|_{C_n}, \tag{7}$$

$$\| \int_{\xi_j}^{\xi_{j+1}} K(x) \int_0^x L(t)f(t)dtdx \|_{C_n} \le k_0 l_0 n^2 \|f\|_{C_n}, \tag{8}$$

$$\|H(x)\int_0^1 K(\xi)\int_0^\xi L(t)f(t)dtd\xi\|_{C_n} \leq h_0k_0l_0n^3\|f\|_{C_n}, \quad x \in [0,1]. \quad (9)$$

Proof The properties (4)–(6), (8) have been proved in [6]. We prove (7) and (9). Let $\phi(x) = col(\phi_1(x), \ldots, \phi_n(x)) = \int_0^x L(t)f(t)dt$. Then, from (5) follows that

$$\begin{aligned}\|\int_0^1 K(x)\int_0^x L(t)f(t)dtdx\|_{C_n} &= \|\int_0^1 K(x)\phi(x)dx\|_{C_n}\\ &\leq k_0n\|\phi(x)\|_{C_n}\\ &= k_0n\|\int_0^x L(t)f(t)dt\|_{C_n}\\ &\leq k_0l_0n^2\|f\|_{C_n}.\end{aligned}$$

We now prove (9). Let $\phi = col\left(\phi_1, \ldots, \phi_n\right) = \int_0^1 K(\xi)\int_0^\xi L(t)f(t)dtd\xi$. Then, from (4) and (7) follows that

$$\begin{aligned}\|H(x)\int_0^1 K(\xi)\int_0^\xi L(t)f(t)dtd\xi\|_{C_n} &= \|H(x)\phi\|_{C_n}\\ &\leq h_0n\|\phi\|_{C_n}\\ &= h_0n\|\int_0^1 K(\xi)\int_0^\xi L(t)f(t)dtd\xi\|_{C_n}\\ &\leq h_0k_0l_0n^3\|f\|_{C_n}.\end{aligned}$$

The lemma is proved. □

3 Main Results

Let the operator P associated with problem (1) be defined as

$$Py = y'(x) - Ay(x) - \sum_{i=0}^m G_i(x)\int_0^1 H_i(t)y(t)dt,$$

$$D(P) = \left\{y(x) \in C_n^1[0,1] : \sum_{i=0}^m A_iy(x_i) + \sum_{j=0}^s B_j\int_{\xi_j}^{\xi_{j+1}} C_j(t)y(t)dt = \mathbf{0}\right\}, \quad (10)$$

where A, A_i, B_j are $n \times n$ constant matrices and $G_i(x), H_i(x), C_j(x)$ are variable $n \times n$ matrices with elements continuous functions on $[0, 1]$; the points x_i, ξ_j satisfy the conditions $0 = x_0 < x_1 < \cdots < x_{m-1} < x_m = 1, 0 = \xi_0 < \xi_1 < \cdots < \xi_s < \xi_{s+1} = 1$. Note that the operator P is an extension of the minimal operator P_0 defined by

$$P_0 y = y'(x) - Ay(x),$$

$$D(P_0) = \left\{ y(x) \in C_n^1[0, 1] : y(x_i) = \mathbf{0}, \int_{\xi_j}^{\xi_{j+1}} C_j(t)y(t)dt = \mathbf{0}, \right.$$

$$\left. \int_0^1 H_i(t)y(t)dt = \mathbf{0}, \ i = 0, \dots, m, \ j = 0, \dots, s \right\}. \tag{11}$$

Moreover, we may write the operator P compactly as

$$Py = y'(x) - Ay(x) - \mathbf{G}z(y),$$

$$D(P) = \left\{ y(x) \in C_n^1[0, 1] : \mathbf{A}y(\mathbf{x}) + \mathbf{B}\psi(y) = \mathbf{0} \right\}, \tag{12}$$

where the composite matrices

$$\mathbf{G} = \left[G_0 \ G_1 \ \dots \ G_m \right], \quad \mathbf{A} = \left[A_0 \ A_1 \ \dots \ A_m \right], \quad \mathbf{B} = \left[B_0 \ B_1 \ \dots \ B_s \right],$$

the compound column vectors

$$z(y) = col\,(z_0(y), z_1(y), \dots, z_m(y)),$$

$$y(\mathbf{x}) = col\,(y(x_0), y(x_1), \dots, y(x_m)),$$

$$\psi(y) = col\,(\psi_0(y), \psi_1(y), \dots, \psi_s(y)),$$

and the $n \times 1$ vectors

$$z_i(y) = \int_0^1 H_i(t)y(t)dt, \quad i = 0, \dots, m,$$

$$\psi_j(y) = \int_{\xi_j}^{\xi_{j+1}} C_j(t)y(t)dt, \quad j = 0, \dots, s.$$

By using (10) or (12), we can express the system (1) equivalently in the elegant operator form

$$Py = f(x), \quad f(x) \in C_n[0, 1]. \tag{13}$$

Theorem 1 below provides the criteria for the existence of a unique solution to the problem (13).

We first consider the $n \times n$ matrix e^{xA} and define the following $n \times n$ matrices

$$\begin{aligned} L_i &= \int_0^1 H_i(t) e^{tA} dt, \\ \Lambda_j &= \int_{\xi_j}^{\xi_{j+1}} C_j(t) e^{tA} dt, \\ \Delta_{ik}(G) &= \int_0^{x_i} e^{(x_i - t)A} G_k(t) dt, \\ V_{ik}(G) &= \int_0^1 H_i(x) \int_0^x e^{(x-t)A} G_k(t) dt dx, \\ W_{jk}(G) &= \int_{\xi_j}^{\xi_{j+1}} C_j(x) \int_0^x e^{(x-t)A} G_k(t) dt dx, \end{aligned}$$

where $i, k = 0, \ldots, m,\ j = 0, \ldots, s$, and the compound matrices

$$\begin{aligned} e^{\mathbf{x}A} &= col \left[e^{x_0 A} \ e^{x_1 A} \ \ldots \ e^{x_m A} \right], \\ L &= col \left[L_0 \ L_1 \ \cdots \ L_m \right], \quad \Lambda = col \left[\Lambda_0 \ \Lambda_1 \ \cdots \ \Lambda_s \right], \\ \Delta_G &= \left[\Delta_{ik}(G) \right], \quad V_G = \left[V_{ik}(G) \right], \quad W_G = \left[W_{jk}(G) \right]. \end{aligned} \tag{14}$$

Theorem 1 *The problem (13) is uniquely solvable on $C_n[0, 1]$ if*

$$\det \mathbf{T} = \det \begin{bmatrix} \mathbf{A}\Delta_G + \mathbf{B}W_G & \mathbf{A}e^{\mathbf{x}A} + \mathbf{B}\Lambda \\ V_G - I & L \end{bmatrix} \neq 0. \tag{15}$$

Proof It suffices to show that $\ker P = \{0\}$ if $\det T \neq 0$. Assume that $\det T \neq 0$. Consider the homogeneous problem $Py = \mathbf{0}$ consisting of the homogeneous equation

$$y'(x) - Ay(x) - \mathbf{G}z(y) = \mathbf{0}, \tag{16}$$

and the boundary conditions

$$\mathbf{A}y(\mathbf{x}) + \mathbf{B}\psi(y) = \mathbf{0}. \tag{17}$$

Let the auxiliary integro-functional equation

$$y(x) = e^{xA}\mathbf{d} + e^{xA} \sum_{i=0}^{m} \int_0^x e^{-tA} G_i(t) dt z_i(y), \tag{18}$$

or in a compact form

$$y(x) = e^{xA}\mathbf{d} + e^{xA}\int_0^x e^{-tA}\mathbf{G}(t)dtz(y), \tag{19}$$

where e^{xA} is a fundamental $n \times n$ matrix to the homogeneous differential equation $y'(x) - Ay(x) = \mathbf{0}$ and $\mathbf{d}$ is an arbitrary column vector with constant coefficients. It is easy to verify that from (19) follows the homogeneous equation (16). Therefore every solution of (19) is also a solution of (16). From (18), we have

$$y(x_i) = e^{x_i A}\mathbf{d} + e^{x_i A}\sum_{k=0}^{m}\int_0^{x_i} e^{-tA}G_k(t)dtz_k(y), \tag{20}$$

$$H_i(x)y(x) = H_i(x)e^{xA}\mathbf{d} + H_i(x)e^{xA}\sum_{k=0}^{m}\int_0^{x} e^{-tA}G_k(t)dtz_k(y), \tag{21}$$

$$C_j(x)y(x) = C_j(x)e^{xA}\mathbf{d} + C_j(x)e^{xA}\sum_{k=0}^{m}\int_0^{x} e^{-tA}G_k(t)dtz_k(y), \tag{22}$$

for $i = 0, \ldots, m,\ j = 0, \ldots, s$. By integrating (21) and (22), we get

$$\begin{aligned}\int_0^1 H_i(x)y(x)dx = &\int_0^1 H_i(x)e^{xA}dx\mathbf{d}\\ &+\sum_{k=0}^{m}\int_0^1 H_i(x)e^{xA}\int_0^x e^{-tA}G_k(t)dtdxz_k(y),\end{aligned} \tag{23}$$

$$\begin{aligned}\int_{\xi_j}^{\xi_{j+1}} C_j(x)y(x)dx = &\int_{\xi_j}^{\xi_{j+1}} C_j(x)e^{xA}dx\mathbf{d}\\ &+\sum_{k=0}^{m}\int_{\xi_j}^{\xi_{j+1}} C_j(x)e^{xA}\int_0^x e^{-tA}G_k(t)dtdxz_k(y).\end{aligned} \tag{24}$$

We rewrite (20), (23), (24) in the compact matrix form

$$y(\mathbf{x}) = e^{\mathbf{x}A}\mathbf{d} + \Delta_G z(y), \tag{25}$$

$$z(y) = L\mathbf{d} + V_G z(y), \tag{26}$$

$$\psi(y) = \Lambda\mathbf{d} + W_G z(y), \tag{27}$$

where the matrices $e^{\mathbf{x}A},\ L,\ \Lambda,\ \Delta_G,\ V_G,\ W_G$ are defined in (14). By utilizing (25) and (27), the boundary conditions in (17) are written as

$$\mathbf{A}\left(e^{\mathbf{x}A}\mathbf{d}+\Delta_G z(y)\right)+\mathbf{B}\left(\Lambda\mathbf{d}+W_G z(y)\right)=\mathbf{0}. \tag{28}$$

From (26) and (28) we obtain the system

$$\begin{bmatrix} \mathbf{A}\Delta_G+\mathbf{B}W_G & \mathbf{A}e^{\mathbf{x}A}+\mathbf{B}\Lambda \\ V_G-I & L \end{bmatrix}\begin{pmatrix} z(y) \\ \mathbf{d} \end{pmatrix}=\begin{pmatrix} \mathbf{0} \\ \mathbf{0} \end{pmatrix}, \tag{29}$$

or

$$\mathbf{T}\begin{pmatrix} z(y) \\ \mathbf{d} \end{pmatrix}=\begin{pmatrix} \mathbf{0} \\ \mathbf{0} \end{pmatrix}, \tag{30}$$

The assumption that $\det\mathbf{T}\neq 0$ implies $z(y)=\mathbf{0},\ \mathbf{d}=\mathbf{0}$. Substituting these values into (19), we obtain $y(x)=\mathbf{0}$. Hence $\ker P=\{0\}$ and the operator P is uniquely solvable. The theorem is proved. □

Remark 1 Note that the system of integro-differential equations (1) for $G_i\equiv[0],\ i=0,\ldots,m$, degenerates to the system of differential equations (2). By setting $G_i=[0],\ i=0,\ldots,m$, into (15), we obtain

$$\begin{aligned}
\det\mathbf{T} &= \det\begin{bmatrix} \mathbf{A}\Delta_G+\mathbf{B}W_G & \mathbf{A}e^{\mathbf{x}A}+\mathbf{B}\Lambda \\ V_G-I & L \end{bmatrix} \\
&\overset{G_i=[0]}{=} \det\begin{bmatrix} [0] & \mathbf{A}e^{\mathbf{x}A}+\mathbf{B}\Lambda \\ -I & L \end{bmatrix} \\
&= \det\left[\mathbf{A}e^{\mathbf{x}A}+\mathbf{B}\Lambda\right] \\
&= \det\left[\sum_{i=0}^{m}A_i e^{x_i A}+\sum_{j=0}^{s}B_j\int_{\xi_j}^{\xi_{j+1}}C_j(x)e^{xA}dx\right]\neq 0,
\end{aligned} \tag{31}$$

which is the sufficient solvability condition for the differential problem (2) derived in [6].

We introduce now the $\mathbf{C}^n$ vectors

$$\begin{aligned}
\phi_i(f) &= \int_0^{x_i} e^{(x_i-t)A}f(t)dt, \\
\nu_i(f) &= \int_0^1 H_i(x)\int_0^x e^{(x-t)A}f(t)dtdx, \\
\omega_j(f) &= \int_{\xi_j}^{\xi_{j+1}} C_j(x)\int_0^x e^{(x-t)A}f(t)dtdx,
\end{aligned}$$

for $i=0,\ldots,m,\ j=0,\ldots,s$, and the combined vectors

$$\phi_f = col(\phi_0(f), \phi_1(f), \ldots, \phi_m(f)), \quad \nu_f = col(\nu_0(f), \ldots, \nu_m(f)),$$
$$\omega_f = col(\omega_0(f), \ldots, \omega_s(f)). \tag{32}$$

Theorem 2 *Let (15) hold true. Then the problem (13) is correct on $C_n[0, 1]$ and its unique solution is given by*

$$y(x) = e^{xA}\int_0^x e^{-tA}f(t)dt - \left[e^{xA}\int_0^x e^{-tA}\mathbf{G}(t)dt \;\; e^{xA}\right]\mathbf{T}^{-1}\begin{pmatrix}\mathbf{A}\phi_f + \mathbf{B}\omega_f\\ \nu_f\end{pmatrix}. \tag{33}$$

Proof The problem (13) encompasses the nonhomogeneous system of integro-differential equations

$$y'(x) - Ay(x) - \mathbf{G}z(y) = f(x), \tag{34}$$

and the boundary conditions

$$\mathbf{A}y(\mathbf{x}) + \mathbf{B}\psi(y) = \mathbf{0}. \tag{35}$$

Take the auxiliary integro-functional equation

$$y(x) = e^{xA}\mathbf{d} + e^{xA}\sum_{i=0}^{m}\int_0^x e^{-tA}G_i(t)dtz_i(y) + e^{xA}\int_0^x e^{-tA}f(t)dt, \tag{36}$$

or in the compact matrix form

$$y(x) = \left[e^{xA}\int_0^x e^{-tA}\mathbf{G}(t)dt \;\; e^{xA}\right]\begin{pmatrix}z(y)\\ \mathbf{d}\end{pmatrix} + e^{xA}\int_0^x e^{-tA}f(t)dt, \tag{37}$$

for every $f(x) \in C_n[0, 1]$; e^{xA} is a fundamental $n \times n$ matrix to the homogeneous differential equation $y'(x) - Ay(x) = \mathbf{0}$ and $\mathbf{d}$ is an arbitrary column vector with constant elements. Observe that differentiation of (37) yields (34). Hence, a solution of (37) is also a solution of (34). From (36), we get

$$y(x_i) = e^{x_iA}\mathbf{d} + e^{x_iA}\sum_{k=0}^{m}\int_0^{x_i} e^{-tA}G_k(t)dtz_k(y) + e^{x_iA}\int_0^{x_i} e^{-tA}f(t)dt, \tag{38}$$

$$H_i(x)y(x) = H_i(x)e^{xA}\mathbf{d} + H_i(x)e^{xA}\sum_{k=0}^{m}\int_0^x e^{-tA}G_k(t)dtz_k(y)$$
$$+H_i(x)e^{xA}\int_0^x e^{-tA}f(t)dt, \tag{39}$$

$$C_j(x)y(x) = C_j(x)e^{xA}\mathbf{d} + C_j(x)e^{xA}\sum_{k=0}^{m}\int_0^x e^{-tA}G_k(t)dtz_k(y)$$
$$+C_j(x)e^{xA}\int_0^x e^{-tA}f(t)dt, \tag{40}$$

for $i = 0, \ldots, m,\ j = 0, \ldots, s$. By integrating (39), (40), we obtain

$$\int_0^1 H_i(x)y(x)dx = \int_0^1 H_i(x)e^{xA}dx\mathbf{d}$$
$$+\sum_{k=0}^{m}\int_0^1 H_i(x)e^{xA}\int_0^x e^{-tA}G_k(t)dtdxz_k(y)$$
$$+\int_0^1 H_i(x)e^{xA}\int_0^x e^{-tA}f(t)dtdx, \tag{41}$$

$$\int_{\xi_j}^{\xi_{j+1}} C_j(x)y(x)dx = \int_{\xi_j}^{\xi_{j+1}} C_j(x)e^{xA}dx\mathbf{d}$$
$$+\sum_{k=0}^{m}\int_{\xi_j}^{\xi_{j+1}} C_j(x)e^{xA}\int_0^x e^{-tA}G_k(t)dtdxz_k(y)$$
$$+\int_{\xi_j}^{\xi_{j+1}} C_j(x)e^{xA}\int_0^x e^{-tA}f(t)dtdx, \tag{42}$$

for $i = 0, \ldots, m,\ j = 0, \ldots, s$. We rewrite (38), (41), (42) in the compact matrix form

$$y(\mathbf{x}) = e^{\mathbf{x}A}\mathbf{d} + \Delta_G z(y) + \phi_f, \tag{43}$$
$$z(y) = L\mathbf{d} + V_G z(y) + \nu_f, \tag{44}$$
$$\psi(y) = \Lambda\mathbf{d} + W_G z(y) + \omega_f, \tag{45}$$

where the matrices $e^{\mathbf{x}A},\ L,\ \Lambda,\ \Delta_G,\ V_G,\ W_G$ are defined in (14) and the vectors $\phi_f,\ \nu_f,\ \omega_f$ are given in (32). By utilizing (43) and (45), the boundary conditions in (35) are recast as

$$\mathbf{A}\left(e^{\mathbf{x}A}\mathbf{d} + \Delta_G z(y) + \phi_f\right) + \mathbf{B}\left(\Lambda\mathbf{d} + W_G z(y) + \omega_f\right) = \mathbf{0}. \tag{46}$$

From (44) and (46), we obtain the system

$$\begin{bmatrix} \mathbf{A}\Delta_G + \mathbf{B}W_G & \mathbf{A}e^{\mathbf{x}A} + \mathbf{B}\Lambda \\ V_G - I & L \end{bmatrix}\begin{pmatrix} z(y) \\ \mathbf{d} \end{pmatrix} = -\begin{pmatrix} \mathbf{A}\phi_f + \mathbf{B}\omega_f \\ \nu_f \end{pmatrix}, \tag{47}$$

or

$$\mathbf{T}\begin{pmatrix} z(y) \\ \mathbf{d} \end{pmatrix} = -\begin{pmatrix} \mathbf{A}\phi_f + \mathbf{B}\omega_f \\ \nu_f \end{pmatrix}. \tag{48}$$

Since $\det \mathbf{T} \neq 0$ by hypothesis, we have

$$\begin{pmatrix} z(y) \\ \mathbf{d} \end{pmatrix} = -\mathbf{T}^{-1}\begin{pmatrix} \mathbf{A}\phi_f + \mathbf{B}\omega_f \\ \nu_f \end{pmatrix}. \tag{49}$$

Substitution of (49) into (37) yields the solution (33) to the problem (34)–(35). Since this solution holds for all $f(x) \in C_n[0, 1]$, then the system (34)–(35) is everywhere solvable. Thus, (33) is the unique solution to the nonhomogeneous problem (13) which can be denoted conveniently as $y(x) = P^{-1}f(x)$. To prove the correctness of the problem (13) it remains to show that the inverse operator P^{-1} is bounded.

Let $r = m + 1$ and write the matrix $\mathbf{T}$ conveniently as

$$\mathbf{T} = \begin{bmatrix} \mathbf{A}\Delta_G + \mathbf{B}W_G & \mathbf{A}e^{\mathbf{x}A} + \mathbf{B}\Lambda \\ V_G - I & L \end{bmatrix} = \begin{bmatrix} \mathbf{T}^{1r} & \mathbf{T}^{11} \\ \mathbf{T}^{rr} & \mathbf{T}^{r1} \end{bmatrix}, \tag{50}$$

where $\mathbf{T}^{1r} = \mathbf{A}\Delta_G + \mathbf{B}W_G$, $\mathbf{T}^{11} = \mathbf{A}e^{\mathbf{x}A} + \mathbf{B}\Lambda$, $\mathbf{T}^{rr} = V_G - I$ and $\mathbf{T}^{r1} = L$. Let also the analogously partitioned matrix

$$\Pi = \mathbf{T}^{-1} = \begin{bmatrix} \Pi^{r1} & \Pi^{rr} \\ \Pi^{11} & \Pi^{1r} \end{bmatrix}, \tag{51}$$

where

$$\Pi^{1r} = \begin{bmatrix} \Pi_0^{1r} \cdots \Pi_m^{1r} \end{bmatrix}, \quad \Pi^{r1} = \begin{bmatrix} \Pi_0^{r1} \\ \vdots \\ \Pi_m^{r1} \end{bmatrix}, \quad \Pi^{rr} = \begin{bmatrix} \Pi_{00}^{rr} & \cdots & \Pi_{0m}^{rr} \\ \vdots & \cdots & \vdots \\ \Pi_{m0}^{rr} & \cdots & \Pi_{mm}^{rr} \end{bmatrix}, \tag{52}$$

and Π^{11}, Π_i^{1r}, Π_i^{r1}, Π_{ik}^{rr}, $i, k = 0, \ldots, m$ are $n \times n$ matrices. Then,

$$\Pi\begin{pmatrix} \mathbf{A}\phi_f + \mathbf{B}\omega_f \\ \nu_f \end{pmatrix} = \begin{pmatrix} \Pi^{r1}\left(\mathbf{A}\phi_f + \mathbf{B}\omega_f\right) + \Pi^{rr}\nu_f \\ \Pi^{11}\left(\mathbf{A}\phi_f + \mathbf{B}\omega_f\right) + \Pi^{1r}\nu_f \end{pmatrix}. \tag{53}$$

Substitution of (53) into solution (33) yields

$$\begin{aligned} y(x) &= \int_0^x e^{(x-t)A} f(t)dt \\ &\quad - \left[\int_0^x e^{(x-t)A}\mathbf{G}(t)dt \;\; e^{xA}\right]\begin{pmatrix} \Pi^{r1}(\mathbf{A}\phi_f + \mathbf{B}\omega_f) + \Pi^{rr}\nu_f \\ \Pi^{11}(\mathbf{A}\phi_f + \mathbf{B}\omega_f) + \Pi^{1r}\nu_f \end{pmatrix} \end{aligned}$$

$$= \int_0^x e^{(x-t)A} f(t)dt$$
$$- \int_0^x e^{(x-t)A}\mathbf{G}(t)dt\Pi^{r1}(\mathbf{A}\phi_f + \mathbf{B}\omega_f) - \int_0^x e^{(x-t)A}\mathbf{G}(t)dt\Pi^{rr}\nu_f$$
$$-e^{xA}\Pi^{11}(\mathbf{A}\phi_f + \mathbf{B}\omega_f) - e^{xA}\Pi^{1r}\nu_f. \tag{54}$$

Let the maxima absolute elements (ae) for each of the following $n \times n$ matrices be denoted by

$$k^{(0)} = \max_{ae}\left[e^{xA}\right], \quad k_j = \max_{ae}\left[B_j C_j(x)e^{xA}\right], \quad l^{(0)} = \max_{ae}\left[e^{-xA}\right],$$
$$l^{(1)} = \max_{ae}\left[\int_0^x \sum_{i=0}^m e^{(x-t)A} G_i(t)\Pi_i^{r1} dt\right], \quad l_i^{(2)} = \max_{ae}\left[A_i e^{(x_i-t)A}\right],$$
$$l_k^{(3)} = \max_{ae}\left[\sum_{i=0}^m \int_0^x e^{(x-t)A} G_i(t)\Pi_{ik}^{rr} dt\right], \quad l^{(4)} = \max_{ae}\left[e^{xA}\Pi^{11}\right],$$
$$\hat{h}_k = \max_{ae}\left[H_k(x)e^{xA}\right], \quad \tilde{h}_i = \max_{ae}\left[e^{xA}\Pi_i^{1r}\right]. \tag{55}$$

Notice that the elements of the above matrices are continuous functions on $[0, 1]$ since the elements of the fundamental matrix e^{xA} and the inverse matrix e^{-xA} are continuous functions.

We now find some estimates for the terms appearing in (54). First, note that $\mathbf{A}\phi_f + \mathbf{B}\omega_f \in \mathbf{C}^n$, since both $A_i\phi_i(f)$ and $B_j\omega_j(f) \in \mathbf{C}^n$, and by the triangle inequality and properties (5), (8), we have

$$\|\mathbf{A}\phi_f + \mathbf{B}\omega_f\|_{C_n} \leq \|\mathbf{A}\phi_f\|_{C_n} + \|\mathbf{B}\omega_f\|_{C_n}$$
$$= \|\sum_{i=0}^m A_i\phi_i(f)\|_{C_n} + \|\sum_{j=0}^s B_j\omega_j(f)\|_{C_n}$$
$$\leq \sum_{i=0}^m \|\int_0^{x_i} A_i e^{(x_i-t)A} f(t)dt\|_{C_n}$$
$$+ \sum_{j=0}^s \|\int_{\xi_j}^{\xi_{j+1}} B_j C_j(x)e^{xA}\int_0^x e^{-tA} f(t)dtdx\|_{C_n}$$
$$\leq \sum_{i=0}^m l_i^{(2)} n\|f\|_{C_n} + \sum_{j=0}^s k_j l^{(0)} n^2\|f\|_{C_n}$$

$$= n\left(\sum_{i=0}^{m} l_i^{(2)} + l^{(0)} n \sum_{j=0}^{s} k_j\right) \|f\|_{C_n}. \tag{56}$$

By means of (6), we obtain

$$\|\int_0^x e^{(x-t)A} f(t)dt\|_{C_n} = \|e^{xA}\int_0^x e^{-tA} f(t)dt\|_{C_n} \leq k^{(0)} l^{(0)} n^2 \|f\|_{C_n}. \tag{57}$$

Utilization of (4) and the relation (56) produces

$$\begin{aligned}
&\|\int_0^x e^{(x-t)A}\mathbf{G}(t)dt\Pi^{r1}(\mathbf{A}\phi_f + \mathbf{B}\omega_f)\|_{C_n}\\
&= \|\int_0^x \sum_{i=0}^{m} e^{(x-t)A} G_i(t)\Pi_i^{r1} dt(\mathbf{A}\phi_f + \mathbf{B}\omega_f)\|_{C_n}\\
&\leq l^{(1)} n\|\mathbf{A}\phi_f + \mathbf{B}\omega_f\|_{C_n}\\
&\leq l_1 n^2\left(\sum_{i=0}^{m} l_i^{(2)} + l^{(0)} n\sum_{j=0}^{s} k_j\right)\|f\|_{C_n}.
\end{aligned} \tag{58}$$

From (4) and (7) follows that

$$\begin{aligned}
&\|\int_0^x e^{(x-t)A}\mathbf{G}(t)dt\Pi^{rr}\nu_f\|_{C_n}\\
&= \|\int_0^x e^{(x-t)A}\left(\sum_{i=0}^{m} G_i(t)\Pi_{i0}^{rr}, \ldots, \sum_{i=0}^{m} G_i(t)\Pi_{im}^{rr}\right) dt col\,(\nu_0(f), \ldots, \nu_m(f))\|_{C_n}\\
&\leq \sum_{k=0}^{m}\|\sum_{i=0}^{m}\int_0^x e^{(x-t)A} G_i(t)\Pi_{ik}^{rr} dt\nu_k(f)\|_{C_n}\\
&\leq \sum_{k=0}^{m} l_k^{(3)} n\|\nu_k(f)\|_{C_n}\\
&= \sum_{k=0}^{m} l_k^{(3)} n\|\int_0^1 H_k(x)e^{xA}\int_0^x e^{-tA} f(t)dtdx\|_{C_n}\\
&\leq \sum_{k=0}^{m} l_k^{(3)} n\hat{h}_k n l^{(0)} n\|f\|_{C_n}\\
&= l^{(0)} n^3\sum_{k=0}^{m} l_k^{(3)}\hat{h}_k\|f\|_{C_n}.
\end{aligned} \tag{59}$$

Further, by using property (4) and the relation (56), we acquire

$$\|e^{xA}\Pi^{11}(\mathbf{A}\phi_f+\mathbf{B}\omega_f)\|_{C_n} \leq l^{(4)}n\|\mathbf{A}\phi_f+\mathbf{B}\omega_F\|_{C_n}$$
$$\leq l^{(4)}n^2\left(\sum_{i=0}^{m}l_i^{(2)}+l^{(0)}n\sum_{j=0}^{s}k_j\right)\|f\|_{C_n}. \quad (60)$$

Finally, by employing (9), we get

$$\|e^{xA}\Pi^{1r}\nu_f\|_{C_n} = \|e^{xA}\sum_{i=0}^{m}\Pi_i^{1r}\nu_i(f)\|_{C_n}$$
$$\leq \sum_{i=0}^{m}\|e^{xA}\Pi_i^{1r}\nu_i(f)\|_{C_n}$$
$$= \sum_{i=0}^{m}\|e^{xA}\Pi_i^{1r}\int_0^1 H_i(\xi)e^{\xi A}\int_0^{\xi}e^{-tA}f(t)dtd\xi\|_{C_n}$$
$$\leq l^{(0)}n^3\sum_{i=0}^{m}\hat{h}_i\tilde{h}_i\|f\|_{C_n}. \quad (61)$$

From (54) and (57)–(60), follows that

$$\|y(x)\|_{C_n} \leq \left[k^{(0)}l^{(0)}n^2+l_1n^2\left(\sum_{i=0}^{m}l_i^{(2)}+l^{(0)}n\sum_{j=0}^{s}k_j\right)+l^{(0)}n^3\sum_{k=0}^{m}l_k^{(3)}\hat{h}_k\right.$$
$$\left.+l^{(4)}n^2\left(\sum_{i=0}^{m}l_i^{(2)}+l^{(0)}n\sum_{j=0}^{s}k_j\right)+l^{(0)}n^3\sum_{i=0}^{m}\hat{h}_i\tilde{h}_i\right]\|f\|_{C_n}$$
$$\leq \gamma\|f\|_{C_n}. \quad (62)$$

where $\gamma > 0$. The last inequality proves the boundedness and correctness of the operator P and problem (13). The theorem is proved. □

4 Conclusions

We have studied a class of nonhomogeneous systems of n linear first-order ordinary Fredholm type integro-differential equations subject to general multipoint and integral boundary constraints. We have established sufficient solvability and uniqueness criteria and we have derived a ready to use exact solution formula. The method proposed requires the knowledge of a fundamental matrix of the corresponding homogeneous system of first-order differential equations. The solution process can be easily implemented to any computer algebra system.

References

1. A.R. Abdullaev, E.A. Skachkova, On a multipoint boundary-value problem for a second-order differential equation. Vestn. Perm. Univ. Ser. Math. Mech. Inf. **2**(25), 5–9 (2014)
2. A.R. Abdullaev, E.A. Skachkova, On one class of multipoint boundary-value problems for a second-order linear functional-differential equation. J. Math. Sci. **230**, 647–650 (2018). https://doi.org/10.1007/s10958-018-3761-9
3. R.P. Agarwal, The numerical solutions of multipoint boundary value problems. J. Comput. Appl. Math. **5**, 17–24 (1979)
4. M.T. Ashordiya, A Criterion for the solvability of a multipoint boundary value problem for a system of generalized ordinary differential equations. Differ. Uravn. **32**, 1303–1311 (1996)
5. M.M. Baiburin, Multipoint value problems for second order differential operator. Vestn. KAZgu. Math. **1**(37), 36–40 (2005) [in Russian]
6. M.M. Baiburin, On multi-point boundary value problems for first order linear differential equations system. Appl. Math. Control Prob. **3**, 16–30 (2018) [in Russian]
7. D.S. Dzhumabaev, On one approach to solve the linear boundary value problems for Fredholm integro-differential equations. J. Comput. Appl. Math. **294**, 342–357 (2016). https://doi.org/10.1016/j.cam.2015.08.023
8. C.P. Gupta, A generalized multi-point boundary value problem for second order ordinary differential equations. Appl. Math. Comput. **89**, 133–146 (1998). https://doi.org/10.1016/S0096-3003(97)81653-0
9. V.A. Il'in, E.I. Moiseev, Nonlocal boundary value problem of the first kind for a Sturm-Liouville operator in its differential and finite difference aspects. Differ. Equ. **23**, 803–810 (1987)
10. A.K. Iskakova, On the correctness of a single operator in $C_n[0, 1]$. Vestn. KAZgu. Math. Mech. Inf. **5**(19), 68–72 (1999) [in Russian]
11. I.T. Kiguradze, Boundary-value problems for systems of ordinary differential equations. J. Math. Sci. **43**, 2259–2339 (1988). https://doi.org/10.1007/BF01100360
12. V.S. Klimov, A multipoint boundary value problem for a system of differential equations. Differ. Uravn. **5**, 1532–1534 (1969) [in Russian]
13. B.K. Kokebaev, M. Otelbaev, A.N. Shynybekov, Questions on extension and restriction of operators. Dokl. Akad. Nauk SSSR **271**, 24–26 (1983)
14. R.O. Oinarov, I.N. Parasidis, Correctly solvable extensions of operators with finite deficiencies in Banach space. Izv. Akad. Kaz. SSR. **5**, 42–46 (1988) [in Russian]
15. T. Ojika, W. Wayne, A numerical method for the solution of multi-point problems for ordinary differential equations with integral constraints. J. Math. Anal. Appl. **72**, 500–511 (1979). https://doi.org/10.1016/0022-247X(79)90243-9
16. M.O. Otelbaev, A.K. Iskakova, On the multipoint value problem for the operator $Ly = y'(t)$ in $C_1[0, 1]$. Vestn. KAZgu. Math. Mech. Inf. **5**(19), 111–115 (1999) [in Russian]
17. I.N. Parasidis, E. Providas, Resolvent operators for some classes of integro-differential equations, in *Mathematical Analysis, Approximation Theory and Their Applications*, ed. by T. Rassias, V. Gupta. Springer Optimization and Its Applications, vol. 111 (Springer, Cham, 2016), pp. 535–558. https://doi.org/10.1007/978-3-319-31281-1
18. I.N. Parasidis, E. Providas, Extension operator method for the exact solution of integro-differential equations, in *Contributions in Mathematics and Engineering*, ed. by P. Pardalos, T. Rassias (Springer, Cham, 2016), pp. 473–496. https://doi.org/10.1007/978-3-319-31317-7

19. I.N. Parasidis, E. Providas, On the exact solution of nonlinear integro-differential equations, in *Applications of Nonlinear Analysis*, ed. by T. Rassias. Springer Optimization and Its Applications, vol. 134 (Springer, Cham, 2018), pp. 591–609. https://doi.org/10.1007/978-3-319-89815-5
20. I. Parassidis, P. Tsekrekos, Correct selfadjoint and positive extensions of nondensely defined symmetric operators. Abstr. Appl. Anal. **2005**(7), 767–790 (2005)
21. I.V. Parkhimovich, Multipoint boundary value problems for linear integro-differential equations in a class of smooth functions. Differ. Uravn. **8**, 549–552 (1972) [in Russian]
22. M.N. Yakovlev, Estimates of the solutions systems of integro-differential equations subject to many-points and integral boundary conditions. Zap. Nauchn. Sem. LOMI **124**, 131–139 (1983) [in Russian]

A Variational Approach to the Financial Problem with Insolvencies and Analysis of the Contagion

Giorgia Cappello, Patrizia Daniele, Sofia Giuffrè, and Antonino Maugeri

Abstract In this chapter we improve some results in literature on the general financial equilibrium problem related to individual entities, called sectors, which invest in financial instruments as assets and as liabilities. Indeed the model, studied in the chapter, takes into account the insolvencies and we analyze how these insolvencies affect the financial problem. For this improved model we describe a variational inequality for which we provide an existence result. Moreover, we study the dual Lagrange problem, in which the Lagrange variables, which represent the deficit and the surplus per unit, appear and an economical indicator is provided. Finally, we perform the contagion by means of the deficit and surplus variables. As expected, the presence of the insolvencies makes it more difficult to reach the financial equilibrium and increases the risk of a negative contagion for all the systems.

1 Introduction

The term "insolvency" is often used to denote that an individual or an organization can no longer meet its financial obligations with its lender. Usually, before getting involved in insolvency proceedings, some informal arrangements with creditors are attempted. Insolvency can be caused by poor cash management, a reduction in cash inflow forecasts or by an increase in expenses.

When insolvent, the credit loans are revoked both at the credit institution concerned and at all the institutions and banks to which the customer has had debts;

G. Cappello · P. Daniele (✉) · A. Maugeri
Department of Mathematics and Computer Science, University of Catania, Catania, Italy
e-mail: giorgia.cappello@unict.it; daniele@dmi.unict.it; maugeri@dmi.unict.it

S. Giuffrè
D.I.I.E.S. "Mediterranea" University of Reggio Calabria, Reggio Calabria, Italy
e-mail: sofia.giuffre@unirc.it

T. M. Rassias, P. M. Pardalos (eds.), *Mathematical Analysis and Applications*, Springer Optimization and Its Applications 154,
https://doi.org/10.1007/978-3-030-31339-5_2

further, it becomes impossible for the client/company to obtain liquidity from other institutions.

In the USA the number of bankruptcies decreased to 23,106 companies in the first quarter of 2018 from 23,157 companies in the fourth quarter of 2017. According to The Guardian, the number of people who went bankrupt in 2017 in the United Kingdom rose to the highest level after the financial crisis, revealing the devastating toll of rising debts for the families. According to the Insolvency Service 99,196 people were declared insolvent in 2017, with an increase of 9.4% with respect to the year before and very close to the peak recorded during the recession. Lots of households (about 59,220 in 2017) are turning to "bankruptcy-lite" debt deals, where individuals reschedule their debts and agree to much lower payments. Italy confirms the unenviable leadership in the ranking of companies in difficulty among the main Western European countries. According to the surveys of Coface, a group at the top in credit insurance, in Italy there are 7.2% of companies in difficulty, in Spain 6.3%, in France 5.7%, and in Germany 4.9%. The percentage takes into account the insolvent companies and those indebted, unprofitable, who struggle to honor the payments at maturity. In Italy, the current levels of insolvency are more than double that of 2007, with one of the worst performances recorded at the European level. In general, the trend of insolvencies at global level is almost stable in 2017. The modest decline that was expected last year, equal to about a -1%, is in fact the weakest result since 2009.

Some financial network models have already been studied in the literature. The first authors to develop a multi-sector, multi-instrument financial equilibrium model using the variational inequality theory were Nagurney et al. [35]. Recently, in [1, 7, 8, 11] more general models have been studied allowing that the data are evolving over time.

In this chapter we improve the previous results, including the insolvencies of the financial institutions.

We obtain such a result, considering in the utility function the presence of the term $\sum_{j=1}^{n} r_j(t)(1-\tau_{ij}(t))c_j(t)(1+h_j(t))y_{ij}(t)$, which represents, by means of the insolvency coefficients $c_j(t)$, the portion of liabilities that are not reimbursed. Since a big number of critic situations have been caused by the fact that the banks or the financial institutions were not able to recover a part of their debts, we focus our attention on this more complete model, deriving the variational formulation, applying the infinite-dimensional duality theory and examining the contagion effect on the economy. In this context, a particular attention is devoted to the problem of the contagion, in order to know when it happens and also to establish how the insolvencies contribute to the occurrence of the contagion. We are able to control the contagion, using the dual Lagrange problem and the dual Lagrange variables, which represent the deficit and the surplus per unit, arising from instrument j. Considering the dual problem, we can examine the financial model both from the *Point of View of the Sectors* and from the *System Point of View* (see Section 3.3) and we can clearly see that liabilities from the point of view of the sectors are investments for the economic system, namely a positive factor, upon which to base the development of the economy. As expected, the presence of the insolvencies, that we are able to quantify, makes it more difficult to reach the financial equilibrium, since reduced income has to balance all the expenditure of the system.

The chapter is organized as follows: in Section 2 we present the detailed financial model, together with the evolutionary variational inequality formulation of the equilibrium conditions, and an existence result is provided; in Section 3 we apply the duality to the general financial equilibrium problem, deriving the Deficit Formula, the Balance Law, and the Liability Formula, we give the dual formulation of the financial problem, we study the regularity of the Lagrange variables, deficit and surplus, and we analyze, by means of these variables, the financial contagion; in Section 4 we provide a numerical financial example and, finally, in Section 5 we summarize our results and conclusions.

It is worth mentioning that the methods applied in this chapter may be used in the study of many other equilibrium problems [4, 5, 10, 19–26, 34].

2 The Financial Model and the Equilibrium Conditions

2.1 *Presentation of the Model*

For the reader's convenience, we present the detailed financial model (see also [1]). We consider a financial economy consisting of m sectors, for example households, domestic business, banks and other financial institutions, as well as state and local governments, with a typical sector denoted by i, and of n instruments, for example mortgages, mutual funds, saving deposits, money market funds, with a typical financial instrument denoted by j, in the time interval $[0, T]$. Let $s_i(t)$ denote the total financial volume held by sector i at time t as assets, and let $l_i(t)$ be the total financial volume held by sector i at time t as liabilities. Further, we allow markets of assets and liabilities to have different investments $s_i(t)$ and $l_i(t)$, respectively. Since we are working in the presence of uncertainty and of risk perspectives, the volumes $s_i(t)$ and $l_i(t)$ held by each sector cannot be considered stable with respect to time and may decrease or increase. For instance, depending on the crisis periods, a sector may decide not to invest on instruments and to buy goods as gold and silver. At time t, we denote the amount of instrument j held as an asset in sector i's portfolio by $x_{ij}(t)$ and the amount of instrument j held as a liability in sector i's portfolio by $y_{ij}(t)$. The assets and liabilities in all the sectors are grouped into the matrices $x(t), y(t) \in \mathbb{R}^{m \times n}$, respectively. At time t we denote the price of instrument j held as an asset and as a liability by $r_j(t)$ and by $(1 + h_j(t))r_j(t)$, respectively, where h_j is a nonnegative function defined into $[0, T]$ and belonging to $L^\infty([0, T], \mathbb{R})$. We introduce the term $h_j(t)$ because the prices of liabilities are generally greater than or equal to the prices of assets. In this manner we describe, in a more realistic way, the behavior of the markets for which the liabilities are more expensive than the assets. We group the instrument prices held as an asset and as a liability into the vectors $r(t) = [r_1(t), r_2(t), \ldots, r_i(t), \ldots, r_n(t)]^T$ and $(1 + h(t))r(t) = [(1 + h_1(t))r_1(t), (1 + h_2(t))r_2(t), \ldots, (1 + h_i(t))r_i(t), \ldots, (1 + h_n(t))r_n(t)]^T$, respectively. In our problem the prices of each instrument appear as unknown variables. Under the assumption of perfect competition, each sector will behave as if it has no influence on the instrument prices or on the behavior of the other sectors, but on the total amount of the investments and the liabilities of each sector.

We choose as a functional setting the very general Lebesgue space

$$L^2([0, T], \mathbb{R}^p) = \left\{ f : [0, T] \to \mathbb{R}^p \text{ measurable} : \int_0^T \|f(t)\|_p^2 dt < +\infty \right\},$$

with the norm

$$\|f\|_{L^2([0,T],\mathbb{R}^p)} = \left(\int_0^T \|f(t)\|_p^2 dt \right)^{\frac{1}{2}}.$$

Then, the set of feasible assets and liabilities for each sector $i = 1, \ldots, m$ becomes

$$P_i = \left\{ (x_i(t), y_i(t)) \in L^2([0, T], \mathbb{R}_+^{2n}) : \right.$$
$$\left. \sum_{j=1}^n x_{ij}(t) = s_i(t), \quad \sum_{j=1}^n y_{ij}(t) = l_i(t) \text{ a.e. in } [0, T] \right\}$$

and the set of all feasible assets and liabilities becomes

$$P = \left\{ (x(t), y(t)) \in L^2([0, T], \mathbb{R}^{2mn}) : (x_i(t), y_i(t)) \in P_i, \ i = 1, \ldots, m \right\}.$$

Now, we introduce the ceiling and the floor price associated with instrument j, denoted by $\overline{r}_j$ and by $\underline{r}_j$, respectively, with $\overline{r}_j(t) > \underline{r}_j(t) \geq 0$, a.e. in $[0, T]$. The floor price $\underline{r}_j(t)$ is determined on the basis of the official interest rate fixed by the central banks, which, in turn, take into account the consumer price inflation. Then the equilibrium prices $r_j^*(t)$ cannot be less than these floor prices. The ceiling price $\overline{r}_j(t)$ derives from the financial need to control the national debt arising from the amount of public bonds and of the rise in inflation. It is a sign of the difficulty on the recovery of the economy. However it should be not overestimated because it produced an availability of money.

In detail, the meaning of the lower and upper bounds is that to each investor a minimal price $\underline{r}_j$ for the assets held in the instrument j is guaranteed, whereas each investor is requested to pay for the liabilities in any case a minimal price $(1+h_j)\underline{r}_j$. Analogously each investor cannot obtain for an asset a price greater than $\overline{r}_j$ and as a liability the price cannot exceed the maximum price $(1 + h_j)\overline{r}_j$.

We denote the given tax rate levied on sector i's net yield on financial instrument j, as τ_{ij}. Assume that the tax rates lie in the interval $[0, 1)$ and belong to

$L^\infty([0, T], \mathbb{R})$. Therefore, the government in this model has the flexibility of levying a distinct tax rate across both sectors and instruments.

We group the instrument ceiling and floor prices into the column vectors $\overline{r}(t) = (\overline{r}_j(t))_{j=1,\dots,n}$ and $\underline{r}(t) = (\underline{r}_j(t))_{j=1,\dots,n}$, respectively, and the tax rates τ_{ij} into the matrix $\tau(t) \in L^2([0, T], \mathbb{R}^{m\times n})$.

The set of feasible instrument prices is:

$$\mathscr{R} = \{r \in L^2([0, T], \mathbb{R}^n) : \underline{r}_j(t) \le r_j(t) \le \overline{r}_j(t), \quad j = 1, \dots, n, \text{ a.e. in } [0, T]\},$$

where $\underline{r}$ and $\overline{r}$ are assumed to belong to $L^2([0, T], \mathbb{R}^n)$.

In order to determine for each sector i the optimal distribution of instruments held as assets and as liabilities, we consider, as usual, the influence due to risk-aversion and the optimality conditions of each sector in the financial economy, namely the desire to maximize the value of the asset holdings while minimizing the value of liabilities. In the current economic situation there is a serious problem caused by the suffering that undermines the whole system. For this reason we intend to address the study of the financial problem in the presence of insolvencies.

Hence, in order to meet this need, we take into account the non-performing loans, introducing the insolvency coefficients $c_j(t)$, $j = 1, \dots, n$. We assume that the insolvency coefficients $c_j(t)$ lie in the interval $[0, 1)$ and belong to $L^\infty([0, T], \mathbb{R})$.

Then, we introduce the utility function $U_i(t, x_i(t), y_i(t), r(t))$, for each sector i, defined as follows:

$$U_i(t, x_i(t), y_i(t), r(t)) = u_i(t, x_i(t), y_i(t))$$

$$+\sum_{j=1}^{n} r_j(t)(1 - \tau_{ij}(t))[x_{ij}(t) - (1 - c_j(t))(1 + h_j(t))y_{ij}(t)],$$

where the term $-u_i(t, x_i(t), y_i(t))$ represents a measure of the risk of the financial agent, the term $\sum_{j=1}^{n} r_j(t)(1 - \tau_{ij}(t))[x_{ij}(t) - (1 + h_j(t))y_{ij}(t)]$ represents the value of the difference between the asset holdings and the value of liabilities, and the term $\sum\limits_{j=1}^{n} r_j(t)(1 - \tau_{ij}(t))c_j(t)(1 + h_j(t))y_{ij}(t)$ represents, by means of the insolvency coefficients $c_j(t)$, the portion of liabilities that are not reimbursed. Such a term appears as a positive contribute for sector i and a loss for the system.

We suppose that the sector's utility function $U_i(t, x_i(t), y_i(t))$ is defined on $[0, T] \times \mathbb{R}^n \times \mathbb{R}^n$, is measurable in t, and is continuous with respect to x_i and y_i. Moreover we assume that $\dfrac{\partial u_i}{\partial x_{ij}}$ and $\dfrac{\partial u_i}{\partial y_{ij}}$ exist and that they are measurable in t and continuous with respect to x_i and y_i. Further, we require that $\forall i = 1, \dots, m$, $\forall j = 1, \dots, n$, and a.e. in $[0, T]$ the following growth conditions hold true:

$$|u_i(t, x, y)| \le \alpha_i(t)\|x\|\|y\|, \quad \forall x, y \in \mathbb{R}^n, \tag{1}$$

and

$$\left|\frac{\partial u_i(t,x,y)}{\partial x_{ij}}\right| \leq \beta_{ij}(t)\|y\|, \qquad \left|\frac{\partial u_i(t,x,y)}{\partial y_{ij}}\right| \leq \gamma_{ij}(t)\|x\|, \tag{2}$$

where $\alpha_i, \beta_{ij}, \gamma_{ij}$ are nonnegative functions of $L^\infty([0,T],\mathbb{R})$. Finally, we suppose that the function $u_i(t,x,y)$ is concave.

An example of measure of the risk aversion is given by a generalization to the evolutionary case of the well-known Markowitz quadratic function based on the variance-covariance matrix denoting the sector's assessment of the standard deviation of prices for each instrument (see [31, 32]). This evolutionary measure of Markowitz type can be refined in such a way that it can incorporate the adjustment in time which depends on the previous equilibrium states.

In Section 2.4 we define a utility function of Markowitz type.

2.2 *The Equilibrium Flows and Prices*

Now, we establish the equilibrium conditions for the prices, which express the equilibration of the total assets, the total liabilities, and the portion of financial transactions per unit F_j employed to cover the expenses of the financial institutions, including possible dividends and manager bonus. Indeed, the equilibrium condition for the price r_j of instrument j is the following:

$$\sum_{i=1}^{m}(1-\tau_{ij}(t))\left[x_{ij}^*(t)-(1-c_j(t))(1+h_j(t))y_{ij}^*(t)\right]+F_j(t)$$

$$\begin{cases} \geq 0 \text{ if } r_j^*(t) = \underline{r}_j(t) \\ = 0 \text{ if } \underline{r}_j(t) < r_j^*(t) < \overline{r}_j(t) \\ \leq 0 \text{ if } r_j^*(t) = \overline{r}_j(t) \end{cases} \tag{3}$$

where (x^*, y^*, r^*) is the equilibrium solution for the investments as assets and as liabilities and for the prices. In other words, the prices are determined taking into account the amount of the supply, the demand of an instrument, and the charges F_j, namely if there is an actual supply excess of an instrument as assets and of the charges F_j in the economy, then its price must be the floor price. If the price of an instrument is greater than the floor price, but not at the ceiling, then the market of that instrument must clear. Finally, if there is an actual demand excess of an instrument as liabilities in the economy, then the price must be at the ceiling.

Now, we can give different but equivalent equilibrium conditions, each of which is useful to illustrate particular features of the equilibrium.

Definition 1 A vector of sector assets, liabilities, and instrument prices $(x^*(t), y^*(t), r^*(t)) \in P \times \mathscr{R}$ is an equilibrium of the dynamic financial model if and only if $\forall i = 1, \dots, m$, $\forall j = 1, \dots, n$, and a.e. in $[0, T]$, it satisfies the system of inequalities

$$-\frac{\partial u_i(t, x^*, y^*)}{\partial x_{ij}} - (1 - \tau_{ij}(t))r_j^*(t) - \mu_i^{(1)*}(t) \geq 0, \tag{4}$$

$$-\frac{\partial u_i(t, x^*, y^*)}{\partial y_{ij}} + (1 - \tau_{ij}(t))(1 - c_j(t))(1 + h_j(t))r_j^*(t) - \mu_i^{(2)*}(t) \geq 0, \tag{5}$$

and equalities

$$x_{ij}^*(t)\Big[- \frac{\partial u_i(t, x^*, y^*)}{\partial x_{ij}} - (1 - \tau_{ij}(t))r_j^*(t) - \mu_i^{(1)*}(t)\Big] = 0, \tag{6}$$

$$y_{ij}^*(t)\Big[-\frac{\partial u_i(t, x^*, y^*)}{\partial y_{ij}} + (1 - \tau_{ij}(t))(1 - c_j(t))(1 + h_j(t))r_j^*(t) - \mu_i^{(2)*}(t)\Big] = 0, \tag{7}$$

where $\mu_i^{(1)*}(t)$, $\mu_i^{(2)*}(t) \in L^2([0, T], \mathbb{R})$ are Lagrange multipliers, and verifies conditions (3) a.e. in $[0, T]$.

We associate with each financial volumes s_i and l_i held by sector i the functions $\mu_i^{(1)*}(t)$ and $\mu_i^{(2)*}(t)$, related, respectively, to the assets and to the liabilities and which represent the "equilibrium disutilities" per unit of sector i. Then, (4) and (6) mean that the financial volume invested in instrument j as assets x_{ij}^* is greater than or equal to zero if the j-th component $-\frac{\partial u_i(t, x^*, y^*)}{\partial x_{ij}} - (1 - \tau_{ij}(t))r_j^*(t)$ of the disutility is equal to $\mu_i^{(1)*}(t)$, whereas if $-\frac{\partial u_i(t, x^*, y^*)}{\partial x_{ij}} - (1 - \tau_{ij}(t))r_j^*(t) > \mu_i^{(1)*}(t)$, then $x_{ij}^*(t) = 0$. The same occurs for the liabilities.

The functions $\mu_i^{(1)*}(t)$ and $\mu_i^{(2)*}(t)$ are the Lagrange multipliers associated a.e. in $[0, T]$ with the constraints $\sum_{j=1}^{n} x_{ij}(t) - s_i(t) = 0$ and $\sum_{j=1}^{n} y_{ij}(t) - l_i(t) = 0$, respectively. They are unknown a priori, but this fact has no influence because we will prove in the following theorem that Definition 1 is equivalent to a variational inequality in which $\mu_i^{(1)*}(t)$ and $\mu_i^{(2)*}(t)$ do not appear (see [1, Theorem 2.1]).

Theorem 1 *A vector $(x^*, y^*, r^*) \in P \times \mathcal{R}$ is a dynamic financial equilibrium if and only if it satisfies the following variational inequality:*

Find $(x^*, y^*, r^*) \in P \times \mathcal{R}$:

$$\sum_{i=1}^{m} \int_0^T \Big\{ \sum_{j=1}^{n} \Big[- \frac{\partial u_i(t, x_i^*(t), y_i^*(t))}{\partial x_{ij}} - (1 - \tau_{ij}(t))r_j^*(t)\Big]$$
$$\times [x_{ij}(t) - x_{ij}^*(t)]$$

$$+\sum_{j=1}^{n}\Big[-\frac{\partial u_i(t,x_i^*(t),y_i^*(t))}{\partial y_{ij}}+(1-\tau_{ij}(t))(1-c_j(t))r_j^*(t)(1+h_j(t))\Big]$$

$$\times[y_{ij}(t)-y_{ij}^*(t)]\Big\}dt$$

$$+\sum_{j=1}^{n}\int_0^T\sum_{i=1}^{m}\Big\{(1-\tau_{ij}(t))\Big[x_{ij}^*(t)-(1-c_j(t))(1+h_j(t))y_{ij}^*(t)\Big]+F_j(t)\Big\}$$

$$\times\big[r_j(t)-r_j^*(t)\big]dt\geq 0,\qquad \forall(x,y,r)\in P\times\mathscr{R}. \tag{8}$$

Remark 1 We would like to explicitly remark that our definition of equilibrium conditions (Definition 1) is equivalent to the equilibrium definition given by a vector $(x^*,y^*,r^*)\in P\times\mathscr{R}$ satisfying (3) and, $\forall i=1,\ldots,m$:

$$\max_{P_i}\int_0^T\Big\{u_i(t,x_i(t),y_i(t))+\sum_{j=1}^{n}(1-\tau_{ij}(t))r_j^*(t)[x_{ij}(t)-(1-c_j(t))(1+h_j(t))y_{ij}(t)]\Big\}dt=$$
$$\int_0^T\Big\{u_i(t,x_i^*(t),y_i^*(t))+\sum_{j=1}^{n}(1-\tau_{ij}(t))r_j^*(t)[x_{ij}^*(t)-(1-c_j(t))(1+h_j(t))y_{ij}^*(t)]\Big\}dt.$$

We prefer to use Definition 1, since it is expressed in terms of equilibrium disutilities.

2.3 Existence Theorem

Now, we would like to give an existence result. First of all, we remind some definitions. Let X be a reflexive Banach space and let $\mathbb{K}$ be a subset of X and X^* be the dual space of X.

Definition 2 A mapping $A:\mathbb{K}\to X^*$ is pseudomonotone in the sense of Brezis (B-pseudomonotone) iff

1. For each sequence u_n weakly converging to u (in short $u_n\rightharpoonup u$) in $\mathbb{K}$ and such that $\limsup_n\langle Au_n,u_n-v\rangle\leq 0$ it results that:

$$\liminf_n\langle Au_n,u_n-v\rangle\geq\langle Au,u-v\rangle,\qquad\forall v\in\mathbb{K}.$$

2. For each $v\in\mathbb{K}$ the function $u\mapsto\langle Au,u-v\rangle$ is lower bounded on the bounded subset of $\mathbb{K}$.

Definition 3 A mapping $A : \mathbb{K} \to X^*$ is hemicontinuous in the sense of Fan (F-hemicontinuous) iff for all $v \in \mathbb{K}$ the function $u \mapsto \langle Au, u - v \rangle$ is weakly lower semicontinuous on $\mathbb{K}$.

The following existence result does not require any kind of monotonicity assumptions.

Theorem 2 *Let $\mathbb{K} \subset X$ be a nonempty closed convex bounded set and let $A : \mathbb{K} \subset E \to X^*$ be B-pseudomonotone or F-hemicontinuous. Then, variational inequality*

$$\langle Au, v - u \rangle \geq 0 \quad \forall v \in \mathbb{K} \tag{9}$$

admits a solution.

In the following subsection we shall present an example of a function, which satisfies the above assumptions.

2.4 An Example of a Markowitz-Type Risk Measure

We generalize and provide an evolutionary Markowitz-type measure of the risk proposed with a memory term. This function is effective, namely an existence theorem for the general financial problem holds (see [17]). In this way we cover a lack, providing the existence of a significant evolutionary measure of the risk. The particular, but significant, example of utility function is:

$$u_i(x_i(t), y_i(t))$$

$$= \begin{bmatrix} x_i(t) \\ y_i(t) \end{bmatrix}^T Q^i \begin{bmatrix} x_i(t) \\ y_i(t) \end{bmatrix} + \int_0^t \begin{bmatrix} x_i(t-z) \\ y_i(t-z) \end{bmatrix}^T Q^i \begin{bmatrix} x_i(t-z) \\ y_i(t-z) \end{bmatrix} dz, \tag{10}$$

where Q^i denotes the sector i's assessment of the standard deviation of prices for each instrument j.

In [17] it has been proven that Markowitz function verifies all the assumptions of the existence theorem, hence a problem with a function like this admits solutions.

3 The Duality for the Financial Equilibrium Problem

In this section we study the duality for the financial equilibrium problem (see also [6]).

To this end, for reader's convenience, we recall here some definitions and results of the infinite dimensional duality theory.

3.1 The New Infinite-Dimensional Duality Theory

In order to obtain the strong duality, we need that some delicate conditions, called "constraints qualification conditions," hold. In the infinite dimensional settings the next assumption, the so-called *Assumption S*, results to be a necessary and sufficient condition for the strong duality (see [3, 9, 12, 13, 33]).

Let $f : S \to \mathbb{R}$, $g : S \to Y$, $h : S \to Z$ be three mappings, where S is a convex subset of a real normed space X, Y is a real normed space ordered by a convex cone C, Z is a real normed space and consider the optimization problem:

$$\begin{cases} f(x_0) = \min_{x \in \mathbb{K}} f(x) \\ x_0 \in \mathbb{K} = \{x \in S : g(x) \in -C, \ h(x) = \theta_Z\}, \end{cases} \tag{11}$$

where θ_Z is the zero element in the space Z.

Its Lagrange dual problem is:

$$\max_{\lambda \in C^*, \, \mu \in Z^*} \inf_{x \in S} [f(x) + \langle \lambda, g(x) \rangle + \langle \mu, h(x) \rangle], \tag{12}$$

where

$$C^* := \{u \in Y^* : \langle u, y \rangle \geq 0, \ \forall y \in C\}$$

is the dual cone of C and Z^* is the dual space of Z. Then, we say that the strong duality holds for problems (11) and (12) if and only if problems (11) and (12) admit a solution and the optimal values coincide.

Some classical results due to Rockafellar [36], Holmes [27], Borwein and Lewis [2] give sufficient conditions in order that the strong duality between problems (11) and (12) holds, which use concepts such as the core, the intrinsic core, the strong quasi-relative interior of C. Such concepts (see [2, 27, 29, 36]) require the nonemptiness of the ordering cone, which defines the cone constraints in convex optimization and variational inequalities. However, the ordering cone of almost all the known problems, stated in infinite dimensional spaces, has the interior (and all the above generalized interior concepts) empty. Hence, the above interior conditions cannot be used to guarantee the strong duality.

Only recently, in [12] the authors introduced a new condition called *S*, which turns out to be a necessary and sufficient condition for the strong duality and really useful in the applications. This condition does not require the nonemptiness of the interior of the ordering cone. This new strong duality theory was then refined in [9, 13, 15, 28, 33].

Now we present in detail these new conditions.

Let us first recall that for a subset $C \subseteq X$ and $x \in X$ the tangent cone to C at x is defined as

$$T_C(x) = \{y \in X : \ y = \lim_{n \to \infty} \lambda_n (x_n - x), \ \lambda_n > 0, \ x_n \in C, \ \lim_{n \to \infty} x_n = x\}.$$

If $x \in clC$ (the closure of C) and C is convex, we have

$$T_C(x) = clcone(C - \{x\}),$$

where the $coneA = \{\lambda x : x \in A, \lambda \in \mathbb{R}^+\}$ denotes the cone hull of a general subset A of the space.

Definition 4 (Assumption S) Given the mappings f, g, h and the set $\mathbb{K}$ as above, we shall say that *Assumption S* is fulfilled at a point $x_0 \in \mathbb{K}$ if it results to be

$$T_{\widetilde{M}}(0, \theta_Y, \theta_Z) \cap \Big( ]-\infty, 0[\times \{\theta_Y\} \times \{\theta_Z\} \Big) = \emptyset, \tag{13}$$

where

$$\widetilde{M} = \{(f(x) - f(x_0) + \alpha, g(x) + y, h(x)) : x \in S \setminus \mathbb{K}, \ \alpha \geq 0, \ y \in C\}.$$

The following theorem holds (see Theorem 1.1 in [13] for the proof):

Theorem 3 *Under the above assumptions on f, g, h, and C, if problem* (11) *is solvable and Assumption S is fulfilled at the extremal solution $x_0 \in \mathbb{K}$, then also problem* (12) *is solvable, the extreme values of both problems are equal, namely, if $(x_0, \lambda^*, \mu^*) \in \mathbb{K} \times C^* \times Z^*$ is the optimal point of problem* (12),

$$\begin{aligned} f(x_0) &= \min_{x \in \mathbb{K}} f(x) = f(x_0) + \langle \lambda^*, g(x_0) \rangle + \langle \mu^*, h(x_0) \rangle \\ &= \max_{\substack{\lambda \in C^* \\ \mu \in Z^*}} \inf_{x \in S} \{ f(x) + \langle \lambda, g(x) \rangle + \langle \mu, h(x) \rangle \} \end{aligned} \tag{14}$$

and, it results to be:

$$\langle \lambda^*, g(x_0) \rangle = 0.$$

3.2 Existence of Lagrange Multipliers

Now, we can apply the infinite-dimensional duality for the financial equilibrium problem expressed by variational inequality (8), which ensures the existence of the Lagrange multipliers. To this end, let us set:

$$f(x, y, r) = \int_0^T \Bigg\{ \sum_{i=1}^m \sum_{j=1}^n \Bigg[-\frac{\partial u_i(t, x^*(t), y^*(t))}{\partial x_{ij}} - (1 - \tau_{ij}(t)) r_j^*(t) \Bigg] \times [x_{ij}(t) - x_{ij}^*(t)]$$

$$+\sum_{i=1}^{m}\sum_{j=1}^{n}\left[-\frac{\partial u_i(t,x^*(t),y^*(t))}{\partial y_{ij}}+(1-\tau_{ij}(t))(1-c_j(t))(1+h_j(t))r_j^*(t)\right]$$
$$\times [y_{ij}(t)-y_{ij}^*(t)]$$
$$+\sum_{j=1}^{n}\left[\sum_{i=1}^{m}(1-\tau_{ij}(t))\left[x_{ij}^*(t)-(1-c_j(t))(1+h_j(t))y_{ij}^*(t)\right]+F_j(t)\right]$$
$$\times\left[r_j(t)-r_j^*(t)\right]\Bigg\}\,dt.$$

Then, the Lagrange functional is

$$\begin{aligned}
&\mathscr{L}(x,y,r,\lambda^{(1)},\lambda^{(2)},\mu^{(1)},\mu^{(2)},\rho^{(1)},\rho^{(2)})=f(x,y,r)\\
&-\sum_{i=1}^{m}\sum_{j=1}^{n}\int_0^T\lambda_{ij}^{(1)}(t)x_{ij}(t)\,dt-\sum_{i=1}^{m}\sum_{j=1}^{n}\int_0^T\lambda_{ij}^{(2)}y_{ij}(t)\,dt\\
&-\sum_{i=1}^{m}\int_0^T\mu_i^{(1)}(t)\left(\sum_{j=1}^{n}x_{ij}(t)-s_i(t)\right)dt\\
&-\sum_{i=1}^{m}\int_0^T\mu_i^{(2)}(t)\left(\sum_{j=1}^{n}y_{ij}(t)-l_i(t)\right)dt\\
&+\sum_{j=1}^{n}\int_0^T\rho_j^{(1)}(t)(\underline{r}_j(t)-r_j(t))\,dt+\sum_{j=1}^{n}\int_0^T\rho_j^{(2)}(t)(r_j(t)-\overline{r}_j(t))\,dt,
\end{aligned}\tag{15}$$

where $(x,y,r)\in L^2([0,T],\mathbb{R}^{2mn+n})$, $\lambda^{(1)},\lambda^{(2)}\in L^2([0,T],\mathbb{R}^{mn}_+)$, $\mu^{(1)},\mu^{(2)}\in L^2([0,T],\mathbb{R}^m)$, $\rho^{(1)},\rho^{(2)}\in L^2([0,T],\mathbb{R}^n_+)$ and $\lambda^{(1)}$, $\lambda^{(2)}$, $\rho^{(1)}$, $\rho^{(2)}$ are the Lagrange multipliers associated, a.e. in $[0,T]$, with the sign constraints $x_i(t)\geq 0$, $y_i(t)\geq 0$, $r_j(t)-\underline{r}_j(t)\geq 0$, $\overline{r}_j(t)-r_j(t)\geq 0$, respectively, whereas the functions $\mu^{(1)}(t)$ and $\mu^{(2)}(t)$ are the Lagrange multipliers associated, a.e. in $[0,T]$, with the equality constraints $\sum_{j=1}^{n}x_{ij}(t)-s_i(t)=0$ and $\sum_{j=1}^{n}y_{ij}(t)-l_i(t)=0$, respectively.

Applying the new strong duality theory, the following theorem holds.

Theorem 4 *Let* $(x^*,y^*,r^*)\in P\times\mathscr{R}$ *be a solution to variational inequality* (8) *and let us consider the associated Lagrange functional* (15). *Then, the strong duality holds and there exist* $\lambda^{(1)*},\lambda^{(2)*}\in L^2([0,T],\mathbb{R}^{mn}_+)$, $\mu^{(1)*},\mu^{(2)*}\in L^2([0,T],\mathbb{R}^m)$, $\rho^{(1)*},\rho^{(2)*}\in L^2([0,T],\mathbb{R}^n_+)$ *such that* $(x^*,y^*,r^*,\lambda^{(1)*},\lambda^{(2)*},\mu^{(1)*},\mu^{(2)*},\rho^{(1)*},\rho^{(2)*})$ *is a saddle point of the Lagrange functional, namely*

$$\begin{aligned} &\mathscr{L}(x^*, y^*, r^*, \lambda^{(1)}, \lambda^{(2)}, \mu^{(1)}, \mu^{(2)}, \rho^{(1)}, \rho^{(2)}) \\ &\leq \mathscr{L}(x^*, y^*, r^*, \lambda^{(1)*}, \lambda^{(2)*}, \mu^{(1)*}, \mu^{(2)*}, \rho^{(1)*}, \rho^{(2)*}) = 0 \\ &\leq \mathscr{L}(x, y, r, \lambda^{(1)*}, \lambda^{(2)*}, \mu^{(1)*}, \mu^{(2)*}, \rho^{(1)*}, \rho^{(2)*}) \end{aligned} \tag{16}$$

$\forall (x, y, r) \in L^2([0,T], \mathbb{R}^{2mn+n})$, $\forall \lambda^{(1)}, \lambda^{(2)} \in L^2([0,T], \mathbb{R}^{mn}_+)$, $\forall \mu^{(1)}, \mu^{(2)} \in L^2([0,T], \mathbb{R}^m)$, $\forall \rho^{(1)}, \rho^{(2)} \in L^2([0,T], \mathbb{R}^n_+)$ *and, a.e. in* $[0,T]$,

$$-\frac{\partial u_i(t, x^*(t), y^*(t))}{\partial x_{ij}} - (1 - \tau_{ij}(t)) r_j^*(t) - \lambda_{ij}^{(1)*}(t) - \mu_i^{(1)*}(t) = 0,$$

$$\forall i = 1, \ldots, m, \ \forall j = 1 \ldots, n;$$

$$-\frac{\partial u_i(t, x^*(t), y^*(t))}{\partial y_{ij}} + (1 - c_j(t))(1 - \tau_{ij}(t))(1 + h_j(t)) r_j^*(t) - \lambda_{ij}^{(2)*}(t) - \mu_i^{(2)*}(t) = 0,$$

$$\forall i = 1, \ldots, m, \ \forall j = 1 \ldots, n;$$

$$\sum_{i=1}^m (1 - \tau_{ij}(t)) \left[x_{ij}^*(t) - (1 - c_j(t))(1 + h_j(t)) y_{ij}^*(t) \right] + F_j(t) + \rho_j^{(2)*}(t) = \rho_j^{(1)*}(t), \tag{17}$$

$$\forall j = 1, \ldots, n;$$

$$\lambda_{ij}^{(1)*}(t) x_{ij}^*(t) = 0, \ \lambda_{ij}^{(2)*}(t) y_{ij}^*(t) = 0, \quad \forall i = 1, \ldots, m, \ \forall j = 1, \ldots, n \tag{18}$$

$$\mu_i^{(1)*}(t) \left(\sum_{j=1}^n x_{ij}^*(t) - s_i(t) \right) = 0, \quad \mu_i^{(2)*}(t) \left(\sum_{j=1}^n y_{ij}^*(t) - l_i(t) \right) = 0, \tag{19}$$

$$\forall i = 1, \ldots, m$$

$$\rho_j^{(1)*}(t)(\underline{r}_j(t) - r_j^*(t)) = 0, \ \rho_j^{(2)*}(t)(r_j^*(t) - \overline{r}_j(t)) = 0, \quad \forall j = 1, \ldots, n. \tag{20}$$

Formula (17) represents the Deficit Formula. Indeed, if $\rho_j^{(1)*}(t)$ is positive, then the prices are minimal and there is a supply excess of instrument j as an asset and of the charge $F_j(t)$, namely the economy is in deficit and, for this reason, $\rho_j^{(1)*}(t)$ is called *the deficit variable* and represents the deficit per unit.

Analogously, if $\rho_j^{(2)*}(t)$ is positive, then the prices are maximal and there is a demand excess of instrument j as a liability, namely there is a surplus in the economy. For this reason $\rho_j^{(2)*}(t)$ is called *the surplus variable* and represents the surplus per unit.

From (17) it is possible to obtain the Balance Law

$$\sum_{i=1}^{m} l_i(t) = \sum_{i=1}^{m} s_i(t) - \sum_{i=1}^{m}\sum_{j=1}^{n} \tau_{ij}(t)\left[x^*_{ij}(t) - y^*_{ij}(t)\right]$$
$$-\sum_{i=1}^{m}\sum_{j=1}^{n}(1-\tau_{ij}(t))h_j(t)y^*_{ij}(t) + \sum_{i=1}^{m}\sum_{j=1}^{n}(1-\tau_{ij}(t))c_j(t)(1+h_j(t))y^*_{ij}(t)$$
$$+\sum_{j=1}^{n} F_j(t) - \sum_{j=1}^{n} \rho_j^{(1)*}(t) + \sum_{j=1}^{n} \rho_j^{(2)*}(t). \tag{21}$$

Finally, assuming that the taxes $\tau_{ij}(t)$, $i = 1,\ldots,m$, $j = 1,\ldots,n$, have a common value $\theta(t)$, the increments $h_j(t)$, $j = 1,\ldots,n$, have a common value $i(t)$, and the insolvency coefficients $c_j(t)$, $j = 1,\ldots,n$, have a common value $c(t)$, otherwise we can consider the average values (see Remark 7.1 in [1]), the significant Liability Formula follows

$$(1-c(t))\sum_{i=1}^{m} l_i(t) = \frac{(1-\theta(t))\sum_{i=1}^{m} s_i(t) + \sum_{j=1}^{n} F_j(t) - \sum_{j=1}^{n} \rho_j^{(1)*}(t) + \sum_{j=1}^{n} \rho_j^{(2)*}(t)}{(1-\theta(t))(1+i(t))}. \tag{22}$$

From (22) we can deduce that in this situation to reach the equilibrium is even more difficult than in the case of absence of insolvencies, because only a portion of liabilities must balance all the expenses.

3.3 The Viewpoints of the Sector and of the System

The financial problem can be considered from two different perspectives: one from the *Point of View of the Sectors*, which try to maximize the utility and a second point of view, that we can call *System Point of View*, which regards the whole equilibrium, namely in respect of the previous laws. For example, from the point of view of the sectors, $l_i(t)$, for $i = 1,\ldots,m$, are liabilities, whereas for the economic system they are investments and, hence, the Liability Formula, from the system point of view, can be called "*Investments Formula*." The system point of view coincides with the dual Lagrange problem (the so-called shadow market) in which $\rho_j^{(1)}(t)$ and $\rho_j^{(2)}(t)$ are the dual multipliers, representing the deficit and the surplus per unit arising from instrument j. Formally, the dual problem is given by

Find $(\rho^{(1)*}, \rho^{(2)*}) \in L^2([0,T], \mathbb{R}^{2n}_+)$ such that

$$\sum_{j=1}^{n}\int_0^T (\rho_j^{(1)}(t) - \rho_j^{(1)*}(t))(\underline{r}_j(t) - r^*_j(t))dt \tag{23}$$

$$+\sum_{j=1}^{n}\int_0^T(\rho_j^{(2)}(t)-\rho_j^{(2)*}(t))(r_j^*(t)-\overline{r}_j(t))dt\le 0,$$

$$\forall(\rho^{(1)},\rho^{(2)})\in L^2([0,T],\mathbb{R}_+^{2n}).$$

Indeed, taking into account inequality (16), we get

$$-\sum_{i=1}^{m}\sum_{j=1}^{n}\int_0^T(\lambda_{ij}^{(1)}(t)-\lambda_{ij}^{(1)*}(t))x_{ij}^*(t)\,dt-\sum_{i=1}^{m}\sum_{j=1}^{n}\int_0^T(\lambda_{ij}^{(2)}-\lambda_{ij}^{(2)*})y_{ij}^*(t)\,dt$$
$$-\sum_{i=1}^{m}\int_0^T(\mu_i^{(1)}(t)-\mu_i^{(1)*}(t))\left(\sum_{j=1}^{n}x_{ij}^*(t)-s_i(t)\right)dt$$
$$-\sum_{i=1}^{m}\int_0^T(\mu_i^{(2)}(t)-\mu_i^{(2)*}(t))\left(\sum_{j=1}^{n}y_{ij}^*(t)-l_i(t)\right)dt$$
$$+\sum_{j=1}^{n}\int_0^T(\rho_j^{(1)}(t)-\rho_j^{(1)*}(t))(\underline{r}_j(t)-r_j^*(t))\,dt$$
$$+\sum_{j=1}^{n}\int_0^T(\rho_j^{(2)}(t)-\rho_j^{(2)*}(t))(r_j^*(t)-\overline{r}_j(t))\,dt\le 0$$

$\forall\lambda^{(1)},\lambda^{(2)}\in L^2([0,T],\mathbb{R}_+^{mn})$, $\mu^{(1)},\mu^{(2)}\in L^2([0,T],\mathbb{R}^m)$, $\rho^{(1)},\rho^{(2)}\in L^2([0,T],\mathbb{R}_+^n)$.

Choosing $\lambda^{(1)}=\lambda^{(1)*}$, $\lambda^{(2)}=\lambda^{(2)*}$, $\mu^{(1)}=\mu^{(1)*}$, $\mu^{(2)}=\mu^{(2)*}$, we obtain the dual problem (23)

Note that, from the *System Point of View*, also the expenses of the institutions $F_j(t)$ are supported from the liabilities of the sectors.

Remark 2 Let us recall that from the Liability Formula we get the following index $E(t)$, called "Evaluation Index," that is very useful for the rating procedure:

$$E(t)=\frac{(1-c(t))\sum_{i=1}^{m}l_i(t)}{\sum_{i=1}^{m}\tilde{s}_i(t)+\sum_{j=1}^{n}\tilde{F}_j(t)},\tag{24}$$

where we set

$$\tilde{s}_i(t)=\frac{s_i(t)}{1+i(t)},\quad \tilde{F}_j(t)=\frac{F_j(t)}{(1+i(t))(1-\theta(t))}.$$

From the Liability Formula we obtain

$$E(t)=1-\frac{\sum_{j=1}^{n}\rho_j^{(1)*}(t)}{(1-\theta(t))(1+i(t))\left(\sum_{i=1}^{m}\tilde{s}_i(t)+\sum_{j=1}^{n}\tilde{F}_j(t)\right)}+\frac{\sum_{j=1}^{n}\rho_j^{(2)*}(t)}{(1-\theta(t))(1+i(t))\left(\sum_{i=1}^{m}\tilde{s}_i(t)+\sum_{j=1}^{n}\tilde{F}_j(t)\right)} \tag{25}$$

If $E(t)$ is greater than or equal to 1, the evaluation of the financial equilibrium is positive (better if $E(t)$ is proximal to 1), whereas if $E(t)$ is less than 1, the evaluation of the financial equilibrium is negative.

The term $(1-c(t))\sum_{i=1}^{m} l_i(t)$ in (24) represents the effective liabilities (or the effective investments from the system point of view). The evaluation index (25) is less than the one in the model in [1], where the insolvency coefficients are not considered, and this means that, in presence of insolvency, it is more difficult to reach the financial equilibrium.

3.4 Regularity Results

In [16] a regularity result of $\rho_j^{(1)*}(t)$, $\rho_j^{(2)*}(t)$, has been proved. Let us set

$$F(t)=[F_1(t),F_2(t),\ldots,F_n(t)]^T;$$

$$\nu=(x,y,r)=\left(\left(x_{ij}\right)_{\substack{i=1,\ldots,m\\ j=1,\ldots,n}},\left(y_{ij}\right)_{\substack{i=1,\ldots,m\\ j=1,\ldots,n}},\left(r_j\right)_{j=1,\ldots,n}\right);$$

$$A(t,\nu)=\left(\left[-\frac{\partial u_i(t,x,y)}{\partial x_{ij}}-(1-\tau_{ij}(t))r_j(t)\right]_{\substack{i=1,\ldots,m\\ j=1,\ldots,n}},\right.$$
$$\left[-\frac{\partial u_i(t,x,y)}{\partial y_{ij}}+(1-\tau_{ij}(t))(1-c_j(t))(1+h_j(t))r_j(t)\right]_{\substack{i=1,\ldots,m\\ j=1,\ldots,n}}, \tag{26}$$

$$\left[\sum_{i=1}^{m}(1-\tau_{ij}(t))\left(x_{ij}(t)-(1-c_j(t))(1+h_j(t))y_{ij}(t)\right)+F_j(t)\right]_{j=1,\dots,n}\Bigg);$$

$$A:\mathscr{K}\to L^2([0,T],\mathbb{R}^{2mn+n}),$$

with

$$\mathscr{K}=P\times\mathscr{R}.$$

Let us note that $\mathscr{K}$ is a convex, bounded, and closed subset of $L^2([0,T],\mathbb{R}^{2mn+n})$. Moreover assumption (2) implies that A is lower semicontinuous along line segments.

The following result holds true (see [16, Theorem 2.4]):

Theorem 5 *Let $A\in C^0([0,T],\mathbb{R}^{2mn+n})$ be strongly monotone in x and y, monotone in r, namely, there exists α such that, for $t\in[0,T]$,*

$$\langle\langle A(t,\nu_1)-A(t,\nu_2),\nu_1-\nu_2\rangle\rangle\geq\alpha(\|x_1-x_2\|^2+\|y_1-y_2\|^2), \tag{27}$$

$\forall\nu_1=(x_1,y_1,r_1),\ \nu_2=(x_2,y_2,r_2)\in\mathbb{R}^{2mn+n}$.

Let $\underline{r}(t)$, $\overline{r}(t)$, $h(t)$, $F(t)=[F_1(t),F_2(t),\dots,F_n(t)]^T$, $C(t)=[c_1(t),c_2(t),\dots,c_n(t)]^T\in C^0([0,T],\mathbb{R}^n_+)$, let $\tau(t)\in C^0([0,T],\mathbb{R}^{mn})$ and let $s,l\in C^0([0,T],\mathbb{R}^m)$, satisfying the following assumption (β):

- *there exists $\delta_1(t)\in L^2([0,T])$ and $c_1\in\mathbb{R}$ such that, for a.a. $t\in[0,T]$:*

$$\|s(t)\|\leq\delta_1(t)+c_1;$$

- *there exists $\delta_2(t)\in L^2([0,T])$ and $c_2\in\mathbb{R}$ such that, for a.a. $t\in[0,T]$:*

$$\|l(t)\|\leq\delta_2(t)+c_2.$$

Then the Lagrange variables, $\rho^{(1)*}(t)$, $\rho^{(2)*}(t)$, which represent the deficit and the surplus per unit, respectively, are continuous too.

3.5 The Contagion Problem

In this section we want to show that it is possible to establish when the economy becomes negative by means of the dual variables $\rho^{(1)*}(t)$, $\rho^{(2)*}(t)$ (see also [14]).

Contagion can be explained as a situation when a crisis in a particular economy or region spreads out and affects others (see [18] for a complete survey on the financial contagion). The Lehman Brothers' failure in the USA is an example of contagion. Fundamental problems in the contagion are to try to know when it can happen, to give a measure of it, and to understand why it occurs. In the particular financial problem we are dealing with, which is based on portfolio flows and investment positions, namely on assets and liabilities of different sectors, we perform the contagion by using the deficit and the surplus variables as well as the balance law. Specifically, we recall that $\rho^{(1)*}(t)$ represents the deficit variable and $\rho^{(2)*}(t)$ represents the surplus variable. For our purpose it is useful to recall also the balance law:

$$\begin{aligned}\sum_{i=1}^{m} l_i(t) - \sum_{i=1}^{m} s_i(t) + \sum_{i=1}^{m}\sum_{j=1}^{n} \tau_{ij}(t)\left[x_{ij}^*(t) - y_{ij}^*(t)\right] \\ + \sum_{i=1}^{m}\sum_{j=1}^{n}(1-\tau_{ij}(t))h_j(t)y_{ij}^*(t) - \sum_{i=1}^{m}\sum_{j=1}^{n}(1-\tau_{ij}(t))c_j(t)(1+h_j(t))y_{ij}^*(t) - \sum_{j=1}^{n} F_j(t) \\ = -\sum_{j=1}^{n}\rho_j^{(1)*}(t) + \sum_{j=1}^{n}\rho_j^{(2)*}(t).\end{aligned} \tag{28}$$

We realize that when the left-hand side is negative, it means that the sum of the liabilities, namely the investments of the system, cannot cover the expenses incurred. The sign of the left-hand side depends on the difference

$$-\sum_{j=1}^{n}\rho_j^{(1)*}(t) + \sum_{j=1}^{n}\rho_j^{(2)*}(t).$$

When such a difference is negative, from (28) it follows that the whole system is at a loss. In this case we say that a negative contagion is determined and we can assume that the insolvencies of individual entities propagate through the entire system. It is sufficient that only one deficit component $\rho_j^{(1)*}(t)$ is very large to obtain, even if the other $\rho_j^{(2)*}(t)$ are lightly positive, a negative balance for the whole system. In addition, if even only one $\rho_j^{(1)*}(t)$ is positive, then for that instrument j all the sectors are already in crisis.

When

$$\sum_{j=1}^{n}\rho_j^{(1)*}(t) > \sum_{j=1}^{n}\rho_j^{(2)*}(t),$$

namely the sum of all the deficit exceeds the sum of all the surplus, we get $E(t) \leq 1$ and, hence, also $E(t)$ is a significant indicator that the financial contagion happens. Causes of contagion are the lack of investments, the financial insolvency, or the excess in the expenses.

4 A Numerical Example

Let us analyze a numerical financial example in which we consider as the risk aversion function an evolutionary measure of Markowitz type, which expresses at each instant $t \in [0, T]$ the risk aversion by means of variance-covariance matrices denoting the sector's assessment of the standard deviation of prices for each instrument.

Let us consider an economy with two sectors and two financial instruments, as shown in Figure 1, and choose as variance-covariance matrices of the two sectors the following ones:

$$Q^1(t) = \begin{bmatrix} 1 & 0 & -0.5t & 0 \\ 0 & 1 & 0 & 0 \\ -0.5t & 0 & 1 & 0 \\ 0 & 0 & 0 & 1 \end{bmatrix}, \quad Q^2(t) = \begin{bmatrix} 1 & 0 & 0 & 0 \\ 0 & 1 & -0.5t & 0 \\ 0 & -0.5t & 1 & 0 \\ 0 & 0 & 0 & 1 \end{bmatrix}.$$

We define the feasible set as follows:

$$\mathbb{K} = \Big\{ (x_{11}(t), x_{12}(t), x_{21}(t), x_{22}(t), y_{11}(t), y_{12}(t), y_{21}(t), y_{22}(t), r_1(t), r_2(t)) \in L^2([0,1], \mathbb{R}^{10}_+) :$$
$$x_{11}(t) + x_{12}(t) = t + 2, \quad x_{21}(t) + x_{22}(t) = 2t + 3, \quad \text{a.e. in } [0,1]$$
$$y_{11}(t) + y_{12}(t) = 2t, \quad y_{21}(t) + y_{22}(t) = 3t, \quad \text{a.e. in } [0,1]$$

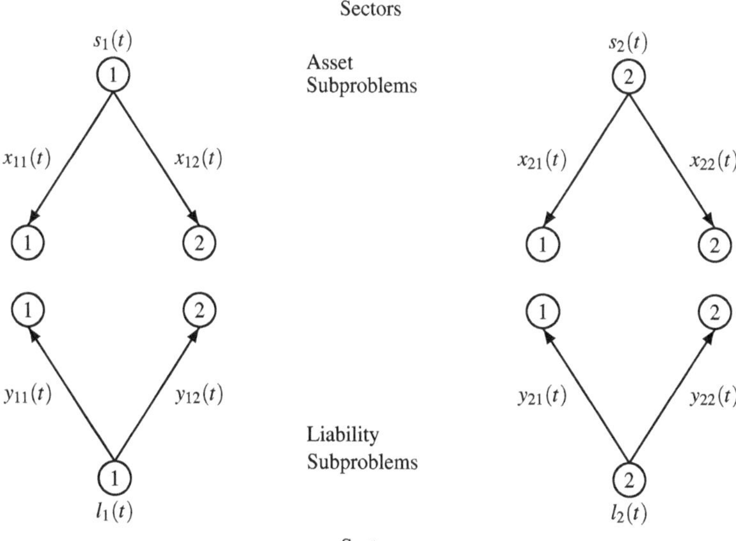

Fig. 1 Two sectors and two financial instruments network a.e. in [0, 1]

$$4t \leq r_1(t) \leq 5t + 12, \ \ t \leq r_2(t) \leq 6t + 5, \ \text{ a.e. in } [0, 1]\Big\}.$$

Let us assume that

$$h_1(t) = \frac{3}{2}t, \text{ and } h_2(t) = \frac{t}{2}.$$

Finally, let us consider

$$\tau_{11}(t) = \frac{t}{2}, \quad \tau_{12}(t) = \frac{3}{4}t, \quad \tau_{21}(t) = \frac{t}{2}, \quad \tau_{22}(t) = \frac{t}{4}$$

and

$$c_1(t) = 0.1 \quad c_2(t) = 0.15.$$

Then, variational inequality (8) becomes:

$$\begin{aligned}
&\int_0^1 \Big\{\Big[2x^*_{11}(t) - ty^*_{11}(t) - \left(1 - \frac{t}{2}\right) r^*_1(t)\Big]\big(x_{11}(t) - x^*_{11}(t)\big) \\
&\quad + \Big[2x^*_{12}(t) - \left(1 - \frac{3}{4}t\right) r^*_2(t)\Big]\big(x_{12}(t) - x^*_{12}(t)\big) \\
&\quad + \Big[2x^*_{21}(t) - \left(1 - \frac{t}{2}\right) r^*_1(t)\Big]\big(x_{21}(t) - x^*_{21}(t)\big) \\
&\quad + \Big[2x^*_{22}(t) - ty^*_{21}(t) - \left(1 - \frac{t}{4}\right) r^*_2(t)\Big]\big(x_{22}(t) - x^*_{22}(t)\big) \\
&\quad + \Big[2y^*_{11}(t) - tx^*_{11}(t) + 0.9\left(1 + \frac{3}{2}t\right)\left(1 - \frac{t}{2}\right) r^*_1(t)\Big]\big(y_{11}(t) - y^*_{11}(t)\big) \\
&\quad + \Big[2y^*_{12}(t) + 0.85\left(1 + \frac{t}{2}\right)\left(1 - \frac{3}{4}t\right) r^*_2(t)\Big]\big(y_{12}(t) - y^*_{12}(t)\big) \\
&\quad + \Big[2y^*_{21}(t) - tx^*_{22}(t) + 0.9\left(1 + \frac{3}{2}t\right)\left(1 - \frac{t}{2}\right) r^*_1(t)\Big]\big(y_{21}(t) - y^*_{21}(t)\big) \\
&\quad + \Big[2y^*_{22}(t) + 0.85\left(1 + \frac{t}{2}\right)\left(1 - \frac{t}{4}\right) r^*_2(t)\Big]\big(y_{22}(t) - y^*_{22}(t)\big) \\
&\quad + \Big[\left(1 - \frac{t}{2}\right) x^*_{11}(t) + \left(1 - \frac{t}{2}\right) x^*_{21}(t) - 0.9\left(1 + \frac{3}{2}t\right) \\
&\quad \Big[\left(1 - \frac{t}{2}\right) y^*_{11}(t) + \left(1 - \frac{t}{2}\right) y^*_{21}(t)\Big] + F_1(t)\Big]\big(r_1(t) - r^*_1(t)\big)
\end{aligned}$$

$$+[\left(1-\frac{3}{4}t\right)x^*_{12}(t)+\left(1-\frac{t}{4}\right)x^*_{22}(t)-0.85\left(1+\frac{t}{2}\right)$$

$$\left[\left(1-\frac{3}{4}t\right)y^*_{12}(t)+\left(1-\frac{t}{4}\right)y^*_{22}(t)\right]+F_2(t)](r_2(t)-r^*_2(t))\Big\}dt\geq 0,$$

$$\forall(x,y,r)\in\mathbb{K}. \tag{29}$$

Using the direct method we get the following solution:

$$\begin{aligned}
&x^\star_{11}(t)=-\frac{381t^4-610t^3-600t^2+3200t+2560}{160(t^2-16)}; && x^\star_{12}(t)=\frac{381t^4-450t^3-280t^2+640t-2560}{160(t^2-16)}\\
&x^\star_{21}(t)=-\frac{-415t^4+222t^3-2120t^2+4480t+3840}{160(t^2-16)}; && x^\star_{22}(t)=\frac{-415t^4+542t^3-1640t^2-640t-3840}{160(t^2-16)}\\
&y^\star_{11}(t)=-\frac{331t^3-410t^2+360t}{40(t^2-16)}; && y^\star_{12}(t)=\frac{411t^3-410t^2-920t}{40(t^2-16)}\\
&y^\star_{21}(t)=-\frac{485t^3-502t^2+760t}{40(t^2-16)}; && y^\star_{22}(t)=\frac{605t^3-502t^2-1160t}{40(t^2-16)}\\
&r^\star_1(t)=4t; && r^\star_2(t)=t
\end{aligned} \tag{30}$$

Since $r^\star_1(t)$ e $r^\star_2(t)$ are the floor prices then $\rho^{(2)}_1(t)=\rho^{(2)}_2(t)=0$. From the Deficit Formula (17) we obtain that:

$$\rho^{(1)}_1(t)=\frac{(2-t)}{160(t^2-16)}[2220.2t^4-799.6t^3+2742.4t^2-1824t-3200]+F_1(t),$$

$$\rho^{(1)}_2(t)=\frac{1}{160(t^2-16)}[(2396.6t^5-2932t^4-17320.8t^3+4259.2t^2+39808t-25600)]+F_2(t).$$

$\rho^{(1)*}_1(t)$ is strictly positive for each $F_1(t)\geq 0$, whereas, for each $F_2(t)$ nonnegative, $\rho^{(1)*}_2(t)$ is positive in the interval $[0,\bar{t}]$ $\bar{t}=0.827636$. In such an interval the solution of the problem is given by (30).

The deficits can be reduced only if $F_1(t)$ and $F_2(t)$ decrease, even if we cannot obtain the financial equilibrium.

In the interval $[\bar{t},1]$ it is possible that the financial equilibrium can be reached obtaining also a surplus. A suggestion in this sense is given by the Evaluation Index, which gives complete information on the behavior of the economy and of the contagion.

Actually we have

$$\theta(t)=\frac{t}{2};\quad i(t)=t;\quad c(t)=0.125;$$

$$\sum_{i=1}^{2}l_i(t)=5t;\quad \sum_{i=1}^{2}s_i(t)=3t+5;$$

$$\sum_{i=1}^{2} \tilde{s}_i(t) = \frac{3t+5}{1+t}; \quad \sum_{j=1}^{2} \tilde{F}_j(t) = \frac{F_1(t)+F_2(t)}{(1+t)(1-\frac{t}{2})}.$$

Thus, the Evaluation Index is:

$$E(t) = \frac{(1-c(t)) \sum_{i=1}^{2} l_i(t)}{\sum_{i=1}^{2} \tilde{s}_i(t) + \sum_{j=1}^{2} \tilde{F}_j(t)} = \frac{(4.375t)(1+t)(2-t)}{(2-t)(5+3t)+2(F_1(t)+F_2(t))}. \tag{31}$$

In the interval $\left[\frac{\sqrt{5721}-11}{70}, 1\right]$ (where $3t^2+11t-40>0$), the economy has a positive average evaluation, if the condition

$$F_1(t)+F_2(t) \leq \frac{2-t}{16}(35t^2+11t-40)$$

is verified.

This result has been obtained considering the average $\theta(t)$ and $i(t)$, however it seems convenient and desirable that the data $\tau_{ij}(t)$ and $h_j(t)$ are not too different.

In our model, which takes into account the insolvencies, the Evaluation index (31) is less than the one obtained in [1], in which the insolvencies are not considered. Then, as expected, in the presence of insolvencies the economy gets worse. If we do not take into account the insolvencies, the Evaluation index (31) coincides with the one in [1].

5 Conclusions

In the chapter, we assessed the influence of the insolvencies on the financial model and on the financial contagion. Our results show that the risk of contagion increases with the presence of insolvencies, with decreasing investments and increasing expenditure. Then, our conclusion is that it is necessary to focus on these three factors, in order to improve the financial equilibrium. The suggestion to the governments, that follows from our analysis, is to reduce the insolvencies, deferring in time the payment of the liabilities, and supporting the sectors.

References

1. A. Barbagallo, P. Daniele, S. Giuffrè, A. Maugeri, Variational approach for a general financial equilibrium problem: The Deficit Formula, the Balance Law and the Liability Formula. A path to the economy recovery. Eur. J. Oper. Res. **237**(1), 231–244 (2014)
2. J.M. Borwein, V. Jeyakumar, A.S. Lewis, M. Wolkowicz, Constrained approximation via convex programming, University of Waterloo (1988). Preprint
3. R.I. Bot, E.R. Csetnek, A. Moldovan, Revisiting some duality theorems via the quasirelative interior in convex optimization. J. Optim. Theory Appl. **139**(1), 67–84 (2008)
4. V. Caruso, P. Daniele, A network model for minimizing the total organ transplant costs. Eur. J. Oper. Res. **266**(2), 652–662 (2018)
5. G. Colajanni, P. Daniele, S. Giuffrè, A. Nagurney, Cybersecurity investments with nonlinear budget constraints and conservation laws: variational equilibrium, marginal expected utilities, and Lagrange multipliers. Int. Trans. Oper. Res. **25**(5), 1415–1714 (2018)
6. G. Colajanni, P. Daniele, S. Giuffrè, A. Maugeri, Nonlinear duality in Banach spaces and applications to finance and elasticity, in *Applications of Nonlinear Analysis*, ed. by Th. M. Rassias. Springer Optimization and Its Applications, vol. 134 (Springer, New York, 2018), pp. 101–139
7. P. Daniele, *Dynamic Networks and Evolutionary Variational Inequalities* (Edward Elgar Publishing, Cheltenham, 2006)
8. P. Daniele, Evolutionary variational inequalities and applications to complex dynamic multi-level models. Transport. Res. Part E **46**, 855–880 (2010)
9. P. Daniele, S. Giuffrè, General infinite dimensional duality and applications to evolutionary network equilibrium problems. Optim. Lett. **1**, 227–243 (2007)
10. P. Daniele, S. Giuffrè, Random variational inequalities and the random traffic equilibrium problem. J. Optim. Theory Appl. **167**(1), 363–381 (2015)
11. P. Daniele, S. Giuffrè, S. Pia, Competitive financial equilibrium problems with policy interventions. J. Ind. Manag. Optim. **1**(1), 39–52 (2005)
12. P. Daniele, S. Giuffrè, G. Idone, A. Maugeri, Infinite dimensional duality and applications. Math. Ann. **339**, 221–239 (2007)
13. P. Daniele, S. Giuffrè, A. Maugeri, Remarks on general infinite dimensional duality with cone and equality constraints. Commun. Appl. Anal. **13**(4), 567–578 (2009)
14. P. Daniele, S. Giuffrè, M. Lorino, A. Maugeri, C. Mirabella, Functional inequalities and analysis of contagion in the financial networks, in *Handbook of Functional Equations - Functional Inequalities*, ed. by Th.M. Rassias. Optimization and Its Applications, vol. 95 (Springer, New York, 2014), pp. 129–146
15. P. Daniele, S. Giuffrè, A. Maugeri, F. Raciti, Duality theory and applications to unilateral problems. J. Optim. Theory Appl. **162**(3), 718–734 (2014)
16. P. Daniele, S. Giuffrè, M. Lorino, Functional inequalities, regularity and computation of the deficit and surplus variables in the financial equilibrium problem. J. Glob. Optim. **65**, 575–596 (2016)
17. P. Daniele, M. Lorino, C. Mirabella, The financial equilibrium problem with a Markowitz-type memory term and adaptive, constraints. J. Optim. Theory Appl. **171**, 276–296 (2016)
18. K. Forbes, The “Big C": Identifying contagion. NBER Working Paper No. 18465, 2012
19. S. Giuffrè, Strong solvability of boundary value contact problems. Appl. Math. Optim. **51**(3), 361–372 (2005)
20. S. Giuffrè, A. Maugeri, New results on infinite dimensional duality in elastic-plastic torsion. Filomat **26**(5), 1029–1036 (2012)
21. S. Giuffrè, A. Maugeri, Lagrange multipliers in elastic-plastic torsion, in *AIP Conference Proceedings*, Rodi, September 2013, vol. 1558, pp. 1801–1804
22. S. Giuffrè, A. Maugeri, A measure-type Lagrange multiplier for the elastic-plastic torsion. Nonlinear Anal. **102**, 23–29 (2014)

23. S. Giuffrè, S. Pia, Weighted traffic equilibrium problem in non pivot Hilbert spaces with long term memory, in *AIP Conference Proceedings*, Rodi, September 2010, vol. 1281, 2010, pp. 282–285
24. S. Giuffrè, G. Idone, A. Maugeri, Duality theory and optimality conditions for generalized complementary problems. Nonlinear Anal. **63**, e1655–e1664 (2005)
25. S. Giuffrè, A. Maugeri, D. Puglisi, Lagrange multipliers in elastic-plastic torsion problem for nonlinear monotone operators. J. Differ. Equ. **259**(3), 817–837 (2015)
26. S. Giuffrè, A. Pratelli, D. Puglisi, Radial solutions and free boundary of the elastic-plastic torsion problem. J. Convex Anal. **25**(2), 529–543 (2018)
27. R.B. Holmes, *Geometric Functional Analysis* (Springer, Berlin, 1975)
28. G. Idone, A. Maugeri, Generalized constraints qualification and infinite dimensional duality. Taiwan. J. Math. **13**, 1711–1722 (2009)
29. V. Jeyakumar, H. Wolkowicz, Generalizations of slater constraint qualification for infinite convex programs. Math. Program. **57**, 85–101 (1992)
30. A. Maugeri, D. Puglisi, On nonlinear strong duality and the infinite dimensional Lagrange multiplier rule. J. Nonlinear Convex Anal. **18**(3), 369–378 (2017)
31. H.M. Markowitz, Portfolio selection. J. Finan. **7**, 77–91 (1952)
32. H.M. Markowitz, *Portfolio Selection: Efficient Diversification of Investments* (Wiley, New York, 1959)
33. A. Maugeri, F. Raciti, Remarks on infinite dimensional duality. J. Global Optim. **46**, 581–588 (2010)
34. A. Maugeri, L. Scrimali, New approach to solve convex infinite-dimensional bilevel problems: application to the pollution emission price problem. J. Optim. Theory Appl. **169**(2), 370–387 (2016)
35. A. Nagurney, J. Dong, M. Hughes, Formulation and computation of general financial equilibrium. Optimization **26**, 339–354 (1992)
36. R.T. Rockafellar, Conjugate duality and optimization, in *Conference board of the Mathematical Science Regional Conference Series in Applied Mathematics*, vol. 16 (Society for Industrial and Applied Mathematics, Philadelphia, 1974)

Fixed Point Theorems for a System of Mappings in Generalized b-Metric Spaces

Stefan Czerwik and Themistocles M. Rassias

Abstract In this paper we present some results of both global and local type on the existence of fixed points for a system of mappings in generalized b-metric spaces. In particular, we obtain a strict generalization of the Banach contraction principle for mappings in ordinary complete metric spaces.

Mathematics Subject Classification (2010) 47H10, 54E99, 46-99

1 Introduction

We shall utilize the idea of a generalized b-metric space (gbms shortly). For details see [1]. Assume that X is a nonempty set. A function $d : X \times X \to [0, \infty]$ is said to be a generalized b-metric on X, iff for $x, y, z \in X$ it holds:

1. $d(x, y) = 0$ iff $x = y$,
2. $d(x, y) = d(y, x)$,
3. $d(x, y) \leqslant s[d(x, z) + d(z, y)]$, where $s \geqslant 1$ is a fixed real constant.

A pair (X, d) is called a generalized b-metric space with a generalized b-metric d.

For any $f : X \to X$, by f^n we denote the n-th iterate of f, defined by

$$f^0(x) = x \text{ for } x \in X, \quad f^{n+1} = f(f^n), \quad n \in \mathbb{N}_0.$$

S. Czerwik (✉)
Institute of Mathematics, Silesian University of Technology, Gliwice, Poland

T. M. Rassias
Department of Mathematics, Zografou Campus, National Technical University of Athens, Athens, Greece
e-mail: trassias@math.ntua.gr

T. M. Rassias, P. M. Pardalos (eds.), *Mathematical Analysis and Applications*, Springer Optimization and Its Applications 154,
https://doi.org/10.1007/978-3-030-31339-5_3

By $\mathbb{N}_0, \mathbb{N}, \mathbb{R}, \mathbb{R}_+$ as usual, we denote the set of all nonnegative integers, the set of all natural numbers, the set of all real numbers, or the set of all nonnegative real numbers, respectively.

One can find more about b-metric spaces, for example, in [4, 5] (see also [13, 14] for related topics).

Let's note also the following (see [1, 6]) theorem, which will be used later on.

Theorem 1 *Let (X, d) be a complete b-metric space and $T : X \to X$ satisfy*

$$d(T(x), T(y)) \leqslant \varphi[d(x, y)], \quad x, y \in X,$$

where $\varphi : R_+ \to R_+$ is a nondecreasing function such that

$$\lim_{n\to\infty} \varphi^n(t) = 0$$

for each $t > 0$. Then T has exactly one fixed point $u \in X$, and

$$\lim_{n\to\infty} d[T^n(x), u] = 0$$

for each $x \in X$.

2 Main Results

2.1 Basic Theorem

In this section we prove the result for a system of mappings in generalized b-metric spaces (we apply the ideas of [7]).

Theorem 2 *Let $(X_i, d_i), i = 1, \ldots, n$ be complete generalized b-metric spaces. Assume that there exist nonnegative real numbers $a_{i,k}, \quad i, k = 1, \ldots, n$ such that the mappings $T_i : X_1 \times \ldots \times X_n \to X_i, \quad i = 1, \ldots, n$ satisfy*

$$d_i(T_i(x_1, \ldots, x_n), T_i(z_1, \ldots, z_n)) \leqslant \sum_{k=1}^{n} a_{i,k} d_k(x_k, z_k), \tag{1}$$

for $x_k, z_k \in X_k, d_k(x_k, z_k) < \infty, k = 1, \ldots, n$.

Moreover, there exists a system of positive real numbers $r_i, i = 1, \ldots, n$ satisfying the inequalities

$$\sum_{i=1}^{n} r_i a_{i,k} < r_k, \quad k = 1, \ldots, n. \tag{2}$$

For any fixed $x^0 \in X = X_1 \times \ldots \times X_n$ *consider the sequence of successive approximations*

$$x_i^{m+1} = T_i(x_1^m, \ldots, x_n^m), \quad m = 0, 1, \ldots, \quad i = 1, \ldots, n. \tag{3}$$

Then either

(A) for any non-negative integer v *there exists an* $i \in \{1, \ldots, n\}$ *such that*

$$d_i(x_i^v, T_i(x_1^v, \ldots, x_n^v)) = \infty,$$

or

(B) there exists a non-negative integer v *such that for every* $i = 1, \ldots, n$,

$$d_i(x_i^v, x_i^{v+1}) < \infty. \tag{4}$$

In (B) the sequence $x^m = (x_1^m, \ldots, x_n^m)$ *given by (3) converges to a fixed point* $u = (u_1, \ldots, u_n) \in X$ *of* $T = (T_1, \ldots, T_n)$*, i.e.*

$$T_i(u_1, \ldots, u_n) = u_i, i = 1, \ldots, n.$$

In the space $K = K_1 \times \ldots \times K_n$ *where*

$$K_i = \{x_i \in X_i : d_i(x_i^v, x_i) < \infty\}, \quad i = 1, \ldots, n, \tag{5}$$

the point u *is the unique fixed point of* T.

Proof In (B) by (4) we get for $i = 1, \ldots, n$,

$$d_i(T_i(x^v), T_i(x^{v+1})) \leqslant \sum_{k=1}^{n} a_{i,k} d_k(x_k^v, x_k^{v+1}) < \infty,$$

and consequently by induction $d_i(x_i^{v+l}, x_i^{v+l+1}) < \infty$ for all $l \in \mathbb{N}_0, \quad i = 1, \ldots, n$.

Consider the number

$$\alpha = \max_k \left\{ \frac{1}{r_k} \sum_{i=1}^{n} r_i a_{i,k} \right\}.$$

Clearly

$$0 \leqslant \alpha < 1 \tag{6}$$

and

$$\sum_{i=1}^{n} r_i a_{i,k} \leqslant \alpha r_k, \quad k = 1, \ldots, n. \tag{7}$$

Now we verify that $T : K \to K$. For if $x \in K$,

$$d_k(x_k^v, T_k(x)) \leqslant s_k[d_k(x_k^v, T_k(x^v)) + d_k(T_k(x^v), T_k(x))]$$
$$\leqslant s_k[d_k(x_k^v, x_k^{v+1}) + \sum_{l=1}^{n} a_{k,l} d_l(x_l^v, x_l)] < \infty.$$

Define

$$D(x, y) := \sum_{i=1}^{n} r_i d_i(x_i, y_i), \quad x, y \in K. \tag{8}$$

We can show that D is a b-metric in K with $s = \max_i(s_i) > 0$. It is also easy to prove that (K, D) is a complete b-metric space (see also [7]).

Now we prove that T is a contraction mapping in K. In fact, for $x, z \in K$ we obtain

$$D(T(x), T(z)) = \sum_{i=1}^{n} r_i d_i(T_i(x), T_i(z))$$
$$\leqslant \sum_{i=1}^{n} r_i \Big[\sum_{k=1}^{n} a_{i,k} d_k(x_k, z_k)\Big]$$
$$\leqslant \sum_{k=1}^{n} \Big(\sum_{i=1}^{n} r_i a_{i,k}\Big) d_k(x_k, z_k)$$
$$\leqslant \sum_{k=1}^{n} \alpha r_k d_k(x_k, z_k) = \alpha D(x, z),$$

which means that

$$D(T(x), T(z)) \leqslant \alpha D(r, z), \quad x, z \in K.$$

Since by (6), $0 \leqslant \alpha < 1$, T is a strict contraction in K.

Eventually, in view of Theorem 1 for $\varphi(t) = \alpha t, \quad t \geqslant 0$, T has in K exactly one fixed point u which is the limit of successive approximations with any initial element from K (and hence T has a fixed point in X). This concludes the proof.

We can also prove the following.

Corollary 1 *Let the assumptions of Theorem 2 be satisfied. If, moreover,*

$$s\alpha < 1, \tag{9}$$

then

$$D(y,u) \leqslant \frac{s}{1-s\alpha} D(y,T(y)), \quad y \in B. \tag{10}$$

Proof For (B) and $y \in K$, $(u = T(u))$ one has

$$\begin{aligned} D(y,u) &\leqslant s[D(y,T(y)) + D(T(y),T(u))] \\ &\leqslant s[D(y,T(y)) + \alpha D(y,u)] \end{aligned}$$

whence

$$D(y,u) \leqslant \frac{s}{1-s\alpha} D(y,T(y)), \quad y \in K.$$

Corollary 2 *Let the assumptions of Corollary 1 be satisfied. If, moreover,* $d_i, i = 1,\ldots,n$, *are continuous (with respect to one variable), then*

$$D(T^m(y),u) \leqslant \frac{s\alpha^m}{1-s\alpha} D(y,T(y)), \quad y \in B. \tag{11}$$

Proof From Theorem 1, for $\varphi(t) = \alpha t, \quad t \geqslant 0, \quad z \in K$ and continuity of D, one gets

$$D(T^m(z),u) \leqslant \sum_{k=0}^{\infty} s^{k+1} \varphi^{m+k}[D(z,T(z))]$$

$$= \sum_{k=0}^{\infty} s^{k+1} \alpha^{m+k} D(z,T(z)) = \frac{s\alpha^m}{1-s\alpha} D(z,T(z)),$$

i.e., the inequality (11), which completes the proof.

Remark 1 A function D (b-metric) may not be continuous (cf. [15]).

Remark 2 From Theorem 2 we get theorem of Diaz, Margolis [8], Luxemburg [10], Banach [2], Matkowski [11], Czerwik [4, 5].

2.2 Local Theorems

First of all we present the local result for a system of mappings in generalized b-metric spaces. Namely, we have

Theorem 3 *Suppose that* (X_i, d_i), $i = 1,\ldots,m$ *are complete generalized b-metric spaces. Assume that there exist non-negative real numbers* $a_{i,k}, \quad i,k = 1,\ldots,n$

and $c > 0$ such that the mappings $T_i : X_1 \times \ldots \times X_n \to X_i, \quad i = 1, \ldots, n$ fulfill the inequalities

$$d_i(T_i(x_1, \ldots, x_n), T_i(z_1, \ldots, z_n)) \leqslant \sum_{k=1}^{n} a_{i,k} d_k(x_k, z_k) \tag{12}$$

for $d_k(x_k, z_k) < c$, and $x_k, z_k \in X_k, \quad k = 1, \ldots, n$. Additionally, let the characteristic roots $d_i, i = 1, \ldots, n$ of the matrix $[a_{i,k}]$ satisfy

$$\alpha = \max\{|d_i| : i = 1, \ldots, n\} < 1. \tag{13}$$

Let $x^0 \in X = X_1 \times \ldots X_n$ be arbitrarily fixed. Consider the sequence of successive approximations (3). Then the following alternative holds: either

(C) for any $v \in \mathbb{N}_0$ there exists an $i \in \{1, \ldots, n\} = A$ such that

$$d_i(x_i^v, T_i(x_i^v, \ldots, x_n^v)) \geqslant c, \tag{14}$$

or

(D) there exists a non-negative integer v such that for every $i = 1, \ldots, n$,

$$d_i(x_i^v, x_i^{v+1}) < c. \tag{15}$$

In the case (D), if, moreover, the numbers in (15) are sufficiently small and (9) holds true, then $T = (T_1, \ldots, T_n)$ has a fixed point $u \in X$.

Proof According to the result of Perron-Frobenius ([9, pp. 354–355]) the number given in (13) is the characteristic root of the matrix $[a_{i,k}]$ with the eigenvector $(r_1, \ldots, r_n), r_i > 0, i = 1, \ldots, n$, i.e.

$$\sum_{k=1}^{n} a_{i,k} r_k = \alpha r_i, \quad i = 1, \ldots, n. \tag{16}$$

Since the equations (16) are homogeneous, also $ar_1, \ldots, ar_n, \quad a > 0$ is a solution of (16). So we can assume that

$$r_i < \frac{c}{2s}, \quad i = 1, \ldots, n. \tag{17}$$

By the assumptions, suppose that

$$d_k(T_k(x^v), x_k^v) \leqslant (1 - \alpha s_k)\frac{r_k}{s_k}, \quad k = 1, \ldots, n. \tag{18}$$

Define

$$B_k := \{z_k \in X_k : d_k(z_k, x_k^v) \leqslant r_k\}, \quad k = 1, \ldots, n. \tag{19}$$

Let $B = B_1 \times \ldots \times B_k$. For $z \in B$, by (12), (17), (19), (16) and (18), we get $(k = 1, \ldots, n)$:

$$\begin{aligned} d_k(T_k(z), x_k^v) &\leqslant s_k[d_k(T_k(z), T_k(x^v)) + d_k(T_k(x^v), x_k^v)] \\ &\leqslant s_k\Big[\sum_{l=1}^{n} a_{k,l} d_l(z_l, x_l^v) + d_k(T_k(x^v), x_k^v)\Big] \\ &\leqslant s_k\Big[\alpha r_k + \frac{r_k}{s_k} - \alpha r_k\Big] = r_k. \end{aligned}$$

Thus $T_k(z) \in B_k, \quad k = 1, \ldots, n$ which means that $T(z) \in B$ for $z \in B$, i.e. $T(B) \subset B$.

Note that also the matrix $[a_{i,k}]^T$ has exactly the same characteristic roots as $[a_{i,k}]$, so by the Perron-Frobenius theorem there exists a system of positive numbers $\xi_i, \quad i = 1, \ldots, n$, which is the solution of the system of equations (inequalities)

$$\sum_{i=1}^{n} \xi_i a_{i,k} \leqslant \alpha \xi_k, \quad k = 1, \ldots, n. \tag{20}$$

Define

$$D(x, y) := \sum_{i=1}^{n} \xi_i d_i(x_i, y_i), \quad x, y \in B \tag{21}$$

and consider the space (B, D). Then (B, D) is a b-metric space with $s = \max_i(s_i)$. For any $x, y \in B$, we have for $k = 1, \ldots, n$ (see (17))

$$\begin{aligned} d_k(x_k, y_k) &\leqslant s_k[d_k(x_k, x_k^v) + d_k(x_k^v, y_k)] \\ &\leqslant s_k[r_k + r_k] < 2s_k \frac{c}{2s} \leqslant c. \end{aligned}$$

Therefore, by (12), (21), and (20), for $x, y \in B$, we can verify that (see the proof of Theorem 2):

$$D(T(x), T(y)) \leqslant \alpha D(x, y) \text{ for } x, y \in B. \tag{22}$$

Consequently for any $x \in B$, $\{T^n(x)\} \subset B$ that is a Cauchy sequence. Indeed, we have

$$D(T(x), T^2(x)) \leqslant \alpha D(x, T(x)),$$

and by the induction principle

$$D(T^m(x), T^{m+1}(x)) \leqslant \alpha^m D(x, T(x)), \quad m \geqslant 1. \tag{23}$$

Consequently, for $m, l \in \mathbb{N}_0$, by (23) and (9) one gets

$$\begin{aligned} D(T^m(x), T^{m+l}(x)) &\leqslant s D(T^m(x), T^{m+1}(x)) + \ldots + s^l D(T^{m+l-1}(x), T^{m+l}(x)) \\ &\leqslant s\alpha^m D(x, T(x)) + \ldots + s^l \alpha^{m+l-1} D(x, T(x)) \\ &\leqslant s\alpha^m [1 + (s\alpha) + \ldots + (s\alpha)^{l-1}] D(x, T(x)) \\ &\leqslant \frac{s\alpha^m}{1 - s\alpha} D(x, T(x)). \end{aligned}$$

Eventually, for $x \in B$ and $m, l \in \mathbb{N}_0$,

$$D(T^m(x), T^{m+l}(x)) \leqslant \frac{s\alpha^m}{1 - s\alpha} D(x, T(x)). \tag{24}$$

Hence, it follows that $\{T^n(x)\}$ is a Cauchy sequence for $x \in B$.

Since $X_k, k = 1, \ldots, n$ are complete, there exist $u_k \in X_k, \quad k = 1, \ldots, n$ such that

$$T_k^m(x) \to u_k, \quad k = 1, \ldots, n \text{ as } m \to \infty.$$

Therefore for $i = 1, \ldots, n$, m sufficiently large and $\epsilon > 0$

$$\begin{aligned} d_i(T_i(u), u_i) &\leqslant s_i [d_i(T_i(u), T_i(x^m)) + d_i(T_i(x^m), u_i)] \\ &\leqslant s_i \Big[\sum_{k=1}^{n} a_{i,k} d_k(u_k, x_k^m) + d_i(x_i^{m+1}, u_i) \Big] \\ &\leqslant s_i \Big[\sum_{k=1}^{n} a_{i,k} \epsilon + \epsilon \Big] \\ &\leqslant s\epsilon \Big[\sum_{k=1}^{n} a_{i,k} + 1 \Big] \to 0 \text{ as } \epsilon \to 0. \end{aligned}$$

Therefore we get for $i = 1, \ldots, n$ the following:

$$d_i(T_i(u), u_i) = 0 \Rightarrow T_i(u) = u_i \Rightarrow T(u) = u,$$

which means that $u \in X$ is a fixed point of T, and this concludes the proof of the theorem.

Remark 3 The point $u = (u_1, \ldots, u_n)$ satisfies the condition

$$d_k(u_k, x_k^v) \leqslant s_k r_k, \quad k = 1, \ldots, n. \tag{25}$$

In fact, we have for $m \in \mathbb{N}$ sufficiently large, $\epsilon > 0$ and $x \in B$:

$$\begin{aligned} d_k(u_k, x_k^v) &\leqslant s_k[d_k(u_k, T_k^m(\alpha) + d_k(T_k^m(\alpha), x_k^v)] \\ &\leqslant s_k[\epsilon + r_k] \to s_k r_k \text{ as } \epsilon \to 0, \end{aligned}$$

whence we get (25).

Remark 4 It is an open question whether this result is true without the assumption that T does not displace the center of the ball B too far.

Remark 5 In B, the mapping T may have at most one fixed point. Indeed, assume, on the contrary, that $u, w \in B$, $u \neq w$, and $T(u) = u, T(w) = w$. Note that (see (17))

$$d_k(u_k, w_k) < c, \quad k = 1, \ldots, n,$$

and hence by (12) and (16), we obtain for $k = 1, \ldots, n$ and

$$\xi = \max_k \Big(\frac{1}{r_k} d_k(u_k, w_k)\Big),$$

$$\begin{aligned} d_k(u_k, w_k) = d_k(T_k(u), T_k(w)) &\leqslant \sum_{l=1}^{n} a_{k,l} d_l(u_l, w_l) \\ \leqslant \sum_{l=1}^{n} a_{k,l} r_l \cdot \frac{1}{r_l} d_l(u_l, w_l) &\leqslant \xi \sum_{l=1}^{n} a_{k,l} r_l \leqslant \xi \alpha r_k. \end{aligned}$$

Similarly,

$$d_k(u_k, w_k) = \sum_{l=1}^{n} a_{k,l} d_l(u_l, w_l) \leqslant \sum_{l=1}^{n} a_{k,l} \xi \alpha r_l \leqslant \xi \alpha^2 r_k.$$

By induction one gets

$$d_k(u_k, w_k) \leqslant \xi \alpha^m r_k, \quad m \in \mathbb{N}, \quad k = 1, \ldots, n.$$

Thus, in view of (13), one has

$$d_k(u_k, w_k) = 0, \quad k = 1, \ldots, n \Rightarrow u = w,$$

and the proof is completed.

Remark 6 For some conditions equivalent to the condition (13) cf. [3, 12].

In the sequel we prove a similar result for systems of mappings in b-metric spaces.

Theorem 4 *Let $(X_i, d_i), i = 1, \ldots, n$ be complete b-metric spaces with $s_i \geqslant 1$, $i = 1, \ldots, n$. Suppose that $a_{i,k} \geqslant 0$, $i, k = 1, \ldots, n$ and the characteristic roots $\lambda_i, i = 1, \ldots, n$ of the matrix $[a_{i,k}]$ satisfy (13). Consider any*

$$x^0 \in X_1 \times \ldots \times X_n = X$$

and set

$$B_k := \{z_k \in X_k : d_k(z_k, x_k^0) \leqslant r_k\}, \quad k = 1, \ldots, n,$$

and $B = B_1 \times \ldots \times B_n$, where $r_k > 0, k = 1, \ldots, n$ satisfy (16).
Assume that $T_k : B \to X_k, \quad k = 1, \ldots, n$ satisfy

$$d_k(T_k(x), T_k(y)) \leqslant \sum_{l=1}^{n} a_{k,l} d_l(x_l, y_l), \quad x, y \in B. \tag{26}$$

If, moreover,

$$d_k(T_k(x^0), x_k^0) \leqslant (1 - \alpha s_k)\frac{r_k}{s_k}, \quad k = 1, \ldots, n, \tag{27}$$

and

$$\alpha s < 1, \quad s = \max_i(s_i), \tag{28}$$

then $T = (T_1, \ldots, T_n)$ has a fixed point in X.

Proof We verify that $T(B) \subset B$. By (26), (16) and (27), it can be carried through exactly as in the proof of Theorem 3.
Next, consider numbers $\xi_k > 0$, $k = 1, \ldots, n$, fulfilling inequalities (20) and the number α satisfying inequality (13).
Consider the b-metric defined by

$$D(x, y) := \sum_{k=1}^{n} \xi_k d_k(x_k, y_k), \quad x, y \in B. \tag{29}$$

Then (B, D) is a b-metric space with $s = \max_i(s_i)$. Also $T : B \to B$ is a contraction: for $x, y \in B$, we have by (29), (26), and (20),

$$D(T(x), T(y)) \leqslant \alpha D(x, y), \quad x, y \in B.$$

Therefore, for any $x \in B$, $\{T^n(x)\}$ is a Cauchy sequence (see (28)). The rest of the proof follows the same arguments as in the proof of Theorem 3.

References

1. H. Aydi, S. Czerwik, Fixed point theorems in generalized b-metric spaces, in *Modern Discrete Mathematics and Analysis: With Applications in Cryptography, Information Systems and Modeling*, ed. by N. Daras, Th. Rassias (Springer, Cham, 2018)
2. S. Banach, Sur les operations dans les ensembles abstrait et leur application aux equations integrals. Fundam. Math. **3**, 133–181 (1922)
3. S. Czerwik, Some inequalities, characteristic roots of a matrix and Edelstein's fixed point theorem. Bull Acad. Polon. Sci. Ser. Sci. Math. Astronom. Phys. **24**, 827–828 (1976)
4. S. Czerwik, Contraction Mappings in b-metric spaces, Acta Math. et Informatica Universitatis Ostraviensis **1**, 5–11 (1993)
5. S. Czerwik, Nonlinear set-valued contraction mappings in B-metric spaces. Atti Sem. Mat. Fis. Univ. Modena **XLVI**, 263–276 (1998)
6. S. Czerwik, *Functional Equations and Inequalities in Several Variables* (World Scientific, Singapore, 2002), pp. 1–410
7. S. Czerwik, K. Król, Fixed point theorems for a system of transformations in generalized metric spaces. Asian-Eur. J. Math. **8**(4), 1–8 (2015)
8. J.B. Diaz, B. Margolis, A fixed point theorem of the alternative, for contractions on a generalized complete metric space. Bull. Am. Math. Soc. **74**, 305–309 (1968)
9. F.R. Gantmacher, *Theory of Matrix* (Nauks, Moscow, 1966) (in Russian)
10. W.A.J. Luxemburg, On the convergence of successive approximations in the theory of ordinary differential equations. Indag. Math. **20**(5), 540–546 (1958)
11. J. Matkowski, Some inequalities and a generalization of Banach's principle, Bull Acad. Polon. Sci. Ser. Sci. Math. Astronom. Phys. **21**, 323–324 (1973)
12. J. Matkowski, Integrable solutions of functional equations. Dissertationes Math. **127**, 5–63 (1975)
13. G.V. Milovanović, M. Th. Rassias (eds.), *Analytic Number Theory*. Approximation Theory and Special Functions (Springer, New York, 2014)
14. J.F. Nash Jr., M. Th. Rassias, *Open Problems in Mathematics* (Springer, New York, 2016)
15. J.R. Roshan, N. Shobkolaei, S. Sedghi, M. Abbas, Common fixed point of four maps in b-metric spaces. Hacettepe J. Math. Stat. **43**(4), 613–624 (2014)

Inequalities and Approximations for the Finite Hilbert Transform: A Survey of Recent Results

Silvestru Sever Dragomir

Abstract In this paper we survey some recent results due to the author concerning various inequalities and approximations for the finite Hilbert transform of a function belonging to several classes of functions, such as: Lipschitzian, monotonic, convex or with the derivative of bounded variation or absolutely continuous. More accurate estimates in the case that the higher order derivatives are absolutely continuous are also provided. Some quadrature rules with error bounds are derived. They can be used in the numerical integration of the finite Hilbert transform and, due to the explicit form of the error bounds, enable the user to predict a priori the accuracy.

1 Introduction

Let $\Omega = (-1, 1)$ where $1 \leq p < \infty$, the usual $\mathfrak{L}^p$-space with respect to the Lebesgue measure λ restricted to the open interval Ω will be denoted by $\mathfrak{L}^p(\Omega)$.

We define a linear operator T (see [24]) from the vector space $\mathfrak{L}^1(\Omega)$ into the vector space of all λ-measurable functions on Ω as follows. Let $f \in \mathfrak{L}^1(\Omega)$. The Cauchy principal value

$$\frac{1}{\pi} PV \int_{-1}^{1} \frac{f(\tau)}{\tau - t} d\tau = \lim_{\varepsilon \to 0+} \left[\int_{-1}^{t-\varepsilon} + \int_{t+\varepsilon}^{1} \right] \frac{f(\tau)}{\pi(\tau - t)} d\tau \tag{1}$$

exists for λ-almost every $t \in \Omega$.

We denote the left-hand side of (1) by $(Tf)(t)$ for each $t \in \Omega$ for which $(Tf)(t)$ exists. The so-defined function Tf, which we call the *finite Hilbert transform* of f, is defined λ-almost everywhere on Ω and is λ-measurable; (see, for example, [2,

S. S. Dragomir (✉)
Mathematics, College of Engineering & Science, Victoria University, Melbourne, VIC, Australia

DST-NRF Centre of Excellence in the Mathematical and Statistical Sciences, School of Computer Science & Applied Mathematics, University of the Witwatersrand, Johannesburg, South Africa
e-mail: sever.dragomir@vu.edu.au

T. M. Rassias, P. M. Pardalos (eds.), *Mathematical Analysis and Applications*, Springer Optimization and Its Applications 154,
https://doi.org/10.1007/978-3-030-31339-5_4

Theorem 8.1.5]). The resulting linear operator T will be called the *finite Hilbert transform operator* or Cauchy kernel operator.

It is known that $\mathfrak{L}^1(\Omega)$ is not invariant under T, namely, $T\left(\mathfrak{L}^1(\Omega)\right) \not\subset \mathfrak{L}^1(\Omega)$ [19, Proof of Theorem 1 (b)].

The following basic results are well known and their proofs may be found in Propositions 8.1.9 and 8.2.1 of [2], respectively.

Theorem 1 (M. Riesz) *Let $1 < p < \infty$. Then $T(\mathfrak{L}^p(\Omega)) \subset \mathfrak{L}^p(\Omega)$ and the linear operator*

$$T_p : f \mapsto Tf,\ f \in \mathfrak{L}^p(\Omega)$$

on $\mathfrak{L}^p(\Omega)$ is continuous.

Theorem 2 (Parseval) *Let $1 < p < \infty$ and $q = \frac{p}{p-1}$. Then*

$$\int_{-1}^{1} (fTg + gTf)\, d\lambda = 0 \tag{2}$$

for every $f \in \mathfrak{L}^p(\Omega)$ and $g \in \mathfrak{L}^q(\Omega)$.

We introduce the following definition.

Definition 1 A function $f : \Omega \to \mathbb{C}$ is said to be *α-Hölder continuous* $(0 < \alpha \leq 1)$ in a subinterval Ω_0 of Ω if there exists a constant $c > 0$, dependent upon Ω_0, such that

$$|f(s) - f(t)| \leq c\,|s - t|^{\alpha},\quad s,\ t \in \Omega_0. \tag{3}$$

A function on Ω is said to be *locally α-Hölder continuous* if it is α-Hölder continuous in every compact subinterval of Ω. We denote by $H^{\alpha}_{loc}(\Omega)$ the space of all locally α-Hölder continuous functions on Ω.

The class of Hölder continuous functions on Ω is independent because the finite Hilbert transform of such a function exists everywhere on Ω (see [18, Section 3.2] or [23, Lemma II.1.1]).

This is in contrast to the λ-almost everywhere existence of the finite Hilbert transform of functions in $\mathfrak{L}^1(\Omega)$.

There are continuous functions $f \in \mathfrak{L}^1(\Omega)$ such that $(Tf)(t)$ does not exist at some point $t \in \Omega$. An example is given by the function f defined by (see [24])

$$f(t) = \begin{cases} 0 & \text{if } -1 < t \leq 0, \\ \frac{1}{\ln t - \ln 2} & \text{if } 0 < t < 1. \end{cases}$$

It readily follows that $(Tf)(0)$ does not exist.

In paper [24] it is proved amongst others the following result.

Theorem 3 (Okada-Elliot [24]) *The space $\mathfrak{L}^p(\Omega) \cap H^\alpha_{loc}(\Omega)$ is invariant under the finite Hilbert transform operator T and the restriction of T to that space is continuous whenever $1 < p < \infty$. This, however, is not true when $p = 1$.*

Allover this paper, we consider the finite Hilbert transform on the open interval (a, b) defined by

$$(Tf)(a, b; t) := \frac{1}{\pi} PV \int_a^b \frac{f(\tau)}{\tau - t} d\tau := \lim_{\varepsilon \to 0+} \left[\int_a^{t-\varepsilon} + \int_{t+\varepsilon}^b \right] \frac{f(\tau)}{\pi(\tau - t)} d\tau$$

for $t \in (a, b)$ and for various classes of functions f for which the above Cauchy principal value integral exists.

For several recent papers devoted to inequalities for the finite Hilbert transform (Tf), see [20–22, 25] and [26].

In this paper we survey some recent results due to the author concerning various inequalities and approximations for the finite Hilbert transform of a function belonging to several classes of functions, such as: Lipschitzian, monotonic, convex or with the derivative of bounded variation or absolutely continuous. More accurate estimates in the case that the higher order derivatives are absolutely continuous, are also provided. Some quadrature rules with error bounds are derived. They can be used in the numerical integration of the finite Hilbert transform and, due to the explicit form of the error bounds, enable the user to predict a priory the accuracy.

2 Inequalities for Some Classes of Functions

2.1 Some Estimates for α-Hölder Continuous Mappings

We say that the function $f : [a, b] \to \mathbb{R}$ is α-H-Hölder continuous on (a, b), if

$$|f(t) - f(s)| \le H |t - s|^\alpha \quad \text{for all } t,\ s \in (a, b), \tag{4}$$

where $\alpha \in (0, 1],\ H > 0$.

The following theorem holds.

Theorem 4 (Dragomir et al. [13]) *If $f : [a, b] \to \mathbb{R}$ is α-H-Hölder continuous on (a, b), then we have the estimate*

$$\begin{aligned} \left| (Tf)(a, b; t) - \frac{f(t)}{\pi} \ln\left(\frac{b - t}{t - a}\right) \right| &\le \frac{H}{\alpha\pi} \left[(t - a)^\alpha + (b - t)^\alpha \right] \\ &\le \frac{2^{1-\alpha}}{\alpha\pi} H (b - a)^\alpha, \end{aligned} \tag{5}$$

for all $t \in (a, b)$.

Proof As for the mapping $f : (a, b) \to \mathbb{R}$, $f(t) = 1$, $t \in (a, b)$, we have

$$\begin{aligned}
(Tf)(a, b; t) &= \frac{1}{\pi} PV \int_a^b \frac{1}{\tau - t} d\tau \\
&= \frac{1}{\pi} \lim_{\varepsilon \to 0+} \left[\int_a^{t-\varepsilon} \frac{1}{\tau - t} d\tau + \int_{t+\varepsilon}^b \frac{1}{\tau - t} d\tau \right] \\
&= \frac{1}{\pi} \lim_{\varepsilon \to 0+} \left[\ln |\tau - t| |_a^{t-\varepsilon} + \ln (\tau - t) |_{t+\varepsilon}^b \right] \\
&= \frac{1}{\pi} \lim_{\varepsilon \to 0+} [\ln \varepsilon - \ln (t - a) + \ln (b - t) - \ln \varepsilon] \\
&= \frac{1}{\pi} \ln \left(\frac{b - t}{t - a} \right), \quad t \in (a, b).
\end{aligned}$$

Then, obviously

$$\begin{aligned}
(Tf)(a, b; t) &= \frac{1}{\pi} PV \int_a^b \frac{f(\tau) - f(t) + f(t)}{\tau - t} d\tau \\
&= \frac{1}{\pi} PV \int_a^b \frac{f(\tau) - f(t)}{\tau - t} d\tau + \frac{f(t)}{\pi} PV \int_a^b \frac{1}{\tau - t} d\tau
\end{aligned}$$

from where we get the equality

$$(Tf)(a, b; t) - \frac{f(t)}{\pi} \ln \left(\frac{b - t}{t - a} \right) = \frac{1}{\pi} PV \int_a^b \frac{f(\tau) - f(t)}{\tau - t} d\tau \tag{6}$$

for all $t \in (a, b)$.

By (6) and by the modulus properties, we have

$$\begin{aligned}
&\left| (Tf)(a, b; t) - \frac{f(t)}{\pi} \ln \left(\frac{b - t}{t - a} \right) \right| \qquad (7) \\
&= \frac{1}{\pi} \left| PV \int_a^b \frac{f(\tau) - f(t)}{\tau - t} d\tau \right| \le \frac{1}{\pi} PV \int_a^b \left| \frac{f(\tau) - f(t)}{\tau - t} \right| d\tau \\
&\le \frac{1}{\pi} PV \int_a^b \frac{|\tau - t|^\alpha}{|\tau - t|} d\tau = \frac{1}{\pi} PV \int_a^b \frac{d\tau}{|\tau - t|^{1-\alpha}}.
\end{aligned}$$

However,

$$PV \int_a^b \frac{d\tau}{|\tau - t|^{1-\alpha}} = \lim_{\varepsilon \to 0+} \left[\int_a^{t-\varepsilon} \frac{d\tau}{(t - \tau)^{1-\alpha}} + \int_{t+\varepsilon}^b \frac{d\tau}{(\tau - t)^{1-\alpha}} \right]$$

$$= \lim_{\varepsilon \to 0+} \left[\frac{-(t-\tau)^{\alpha}}{\alpha} \Big|_{a}^{t-\varepsilon} + \frac{(\tau - t)^{\alpha}}{\alpha} \Big|_{t+\varepsilon}^{b} \right]$$

$$= \lim_{\varepsilon \to 0+} \left[\frac{(t-a)^{\alpha} - \varepsilon^{\alpha}}{\alpha} + \frac{(b-t)^{\alpha} - \varepsilon^{\alpha}}{\alpha} \right]$$

$$= \frac{(t-a)^{\alpha} + (b-t)^{\alpha}}{\alpha}.$$

Using (7), we get the first inequality in (5).

Consider the mapping $\phi(t) := (t-a)^{\alpha} + (b-t)^{\alpha}$, $\alpha \in (0,1]$, $t \in [a,b]$. Then, obviously

$$\phi'(t) = \frac{\alpha \left[(b-t)^{1-\alpha} - (t-a)^{1-\alpha} \right]}{[(t-a)(b-t)]^{1-\alpha}}.$$

We observe that $\phi'(t) = 0$ iff $t = \frac{a+b}{2}$ and $\phi'(t) > 0$ if $t \in \left(a, \frac{a+b}{2}\right)$ and $\phi'(t) < 0$ if $t \in \left(\frac{a+b}{2}, b\right)$, which shows that

$$\max_{t \in [a,b]} \phi(t) = \phi\left(\frac{a+b}{2}\right) = 2\left(\frac{b-a}{2}\right)^{\alpha} = 2^{1-\alpha}(b-a)^{\alpha}$$

and the last part of (5) is proved.

The following two corollaries are natural.

Corollary 1 *Let $f : [a,b] \to \mathbb{R}$ be an L-Lipschitzian mapping on $[a,b]$, i.e., f satisfies the condition*

$$|f(t) - f(s)| \le L|t-s| \text{ for all } t, s \in [a,b], \ (L > 0). \tag{8}$$

Then we have the inequality

$$\left| (Tf)(a,b;t) - \frac{f(t)}{\pi} \ln\left(\frac{b-t}{t-a}\right) \right| \le \frac{L(b-a)}{\pi} \tag{9}$$

for all $t \in (a,b)$.

Corollary 2 *Let $f : [a,b] \to \mathbb{R}$ be an absolutely continuous mapping on $[a,b]$. If $f' \in L_{\infty}[a,b]$, then, for all $t \in (a,b)$, we have*

$$\left| (Tf)(a,b;t) - \frac{f(t)}{\pi} \ln\left(\frac{b-t}{t-a}\right) \right| \le \frac{\|f'\|_{\infty}(b-a)}{\pi}, \tag{10}$$

where $\|f'\|_{\infty} = \text{essup}_{t \in (a,b)} |f'(t)| < \infty$.

The following result also holds.

Corollary 3 *Let $f : [a, b] \to \mathbb{R}$ be an absolutely continuous mapping on $[a, b]$ whose derivative $f' \in L_p[a, b]$, $p \in (1, \infty)$. Then for all $t \in (a, b)$ we have*

$$\left|(Tf)(a, b; t) - \frac{f(t)}{\pi} \ln\left(\frac{b-t}{t-a}\right)\right| \tag{11}$$

$$\leq \frac{p}{(p-1)\pi}\left[(t-a)^{\frac{p-1}{p}} + (b-t)^{\frac{p-1}{p}}\right]\left\|f'\right\|_p \leq \frac{2^{\frac{1}{p}} p}{(p-1)\pi}(b-a)^{\frac{p-1}{p}}\left\|f'\right\|_p,$$

where, as is usually the case, $\left\|f'\right\|_p := \left(\int_a^b \left|f'(t)\right|^p dt\right)^{\frac{1}{p}} < \infty$.

Proof As f is absolutely continuous on $[a, b]$, we can state that

$$\begin{aligned} |f(x) - f(y)| &= \left|\int_x^y f'(t)\, dt\right| \leq \left|\int_x^y \left|f'(t)\right| dt\right| \qquad (12)\\ &\leq \left|\int_x^y dt\right|^{\frac{1}{q}} \left|\int_x^y \left|f'(t)\right|^p dt\right|^{\frac{1}{p}} \quad \text{(by Hölder's inequality)}\\ &\leq |y-x|^{\frac{1}{q}} \left(\int_a^b \left|f'(t)\right|^p dt\right)^{\frac{1}{p}} = |y-x|^{\frac{1}{q}} \left\|f'\right\|_p, \end{aligned}$$

where $p > 1$, $\frac{1}{p} + \frac{1}{q} = 1$.

Thus, f is α-H-Hölder continuous with $\alpha = \frac{1}{q} = \frac{p-1}{p} \in (0, 1)$ and $H = \left\|f'\right\|_p$. Applying Theorem 4 we get the desired result (11).

The particular case for euclidean norms may be useful.

Corollary 4 *If f is absolutely continuous on $[a, b]$ and $f' \in L_2[a, b]$, then for all $t \in (a, b)$ we have*

$$\left|(Tf)(a, b; t) - \frac{f(t)}{\pi} \ln\left(\frac{b-t}{t-a}\right)\right| \tag{13}$$

$$\leq \frac{2}{\pi}\left(\sqrt{t-a} + \sqrt{b-t}\right)\left\|f'\right\|_2 \leq \frac{2\sqrt{2}}{\pi}\sqrt{b-a}\left\|f'\right\|_2.$$

2.2 Some Results for Monotonic Functions

The following result holds.

Theorem 5 (Dragomir et al. [13]) *Let $f : [a, b] \to \mathbb{R}$ be a monotonic nondecreasing (nonincreasing) function on $[a, b]$. If the finite Hilbert transform $(Tf)(a, b, \cdot)$ exists in every $t \in (a, b)$, then*

$$(Tf)(a,b;t) \geq (\leq) \frac{1}{\pi} f(t) \ln\left(\frac{b-t}{t-a}\right) \tag{14}$$

for all $t \in (a,b)$.

Proof Using the equality (6) we have

$$\begin{aligned}
&(Tf)(a,b;t) - \frac{f(t)}{\pi} \ln\left(\frac{b-t}{t-a}\right) \qquad (15)\\
&= \frac{1}{\pi} PV \int_a^b \frac{f(\tau)-f(t)}{\tau - t} d\tau \\
&= \frac{1}{\pi} \lim_{\varepsilon \to 0+} \left[\int_a^{t-\varepsilon} \frac{f(\tau)-f(t)}{\tau - t} d\tau + \int_{t+\varepsilon}^b \frac{f(\tau)-f(t)}{\tau - t} d\tau\right].
\end{aligned}$$

If we assume that f is nondecreasing, then for both $\tau \in [a, t-\varepsilon]$ and $\tau \in [t+\varepsilon, b]$ we have

$$\frac{f(\tau)-f(t)}{\tau - t} \geq 0$$

which shows that the right side of (15) is positive and hence the inequality (14) holds.

The following result can be useful in practice.

Corollary 5 *Let* $f : [a,b] \to \mathbb{R}$ *and* $e : [a,b] \to \mathbb{R}$, $e(t) = t$ *such that* $f - me$, $Me - f$ *are monotonic nondecreasing, where* m, M *are given real numbers. If* $(Tf)(a,b,\cdot)$ *exists in every point* $t \in (a,b)$, *then we have the inequality*

$$\frac{(b-a)m}{\pi} \leq (Tf)(a,b;t) - \frac{1}{\pi} f(t) \ln\left(\frac{b-t}{t-a}\right) \leq \frac{(b-a)M}{\pi} \tag{16}$$

for all $t \in (a,b)$.

Proof We simply apply Theorem 5 for the mappings $f - me$ and $Me - f$ which are monotonic on $[a,b]$.

For example, for the first mapping $f - me$ we get

$$T(f-me)(a,b;t) \geq \frac{1}{\pi}[f(t) - mt] \ln\left(\frac{b-t}{t-a}\right) \tag{17}$$

for all $t \in (a,b)$.

However,

$$T(f-me)(a,b;t) = T(f)(a,b;t) - mT(e)(a,b;t)$$

and as

$$\begin{aligned}
T(e)(a,b;t) &= \frac{1}{\pi}\lim_{\varepsilon\to 0+}\left[\int_a^{t-\varepsilon}\frac{\tau}{\tau-t}d\tau+\int_{t+\varepsilon}^{b}\frac{\tau}{\tau-t}d\tau\right]\\
&= \frac{1}{\pi}\lim_{\varepsilon\to 0+}\left[\int_a^{t-\varepsilon}\left(1+\frac{t}{\tau-t}\right)d\tau+\int_{t+\varepsilon}^{b}\left(1+\frac{t}{\tau-t}\right)d\tau\right]\\
&= \frac{1}{\pi}\lim_{\varepsilon\to 0+}\left[t-\varepsilon-a+\int_a^{t-\varepsilon}\frac{t}{\tau-t}d\tau+b-t-\varepsilon+\int_{t+\varepsilon}^{b}\frac{t}{\tau-t}d\tau\right]\\
&= \frac{1}{\pi}\left(b-a+tT(1)(a,b;t)\right)\\
&= \frac{b-a+tT(1)(a,b;t)}{\pi}=\frac{1}{\pi}\left[b-a+t\ln\left(\frac{b-t}{t-a}\right)\right]
\end{aligned}$$

then by (17) we get

$$T(f)(a,b;t)-\frac{(b-a)m}{\pi}-\frac{mt}{\pi}\ln\left(\frac{b-t}{t-a}\right)\geq\frac{1}{\pi}f(t)\ln\left(\frac{b-t}{t-a}\right)-\frac{mt}{\pi}\ln\left(\frac{b-t}{t-a}\right)$$

and the first inequality in (16) is obtained.

The second inequality goes likewise by applying Theorem 5 to the mapping $Me-f$.

Remark 1 If the mapping is differentiable on (a,b), the condition that $f-me$, $Me-f$ are monotonic nondecreasing is equivalent with the following more practical condition

$$m\leq f'(t)\leq M \text{ for all } t\in(a,b). \tag{18}$$

Remark 2 From (16) we may deduce the following approximation result

$$\left|(Tf)(a,b;t)-\frac{1}{\pi}f(t)\ln\left(\frac{b-t}{t-a}\right)-\frac{M+m}{2\pi}(b-a)\right|\leq\frac{M-m}{2\pi}(b-a). \tag{19}$$

for all $t\in(a,b)$.

The above procedure for estimating the finite Hilbert transform can be extended as follows.

Theorem 6 (Dragomir et al. [13]) *Let $f, g:[a,b]\to\mathbb{R}$ and $\gamma, \Gamma\in\mathbb{R}$ be such that $f-\gamma g$, $\Gamma g-f$ are monotonic on $[a,b]$. If $(Tf)(a,b,\cdot)$ and $(Tg)(a,b,\cdot)$ exist in every point $t\in(a,b)$, then we have the inequality:*

$$\gamma\left[T(g)(a,b;t)-\frac{1}{\pi}g(t)\ln\left(\frac{b-t}{t-a}\right)\right] \tag{20}$$

$$\leq (Tf)(a,b;t) - \frac{1}{\pi} f(t) \ln\left(\frac{b-t}{t-a}\right)$$
$$\leq \Gamma \left[T(g)(a,b;t) - \frac{1}{\pi} g(t) \ln\left(\frac{b-t}{t-a}\right) \right]$$

for all $t \in (a,b)$.

Proof As above, we apply Theorem 5 for the monotonic nondecreasing function $f - \gamma g$ to get

$$T(f - \gamma g)(a,b;t) \geq \frac{1}{\pi} [f(t) - \gamma g(t)] \ln\left(\frac{b-t}{t-a}\right) \tag{21}$$

and as, by the linearity of T, we have

$$T(f - \gamma g)(a,b;t) = T(f)(a,b;t) - \gamma T(g)(a,b;t),$$

then, by (21) we obtain the first inequality in (20).

The second inequality goes likewise and we omit the details.

Remark 3 If we assume that the mappings f, g are differentiable on (a,b), $g'(t) > 0$ on (a,b) and

$$\gamma = \inf_{t \in (a,b)} \frac{f'(t)}{g'(t)}, \quad \Gamma = \sup_{t \in (a,b)} \frac{f'(t)}{g'(t)},$$

then the inequality (20) holds.

2.3 Some Results for Convex Functions

Now, if we assume that the mapping $f : (a,b) \to \mathbb{R}$ is convex on (a,b), then it is locally Lipschitzian on (a,b) and then the finite Hilbert transform of f exists in every point $t \in (a,b)$.

The following result holds.

Theorem 7 (Dragomir et al. [13]) *Let* $f : (a,b) \to \mathbb{R}$ *be a convex mapping on* (a,b). *Then we have*

$$\frac{1}{\pi} \left[f(t) \ln\left(\frac{b-t}{t-a}\right) + f(t) - f(a) + l(t)(b-t) \right] \tag{22}$$
$$\leq (Tf)(a,b;t)$$

$$\leq \frac{1}{\pi}\left[f(t)\ln\left(\frac{b-t}{t-a}\right)+f(b)-f(t)+l(t)(t-a)\right],$$

where $l(s)\in\left[f'_{-}(s),f'_{+}(s)\right]$, $s\in(a,b)$.

Proof By the convexity of f we can state that for all $a\leq c<d\leq b$ we have

$$\frac{f(d)-f(c)}{d-c}\geq l(c), \tag{23}$$

where $l(c)\in\left[f'_{-}(c),f'_{+}(c)\right]$.

Using (23), we have

$$\int_a^{t-\varepsilon}\frac{f(t)-f(\tau)}{t-\tau}d\tau\geq\int_a^{t-\varepsilon}l(\tau)\,d\tau \tag{24}$$

and

$$\int_{t+\varepsilon}^{b}\frac{f(\tau)-f(t)}{\tau-t}d\tau\geq\int_{t+\varepsilon}^{b}l(t)\,d\tau=l(t)(b-t-\varepsilon) \tag{25}$$

and then, by adding (24) and (25), we get

$$\begin{aligned}&\lim_{\varepsilon\to0+}\left[\int_a^{t-\varepsilon}\frac{f(t)-f(\tau)}{t-\tau}d\tau+\int_{t+\varepsilon}^{b}\frac{f(\tau)-f(t)}{\tau-t}d\tau\right]\\&\geq\lim_{\varepsilon\to0+}\left[\int_a^{t-\varepsilon}l(\tau)\,d\tau+l(t)(b-t-\varepsilon)\right]\\&=\int_a^{t}l(\tau)\,d\tau+l(t)(b-t)=f(t)-f(a)+l(t)(b-t)\end{aligned}$$

Consequently, we have

$$PV\int_a^{b}\frac{f(\tau)-f(t)}{\tau-t}d\tau\geq f(t)-f(a)+l(t)(b-t)$$

and by the identity (6), we deduce the first inequality in (22).

Similarly, by the convexity of f we have for $a\leq c<d\leq b$

$$l(d)\geq\frac{f(d)-f(c)}{d-c}, \tag{26}$$

where $l(d)\in\left[f'_{-}(d),f'_{+}(d)\right]$.

Using (26) we may state

$$\int_a^{t-\varepsilon} \frac{f(t)-f(\tau)}{t-\tau} d\tau \le \int_a^{t-\varepsilon} l(t)\, d\tau = l(t)(t-\varepsilon-a)$$

and

$$\int_{t+\varepsilon}^{b} \frac{f(\tau)-f(t)}{\tau-t} d\tau \le \int_{t+\varepsilon}^{b} l(\tau)\, d\tau.$$

Now, in the same manner as that employed above, we may obtain the second part of (22).

Corollary 6 *Let $f:(a,b)\to\mathbb{R}$ be a differentiable convex function on (a,b). Then we have the inequality*

$$\begin{aligned} &\frac{1}{\pi}\left[f(t)\ln\left(\frac{b-t}{t-a}\right)+f(t)-f(a)+f'(t)(b-t)\right] \\ &\le (Tf)(a,b;t) \\ &\le \frac{1}{\pi}\left[f(t)\ln\left(\frac{b-t}{t-a}\right)+f(b)-f(t)+f'(t)(t-a)\right] \end{aligned} \tag{27}$$

for all $t\in(a,b)$.

3 Inequalities of Trapezoid Type

3.1 Trapezoid Type Inequalities

The following theorem holds.

Theorem 8 (Dragomir et al. [14]) *Let $f:[a,b]\to\mathbb{R}$ be such that $f':(a,b)\to\mathbb{R}$ is absolutely continuous on (a,b). Then we have the bounds*

$$\left|T(f)(a,b;t)-\frac{f(t)}{\pi}\ln\left(\frac{b-t}{t-a}\right)-\frac{1}{2\pi}\left[f(b)-f(a)+f'(t)(b-a)\right]\right|$$

$$\le \begin{cases} \frac{\|f''\|_\infty}{4\pi}\left[\frac{(b-a)^2}{4}+\left(t-\frac{a+b}{2}\right)^2\right], & \text{if } f''\in L_\infty[a,b]; \\ \frac{q\|f''\|_p}{2\pi(q+1)^{\frac{q+1}{q}}}\left[(t-a)^{\frac{q+1}{q}}+(b-t)^{\frac{q+1}{q}}\right], & \text{if } f''\in L_p[a,b], \\ & p>1,\ \frac{1}{p}+\frac{1}{q}=1; \\ \frac{1}{2\pi}\left\|f''\right\|_1(b-a) \end{cases} \tag{28}$$

$$\leq \begin{cases} \frac{\|f''\|_{\infty}(b-a)^2}{8\pi}, & \text{if } f'' \in L_{\infty}[a,b]; \\ \frac{q\|f''\|_p (b-a)^{\frac{q+1}{q}}}{2\pi(q+1)^{\frac{q+1}{q}}}, & \text{if } f'' \in L_p[a,b], \\ & p>1,\ \frac{1}{p}+\frac{1}{q}=1; \\ \frac{1}{2\pi}\left\|f''\right\|_1(b-a) \end{cases}$$

for all $t \in (a,b)$, *where* $\|\cdot\|_p$ *are the usual Lebesgue norms in* $L_p[a,b]$ $(1 \leq p \leq \infty)$.

Proof We start with the following elementary identity which can be proved using the integration by parts formula

$$\int_{\alpha}^{\beta} g(u)\,du = \frac{g(\alpha)+g(\beta)}{2}(\beta-\alpha) + \int_{\alpha}^{\beta}\left(\frac{\alpha+\beta}{2}-u\right)g'(u)\,du, \tag{29}$$

provided that g is absolutely continuous on the interval $[\alpha,\beta]$ if $\alpha \leq \beta$ (or $[\beta,\alpha]$ if $\beta \leq \alpha$).

As for the mapping $f:(a,b)\to\mathbb{R}$, $f(t)=1$, $t\in(a,b)$, we have

$$(Tf)(a,b;t) = \frac{1}{\pi}\ln\left(\frac{b-t}{t-a}\right),\ t\in(a,b)$$

then obviously

$$\begin{aligned}(Tf)(a,b;t) &= \frac{1}{\pi}PV\int_a^b \frac{f(\tau)-f(t)+f(t)}{\tau-t}d\tau \\ &= \frac{1}{\pi}PV\int_a^b \frac{f(\tau)-f(t)}{\tau-t}d\tau + \frac{f(t)}{\pi}PV\int_a^b \frac{d\tau}{\tau-t}\end{aligned}$$

from where we get the equality

$$(Tf)(a,b;t) - \frac{f(t)}{\pi}\ln\left(\frac{b-t}{t-a}\right) = \frac{1}{\pi}PV\int_a^b \frac{f(\tau)-f(t)}{\tau-t}d\tau. \tag{30}$$

Using (29), we obtain

$$\begin{aligned}&PV\int_a^b \frac{f(\tau)-f(t)}{\tau-t}d\tau \\ &= PV\int_a^b \frac{\int_t^{\tau} f'(u)\,du}{\tau-t}d\tau\end{aligned}$$

$$= PV\int_a^b \frac{\frac{f'(\tau)+f'(t)}{2}(\tau-t)+\int_t^\tau\left(\frac{t+\tau}{2}-u\right)f''(u)\,du}{\tau-t}d\tau$$

$$= \frac{1}{2}\int_a^b\left[f'(\tau)+f'(t)\right]d\tau + PV\int_a^b\frac{\int_t^\tau\left(\frac{t+\tau}{2}-u\right)f''(u)\,du}{\tau-t}d\tau$$

$$= \frac{1}{2}\left[f(b)-f(a)+f'(t)(b-a)\right]+PV\int_a^b\frac{\int_t^\tau\left(\frac{t+\tau}{2}-u\right)f''(u)\,du}{\tau-t}d\tau$$

and then, by (30), we can state the identity

$$T(f)(a,b;t)-\frac{f(t)}{\pi}\ln\left(\frac{b-t}{t-a}\right)-\frac{1}{2\pi}\left[f(b)-f(a)+f'(t)(b-a)\right] \tag{31}$$

$$=\frac{1}{\pi}PV\int_a^b\frac{\int_t^\tau\left(\frac{t+\tau}{2}-u\right)f''(u)\,du}{\tau-t}d\tau.$$

Using the property of modulus

$$\left|(Tf)(a,b;t)-\frac{f(t)}{\pi}\ln\left(\frac{b-t}{t-a}\right)-\frac{1}{2\pi}\left[f(b)-f(a)+f'(t)(b-a)\right]\right| \tag{32}$$

$$\le\frac{1}{\pi}PV\int_a^b\left|\frac{\int_t^\tau\left(\frac{t+\tau}{2}-u\right)f''(u)\,du}{\tau-t}\right|d\tau =: A(a,b;t).$$

Now, it is obvious that

$$\left|\int_t^\tau\left(\frac{t+\tau}{2}-u\right)f''(u)\,du\right| \le \operatorname*{essup}_{u\in[a,b]}\left|f''(u)\right|\left|\int_t^\tau\left|\frac{t+\tau}{2}-u\right|du\right|$$

$$=\left\|f''\right\|_\infty\frac{|t-\tau|^2}{4},\quad \text{for all } t,\tau\in(a,b).$$

Then

$$A(a,b;t)\le\frac{1}{\pi}PV\int_a^b\frac{\left\|f''\right\|_\infty|t-\tau|}{4}d\tau$$

$$=\frac{\left\|f''\right\|_\infty}{4}\lim_{\varepsilon\to0+}\left[\int_a^{t-\varepsilon}(t-\tau)\,d\tau+\int_{t+\varepsilon}^b(\tau-t)\,d\tau\right]$$

$$=\frac{\left\|f''\right\|_\infty}{4}\lim_{\varepsilon\to0+}\left[\frac{(t-a)^2}{2}+\frac{\varepsilon^2}{2}+\frac{(b-t)^2}{2}-\frac{\varepsilon^2}{2}\right]$$

$$= \frac{\|f''\|_\infty}{4\pi} \cdot \frac{(t-a)^2 + (b-t)^2}{2}$$
$$= \frac{\|f''\|_\infty}{4\pi} \left[\left(t - \frac{a+b}{2}\right)^2 + \frac{(b-a)^2}{4} \right].$$

Using Hölder's integral inequality, we can state for $p > 1$, $\frac{1}{p} + \frac{1}{q} = 1$ that

$$\left| \int_t^\tau \left(\frac{t+\tau}{2} - u \right) f''(u)\, du \right|$$
$$\leq \left| \int_t^\tau |f''(u)|^p\, du \right|^{\frac{1}{p}} \left| \int_t^\tau \left| \frac{t+\tau}{2} - u \right|^q du \right|^{\frac{1}{q}}$$
$$\leq \left(\int_a^b |f''(u)|^p\, du \right)^{\frac{1}{p}} \left| \int_t^{\frac{t+\tau}{2}} \left(\frac{t+\tau}{2} - u \right)^q du + \int_{\frac{t+\tau}{2}}^\tau \left(u - \frac{t+\tau}{2} \right)^q du \right|^{\frac{1}{q}}$$
$$= \|f''\|_p \frac{|t-\tau|^{\frac{q+1}{q}}}{2(q+1)^{\frac{1}{q}}} \quad \text{for all } t, \tau \in (a,b).$$

Then

$$A(a,b;t) \leq \frac{1}{\pi} PV \int_a^b \frac{\|f''\|_p |t-\tau|^{q^{-1}}}{2(q+1)^{\frac{1}{q}}} d\tau$$
$$= \frac{\|f''\|_p}{2\pi (q+1)^{\frac{1}{q}}} PV \int_a^b |t-\tau|^{q^{-1}}\, d\tau$$
$$= \frac{\|f''\|_p}{2\pi (q+1)^{\frac{1}{q}}} \lim_{\varepsilon \to 0+} \left[\int_a^{t-\varepsilon} (t-\tau)^{q^{-1}}\, d\tau + \int_{t+\varepsilon}^b (t-\tau)^{q^{-1}}\, d\tau \right]$$

$$= \frac{\|f''\|_p}{2\pi (q+1)^{\frac{1}{q}}} \lim_{\varepsilon \to 0+} \left[\frac{(t-a)^{q^{-1}+1} - \varepsilon^{q^{-1}+1}}{q^{-1}+1} + \frac{(b-t)^{q^{-1}+1} - \varepsilon^{q^{-1}+1}}{q^{-1}+1} \right]$$
$$= \frac{\|f''\|_p}{2\pi (q+1)^{\frac{1}{q}}} \left[\frac{(t-a)^{q^{-1}+1} + (b-t)^{q^{-1}+1}}{q^{-1}+1} \right]$$
$$= \frac{q \|f''\|_p \left[(t-a)^{\frac{q+1}{q}} + (b-t)^{\frac{q+1}{q}} \right]}{2\pi (q+1)^{\frac{q+1}{q}}}$$

and the second bound in (28) is proved.

Finally, we observe that

$$\left|\int_t^\tau \left(\frac{t+\tau}{2}-u\right) f''(u)\,du\right| \le \sup_{u\in[t,\tau]}\left|\frac{t+\tau}{2}-u\right|\left|\int_t^\tau |f''(u)|\,du\right|$$
$$\le \frac{|t-\tau|}{2}\int_a^b |f''(u)|\,du = \frac{|t-\tau|}{2}\left\|f''\right\|_1.$$

Consequently,

$$A(a,b;t) \le \frac{1}{\pi} PV \int_a^b \frac{1}{2}\left\|f''\right\|_1 d\tau = \frac{1}{2\pi}\left\|f''\right\|_1 (b-a)$$

and the theorem is proved.

Remark 4 It is obvious that for small intervals (a,b), the approximation provided by (28) is accurate.

The best inequality we can get from the first and second part of (28) is the one for $t=\frac{a+b}{2}$, and thus we can state the following corollary.

Corollary 7 *With the assumptions of Theorem 8, we have*

$$\left|(Tf)\left(a,b;\frac{a+b}{2}\right)-\frac{1}{2\pi}\left[f(b)-f(a)+f'\left(\frac{a+b}{2}\right)(b-a)\right]\right| \tag{33}$$
$$\le \begin{cases} \frac{\|f''\|_\infty (b-a)^2}{16\pi} & \text{if } f''\in L_\infty[a,b]; \\ \frac{q\|f''\|_p}{2^{\frac{q+1}{q}}(q+1)^{\frac{q+1}{q}}\pi}(b-a)^{\frac{q+1}{q}}, & \text{if } f''\in L_p[a,b],\ p>1,\ \frac{1}{p}+\frac{1}{q}=1. \end{cases}$$

The following result also holds.

Theorem 9 (Dragomir et al. [14]) *Let $f:[a,b]\to\mathbb{R}$ be such that $f'':(a,b)\to\mathbb{R}$ is absolutely continuous on (a,b). Then we have the bounds*

$$\left|(Tf)(a,b;t)-\frac{f(t)}{\pi}\ln\left(\frac{b-t}{t-a}\right)-\frac{1}{2\pi}\left[f(b)-f(a)+f'(t)(b-a)\right]\right| \tag{34}$$

$$\le \begin{cases} \frac{\|f'''\|_\infty (b-a)}{12\pi}\left[\frac{(b-a)^2}{12}+\left(t-\frac{a+b}{2}\right)^2\right], & \text{if } f'''\in L_\infty[a,b]; \\ \frac{q\|f'''\|_p [B(q+1,q+1)]^{\frac{1}{q}}}{2\pi(2q+1)}\left[(b-t)^{2+\frac{1}{q}}+(t-a)^{2+\frac{1}{q}}\right], & \text{if } f'''\in L_p[a,b], \\ & p>1,\ \frac{1}{p}+\frac{1}{q}=1; \\ \frac{\|f'''\|_1}{8\pi}\left[\frac{(b-a)^2}{4}+\left(t-\frac{a+b}{2}\right)^2\right] \end{cases}$$

$$\leq \begin{cases} \frac{\|f'''\|_\infty (b-a)^3}{36\pi}, & \text{if } f''' \in L_\infty[a,b]; \\ \frac{q\|f'''\|_p [B(q+1,q+1)]^{\frac{1}{q}}}{2\pi(2q+1)} (b-a)^{2+\frac{1}{q}}, & \text{if } f''' \in L_p[a,b], \\ & p>1,\ \frac{1}{p}+\frac{1}{q}=1; \\ \frac{\|f'''\|_1}{16\pi}(b-a)^2 \end{cases}$$

for all $t \in (a, b)$, *where* $\|\cdot\|_p$ *are the usual p-norms and* $B(\cdot, \cdot)$ *is Euler's Beta mapping*

$$B(\alpha, \beta) = \int_0^1 t^{\alpha-1}(1-t)^{\beta-1}\, dt, \quad \alpha, \beta > 0. \tag{35}$$

Proof Using the integration by parts formula, we obtain the equality:

$$\int_\alpha^\beta g(u)\, du = \frac{g(\alpha)+g(\beta)}{2}(\beta-\alpha) - \frac{1}{2}\int_\alpha^\beta (u-\alpha)(\beta-u)\, g''(u)\, du, \tag{36}$$

where g is such that g' is absolutely continuous on $[\alpha, \beta]$ (if $\alpha < \beta$), or on $[\beta, \alpha]$ (if $\beta < \alpha$).

By a similar procedure to that in Theorem 8, we have

$$(Tf)(a, b; t) - \frac{f(t)}{\pi}\ln\left(\frac{b-t}{t-a}\right) - \frac{1}{2\pi}\left[f(b) - f(a) + f'(t)(b-a)\right] \tag{37}$$

$$= -\frac{1}{2\pi} PV \int_a^b \frac{\int_t^\tau (u-t)(\tau-u) f'''(u)\, du}{\tau - t} d\tau.$$

Using the property of modulus, we have

$$\left|(Tf)(a, b; t) - \frac{f(t)}{\pi}\ln\left(\frac{b-t}{t-a}\right) - \frac{1}{2\pi}\left[f(b) - f(a) + f'(t)(b-a)\right]\right|$$

$$\leq \frac{1}{2\pi} PV \int_a^b \left|\frac{\int_t^\tau (u-t)(\tau-u) f'''(u)\, du}{\tau - t}\right| d\tau =: B(a, b; t).$$

Firstly, let us observe that

$$\left|\int_t^\tau (u-t)(\tau-u) f'''(u)\, du\right| \leq \sup_{u\in[t,\tau]} \left|f'''(u)\right| \left|\int_t^\tau |u-t|\,|\tau-u|\, du\right|$$

$$\leq \left\|f'''\right\|_\infty \frac{|t-\tau|^3}{6}$$

and then

$$B(a,b;t) \leq \frac{\|f'''\|_\infty}{12\pi} PV \int_a^b |t-\tau|^2 d\tau$$
$$= \frac{\|f'''\|_\infty}{12\pi} \cdot \frac{(b-t)^3 + (t-a)^3}{3}$$
$$= \frac{\|f'''\|_\infty (b-a)}{12\pi} \left[\frac{(b-a)^2}{12} + \left(t - \frac{a+b}{2} \right)^2 \right],$$

which proves the first part of (34).

For the second part, we apply Hölder's integral inequality to obtain

$$\left| \int_t^\tau (u-t)(\tau-u) f'''(u) du \right| \leq \left| \int_t^\tau |u-t|^q |\tau-u|^q du \right|^{\frac{1}{q}} \left| \int_t^\tau |f'''(u)|^p du \right|^{\frac{1}{p}}$$
$$\leq \|f'''\|_p |t-\tau|^{2+\frac{1}{q}} [B(q+1,q+1)]^{\frac{1}{q}}$$

for all $t, \tau \in (a,b)$.

Indeed, if we assume that $\tau > t$, then, by $u = (1-s)t + s\tau$, $s \in [0,1]$, we get

$$\int_t^\tau |u-t|^q |\tau-u|^q du = \int_t^\tau (u-t)^q (\tau-u)^q du$$
$$= (\tau-t)^{2q+1} \int_0^1 s^q (1-s)^q ds$$
$$= (\tau-t)^{2q+1} B(q+1,q+1).$$

Consequently, we have

$$B(a,b;t) \leq \frac{\|f'''\|_p [B(q+1,q+1)]^{\frac{1}{q}}}{2\pi} PV \int_a^b |t-\tau|^{1+\frac{1}{q}} d\tau$$
$$= \frac{\|f'''\|_p [B(q+1,q+1)]^{\frac{1}{q}}}{2\pi} \cdot \frac{(b-t)^{2+\frac{1}{q}} + (t-a)^{2+\frac{1}{q}}}{1+\frac{1}{q}}$$
$$= \frac{q\|f'''\|_p [B(q+1,q+1)]^{\frac{1}{q}}}{2\pi(2q+1)} \left[(b-t)^{2+\frac{1}{q}} + (t-a)^{2+\frac{1}{q}} \right]$$

and the second part of (34) holds.

For the last part, we observe that

$$\left|\int_t^\tau (u-t)(\tau-u) f'''(u)\, du\right| \le \max_u |(u-t)(\tau-u)| \int_t^\tau |f'''(u)|\, du$$
$$\le \|f'''\|_1 \frac{(t-\tau)^2}{4}$$

since a simple calculation shows that

$$\max_{\substack{u\in[t,\tau]\\ (u\in[\tau,t])}} |(u-t)(\tau-u)| = \frac{(t-\tau)^2}{4}.$$

Thus, we can write the following inequality

$$\begin{aligned} B(a,b;t) &\le \frac{\|f'''\|_1}{8\pi} PV \int_a^b |t-\tau|\, d\tau \\ &= \frac{\|f'''\|_1}{8\pi} \cdot \frac{(b-t)^2+(t-a)^2}{2} \\ &= \frac{\|f'''\|_1}{8\pi} \left[\frac{(b-a)^2}{4} + \left(t - \frac{a+b}{2}\right)^2\right] \end{aligned}$$

and the theorem is proved.

Remark 5 It is obvious that if $(b-a) \to 0$, then (34) provides an accurate approximation for the finite Hilbert transform.

Taking into account the fact that all the mappings depending on t from the right-hand side of (34) are convex on the interval (a,b), it is obvious that the best inequality from (34) is that one for which $t = \frac{a+b}{2}$.

Corollary 8 *Let f be as in Theorem 9. Then we have the inequality*

$$\left|(Tf)\left(a,b;\frac{a+b}{2}\right) - \frac{1}{2\pi}\left[f(b) - f(a) + f'\left(\frac{a+b}{2}\right)(b-a)\right]\right| \tag{38}$$

$$\le \begin{cases} \frac{\|f'''\|_\infty (b-a)^3}{144\pi}, & \text{if } f''' \in L_\infty[a,b]; \\ \frac{q\|f'''\|_p [B(q+1,q+1)]^{\frac{1}{q}}}{2^{2+\frac{1}{q}}\pi(2q+1)} (b-a)^{2+\frac{1}{q}}, & \text{if } f''' \in L_p[a,b], \\ & p>1,\ \frac{1}{p}+\frac{1}{q}=1; \\ \frac{\|f'''\|_1 (b-a)^2}{32\pi}. \end{cases}$$

3.2 Agarwal-Dragomir Type Inequalities

In [1], R. P. Agarwal and S. S. Dragomir proved the following trapezoid type inequality

$$\begin{aligned} &\left| \frac{1}{b-a} \int_a^b g(x)\,dx - \frac{g(a)+g(b)}{2} \right| \\ &\leq \frac{[g(b)-g(a)-m(b-a)][M(b-a)-g(b)+g(a)]}{2(M-m)(b-a)} \\ &\leq \frac{(M-m)(b-a)}{8}, \end{aligned} \tag{39}$$

provided that $g : [a, b] \to \mathbb{R}$ is absolutely continuous on $[a, b]$ and $M = \sup_{x\in(a,b)} g'(x) < \infty$, $m = \inf_{x\in(a,b)} g'(x) > -\infty$ and $M > m$.

Using the above inequality, we can state and prove the following theorem.

Theorem 10 (Dragomir et al. [14]) *Let $f : [a, b] \to \mathbb{R}$ be such that $f' : (a, b) \to \mathbb{R}$ is absolutely continuous on (a, b) and*

$$\Gamma = \sup_{t\in(a,b)} f''(t) < \infty, \quad \gamma = \inf_{t\in(a,b)} f''(t) > -\infty, \quad \Gamma > \gamma. \tag{40}$$

Then we have the bound:

$$\begin{aligned} &\left| (Tf)(a, b; t) - \frac{f(t)}{\pi} \ln\left(\frac{b-t}{t-a}\right) - \frac{1}{2\pi}\left[f(b) - f(a) + f'(t)(b-a)\right] \right| \\ &\leq \frac{(\Gamma-\gamma)}{8\pi}\left[\left(t - \frac{a+b}{2}\right)^2 + \frac{(b-a)^2}{4}\right] \leq \frac{(\Gamma-\gamma)(b-a)^2}{16\pi}, \end{aligned} \tag{41}$$

for all $t \in (a, b)$.

Proof Applying the inequality (39) written for $f'(\cdot)$ in the following version:

$$\left| \frac{1}{\tau-t} \int_t^\tau f'(u)\,du - \frac{f'(\tau)+f'(t)}{2} \right| \leq \frac{(\Gamma-\gamma)|t-\tau|}{8},$$

we can state the inequality

$$\left| \frac{f(\tau)-f(t)}{\tau-t} - \frac{f'(\tau)+f'(t)}{2} \right| \leq \frac{(\Gamma-\gamma)|t-\tau|}{8} \tag{42}$$

for all $t, \tau \in [a, b]$, $t \neq \tau$.

The following property of the Cauchy-Principal Value follows by the properties of integral, modulus and limit,

$$\left|PV\int_{a}^{b}A(t,s)\,ds\right|\leq PV\int_{a}^{b}|A(t,s)|\,ds, \tag{43}$$

holds, assuming that the PV involved exist for all $t\in(a,b)$.

Using (31) and (32), we may write

$$\begin{aligned}&\left|PV\int_{a}^{b}\frac{f(\tau)-f(t)}{\tau-t}d\tau-PV\int_{a}^{b}\frac{f'(\tau)+f'(t)}{2}d\tau\right| \\ &\leq PV\int_{a}^{b}\left|\frac{f(\tau)-f(t)}{\tau-t}-\frac{f'(\tau)+f'(t)}{2}\right|d\tau \\ &\leq PV\int_{a}^{b}\frac{(\Gamma-\gamma)|t-\tau|}{8}d\tau,\end{aligned} \tag{44}$$

and as

$$\frac{1}{\pi}PV\int_{a}^{b}\frac{f(\tau)-f(t)}{\tau-t}d\tau=(Tf)(a,b;t)-\frac{f(t)}{\pi}\ln\left(\frac{b-t}{t-a}\right),$$

$$\frac{1}{\pi}PV\int_{a}^{b}\frac{f'(\tau)+f'(t)}{2}d\tau=\frac{1}{2\pi}\left[f(b)-f(a)+f'(t)(b-a)\right]$$

and

$$\frac{1}{\pi}PV\int_{a}^{b}|t-\tau|\,d\tau=\frac{(t-a)^2+(b-t)^2}{2\pi}=\frac{1}{\pi}\left[\left(t-\frac{a+b}{2}\right)^2+\frac{(b-a)^2}{4}\right]$$

then by (44) we deduce the desired inequality (41).

It is obvious that the best inequality we can get from (41) is that one for which we have $t=\frac{a+b}{2}$, obtaining the following result.

Corollary 9 *Under the assumptions of Theorem 10, we have the inequality:*

$$\begin{aligned}&\left|(Tf)\left(a,b;\frac{a+b}{2}\right)-\frac{1}{2\pi}\left[f(b)-f(a)+f'\left(\frac{a+b}{2}\right)(b-a)\right]\right| \\ &\leq\frac{(\Gamma-\gamma)(b-a)^2}{32\pi}.\end{aligned} \tag{45}$$

3.3 Compounding Trapezoid Type Inequalities

The following inequality concerning the trapezoid inequality for absolutely continuous functions holds (see, for example, [9, p. 32]).

Lemma 1 *Let $u : [a, b] \to \mathbb{R}$ be an absolutely continuous function on $[a, b]$. Then one has the inequalities:*

$$\left| \int_a^b u(s)\, ds - \frac{u(a) + u(b)}{2}(b - a) \right| \tag{46}$$

$$\leq \begin{cases} \dfrac{(b-a)^2}{4} \left\| u' \right\|_{[a,b],\infty} & \text{if } u' \in L_\infty[a, b]; \\[2ex] \dfrac{(b-a)^{1+\frac{1}{q}}}{2(q+1)^{\frac{1}{q}}} \left\| u' \right\|_{[a,b],p} & \text{if } u' \in L_p[a, b], \\ & p > 1,\ \frac{1}{p} + \frac{1}{q} = 1; \\ \dfrac{(b-a)}{2} \left\| u' \right\|_{[a,b],1}. \end{cases}$$

A simple proof of this fact can be done by using the identity

$$\int_a^b u(s)\, ds - \frac{u(a) + u(b)}{2}(b - a) = -\int_a^b \left(s - \frac{a+b}{2} \right) u'(s)\, ds, \tag{47}$$

and we omit the details.

The following lemma holds.

Lemma 2 *Let $u : [a, b] \to \mathbb{R}$ be an absolutely continuous function on $[a, b]$. Then for any $t, \tau \in [a, b]$, $t \neq \tau$ and $n \in \mathbb{N}$, $n \geq 1$, we have the inequality:*

$$\left| \frac{1}{\tau - t} \int_t^\tau u(s)\, ds - \frac{1}{2n} \sum_{i=0}^{n-1} \left[u\left(t + i \cdot \frac{\tau - t}{n} \right) + u\left(t + (i+1) \cdot \frac{\tau - t}{n} \right) \right] \right| \tag{48}$$

$$\leq \begin{cases} \dfrac{|\tau - t|}{4n} \left\| u' \right\|_{[t,\tau],\infty} & \text{if } u' \in L_\infty[a, b]; \\[2ex] \dfrac{|\tau - t|^{\frac{1}{q}}}{2(q+1)^{\frac{1}{q}} n} \left\| u' \right\|_{[t,\tau],p} & \text{if } u' \in L_p[a, b], \\ & p > 1,\ \frac{1}{p} + \frac{1}{q} = 1; \\ \dfrac{1}{2n} \left\| u' \right\|_{[t,\tau],1}; \end{cases}$$

where

$$\left\|u'\right\|_{[t,\tau],\infty} := \underset{\substack{s\in[t,\tau] \\ (s\in[\tau,t])}}{\text{essup}} \left|u'(s)\right|$$

and

$$\left\|u'\right\|_{[t,\tau],p} := \left|\int_t^\tau \left|u'(s)\right|^p ds\right|^{\frac{1}{p}}$$

for $p \geq 1$.

Proof Consider the equidistant division of $[t, \tau]$ (if $t < \tau$) or $[\tau, t]$ (if $\tau < t$) given by

$$E_n : x_i = t + i \cdot \frac{\tau - t}{n}, \quad i = \overline{0, n}.$$

If we apply the inequality (46) on the interval $[x_i, x_{i+1}]$, we may write that

$$\left|\int_{x_i}^{x_{i+1}} u(s)\,ds - \frac{u\left(t + i \cdot \frac{\tau-t}{n}\right) + u\left(t + (i+1) \cdot \frac{\tau-t}{n}\right)}{2} \cdot \frac{\tau - t}{n}\right|$$

$$\leq \begin{cases} \dfrac{(\tau - t)^2}{4n^2} \left\|u'\right\|_{[x_i,x_{i+1}],\infty} & \text{if } u' \in L_\infty[a,b]; \\[2ex] \dfrac{|\tau - t|^{1+\frac{1}{q}}}{2n^{1+\frac{1}{q}} (q+1)^{\frac{1}{q}}} \left\|u'\right\|_{[x_i,x_{i+1}],p} & \text{if } u' \in L_p[a,b], \\ & \quad p > 1,\ \frac{1}{p} + \frac{1}{q} = 1; \\[2ex] \dfrac{|\tau - t|}{2n} \left\|u'\right\|_{[x_i,x_{i+1}],1}; \end{cases}$$

from where we get

$$\left|\frac{1}{\tau - t}\int_{x_i}^{x_{i+1}} u(s)\,ds - \frac{1}{2n}\left[u\left(t + i \cdot \frac{\tau - t}{n}\right) + u\left(t + (i+1) \cdot \frac{\tau - t}{n}\right)\right]\right|$$

$$\leq \begin{cases} \dfrac{|\tau - t|}{4n^2} \left\|u'\right\|_{[x_i,x_{i+1}],\infty} \\[2ex] \dfrac{|\tau - t|^{\frac{1}{q}}}{2n^{1+\frac{1}{q}} (q+1)^{\frac{1}{q}}} \left\|u'\right\|_{[x_i,x_{i+1}],p} \\[2ex] \dfrac{1}{2n} \left\|u'\right\|_{[x_i,x_{i+1}],1}. \end{cases}$$

Summing over i from 0 to $n-1$ and using the generalized triangle inequality, we may write

$$\left|\frac{1}{\tau-t}\int_t^\tau u(s)\,ds-\frac{1}{2n}\sum_{i=0}^{n-1}\left[u\left(t+i\cdot\frac{\tau-t}{n}\right)+u\left(t+(i+1)\cdot\frac{\tau-t}{n}\right)\right]\right|$$

$$\leq\sum_{i=0}^{n-1}\left|\frac{1}{\tau-t}\int_{x_i}^{x_{i+1}} u(s)\,ds-\frac{1}{2n}\left[u\left(t+i\cdot\frac{\tau-t}{n}\right)+u\left(t+(i+1)\cdot\frac{\tau-t}{n}\right)\right]\right|$$

$$\leq\begin{cases}\dfrac{|\tau-t|}{4n^2}\displaystyle\sum_{i=0}^{n-1}\left\|u'\right\|_{[x_i,x_{i+1}],\infty}\\[2ex]\dfrac{|\tau-t|^{\frac{1}{q}}}{2n^{1+\frac{1}{q}}(q+1)^{\frac{1}{q}}}\displaystyle\sum_{i=0}^{n-1}\left\|u'\right\|_{[x_i,x_{i+1}],p}\\[2ex]\dfrac{1}{2n}\displaystyle\sum_{i=0}^{n-1}\left\|u'\right\|_{[x_i,x_{i+1}],1}.\end{cases}$$

However,

$$\sum_{i=0}^{n-1}\left\|u'\right\|_{[x_i,x_{i+1}],\infty}\leq n\left\|u'\right\|_{[t,\tau],\infty},$$

$$\sum_{i=0}^{n-1}\left\|u'\right\|_{[x_i,x_{i+1}],p}=\sum_{i=0}^{n-1}\left|\int_{x_i}^{x_{i+1}}\left|u'(s)\right|^p ds\right|^{\frac{1}{p}}$$

$$\leq n^{\frac{1}{q}}\left[\left(\sum_{i=0}^{n-1}\left|\int_{x_i}^{x_{i+1}}\left|u'(s)\right|^p ds\right|^{\frac{1}{p}}\right)^p\right]^{\frac{1}{p}}$$

$$=n^{\frac{1}{q}}\left|\int_t^\tau\left|u'(s)\right|^p ds\right|^{\frac{1}{p}}=n^{\frac{1}{q}}\left\|u'\right\|_{[t,\tau],p}$$

and

$$\sum_{i=0}^{n-1} \|u'\|_{[x_i,x_{i+1}],1} \le \left|\sum_{i=0}^{n-1} \int_{x_i}^{x_{i+1}} |u'(s)|\, ds\right| = \left|\int_t^\tau |u'(s)|\, ds\right| = \|u'\|_{[t,\tau],1}$$

and the lemma is proved.

The following theorem in approximating the Hilbert transform of a differentiable function whose derivative is absolutely continuous holds.

Theorem 11 (Dragomir et al. [16]) *Let $f : [a,b] \to \mathbb{R}$ be a differentiable function such that its derivative f' is absolutely continuous on $[a,b]$. If*

$$T_n(f;t) = \frac{f(b) - f(a) + f'(t)(b-a)}{2n\pi} + \frac{b-a}{n\pi} \sum_{i=1}^{n-1} \left[f; t - \frac{t-a}{n} \cdot i, t + \frac{b-t}{n} \cdot i\right], \tag{49}$$

then we have the estimate

$$\left|(Tf)(a,b;t) - \frac{f(t)}{\pi} \ln\left(\frac{b-t}{t-a}\right) - T_n(f;t)\right| \tag{50}$$

$$\le \begin{cases} \frac{1}{4\pi n}\left[\frac{1}{4}(b-a)^2 + \left(t - \frac{a+b}{2}\right)^2\right] \|f''\|_{[a,b],\infty} & \text{if } f'' \in L_\infty[a,b]; \\ \frac{q}{2\pi n (q+1)^{1+\frac{1}{q}}}\left[(t-a)^{1+\frac{1}{q}} + (b-t)^{1+\frac{1}{q}}\right] \|f''\|_{[a,b],p} & \text{if } f'' \in L_p[a,b], \\ & p > 1,\ \frac{1}{p} + \frac{1}{q} = 1; \\ \frac{1}{2\pi n}\left[\frac{1}{2}(b-a) + \left|t - \frac{a+b}{2}\right|\right] \|f''\|_{[a,b],1} \end{cases}$$

$$\le \begin{cases} \frac{1}{8\pi n}(b-a)^2 \|f''\|_{[a,b],\infty} & \text{if } f'' \in L_\infty[a,b]; \\ \frac{q}{2\pi n (q+1)^{1+\frac{1}{q}}}(b-a)^{1+\frac{1}{q}} \|f''\|_{[a,b],p} & \text{if } f'' \in L_p[a,b], \\ & p > 1,\ \frac{1}{p} + \frac{1}{q} = 1; \\ \frac{1}{2\pi n}(b-a) \|f''\|_{[a,b],1} \end{cases}$$

for all $t \in (a,b)$, where $[f;c,d]$ denotes the divided difference

$$[f;c,d] := \frac{f(c) - f(d)}{c-d}.$$

Proof Applying Lemma 2 for the function f', we may write that

$$\left| \frac{f(\tau) - f(t)}{\tau - t} - \frac{1}{2n} \left[f'(t) + \sum_{i=1}^{n-1} f'\left(t + i \cdot \frac{\tau - t}{n}\right) \right. \right. \tag{51}$$

$$\left. \left. + \sum_{i=0}^{n-2} f'\left(t + (i+1) \cdot \frac{\tau - t}{n}\right) + f'(\tau) \right] \right|$$

$$\leq \begin{cases} \dfrac{(\tau - t)}{4n} \left\| f'' \right\|_{[t,\tau],\infty} & \text{if } f'' \in L_{\infty}[a,b]; \\ \dfrac{|\tau - t|^{\frac{1}{q}}}{2(q+1)^{\frac{1}{q}} n} \left\| f'' \right\|_{[t,\tau],p} & \text{if } f'' \in L_p[a,b], \\ & p > 1,\ \frac{1}{p} + \frac{1}{q} = 1; \\ \dfrac{1}{2n} \left\| f'' \right\|_{[t,\tau],1}. \end{cases}$$

However,

$$\sum_{i=1}^{n-1} f'\left(t + i \cdot \frac{\tau - t}{n}\right) = \sum_{i=0}^{n-2} f'\left(t + (i+1) \cdot \frac{\tau - t}{n}\right)$$

and then, by (51), we may write

$$\left| \frac{f(\tau) - f(t)}{\tau - t} - \left[\frac{f'(t) + f'(\tau)}{2n} + \frac{1}{n} \sum_{i=1}^{n-1} f'\left(t + i \cdot \frac{\tau - t}{n}\right) \right] \right| \tag{52}$$

$$\leq \begin{cases} \dfrac{|\tau - t|}{4n} \left\| f'' \right\|_{[t,\tau],\infty} \\ \dfrac{|\tau - t|^{\frac{1}{q}}}{2(q+1)^{\frac{1}{q}} n} \left\| f'' \right\|_{[t,\tau],p} \\ \dfrac{1}{2n} \left\| f'' \right\|_{[t,\tau],1} \end{cases}$$

for any $t, \tau \in [a,b]$, $t \neq \tau$.

Consequently, we have

$$\left|\frac{1}{\pi}PV\int_a^b \frac{f(\tau)-f(t)}{\tau-t}d\tau - \frac{1}{\pi}PV\int_a^b\left[\frac{f'(t)+f'(\tau)}{2n}\right.\right. \tag{53}$$

$$\left.\left. + \frac{1}{n}\sum_{i=1}^{n-1} f'\left(t+i\cdot\frac{\tau-t}{n}\right)\right]d\tau\right|$$

$$\leq \begin{cases} \dfrac{1}{4\pi n}PV\displaystyle\int_a^b |\tau-t|\,\|f''\|_{[t,\tau],\infty}\,d\tau, \\[2ex] \dfrac{1}{2\pi(q+1)^{\frac{1}{q}}n}PV\displaystyle\int_a^b |\tau-t|^{\frac{1}{q}}\,\|f''\|_{[t,\tau],p}\,d\tau, \\[2ex] \dfrac{1}{2\pi n}PV\displaystyle\int_a^b \|f''\|_{[t,\tau],1}\,d\tau. \end{cases}$$

Since

$$PV\int_a^b\left[\frac{f'(t)+f'(\tau)}{2n}+\frac{1}{n}\sum_{i=1}^{n-1} f'\left(t+i\cdot\frac{\tau-t}{n}\right)\right]d\tau$$

$$= \lim_{\varepsilon\to 0+}\left(\int_a^{t-\varepsilon}+\int_{t+\varepsilon}^b\right)\left(\frac{f'(t)+f'(\tau)}{2n}+\frac{1}{n}\sum_{i=1}^{n-1} f'\left(t+i\cdot\frac{\tau-t}{n}\right)\right)d\tau$$

$$= \frac{f'(t)(b-a)+f(b)-f(a)}{2n}$$

$$+\frac{1}{n}\sum_{i=1}^{n-1}\left[\lim_{\varepsilon\to 0+}\left(\int_a^{t-\varepsilon}+\int_{t+\varepsilon}^b\right)\left(f'\left(t+i\cdot\frac{\tau-t}{n}\right)\right)d\tau\right]$$

$$= \frac{f'(t)(b-a)+f(b)-f(a)}{2n}$$

$$+\frac{1}{n}\sum_{i=1}^{n-1}\left[\lim_{\varepsilon\to 0+}\left[\frac{n}{i}\cdot f\left(t+i\cdot\frac{\tau-t}{n}\right)\Big|_a^{t-\varepsilon}+\frac{n}{i}\cdot f\left(t+i\cdot\frac{\tau-t}{n}\right)\Big|_{t+\varepsilon}^b\right]\right]$$

$$= \frac{f'(t)(b-a) + f(b) - f(a)}{2n} + \frac{1}{n}\sum_{i=1}^{n-1}\frac{n}{i}\left[f\left(t + i\cdot\frac{b-t}{n}\right) - f\left(t + i\cdot\frac{a-t}{n}\right)\right]$$

$$= \frac{f'(t)(b-a) + f(b) - f(a)}{2n} + \frac{b-a}{n}\sum_{i=1}^{n-1}\left[f; t + i\cdot\frac{b-t}{n}, t + i\cdot\frac{a-t}{n}\right],$$

and

$$\begin{aligned} PV\int_a^b |\tau - t| \left\|f''\right\|_{[t,\tau],\infty} d\tau &\le \left\|f''\right\|_{[a,b],\infty} PV\int_a^b |\tau - t|\, d\tau \\ &= \left\|f''\right\|_{[a,b],\infty} \frac{(t-a)^2 + (b-t)^2}{2} \\ &= \left\|f''\right\|_{[a,b],\infty}\left[\frac{1}{4}(b-a)^2 + \left(t - \frac{a+b}{2}\right)^2\right], \end{aligned}$$

$$\begin{aligned} PV\int_a^b |\tau - t|^{\frac{1}{q}} \left\|f''\right\|_{[t,\tau],p} d\tau &\le \left\|f''\right\|_{[a,b],p} PV\int_a^b |\tau - t|^{\frac{1}{q}} d\tau \\ &= \left\|f''\right\|_{[a,b],p}\left[\frac{(t-a)^{1+\frac{1}{q}} + (b-t)^{1+\frac{1}{q}}}{1+\frac{1}{q}}\right] \\ &= \frac{q\left\|f''\right\|_{[a,b],p}}{q+1}\left[(t-a)^{1+\frac{1}{q}} + (b-t)^{1+\frac{1}{q}}\right] \end{aligned}$$

and

$$\begin{aligned} PV\int_a^b \left\|f''\right\|_{[t,\tau],1} d\tau &= PV\left[\int_a^t \left\|f''\right\|_{[\tau,t],1} d\tau + \int_\tau^b \left\|f''\right\|_{[t,\tau],1} d\tau\right] \\ &\le \left\|f''\right\|_{[a,t],1}(t-a) + \left\|f''\right\|_{[t,b],1}(b-t) \\ &\le \max(t-a, b-t)\left[\left\|f''\right\|_{[a,t],1} + \left\|f''\right\|_{[t,b],1}\right] \\ &= \left[\frac{1}{2}(b-a) + \left|t - \frac{a+b}{2}\right|\right]\left\|f''\right\|_{[a,b],1} \end{aligned}$$

then, by (53) we get

$$\left| \frac{1}{\pi} PV \int_a^b \frac{f(\tau) - f(t)}{\tau - t} d\tau - \frac{f'(t)(b-a) + f(b) - f(a)}{2n\pi} \right. \tag{54}$$

$$\left. - \frac{b-a}{n\pi} \sum_{i=1}^{n-1} \left[f; t - \frac{t-a}{n} \cdot i, t + \frac{b-t}{n} \cdot i \right] \right|$$

$$\leq \begin{cases} \dfrac{\|f''\|_{[a,b],\infty}}{4\pi n} \left[\frac{1}{4}(b-a)^2 + \left(t - \frac{a+b}{2}\right)^2 \right] & \text{if } f'' \in L_\infty [a,b]; \\ \dfrac{q \|f''\|_{[a,b],p}}{2\pi (q+1)^{1+\frac{1}{q}} n} \left[(t-a)^{1+\frac{1}{q}} + (b-t)^{1+\frac{1}{q}} \right] & \text{if } f'' \in L_p [a,b], \\ & p > 1, \ \frac{1}{p} + \frac{1}{q} = 1; \\ \dfrac{\|f''\|_{[a,b],1}}{2\pi n} \left[\frac{1}{2}(b-a) + \left| t - \frac{a+b}{2} \right| \right]. \end{cases}$$

On the other hand, as for the function $f_0 : (a,b) \to \mathbb{R}$, $f_0(t) = 1$, we have

$$(Tf_0)(a,b;t) = \frac{1}{\pi} \ln \left(\frac{b-t}{t-a} \right), \quad t \in (a,b)$$

then obviously

$$\begin{aligned} (Tf)(a,b;t) &= \frac{1}{\pi} PV \int_a^b \frac{f(\tau) - f(t) + f(t)}{\tau - t} d\tau \\ &= \frac{1}{\pi} PV \int_a^b \frac{f(\tau) - f(t)}{\tau - t} d\tau + \frac{f(t)}{\pi} PV \int_a^b \frac{d\tau}{\tau - t}, \end{aligned}$$

from where we get the equality:

$$(Tf)(a,b;t) - \frac{f(t)}{\pi} \ln \left(\frac{b-t}{t-a} \right) = \frac{1}{\pi} PV \int_a^b \frac{f(\tau) - f(t)}{\tau - t} d\tau. \tag{55}$$

Finally, using (54) and (55), we deduce (50).

Before we proceed with another estimate of the remainder in approximating the Hilbert transform for functions whose second derivatives are absolutely continuous, we need the following lemma (see, for example, [9, p. 39]).

Lemma 3 *Let $u : [a,b] \to \mathbb{R}$ be a function such that its derivative is absolutely continuous on $[a,b]$. Then one has the inequalities*

$$\left| \int_a^b u(s)\,ds - \frac{u(a)+u(b)}{2}(b-a) \right| \tag{56}$$

$$\leq \begin{cases} \dfrac{(b-a)^3}{12} \left\| u'' \right\|_{[a,b],\infty} & \textit{if } u'' \in L_\infty[a,b]; \\ \dfrac{(b-a)^{2+\frac{1}{q}}}{2} \left[B(q+1,q+1) \right]^{\frac{1}{q}} \left\| u'' \right\|_{[a,b],p} & \textit{if } u'' \in L_p[a,b], \\ & p>1,\ \frac{1}{p}+\frac{1}{q}=1; \\ \dfrac{(b-a)^2}{8} \left\| u'' \right\|_{[a,b],1}. \end{cases}$$

where $B(\cdot,\cdot)$ is the Beta function

$$B(\alpha,\beta) := \int_0^1 t^{\alpha-1}(1-t)^{\beta-1}\,dt, \quad \alpha,\beta>0.$$

A simple proof of the fact can be done by the use of the following identity:

$$\int_a^b u(s)\,ds - \frac{u(a)+u(b)}{2}(b-a) = -\frac{1}{2}\int_a^b (b-s)(s-a)\,u''(s)\,ds, \tag{57}$$

and we omit the details.

The following lemma also holds.

Lemma 4 *Let $u:[a,b]\to\mathbb{R}$ be a differentiable function such that $u':[a,b]\to\mathbb{R}$ is absolutely continuous on $[a,b]$. Then for any $t,\tau\in[a,b]$, $t\neq\tau$ and $n\in\mathbb{N}$, $n\geq 1$, we have the inequality:*

$$\left| \frac{1}{\tau-t}\int_t^\tau u(s)\,ds - \frac{1}{2n}\sum_{i=0}^{n-1}\left[u\left(t+i\cdot\frac{\tau-t}{n}\right) + u\left(t+(i+1)\cdot\frac{\tau-t}{n}\right)\right] \right| \tag{58}$$

$$\leq \begin{cases} \dfrac{|\tau-t|^2}{12n^2} \left\| u'' \right\|_{[t,\tau],\infty} & \textit{if } u'' \in L_\infty[a,b]; \\ \dfrac{|\tau-t|^{1+\frac{1}{q}}}{2n^2} \left[B(q+1,q+1) \right]^{\frac{1}{q}} \left\| u'' \right\|_{[t,\tau],p} & \textit{if } u'' \in L_p[a,b], \\ & p>1,\ \frac{1}{p}+\frac{1}{q}=1; \\ \dfrac{|\tau-t|}{8n^2} \left\| u'' \right\|_{[t,\tau],1}, \end{cases}$$

where $B(\cdot,\cdot)$ is the Beta function.

Proof Consider the equidistant division of $[t,\tau]$ (or $[\tau,t]$)

$$E_n : x_i = t + i \cdot \frac{\tau - t}{n}, \quad i = \overline{0,n}.$$

If we apply the inequality (56), we may state that

$$\left| \int_{x_i}^{x_{i+1}} u(s)\,ds - \frac{u\left(t + i \cdot \frac{\tau-t}{n}\right) + u\left(t + (i+1) \cdot \frac{\tau-t}{n}\right)}{2} \cdot \frac{\tau - t}{n} \right|$$

$$\leq \begin{cases} \dfrac{|\tau-t|^3}{12n^3} \left\|u''\right\|_{[x_i,x_{i+1}],\infty} & \text{if } u'' \in L_\infty[a,b]; \\ \dfrac{|\tau-t|^{2+\frac{1}{q}}}{2n^{2+\frac{1}{q}}} \left[B(q+1,q+1)\right]^{\frac{1}{q}} \left\|u''\right\|_{[x_i,x_{i+1}],p} & \text{if } u'' \in L_p[a,b], \\ & p>1,\ \frac{1}{p}+\frac{1}{q}=1; \\ \dfrac{|\tau-t|^2}{8n^2} \left\|u''\right\|_{[x_i,x_{i+1}],1}. \end{cases}$$

Dividing by $|\tau - t| > 0$ and using a similar argument to the one in Lemma 2, we conclude that the desired inequality (58) holds.

The following theorem in approximating the Hilbert transform of a twice differentiable function whose second derivative f'' is absolutely continuous also holds.

Theorem 12 (Dragomir et al. [16]) *Let $f : [a,b] \to \mathbb{R}$ be a twice differentiable function such that the second derivative f'' is absolutely continuous on $[a,b]$. Then*

$$\left| (Tf)(a,b;t) - \frac{f(t)}{\pi} \ln\left(\frac{b-t}{t-a}\right) - T_n(f;t) \right| \tag{59}$$

$$\leq \begin{cases} \dfrac{1}{12n^2\pi} \left[\dfrac{(b-a)^2}{12} + \left(t - \dfrac{a+b}{2}\right)^2 \right] (b-a) \left\|f'''\right\|_{[a,b],\infty} \\ \qquad \text{if } f''' \in L_\infty[a,b]; \\ \dfrac{q\left[B(q+1,q+1)\right]^{\frac{1}{q}}}{2(2q+1)n^2\pi} \left[(t-a)^{2+\frac{1}{q}} + (b-t)^{2+\frac{1}{q}} \right] \left\|f'''\right\|_{[a,b],p} \\ \qquad \text{if } f''' \in L_p[a,b],\ p>1,\ \frac{1}{p}+\frac{1}{q}=1; \\ \dfrac{1}{8\pi n^2} \left[\dfrac{(b-a)^2}{4} + \left(t - \dfrac{a+b}{2}\right)^2 \right] \left\|f'''\right\|_{[a,b],1} \end{cases}$$

$$\leq \begin{cases} \dfrac{(b-a)^3}{36\pi n^2} \left\| f''' \right\|_{[a,b],\infty} & \text{if } f''' \in L_\infty [a,b]; \\ \dfrac{q\left[B\left(q+1,q+1\right)\right]^{\frac{1}{q}}(b-a)^{2+\frac{1}{q}}}{2\pi\left(2q+1\right)n^2} \left\| f''' \right\|_{[a,b],p} & \text{if } f''' \in L_p [a,b], \\ & p>1,\ \frac{1}{p}+\frac{1}{q}=1; \\ \dfrac{\left\| f''' \right\|_{[a,b],1}}{16\pi n^2}(b-a) \end{cases}$$

for all $t \in (a,b)$, where $T_n(f;t)$ is defined by (48).

Proof Applying Lemma 4 for the function f'', we may write that (see also Theorem 11)

$$\left| \frac{f(\tau)-f(t)}{\tau - t} - \left[\frac{f'(t)+f'(\tau)}{2n} + \frac{1}{n}\sum_{i=0}^{n-2} f'\left(t + i\cdot\frac{\tau - t}{n}\right)\right]\right| \tag{60}$$

$$\leq \begin{cases} \dfrac{|\tau-t|^2}{12n^2} \left\| f''' \right\|_{[t,\tau],\infty} & \text{if } f''' \in L_\infty [a,b]; \\ \dfrac{|\tau-t|^{1+\frac{1}{q}}}{2n^2}\left[B\left(q+1,q+1\right)\right]^{\frac{1}{q}} \left\| f''' \right\|_{[t,\tau],p} & \text{if } f''' \in L_p [a,b], \\ & p>1,\ \frac{1}{p}+\frac{1}{q}=1; \\ \dfrac{|\tau-t|}{8n^2} \left\| f''' \right\|_{[t,\tau],1}. \end{cases}$$

for all $t, \tau \in [a,b]$, $t \neq \tau$.

Consequently, we may write

$$\left| \frac{1}{\pi} PV \int_a^b \frac{f(\tau)-f(t)}{\tau - t} d\tau - \frac{1}{\pi} PV \int_a^b \left[\frac{f'(t)+f'(\tau)}{2n} \right.\right. \tag{61}$$

$$\left.\left. + \frac{1}{n}\sum_{i=1}^{n-1} f'\left(t + i\cdot\frac{\tau - t}{n}\right)\right] d\tau \right|$$

$$\leq \begin{cases} \frac{1}{12n^2\pi} PV \int_a^b |\tau - t|^2 \left\| f''' \right\|_{[t,\tau],\infty} d\tau, \\ \\ \frac{[B(q+1,q+1)]^{\frac{1}{q}}}{2n^2} PV \int_a^b |\tau - t|^{1+\frac{1}{q}} \left\| f''' \right\|_{[t,\tau],p} d\tau, \\ \\ \frac{1}{8n^2} PV \int_a^b |\tau - t| \left\| f''' \right\|_{[t,\tau],1} d\tau. \end{cases}$$

Since

$$\begin{aligned} & PV \int_a^b |\tau - t|^2 \left\| f''' \right\|_{[t,\tau],\infty} d\tau \\ & \leq \left\| f''' \right\|_{[a,b],\infty} PV \int_a^b |\tau - t|^2 d\tau = \left\| f''' \right\|_{[a,b],\infty} \left[\frac{(t-a)^3 + (b-t)^3}{3} \right] \\ & = \left\| f''' \right\|_{[a,b],\infty} \left[\frac{(b-a)^2}{12} + \left(t - \frac{a+b}{2} \right)^2 \right] (b-a), \end{aligned}$$

$$\begin{aligned} PV \int_a^b |\tau - t|^{1+\frac{1}{q}} \left\| f''' \right\|_{[t,\tau],p} d\tau & \leq \left\| f''' \right\|_{[a,b],p} PV \int_a^b |\tau - t|^{1+\frac{1}{q}} d\tau \\ & = \left\| f''' \right\|_{[a,b],p} \frac{(b-t)^{2+\frac{1}{q}} + (t-a)^{2+\frac{1}{q}}}{2+\frac{1}{q}} \\ & = \frac{q \left\| f''' \right\|_{[a,b],p}}{2q+1} \left[(b-t)^{2+\frac{1}{q}} + (t-a)^{2+\frac{1}{q}} \right] \end{aligned}$$

and

$$\begin{aligned} & PV \int_a^b |\tau - t| \left\| f''' \right\|_{[t,\tau],1} d\tau \\ & \leq \left\| f''' \right\|_{[a,b],1} PV \int_a^b |\tau - t| d\tau = \frac{(t-a)^2 + (b-t)^2}{2} \left\| f''' \right\|_{[a,b],1} \\ & = \left[\frac{(b-a)^2}{4} + \left(t - \frac{a+b}{2} \right)^2 \right] \left\| f''' \right\|_{[a,b],1}. \end{aligned}$$

Then by (61), we deduce the first part of (59).

4 Inequalities of Midpoint Type

4.1 Midpoint Type Inequalities

The following result holds.

Theorem 13 (Dragomir et al. [15]) *Assume that the function* $f : [a, b] \to \mathbb{R}$ *is such that* $f' : [a, b] \to \mathbb{R}$ *is absolutely continuous on* $[a, b]$. *Then we have the inequality:*

$$\left| T(f)(a, b; t) - \frac{f(t)}{\pi} \ln\left(\frac{b-t}{t-a}\right) - \frac{2}{\pi}\left[f\left(\frac{b+t}{2}\right) - f\left(\frac{t+a}{2}\right)\right]\right| \tag{62}$$

$$\leq \begin{cases} \frac{\|f''\|_\infty}{4\pi}\left[\left(t - \frac{a+b}{2}\right)^2 + \frac{(b-a)^2}{4}\right] & \text{if } f'' \in L_\infty[a, b]; \\ \frac{q\|f''\|_p}{2\pi(q+1)^{1+\frac{1}{q}}}\left[(t-a)^{1+\frac{1}{q}} + (b-t)^{1+\frac{1}{q}}\right] & \text{if } f'' \in L_p[a, b], \\ & p > 1,\ \frac{1}{p} + \frac{1}{q} = 1; \\ \frac{\|f''\|_1}{2\pi}(b-a), \end{cases}$$

$$\leq \begin{cases} \frac{\|f''\|_\infty}{8\pi}(b-a)^2 & \text{if } f'' \in L_\infty[a, b]; \\ \frac{q\|f''\|_p}{\pi(q+1)^{1+\frac{1}{q}}}(b-a)^{1+\frac{1}{q}} & \text{if } f'' \in L_p[a, b], \\ & p > 1,\ \frac{1}{p} + \frac{1}{q} = 1; \\ \frac{1}{2\pi}\|f''\|_1(b-a), \end{cases}$$

for any $t \in (a, b)$. *The* $\|\cdot\|_p$, $p \in [1, \infty]$ *denote the usual norms, i.e.,*

$$\|g\|_\infty := \operatorname*{essup}_{t\in[a,b]} |g(t)| \quad \text{if } g \in L_\infty[a, b]$$

and

$$\|g\|_p := \left(\int_a^b |g(t)|^p\, dt\right)^{\frac{1}{p}} \quad \text{if } g \in L_p[a, b],\ p \geq 1.$$

Proof As for the mapping $f_0 : (a, b) \to \mathbb{R}$, $f_0(t) = 1, t \in (a, b)$, we have

$$T(f_0)(a, b; t) = \frac{1}{\pi} \ln\left(\frac{b-t}{t-a}\right), \ t \in (a, b),$$

then, obviously

$$\begin{aligned}(Tf)(a, b; t) &= \frac{1}{\pi} PV \int_a^b \frac{f(\tau) - f(t) + f(t)}{\tau - t} d\tau \\ &= \frac{1}{\pi} PV \int_a^b \frac{f(\tau) - f(t)}{\tau - t} d\tau + \frac{f(t)}{\pi} PV \int_a^b \frac{d\tau}{\tau - t},\end{aligned}$$

from where we get the identity

$$(Tf)(a, b; t) - \frac{f(t)}{\pi} \ln\left(\frac{b-t}{t-a}\right) = \frac{1}{\pi} PV \int_a^b \frac{f(\tau) - f(t)}{\tau - t} d\tau. \tag{63}$$

If we use the known identity, which can easily be proved using the integration by parts formula,

$$\int_\alpha^\beta g(u)\, du = g\left(\frac{\alpha+\beta}{2}\right)(\beta - \alpha) + \int_\alpha^\beta K(u)\, g'(u)\, du, \tag{64}$$

where

$$K(u) := \begin{cases} u - \alpha \text{ if } u \in \left[\alpha, \frac{\alpha+\beta}{2}\right] \\ \\ u - \beta \text{ if } u \in \left(\frac{\alpha+\beta}{2}, \beta\right] \end{cases}$$

and g is absolutely continuous on $[a, b]$, we may write

$$\begin{aligned} PV &\int_a^b \frac{f(\tau) - f(t)}{\tau - t} d\tau = PV \int_a^b \frac{\int_t^\tau f'(u)\, du}{\tau - t} d\tau \\ &= PV \int_a^b \left[\frac{f'\left(\frac{\tau+t}{2}\right)(\tau - t) + \int_t^\tau K(u)\, f''(u)\, du}{\tau - t}\right] d\tau \\ &= PV \int_a^b f'\left(\frac{\tau+t}{2}\right) d\tau + PV \int_a^b \left(\frac{1}{\tau - t} \int_t^\tau K(u)\, f''(u)\, du\right) d\tau \\ &= 2\left[f\left(\frac{b+t}{2}\right) - f\left(\frac{a+t}{2}\right)\right] + PV \int_a^b \left(\frac{1}{\tau - t} \int_t^\tau K(u)\, f''(u)\, du\right) d\tau. \end{aligned}$$

Consequently, by (63), we obtain the identity

$$(Tf)(a,b;t) - \frac{f(t)}{\pi}\ln\left(\frac{b-t}{t-a}\right) - \frac{2}{\pi}\left[f\left(\frac{b+t}{2}\right) - f\left(\frac{t+a}{2}\right)\right] \tag{65}$$
$$= \frac{1}{\pi}PV\int_a^b\left(\frac{1}{\tau-t}\int_t^\tau K(u)f''(u)\,du\right)d\tau,$$

where

$$K(u) = \begin{cases} u-t & \text{if } u\in\left[t,\frac{\tau+t}{2}\right] \\ u-\tau & \text{if } u\in\left(\frac{\tau+t}{2},\tau\right] \end{cases}.$$

Using the properties of modulus, we get, by (65), that

$$\left|(Tf)(a,b;t) - \frac{f(t)}{\pi}\ln\left(\frac{b-t}{t-a}\right) - \frac{2}{\pi}\left[f\left(\frac{b+t}{2}\right) - f\left(\frac{t+a}{2}\right)\right]\right| \tag{66}$$
$$\le \frac{1}{\pi}PV\int_a^b\left|\frac{1}{\tau-t}\int_t^\tau K(u)f''(u)\,du\right|d\tau =: D(a,b;t).$$

Now, it is obvious that

$$\begin{aligned}\left|\int_t^\tau K(u)f''(u)\,du\right| &\le \sup_{u\in[a,b]}\left|f''(u)\right|\left|\int_t^\tau K(u)\,du\right| \\ &= \left\|f''\right\|_\infty\left|\int_t^{\frac{\tau+t}{2}}(u-t)\,du + \int_{\frac{\tau+t}{2}}^\tau (t-u)\,du\right| \\ &= \left\|f''\right\|_\infty\frac{(t-\tau)^2}{4}.\end{aligned}$$

Then

$$\begin{aligned}D(a,b;t) &\le \frac{1}{4\pi}\left\|f''\right\|_\infty PV\int_a^b |t-\tau|\,d\tau \\ &= \frac{\left\|f''\right\|_\infty}{4\pi}\cdot\frac{(t-a)^2+(b-t)^2}{2} \\ &= \frac{\left\|f''\right\|_\infty}{4\pi}\left[\left(t-\frac{a+b}{2}\right)^2 + \frac{(b-a)^2}{4}\right].\end{aligned}$$

Using Hölder's integral equality, we have

$$\begin{aligned}\left|\int_t^\tau K(u) f''(u)\,du\right| &\le \left|\int_t^\tau |f''(u)|^p du\right|^{\frac{1}{p}} \left|\int_t^\tau |K(u)|^q du\right|^{\frac{1}{q}} \\ &\le \|f''\|_p \left|\int_t^\tau |K(u)|^q du\right|^{\frac{1}{q}} \\ &= \|f''\|_p \left|\int_t^{\frac{\tau+t}{2}} (u-t)^q du + \int_{\frac{\tau+t}{2}}^\tau (t-u)^q du\right|^{\frac{1}{q}} \\ &= \|f''\|_p \left[\frac{|\tau-t|^{q+1}}{2^q(q+1)}\right]^{\frac{1}{q}} = \frac{\|f''\|_p |t-\tau|^{1+\frac{1}{q}}}{2(q+1)^{\frac{1}{q}}}\end{aligned}$$

for all $t, \tau \in (a,b)$.

Then

$$\begin{aligned}D(a,b;t) &\le \frac{1}{\pi}\|f''\|_p PV \int_a^b \frac{|t-\tau|^{\frac{1}{q}}}{2(q+1)^{\frac{1}{q}}} d\tau \\ &= \frac{q\|f''\|_p \left[(t-a)^{1+\frac{1}{q}} + (b-t)^{1+\frac{1}{q}}\right]}{2\pi (q+1)^{1+\frac{1}{q}}}\end{aligned}$$

proving the second part of the first inequality in (62).

Finally, we observe that

$$\left|\int_t^\tau K(u) f''(u)\,du\right| \le \sup_{u\in[t,\tau]} |K(u)| \left|\int_t^\tau |f''(u)|\,du\right| = \frac{\|f''\|_1}{2\pi}|t-\tau|$$

and then

$$D(a,b;t) \le \frac{1}{2\pi}\|f''\|_1 PV \int_a^b d\tau = \frac{1}{2\pi}\|f''\|_1 (b-a),$$

proving the last part of the second inequality in (62).

The last part of (62) is obvious.

The best inequality we can get from (62) is embodied in the following corollary.

Corollary 10 *With the assumptions in Theorem 13, we have*

$$\left|(Tf)\left(a,b;\frac{a+b}{2}\right) - \frac{2}{\pi}\left[f\left(\frac{a+3b}{4}\right) - f\left(\frac{3a+b}{4}\right)\right]\right| \tag{67}$$

$$\leq \begin{cases} \frac{1}{16\pi} \left\| f'' \right\|_{\infty} (b-a)^2 & \text{if } f'' \in L_{\infty}[a,b]; \\ \frac{q}{2^{1+\frac{1}{q}} \pi (q+1)^{1+\frac{1}{q}}} \left\| f'' \right\|_p (b-a)^{1+\frac{1}{q}} & \text{if } f'' \in L_p[a,b], \\ & p>1,\ \frac{1}{p}+\frac{1}{q}=1; \end{cases}$$

Remark 6 It is also obvious that if $b-a \to 0$, then both the inequalities (62) and (67) provide accurate approximations.

4.2 Other Midpoint Type Inequalities

The following result holds.

Theorem 14 (Dragomir et al. [15]) *Assume that the function $f:[a,b]\to\mathbb{R}$ is such that $f'':[a,b]\to\mathbb{R}$ is absolutely continuous on $[a,b]$. Then we have the inequalities:*

$$\left| (Tf)(a,b;t) - \frac{f(t)}{\pi} \ln\left(\frac{b-t}{t-a}\right) - \frac{2}{\pi}\left[f\left(\frac{b+t}{2}\right) - f\left(\frac{t+a}{2}\right)\right]\right| \tag{68}$$

$$\leq \begin{cases} \frac{\|f'''\|_{\infty}}{24\pi}(b-a)\left[\frac{(b-a)^2}{12}+\left(t-\frac{a+b}{2}\right)^2\right] & \text{if } f''' \in L_{\infty}[a,b]; \\ \frac{q\|f'''\|_p}{8\pi(2q+1)^{1+\frac{1}{q}}}\left[(b-t)^{2+\frac{1}{q}}+(t-a)^{2+\frac{1}{q}}\right] & \text{if } f''' \in L_p[a,b], \\ & p>1,\ \frac{1}{p}+\frac{1}{q}=1; \\ \frac{\|f'''\|_1}{8\pi}\left[\frac{(b-a)^2}{4}+\left(t-\frac{a+b}{2}\right)^2\right], & \end{cases}$$

$$\leq \begin{cases} \frac{\|f'''\|_{\infty}(b-a)^3}{72\pi} & \text{if } f''' \in L_{\infty}[a,b]; \\ \frac{q\|f'''\|_p(b-a)^{2+\frac{1}{q}}}{8\pi(2q+1)^{1+\frac{1}{q}}} & \text{if } f''' \in L_p[a,b], \\ & p>1,\ \frac{1}{p}+\frac{1}{q}=1; \\ \frac{\|f'''\|_1(b-a)^2}{16\pi}. & \end{cases}$$

Proof If we use the identity (63) and the following identity, which can be proved by applying the integration by parts formula twice,

$$\int_{\alpha}^{\beta} g(u)\,du = g\left(\frac{\alpha+\beta}{2}\right)(\beta-\alpha) + \frac{1}{2}\int_{\alpha}^{\beta} L(u)\,g''(u)\,du,$$

where

$$L(u) := \begin{cases} (u-\alpha)^2 \text{ if } u \in \left[\alpha, \frac{\alpha+\beta}{2}\right] \\ \\ (u-\beta)^2 \text{ if } u \in \left(\frac{\alpha+\beta}{2}, \beta\right] \end{cases}$$

and g is such that g' is absolutely continuous on $[\alpha, \beta]$, then we get

$$\begin{aligned} &(Tf)(a,b;t) - \frac{f(t)}{\pi}\ln\left(\frac{b-t}{t-a}\right) \\ &= \frac{1}{\pi} PV \int_a^b \left[\frac{f'\left(\frac{\tau+t}{2}\right)(\tau-t)\frac{1}{2}\int_t^{\tau} L(u)\,f'''(u)\,du}{\tau-t}\right] d\tau \\ &= \frac{2}{\pi}\left[f\left(\frac{b+t}{2}\right) - f\left(\frac{t+a}{2}\right)\right] + \frac{1}{2\pi} PV \int_a^b \left[\frac{1}{\tau-t}\int_t^{\tau} L(u)\,f'''(u)\,du\right] d\tau. \end{aligned}$$

Consequently, we have the identity:

$$\begin{aligned} &(Tf)(a,b;t) - \frac{f(t)}{\pi}\ln\left(\frac{b-t}{t-a}\right) - \frac{2}{\pi}\left[f\left(\frac{b+t}{2}\right) - f\left(\frac{t+a}{2}\right)\right] \\ &= \frac{1}{2\pi} PV \int_a^b \left[\frac{1}{\tau-t}\int_t^{\tau} L(u)\,f'''(u)\,du\right] d\tau, \end{aligned} \tag{69}$$

where

$$L(u) = \begin{cases} (u-t)^2 \text{ if } u \in \left[t, \frac{\tau+t}{2}\right] \\ \\ (u-\tau)^2 \text{ if } u \in \left(\frac{\tau+t}{2}, \tau\right] \end{cases}.$$

Using the modulus properties, we may write, by (69), that

$$\begin{aligned} &\left|(Tf)(a,b;t) - \frac{f(t)}{\pi}\ln\left(\frac{b-t}{t-a}\right) - \frac{2}{\pi}\left[f\left(\frac{b+t}{2}\right) - f\left(\frac{t+a}{2}\right)\right]\right| \\ &\le \frac{1}{2\pi} PV \int_a^b \left|\frac{1}{\tau-t}\int_t^{\tau} |L(u)|\left|f'''(u)\right| du\right| d\tau =: E(a,b;t). \end{aligned} \tag{70}$$

Now, observe that

$$\left|\int_t^\tau |L(u)|\left|f'''(u)\right|du\right| \le \left\|f'''\right\|_\infty \left|\int_t^{\frac{\tau+t}{2}} (u-t)^2\,du + \int_{\frac{\tau+t}{2}}^\tau (t-u)^2\,du\right|$$
$$= \frac{\left\|f'''\right\|_\infty}{12}|t-\tau|^3$$

and then

$$E(a,b;t) \le \frac{\left\|f'''\right\|_\infty}{24\pi}\int_a^b (t-\tau)^2\,d\tau = \frac{\left\|f'''\right\|_\infty}{24\pi}\cdot\frac{(b-t)^3+(t-a)^3}{3}$$
$$= \frac{\left\|f'''\right\|_\infty}{24\pi}\left[\frac{(b-a)^2}{12}+\left(t-\frac{a+b}{2}\right)^2\right](b-a),$$

giving the first part of the first inequality in (68).

Using Hölder's inequality, we may write that

$$\left|\int_t^\tau |L(u)|\left|f'''(u)\right|du\right| \le \left\|f'''\right\|_p \left|\int_t^{\frac{\tau+t}{2}} |u-t|^{2q}\,du + \int_{\frac{\tau+t}{2}}^\tau |t-u|^{2q}\,du\right|^{\frac{1}{q}}$$
$$= \left\|f'''\right\|_p\left[\frac{2\cdot\left|\frac{t-\tau}{2}\right|^{2q+1}}{2q+1}\right]^{\frac{1}{q}} = \frac{1}{4(2q+1)^{\frac{1}{q}}}\left\|f'''\right\|_p|t-\tau|^{2+\frac{1}{q}}$$

and then

$$E(a,b;t) \le \frac{\left\|f'''\right\|_p}{8\pi(2q+1)^{\frac{1}{q}}}PV\int_a^b |t-\tau|^{1+\frac{1}{q}}\,d\tau$$
$$= \frac{\left\|f'''\right\|_p}{8\pi(2q+1)^{\frac{1}{q}}}\cdot\frac{(b-t)^{2+\frac{1}{q}}+(t-a)^{2+\frac{1}{q}}}{\frac{2q+1}{q}}$$
$$= \frac{q\left\|f'''\right\|_p}{8\pi(2q+1)^{\frac{1}{q}}}\left[(b-t)^{2+\frac{1}{q}}+(t-a)^{2+\frac{1}{q}}\right],$$

which proves the second part of the first inequality in (68).

Finally,

$$\left|\int_t^\tau |L(u)|\left|f'''(u)\right|du\right| \le \sup_{u\in[t,\tau]}|L(u)|\left\|f'''\right\|_1 = \frac{|t-\tau|}{4}\left\|f'''\right\|_1,$$

giving

$$E(a,b;t) \le \frac{\|f'''\|_1}{8\pi} PV \int_a^b |t-\tau|\,d\tau = \frac{\|f'''\|_1}{8\pi} \cdot \frac{(b-t)^2+(t-a)^2}{2}$$
$$= \frac{\|f'''\|_1}{8\pi}\left[\frac{(b-a)^2}{4} + \left(t - \frac{a+b}{2}\right)^2\right],$$

which proves the last part of the first inequality in (68).

The best inequality we may obtain from (68) is embodied in the following corollary.

Corollary 11 *With the assumptions of Theorem 14, we have*

$$\left|(Tf)\left(a,b;\frac{a+b}{2}\right) - \frac{2}{\pi}\left[f\left(\frac{a+3b}{4}\right) - f\left(\frac{3a+b}{4}\right)\right]\right| \tag{71}$$

$$\le \begin{cases} \dfrac{\|f'''\|_\infty (b-a)^3}{288\pi} & \text{if } f''' \in L_\infty[a,b]; \\[2ex] \dfrac{q\,\|f'''\|_p (b-a)^{2+\frac{1}{q}}}{16 \cdot 2^{\frac{1}{q}} \pi (2q+1)^{\frac{1}{q}}} & \text{if } f''' \in L_p[a,b], \\ & \quad p>1,\ \frac{1}{p}+\frac{1}{q}=1; \\ \dfrac{\|f'''\|_1 (b-a)^2}{32\pi}. \end{cases}$$

4.3 Compounding Midpoint Type Inequalities

Before we point out the quadrature formula for the finite Hilbert transform, we need the following two technical lemmas:

Lemma 5 *Let $u : [a,b] \to \mathbb{R}$ be an absolutely continuous function on $[a,b]$. Then we have the midpoint inequalities:*

$$\left|\int_a^b u(s)\,ds - u\left(\frac{a+b}{2}\right)(b-a)\right| \tag{72}$$

$$\leq \begin{cases} \frac{(b-a)^2}{4} \left\| u' \right\|_{[a,b],\infty} & \text{if } u' \in L_\infty [a,b]; \\ \frac{(b-a)^{1+\frac{1}{q}}}{2(q+1)^{\frac{1}{q}}} \left\| u' \right\|_{[a,b],p} & \text{if } u' \in L_p [a,b], \\ & p > 1,\ \frac{1}{p} + \frac{1}{q} = 1; \\ \frac{(b-a)}{2} \left\| u' \right\|_{[a,b],1} . \end{cases}$$

A simple proof of this fact can be done by using the identity (see, for example, [9, p. 34]):

$$\int_a^b u(s)\,ds - u\left(\frac{a+b}{2}\right)(b-a) \tag{73}$$
$$= -\int_a^{\frac{a+b}{2}} (s-a) f'(s)\,ds + \int_{\frac{a+b}{2}}^b (s-b) f'(s)\,ds.$$

We omit the details.

The following lemma also holds.

Lemma 6 *Let $u : [a,b] \to \mathbb{R}$ be an absolutely continuous function on $[a,b]$. Then for any $t, \tau \in [a,b]$, $t \neq \tau$ and $n \in \mathbb{N}$, $n \geq 1$, we have the inequality:*

$$\left| \frac{1}{\tau - t} \int_t^\tau u(s)\,ds - \frac{1}{n} \sum_{i=0}^{n-1} u\left(t + \left(i + \frac{1}{2}\right) \frac{\tau - t}{n}\right) \right| \tag{74}$$

$$\leq \begin{cases} \frac{|\tau - t|}{4n} \left\| u' \right\|_{[t,\tau],\infty} & \text{if } u' \in L_\infty [a,b]; \\ \frac{|\tau - t|^{\frac{1}{q}}}{2(q+1)^{\frac{1}{q}} n} \left\| u' \right\|_{[t,\tau],p} & \text{if } u' \in L_p [a,b], \\ & p > 1,\ \frac{1}{p} + \frac{1}{q} = 1; \\ \frac{1}{2n} \left\| u' \right\|_{[t,\tau],1} , \end{cases}$$

where

$$\left\| u' \right\|_{[t,\tau],\infty} := \underset{\substack{t \in [t,\tau] \\ (t \in [\tau,t])}}{\text{essup}} \left| u'(t) \right|$$

and

$$\|u'\|_{[t,\tau],p} := \left| \int_t^{\tau} |u'(s)|^p \, ds \right|^{\frac{1}{p}}$$

for $p \geq 1$.

Proof Consider the equidistant division of $[t, \tau]$ (if $t < \tau$) or $[\tau, t]$ (if $\tau < t$) given by

$$E_n : x_i = t + i \cdot \frac{\tau - t}{n}, \quad i = \overline{0, n}.$$

If we apply the inequality (72) on the interval $[x_i, x_{i+1}]$, we may write that

$$\left| \int_{x_i}^{x_{i+1}} u(s) \, ds - \frac{\tau - t}{n} \cdot u\left(t + \left(i + \frac{1}{2}\right) \frac{\tau - t}{n}\right) \right|$$

$$\leq \begin{cases} \dfrac{(\tau - t)^2}{4n^2} \|u'\|_{[x_i, x_{i+1}],\infty} & \text{if } u' \in L_\infty [a, b]; \\[2ex] \dfrac{|\tau - t|^{1+\frac{1}{q}}}{2n^{1+\frac{1}{q}} (q + 1)^{\frac{1}{q}}} \|u'\|_{[x_i, x_{i+1}],p} & \text{if } u' \in L_p [a, b], \\ & p > 1, \ \frac{1}{p} + \frac{1}{q} = 1; \\[2ex] \dfrac{|\tau - t|}{2n} \|u'\|_{[x_i, x_{i+1}],1}, \end{cases}$$

from where we get

$$\left| \frac{1}{\tau - t} \int_{x_i}^{x_{i+1}} u(s) \, ds - \frac{1}{n} \cdot u\left(t + \left(i + \frac{1}{2}\right) \frac{\tau - t}{n}\right) \right|$$

$$\leq \begin{cases} \dfrac{|\tau - t|}{4n^2} \|u'\|_{[x_i, x_{i+1}],\infty}; \\[2ex] \dfrac{|\tau - t|^{\frac{1}{q}}}{2n^{1+\frac{1}{q}} (q + 1)^{\frac{1}{q}}} \|u'\|_{[x_i, x_{i+1}],p}; \\[2ex] \dfrac{1}{2n} \|u'\|_{[x_i, x_{i+1}],1}, \end{cases}$$

for all $i = \overline{0, n - 1}$.

Summing over i from 0 to $n-1$ and using the generalized triangle inequality, we may write

$$\left| \frac{1}{\tau - t} \int_t^\tau u(s)\, ds - \frac{1}{n} \sum_{i=0}^{n-1} u\left(t + \left(i + \frac{1}{2} \right) \frac{\tau - t}{n} \right) \right|$$

$$\leq \sum_{i=0}^{n-1} \left| \frac{1}{\tau - t} \int_{x_i}^{x_{i+1}} u(s)\, ds - \frac{1}{n} \cdot u\left(t + \left(i + \frac{1}{2} \right) \frac{\tau - t}{n} \right) \right|$$

$$\leq \begin{cases} \frac{|\tau - t|}{4n^2} \sum_{i=0}^{n-1} \left\| u' \right\|_{[x_i, x_{i+1}], \infty}; \\ \\ \frac{|\tau - t|^{\frac{1}{q}}}{2n^{1+\frac{1}{q}} (q+1)^{\frac{1}{q}}} \sum_{i=0}^{n-1} \left\| u' \right\|_{[x_i, x_{i+1}], p}; \\ \\ \frac{1}{2n} \sum_{i=0}^{n-1} \left\| u' \right\|_{[x_i, x_{i+1}], 1}. \end{cases}$$

However,

$$\sum_{i=0}^{n-1} \left\| u' \right\|_{[x_i, x_{i+1}], \infty} \leq n \left\| u' \right\|_{[t, \tau], \infty},$$

$$\begin{aligned} \sum_{i=0}^{n-1} \left\| u' \right\|_{[x_i, x_{i+1}], p} &\leq \sum_{i=0}^{n-1} \left| \int_{x_i}^{x_{i+1}} \left| u'(s) \right|^p ds \right|^{\frac{1}{p}} \\ &\leq n^{\frac{1}{q}} \left[\left(\sum_{i=0}^{n-1} \left| \int_{x_i}^{x_{i+1}} \left| u'(s) \right|^p ds \right|^{\frac{1}{p}} \right)^p \right]^{\frac{1}{p}} \\ &= n^{\frac{1}{q}} \left| \int_t^\tau \left| u'(s) \right|^p ds \right|^{\frac{1}{p}} = n^{\frac{1}{q}} \left\| u' \right\|_{[t, \tau], p} \end{aligned}$$

and

$$\sum_{i=0}^{n-1} \left\| u' \right\|_{[x_i, x_{i+1}], 1} = \sum_{i=0}^{n-1} \left| \int_{x_i}^{x_{i+1}} \left| u'(s) \right| ds \right| = \left| \int_t^\tau \left| u'(s) \right| ds \right| = \left\| u' \right\|_{[t, \tau], 1}$$

and the lemma is proved.

The following theorem in approximating the Hilbert transform of a differentiable function whose derivative f' is absolutely continuous holds.

Theorem 15 (Dragomir et al. [17]) *Let $f : [a,b] \to \mathbb{R}$ be a differentiable function such that its derivative f' is absolutely continuous on $[a,b]$. If*

$$M_n(f;t) = \frac{1}{n\pi}(b-a)\sum_{i=0}^{n-1}\left[f; t + \left(i+\frac{1}{2}\right)\frac{b-t}{n}, t - \left(i+\frac{1}{2}\right)\frac{t-a}{n}\right] \tag{75}$$

then we have the estimate:

$$\left|(Tf)(a,b;t) - \frac{f(t)}{\pi}\ln\left(\frac{b-t}{t-a}\right) - M_n(f;t)\right| \tag{76}$$

$$\leq \begin{cases} \frac{1}{4\pi n}\left[\frac{1}{4}(b-a)^2 + \left(t-\frac{a+b}{2}\right)^2\right]\|f''\|_{[a,b],\infty} & \text{if } f'' \in L_\infty[a,b]; \\ \frac{q}{2\pi n(q+1)^{\frac{1}{q}}}\left[(t-a)^{1+\frac{1}{q}} + (b-t)^{1+\frac{1}{q}}\right]\|f''\|_{[a,b],p} & \text{if } f'' \in L_p[a,b], \\ & p>1,\ \frac{1}{p}+\frac{1}{q}=1; \\ \frac{1}{2\pi n}\left[\frac{1}{2}(b-a) + \left|t-\frac{a+b}{2}\right|\right]\|f''\|_{[a,b],1}; \end{cases}$$

$$\leq \begin{cases} \frac{1}{8\pi n}(b-a)^2\|f''\|_{[a,b],\infty} & \text{if } f'' \in L_\infty[a,b]; \\ \frac{q}{2\pi n(q+1)^{\frac{1}{q}}}(b-a)^{1+\frac{1}{q}}\|f''\|_{[a,b],p} & \text{if } f'' \in L_p[a,b], \\ & p>1,\ \frac{1}{p}+\frac{1}{q}=1; \\ \frac{1}{2\pi n}(b-a)\|f''\|_{[a,b],1}; \end{cases}$$

for all $t \in (a,b)$, where $[f; c, d]$ denotes the divided difference

$$[f; c, d] := \frac{f(c) - f(d)}{c-d}.$$

Proof Applying Lemma 6 for the function f', we may write that

$$\left|\frac{f(\tau)-f(t)}{\tau - t} - \frac{1}{n}\sum_{i=0}^{n-1} f'\left(t + \left(i+\frac{1}{2}\right)\frac{(\tau - t)}{n}\right)\right|$$

$$\leq \begin{cases} \dfrac{|\tau - t|}{4n} \left\| f'' \right\|_{[t,\tau],\infty} \\[2ex] \dfrac{|\tau - t|^{\frac{1}{q}}}{2(q+1)^{\frac{1}{q}} n} \left\| f'' \right\|_{[t,\tau],p} \\[2ex] \dfrac{1}{2n} \left\| f'' \right\|_{[t,\tau],1}, \end{cases}$$

for any $t, \tau \in [a,b]$, $t \neq \tau$.

Consequently, we have

$$\left| \frac{1}{\pi} PV \int_a^b \frac{f(\tau) - f(t)}{\tau - t} d\tau - \frac{1}{n\pi} \sum_{i=0}^{n-1} PV \int_a^b f'\left(t + \left(i + \frac{1}{2}\right) \frac{\tau - t}{n}\right) d\tau \right| \tag{77}$$

$$\leq \begin{cases} \dfrac{1}{4\pi n} PV \displaystyle\int_a^b |\tau - t| \left\| f'' \right\|_{[t,\tau],\infty} d\tau & \text{if } f'' \in L_\infty[a,b]; \\[2ex] \dfrac{1}{2\pi (q+1)^{\frac{1}{q}} n} PV \displaystyle\int_a^b |\tau - t|^{\frac{1}{q}} \left\| f'' \right\|_{[t,\tau],p} d\tau & \text{if } f'' \in L_p[a,b], \\ & p > 1,\ \frac{1}{p} + \frac{1}{q} = 1; \\[2ex] \dfrac{1}{2\pi n} PV \displaystyle\int_a^b \left\| f'' \right\|_{[t,\tau],1} d\tau. \end{cases}$$

Since

$$\begin{aligned} &PV \int_a^b f'\left(t + \left(i + \frac{1}{2}\right) \frac{\tau - t}{n}\right) d\tau \\ &= \lim_{\varepsilon \to 0+} \left(\int_a^{t-\varepsilon} + \int_{t+\varepsilon}^b \right) \left(f'\left(t + \left(i + \frac{1}{2}\right) \frac{\tau - t}{n}\right)\right) d\tau \\ &= \lim_{\varepsilon \to 0+} \left[\frac{n}{i + \frac{1}{2}} f\left(t + \left(i + \frac{1}{2}\right) \frac{\tau - t}{n}\right)\Bigg|_a^{t-\varepsilon} + \frac{n}{i + \frac{1}{2}} f\left(t + \left(i + \frac{1}{2}\right) \frac{\tau - t}{n}\right)\Bigg|_{t+\varepsilon}^b \right] \\ &= \frac{n}{i + \frac{1}{2}} \left[f(t) - f\left(t + \left(i + \frac{1}{2}\right) \frac{a - t}{n}\right) + f\left(t + \left(i + \frac{1}{2}\right) \frac{b - t}{n}\right) - f(t) \right] \\ &= \frac{n}{i + \frac{1}{2}} \left[f\left(t + \left(i + \frac{1}{2}\right) \frac{b - t}{n}\right) - f\left(t + \left(i + \frac{1}{2}\right) \frac{a - t}{n}\right) \right] \\ &= (b - a) \left[f; t + \left(i + \frac{1}{2}\right) \frac{b - t}{n}, t - \left(i + \frac{1}{2}\right) \frac{t - a}{n} \right] \end{aligned}$$

and

$$PV\int_a^b |\tau - t| \, \|f''\|_{[t,\tau],\infty} \, d\tau \le \|f''\|_{[a,b],\infty} \, PV\int_a^b |\tau - t| \, d\tau$$
$$= \|f''\|_{[a,b],\infty}\left[\frac{1}{4}(b-a)^2 + \left(t - \frac{a+b}{2}\right)^2\right],$$

$$PV\int_a^b |\tau - t|^{\frac{1}{q}} \, \|f''\|_{[t,\tau],p} \, d\tau \le \|f''\|_{[a,b],p} \, PV\int_a^b |\tau - t|^{\frac{1}{q}} \, d\tau$$
$$= \frac{q\,\|f''\|_{[a,b],p}}{q+1}\left[(t-a)^{1+\frac{1}{q}} + (b-t)^{1+\frac{1}{q}}\right]$$

and

$$PV\int_a^b \|f''\|_{[t,\tau],1} \, d\tau = PV\left[\int_a^t \|f''\|_{[\tau,t],1} \, d\tau + \int_t^b \|f''\|_{[t,\tau],1} \, d\tau\right]$$
$$\le (t-a)\,\|f''\|_{[a,t],1} + (b-t)\,\|f''\|_{[t,b],1}$$
$$\le \max(t-a, b-t)\left[\|f''\|_{[a,t],1} + \|f''\|_{[t,b],1}\right]$$
$$= \left(\frac{1}{2}(b-a) + \left|t - \frac{a+b}{2}\right|\right)\|f''\|_{[a,b],1}$$

then by (77) we obtain

$$\left|\frac{1}{\pi}PV\int_a^b \frac{f(\tau) - f(t)}{\tau - t} d\tau - M_n(f;t)\right| \tag{78}$$

$$\le \begin{cases} \dfrac{1}{4\pi n}\left[\dfrac{1}{4}(b-a)^2 + \left(t - \dfrac{a+b}{2}\right)^2\right]\|f''\|_{[a,b],\infty} & \text{if } f'' \in L_\infty[a,b]; \\[2ex] \dfrac{q}{2\pi n (q+1)^{\frac{1}{q}}}\left[(t-a)^{1+\frac{1}{q}} + (b-t)^{1+\frac{1}{q}}\right]\|f''\|_{[a,b],p} & \text{if } f'' \in L_p[a,b], \\ & p>1,\ \frac{1}{p}+\frac{1}{q}=1; \\[2ex] \dfrac{1}{2\pi n}\left[\dfrac{1}{2}(b-a) + \left|t - \dfrac{a+b}{2}\right|\right]\|f''\|_{[a,b],1}. & \end{cases}$$

On the other hand, as for the function $f_0 : (a,b) \to \mathbb{R}$, $f_0(t) = 1$, we have

$$(Tf_0)(a,b;t) = \frac{1}{\pi}\ln\left(\frac{b-t}{t-a}\right), \quad t \in (a,b),$$

then obviously

$$
\begin{aligned}
(Tf)(a,b;t) &= \frac{1}{\pi} PV \int_a^b \frac{f(\tau) - f(t) + f(t)}{\tau - t} d\tau \\
&= \frac{1}{\pi} PV \int_a^b \frac{f(\tau) - f(t)}{\tau - t} d\tau + \frac{f(t)}{\pi} PV \int_a^b \frac{d\tau}{\tau - t}
\end{aligned}
$$

from where we get the equality:

$$
(Tf)(a,b;t) - \frac{f(t)}{\pi} \ln\left(\frac{b-t}{t-a}\right) = \frac{1}{\pi} PV \int_a^b \frac{f(\tau) - f(t)}{\tau - t} d\tau. \tag{79}
$$

Finally, using (78) and (79), we deduce (76).

Before we go further and point out another estimate of the remainder in approximating the Hilbert transform for functions whose second derivatives are absolutely continuous, we need the following lemma.

Lemma 7 *Let $u : [a,b] \to \mathbb{R}$ be a function such that its derivative is absolutely continuous on $[a,b]$. Then one has the inequalities*

$$
\left| \int_a^b u(s)\,ds - u\left(\frac{a+b}{2}\right)(b-a) \right| \tag{80}
$$

$$
\leq \begin{cases}
\dfrac{(b-a)^3}{24} \left\| u'' \right\|_{[a,b],\infty} & \text{if } u'' \in L_\infty[a,b]; \\[2ex]
\dfrac{|b-a|^{2+\frac{1}{q}}}{8(2q+1)^{\frac{1}{q}}} \left\| u'' \right\|_{[a,b],p} & \text{if } u'' \in L_p[a,b], \\
& p > 1,\ \frac{1}{p} + \frac{1}{q} = 1; \\[2ex]
\dfrac{(b-a)^2}{8} \left\| u'' \right\|_{[a,b],1}.
\end{cases}
$$

A simple proof of this inequality may be done by using the identity:

$$
\begin{aligned}
&\int_a^b u(s)\,ds - u\left(\frac{a+b}{2}\right)(b-a) \\
&= \frac{1}{2} \int_a^{\frac{a+b}{2}} (s-a)^2 f''(s)\,ds + \frac{1}{2} \int_{\frac{a+b}{2}}^b (b-s)^2 f''(s)\,ds.
\end{aligned} \tag{81}
$$

We omit the details.

The following lemma also holds.

Lemma 8 *Let $u : [a, b] \to \mathbb{R}$ be a differentiable function such that $u' : [a, b] \to \mathbb{R}$ is absolutely continuous on $[a, b]$. Then for any $t, \tau \in [a, b]$, $t \in \tau$ and $n \in \mathbb{N}$, $n \geq 1$, we have the inequality:*

$$\left| \frac{1}{\tau - t} \int_t^\tau u(s)\, ds - \frac{1}{n} \sum_{i=0}^{n-1} u\left(t + \left(i + \frac{1}{2}\right) \cdot \frac{\tau - t}{n}\right) \right| \tag{82}$$

$$\leq \begin{cases} \dfrac{(\tau - t)^2}{24n^2} \left\| u'' \right\|_{[t,\tau],\infty} & \text{if } u'' \in L_\infty[a, b]; \\[2ex] \dfrac{|\tau - t|^{1+\frac{1}{q}}}{8(2q + 1)^{\frac{1}{q}} n^2} \left\| u'' \right\|_{[t,\tau],p} & \text{if } u'' \in L_p[a, b], \\ & p > 1,\ \frac{1}{p} + \frac{1}{q} = 1; \\[2ex] \dfrac{|\tau - t|}{8n^2} \left\| u'' \right\|_{[t,\tau],1}. \end{cases}$$

Proof Consider the equidistant division of $[t, \tau]$ (or $[\tau, t]$)

$$E_n : x_i = t + i \cdot \frac{\tau - t}{n}, \quad i = \overline{0, n}.$$

If we apply the inequality (80), we may state that

$$\left| \int_{x_i}^{x_{i+1}} u(s)\, ds - \frac{\tau - t}{n} \cdot u\left(t + \left(i + \frac{1}{2}\right) \frac{\tau - t}{n}\right) \right|$$

$$\leq \begin{cases} \dfrac{(\tau - t)^3}{24n^3} \left\| u'' \right\|_{[x_i,x_{i+1}],\infty} & \text{if } u'' \in L_\infty[a, b]; \\[2ex] \dfrac{|\tau - t|^{2+\frac{1}{q}}}{8(2q + 1)^{\frac{1}{q}} n^{2+\frac{1}{q}}} \left\| u'' \right\|_{[x_i,x_{i+1}],p} & \text{if } u'' \in L_p[a, b], \\ & p > 1,\ \frac{1}{p} + \frac{1}{q} = 1; \\[2ex] \dfrac{|\tau - t|^2}{8n^2} \left\| u'' \right\|_{[x_i,x_{i+1}],1}. \end{cases}$$

Dividing by $|\tau - t| > 0$ and using a similar argument to the one in Lemma 6, we conclude that the desired inequality (82) holds.

The following theorem also holds.

Theorem 16 (Dragomir et al. [17]) *Let $f : [a, b] \to \mathbb{R}$ be a twice differentiable function such that the second derivative f'' is absolutely continuous on $[a, b]$. Then*

$$\left| (Tf)(a, b; t) - \frac{f(t)}{\pi} \ln\left(\frac{b-t}{t-a}\right) - M_n(f; t) \right| \tag{83}$$

$$\leq \begin{cases} \dfrac{1}{24\pi n^2}\left[\dfrac{(b-a)^2}{12} + \left(t - \dfrac{a+b}{2}\right)^2\right](b-a)\left\|f'''\right\|_{[a,b],\infty} & \text{if } f''' \in L_\infty[a,b]; \\[2ex] \dfrac{1}{8(2q+1)^{\frac{1}{q}}\pi n^2}\left[(t-a)^{2+\frac{1}{q}} + (b-t)^{2+\frac{1}{q}}\right]\left\|f'''\right\|_{[a,b],p} & \text{if } f''' \in L_p[a,b], \\ & p > 1,\ \frac{1}{p} + \frac{1}{q} = 1; \\[1ex] \dfrac{1}{8\pi n^2}\left[\dfrac{(b-a)^2}{4} + \left(t - \dfrac{a+b}{2}\right)^2\right]\left\|f'''\right\|_{[a,b],1}; \end{cases}$$

$$\leq \begin{cases} \dfrac{1}{72\pi n^2}(b-a)^3\left\|f'''\right\|_{[a,b],\infty} & \text{if } f''' \in L_\infty[a,b]; \\[2ex] \dfrac{1}{8(2q+1)^{\frac{1}{q}}\pi n^2}(b-a)^{2+\frac{1}{q}}\left\|f'''\right\|_{[a,b],p} & \text{if } f''' \in L_p[a,b], \\ & p > 1,\ \frac{1}{p} + \frac{1}{q} = 1; \\[1ex] \dfrac{\left\|f'''\right\|_{[a,b],1}}{16\pi n^2}(b-a)^2. \end{cases}$$

Proof Applying Lemma 8 for the function f'', we may write that (see also Theorem 15)

$$\left| \frac{f(\tau) - f(t)}{\tau - t} - \frac{1}{n}\sum_{i=0}^{n-1} f'\left(t + \left(i + \frac{1}{2}\right)\frac{\tau - t}{n}\right) \right| \tag{84}$$

$$\leq \begin{cases} \dfrac{(\tau-t)^2}{24n^2}\left\|f'''\right\|_{[t,\tau],\infty} & \text{if } f''' \in L_\infty[a,b]; \\[2ex] \dfrac{|\tau-t|^{1+\frac{1}{q}}}{8(2q+1)^{\frac{1}{q}}n^2}\left\|f'''\right\|_{[t,\tau],p} & \text{if } f''' \in L_p[a,b], \\ & p > 1,\ \frac{1}{p} + \frac{1}{q} = 1; \\[1ex] \dfrac{|\tau-t|}{8n^2}\left\|f'''\right\|_{[t,\tau],1}, \end{cases}$$

for all $t, \tau \in [a, b]$, $t \neq \tau$.

Consequently, by (84), we may write that

$$\left|\frac{1}{\pi}PV\int_a^b \frac{f(\tau)-f(t)}{\tau-t}d\tau - \frac{1}{\pi n}\sum_{i=0}^{n-1} PV\int_a^b f'\left(t+\left(i+\frac{1}{2}\right)\frac{\tau-t}{n}\right)d\tau\right|$$

$$\leq \begin{cases} \dfrac{1}{24\pi n^2}PV\displaystyle\int_a^b |\tau-t|^2 \left\|f'''\right\|_{[t,\tau],\infty} d\tau \\ \dfrac{1}{8\pi(2q+1)^{\frac{1}{q}} n^2}PV\displaystyle\int_a^b |\tau-t|^{1+\frac{1}{q}} \left\|f'''\right\|_{[t,\tau],p} d\tau \\ \dfrac{1}{8\pi n^2}PV\displaystyle\int_a^b |\tau-t| \left\|f'''\right\|_{[t,\tau],1} d\tau. \end{cases} \tag{85}$$

Since

$$\begin{aligned} PV\int_a^b |\tau-t|^2 \left\|f'''\right\|_{[t,\tau],\infty} d\tau &\leq \left\|f'''\right\|_{[a,b],\infty} PV\int_a^b |\tau-t|^2 d\tau \\ &= \left\|f'''\right\|_{[a,b],\infty}\left[\frac{(t-a)^3+(b-t)^3}{3}\right] \\ &= \left\|f'''\right\|_{[a,b],\infty}\left[\frac{(b-a)^2}{12}+\left(t-\frac{a+b}{2}\right)^2\right](b-a), \end{aligned}$$

$$\begin{aligned} PV\int_a^b |\tau-t|^{1+\frac{1}{q}} \left\|f'''\right\|_{[t,\tau],p} d\tau &\leq \left\|f'''\right\|_{[a,b],p} PV\int_a^b |\tau-t|^{1+\frac{1}{q}} d\tau \\ &= \frac{q\left\|f'''\right\|_{[a,b],p}}{2q+1}\left[(t-a)^{2+\frac{1}{q}}+(b-t)^{2+\frac{1}{q}}\right] \end{aligned}$$

and

$$\begin{aligned} PV\int_a^b |\tau-t| \left\|f'''\right\|_{[t,\tau],1} d\tau &\leq \left\|f'''\right\|_{[a,b],1} PV\int_a^b |\tau-t| d\tau \\ &= \left\|f'''\right\|_{[a,b],1}\frac{(t-a)^2+(b-t)^2}{2} \\ &= \left\|f'''\right\|_{[a,b],1}\left[\frac{(b-a)^2}{4}+\left(t-\frac{a+b}{2}\right)^2\right], \end{aligned}$$

then, as in Theorem 15, by (85) we deduce the first part of (83). The second part is obvious.

5 Estimates for Derivatives of Bounded Variation

5.1 Some Integral Inequalities

We start with the following lemma proved in [3] dealing with an Ostrowski type inequality for functions of bounded variation.

Lemma 9 *Let $u : [a,b] \to \mathbb{R}$ be a function of bounded variation on $[a,b]$. Then, for all $x \in [a,b]$, we have the inequality:*

$$\left| u(x)(b-a) - \int_a^b u(t)\,dt \right| \leq \left[\frac{1}{2}(b-a) + \left| x - \frac{a+b}{2} \right| \right] \bigvee_a^b (u), \tag{86}$$

where $\bigvee_a^b (u)$ denotes the total variation of u on $[a,b]$.

The constant $\frac{1}{2}$ is the possible one.

Proof For the sake of completeness and since this result will be essentially used in what follows, we give here a short proof.

Using the integration by parts formula for the Riemann-Stieltjes integral we have

$$\int_a^x (t-a)\,du(t) = u(x)(x-a) - \int_a^x u(t)\,dt$$

and

$$\int_x^b (t-b)\,du(t) = u(x)(b-x) - \int_x^b u(t)\,dt.$$

If we add the above two equalities, we get

$$u(x)(b-a) - \int_a^b u(t)\,dt = \int_a^x (t-a)\,du(t) + \int_x^b (t-b)\,du(t) \tag{87}$$

for any $x \in [a,b]$.

If $p : [c,d] \to \mathbb{R}$ is continuous on $[c,d]$ and $v : [c,d] \to \mathbb{R}$ is of bounded variation on $[c,d]$, then:

$$\left| \int_c^d p(x)\,dv(x) \right| \leq \sup_{x \in [c,d]} |p(x)| \bigvee_c^d (u). \tag{88}$$

Using (87) and (88), we deduce

$$\begin{aligned}\left|u(x)(b-a)-\int_a^b u(t)\,dt\right| &\le \left|\int_a^x (t-a)\,du(t)\right|+\left|\int_x^b (t-b)\,du(t)\right|\\ &\le (x-a)\bigvee_a^x(u)+(b-x)\bigvee_x^b(u)\\ &\le \max\{x-a,b-x\}\left[\bigvee_a^x(u)+\bigvee_x^b(u)\right]\\ &=\left[\frac{1}{2}(b-a)+\left|x-\frac{a+b}{2}\right|\right]\bigvee_a^b(u)\end{aligned}$$

and the inequality (86) is proved.

Now, assume that the inequality (87) holds with a constant $c>0$, i.e.,

$$\left|u(x)(b-a)-\int_a^b u(t)\,dt\right|\le\left[c(b-a)+\left|x-\frac{a+b}{2}\right|\right]\bigvee_a^b(u) \tag{89}$$

for all $x\in[a,b]$.

Consider the function $u_0:[a,b]\to\mathbb{R}$ given by

$$u_0(x)=\begin{cases}0 \text{ if } x\in[a,b]\setminus\{\frac{a+b}{2}\}\\ \\ 1 \text{ if } x=\frac{a+b}{2}.\end{cases}$$

Then u_0 is of bounded variation on $[a,b]$ and

$$\bigvee_a^b(u_0)=2,\quad \int_a^b u_0(t)\,dt=0.$$

If we apply (89) for u_0 and choose $x=\frac{a+b}{2}$, then we get $2c\ge 1$ which implies that $c\ge\frac{1}{2}$ showing that $\frac{1}{2}$ is the best possible constant in (86).

The best inequality we can get from (86) is the following midpoint inequality.

Corollary 12 *With the assumptions in Lemma 9, we have*

$$\left|u\left(\frac{a+b}{2}\right)(b-a)-\int_a^b u(t)\,dt\right|\le\frac{1}{2}(b-a)\bigvee_a^b(u). \tag{90}$$

The constant $\frac{1}{2}$ is best possible.

Using the above Ostrowski type inequality we may point out the following result in estimating the finite Hilbert transform.

Theorem 17 (Dragomir [4]) *Let $f : [a,b] \to \mathbb{R}$ be a function such that its derivative $f' : [a,b] \to \mathbb{R}$ is of bounded variation on $[a,b]$. Then we have the inequality:*

$$\left|(Tf)(a,b;t) - \frac{f(t)}{\pi}\ln\left(\frac{b-t}{t-a}\right) - \frac{b-a}{\pi}[f;\lambda t + (1-\lambda)b, \lambda t + (1-\lambda)a]\right| \tag{91}$$

$$\leq \frac{1}{\pi}\left[\frac{1}{2} + \left|\lambda - \frac{1}{2}\right|\right]\left[\frac{1}{2}(b-a) + \left|t - \frac{a+b}{2}\right|\right]\bigvee_a^b (f'),$$

for any $t \in (a,b)$ and $\lambda \in [0,1)$, where $[f;\alpha,\beta]$ is the divided difference, i.e.,

$$[f;\alpha,\beta] := \frac{f(\alpha) - f(\beta)}{\alpha - \beta}.$$

Proof Since f' is bounded on $[a,b]$, it follows that f is Lipschitzian on $[a,b]$ and thus the finite Hilbert transform exists everywhere in (a,b).

As for the function $f_0 : (a,b) \to \mathbb{R}$, $f_0(t) = 1$, $t \in (a,b)$, we have

$$(Tf_0)(a,b;t) = \frac{1}{\pi}\ln\left(\frac{b-t}{t-a}\right), \quad t \in (a,b),$$

then obviously

$$(Tf)(a,b;t) - \frac{f(t)}{\pi}\ln\left(\frac{b-t}{t-a}\right) = \frac{1}{\pi}PV\int_a^b \frac{f(\tau) - f(t)}{\tau - t}d\tau. \tag{92}$$

Now, if we choose in (86), $u = f'$, $x = \lambda c + (1-\lambda)d$, $\lambda \in [0,1]$, then we get

$$\left|f(d) - f(c) - (d-c)f'(\lambda c + (1-\lambda)d)\right|$$
$$\leq \left[\frac{1}{2}|d-c| + \left|\lambda c + (1-\lambda)d - \frac{c+d}{2}\right|\right]\left|\bigvee_c^d (f')\right|$$

where $c, d \in (a,b)$, which is equivalent to

$$\left|\frac{f(d) - f(c)}{d-c} - f'(\lambda c + (1-\lambda)d)\right| \leq \left[\frac{1}{2} + \left|\lambda - \frac{1}{2}\right|\right]\left|\bigvee_c^d (f')\right| \tag{93}$$

for any $c, d \in (a,b)$, $c \neq d$.

Using (93), we may write

$$\begin{aligned} & \left| \frac{1}{\pi} PV \int_a^b \frac{f(\tau) - f(t)}{\tau - t} d\tau - \frac{1}{\pi} PV \int_a^b f'(\lambda t + (1-\lambda)\tau) d\tau \right| \qquad (94) \\ & \leq \frac{1}{\pi} \left[\frac{1}{2} + \left| \lambda - \frac{1}{2} \right| \right] PV \int_a^b \left| \bigvee_\tau^t (f') \right| dt \\ & = \frac{1}{\pi} \left[\frac{1}{2} + \left| \lambda - \frac{1}{2} \right| \right] \left[\int_a^t \left(\bigvee_\tau^t (f') \right) dt + \int_t^b \left(\bigvee_t^\tau (f') \right) dt \right] \\ & \leq \frac{1}{\pi} \left[\frac{1}{2} + \left| \lambda - \frac{1}{2} \right| \right] \left[(t-a) \bigvee_a^t (f') + (b-t) \bigvee_t^b (f') \right] \\ & \leq \frac{1}{\pi} \left[\frac{1}{2} + \left| \lambda - \frac{1}{2} \right| \right] \left[\frac{1}{2}(b-a) + \left| t - \frac{a+b}{2} \right| \right] \bigvee_a^b (f'). \end{aligned}$$

Since (for $\lambda \neq 1$)

$$\begin{aligned} & \frac{1}{\pi} PV \int_a^b f'(\lambda t + (1-\lambda)\tau) d\tau \\ & = \frac{1}{\pi} \lim_{\varepsilon \to 0+} \left[\int_a^{t-\varepsilon} + \int_{t+\varepsilon}^b \right] \left(f'(\lambda t + (1-\lambda)\tau) d\tau \right) \\ & = \frac{1}{\pi} \lim_{\varepsilon \to 0+} \left[\frac{1}{1-\lambda} f(\lambda t + (1-\lambda)\tau) \Big|_a^{t-\varepsilon} + \frac{1}{1-\lambda} f(\lambda t + (1-\lambda)\tau) \Big|_{t+\varepsilon}^b \right] \\ & = \frac{1}{\pi} \cdot \frac{f(t) - f(\lambda t + (1-\lambda)a) + f(\lambda t + (1-\lambda)b) - f(t)}{1-\lambda} \\ & = \frac{b-a}{\pi} [f; \lambda t + (1-\lambda)b, \lambda t + (1-\lambda)a]. \end{aligned}$$

Using (94) and (92), we deduce the desired result (91).

It is obvious that the best inequality we can get from (91) is the one for $\lambda = \frac{1}{2}$. Thus, we may state the following corollary.

Corollary 13 *With the assumptions of Theorem 17, we have*

$$\begin{aligned} & \left| (Tf)(a,b;t) - \frac{f(t)}{\pi} \ln\left(\frac{b-t}{t-a} \right) - \frac{b-a}{\pi} \left[f; \frac{t+b}{2}, \frac{a+t}{2} \right] \right| \qquad (95) \\ & \leq \frac{1}{2\pi} \left[\frac{1}{2}(b-a) + \left| t - \frac{a+b}{2} \right| \right] \bigvee_a^b (f'). \end{aligned}$$

The above Theorem 17 may be used to point out some interesting inequalities for the functions for which the finite Hilbert transforms $(Tf)(a, b; t)$ can be expressed in terms of special functions.

For instance, we have

1. Assume that $f : [a, b] \subset (0, \infty) \to \mathbb{R}$, $f(x) = \frac{1}{x}$. Then

$$(Tf)(a, b; t) = \frac{1}{\pi t} \ln \left[\frac{(b-t)a}{(t-a)b} \right], \quad t \in (a, b),$$

$$\begin{aligned} &\frac{b-a}{\pi} \cdot [f; \lambda t + (1-\lambda) b, \lambda t + (1-\lambda) a] \\ &= -\frac{1}{\pi} \cdot \frac{b-a}{[\lambda t + (1-\lambda) b][\lambda t + (1-\lambda) a]}, \end{aligned}$$

$$\bigvee_a^b (f') = \int_a^b |f''(t)| \, dt = \frac{b^2 - a^2}{a^2 b^2}.$$

Using the inequality (91) we may write that

$$\begin{aligned} &\left| \frac{1}{\pi t} \ln \left[\frac{(b-t)a}{(t-a)b} \right] - \frac{1}{\pi t} \ln \left(\frac{b-t}{t-a} \right) + \frac{b-a}{\pi [\lambda t + (1-\lambda) b][\lambda t + (1-\lambda) a]} \right| \\ &\leq \frac{1}{\pi} \left[\frac{1}{2} + \left| \lambda - \frac{1}{2} \right| \right] \left[\frac{1}{2}(b-a) + \left| t - \frac{a+b}{2} \right| \right] \cdot \frac{b^2 - a^2}{a^2 b^2} \end{aligned}$$

which is equivalent to

$$\begin{aligned} &\left| \frac{b-a}{[\lambda t + (1-\lambda) b][\lambda t + (1-\lambda) a]} - \frac{1}{t} \ln \left(\frac{b}{a} \right) \right| \\ &\leq \left[\frac{1}{2} + \left| \lambda - \frac{1}{2} \right| \right] \left[\frac{1}{2}(b-a) + \left| t - \frac{a+b}{2} \right| \right] \cdot \frac{b^2 - a^2}{a^2 b^2}. \end{aligned} \tag{96}$$

If we use the notations

$$L(a, b) := \frac{b-a}{\ln b - \ln a} \quad \text{(the logarithmic mean)}$$

$$A_\lambda(x, y) := \lambda x + (1-\lambda) y \quad \text{(the weighted arithmetic mean)}$$

$$G(a, b) := \sqrt{ab} \quad \text{(the geometric mean)}$$

$$A(a, b) := \frac{a+b}{2} \quad \text{(the arithmetic mean)}$$

then by (96) we deduce

$$\left|\frac{1}{A_\lambda(t,b)A_\lambda(t,a)}-\frac{1}{tL(a,b)}\right| \leq\left[\frac{1}{2}+\left|\lambda-\frac{1}{2}\right|\right]\left[\frac{1}{2}(b-a)+|t-A(a,b)|\right]\frac{2A(a,b)}{G^4(a,b)},$$

giving the following proposition:

Proposition 1 *With the above assumption, we have*

$$|tL(a,b)-A_\lambda(t,b)A_\lambda(t,a)| \leq\frac{2A(a,b)}{G^4(a,b)}\left[\frac{1}{2}+\left|\lambda-\frac{1}{2}\right|\right]\left[\frac{1}{2}(b-a)+\left|t-\frac{a+b}{2}\right|\right]tA_\lambda(t,b)A_\lambda(t,a)L(a,b) \tag{97}$$

for any $t\in(a,b)$, $\lambda\in[0,1)$.

In particular, for $t=A(a,b)$ and $\lambda=\frac{1}{2}$, we get

$$\left|A(a,b)L(a,b)-\frac{(A(a,b)+a)(A(a,b)+b)}{4}\right| \leq\frac{1}{2}\cdot\frac{A^2(a,b)}{G^4(a,b)}\cdot\frac{(A(a,b)+a)(A(a,b)+b)}{4}L(a,b). \tag{98}$$

2. Assume that $f:[a,b]\subset\mathbb{R}\to\mathbb{R}$, $f(x)=\exp(x)$. Then

$$(Tf)(a,b;t)=\frac{\exp(t)}{\pi}[Ei(b-t)-Ei(a-t)],$$

where

$$Ei(z):=PV\int_{-\infty}^{z}\frac{\exp(t)}{t}dt,\quad z\in\mathbb{R}.$$

Also, we have

$$\frac{b-a}{\pi}[\exp;\lambda t+(1-\lambda)b,\lambda t+(1-\lambda)a] =\frac{1}{\pi}\cdot\frac{\exp(\lambda t+(1-\lambda)b)-\exp(\lambda t+(1-\lambda)a)}{1-\lambda},$$

$$\bigvee_a^b(f')=\int_a^b|f''(t)|\,dt=\exp(b)-\exp(a).$$

Using the inequality (91) we may write

$$\left| \exp(t) \left[Ei(b-t) - Ei(a-t) - \ln\left(\frac{b-t}{t-a}\right) \right] - \frac{\exp(\lambda t + (1-\lambda) b) - \exp(\lambda t + (1-\lambda) a)}{1-\lambda} \right| \le \left[\frac{1}{2} + \left| \lambda - \frac{1}{2} \right| \right] \left[\frac{1}{2}(b-a) + \left| t - \frac{a+b}{2} \right| \right] [\exp(b) - \exp(a)] \tag{99}$$

for any $t \in (a, b)$.

The reader may get other similar inequalities for special functions if appropriate examples of functions f are chosen.

5.2 A Quadrature Formula for Equidistant Divisions

The following lemma is of interest in itself.

Lemma 10 *Let $u : [a, b] \to \mathbb{R}$ be a function of bounded variation on $[a, b]$. Then for all $n \ge 1$, $\lambda_i \in [0, 1)$ $(i = 0, \dots, n-1)$ and $t, \tau \in [a, b]$ with $t \ne \tau$, we have the inequality:*

$$\left| \frac{1}{\tau - t} \int_t^\tau u(s)\, ds - \frac{1}{n} \sum_{i=0}^{n-1} u\left[t + (i + 1 - \lambda_i) \frac{\tau - t}{n} \right] \right| \le \frac{1}{n} \left[\frac{1}{2} + \max_{i=\overline{0,n-1}} \left| \lambda_i - \frac{1}{2} \right| \right] \left| \bigvee_t^\tau (u) \right|. \tag{100}$$

Proof Consider the equidistant division of $[t, \tau]$ (if $t < \tau$) or $[\tau, t]$ (if $\tau < t$) given by

$$E_n : x_i = t + i \cdot \frac{\tau - t}{n}, \quad i = \overline{0, n}. \tag{101}$$

Then the points $\xi_i = \lambda_i \left[t + i \cdot \frac{\tau - t}{n} \right] + (1 - \lambda_i) \left[t + (i+1) \cdot \frac{\tau - t}{n} \right]$ $\left(\lambda_i \in [0, 1],\ i = \overline{0, n-1}\right)$ are between x_i and x_{i+1}. We observe that we may write for simplicity $\xi_i = t + (i + 1 - \lambda_i) \frac{\tau - t}{n}$ $\left(i = \overline{0, n-1}\right)$. We also have

$$\xi_i - \frac{x_i + x_{i+1}}{2} = \frac{\tau - t}{2n}(1 - 2\lambda_i), \quad \xi_i - x_i = (1 - \lambda_i) \frac{\tau - t}{n}$$

and

$$x_{i+1} - \xi_i = \lambda_i \cdot \frac{\tau - t}{n}$$

for any $i = \overline{0, n-1}$.

If we apply the inequality (86) on the interval $[x_i, x_{i+1}]$ and the intermediate point ξ_i $\left(i = \overline{0, n-1}\right)$, then we may write that

$$\left|\frac{\tau - t}{n} u\left(t + (i + 1 - \lambda_i)\frac{\tau - t}{n}\right) - \int_{x_i}^{x_{i+1}} u(s)\,ds\right| \tag{102}$$
$$\leq \left[\frac{1}{2} \cdot \frac{|\tau - t|}{n} + \left|\frac{\tau - t}{2n}(1 - 2\lambda_i)\right|\right] \left|\bigvee_{x_i}^{x_{i+1}}(u)\right|.$$

Summing, we get

$$\left|\int_t^{\tau} u(s)\,ds - \frac{\tau - t}{n}\sum_{i=0}^{n-1} u\left[t + (i + 1 - \lambda_i)\frac{\tau - t}{n}\right]\right|$$
$$\leq \frac{|\tau - t|}{2n}\sum_{i=0}^{n-1}[1 + |1 - 2\lambda_i|]\left|\bigvee_{x_i}^{x_{i+1}}(u)\right|$$
$$= \frac{|\tau - t|}{n}\left[\frac{1}{2} + \max_{i=\overline{0,n-1}}\left|\lambda_i - \frac{1}{2}\right|\right]\left|\bigvee_t^{\tau}(u)\right|,$$

which is equivalent to (100).

We may now state the following theorem in approximating the finite Hilbert transform of a differentiable functions with the derivative of bounded variation on $[a, b]$.

Theorem 18 (Dragomir [4]) *Let $f : [a, b] \to \mathbb{R}$ be a differentiable function such that its derivative f' is of bounded variation on $[a, b]$. If $\boldsymbol{\lambda} = (\lambda_i)_{i=\overline{0,n-1}}$, $\lambda_i \in [0, 1)$ $\left(i = \overline{0, n-1}\right)$ and*

$$S_n(f; \boldsymbol{\lambda}, t) := \frac{b - a}{\pi n}\sum_{i=0}^{n-1}\left[f; (i + 1 - \lambda_i)\frac{b - t}{n} + t, (i + 1 - \lambda_i)\frac{a - t}{n} + t\right], \tag{103}$$

then we have the estimate:

$$\left|(Tf)(a, b; t) - \frac{f(t)}{\pi}\ln\left(\frac{b - t}{t - a}\right) - S_n(f; \boldsymbol{\lambda}, t)\right| \tag{104}$$

$$\leq \frac{b-a}{n\pi}\left[\frac{1}{2}+\max_{i=\overline{0,n-1}}\left|\lambda_i-\frac{1}{2}\right|\right]\left[\frac{1}{2}(b-a)+\left|t-\frac{a+b}{2}\right|\right]\bigvee_a^b(f')$$

$$\leq \frac{b-a}{n\pi}\bigvee_a^b(f').$$

Proof Applying Lemma 10 for the function f', we may write that

$$\left|\frac{f(\tau)-f(t)}{\tau-t}-\frac{1}{n}\sum_{i=0}^{n-1}f'\left[t+(i+1-\lambda_i)\frac{\tau-t}{n}\right]\right| \tag{105}$$

$$\leq \frac{1}{n}\left[\frac{1}{2}+\max_{i=\overline{0,n-1}}\left|\lambda_i-\frac{1}{2}\right|\right]\left|\bigvee_t^\tau(f')\right|$$

for any $t,\tau\in[a,b]$, $t\neq\tau$.

Consequently, we have

$$\left|\frac{1}{\pi}PV\int_a^b\frac{f(\tau)-f(t)}{\tau-t}d\tau-\frac{1}{\pi n}\sum_{i=0}^{n-1}PV\int_a^b f'\left[t+(i+1-\lambda_i)\frac{\tau-t}{n}\right]d\tau\right| \tag{106}$$

$$\leq \frac{1}{n\pi}\left[\frac{1}{2}+\max_{i=\overline{0,n-1}}\left|\lambda_i-\frac{1}{2}\right|\right]PV\int_a^b\left|\bigvee_t^\tau(f')\right|d\tau$$

$$\leq \frac{1}{n\pi}\left[\frac{1}{2}+\max_{i=\overline{0,n-1}}\left|\lambda_i-\frac{1}{2}\right|\right]\left[\frac{1}{2}(b-a)+\left|t-\frac{a+b}{2}\right|\right]\bigvee_a^b(f').$$

On the other hand

$$PV\int_a^b f'\left[t+(i+1-\lambda_i)\frac{\tau-t}{n}\right]d\tau \tag{107}$$

$$=\lim_{\varepsilon\to0+}\left[\int_a^{t-\varepsilon}+\int_{t+\varepsilon}^b\right]\left(f'\left[t+(i+1-\lambda_i)\frac{\tau-t}{n}\right]d\tau\right)$$

$$=\lim_{\varepsilon\to0+}\left[\frac{n}{i+1-\lambda_i}f\left(t+(i+1-\lambda_i)\frac{\tau-t}{n}\right)\Big|_a^{t-\varepsilon}\right.$$

$$\left.+\frac{n}{i+1-\lambda_i}f\left(t+(i+1-\lambda_i)\frac{\tau-t}{n}\right)\Big|_{t+\varepsilon}^b\right]$$

$$= \frac{n}{i+1-\lambda_i}\left[f\left(t+(i+1-\lambda_i)\frac{b-t}{n}\right)-f\left(t+(i+1-\lambda_i)\frac{a-t}{n}\right)\right]$$
$$= (b-a)\left[f; t+(i+1-\lambda_i)\frac{b-t}{n}, (i+1-\lambda_i)\frac{a-t}{n}+t\right].$$

Since (see, for example, (92)),

$$(Tf)(a,b;t) = \frac{1}{\pi}PV\int_a^b \frac{f(\tau)-f(t)}{\tau-t}d\tau + \frac{f(t)}{\pi}\ln\left(\frac{b-t}{t-a}\right)$$

for $t \in (a,b)$, then by (106) and (107) we deduce the desired estimate (104).

Remark 7 For $n = 1$, we recapture the inequality (91).

Corollary 14 *With the assumptions of Theorem 18, we have*

$$(Tf)(a,b;t) = \frac{f(t)}{\pi}\ln\left(\frac{b-t}{t-a}\right) + \lim_{n\to\infty} S_n(f;\boldsymbol{\lambda},t) \tag{108}$$

uniformly by rapport of $t \in (a,b)$ and $\boldsymbol{\lambda}$ with $\lambda_i \in [0,1)$ $(i \in \mathbb{N})$.

Remark 8 If one needs to approximate the finite Hilbert transform $(Tf)(a,b;t)$ in terms of

$$\frac{f(t)}{\pi}\ln\left(\frac{b-t}{t-a}\right) + S_n(f;\boldsymbol{\lambda},t)$$

with the accuracy $\varepsilon > 0$ (ε small), then the theoretical minimal number n_ε to be chosen is

$$n_\varepsilon := \left[\frac{b-a}{\varepsilon\pi}\bigvee_a^b (f')\right] + 1 \tag{109}$$

where $[\alpha]$ is the integer part of α.

It is obvious that the best inequality we can get in (104) is for $\lambda_i = \frac{1}{2}$ $\left(i = \overline{0, n-1}\right)$ obtaining the following corollary.

Corollary 15 *Let f be as in Theorem 18. Define*

$$M_n(f;t) := \frac{b-a}{\pi n}\sum_{i=0}^{n-1}\left[f; \left(i+\frac{1}{2}\right)\frac{b-t}{n}+t, \left(i+\frac{1}{2}\right)\frac{a-t}{n}+t\right]. \tag{110}$$

Then we have the estimate

$$\left|(Tf)(a,b;t)-\frac{f(t)}{\pi}\ln\left(\frac{b-t}{t-a}\right)-M_n(f;t)\right| \tag{111}$$
$$\leq \frac{b-a}{2n\pi}\left[\frac{1}{2}(b-a)+\left|t-\frac{a+b}{2}\right|\right]\bigvee_a^b(f')$$

for any $t\in(a,b)$.

5.3 A More General Quadrature Formula

We may state the following lemma.

Lemma 11 *Let* $u:[a,b]\to\mathbb{R}$ *be a function of bounded variation on* $[a,b]$, $0=\mu_0<\mu_1<\cdots<\mu_{n-1}<\mu_n=1$ *and* $\nu_i\in\left[\mu_i,\mu_{i+1}\right]$, $i=\overline{0,n-1}$. *Then for any* $t,\tau\in[a,b]$ *with* $t\neq\tau$, *we have the inequality:*

$$\left|\frac{1}{\tau-t}\int_t^\tau u(s)\,ds-\sum_{i=0}^{n-1}\left(\mu_{i+1}-\mu_i\right)u\left[(1-\nu_i)t+\nu_i\tau\right]\right| \tag{112}$$
$$\leq\left[\frac{1}{2}\Delta_n(\boldsymbol{\mu})+\max_{i=\overline{0,n-1}}\left|\nu_i-\frac{\mu_i+\mu_{i+1}}{2}\right|\right]\left|\bigvee_t^\tau(u)\right|,$$

where $\Delta_n(\boldsymbol{\mu}):=\max_{i=\overline{0,n-1}}\left(\mu_{i+1}-\mu_i\right)$.

Proof Consider the division of $[t,\tau]$ (if $t<\tau$) or $[\tau,t]$ (if $\tau<t$) given by

$$I_n: x_i:=\left(1-\mu_i\right)t+\mu_i\tau\quad\left(i=\overline{0,n}\right). \tag{113}$$

Then the points $\xi_i:=(1-\nu_i)t+\nu_i\tau\ \left(i=\overline{0,n-1}\right)$ are between x_i and x_{i+1}. We have

$$x_{i+1}-x_i=\left(\mu_{i+1}-\mu_i\right)(\tau-t)\quad\left(i=\overline{0,n-1}\right)$$

and

$$\xi_i-\frac{x_i+x_{i+1}}{2}=\left(\nu_i-\frac{\mu_i+\mu_{i+1}}{2}\right)(\tau-t)\quad\left(i=\overline{0,n-1}\right).$$

Applying the inequality (86) on $[x_i,x_{i+1}]$ with the intermediate points ξ_i $\left(i=\overline{0,n-1}\right)$, we get

$$\left|\int_{x_i}^{x_{i+1}} u(s)\,ds - \left(\mu_{i+1}-\mu_i\right)(\tau - t)\,u\left[(1-\nu_i)t + \nu_i\tau\right]\right|$$
$$\leq \left[\frac{1}{2}\left(\mu_{i+1}-\mu_i\right)|\tau - t| + |\tau - t|\left|\nu_i - \frac{\mu_i+\mu_{i+1}}{2}\right|\right]\left|\bigvee_{x_i}^{x_{i+1}}(u)\right|$$

for any $i=\overline{0,n-1}$. Summing over i, using the generalized triangle inequality and dividing by $|t-\tau|>0$, we obtain

$$\left|\frac{1}{\tau - t}\int_a^b u(s)\,ds - \sum_{i=0}^{n-1}\left(\mu_{i+1}-\mu_i\right)u\left[(1-\nu_i)t+\nu_i\tau\right]\right|$$
$$\leq \sum_{i=0}^{n-1}\left[\frac{1}{2}\left(\mu_{i+1}-\mu_i\right) + \left|\nu_i - \frac{\mu_i+\mu_{i+1}}{2}\right|\right]\left|\bigvee_{x_i}^{x_{i+1}}(u)\right|$$
$$\leq \left[\frac{1}{2}\Delta_n(\boldsymbol{\mu}) + \max_{i=\overline{0,n-1}}\left|\nu_i - \frac{\mu_i+\mu_{i+1}}{2}\right|\right]\left|\bigvee_{t}^{\tau}(u)\right|$$

and the inequality (112) is proved.

The following theorem holds.

Theorem 19 (Dragomir [4]) *Let $f:[a,b]\to\mathbb{R}$ be a differentiable function such that its derivative f' is of bounded variation on $[a,b]$. If $0=\mu_0<\mu_1<\cdots<\mu_{n-1}<\mu_n=1$ and $\nu_i\in\left[\mu_i,\mu_{i+1}\right]$, $\left(i=\overline{0,n-1}\right)$, then*

$$(Tf)(a,b;t) = \frac{f(t)}{\pi}\ln\left(\frac{b-t}{t-a}\right) + \frac{1}{\pi}Q_n(\boldsymbol{\mu},\boldsymbol{\nu},t) + W_n(\boldsymbol{\mu},\boldsymbol{\nu},t) \tag{114}$$

for any $t\in(a,b)$, where

$$\begin{aligned} &Q_n(\boldsymbol{\mu},\boldsymbol{\nu},t) \\ &:= \mu_1 f'(t)(b-a) + (b-a)\sum_{i=1}^{n-2}\Big\{\left(\mu_{i+1}-\mu_i\right) \\ &\times\left[f;(1-\nu_i)t+\nu_i b,(1-\nu_i)t+\nu_i a\right]\Big\} + \left(1-\mu_{n-1}\right)\left[f(b)-f(a)\right] \end{aligned} \tag{115}$$

if $\nu_0=0$, $\nu_{n-1}=1$,

$$\begin{aligned} Q_n(\boldsymbol{\mu},\boldsymbol{\nu},t) &:= \mu_1 f'(t)(b-a) + (b-a)\sum_{i=1}^{n-1}\left(\mu_{i+1}-\mu_i\right) \\ &\times\left[f;(1-\nu_i)t+\nu_i b,(1-\nu_i)t+\nu_i a\right] \end{aligned} \tag{116}$$

if $\nu_0 = 0,\ \nu_{n-1} < 1,$

$$Q_n(\boldsymbol{\mu}, \boldsymbol{\nu}, t) := (b-a)\sum_{i=1}^{n-2}\left(\mu_{i+1} - \mu_i\right)$$

$$\times\left[f; (1-\nu_i)t + \nu_i b, (1-\nu_i)t + \nu_i a\right] + \left(1 - \mu_{n-1}\right)\left[f(b) - f(a)\right] \tag{117}$$

if $\nu_0 > 0,\ \nu_{n-1} = 1$ *and*

$$Q_n(\boldsymbol{\mu}, \boldsymbol{\nu}, t) := (b-a)\sum_{i=1}^{n-1}\left(\mu_{i+1} - \mu_i\right)\left[f; (1-\nu_i)t + \nu_i b, (1-\nu_i)t + \nu_i a\right] \tag{118}$$

if $\nu_0 > 0,\ \nu_{n-1} < 1.$
In all cases, the remainder satisfies the estimate:

$$|W_n(\boldsymbol{\mu}, \boldsymbol{\nu}, t)| \le \frac{1}{\pi}\left[\frac{1}{2}\Delta_n(\boldsymbol{\mu}) + \max_{i=0,n-1}\left|\nu_i - \frac{\mu_i + \mu_{i+1}}{2}\right|\right] \tag{119}$$

$$\times\left[\frac{1}{2}(b-a) + \left|t - \frac{a+b}{2}\right|\right]\bigvee_a^b (f')$$

$$\le \frac{1}{\pi}\Delta_n(\boldsymbol{\mu})\left[\frac{1}{2}(b-a) + \left|t - \frac{a+b}{2}\right|\right]\left|\bigvee_a^b (f')\right|$$

$$\le \frac{1}{\pi}\Delta_n(\boldsymbol{\mu})(b-a)\bigvee_a^b (f').$$

Proof If we apply Lemma 11 for the function f', we may write that

$$\left|\frac{f(\tau) - f(t)}{\tau - t} - \sum_{i=0}^{n-1}\left(\mu_{i+1} - \mu_i\right) f'\left[(1-\nu_i)t + \nu_i \tau\right]\right|$$

$$\le \left[\frac{1}{2}\Delta_n(\boldsymbol{\mu}) + \max_{i=0,n-1}\left|\nu_i - \frac{\mu_i + \mu_{i+1}}{2}\right|\right]\left|\bigvee_t^\tau (f')\right|$$

for any $t, \tau \in [a, b],\ t \neq \tau.$

Taking the PV in both sides, we may write that

$$\left|\frac{1}{\pi}PV\int_a^b \frac{f(\tau)-f(t)}{\tau-t}d\tau\right. \tag{120}$$
$$\left.-\frac{1}{\pi}PV\int_a^b\left(\sum_{i=0}^{n-1}\left(\mu_{i+1}-\mu_i\right)f'\left[(1-\nu_i)t+\nu_i\tau\right]\right)d\tau\right|$$
$$\leq\frac{1}{\pi}\left[\frac{1}{2}\Delta_n(\boldsymbol{\mu})+\max_{i=0,n-1}\left|\nu_i-\frac{\mu_i+\mu_{i+1}}{2}\right|\right]PV\int_a^b\left|\bigvee_t^\tau(f')\right|d\tau.$$

If $\nu_0=0,\ \nu_{n-1}=1$, then

$$PV\int_a^b\left(\sum_{i=0}^{n-1}\left(\mu_{i+1}-\mu_i\right)f'\left[(1-\nu_i)t+\nu_i\tau\right]\right)d\tau$$
$$=PV\int_a^b\mu_1 f'(t)\,d\tau+\sum_{i=1}^{n-2}\left(\mu_{i+1}-\mu_i\right)PV\int_a^b f'\left[(1-\nu_i)t+\nu_i\tau\right]d\tau$$
$$+\left(1-\mu_{n-1}\right)PV\int_a^b f'(\tau)\,d\tau$$
$$=\mu_1 f'(t)(b-a)+(b-a)\sum_{i=1}^{n-2}\left(\mu_{i+1}-\mu_i\right)\left[f;(1-\nu_i)t+\nu_i b,(1-\nu_i)t+\nu_i a\right]$$
$$+\left(1-\mu_{n-1}\right)\left[f(b)-f(a)\right].$$

If $\nu_0=0,\ \nu_{n-1}<1$, then

$$PV\int_a^b\left(\sum_{i=0}^{n-1}\left(\mu_{i+1}-\mu_i\right)f'\left[(1-\nu_i)t+\nu_i\tau\right]\right)d\tau$$
$$=\mu_1 f'(t)(b-a)+(b-a)\sum_{i=1}^{n-1}\left(\mu_{i+1}-\mu_i\right)\left[f;(1-\nu_i)t+\nu_i b,(1-\nu_i)t+\nu_i a\right].$$

If $\nu_0>0,\ \nu_{n-1}=1$, then

$$PV\int_a^b\left(\sum_{i=0}^{n-1}\left(\mu_{i+1}-\mu_i\right)f'\left[(1-\nu_i)t+\nu_i\tau\right]\right)d\tau$$
$$=(b-a)\sum_{i=1}^{n-2}\left(\mu_{i+1}-\mu_i\right)\left[f;(1-\nu_i)t+\nu_i b,(1-\nu_i)t+\nu_i a\right]+\left(1-\mu_{n-1}\right)\left[f(b)-f(a)\right].$$

and, finally, if $\nu_0>0,\ \nu_{n-1}<1$, then

$$PV\int_a^b \left(\sum_{i=0}^{n-1} (\mu_{i+1}-\mu_i)\, f'[(1-\nu_i)t+\nu_i\tau]\right) d\tau$$
$$= (b-a)\sum_{i=1}^{n-1} (\mu_{i+1}-\mu_i)\,[f;(1-\nu_i)t+\nu_i b,(1-\nu_i)t+\nu_i a].$$

Since

$$PV\int_a^b \left|\bigvee_t^\tau (f')\right| d\tau \le \left[\frac{1}{2}(b-a)+\left|t-\frac{a+b}{2}\right|\right]\bigvee_a^b (f')$$

and

$$(Tf)(a,b;t) = \frac{1}{\pi}PV\int_a^b \frac{f(\tau)-f(t)}{\tau-t}d\tau + \frac{f(t)}{\pi}\ln\left(\frac{b-t}{t-a}\right),$$

then by (120) we deduce (114).

6 Estimates for Absolutely Continuous Derivatives

6.1 Ostrowski Type Inequalities

For the sake of completeness, we state and prove the following lemma providing some Ostrowski type inequalities for absolutely continuous functions (see [10, 11] and [12]).

Lemma 12 *Let $u:[a,b]\to\mathbb{R}$ be an absolutely continuous function on $[a,b]$. Then we have*

$$\left|u(x)(b-a)-\int_a^b u(t)\,dt\right| \tag{121}$$

$$\le \begin{cases} \left[\frac{1}{4}(b-a)^2+\left(x-\frac{a+b}{2}\right)^2\right]\|u'\|_{[a,b],\infty} & \text{if } u'\in L_\infty[a,b]; \\ \frac{1}{(q+1)^{\frac{1}{q}}}\left[(x-a)^{1+\frac{1}{q}}+(b-x)^{1+\frac{1}{q}}\right]\|u'\|_{[a,b],p} & \text{if } u'\in L_p[a,b], \\ & p>1,\ \frac{1}{p}+\frac{1}{q}=1; \\ \left[\frac{1}{2}(b-a)+\left|x-\frac{a+b}{2}\right|\right]\|u'\|_{[a,b],1}, & \text{if } u'\in L[a,b] \end{cases}$$

where $\|\cdot\|_r$ $(r \in [1, \infty])$ *are the usual Lebesgue norms, i.e., for* $c < d$

$$\|h\|_{[c,d],\infty} := ess \sup_{t\in[c,d]} |h(t)|$$

and

$$\|h\|_{[c,d],r} := \left(\int_c^d |h(t)|^r \, dt \right)^{\frac{1}{r}}, \quad r \geq 1.$$

Proof Using the integration by parts formula, we have

$$\int_a^x (t-a) u'(t) \, dt = u(x)(x-a) - \int_a^x u(t) \, dt$$

and

$$\int_x^b (t-b) u'(t) \, dt = u(x)(b-x) - \int_x^b u(t) \, dt.$$

If we add the above two equalities, we get

$$u(x)(b-a) - \int_a^b u(t) \, dt = \int_a^x (t-a) u'(t) \, dt + \int_x^b (t-b) u'(t) \, dt \quad (122)$$

for any $x \in [a, b]$.

Taking the modulus, we have

$$\left| u(x)(b-a) - \int_a^b u(t) \, dt \right| \leq \int_a^x (t-a) \left| u'(t) \right| dt + \int_x^b (t-b) \left| u'(t) \right| dt$$
$$=: M(x).$$

Now, it is obvious that

$$M(x) \leq \left\| u' \right\|_{[a,x],\infty} \int_a^x (t-a) \, dt + \left\| u' \right\|_{[x,b],\infty} \int_x^b (b-t) \, dt$$
$$= \left\| u' \right\|_{[a,x],\infty} \cdot \frac{(x-a)^2}{2} + \left\| u' \right\|_{[x,b],\infty} \cdot \frac{(b-x)^2}{2}$$
$$\leq \left\| u' \right\|_{[a,b],\infty} \left[\frac{(x-a)^2 + (b-x)^2}{2} \right]$$

$$= \left\|u'\right\|_{[a,b],\infty}\left[\frac{1}{4}(b-a)^2+\left(x-\frac{a+b}{2}\right)^2\right],$$

proving the first part of (121).

Using Hölder's integral inequality, we may write

$$\begin{aligned}
M(x) &\le \left\|u'\right\|_{[a,x],p}\left(\int_a^x (t-a)^q\,dt\right)^{\frac{1}{q}}+\left\|u'\right\|_{[x,b],p}\left(\int_x^b (b-t)^q\,dt\right)^{\frac{1}{q}}\\
&= \left\|u'\right\|_{[a,x],p}\cdot\left[\frac{(x-a)^{q+1}}{q+1}\right]^{\frac{1}{q}}+\left\|u'\right\|_{[x,b],p}\cdot\left[\frac{(b-x)^{q+1}}{q+1}\right]^{\frac{1}{q}}\\
&\le \left\|u'\right\|_{[a,b],p}\frac{1}{(q+1)^{\frac{1}{q}}}\left[(x-a)^{1+\frac{1}{q}}+(b-x)^{1+\frac{1}{q}}\right],
\end{aligned}$$

proving the second part of (121).

Finally, we observe that

$$\begin{aligned}
M(x) &\le (x-a)\left\|u'\right\|_{[a,x],1}+(b-x)\left\|u'\right\|_{[x,b],1}\\
&\le \max\{x-a,b-x\}\left[\left\|u'\right\|_{[a,x],1}+\left\|u'\right\|_{[x,b],1}\right]\\
&= \left[\frac{1}{2}(b-a)+\left|x-\frac{a+b}{2}\right|\right]\left\|u'\right\|_{[a,b],1}
\end{aligned}$$

and the lemma is proved.

The best inequalities we can get from (121) are embodied in the following corollary.

Corollary 16 *With the assumptions of Lemma 12, we have*

$$\left|u\left(\frac{a+b}{2}\right)(b-a)-\int_a^b u(t)\,dt\right| \tag{123}$$

$$\le \begin{cases} \frac{1}{4}(b-a)^2\left\|u'\right\|_{[a,b],\infty} & \text{if } u'\in L_\infty[a,b];\\[2ex] \frac{1}{2(q+1)^{\frac{1}{q}}}(b-a)^{1+\frac{1}{q}}\left\|u'\right\|_{[a,b],p} & \text{if } u'\in L_p[a,b],\\ & p>1,\ \frac{1}{p}+\frac{1}{q}=1;\\[1ex] \frac{1}{2}(b-a)\left\|u'\right\|_{[a,b],1}\,. & \text{if } u'\in L[a,b]\end{cases}$$

The following theorem providing an estimate for the finite Hilbert transform holds.

Theorem 20 (Dragomir [5]) *Let $f : [a,b] \to \mathbb{R}$ be a function so that its derivative $f' : [a,b] \to \mathbb{R}$ is absolutely continuous on $[a,b]$. Then we have the inequalities*

$$\left|(Tf)(a,b;t) - \frac{f(t)}{\pi}\ln\left(\frac{b-t}{t-a}\right) - \frac{b-a}{\pi}[f;\lambda t + (1-\lambda)b, \lambda t + (1-\lambda)a]\right| \tag{124}$$

$$\leq \begin{cases} \frac{1}{\pi}\left[\frac{1}{4} + \left(\lambda - \frac{1}{2}\right)^2\right] \\ \quad \times \left[\frac{1}{4}(b-a)^2 + \left(t - \frac{a+b}{2}\right)^2\right]\|f''\|_{[a,b],\infty} & \text{if } f'' \in L_\infty[a,b]; \\ \frac{1}{\pi}\frac{q}{(q+1)^{1+\frac{1}{q}}}\left[\lambda^{1+\frac{1}{q}} + (1-\lambda)^{1+\frac{1}{q}}\right] \\ \quad \times \left[(t-a)^{1+\frac{1}{q}} + (b-t)^{1+\frac{1}{q}}\right]\|f''\|_{[a,b],p} & \text{if } f'' \in L_p[a,b], \\ & p > 1,\ \frac{1}{p} + \frac{1}{q} = 1; \\ \frac{1}{\pi}\left[\frac{1}{2} + \left|\lambda - \frac{1}{2}\right|\right]\left[\frac{1}{2}(b-a) + \left|t - \frac{a+b}{2}\right|\right]\|f''\|_{[a,b],1}, & \text{if } f'' \in L[a,b] \end{cases}$$

for any $t \in (a,b)$ and $\lambda \in [0,1)$, where $[f;\alpha,\beta]$ is the divided difference, i.e.,

$$[f;\alpha,\beta] = \frac{f(\alpha) - f(\beta)}{\alpha - \beta}.$$

Proof Since f' is bounded on $[a,b]$, it follows that f is Lipschitzian on $[a,b]$ and thus the finite Hilbert transform exists everywhere in (a,b). As for the function $f_0 : (a,b) \to \mathbb{R}$, $f_0(t) = 1$, $t \in (a,b)$, we have

$$(Tf_0)(a,b;t) = \frac{1}{\pi}\ln\left(\frac{b-t}{t-a}\right),\ t \in (a,b),$$

then, obviously,

$$\begin{aligned}(Tf)(a,b;t) &= \frac{1}{\pi}PV\int_a^b \frac{f(\tau) - f(t) + f(t)}{\tau - t}d\tau \\ &= \frac{1}{\pi}PV\int_a^b \frac{f(\tau) - f(t)}{\tau - t}d\tau + \frac{f(t)}{\pi}PV\int_a^b \frac{d\tau}{\tau - t},\end{aligned}$$

from where we get the identity

$$(Tf)(a,b;t) - \frac{f(t)}{\pi}\ln\left(\frac{b-t}{t-a}\right) = \frac{1}{\pi}PV\int_a^b \frac{f(\tau) - f(t)}{\tau - t}d\tau. \tag{125}$$

Now, if we choose in (121), $u = f'$, $x = \lambda c + (1-\lambda) d$, $\lambda \in [0,1]$, $c, d \in [a,b]$ then we get

$$\left| f(d) - f(c) - (d-c) f'(\lambda c + (1-\lambda) d) \right|$$
$$\leq \begin{cases} (d-c)^2 \left[\frac{1}{4} + \left(\lambda - \frac{1}{2} \right)^2 \right] \|f''\|_{[c,d],\infty} & \text{if } f'' \in L_\infty[a,b]; \\ \frac{|d-c|^{1+\frac{1}{q}}}{(q+1)^{\frac{1}{q}}} \left[\lambda^{1+\frac{1}{q}} + (1-\lambda)^{1+\frac{1}{q}} \right] \left| \|f''\|_{[c,d],p} \right| & \text{if } f'' \in L_p[a,b], \\ & p > 1,\ \frac{1}{p} + \frac{1}{q} = 1; \\ |d-c| \left[\frac{1}{2} + \left| \lambda - \frac{1}{2} \right| \right] \left| \|f''\|_{[c,d],1} \right|, \end{cases}$$

which is equivalent to

$$\left| \frac{f(d) - f(c)}{d-c} - f'(\lambda c + (1-\lambda) d) \right| \tag{126}$$
$$\leq \begin{cases} |d-c| \left[\frac{1}{4} + \left(\lambda - \frac{1}{2} \right)^2 \right] \|f''\|_{[c,d],\infty} & \text{if } f'' \in L_\infty[a,b]; \\ \frac{|d-c|^{\frac{1}{q}}}{(q+1)^{\frac{1}{q}}} \left[\lambda^{1+\frac{1}{q}} + (1-\lambda)^{1+\frac{1}{q}} \right] \left| \|f''\|_{[c,d],p} \right| & \text{if } f'' \in L_p[a,b], \\ & p > 1,\ \frac{1}{p} + \frac{1}{q} = 1; \\ \left[\frac{1}{2} + \left| \lambda - \frac{1}{2} \right| \right] \left| \|f''\|_{[c,d],1} \right|. \end{cases}$$

Using (126), we may write

$$\left| \frac{1}{\pi} PV \int_a^b \frac{f(\tau) - f(t)}{\tau - t} d\tau - \frac{1}{\pi} PV \int_a^b f'(\lambda t + (1-\lambda)\tau) \, d\tau \right| \tag{127}$$
$$\leq \begin{cases} \frac{1}{\pi} \left[\frac{1}{4} + \left(\lambda - \frac{1}{2} \right)^2 \right] PV \int_a^b |t-\tau| \left| \|f''\|_{[t,\tau],\infty} \right| d\tau \\ \frac{1}{\pi} \cdot \frac{1}{(q+1)^{\frac{1}{q}}} \left[\lambda^{1+\frac{1}{q}} + (1-\lambda)^{1+\frac{1}{q}} \right] PV \int_a^b |t-\tau|^{\frac{1}{q}} \left| \|f''\|_{[t,\tau],p} \right| d\tau \\ \frac{1}{\pi} \left[\frac{1}{2} + \left| \lambda - \frac{1}{2} \right| \right] PV \int_a^b \left| \|f''\|_{[t,\tau],1} \right| d\tau \end{cases}$$

$$\leq \begin{cases} \frac{1}{\pi}\left[\frac{1}{4}+\left(\lambda-\frac{1}{2}\right)^2\right]\left\|f''\right\|_{[a,b],\infty} PV\int_a^b |t-\tau|\, d\tau \\ \frac{1}{\pi}\cdot\frac{1}{(q+1)^{\frac{1}{q}}}\left[\lambda^{1+\frac{1}{q}}+(1-\lambda)^{1+\frac{1}{q}}\right]\left\|f''\right\|_{[a,b],p} PV\int_a^b |t-\tau|^{\frac{1}{q}}\, d\tau \\ \frac{1}{\pi}\left[\frac{1}{2}+\left|\lambda-\frac{1}{2}\right|\right] PV\left[\int_a^t \left\|f''\right\|_{[\tau,t],1} d\tau+\int_t^b \left\|f''\right\|_{[t,\tau],1} d\tau\right] \end{cases}$$

$$\leq \begin{cases} \frac{1}{\pi}\left[\frac{1}{4}+\left(\lambda-\frac{1}{2}\right)^2\right]\left\|f''\right\|_{[a,b],\infty}\left[\frac{1}{4}(b-a)^2+\left(t-\frac{a+b}{2}\right)^2\right] \\ \frac{1}{\pi}\cdot\frac{1}{(q+1)^{\frac{1}{q}}}\left[\lambda^{1+\frac{1}{q}}+(1-\lambda)^{1+\frac{1}{q}}\right]\left\|f''\right\|_{[a,b],p} \\ \qquad \times\frac{q}{(q+1)}\left[(t-a)^{1+\frac{1}{q}}+(b-t)^{1+\frac{1}{q}}\right] \\ \frac{1}{\pi}\left[\frac{1}{2}+\left|\lambda-\frac{1}{2}\right|\right]\left[\frac{1}{2}(b-a)+\left|t-\frac{a+b}{2}\right|\right]\left\|f''\right\|_{[a,b],1}. \end{cases}$$

Since (note that $\lambda \neq 1$)

$$\begin{aligned} &\frac{1}{\pi} PV\int_a^b f'(\lambda t+(1-\lambda)\tau)\, d\tau \\ &= \frac{1}{\pi}\lim_{\varepsilon\to 0+}\left[\int_a^{t-\varepsilon}+\int_{t+\varepsilon}^b\right]\left(f'(\lambda t+(1-\lambda)\tau)\, d\tau\right) \\ &= \frac{1}{\pi}\lim_{\varepsilon\to 0+}\left[\frac{1}{1-\lambda}f(\lambda t+(1-\lambda)\tau)\Big|_a^{t-\varepsilon}+\frac{1}{1-\lambda}f(\lambda t+(1-\lambda)\tau)\Big|_{t+\varepsilon}^b\right] \\ &= \frac{1}{\pi}\frac{f(t)-f(\lambda t+(1-\lambda)a)+f(\lambda t+(1-\lambda)a)-f(t)}{1-\lambda} \\ &= \frac{b-a}{\pi}\left[f;\lambda t+(1-\lambda)b,\lambda t+(1-\lambda)a\right], \end{aligned}$$

then by (125) and (127) we deduce the desired inequality (124).

The best inequality one may obtain from (124) is embodied in the following corollary.

Corollary 17 *With the assumptions of Theorem 20, one has the inequality*

$$\left|(Tf)(a,b;t)-\frac{f(t)}{\pi}\ln\left(\frac{b-t}{t-a}\right)-\frac{b-a}{\pi}\left[f;\frac{t+b}{2},\frac{a+t}{2}\right]\right|$$
$$\leq\begin{cases}\frac{1}{4\pi}\left[\frac{1}{4}(b-a)^2+\left(t-\frac{a+b}{2}\right)^2\right]\left\|f''\right\|_{[a,b],\infty} & \text{if } f''\in L_\infty[a,b];\\[2ex] \frac{1}{\pi}\frac{q}{2^{\frac{p}{q}}(q+1)^{1+\frac{1}{q}}}\left[(t-a)^{1+\frac{1}{q}}+(b-t)^{1+\frac{1}{q}}\right]\left\|f''\right\|_{[a,b],p} & \text{if } f''\in L_p[a,b],\\ & p>1,\ \frac{1}{p}+\frac{1}{q}=1;\\[2ex] \frac{1}{2\pi}\left[\frac{1}{2}(b-a)+\left|t-\frac{a+b}{2}\right|\right]\left\|f''\right\|_{[a,b],1}, & \text{if } f''\in L[a,b];\end{cases} \tag{128}$$

for any $t\in(a,b)$.

6.2 A Quadrature Formula

The following lemma is of interest in itself.

Lemma 13 *Let* $u:[a,b]\to\mathbb{R}$ *be an absolutely continuous function on* $[a,b]$. *Then for all* $n\geq 1$, $\lambda_i\in[0,1)$ $(i=0,\ldots,n-1)$ *and* $t,\tau\in[a,b]$ *with* $t\neq\tau$, *we have the inequality*

$$\left|\frac{1}{\tau-t}\int_t^\tau u(s)\,ds-\frac{1}{n}\sum_{i=0}^{n-1}u\left[t+(i+1-\lambda_i)\frac{\tau-t}{n}\right]\right| \tag{129}$$

$$\leq\begin{cases}\frac{|t-\tau|}{n}\left[\frac{1}{4}+\frac{1}{n}\sum_{i=0}^{n-1}\left(\lambda_i-\frac{1}{2}\right)^2\right]\left\|u'\right\|_{[t,\tau],\infty};\\[2ex] \frac{|t-\tau|^{\frac{1}{q}}}{(q+1)^{\frac{1}{q}}n}\left[\frac{1}{n}\sum_{i=0}^{n-1}\left(\lambda_i^{1+\frac{1}{q}}+(1-\lambda_i)^{1+\frac{1}{q}}\right)^q\right]^{\frac{1}{q}}\left\|u'\right\|_{[t,\tau],p}\\ \qquad p>1,\ \frac{1}{p}+\frac{1}{q}=1;\\[2ex] \frac{1}{n}\left[\frac{1}{2}+\max\left|\lambda_i-\frac{1}{2}\right|\right]\left\|u'\right\|_{[t,\tau],1};\end{cases}$$

$$\leq \begin{cases} \dfrac{|t-\tau|}{2n} \left\| u' \right\|_{[t,\tau],\infty}; \\ \\ \dfrac{|t-\tau|^{\frac{1}{q}}}{(q+1)^{\frac{1}{q}} n} \left\| u' \right\|_{[t,\tau],p} \ p > 1, \ \frac{1}{p}+\frac{1}{q}=1; \\ \\ \dfrac{1}{n} \left\| u' \right\|_{[t,\tau],1}, \end{cases}$$

where

$$\left\| u' \right\|_{[t,\tau],p} := \left| \int_t^\tau \left| u'(s) \right|^p ds \right|^{\frac{1}{p}}, \quad p \geq 1$$

and

$$\left\| u' \right\|_{[t,\tau],\infty} := ess \sup_{\substack{s\in[t,\tau] \\ (s\in[\tau,t])}} \left| u'(s) \right|.$$

Proof Consider the equidistant division of $[t, \tau]$ (if $t < \tau$) or $[\tau, t]$ (if $\tau < t$) given by

$$E_n : x_i = t + i \cdot \frac{\tau - t}{n}, \quad i = \overline{0, n}. \tag{130}$$

Then the points $\xi_i := \lambda_i \left[t + i \cdot \frac{\tau-t}{n}\right] + (1-\lambda_i)\left[t + (i+1) \cdot \frac{\tau-t}{n}\right]$ $(\lambda_i \in [0, 1),$ $i = \overline{0, n-1})$ are between x_i and x_{i+1}. We observe that we may write for simplicity $\xi_i = t + (i + 1 - \lambda_i) \frac{\tau-t}{n}$ $\left(i = \overline{0, n-1}\right)$. We also have

$$\xi_i - \frac{x_i + x_{i+1}}{2} = \frac{\tau - t}{n}\left(\frac{1}{2} - \lambda_i\right); \ \xi_i - x_i = (1-\lambda_i)\frac{\tau - t}{n}$$

and

$$x_{i+1} - \xi_i = \lambda_i \cdot \frac{\tau - t}{n}$$

for any $i = \overline{0, n-1}$.

If we apply the inequality (121) on the interval $[x_i, x_{i+1}]$ and the intermediate points ξ_i $\left(i = \overline{0, n-1}\right)$, then we may write that

$$\left| \frac{\tau - t}{n} u\left[t + (i+1-\lambda_i)\frac{\tau - t}{n}\right] - \int_{x_i}^{x_{i+1}} u(s)\, ds \right| \tag{131}$$

$$\leq \begin{cases} \left[\frac{1}{4}\frac{(t-\tau)^2}{n^2} + \frac{(t-\tau)^2}{4n^2}(1-2\lambda_i)^2\right]\|u'\|_{[x_i,x_{i+1}],\infty} & \text{if } u' \in L_\infty[a,b], \\ \frac{1}{(q+1)^{\frac{1}{q}}}\left[\frac{|t-\tau|^{1+\frac{1}{q}}}{n^{1+\frac{1}{q}}}\lambda_i^{1+\frac{1}{q}} + \frac{|t-\tau|^{1+\frac{1}{q}}}{n^{1+\frac{1}{q}}}(1-\lambda_i)^{1+\frac{1}{q}}\right]\|u'\|_{[x_i,x_{i+1}],p} & \text{if } u' \in L_p[a,b], \\ & p>1,\ \frac{1}{p}+\frac{1}{q}=1; \\ \left[\frac{1}{2}\frac{|\tau-t|}{n} + \frac{|\tau-t|}{n}\left|\frac{1}{2}-\lambda_i\right|\right]\|u'\|_{[t,\tau],1}; \end{cases}$$

Summing (123), we get

$$\left|\int_t^\tau u(s)\,ds - \frac{\tau-t}{n}\sum_{i=0}^{n-1} u\left[t+(i+1-\lambda_i)\frac{\tau-t}{n}\right]\right| \tag{132}$$

$$\leq \begin{cases} \frac{(t-\tau)^2}{n^2}\sum_{i=0}^{n-1}\left[\frac{1}{4}+\left(\lambda_i-\frac{1}{2}\right)^2\right]\|u'\|_{[x_i,x_{i+1}],\infty}; \\ \frac{|t-\tau|^{1+\frac{1}{q}}}{(q+1)^{\frac{1}{q}}\,n^{1+\frac{1}{q}}}\sum_{i=0}^{n-1}\left[\lambda_i^{1+\frac{1}{q}}+(1-\lambda_i)^{1+\frac{1}{q}}\right]\|u'\|_{[x_i,x_{i+1}],p} \\ \qquad p>1,\ \frac{1}{p}+\frac{1}{q}=1; \\ \frac{|t-\tau|}{n}\sum_{i=0}^{n-1}\left[\frac{1}{2}+\left|\lambda_i-\frac{1}{2}\right|\right]\|u'\|_{[x_i,x_{i+1}],1}. \end{cases}$$

However,

$$\begin{aligned} &\sum_{i=0}^{n-1}\left[\frac{1}{4}+\left(\lambda_i-\frac{1}{2}\right)^2\right]\|u'\|_{[x_i,x_{i+1}],\infty} \\ &= \|u'\|_{[t,\tau],\infty}\left[\frac{1}{4}n+\sum_{i=0}^{n-1}\left(\lambda_i-\frac{1}{2}\right)^2\right], \end{aligned} \tag{133}$$

$$\begin{aligned} &\sum_{i=0}^{n-1}\left[\lambda_i^{1+\frac{1}{q}}+(1-\lambda_i)^{1+\frac{1}{q}}\right]\|u'\|_{[x_i,x_{i+1}],p} \\ &\leq \left(\sum_{i=0}^{n-1}\left[\lambda_i^{1+\frac{1}{q}}+(1-\lambda_i)^{1+\frac{1}{q}}\right]^q\right)^{\frac{1}{q}}\left(\sum_{i=0}^{n-1}\|u'\|_{[x_i,x_{i+1}],p}^p\right)^{\frac{1}{p}} \\ &= \|u'\|_{[t,\tau],p}\left[\sum_{i=0}^{n-1}\left(\lambda_i^{1+\frac{1}{q}}+(1-\lambda_i)^{1+\frac{1}{q}}\right)^q\right]^{\frac{1}{q}} \end{aligned} \tag{134}$$

and

$$\sum_{i=0}^{n-1}\left[\frac{1}{2}+\left|\lambda_i-\frac{1}{2}\right|\right]\left\|u'\right\|_{[x_i,x_{i+1}],1} \tag{135}$$

$$\leq\left[\frac{1}{2}+\max\left|\lambda_i-\frac{1}{2}\right|\right]\sum_{i=0}^{n-1}\left\|u'\right\|_{[x_i,x_{i+1}],1}=\left[\frac{1}{2}+\max\left|\lambda_i-\frac{1}{2}\right|\right]\left\|u'\right\|_{[t,\tau],1}.$$

Now, using (132)–(135), we deduce the first part of (129).

The second part is obvious.

We may now state the following theorem in approximating the finite Hilbert transform of a differentiable function whose derivative is absolutely continuous.

Theorem 21 (Dragomir [5]) *Let $f:[a,b]\to\mathbb{R}$ be a differentiable function so that its derivative f' is absolutely continuous on $[a,b]$. If $\lambda=(\lambda_i)_{i=\overline{0,n-1}}$, $\lambda_i\in[0,1)$ $\left(i=\overline{0,n-1}\right)$ and*

$$S_n(f;\lambda,t):=\frac{b-a}{n\pi}\sum_{i=0}^{n-1}\left[f;t+(i+1-\lambda_i)\frac{b-t}{n},t-(i+1-\lambda_i)\frac{t-a}{n}\right] \tag{136}$$

then we have

$$(Tf)(a,b;t)=\frac{f(t)}{\pi}\ln\left(\frac{b-t}{t-a}\right)+S_n(f;\lambda,t)+R_n(f;\lambda,t) \tag{137}$$

and the remainder $R_n(f;\lambda,t)$ satisfies the estimate:

$$|R_n(f;\lambda,t)| \tag{138}$$

$$\leq\frac{1}{\pi}\times\begin{cases}\frac{1}{n}\left[\frac{1}{4}+\frac{1}{n}\sum_{i=0}^{n-1}\left(\lambda_i-\frac{1}{2}\right)^2\right] & \\ \quad\times\left[\frac{1}{4}(b-a)^2+\left(t-\frac{a+b}{2}\right)^2\right]\left\|f''\right\|_{[a,b],\infty} & \text{if } f''\in L_\infty[a,b];\\ \frac{1}{n}\cdot\frac{q}{(q+1)^{\frac{1}{q}+1}}\left[\frac{1}{n}\sum_{i=0}^{n-1}\left(\lambda_i^{1+\frac{1}{q}}+(1-\lambda_i)^{1+\frac{1}{q}}\right)^q\right]^{\frac{1}{q}} & \\ \quad\times\left[(t-a)^{1+\frac{1}{q}}+(b-t)^{1+\frac{1}{q}}\right]\left\|f''\right\|_{[a,b],p} & \text{if } f''\in L_p[a,b],\\ & p>1,\ \frac{1}{p}+\frac{1}{q}=1;\\ \frac{1}{n}\left[\frac{1}{2}+\max\left|\lambda_i-\frac{1}{2}\right|\right]\left[\frac{1}{2}(b-a)+\left|t-\frac{a+b}{2}\right|\right]\left\|f''\right\|_{[a,b],1}, & \end{cases}$$

$$\leq \frac{1}{\pi}\times\begin{cases}\frac{1}{4n}\left\|f''\right\|_{[a,b],\infty}(b-a)^2 & \text{if } f''\in L_\infty[a,b];\\ \frac{q}{(q+1)^{\frac{1}{q}+1}n}(b-a)^{1+\frac{1}{q}}\left\|f''\right\|_{[a,b],p} & \text{if } f''\in L_p[a,b],\ p>1,\ \frac{1}{p}+\frac{1}{q}=1;\\ \frac{1}{n}(b-a)\left\|f''\right\|_{[a,b],1}, & \end{cases}$$

Proof Applying Lemma 13 for the function f', we may write that

$$\left|\frac{f(t)-f(\tau)}{\tau-t}-\frac{1}{n}\sum_{i=0}^{n-1}f'\left[t+(i+1-\lambda_i)\frac{\tau-t}{n}\right]\right| \tag{139}$$

$$\leq\begin{cases}\frac{|t-\tau|}{n}\left[\frac{1}{4}+\frac{1}{n}\sum_{i=0}^{n-1}\left(\lambda_i-\frac{1}{2}\right)^2\right]\left\|f''\right\|_{[t,\tau],\infty};\\ \frac{|t-\tau|^{\frac{1}{q}}}{(q+1)^{\frac{1}{q}}n}\left[\frac{1}{n}\sum_{i=0}^{n-1}\left(\lambda_i^{1+\frac{1}{q}}+(1-\lambda_i)^{1+\frac{1}{q}}\right)^q\right]^{\frac{1}{q}}\left\|f''\right\|_{[t,\tau],p}\\ \frac{1}{n}\left[\frac{1}{2}+\max\left|\lambda_i-\frac{1}{2}\right|\right]\left\|f''\right\|_{[t,\tau],1}\end{cases}$$

for any $t,\tau\in[a,b]$, $t\neq\tau$.

Taking the PV, we may write

$$\left|\frac{1}{\pi}PV\int_a^b\frac{f(t)-f(\tau)}{\tau-t}d\tau-\frac{1}{n\pi}\sum_{i=0}^{n-1}PV\int_a^b f'\left[t+(i+1-\lambda_i)\frac{\tau-t}{n}\right]d\tau\right| \tag{140}$$

$$\leq\frac{1}{\pi}\times\begin{cases}\frac{1}{n}\left[\frac{1}{4}+\frac{1}{n}\sum_{i=0}^{n-1}\left(\lambda_i-\frac{1}{2}\right)^2\right]PV\int_a^b|t-\tau|\left\|f''\right\|_{[t,\tau],\infty}d\tau;\\ \frac{1}{n(q+1)^{\frac{1}{q}}}\left[\frac{1}{n}\sum_{i=0}^{n-1}\left(\lambda_i^{1+\frac{1}{q}}+(1-\lambda_i)^{1+\frac{1}{q}}\right)^q\right]^{\frac{1}{q}}PV\int_a^b|t-\tau|^{\frac{1}{q}}\left\|f''\right\|_{[t,\tau],p}d\tau\\ \frac{1}{n}\left[\frac{1}{2}+\max\left|\lambda_i-\frac{1}{2}\right|\right]PV\int_a^b\left\|f''\right\|_{[t,\tau],1}d\tau.\end{cases}$$

However,

$$PV\int_a^b |t-\tau| \left\|f''\right\|_{[t,\tau],\infty} d\tau \le \left\|f''\right\|_{[a,b],\infty} PV\int_a^b |t-\tau|\, d\tau$$
$$= \left\|f''\right\|_{[a,b],\infty}\left[\frac{(t-a)^2+(b-t)^2}{2}\right],$$

$$PV\int_a^b |t-\tau|^{\frac{1}{q}} \left\|f''\right\|_{[t,\tau],p} d\tau \le \left\|f''\right\|_{[a,b],p} PV\int_a^b |t-\tau|^{\frac{1}{q}} d\tau$$
$$= \left\|f''\right\|_{[a,b],p}\left[\frac{(t-a)^{\frac{1}{q}+1}+(b-t)^{\frac{1}{q}+1}}{\frac{1}{q}+1}\right]$$
$$= \frac{q}{(q+1)}\left[(t-a)^{\frac{1}{q}+1}+(b-t)^{\frac{1}{q}+1}\right]\left\|f''\right\|_{[a,b],p},$$

$$PV\int_a^b \left\|f''\right\|_{[t,\tau],1} d\tau = PV\left[\int_a^t \left\|f''\right\|_{[\tau,t],1} d\tau + \int_t^b \left\|f''\right\|_{[t,\tau],1} d\tau\right]$$
$$\le \max\{t-a, b-t\}\left\|f''\right\|_{[a,b],1}$$
$$= \left[\frac{1}{2}(b-a)+\left|t-\frac{a+b}{2}\right|\right]\left\|f''\right\|_{[a,b],1}$$

and using the inequality (140) we obtain the desired estimate (138).

The following particular case which may be easily numerically implemented holds.

Corollary 18 *Let f be as in Theorem 21. Define*

$$S_{M,n}(f;t) := \frac{b-a}{n\pi}\sum_{i=0}^{n-1}\left[f; t+\left(i+\frac{1}{2}\right)\frac{b-t}{n}, t-\left(i+\frac{1}{2}\right)\frac{t-a}{n}\right].$$

Then we have the representation:

$$(Tf)(a,b;t) = \frac{f(t)}{\pi}\ln\left(\frac{b-t}{t-a}\right) + S_{M,n}(f;t) + R_{M,n}(f;t)$$

and the remainder $R_{M,n}(f;t)$ satisfies the estimate

$$\left|R_{M,n}(f;t)\right| \tag{141}$$

$$\leq \frac{1}{\pi n} \times \begin{cases} \frac{1}{4}\left[\frac{1}{4}(b-a)^2 + \left(t - \frac{a+b}{2}\right)^2\right] \|f''\|_{[a,b],\infty} & \text{if } f'' \in L_\infty [a,b]; \\ \frac{1}{2^{\frac{1}{q}}(q+1)^{\frac{1}{q}+1}} \left[(t-a)^{1+\frac{1}{q}} + (b-t)^{1+\frac{1}{q}}\right] \|f''\|_{[a,b],p} & \text{if } f'' \in L_p [a,b], \\ & p>1,\ \frac{1}{p}+\frac{1}{q}=1; \\ \frac{1}{2}\left[\frac{1}{2}(b-a) + \left|t - \frac{a+b}{2}\right|\right] \|f''\|_{[a,b],1}, \end{cases}$$

for any $t \in (a,b)$.

7 Inequalities for Convex Derivatives

7.1 An Inequality on the Interval (a, b)

The following result holds.

Theorem 22 (Dragomir [6]) *Assume that the differentiable function* $f : (a,b) \to \mathbb{R}$ *is such that* f' *is convex on* (a,b)*. Then the Hilbert transform* $(Tf)(a,b;\cdot)$ *exists in every point* $t \in (a,b)$ *and*

$$\begin{aligned} &\frac{2}{\pi}\left[f\left(\frac{t+b}{2}\right) - f\left(\frac{t+a}{2}\right)\right] + \frac{f(t)}{\pi}\ln\left(\frac{b-t}{t-a}\right) \\ &\leq (Tf)(a,b;t) \\ &\leq \frac{1}{2\pi}\left[f(b) - f(a) + (b-a)f'(t)\right] + \frac{f(t)}{\pi}\ln\left(\frac{b-t}{t-a}\right) \end{aligned} \tag{142}$$

for any $t \in (a,b)$.

Proof The existence of the Hilbert transform in each point $t \in (a,b)$ follows by the fact that f is locally Lipschitzian on (a,b).

Since f' is convex, we have, by the Hermite-Hadamard inequality, that

$$f'\left(\frac{t+\tau}{2}\right) \leq \frac{1}{\tau - t}\int_t^\tau f'(u)\,du \leq \frac{f'(t)+f'(\tau)}{2} \tag{143}$$

for all $t, \tau \in (a,b)$, $t \neq \tau$, giving

$$f'\left(\frac{t+\tau}{2}\right) \leq \frac{f(\tau)-f(t)}{\tau - t} \leq \frac{f'(t)+f'(\tau)}{2} \tag{144}$$

for all $t, \tau \in (a,b)$, $t \neq \tau$.

Applying the PV in t, i.e., $\lim_{\varepsilon\to 0+}\left(\int_a^{t-\varepsilon}+\int_{t+\varepsilon}^b\right)(\cdot)$, we get

$$PV\int_a^b f'\left(\frac{t+\tau}{2}\right)d\tau \le PV\int_a^b \frac{f(\tau)-f(t)}{\tau-t}d\tau \le PV\int_a^b \frac{f'(t)+f'(\tau)}{2}d\tau. \tag{145}$$

Since

$$\begin{aligned}
&\lim_{\varepsilon\to 0+}\left(\int_a^{t-\varepsilon}+\int_{t+\varepsilon}^b\right)\left(f'\left(\frac{t+\tau}{2}\right)d\tau\right)\\
&=\lim_{\varepsilon\to 0+}\left[\int_a^{t-\varepsilon} f'\left(\frac{t+\tau}{2}\right)d\tau+\int_{t+\varepsilon}^b f'\left(\frac{t+\tau}{2}\right)d\tau\right]\\
&=\lim_{\varepsilon\to 0+} 2\left[\left(f\left(\frac{2t-\varepsilon}{2}\right)-f\left(\frac{t+a}{2}\right)\right)+\left(f\left(\frac{t+b}{2}\right)-f\left(\frac{2t+\varepsilon}{2}\right)\right)\right]\\
&=2\left[f\left(\frac{t+b}{2}\right)-f\left(\frac{t+a}{2}\right)\right]
\end{aligned}$$

and

$$\begin{aligned}
&\lim_{\varepsilon\to 0+}\left(\int_a^{t-\varepsilon}+\int_{t+\varepsilon}^b\right)\left[\frac{f'(t)+f'(\tau)}{2}d\tau\right]\\
&=\frac{1}{2}\lim_{\varepsilon\to 0+}\left[f'(t)(t-\varepsilon-a)+f'(t)(b-t-\varepsilon)+f(t-\varepsilon)-f(a)+f(b)-f(t+\varepsilon)\right]\\
&=\frac{1}{2}\left[f(b)-f(a)+(b-a)f'(t)\right],
\end{aligned}$$

then by (145), we may state that:

$$\begin{aligned}
\frac{2}{\pi}\left[f\left(\frac{t+b}{2}\right)-f\left(\frac{t+a}{2}\right)\right] &\le \frac{1}{\pi}PV\int_a^b \frac{f(\tau)-f(t)}{\tau-t}d\tau \\
&\le \frac{1}{2\pi}\left[f(b)-f(a)+(b-a)f'(t)\right]
\end{aligned} \tag{146}$$

for all $t\in(a,b)$.

As for the function $f_0(t)=1$, $t\in(a,b)$, we have

$$(Tf)(a,b;t)=\frac{1}{\pi}\ln\left(\frac{b-t}{t-a}\right),\quad t\in(a,b),$$

then obviously

$$(Tf_0)(a,b;t) = \frac{1}{\pi} PV \int_a^b \frac{f(\tau) - f(t) + f(t)}{\tau - t} d\tau \tag{147}$$
$$= \frac{1}{\pi} PV \int_a^b \frac{f(\tau) - f(t)}{\tau - t} d\tau + \frac{f(t)}{\pi} \ln\left(\frac{b-t}{t-a}\right)$$

for any $t \in (a,b)$.

Finally, by (146) and (147), we may obtain (155).

The inequality (142) in Theorem 22 may be used to obtain different analytic inequalities for functions $f : [a,b] \to \mathbb{R}$ whose derivatives are convex on (a,b) and the Hilbert transform $(Tf)(a,b;\cdot)$ is known.

For example, the following proposition holds.

Proposition 2 (Dragomir [6]) *For any $a, b \in \mathbb{R}$, $a < b$ and $t \in (a,b)$, we have the inequality:*

$$\ln\left(\frac{b-t}{t-a}\right) + 2\left(e^{\frac{b-t}{2}} - e^{\frac{a-t}{2}}\right) \tag{148}$$
$$\le E_i(b-t) - E_i(a-t)$$
$$\le \ln\left(\frac{b-t}{t-a}\right) + \frac{1}{2}\left[e^{b-t} - e^{a-t} + (b-a)\right],$$

where E_i is defined in (149).

Proof If we consider the function $f(t) = e^t$, $t \in (a,b)$, then f' is convex on (a,b),

$$(Tf)(a,b;t) = \frac{e^t}{\pi}\left[E_i(b-t) - E_i(a-t)\right],$$

where E_i is defined by

$$E_i(z) := PV \int_{-\infty}^{z} \frac{e^t}{t} dt, \quad z \in \mathbb{R}, \tag{149}$$

$$\frac{2}{\pi}\left[f\left(\frac{t+b}{2}\right) - f\left(\frac{t+a}{2}\right)\right] + \frac{f(t)}{\pi} \ln\left(\frac{b-t}{t-a}\right)$$
$$= \frac{2}{\pi}\left[e^{\frac{t+b}{2}} - e^{\frac{t+a}{2}}\right] + \frac{e^t}{\pi} \ln\left(\frac{b-t}{t-a}\right)$$

and

$$\frac{1}{2\pi}\left[f(b)-f(a)+(b-a)f'(t)\right]+\frac{f(t)}{\pi}\ln\left(\frac{b-t}{t-a}\right)$$
$$=\frac{1}{2\pi}\left[e^{b}-e^{a}+(b-a)e^{t}\right]+\frac{e^{t}}{\pi}\ln\left(\frac{b-t}{t-a}\right).$$

Using (142) and dividing by e^t, we deduce (148).

The following inequality also holds.

Proposition 3 (Dragomir [6]) *For any $x>0$, we have the inequality*

$$2\left(e^{\frac{1}{2}x}-e^{-\frac{1}{2}x}\right)\le E_i(x)-E_i(-x)\le\frac{1}{2}\left[e^{x}-e^{-x}+2x\right]. \tag{150}$$

Proof If in (148) we put $t=\frac{a+b}{2}$, then we deduce

$$2\left(e^{\frac{b-a}{4}}-e^{-\frac{b-a}{4}}\right)\le E_i\left(\frac{b-a}{2}\right)-E_i\left(-\frac{b-a}{2}\right)$$
$$\le\frac{1}{2}\left[e^{\frac{b-a}{2}}-e^{-\frac{b-a}{2}}+b-a\right].$$

If we denote $x:=\frac{b-a}{2}$, then we get

$$2\left(e^{\frac{1}{2}x}-e^{-\frac{1}{2}x}\right)\le E_i(x)-E_i(-x)\le\frac{1}{2}\left[e^{x}-e^{-x}+2x\right]. \tag{151}$$

If we choose another function, for instance, $f:(a,b)\subset(0,\infty)\to\mathbb{R}$, $f(t)=-\frac{1}{t}$, then obviously f' is convex on (a,b), and we may state the following result as well.

Proposition 4 (Dragomir [6]) *For any $0<a<b$ and $t\in(a,b)$, we have the inequality:*

$$\frac{2tG^2}{G^2+t^2}\le L\le\frac{t^2+2At+G^2}{4t},\quad \textit{for any } t\in(a,b), \tag{152}$$

where $A=\frac{a+b}{2}$, $G=\sqrt{ab}$ and $L=\frac{b-a}{\ln b-\ln a}$ (the logarithmic mean).

Proof For the function $f:(a,b)\to\mathbb{R}$, $f(t)=-\frac{1}{t}$, we have

$$(Tf)(a,b;t)=\frac{1}{\pi t}\left[\ln\left(\frac{b}{a}\right)-\ln\left(\frac{b-t}{t-a}\right)\right],$$

$$\frac{2}{\pi}\left[f\left(\frac{t+b}{2}\right)-f\left(\frac{t+a}{2}\right)\right]+\frac{f(t)}{\pi}\ln\left(\frac{b-t}{t-a}\right)$$
$$=\frac{4}{\pi}\cdot\frac{b-a}{(t+a)(t+b)}-\frac{1}{\pi t}\ln\left(\frac{b-t}{t-a}\right),$$

and

$$\frac{1}{2\pi}\left[f(b)-f(a)+(b-a)f'(t)\right]+\frac{f(t)}{\pi}\ln\left(\frac{b-t}{t-a}\right)$$
$$=\frac{b-a}{2\pi}\left[\frac{1}{ab}+\frac{1}{t^2}\right]-\frac{1}{\pi t}\ln\left(\frac{b-t}{t-a}\right).$$

Now, if we use (142), we may write

$$\frac{4}{\pi}\cdot\frac{b-a}{(t+a)(t+b)}-\frac{1}{\pi t}\ln\left(\frac{b-t}{t-a}\right)\leq\frac{1}{\pi t}\ln\left(\frac{b}{a}\right)-\frac{1}{\pi t}\ln\left(\frac{b-t}{t-a}\right)$$
$$\leq\frac{b-a}{2\pi}\left(\frac{t^2+ab}{abt^2}\right)-\frac{1}{\pi t}\ln\left(\frac{b-t}{t-a}\right),$$

which is equivalent to

$$\frac{4t}{(t+a)(t+b)}\leq\frac{\ln b-\ln a}{b-a}\leq\frac{t^2+ab}{2tab}.$$

Using the fact that $L:=\frac{b-a}{\ln b-\ln a}$, we deduce (152).

Corollary 19 *We have the inequality*

$$G\leq L\leq\frac{G+A}{2}. \tag{153}$$

Remark 9 The first inequality is a well-known result as the following sequence of inequalities hold:

$$G\leq L\leq I\leq A.$$

The second inequality is equivalent with

$$L(a,b)\leq\left[A\left(\sqrt{a},\sqrt{b}\right)\right]^2, \tag{154}$$

which is interesting in itself.

7.2 An Inequality on an Equidistant Division of (a, b)

The following lemma is interesting in itself.

Lemma 14 *Let $g : [a, b] \to \mathbb{R}$ be a convex function. Then for $n \geq 1$ and $t, \tau \in [a, b]$, $t \neq \tau$, we have the inequality:*

$$\begin{aligned} &\frac{1}{n}\sum_{i=0}^{n-1} g\left[t + \left(i + \frac{1}{2}\right) \cdot \frac{t-\tau}{n}\right] \\ &\leq \frac{1}{\tau - t}\int_t^\tau g(u)\,du \\ &\leq \frac{1}{2n}\sum_{i=0}^{n-1}\left[g\left(t + i \cdot \frac{\tau - t}{n}\right) + g\left(t + (i+1) \cdot \frac{\tau - t}{n}\right)\right]. \end{aligned} \tag{155}$$

Proof Consider the equidistant partitioning of $[t, \tau]$ (if $t < \tau$) or $[\tau, t]$ (if $\tau < t$) given by

$$E_n : x_i = t + i \cdot \frac{\tau - t}{n}, \quad i = \overline{0, n}. \tag{156}$$

Then, applying the Hermite-Hadamard inequality, we may write that:

$$g\left(\frac{x_i + x_{i+1}}{2}\right) \leq \frac{1}{x_{i+1} - x_i}\int_{x_i}^{x_{i+1}} g(u)\,du \leq \frac{g(x_i) + g(x_{i+1})}{2}$$

i.e.,

$$\begin{aligned} g\left(t + \left(i + \frac{1}{2}\right) \cdot \frac{t - \tau}{n}\right) &\leq \frac{n}{\tau - t}\int_{x_i}^{x_{i+1}} g(u)\,du \\ &\leq \frac{1}{2}\left[g\left(t + i \cdot \frac{\tau - t}{n}\right) + g\left(t + (i+1) \cdot \frac{\tau - t}{n}\right)\right]. \end{aligned}$$

Dividing by n and summing over i from 0 to $n - 1$, we deduce the desired inequality (155).

The following generalization of Theorem 22 holds.

Theorem 23 (Dragomir [6]) *Assume that $f : (a, b) \to \mathbb{R}$ fulfills the hypothesis of Theorem 22. Then for all $n \geq 1$, we have the double inequality:*

$$\frac{b-a}{n\pi}\sum_{i=0}^{n-1}\left[f;t-\left(i+\frac{1}{2}\right)\cdot\frac{t-a}{n},t+\left(i+\frac{1}{2}\right)\cdot\frac{b-t}{n}\right]+\frac{f(t)}{\pi}\ln\left(\frac{b-t}{t-a}\right) \tag{157}$$

$$\leq (Tf)(a,b;t)$$

$$\leq \frac{f(b)-f(a)+f'(t)(b-a)}{2n\pi}+\frac{b-a}{n\pi}\sum_{i=1}^{n-1}\left[f;t-i\cdot\frac{t-a}{n},t+i\cdot\frac{b-t}{n}\right]$$

$$+\frac{f(t)}{\pi}\ln\left(\frac{b-t}{t-a}\right)$$

for any $t\in(a,b)$, where $[f;c,d]$ denotes the divided difference $\frac{f(c)-f(d)}{c-d}$.

Proof If we write the inequality (155) for f', then we have

$$\frac{1}{n}\sum_{i=0}^{n-1}f'\left[t+\left(i+\frac{1}{2}\right)\cdot\frac{\tau-t}{n}\right]$$

$$\leq \frac{f(\tau)-f(t)}{\tau-t}$$

$$\leq \frac{1}{2n}\sum_{i=0}^{n-1}\left[f'\left(t+i\cdot\frac{\tau-t}{n}\right)+f'\left(t+(i+1)\cdot\frac{\tau-t}{n}\right)\right]$$

$$=\frac{1}{2n}\left[f'(t)+\sum_{i=1}^{n-1}f'\left(t+i\cdot\frac{\tau-t}{n}\right)+\sum_{i=0}^{n-2}f'\left(t+(i+1)\cdot\frac{\tau-t}{n}\right)+f'(\tau)\right]$$

$$=\frac{1}{2n}\left[f'(t)+f'(\tau)+2\sum_{i=1}^{n-1}f'\left(t+i\cdot\frac{\tau-t}{n}\right)\right], \tag{158}$$

since it is obvious that

$$\sum_{i=1}^{n-1}f'\left(t+i\cdot\frac{\tau-t}{n}\right)=\sum_{i=0}^{n-2}f'\left(t+(i+1)\cdot\frac{\tau-t}{n}\right).$$

Applying the PV over t, i.e., $\lim_{\varepsilon\to 0+}\left(\int_a^{t-\varepsilon}+\int_{t+\varepsilon}^b\right)$ to the inequality (158), we deduce

$$\frac{1}{n}\sum_{i=1}^{n-1}PV\int_a^b f'\left[t+\left(i+\frac{1}{2}\right)\cdot\frac{\tau-t}{n}\right]d\tau \tag{159}$$

$$\leq PV\int_a^b \frac{f(\tau)-f(t)}{\tau-t}d\tau$$
$$\leq \frac{1}{2n}PV\int_a^b \left[f'(t)+f'(\tau)+2\sum_{i=1}^{n-1} f'\left(t+i\cdot\frac{\tau-t}{n}\right)\right]d\tau.$$

Now, as

$$PV\int_a^b f'\left[t+\left(i+\frac{1}{2}\right)\cdot\frac{\tau-t}{n}\right]d\tau$$
$$=\lim_{\varepsilon\to 0+}\left(\int_a^{t-\varepsilon}+\int_{t+\varepsilon}^b\right)\left(f'\left[t+\left(i+\frac{1}{2}\right)\cdot\frac{\tau-t}{n}\right]d\tau\right)$$
$$=\lim_{\varepsilon\to 0+}\frac{n}{i+\frac{1}{2}}\left[f\left(t-\left(i+\frac{1}{2}\right)\cdot\frac{\varepsilon}{n}\right)-f\left(t+\left(i+\frac{1}{2}\right)\cdot\frac{a-t}{n}\right)\right.$$
$$\left.+f\left(t+\left(i+\frac{1}{2}\right)\cdot\frac{b-t}{n}\right)-f\left(t+\left(i+\frac{1}{2}\right)\cdot\frac{\varepsilon}{n}\right)\right]$$
$$=\frac{n}{i+\frac{1}{2}}\left[f\left(t+\left(i+\frac{1}{2}\right)\cdot\frac{b-t}{n}\right)-f\left(t+\left(i+\frac{1}{2}\right)\cdot\frac{a-t}{n}\right)\right]$$
$$=(b-a)\left[f;t-\left(i+\frac{1}{2}\right)\cdot\frac{t-a}{n},t+\left(i+\frac{1}{2}\right)\cdot\frac{b-t}{n}\right],$$

and

$$PV\int_a^b \left[f'(t)+f'(\tau)+2\sum_{i=1}^{n-1} f'\left(t+i\cdot\frac{\tau-t}{n}\right)\right]d\tau$$
$$=\lim_{\varepsilon\to 0+}\left(\int_a^{t-\varepsilon}+\int_{t+\varepsilon}^b\right)\left[f'(t)+f'(\tau)+2\sum_{i=1}^{n-1} f'\left(t+i\cdot\frac{\tau-t}{n}\right)\right]d\tau$$
$$=\lim_{\varepsilon\to 0+}\left[f'(t)(t-\varepsilon-a)+f'(t)(b-t-\varepsilon)+f(t-\varepsilon)-f(a)+f(b)-f(t+\varepsilon)\right.$$
$$\left.+2\sum_{i=1}^{n-1}\frac{n}{i}\left[f\left(t-\frac{i\varepsilon}{n}\right)-f\left(t+i\cdot\frac{a-t}{n}\right)+f\left(t+i\cdot\frac{b-t}{n}\right)-f\left(t+\frac{i\varepsilon}{n}\right)\right]\right]$$
$$=f(b)-f(a)+f'(t)(b-a)+2(b-a)\sum_{i=1}^{n-1}\left[f;t-i\cdot\frac{t-a}{n},t+i\cdot\frac{b-t}{n}\right],$$

then by (159) we deduce

$$\frac{b-a}{n}\sum_{i=0}^{n-1}\left[f;t-\left(i+\frac{1}{2}\right)\cdot\frac{t-a}{n},t+\left(i+\frac{1}{2}\right)\cdot\frac{b-t}{n}\right] \tag{160}$$

$$\leq PV\int_a^b \frac{f(\tau)-f(t)}{\tau-t}d\tau$$

$$\leq \frac{f(b)-f(a)+f'(t)(b-a)}{2n}+\frac{b-a}{n}\sum_{i=1}^{n-1}\left[f;t-i\cdot\frac{t-a}{n},t+i\cdot\frac{b-t}{n}\right].$$

Using the identity (147) and the inequality (160), we obtain the desired result (157).

7.3 The Case of Non-equidistant Partitioning

The following lemma holds.

Lemma 15 *Let $g:[a,b]\to\mathbb{R}$ be a convex function on $[a,b]$ and $t,\tau\in[a,b]$ with $t\neq\tau$. If $0=\lambda_0<\lambda_1<\cdots<\lambda_{n-1}<\lambda_n=1$, then we have the inequality:*

$$\sum_{i=0}^{n-1}(\lambda_{i+1}-\lambda_i)\,g\left[\left(1-\frac{\lambda_i+\lambda_{i+1}}{2}\right)t+\frac{\lambda_i+\lambda_{i+1}}{2}\cdot\tau\right] \tag{161}$$

$$\leq \frac{1}{\tau-t}\int_t^\tau g(u)\,du$$

$$\leq \frac{1}{2}\sum_{i=0}^{n-1}(\lambda_{i+1}-\lambda_i)\left\{g\left[(1-\lambda_i)t+\lambda_i\tau\right]+g\left[(1-\lambda_{i+1})t+\lambda_{i+1}\tau\right]\right\}.$$

Proof Consider the partitioning of $[t,\tau]$ (if $t<\tau$) or $[\tau,t]$ (if $\tau<t$) given by

$$I_n: x_i=(1-\lambda_i)t+\lambda_i\tau,\quad \left(i=\overline{0,n}\right).$$

Then, obviously,

$$\frac{x_i+x_{i+1}}{2}=\left(1-\frac{\lambda_i+\lambda_{i+1}}{2}\right)t+\frac{\lambda_i+\lambda_{i+1}}{2}\cdot\tau,\quad \left(i=\overline{0,n-1}\right)$$

and

$$x_{i+1}-x_i=(\tau-t)(\lambda_{i+1}-\lambda_i),\quad \left(i=\overline{0,n-1}\right).$$

Applying the Hermite-Hadamard inequality on $[x_i,x_{i+1}]$ $\left(i=\overline{0,n-1}\right)$, we may write that

$$g\left[\left(1-\frac{\lambda_i+\lambda_{i+1}}{2}\right)t+\frac{\lambda_i+\lambda_{i+1}}{2}\cdot\tau\right]$$
$$\leq \frac{1}{(\tau-t)(\lambda_{i+1}-\lambda_i)}\int_{x_i}^{x_{i+1}} g(u)\,du$$
$$\leq \frac{1}{2}\{g[(1-\lambda_i)t+\lambda_i\tau]+g[(1-\lambda_{i+1})t+\lambda_{i+1}\tau]\}$$

for any $i=\overline{0,n-1}$.

If we multiply with $\lambda_{i+1}-\lambda_i>0$ and sum over i from 0 to $n-1$, we deduce the desired inequality (161).

The following theorem holds.

Theorem 24 (Dragomir [6]) *Assume that $f:(a,b)\to\mathbb{R}$ fulfills the hypothesis of Theorem 22. Then for all $n\geq 1$, and $0=\lambda_0<\lambda_1<\cdots<\lambda_{n-1}<\lambda_n=1$, we have the inequality*

$$\frac{f(t)}{\pi}\ln\left(\frac{b-t}{t-a}\right)+\frac{b-a}{\pi}\sum_{i=0}^{n-1}(\lambda_{i+1}-\lambda_i)$$
$$\times\left[f;\left(1-\frac{\lambda_i+\lambda_{i+1}}{2}\right)t+\frac{\lambda_i+\lambda_{i+1}}{2}\cdot b,\left(1-\frac{\lambda_i+\lambda_{i+1}}{2}\right)t+\frac{\lambda_i+\lambda_{i+1}}{2}\cdot a\right]$$
$$\leq (Tf)(a,b;t)$$
$$\leq \frac{1}{2\pi}\left\{\lambda_1(b-a)f'(t)+(1-\lambda_{n-1})[f(b)-f(a)]\right\}$$
$$+\frac{b-a}{2\pi}\sum_{i=0}^{n-1}(\lambda_{i+1}-\lambda_i)[f;(1-\lambda_i)t+\lambda_i b,(1-\lambda_i)t+\lambda_i a]+\frac{f(t)}{\pi}\ln\left(\frac{b-t}{t-a}\right) \quad (162)$$

for any $t\in(a,b)$.

Proof If we write the inequality (161) for f', then we have

$$\sum_{i=0}^{n-1}(\lambda_{i+1}-\lambda_i)f'\left[\left(1-\frac{\lambda_i+\lambda_{i+1}}{2}\right)t+\frac{\lambda_i+\lambda_{i+1}}{2}\cdot\tau\right] \quad (163)$$
$$\leq \frac{f(\tau)-f(t)}{\tau-t}$$
$$\leq \frac{1}{2}\sum_{i=0}^{n-1}(\lambda_{i+1}-\lambda_i)\left\{f'[(1-\lambda_i)t+\lambda_i\tau]+f'[(1-\lambda_{i+1})t+\lambda_{i+1}\tau]\right\}$$

$$= \frac{1}{2}\left[\lambda_1 f'(t) + \sum_{i=1}^{n-1}(\lambda_{i+1} - \lambda_i) f'[(1-\lambda_i)t + \lambda_i \tau]\right.$$

$$\left. + \sum_{i=0}^{n-2}(\lambda_{i+1} - \lambda_i) f'[(1-\lambda_{i+1})t + \lambda_{i+1}\tau] + (1-\lambda_{n-1}) f'(\tau)\right]$$

$$= \frac{1}{2}\left[\lambda_1 f'(t) + \sum_{i=1}^{n-1}(\lambda_{i+1} - \lambda_i) f'[(1-\lambda_i)t + \lambda_i \tau]\right.$$

$$\left. + \sum_{i=1}^{n-1}(\lambda_i - \lambda_{i-1}) f'[(1-\lambda_i)t + \lambda_i \tau] + (1-\lambda_{n-1}) f'(\tau)\right]$$

$$= \frac{1}{2}\left[\lambda_1 f'(t) + \sum_{i=1}^{n-1}(\lambda_{i+1} - \lambda_i) f'[(1-\lambda_i)t + \lambda_i \tau] + (1-\lambda_{n-1}) f'(\tau)\right].$$

Applying the PV over t, i.e., $\lim_{\varepsilon \to 0+}\left(\int_a^{t-\varepsilon} + \int_{t+\varepsilon}^b\right)$ to the inequality (163), we deduce

$$\sum_{i=0}^{n-1}(\lambda_{i+1} - \lambda_i)\, PV \int_a^b f'\left[\left(1 - \frac{\lambda_i + \lambda_{i+1}}{2}\right)t + \frac{\lambda_i + \lambda_{i+1}}{2}\cdot \tau\right] d\tau \tag{164}$$

$$\leq PV \int_a^b \frac{f(\tau) - f(t)}{\tau - t} d\tau$$

$$\leq \frac{1}{2}\left[\lambda_1 (b-a) f'(t) + \sum_{i=1}^{n-1}(\lambda_{i+1} - \lambda_i)\, PV \int_a^b f'[(1-\lambda_i)t + \lambda_i \tau]\, d\tau\right.$$

$$\left. + (1-\lambda_{n-1})(f(b) - f(a))\right].$$

Since

$$PV \int_a^b f'\left[\left(1 - \frac{\lambda_i + \lambda_{i+1}}{2}\right)t + \frac{\lambda_i + \lambda_{i+1}}{2}\cdot \tau\right] d\tau$$

$$= \frac{2}{\lambda_i + \lambda_{i+1}}\left(f\left[\left(1 - \frac{\lambda_i + \lambda_{i+1}}{2}\right)t + \frac{\lambda_i + \lambda_{i+1}}{2}\cdot b\right]\right.$$

$$\left. - f\left[\left(1 - \frac{\lambda_i + \lambda_{i+1}}{2}\right)t + \frac{\lambda_i + \lambda_{i+1}}{2}\cdot a\right]\right)$$

$$= (b-a)\left[f; \left(1 - \frac{\lambda_i + \lambda_{i+1}}{2}\right)t + \frac{\lambda_i + \lambda_{i+1}}{2}\cdot b, \left(1 - \frac{\lambda_i + \lambda_{i+1}}{2}\right)t + \frac{\lambda_i + \lambda_{i+1}}{2}\cdot a\right]$$

and

$$PV\int_a^b f'[(1-\lambda_i)t+\lambda_i\tau]\,d\tau=(b-a)[f;(1-\lambda_i)t+\lambda_i b,(1-\lambda_i)t+\lambda_i a],$$

then by (164) we deduce the desired inequality (162).

Remark 10 It is obvious that for $\lambda_i = \frac{i}{n}\ \left(i=\overline{0,n}\right)$, we recapture the inequality (157).

The following corollary also holds.

Corollary 20 *Assume that $f:(a,b)\to\mathbb{R}$ fulfills the hypothesis of Theorem 22. Then for $n\geq 1$ we have*

$$\begin{aligned}
&\frac{f(t)}{\pi}\ln\left(\frac{b-t}{t-a}\right)\\
&\quad+\frac{b-a}{\pi}\left[\frac{1}{2^{n-1}}\left[f;\left(1-\frac{1}{2^n}\right)t+\frac{1}{2^n}b,\left(1-\frac{1}{2^n}\right)t+\frac{1}{2^n}a\right]\right]\\
&+\frac{b-a}{\pi}\sum_{i=1}^{n-1}\frac{1}{2^{n-1}}\left[f;\left(1-\frac{3}{2^{n-i}}\right)t+\frac{3}{2^{n-i}}b,\left(1-\frac{3}{2^{n-i}}\right)t+\frac{3}{2^{n-i}}a\right]\\
&\qquad\leq (Tf)(a,b;t)\\
&\qquad\leq \frac{1}{2\pi}\left\{\frac{(b-a)f'(t)}{2^{n-1}}+\frac{1}{2}[f(b)-f(a)]\right\}\\
&+\frac{b-a}{2\pi}\left(\frac{1}{2^{n-2}}-1\right)\left[f;\left(1-\frac{1}{2^{n-1}}\right)t+\frac{1}{2^{n-1}}b,\left(1-\frac{1}{2^{n-1}}\right)t+\frac{1}{2^{n-1}}a\right]\\
&+3\cdot\frac{b-a}{2\pi}\sum_{i=2}^{n-1}\frac{1}{2^{n-i+1}}\left[f;\left(1-\frac{1}{2^{n-i}}\right)t+\frac{1}{2^{n-i}}b,\left(1-\frac{1}{2^{n-i}}\right)t+\frac{1}{2^{n-i}}a\right]\\
&\qquad\qquad+\frac{f(t)}{\pi}\ln\left(\frac{b-t}{t-a}\right), \qquad (165)
\end{aligned}$$

for any $t\in(a,b)$.

The proof follows by Theorem 24 applied for $\lambda_0=0$, $\lambda_i=\frac{2^i}{2^n}$, $i=\overline{1,n}$. We omit the details.

8 Inequalities for Products

8.1 Some Basic Inequalities

The following lemma holds.

Lemma 16 (Dragomir [7]) *If f and g are locally Hölder continuous on $[a,b]$, then fg is also locally Hölder continuous on $[a,b]$ and*

$$\begin{aligned} &T(fg)(a,b;t) \\ &= f(t)T(g)(a,b;t) + g(t)T(f)(a,b;t) \\ &- \frac{1}{\pi} f(t) g(t) \ln\left(\frac{b-t}{t-a}\right) + \frac{1}{\pi} PV \int_a^b \frac{(f(\tau)-f(t))(g(\tau)-g(t))}{\tau - t} d\tau \end{aligned} \tag{166}$$

for any $t \in (a,b)$.

Proof Assume that for a subinterval $[c,d] \subseteq [a,b]$, we have

$$|f(s) - f(u)| \le L_1 |s-u|^{r_1} \quad \text{for any } s,u \in [c,d]; \tag{167}$$

$$|g(s) - g(u)| \le L_2 |s-u|^{r_2} \quad \text{for any } s,u \in [c,d]. \tag{168}$$

Then

$$\begin{aligned} |f(s)g(s) - f(u)g(u)| &= |f(s)g(s) - f(s)g(u) + f(s)g(u) - f(u)g(u)| \\ &\le |f(s)| |g(s) - g(u)| + |g(u)| |f(s) - f(u)| \\ &\le M_1 L_1 |s-u|^{r_1} + M_2 L_2 |s-u|^{r_2} \\ &\le |s-u|^r \left[M_1 L_1 |s-u|^{r_1 - r} + M_2 L_2 |s-u|^{r_2 - r}\right] \\ &\le |s-u|^r \left[M_1 L_1 |d-c|^{r_1 - r} + M_2 L_2 |d-c|^{r_2 - r}\right] \\ &= M |s-u|^r \end{aligned}$$

where

$$M_1 := \sup_{s \in [c,d]} |f(s)|, \quad M_2 := \sup_{u \in [c,d]} |g(u)|, \quad r = \min(r_1, r_2),$$

and

$$M = M_1 L_1 |d-c|^{r_1 - r} + M_2 L_2 |d-c|^{r_2 - r},$$

proving that fg is locally Hölder continuous on $[a,b]$.

Now, for any $t,\tau\in[a,b]$, we may write that

$$(f(\tau)-f(t))(g(\tau)-g(t))=f(\tau)g(\tau)+f(t)g(t)-f(t)g(\tau)-f(\tau)g(t)$$

giving

$$\frac{f(\tau)g(\tau)}{\tau-t}=f(t)\cdot\frac{g(\tau)}{\tau-t}+g(t)\cdot\frac{f(\tau)}{\tau-t}-\frac{f(t)g(t)}{\tau-t}+\frac{(f(\tau)-f(t))(g(\tau)-g(t))}{\tau-t}$$

for any $t,\tau\in[a,b]$, $t\neq\tau$.

Consequently,

$$\begin{aligned}
&T(fg)(a,b;t)\\
&=\frac{1}{\pi}PV\int_a^b\frac{f(\tau)g(\tau)}{\tau-t}d\tau\\
&=\frac{1}{\pi}f(t)\,PV\int_a^b\frac{g(\tau)}{\tau-t}d\tau+g(t)\frac{1}{\pi}PV\int_a^b\frac{f(\tau)}{\tau-t}d\tau\\
&-\frac{1}{\pi}f(t)g(t)\,PV\int_a^b\frac{d\tau}{\tau-t}+\frac{1}{\pi}PV\int_a^b\frac{(f(\tau)-f(t))(g(\tau)-g(t))}{\tau-t}d\tau\\
&=f(t)T(g)(a,b;t)+g(t)T(f)(a,b;t)\\
&-\frac{f(t)g(t)}{\pi}\ln\left(\frac{b-t}{t-a}\right)+\frac{1}{\pi}PV\int_a^b\frac{(f(\tau)-f(t))(g(\tau)-g(t))}{\tau-t}d\tau
\end{aligned}$$

for any $t\in(a,b)$, and the identity (166) is proved.

Theorem 25 (Dragomir [7]) *Assume that f is of L_1-r_1-Hölder type and g is of L_2-r_2-Hölder type on $[a,b]$, where $L_1,L_2>0$, $r_1,r_2\in(0,1]$. Then we have the inequality:*

$$\begin{aligned}
&\Big|T(fg)(a,b;t)-f(t)T(g)(a,b;t)\\
&\qquad-g(t)T(f)(a,b;t)+\frac{1}{\pi}f(t)g(t)\ln\left(\frac{b-t}{t-a}\right)\Big|\\
&\leq\frac{L_1L_2}{\pi(r_1+r_2)}\left[(b-t)^{r_1+r_2}+(t-a)^{r_1+r_2}\right]\leq\frac{2L_1L_2(b-a)^{r_1+r_2}}{\pi(r_1+r_2)}
\end{aligned}\tag{169}$$

for any $t\in(a,b)$.

Proof Taking the modulus in (166), we may write

$$\begin{aligned}&\Big|T(fg)(a,b;t)-f(t)T(g)(a,b;t)\\&\qquad -g(t)T(f)(a,b;t)+\frac{1}{\pi}f(t)g(t)\ln\left(\frac{b-t}{t-a}\right)\Big|\\&\le\frac{1}{\pi}PV\int_a^b\left|\frac{(f(\tau)-f(t))(g(\tau)-g(t))}{\tau-t}\right|d\tau\le\frac{1}{\pi}PV\int_a^b L_1L_2|\tau-t|^{r_1+r_2-1}d\tau\\&=\frac{L_1L_2}{\pi}\left[\frac{(b-t)^{r_1+r_2}+(t-a)^{r_1+r_2}}{r_1+r_2}\right]\end{aligned}$$

and the first part of inequality (169) is proved. The second part is obvious.

The best inequality we can get from (169) is embodied in the following corollary.

Corollary 21 *With the assumptions in Theorem 25, we have*

$$\begin{aligned}&\Big|T(fg)\left(a,b;\frac{a+b}{2}\right)-f\left(\frac{a+b}{2}\right)T(g)\left(a,b;\frac{a+b}{2}\right)\\&\qquad -g\left(\frac{a+b}{2}\right)T(f)\left(a,b;\frac{a+b}{2}\right)\Big|\\&\le\frac{L_1L_2(b-a)^{r_1+r_2}}{\pi(r_1+r_2)2^{r_1+r_2-1}}.\end{aligned}\tag{170}$$

The following corollary also holds.

Corollary 22 *If f and g are Lipschitzian with the constants K_1 and K_2, then we have the inequality*

$$\begin{aligned}&\left|T(fg)(a,b;t)-f(t)T(g)(a,b;t)-g(t)T(f)(a,b;t)+\frac{1}{\pi}f(t)g(t)\ln\left(\frac{b-t}{t-a}\right)\right|\\&\le\frac{K_1K_2}{\pi}\left[\frac{1}{4}(b-a)^2+\left(t-\frac{a+b}{2}\right)^2\right]\le\frac{K_1K_2}{2\pi}(b-a)^2\end{aligned}\tag{171}$$

for any $t\in(a,b)$. In particular, for $t=\frac{a+b}{2}$, we have

$$\begin{aligned}&\left|T(fg)\left(a,b;\frac{a+b}{2}\right)-f\left(\frac{a+b}{2}\right)T(g)\left(a,b;\frac{a+b}{2}\right)-g\left(\frac{a+b}{2}\right)T(f)\left(a,b;\frac{a+b}{2}\right)\right|\\&\qquad\le\frac{K_1K_2}{4\pi}(b-a)^2.\end{aligned}\tag{172}$$

8.2 Further Estimates

The following theorem also holds.

Theorem 26 (Dragomir [7]) *Assume that f and g are absolutely continuous on $[a, b]$. Then we have the inequality:*

$$\left|T(fg)(a,b;t) - f(t)T(g)(a,b;t) - g(t)T(f)(a,b;t) + \frac{1}{\pi}f(t)g(t)\ln\left(\frac{b-t}{t-a}\right)\right| \tag{173}$$

$$\leq \frac{1}{\pi}\times\begin{cases}
\left[\frac{1}{4}(b-a)^2 + \left(t - \frac{a+b}{2}\right)^2\right]\left\|f'\right\|_{[a,b],\infty}\left\|g'\right\|_{[a,b],\infty} \\
\qquad \text{if } f' \in L_\infty[a,b],\ g' \in L_\infty[a,b]; \\
\frac{\delta}{\delta+1}\left[(b-t)^{1+\frac{1}{\delta}} + (t-a)^{1+\frac{1}{\delta}}\right]\left\|f'\right\|_{[a,b],\infty}\left\|g'\right\|_{[a,b],\gamma} \\
\qquad \text{if } f' \in L_\infty[a,b],\ g' \in L_\gamma[a,b],\ \gamma > 1,\ \frac{1}{\gamma} + \frac{1}{\delta} = 1; \\
(b-a)\left\|f'\right\|_{[a,b],\infty}\left\|g'\right\|_{[a,b],1} \qquad \text{if } f' \in L_\infty[a,b],\ g' \in L_1[a,b]; \\
\frac{\beta}{\beta+1}\left[(b-t)^{1+\frac{1}{\beta}} + (t-a)^{1+\frac{1}{\beta}}\right]\left\|f'\right\|_{[a,b],\alpha}\left\|g'\right\|_{[a,b],\infty} \\
\qquad \text{if } f' \in L_\alpha[a,b],\ \alpha > 1,\ \frac{1}{\alpha} + \frac{1}{\beta} = 1, \text{ and } g' \in L_\infty[a,b]; \\
\frac{\beta\delta}{\beta+\delta}\left[(b-t)^{\frac{\delta+\beta}{\beta+\delta}} + (t-a)^{\frac{\delta+\beta}{\beta+\delta}}\right]\left\|f'\right\|_{[a,b],\alpha}\left\|g'\right\|_{[a,b],\gamma} \\
\text{if } f' \in L_\alpha[a,b],\ \alpha > 1,\ \frac{1}{\alpha} + \frac{1}{\beta} = 1, \text{ and } g' \in L_\gamma[a,b],\ \gamma > 1,\ \frac{1}{\gamma} + \frac{1}{\delta} = 1; \\
\beta\left[(b-t)^{\frac{1}{\beta}} + (t-a)^{\frac{1}{\beta}}\right]\left\|f'\right\|_{[a,b],\alpha}\left\|g'\right\|_{[a,b],1} \\
\qquad \text{if } f' \in L_\alpha[a,b],\ \alpha > 1,\ \frac{1}{\alpha} + \frac{1}{\beta} = 1, \text{ and } g' \in L_1[a,b]; \\
(b-a)\left\|f'\right\|_{[a,b],1}\left\|g'\right\|_{[a,b],\infty} \qquad \text{if } f' \in L_1[a,b], g' \in L_\infty[a,b]; \\
\delta\left[(b-t)^{1+\frac{1}{\delta}} + (t-a)^{1+\frac{1}{\delta}}\right]\left\|f'\right\|_{[a,b],1}\left\|g'\right\|_{[a,b],\gamma} \\
\qquad \text{if } f' \in L_1[a,b],\ g' \in L_\gamma[a,b],\ \gamma > 1,\ \frac{1}{\gamma} + \frac{1}{\delta} = 1;
\end{cases}$$

Proof Since f and g are absolutely continuous on $[a, b]$, we may write that

$$f(\tau) - f(t) = \int_t^\tau f'(u)\,du \text{ and } g(\tau) - g(t) = \int_t^\tau g'(u)\,du$$

which implies:

$$|f(\tau) - f(t)| \le \begin{cases} \|f'\|_{[\tau,t],\infty} |\tau - t| & \text{if } f' \in L_\infty [a, b]; \\ \|f'\|_{[\tau,t],\alpha} |\tau - t|^{\frac{1}{\beta}} & \text{if } f' \in L_\alpha [a, b], \\ & \alpha > 1, \ \frac{1}{\alpha} + \frac{1}{\beta} = 1, \\ \|f'\|_{[\tau,t],1} \end{cases} \tag{174}$$

and

$$|g(\tau) - g(t)| \le \begin{cases} \|g'\|_{[\tau,t],\infty} |\tau - t| & \text{if } g' \in L_\infty [a, b]; \\ \|g'\|_{[\tau,t],\gamma} |\tau - t|^{\frac{1}{\delta}} & \text{if } g' \in L_\gamma [a, b], \\ & \gamma > 1, \ \frac{1}{\gamma} + \frac{1}{\delta} = 1; \\ \|g'\|_{[\tau,t],1}. \end{cases} \tag{175}$$

Using the identity (167), we get

$$\begin{aligned} &\Big| T(fg)(a, b; t) - f(t)T(g)(a, b; t) \\ &\qquad - g(t)T(f)(a, b; t) + \frac{1}{\pi} f(t)g(t)\ln\left(\frac{b-t}{t-a}\right) \Big| \\ &\le \frac{1}{\pi} PV \int_a^b \left| \frac{(f(\tau) - f(t))(g(\tau) - g(t))}{\tau - t} \right| d\tau =: I. \end{aligned} \tag{176}$$

Then we have, by using (174) or (175), that

$$I \leq \frac{1}{\pi} \times \begin{cases} PV \int_a^b \left\| f' \right\|_{[\tau,t],\infty} \left\| g' \right\|_{[\tau,t],\infty} |\tau - t| \, d\tau \\ PV \int_a^b \left\| f' \right\|_{[\tau,t],\infty} \left\| g' \right\|_{[\tau,t],\alpha} |\tau - t|^{\frac{1}{\delta}} \, d\tau \\ PV \int_a^b \left\| f' \right\|_{[\tau,t],\infty} \left\| g' \right\|_{[\tau,t],1} \, d\tau \\ PV \int_a^b \left\| f' \right\|_{[\tau,t],\alpha} \left\| g' \right\|_{[\tau,t],\infty} |\tau - t|^{\frac{1}{\beta}} \, d\tau \\ PV \int_a^b \left\| f' \right\|_{[\tau,t],\alpha} \left\| g' \right\|_{[\tau,t],\gamma} |\tau - t|^{\frac{1}{\beta} + \frac{1}{\delta} - 1} \, d\tau \\ PV \int_a^b \left\| f' \right\|_{[\tau,t],\alpha} \left\| g' \right\|_{[\tau,t],1} |\tau - t|^{\frac{1}{\beta} - 1} \, d\tau \\ PV \int_a^b \left\| f' \right\|_{[\tau,t],1} \left\| g' \right\|_{[\tau,t],\infty} \, d\tau \\ PV \int_a^b \left\| f' \right\|_{[\tau,t],1} \left\| g' \right\|_{[\tau,t],\gamma} |\tau - t|^{\frac{1}{\delta} - 1} \, d\tau \\ PV \int_a^b \left\| f' \right\|_{[\tau,t],1} \left\| g' \right\|_{[\tau,t],1} |\tau - t|^{-1} \, d\tau. \end{cases} \tag{177}$$

However,

$$\begin{aligned} & PV \int_a^b \left\| f' \right\|_{[\tau,t],\infty} \left\| g' \right\|_{[\tau,t],\infty} |\tau - t| \, d\tau \\ & \leq \left\| f' \right\|_{[a,b],\infty} \left\| g' \right\|_{[a,b],\infty} \left[\frac{(b-t)^2 + (t-a)^2}{2} \right] \\ & = \left\| f' \right\|_{[a,b],\infty} \left\| g' \right\|_{[a,b],\infty} \left[\frac{1}{4}(b-a)^2 + \left(t - \frac{a+b}{2} \right)^2 \right], \end{aligned}$$

$$PV \int_a^b \left\| f' \right\|_{[\tau,t],\infty} \left\| g' \right\|_{[\tau,t],\alpha} |\tau - t|^{\frac{1}{\delta}} \, d\tau$$

$$\leq \left\|f'\right\|_{[a,b],\infty} \left\|g'\right\|_{[a,b],\gamma} \left[\frac{(b-t)^{1+\frac{1}{\delta}} + (t-a)^{1+\frac{1}{\delta}}}{\frac{1}{\delta}+1}\right]$$

$$= \frac{\delta}{\delta+1} \left\|f'\right\|_{[a,b],\infty} \left\|g'\right\|_{[a,b],\gamma} \left[(b-t)^{1+\frac{1}{\delta}} + (t-a)^{1+\frac{1}{\delta}}\right],$$

$$PV \int_a^b \left\|f'\right\|_{[\tau,t],\infty} \left\|g'\right\|_{[\tau,t],1} d\tau \leq \left\|f'\right\|_{[a,b],\infty} \left\|g'\right\|_{[a,b],1} (b-a),$$

$$PV \int_a^b \left\|f'\right\|_{[\tau,t],\alpha} \left\|g'\right\|_{[\tau,t],\infty} |\tau - t|^{\frac{1}{\beta}} d\tau$$

$$\leq \left\|f'\right\|_{[a,b],\alpha} \left\|g'\right\|_{[a,b],\infty} \cdot \frac{\beta}{\beta+1} \left[(b-t)^{\frac{1}{\beta}+1} + (t-a)^{\frac{1}{\beta}+1}\right],$$

$$PV \int_a^b \left\|f'\right\|_{[\tau,t],\alpha} \left\|g'\right\|_{[\tau,t],\gamma} |\tau - t|^{\frac{1}{\beta}+\frac{1}{\delta}-1} d\tau$$

$$\leq \left\|f'\right\|_{[a,b],\alpha} \left\|g'\right\|_{[a,b],\gamma} \frac{1}{\frac{1}{\beta}+\frac{1}{\delta}} \left[(b-t)^{\frac{1}{\beta}+\frac{1}{\delta}} + (t-a)^{\frac{1}{\beta}+\frac{1}{\delta}}\right]$$

$$= \frac{\beta\delta}{\beta+\delta} \left\|f'\right\|_{[a,b],\alpha} \left\|g'\right\|_{[a,b],\gamma} \left[(b-t)^{\frac{\delta+\beta}{\beta+\delta}} + (t-a)^{\frac{\delta+\beta}{\beta+\delta}}\right],$$

$$PV \int_a^b \left\|f'\right\|_{[\tau,t],\alpha} \left\|g'\right\|_{[\tau,t],1} |\tau - t|^{\frac{1}{\beta}-1} d\tau$$

$$\leq \left\|f'\right\|_{[a,b],\alpha} \left\|g'\right\|_{[a,b],1} \beta \left[(b-t)^{\frac{1}{\beta}} + (t-a)^{\frac{1}{\beta}}\right],$$

$$PV \int_a^b \left\|f'\right\|_{[\tau,t],1} \left\|g'\right\|_{[\tau,t],\infty} d\tau \leq (b-a) \left\|f'\right\|_{[a,b],1} \left\|g'\right\|_{[a,b],\infty}$$

and

$$PV \int_a^b \left\|f'\right\|_{[\tau,t],1} \left\|g'\right\|_{[\tau,t],\gamma} |\tau - t|^{\frac{1}{\delta}-1} d\tau$$

$$\leq \left\|f'\right\|_{[a,b],1} \left\|g'\right\|_{[a,b],\gamma} \delta \left[(b-t)^{1+\frac{1}{\delta}} + (t-a)^{1+\frac{1}{\delta}}\right].$$

For the last inequality we cannot point out a bound as above.

Using (176) and (177), we deduce the desired inequality (173).

The following lemma also holds.

Lemma 17 (Dragomir [7]) *Let $f : [a, b] \to \mathbb{R}$ be locally Hölder continuous on $[a, b]$ and $g : [a, b] \to \mathbb{R}$ so that g' is absolutely continuous on $[a, b]$. Then we have the identity:*

$$\begin{aligned} &T(fg)(a,b;t) \qquad (178)\\ &= f(t)T(g)(a,b;t) + g(t)T(f)(a,b;t) - \frac{1}{\pi}f(t)g(t)\ln\left(\frac{b-t}{t-a}\right)\\ &+ \frac{1}{\pi}\left[\int_a^b f(\tau)d\tau - (b-a)f(t)\right]g'(t)\\ &- \frac{1}{\pi}PV\int_a^b \frac{f(\tau)-f(t)}{\tau - t}\left(\int_t^\tau (u-\tau)g''(u)du\right)d\tau \end{aligned}$$

for any $t \in (a, b)$.

Proof We use the following identity:

$$\int_\alpha^\beta \varphi(u)du = \varphi(\alpha)(\beta-\alpha) - \int_\alpha^\beta (u-\beta)\varphi'(u)du$$

which holds for any absolutely continuous function $\varphi : [\alpha, \beta] \to \mathbb{R}$.

Then we have

$$\begin{aligned} &\frac{1}{\pi}PV\int_a^b [f(\tau)-f(t)]\left[\frac{1}{\tau-t}(g(\tau)-g(t))\right]d\tau\\ &= \frac{1}{\pi}PV\int_a^b [f(\tau)-f(t)]\left[\frac{1}{\tau-t}\int_t^\tau g'(u)du\right]d\tau\\ &= \frac{1}{\pi}PV\int_a^b [f(\tau)-f(t)]\left[g'(t) - \frac{1}{\tau-t}\int_t^\tau (u-\tau)g''(u)du\right]d\tau\\ &= \frac{1}{\pi}\left[g'(t)\int_a^b f(\tau)d\tau - (b-a)f(t)g'(t)\right.\\ &\left. - PV\int_a^b \frac{f(\tau)-f(t)}{\tau-t}\left(\int_t^\tau (u-\tau)g''(u)du\right)d\tau\right]\\ &= \frac{1}{\pi}g'(t)\int_a^b f(\tau)d\tau - \frac{1}{\pi}(b-a)f(t)g'(t)\\ &- \frac{1}{\pi}PV\int_a^b \frac{f(\tau)-f(t)}{\tau-t}\left(\int_t^\tau (u-\tau)g''(u)du\right)d\tau. \end{aligned}$$

Using (166), we deduce (178).

The following theorem holds.

Theorem 27 (Dragomir [7]) *Assume that* $f : [a, b] \to \mathbb{R}$ *is of* $H - r$*-Hölder type and* $g : [a, b] \to \mathbb{R}$ *is such that* g' *is absolutely continuous on* $[a, b]$. *Then we have the inequality:*

$$\left| T(fg)(a, b; t) - f(t) T(g)(a, b; t) - g(t) T(f)(a, b; t) \right. \tag{179}$$

$$\left. + \frac{1}{\pi} f(t) g(t) \ln\left(\frac{b-t}{t-a}\right) - \frac{1}{\pi}\left[\int_a^b f(\tau)\, d\tau - (b-a) f(t)\right] g'(t) \right|$$

$$\leq \frac{H}{\pi} \begin{cases} \frac{1}{2(r+2)}\left[(b-t)^{r+2} + (t-a)^{r+2}\right] \left\|g''\right\|_{[a,b],\infty} & \text{if } g'' \in L_\infty[a, b]; \\ \frac{q}{(rq+q+1)(q+1)^{\frac{1}{q}}}\left[(b-t)^{r+\frac{1}{q}+1} + (t-a)^{r+\frac{1}{q}+1}\right] \left\|g''\right\|_{[a,b],p} & \text{if } g'' \in L_p[a, b], \\ & p > 1,\ \frac{1}{p} + \frac{1}{q} = 1; \\ \frac{1}{(r+1)}\left[(b-t)^{r+1} + (t-a)^{r+1}\right] \left\|g''\right\|_{[a,b],1}. \end{cases}$$

Proof Using the identity (178), we deduce that the left side in (179) is upper bounded by

$$I := \frac{1}{\pi} PV \int_a^b |f(\tau) - f(t)| \frac{1}{|\tau - t|} \left| \int_t^\tau (u - \tau) g''(u)\, du \right| d\tau$$
$$\leq \frac{H}{\pi} PV \int_a^b |\tau - t|^{r-1} \left| \int_t^\tau (u - \tau) g''(u)\, du \right| d\tau =: J.$$

We observe that

$$\left| \int_t^\tau (u - \tau) g''(u)\, du \right| \leq \left\|g''\right\|_{[t,\tau],\infty} \frac{(\tau - t)^2}{2}$$

if $g'' \in L_\infty[a, b]$,

$$\left| \int_t^\tau (u - \tau) g''(u)\, du \right| \leq \left\|g''\right\|_{[t,\tau],p} \left| \int_t^\tau |t - \tau|^q\, d\tau \right|^{\frac{1}{q}} = \left\|g''\right\|_{[t,\tau],p} \frac{|t - \tau|^{\frac{q+1}{q}}}{(q+1)^{\frac{1}{q}}}$$

if $g'' \in L_p[a, b]$ and, finally,

$$\left| \int_t^\tau (u - \tau) g''(u)\, du \right| \leq |t - \tau| \left\|g''\right\|_{[t,\tau],1}.$$

Consequently, we have

$$J \le \frac{H}{\pi} \times \begin{cases} PV\displaystyle\int_a^b |\tau - t|^{r-1} \cdot \frac{(\tau-t)^2}{2} \cdot \left\|g''\right\|_{[t,\tau],\infty} d\tau \\ PV\displaystyle\int_a^b \frac{|\tau - t|^{r-1} \cdot |t-\tau|^{1+\frac{1}{q}}}{(q+1)^{\frac{1}{q}}} \left\|g''\right\|_{[t,\tau],p} d\tau \\ PV\displaystyle\int_a^b |\tau - t|^{r-1} \cdot |t-\tau| \left\|g''\right\|_{[t,\tau],1} d\tau \end{cases}$$

$$\le \frac{H}{\pi} \times \begin{cases} \dfrac{1}{2} \left\|g''\right\|_{[a,b],\infty} \left[\dfrac{(b-t)^{r+2} + (t-a)^{r+2}}{r+2}\right] \\ \dfrac{1}{(q+1)^{\frac{1}{q}}} \left\|g''\right\|_{[a,b],p} \left[\dfrac{(b-t)^{r+\frac{1}{q}+1} + (t-a)^{r+\frac{1}{q}+1}}{r+\frac{1}{q}+1}\right] \\ \left\|g''\right\|_{[a,b],1} \cdot \left[\dfrac{(b-t)^{r+1} + (t-a)^{r+1}}{r+1}\right], \end{cases}$$

which proves the inequality (179).

The following lemma also holds.

Lemma 18 *Assume that f and g are as in Lemma 17. Then we have the identity:*

$$\begin{aligned} &T(fg)(a,b;t) \qquad (180)\\ &= f(t)\,T(g)(a,b;t) + g(t)\,T(f)(a,b;t) - \frac{1}{\pi} f(t)\,g(t) \ln\left(\frac{b-t}{t-a}\right) \\ &+ \frac{1}{\pi}\left[\int_a^b f(\tau)\,g'(\tau)\,d\tau - [g(b)-g(a)]\,f(t)\right] \\ &- \frac{1}{\pi} PV \int_a^b \frac{f(\tau)-f(t)}{\tau - t} \left(\int_t^\tau (u-t)\,g''(u)\,du\right) d\tau \end{aligned}$$

for any $t \in (a,b)$.

Proof In this case, we use the following identity:

$$\int_\alpha^\beta \varphi(u)\,du = \varphi(\beta)(\beta-\alpha) - \int_\alpha^\beta (u-\alpha)\,\varphi'(u)\,du$$

which holds for any absolutely continuous function $\varphi : [\alpha,\beta] \to \mathbb{R}$.

Then, as above, we have

$$\frac{1}{\pi}PV\int_a^b [f(\tau)-f(t)]\left[\frac{1}{\tau-t}(g(\tau)-g(t))\right]d\tau$$
$$=\frac{1}{\pi}PV\int_a^b [f(\tau)-f(t)]\left[\frac{1}{\tau-t}\int_t^\tau g'(u)\,du\right]d\tau$$

$$=\frac{1}{\pi}PV\int_a^b [f(\tau)-f(t)]\left[g'(t)-\frac{1}{\tau-t}\int_t^\tau (u-\tau)\,g''(u)\,du\right]d\tau$$
$$=\frac{1}{\pi}\left[\int_a^b f(\tau)\,g'(\tau)\,d\tau - f(t)\,PV\int_a^b g'(\tau)\,d\tau\right.$$
$$\left.-\;PV\int_a^b \frac{f(\tau)-f(t)}{\tau-t}\left(\int_t^\tau (u-t)\,g''(u)\,du\right)d\tau\right]$$
$$=\frac{1}{\pi}\left[\int_a^b f(\tau)\,g'(\tau)\,d\tau - [g(b)-g(a)]\,f(t)\right.$$
$$\left.-\;PV\int_a^b \frac{f(\tau)-f(t)}{\tau-t}\left(\int_t^\tau (u-t)\,g''(u)\,du\right)d\tau\right],$$

proving the identity (180).

The following result also holds.

Theorem 28 (Dragomir [7]) *With the assumptions in Theorem 27, we have*

$$\left|T(fg)(a,b;t)-f(t)\,T(g)(a,b;t)-g(t)\,T(f)(a,b;t)\right. \tag{181}$$
$$\left.+\frac{1}{\pi}f(t)\,g(t)\ln\left(\frac{b-t}{t-a}\right)-\frac{1}{\pi}\left[\int_a^b f(\tau)\,g'(\tau)\,d\tau-[g(b)-g(a)]\,f(t)\right]\right|$$
$$\leq\frac{H}{\pi}\begin{cases}\frac{1}{2(r+2)}\left[(b-t)^{r+2}+(t-a)^{r+2}\right]\left\|g''\right\|_{[a,b],\infty} & \text{if } g''\in L_\infty[a,b];\\[2ex] \frac{q}{(rq+q+1)(q+1)^{\frac{1}{q}}}\left[(b-t)^{r+\frac{1}{q}+1}+(t-a)^{r+\frac{1}{q}+1}\right]\left\|g''\right\|_{[a,b],p} & \text{if } g''\in L_p[a,b],\\ & p>1,\ \frac{1}{p}+\frac{1}{q}=1;\\[2ex] \frac{1}{(r+1)}\left[(b-t)^{r+1}+(t-a)^{r+1}\right]\left\|g''\right\|_{[a,b],1}. \end{cases}$$

Proof The proof follows in a similar manner to the one in Theorem 27 by the use of Lemma 18. We omit the details.

9 Estimates via Taylor's Expansion

9.1 Inequalities on the Whole Interval $[a, b]$

The following result holds.

Theorem 29 (Dragomir [8]) *Let $f : [a, b] \to \mathbb{R}$ be such that $f^{(n-1)}$ $(n \geq 1)$ is absolutely continuous on $[a, b]$. Then we have the bounds:*

$$\left| (Tf)(a, b; t) - f(t) \ln\left(\frac{b-t}{t-a}\right) - \sum_{k=1}^{n-1} \frac{f^{(k)}(t)}{k!} \cdot \left[\frac{(b-t)^k + (-1)^{k+1}(t-a)^k}{k}\right] \right| \tag{182}$$

$$\leq \begin{cases} \dfrac{\left\| f^{(n)} \right\|_{[a,b],\infty}}{n \cdot n!} \left[(b-t)^n + (t-a)^n\right], & \text{if } f^{(n)} \in L_\infty[a, b]; \\[2ex] \dfrac{q \left\| f^{(n)} \right\|_{[a,b],p} \left[(b-t)^{n-1+\frac{1}{q}} + (t-a)^{n-1+\frac{1}{q}}\right]}{(n-1)!\left[(n-1)q+1\right]^{1+\frac{1}{q}}}, & \text{if } f^{(n)} \in L_p[a, b], \\ & p > 1,\ \frac{1}{p} + \frac{1}{q} = 1; \\[2ex] \dfrac{\left\| f^{(n)} \right\|_{[a,b],1}}{(n-1) \cdot (n-1)!} \left[(b-t)^{n-1} + (t-a)^{n-1}\right], \end{cases}$$

$$\leq \begin{cases} \dfrac{(b-a)^n}{n \cdot n!} \left\| f^{(n)} \right\|_{[a,b],\infty}, & \text{if } f^{(n)} \in L_\infty[a, b]; \\[2ex] \dfrac{q(b-a)^{n-1+\frac{1}{q}}}{(n-1)!\left[(n-1)q+1\right]^{1+\frac{1}{q}}} \left\| f^{(n)} \right\|_{[a,b],p}, & \text{if } f^{(n)} \in L_p[a, b], \\ & p > 1,\ \frac{1}{p} + \frac{1}{q} = 1; \\[2ex] \dfrac{1}{(n-1) \cdot (n-1)!} (b-a)^{n-1} \left\| f^{(n)} \right\|_{[a,b],1}, \end{cases}$$

for any $t \in (a, b)$.

Proof Start with Taylor's formula for a function $g : I \to \mathbb{R}$ (I is a compact interval) with the property that $g^{(n-1)}$ $(n \geq 1)$ is absolutely continuous on I, then we have

$$g(x) = \sum_{k=0}^{n-1} \frac{(x-a)^k}{k!} g^{(k)}(a) + \frac{1}{(n-1)!} \int_a^x (x-t)^{n-1} g^{(n)}(t)\, dt,$$

where $a, x \in \mathring{I}$ ($\mathring{I}$ is the interior of I). This implies that

$$\left| g(x) - \sum_{k=0}^{n-1} \frac{(x-a)^k}{k!} g^{(k)}(a) \right| \leq \frac{1}{(n-1)!} \left| \int_a^x |x-t|^{n-1} \left| g^{(n)}(t) \right| dt \right|$$
$$=: \frac{1}{(n-1)!} \cdot M(x)$$

for any $a, x \in \mathring{I}$.

Before we estimate $M(x)$, let us introduce the following notations

$$\|h\|_{[a,x],p} := \left| \int_a^x |h(t)|^p \, dt \right|^{\frac{1}{p}} \quad \text{if } p \geq 1$$

and

$$\|h\|_{[a,x],\infty} := ess \sup_{\substack{t\in[a,x] \\ t\in([x,a])}} |h(t)|,$$

where $a, x \in \mathring{I}$.

It is obvious now that

$$M(x) \leq \sup_{\substack{t\in[a,x] \\ (t\in[x,a])}} \left| g^{(n)}(t) \right| \left| \int_a^x |x-t|^{n-1} \, dt \right| = \left\| g^{(n)} \right\|_{[a,x],\infty} \frac{|x-a|^n}{n!}$$

for any $a, x \in \mathring{I}$.

Using Hölder's integral inequality, we may state that

$$M(x) \leq \left| \int_a^x \left| g^{(n)}(t) \right|^p dt \right|^{\frac{1}{p}} \left| \int_a^x |x-t|^{(n-1)q} \, dt \right|^{\frac{1}{q}} = \left\| g^{(n)} \right\|_{[a,x],p} \frac{|x-a|^{n-1+\frac{1}{q}}}{[(n-1)q+1]^{\frac{1}{q}}}$$

for any $a, x \in \mathring{I}$.

Also, we observe that

$$M(x) \leq |x-a|^{n-1} \left| \int_a^x \left| g^{(n)}(t) \right| dt \right| = \left\| g^{(n)} \right\|_{[a,x],1} |x-a|^{n-1}$$

for all $a, x \in \mathring{I}$.

In conclusion, we may state the following inequality which will be used in the sequel

$$\left|g(x)-\sum_{k=0}^{n-1}\frac{(x-a)^k}{k!}g^{(k)}(a)\right| \tag{183}$$

$$\leq\begin{cases}\dfrac{|x-a|^n}{n!}\left\|g^{(n)}\right\|_{[a,x],\infty} & \text{if } g^{(n)}\in L_\infty(\mathring{I});\\[2ex] \dfrac{|x-a|^{n-1+\frac{1}{q}}}{(n-1)!\left[(n-1)q+1\right]^{\frac{1}{q}}}\left\|g^{(n)}\right\|_{[a,x],p} & \text{if } g^{(n)}\in L_p(\mathring{I}),\\ & p>1,\ \frac{1}{p}+\frac{1}{q}=1;\\[2ex] \dfrac{|x-a|^{n-1}}{(n-1)!}\left\|g^{(n)}\right\|_{[a,x],1}\end{cases}$$

for any $a,x\in\mathring{I}$.

Now, let us note for the function $f_0:[a,b]\to\mathbb{R}$, $f_0(t)=1$, we have that

$$(Tf_0)(a,b;t)=\ln\left(\frac{b-t}{t-a}\right),\ t\in(a,b),$$

and then

$$\begin{aligned}(Tf)(a,b;t)&=PV\int_a^b\frac{f(\tau)-f(t)+f(t)}{\tau-t}d\tau\\&=PV\int_a^b\frac{f(\tau)-f(t)}{\tau-t}d\tau+f(t)\ln\left(\frac{b-t}{t-a}\right),\end{aligned}$$

giving the equality

$$(Tf)(a,b;t)-f(t)\ln\left(\frac{b-t}{t-a}\right)=PV\int_a^b\frac{f(\tau)-f(t)}{\tau-t}d\tau. \tag{184}$$

Writing (183) for $g=f$, $x=\tau$, $a=t$, we get

$$\left|\frac{f(\tau)-f(t)}{\tau-t}-\sum_{k=1}^{n-1}\frac{(\tau-t)^{k-1}}{k!}f^{(k)}(t)\right| \tag{185}$$

$$\leq \begin{cases} \frac{|\tau - t|^{n-1}}{n!} \left\| f^{(n)} \right\|_{[t,\tau],\infty} & \text{if } f^{(n)} \in L_\infty [a,b] ; \\ \frac{|\tau - t|^{n-2+\frac{1}{q}}}{(n-1)! \left[(n-1) q + 1\right]^{\frac{1}{q}}} \left\| f^{(n)} \right\|_{[t,\tau],p} & \text{if } f^{(n)} \in L_p [a,b] , \\ & p > 1, \ \frac{1}{p} + \frac{1}{q} = 1; \\ \frac{|\tau - t|^{n-2}}{(n-1)!} \left\| f^{(n)} \right\|_{[t,\tau],1} \end{cases}$$

for any $t, \tau \in (a,b)$, $t \neq \tau$.

If we take the PV in (185), then we may write

$$\left| PV \int_a^b \frac{f(\tau) - f(t)}{\tau - t} d\tau - \sum_{k=1}^{n-1} \frac{f^{(k)}(t)}{k!} PV \int_a^b (\tau - t)^{k-1} d\tau \right| \tag{186}$$

$$\leq PV \int_a^b \left| \frac{f(\tau) - f(t)}{\tau - t} - \sum_{k=1}^{n-1} \frac{(\tau - t)^{k-1}}{k!} f^{(k)}(t) \right| d\tau$$

$$\leq \begin{cases} \frac{1}{n!} PV \int_a^b |\tau - t|^{n-1} \left\| f^{(n)} \right\|_{[t,\tau],\infty} d\tau \\ \frac{1}{(n-1)! \left[(n-1) q + 1\right]^{\frac{1}{q}}} PV \int_a^b |\tau - t|^{n-2+\frac{1}{q}} \left\| f^{(n)} \right\|_{[t,\tau],p} d\tau \\ \frac{1}{(n-1)!} PV \int_a^b |\tau - t|^{n-2} \left\| f^{(n)} \right\|_{[t,\tau],1} d\tau \end{cases}$$

$$\leq \begin{cases} \frac{1}{n!} \left\| f^{(n)} \right\|_{[a,b],\infty} PV \int_a^b |\tau - t|^{n-1} d\tau \\ \frac{1}{(n-1)! \left[(n-1) q + 1\right]^{\frac{1}{q}}} \left\| f^{(n)} \right\|_{[a,b],p} PV \int_a^b |\tau - t|^{n-2+\frac{1}{q}} d\tau \\ \frac{1}{(n-1)!} \left\| f^{(n)} \right\|_{[a,b],1} PV \int_a^b |\tau - t|^{n-2} d\tau . \end{cases}$$

However,

$$PV\int_a^b |\tau - t|^{n-1}\,d\tau = \frac{1}{n}\left[(b-t)^n + (t-a)^n\right],$$

$$PV\int_a^b |\tau - t|^{n-2+\frac{1}{q}}\,d\tau = \frac{q}{[(n-1)q+1]}\left[(b-t)^{n-1+\frac{1}{q}} + (t-a)^{n-1+\frac{1}{q}}\right],$$

$$PV\int_a^b |\tau - t|^{n-2}\,d\tau = \frac{1}{n-1}\left[(b-t)^{n-1} + (t-a)^{n-1}\right]$$

and

$$PV\int_a^b (\tau - t)^{k-1}\,d\tau = \frac{1}{k}\left[(b-t)^k + (-1)^{k+1}(t-a)^k\right]$$

and then by (186) we deduce the desired result (182).

It is obvious that the best inequality one would deduce from (182) is the one for $t = \frac{a+b}{2}$, getting the following corollary.

Corollary 23 *With the assumptions of Theorem 29, we have*

$$\left|(Tf)\left(a,b;\frac{a+b}{2}\right) - \sum_{k=1}^{n-1}\frac{(b-a)^k}{2^k\cdot k\cdot k!}\left[1+(-1)^{k+1}\right]f^{(k)}\left(\frac{a+b}{2}\right)\right|$$

$$\leq \begin{cases} \dfrac{(b-a)^n}{2^{n-1}\cdot n\cdot n!}\left\|f^{(n)}\right\|_{[a,b],\infty}, & \text{if } f^{(n)}\in L_\infty[a,b]; \\ \dfrac{q(b-a)^{n-1+\frac{1}{q}}}{2^{n-2+\frac{1}{q}}(n-1)!\,[(n-1)q+1]^{1+\frac{1}{q}}}\left\|f^{(n)}\right\|_{[a,b],p}, & \text{if } f^{(n)}\in L_p[a,b], \\ & p>1,\ \frac{1}{p}+\frac{1}{q}=1; \\ \dfrac{(b-a)^{n-1}}{2^{n-2}\cdot(n-1)\cdot(n-1)!}\left\|f^{(n)}\right\|_{[a,b],1}. & \end{cases} \tag{187}$$

It is important to note that for small intervals, we basically have the following representation:

Corollary 24 *Assume that $f \in C^\infty[a,b]$ and $0 < b-a \leq 1$. Then*

$$(Tf)(a,b;t) = f(t)\ln\left(\frac{b-t}{t-a}\right) + \sum_{k=1}^{\infty}\frac{f^{(k)}(t)}{k!}\cdot\left[\frac{(b-t)^k + (-1)^{k+1}(t-a)^k}{k}\right]$$

and the convergence is uniform on $[a,b]$.

9.2 The Composite Case

The following lemma holds.

Lemma 19 *Let $g : [a, b] \to \mathbb{R}$ be such that $g^{(n-1)}$ $(n \geq 1)$ is absolutely continuous on $[a, b]$. Then for any $m \in \mathbb{N}$, $m \geq 1$, we have the inequality:*

$$\left| \frac{1}{b-a} \int_a^b g(u)\, du - \sum_{i=0}^{m-1} \sum_{k=1}^{n} \frac{(b-a)^{k-1}}{m^k k!} \cdot g^{(k-1)}\left(a + i \cdot \frac{b-a}{m}\right) \right| \tag{188}$$

$$\leq \begin{cases} \dfrac{(b-a)^n}{m^n (n+1)!} \left\| g^{(n)} \right\|_{[a,b],\infty}, & \textit{if } g^{(n)} \in L_\infty [a,b]; \\[2ex] \dfrac{(b-a)^{n-1+\frac{1}{q}}}{m^n n! (nq+1)^{\frac{1}{q}}} \left\| g^{(n)} \right\|_{[a,b],p}, & \textit{if } g^{(n)} \in L_p [a,b], \\ & p > 1,\ \frac{1}{p} + \frac{1}{q} = 1; \\ \dfrac{(b-a)^{n-1}}{m^n n!} \left\| g^{(n)} \right\|_{[a,b],1}. \end{cases}$$

Proof Write Taylor's formula with the integral remainder for $\varphi(x) = \int_\alpha^x g(u)\, du$ and then choose $x = \beta$, to get

$$\left| \int_\alpha^\beta g(u)\, du - \sum_{k=1}^{n} \frac{(\beta-\alpha)^k}{k!} g^{(k-1)}(\alpha) \right| \tag{189}$$

$$\leq \begin{cases} \dfrac{|\beta-\alpha|^{n+1}}{(n+1)!} \left\| g^{(n)} \right\|_{[\alpha,\beta],\infty}, & \text{if } g^{(n)} \in L_\infty [a,b]; \\[2ex] \dfrac{|\beta-\alpha|^{n+\frac{1}{q}}}{n! (nq+1)^{\frac{1}{q}}} \left\| g^{(n)} \right\|_{[\alpha,\beta],p}, & \text{if } g^{(n)} \in L_p [a,b], \\ & p > 1,\ \frac{1}{p} + \frac{1}{q} = 1; \\ \dfrac{|\beta-\alpha|^n}{n!} \left\| g^{(n)} \right\|_{[\alpha,\beta],1} \end{cases}$$

for any $\alpha, \beta \in [a, b]$.

Now, if we consider the division

$$I_n : x_i = a + i \cdot \frac{b-a}{m}, \quad i = \overline{0, m},$$

and apply (189) on the intervals $[x_{i,}x_{i+1}]$ $(i=\overline{0,m-1})$, we can write

$$\left|\int_{x_{i,}}^{x_{i+1}} g(u)\,du - \sum_{k=1}^{n} \frac{(b-a)^k}{m^k k!} \cdot g^{(k-1)}\left(a+i\cdot\frac{b-a}{m}\right)\right|$$

$$\leq \begin{cases} \dfrac{(b-a)^{n+1}}{m^{n+1}(n+1)!} \left\|g^{(n)}\right\|_{[x_i,x_{i+1}],\infty}, \\[2ex] \dfrac{(b-a)^{n+\frac{1}{q}}}{m^{n+\frac{1}{q}} n!(nq+1)^{\frac{1}{q}}} \left\|g^{(n)}\right\|_{[x_i,x_{i+1}],p}, \\[2ex] \dfrac{(b-a)^n}{m^n n!} \left\|g^{(n)}\right\|_{[x_i,x_{i+1}],1}. \end{cases}$$

Summing over i from 0 to $m-1$ and using the generalized triangle inequality, we deduce (189).

The following main result holds.

Theorem 30 (Dragomir [8]) *Let $f:[a,b]\to\mathbb{R}$ be such that $f^{(n)}$ $(n\geq 0)$ is absolutely continuous on $[a,b]$. Then for any $m\in\mathbb{N}$, $m\geq 1$, we have*

$$(Tf)(a,b;t) = f(t)\ln\left(\frac{b-t}{t-a}\right) + A_{n,m}(f,t) + R_{n,m}(f,t),$$

where

$$\begin{aligned} &A_{n,m}(f,t) \qquad (190)\\ &= \sum_{k=1}^{n} \frac{f^{(k)}(t)}{m^k k!}\cdot\left[\frac{(b-t)^k+(-1)^{k+1}(t-a)^k}{k}\right] + (b-a)\sum_{i=1}^{m-1}\sum_{k=1}^{n}\frac{1}{m^k k!} \\ &\times\left\{\sum_{\nu=1}^{k-1}(-1)^{\nu-1}(k-1)\cdots(k-\nu)\left(\frac{m}{i}\right)^{\nu-1}\right. \\ &\times\left[f^{(k-\nu)};t+\frac{i}{m}(b-t),t-\frac{i}{m}(t-a)\right] \\ &\left.+(-1)^{k-1}\left(\frac{m}{i}\right)^{k-1}(k-1)!\left[f;t+\frac{i}{m}(b-t),t-\frac{i}{m}(t-a)\right]\right\} \end{aligned}$$

and the remainder $R_{n,m}(f,t)$ satisfies the estimate

$$\left|R_{n,m}(f,t)\right| \tag{191}$$

$$\leq \begin{cases} \dfrac{\left\|f^{(n+1)}\right\|_{[a,b],\infty}}{m^n(n+1)!\cdot(n+1)}\left[(b-t)^{n+1}+(t-a)^{n+1}\right], \ \textit{if} \ f^{(n+1)} \in L_\infty[a,b]; \\ \\ \dfrac{q\left[(b-t)^{n+\frac{1}{q}}+(t-a)^{n+\frac{1}{q}}\right]}{m^n n!(nq+1)^{1+\frac{1}{q}}}\left\|f^{(n+1)}\right\|_{[a,b],p}, \quad \textit{if} \ f^{(n+1)} \in L_p[a,b], \\ \qquad\qquad\qquad\qquad\qquad\qquad\qquad p>1, \ \frac{1}{p}+\frac{1}{q}=1; \\ \\ \dfrac{\left\|f^{(n+1)}\right\|_{[a,b],1}}{m^n n!\cdot n}\left[(b-t)^n+(t-a)^n\right]; \end{cases}$$

$$\leq \begin{cases} \dfrac{(b-a)^{n+1}}{m^n(n+1)\cdot(n+1)!}, \left\|f^{(n+1)}\right\|_{[a,b],\infty} \ \textit{if} \ f^{(n+1)} \in L_\infty[a,b]; \\ \\ \dfrac{q(b-a)^{n+\frac{1}{q}}}{m^n n!(nq+1)^{1+\frac{1}{q}}}\left\|f^{(n+1)}\right\|_{[a,b],p}, \quad \textit{if} \ f^{(n+1)} \in L_p[a,b], \\ \qquad\qquad\qquad\qquad\qquad p>1, \ \frac{1}{p}+\frac{1}{q}=1; \\ \\ \dfrac{(b-a)^n}{m^n n!\cdot n}\left\|f^{(n+1)}\right\|_{[a,b],1}; \end{cases}$$

Proof We have (see (184)) that:

$$(Tf)(a,b;t)-f(t)\ln\left(\frac{b-t}{t-a}\right)=PV\int_a^b \frac{f(\tau)-f(t)}{\tau-t}d\tau.$$

If we write the inequality (188) for $g=f'$, we get

$$\left|\frac{f(t)-f(\tau)}{\tau-t}-\sum_{i=0}^{m-1}\sum_{k=1}^{n}\frac{(\tau-t)^{k-1}}{m^k k!}\cdot f^{(k)}\left(t+i\cdot\frac{\tau-t}{m}\right)\right| \tag{192}$$

$$\leq \begin{cases} \dfrac{|\tau-t|^n}{m^n(n+1)!}\left\|f^{(n+1)}\right\|_{[t,\tau],\infty}, \quad \text{if} \ f^{(n+1)} \in L_\infty[a,b]; \\ \\ \dfrac{|\tau-t|^{n-1+\frac{1}{q}}}{m^n n!(nq+1)^{\frac{1}{q}}}\left\|f^{(n+1)}\right\|_{[t,\tau],p}, \ \text{if} \ f^{(n+1)} \in L_p[a,b], \\ \qquad\qquad\qquad\qquad\qquad p>1, \ \frac{1}{p}+\frac{1}{q}=1; \\ \\ \dfrac{|\tau-t|^{n-1}}{m^n n!}\left\|f^{(n+1)}\right\|_{[t,\tau],1}. \end{cases}$$

If we apply PV to (192), we may write that:

$$\left| PV \int_a^b \frac{f(t) - f(\tau)}{\tau - t} d\tau \right. \tag{193}$$
$$\left. - \sum_{i=0}^{m-1} \sum_{k=1}^{n} PV \int_a^b \frac{(\tau - t)^{k-1}}{m^k k!} \cdot f^{(k)}\left(t + i \cdot \frac{\tau - t}{m}\right) d\tau \right|$$

$$\leq \begin{cases} \frac{1}{m^n (n+1)!} PV \int_a^b |\tau - t|^n \left\| f^{(n+1)} \right\|_{[t,\tau],\infty} d\tau, \\ \frac{1}{m^n n! (nq+1)^{\frac{1}{q}}} PV \int_a^b |\tau - t|^{n-1+\frac{1}{q}} \left\| f^{(n+1)} \right\|_{[t,\tau],p} d\tau, \\ \frac{1}{m^n n!} PV \int_a^b |\tau - t|^{n-1} \left\| f^{(n+1)} \right\|_{[t,\tau],1} d\tau, \end{cases}$$

$$\leq \begin{cases} \frac{1}{m^n (n+1)!} \left\| f^{(n+1)} \right\|_{[a,b],\infty} PV \int_a^b |\tau - t|^n d\tau, \\ \frac{1}{m^n n! (nq+1)^{\frac{1}{q}}} \left\| f^{(n+1)} \right\|_{[a,b],p} PV \int_a^b |\tau - t|^{n-1+\frac{1}{q}} d\tau, \\ \frac{1}{m^n n!} \left\| f^{(n+1)} \right\|_{[a,b],1} PV \int_a^b |\tau - t|^{n-1} d\tau, \end{cases}$$

$$\leq \begin{cases} \frac{1}{m^n (n+1)!} \left\| f^{(n+1)} \right\|_{[a,b],\infty} \left[\frac{(b-t)^{n+1} + (t-a)^{n+1}}{n+1} \right], \\ \frac{1}{m^n n! (nq+1)^{\frac{1}{q}}} \left\| f^{(n+1)} \right\|_{[a,b],p} \left[\frac{(b-t)^{n+\frac{1}{q}} + (t-a)^{n+\frac{1}{q}}}{n+\frac{1}{q}} \right], \\ \frac{1}{m^n n!} \left\| f^{(n+1)} \right\|_{[a,b],1} \left[\frac{(b-t)^n + (t-a)^n}{n} \right]. \end{cases}$$

Now, let us denote

$$I_{i,k} := PV \int_a^b (\tau - t)^{k-1} f^{(k)}\left(t + \frac{i}{m}(\tau - t)\right) d\tau,$$

where $i = 0, \ldots, m-1$, $k = 1, \ldots, n$.

For $i=0$, we have

$$I_{0,k} := PV \int_a^b (\tau - t)^{k-1} f^{(k)}(t)\, d\tau = f^{(k)}(t) \cdot \frac{(b-t)^k + (-1)^{k+1}(t-a)^k}{k}$$

for any $k = 1, \ldots, n$.

For $k = 1, \ldots, n$ and $i = 1, \ldots, m-1$, we have

$$\begin{aligned} I_{i,k} &= \lim_{\varepsilon \to 0+} \left[\int_a^{t-\varepsilon} (\tau - t)^{k-1} f^{(k)} \left[t + \frac{i}{m}(\tau - t) \right] d\tau \right. \\ &\quad \left. + \int_{t+\varepsilon}^b (\tau - t)^{k-1} f^{(k)} \left[t + \frac{i}{m}(\tau - t) \right] d\tau \right] \end{aligned} \tag{194}$$

$$\begin{aligned} &= \lim_{\varepsilon \to 0+} \left[\frac{m}{i} f^{(k-1)} \left[t + \frac{i}{m}(\tau - t) \right] (\tau - t)^{k-1} \Big|_a^{t-\varepsilon} \right. \\ &\quad - \frac{m}{i} \int_a^{t-\varepsilon} (k-1)(\tau - t)^{k-2} f^{(k-1)} \left[t + \frac{i}{m}(\tau - t) \right] d\tau \\ &\quad + \frac{m}{i} f^{(k-1)} \left[t + \frac{i}{m}(\tau - t) \right] (\tau - t)^{k-1} \Big|_{t+\varepsilon}^{b} \\ &\quad \left. - \frac{m}{i} \int_{t+\varepsilon}^b (k-1)(\tau - t)^{k-2} f^{(k-1)} \left[t + \frac{i}{m}(\tau - t) \right] d\tau \right] \\ &= \frac{m}{i} \left[f^{(k-1)} \left[t + \frac{i}{m}(b-t) \right] - f^{(k-1)} \left[t - \frac{i}{m}(t-a) \right] \right] \\ &\quad - \frac{m}{i}(k-1)\, PV \int_a^b (\tau - t)^{k-2} f^{(k-1)} \left(t + \frac{i}{m}(\tau - t) \right) d\tau \\ &= (b-a) \left[f^{(k-1)}; t + \frac{i}{m}(b-t), t - \frac{i}{m}(t-a) \right] - \frac{m}{i}(k-1) I_{i,k-1} \end{aligned}$$

for any $k = 2, \ldots, n$.

For $k=1$, we have

$$\begin{aligned} I_{i,1} &= PV \int_a^b f^{(1)} \left(t + \frac{i}{m}(\tau - t) \right) d\tau \\ &= \frac{m}{i} \left[f \left[t + \frac{i}{m}(b-t) \right] - f \left[t - \frac{i}{m}(t-a) \right] \right] \\ &= (b-a) \left[f; t + \frac{i}{m}(b-t), t - \frac{i}{m}(t-a) \right]. \end{aligned}$$

Using the recursive relation (194), we may write

$$
\begin{aligned}
& I_{i,k} \qquad (195) \\
&= (b-a)\left[f^{(k-1)}; t+\frac{i}{m}(b-t), t-\frac{i}{m}(t-a)\right] - \left(\frac{m}{i}\right)(k-1) I_{i,k-1} \\
&= (b-a)\left[f^{(k-1)}; t+\frac{i}{m}(b-t), t-\frac{i}{m}(t-a)\right] - \left(\frac{m}{i}\right)(k-1) \\
&\times \left[(b-a)\left[f^{(k-2)}; t+\frac{i}{m}(b-t), t-\frac{i}{m}(t-a)\right] - \left(\frac{m}{i}\right)(k-2) I_{i,k-2}\right]
\end{aligned}
$$

$$
\begin{aligned}
&= (b-a)\left[f^{(k-1)}; t+\frac{i}{m}(b-t), t-\frac{i}{m}(t-a)\right] \\
&- (b-a)\left(\frac{m}{i}\right)(k-1)\left[f^{(k-2)}; t+\frac{i}{m}(b-t), t-\frac{i}{m}(t-a)\right] \\
&+ \left(\frac{m}{i}\right)^2 (k-1)(k-2) I_{i,k-2}
\end{aligned}
$$

$$
\begin{aligned}
&= \cdots\cdots\cdots\cdots = \\
&= (b-a)\sum_{\nu=1}^{k-1}(-1)^{\nu-1}(k-1)\cdots(k-\nu)\left(\frac{m}{i}\right)^{\nu-1} \\
&\times \left[f^{(k-\nu)}; t+\frac{i}{m}(b-t), t-\frac{i}{m}(t-a)\right] + (-1)^{k-1}\left(\frac{m}{i}\right)^{k-1}(k-1)! I_{i,1} \\
&= (b-a)\left[\sum_{\nu=1}^{k-1}(-1)^{\nu-1}(k-1)\cdots(k-\nu)\left(\frac{m}{i}\right)^{\nu-1}\right. \\
&\times \left[f^{(k-\nu)}; t+\frac{i}{m}(b-t), t-\frac{i}{m}(t-a)\right] \\
&+ (-1)^{k-1}\left(\frac{m}{i}\right)^{k-1}(k-1)!\left[f; t+\frac{i}{m}(b-t), t-\frac{i}{m}(t-a)\right].
\end{aligned}
$$

Replacing $I_{i,k}$ in (193), we deduce the estimate (191) with $A_{m,n}$ as defined by (190). The theorem is thus proved.

References

1. R.P. Agarwal, S.S. Dragomir, An application of Hayashi's inequality for differentiable functions. Comput. Math. Appl. **32**(6), 95–99 (1996)
2. P.L. Butzer, R.J. Nessel, *Fourier Analysis and Approximation Theory*, vol. 1 (Birkhaüser, Basel, 1977)
3. S.S. Dragomir, On the Ostrowski integral inequality for mappings of bounded variation and applications. Math. Inequal. Appl. **3**(1), 59–66 (2001)
4. S.S. Dragomir, Approximating the finite Hilbert transform via an Ostrowski type inequality for functions of bounded variation. J. Inequal. Pure Appl. Math. **3**(4), Article 51, 19 pp. (2002)
5. S.S. Dragomir, Approximating the finite Hilbert transform via Ostrowski type inequalities for absolutely continuous functions. Bull. Korean Math. Soc. **39**(4), 543–559 (2002)
6. S.S. Dragomir, Inequalities for the Hilbert transform of functions whose derivatives are convex. J. Korean Math. Soc. **39**(5), 709–729 (2002)
7. S.S. Dragomir, Some inequalities for the finite Hilbert transform of a product. Commun. Korean Math. Soc. **18**(1), 39–57 (2003)
8. S.S. Dragomir, Sharp error bounds of a quadrature rule with one multiple node for the finite Hilbert transform in some classes of continuous differentiable functions. Taiwan. J. Math. **9**(1), 95–109 (2005)
9. S.S. Dragomir, C.E.M. Pearce, Selected Topics on Hermite-Hadamard Inequalities and Applications, RGMIA Monographs, Victoria University, 2000. http://rgmia.org/monographs/hermite_hadamard.html
10. S.S. Dragomir, S. Wang, Applications of Ostrowski's inequality to the estimation of error bounds for some special means and to some numerical quadrature rules. Comput. Math. Appl. **33**, 15–20 (1997)
11. S.S. Dragomir, S. Wang, A new inequality of Ostrowski type in L_1 norm and applications to some special means and some numerical quadrature rules. Tamkang J. Math. **28**, 239–244 (1997)
12. S.S. Dragomir, S. Wang, A new inequality of Ostrowski's type in L_p norm and applications to some numerical quadrature rules. Indian J. Math. **40**(3), 245–304 (1998)
13. N.M. Dragomir, S.S. Dragomir, P.M. Farrell, Some inequalities for the finite Hilbert transform, in *Inequality Theory and Applications*, vol. I (Nova Science, Huntington, 2001), pp. 113–122
14. N.M. Dragomir, S.S. Dragomir, P.M. Farrell, Approximating the finite Hilbert transform via trapezoid type inequalities. Comput. Math. Appl. **43**(10–11), 1359–1369 (2002)
15. N.M. Dragomir, S.S. Dragomir, P.M. Farrell, G.W. Baxter, On some new estimates of the finite Hilbert transform. Libertas Math. **22**, 65–75 (2002)
16. N.M. Dragomir, S.S. Dragomir, P.M. Farrell, G.W. Baxter, A quadrature rule for the finite Hilbert transform via trapezoid type inequalities. J. Appl. Math. Comput. **13**(1–2), 67–84 (2003)
17. N.M. Dragomir, S.S. Dragomir, P.M. Farrell, G.W. Baxter, A quadrature rule for the finite Hilbert transform via midpoint type inequalities, in *Fixed Point Theory and Applications*, vol. 5 (Nova Science, Hauppauge, 2004), pp. 11–22
18. F.D. Gakhov, *Boundary Value Problems* (English translation) (Pergamon Press, Oxford, 1966)
19. H. Kober, A note on Hilbert's operator. Bull. Am. Math. Soc. **48**, 421–427 (1942)
20. W. Liu, X. Gao, Approximating the finite Hilbert transform via a companion of Ostrowski's inequality for function of bounded variation and applications. Appl. Math. Comput. **247**, 373–385 (2014)
21. W. Liu, N. Lu, Approximating the finite Hilbert transform via Simpson type inequalities and applications. Politehn. Univ. Bucharest Sci. Bull. Ser. A Appl. Math. Phys. **77**(3), 107–122 (2015)
22. W. Liu, X. Gao, Y. Wen, Approximating the finite Hilbert transform via some companions of Ostrowski's inequalities. Bull. Malays. Math. Sci. Soc. **39**(4), 1499–1513 (2016)

23. S.G. Mikhlin, S. Prössdorf, *Singular Integral Operators* (English translation) (Springer, Berlin, 1986)
24. S. Okada, D. Elliot, Hölder continuous functions and the finite Hilbert transform. Math. Nachr. **163**, 250–272 (1994)
25. S. Wang, X. Gao, N. Lu, A quadrature formula in approximating the finite Hilbert transform via perturbed trapezoid type inequalities. J. Comput. Anal. Appl. **22**(2), 239–246 (2017)
26. S. Wang, N. Lu, X. Gao, A quadrature rule for the finite Hilbert transform via Simpson type inequalities and applications. J. Comput. Anal. Appl. **22**(2), 229–238 (2017)

On Hyperstability of the Two-Variable Jensen Functional Equation on Restricted Domain

Iz-iddine EL-Fassi

Abstract We present a method that allows to study approximate solutions to the two-variable Jensen functional equation

$$2f\Big(\frac{x+z}{2}, \frac{y+w}{2}\Big) = f(x, y) + f(z, w)$$

on a restricted domain. Namely, we show that (under some weak natural assumptions) functions that satisfy the equation approximately (in some sense) must be actually solutions to it. The method is based on a quite recent fixed point theorem in some functions spaces and can be applied to various similar equations in many variables. Our outcomes are connected with the well-known issues of Ulam stability and hyperstability.

2010 Mathematics Subject Classifications Primary 39B82, 39B62; Secondary 47H14, 47H10

1 Introduction

In this paper, $\mathbb{N}$, $\mathbb{R}$, and $\mathbb{R}_+$ denote the sets of all positive integers, real numbers, and non-negative real numbers, respectively; $\mathbb{N}_0 := \mathbb{N} \cup \{0\}$. Moreover, X and Y always stand for normed spaces. The next definition describes the notion of hyperstability that we apply here (A^B denotes the family of all functions mapping a set $B \neq \emptyset$ into a set $A \neq \emptyset$).

Definition 1 Let A be a nonempty set, (Z, d) be a metric space, $\chi : A^n \to \mathbb{R}_+$, $B \subset A^n$ be nonempty, and $\mathscr{F}_1, \mathscr{F}_2$ map a nonempty $\mathscr{D} \subset Z^A$ into Z^{A^n}. We say that the conditional equation

I.-i. EL-Fassi (✉)
Department of Mathematics, Faculty of Sciences and Techniques, Sidi Mohamed ben abdellah University, Fez, Morocco

T. M. Rassias, P. M. Pardalos (eds.), *Mathematical Analysis and Applications*, Springer Optimization and Its Applications 154,
https://doi.org/10.1007/978-3-030-31339-5_5

$$\mathscr{F}_1\varphi(x_1,\ldots,x_n) = \mathscr{F}_2\varphi(x_1,\ldots,x_n), \qquad (x_1,\ldots,x_n) \in B, \tag{1}$$

is χ-hyperstable provided every $\varphi_0 \in \mathscr{D}$, satisfying

$$d\big(\mathscr{F}_1\varphi_0(x_1,\ldots,x_n), \mathscr{F}_2\varphi_0(x_1,\ldots,x_n)\big) \le \chi(x_1,\ldots,x_n), \qquad (x_1,\ldots,x_n) \in B, \tag{2}$$

is a solution to (1).

That notion is strictly connected with the well-known issue of Ulam's stability for various (e.g., difference, differential, functional, integral, operator) equations. Let us recall that the study of such problems was motivated by the following question of Ulam (cf. [24, 39]) asked in 1940.

Ulam's question *Let* $(G_1,\cdot)$ *and* $(G_2,\cdot)$ *be two groups and* $d : G_2 \times G_2 \to [0,\infty)$ *be a metric. Given* $\epsilon > 0$, *does there exist* $\delta > 0$ *such that if a mapping* $g : G_1 \to G_2$ *satisfies the inequality*

$$d(g(xy), g(x)g(y)) \le \delta$$

for all $x, y \in G_1$, *then there is a homomorphism* $h : G_1 \to G_2$ *with*

$$d(g(x), h(x)) \le \epsilon$$

for all $x \in G$ *?*

In 1941, Hyers [24] solved the well-known Ulam stability problem for additive mappings subject to the Hyers condition on approximately additive mappings. The following theorem is the most classical result concerning the Hyers-Ulam stability of the Cauchy equation

$$f(x+y) = f(x) + f(y), \qquad x, y \in X. \tag{3}$$

Theorem 1 *Let* $f : X \to Y$ *satisfy the inequality*

$$\|f(x+y) - f(x) - f(y)\| \le \theta(\|x\|^p + \|y\|^p) \tag{4}$$

for all $x, y \in X\backslash\{0\}$, *where* θ *and* p *are real constants with* $\theta > 0$ *and* $p \neq 1$. *Then the following two statements are valid.*

(a) *If* $p \ge 0$ *and* Y *is complete, then there exists a unique solution* $T : X \to Y$ *of* (3) *such that*

$$\|f(x) - T(x)\| \le \frac{\theta}{|1-2^{p-1}|}\|x\|^p, \qquad x \in X\backslash\{0\}. \tag{5}$$

(b) *If* $p < 0$, *then* f *is additive, i.e.,* (3) *holds.*

Note that Theorem 1 reduces to the first result of stability due to Hyers [24] if $p = 0$, Aoki [3] for $0 < p < 1$ (see also Th.M. Rassias' paper [35] in which it is proved for the first time the stability of the linear mapping). Afterward, Gajda [22] obtained this result for $p > 1$ and gave an example to show that Theorem 1 fails whenever $p = 1$. Also, Rassias [36] proved Theorem 1 for $p < 0$ (see [38, page 326] and [7]). Now, it is well known that the statement (b) is valid, i.e., f must be additive in that case, which has been proved for the first time in [32] and next in [8] on the restricted domain. For related results, concerning stability of the homomorphism equation on restricted domains, we refer to [1, 13–16, 25, 26, 29, 30, 34, 37, 38].

We say that a function $f : X \to Y$ satisfies the Jensen equation if

$$2f\Big(\frac{x+y}{2}\Big) = f(x) + f(y), \qquad x, y \in X. \tag{6}$$

The stability of the Jensen equation has been investigated at first by Kominek [31]. In 2006, Bae and Park [4] obtained the generalized Hyers-Ulam stability of a bi-Jensen function. Moreover, the stability problem for the bi-Jensen functional equation was discussed by a number of authors (see [27, 28]).

Recently Aghajani and Zahedi [2] investigated stability of the two-variable Jensen functional equation of the following form:

$$2f\Big(\frac{x+z}{2}, \frac{y+w}{2}\Big) = f(x, y) + f(z, w), \qquad x, y, z, w \in X. \tag{7}$$

The term hyperstability was used for the first time probably in [33]; however, it seems that the first hyperstability result was published in [6] and concerned the ring homomorphisms. For further information concerning the notion of hyperstability we refer to the survey paper [11] (for recent related results see, e.g., [5, 8–10, 17–21, 23]).

The purpose of this work is to prove hyperstability results for the equation of the form (7) on restricted domains, that is some conditional versions of that equation. The method is based on a quite recent fixed point theorem in some functions spaces from [12]. In the same way, we can study approximate solutions on restricted domains to various functional equations (in many variables) that are sufficiently similar to (7).

Let U be a nonempty subset of X. We say that a function $f : U^2 \to Y$ fulfills equation (7) on U (or is a solution to (7) on U) provided

$$2f\Big(\frac{x+z}{2}, \frac{y+w}{2}\Big) = f(x, y) + f(z, w), \tag{8}$$

$$x, y, z, w \in U, \quad \frac{x+z}{2}, \frac{y+w}{2} \in U;$$

if $U = X$, then we simply say that f fulfills (or is a solution to) Equation (7).

We consider functions $f : U^2 \to Y$ fulfilling (8) approximately, i.e., satisfying the inequality

$$\left\| 2f\left(\frac{x+z}{2}, \frac{y+w}{2}\right) - f(x,y) - f(z,w) \right\| \leq \gamma(x,y,z,w), \tag{9}$$

$$x, y, z, w \in U, \quad \frac{x+z}{2}, \frac{y+w}{2} \in U,$$

with a given $\gamma : U^4 \to \mathbb{R}_+$. We prove that, for some natural particular forms of γ (and under some additional assumptions on U), the conditional functional equation (8) is γ-hyperstable in the class of functions $f : U^2 \to Y$, i.e., each $f : U^2 \to Y$ satisfying inequality (9) with such γ must fulfill Equation (8).

2 Auxiliary Results

One of the methods of proof is based on a fixed point result that can be derived from [12]. To present it we need the following three hypothesis:

(H1) W is a nonempty set, Y is a Banach space, $f_1, \ldots . f_k : W \to W$ and $L_1, \ldots . L_k : W \to \mathbb{R}_+$ are given.

(H2) $\mathscr{T} : Y^W \to Y^W$ is an operator satisfying the inequality

$$\|\mathscr{T}\xi(x) - \mathscr{T}\mu(x)\| \leq \sum_{i=1}^{k} L_i(x) \|\xi(f_i(x)) - \mu(f_i(x))\|, \quad \xi, \mu \in Y^W, \ x \in W.$$

(H3) $\Lambda : \mathbb{R}_+{}^W \to \mathbb{R}_+{}^W$ is a linear operator defined by

$$\Lambda\delta(x) := \sum_{i=1}^{k} L_i(x)\delta(f_i(x)), \qquad \delta \in \mathbb{R}_+{}^W, \ x \in W.$$

The mentioned fixed point theorem is stated in [12] as follows.

Theorem 2 *Let hypotheses* **(H1)–(H3)** *be valid and functions* $\varepsilon : W \to \mathbb{R}_+$ *and* $\varphi : W \to Y$ *fulfill the following two conditions:*

$$\|\mathscr{T}\varphi(x) - \varphi(x)\| \leq \varepsilon(x), \qquad x \in W,$$

$$\varepsilon^*(x) := \sum_{n=0}^{\infty} \Lambda^n \varepsilon(x) < \infty, \qquad x \in W.$$

Then, there exists a unique fixed point ψ of $\mathcal{T}$ with

$$\|\varphi(x) - \psi(x)\| \leq \varepsilon^*(x), \qquad x \in W.$$

Moreover

$$\psi(x) = \lim_{n\to\infty} \mathcal{T}^n \varphi(x)$$

for all $x \in W$.

3 Hyperstability Results for Equation (8)

The following theorems are the main results in this paper and concern the γ-hyperstability of (8). Namely, for

$$\gamma(x, y, z, w) = c\|x\|^p \|y\|^q \|z\|^r \|w\|^s,$$

with suitable $c, p, q, r, s \in \mathbb{R}$, and

$$\gamma(x, y, z, w) = c(\|x\|^{p_1} + \|y\|^{p_2} + \|z\|^{p_3} + \|w\|^{p_4})^t$$

with suitable $c, p_1, p_2, p_3, p_4, t \in \mathbb{R}$, under some additional assumptions on nonempty $U \subset X$, we show that the conditional functional equation (8) is γ-hyperstable in the class of functions f mapping U^2 to a normed space.

In the remaining part of the paper, X and Y are normed spaces, $X_0 := X\setminus\{0\}$, and $\mathbb{N}_{m_0}$ denotes the set of all positive integers greater than or equal to a given $m_0 \in \mathbb{N}$.

Theorem 3 *Assume that $U \subset X_0$ is nonempty and there is $m_0 \in \mathbb{N}$, $m_0 > 3$, with*

$$-x, nx \in U, \qquad x \in U,\ n \in \mathbb{N},\ n \geq m_0 - 1. \tag{10}$$

If $f : U \times U \to Y$ satisfies

$$\left\|2f\left(\frac{x+z}{2}, \frac{y+w}{2}\right) - f(x, y) - f(z, w)\right\| \leq c\|x\|^p \|y\|^q \|z\|^r \|w\|^s, \tag{11}$$

$$x, y, z, w \in U, \quad \frac{x+z}{2}, \frac{y+w}{2} \in U,$$

with some $c \geq 0$ and $p, q, r, s \in \mathbb{R}$ such that $p+r < 0$ or $q+s < 0$, then (8) *holds.*

Proof Without loss of generality we can assume that Y is complete, because if this is not the case, then we can simply replace Y by its completion. Assume that $p + r < 0$ (the case $q + s < 0$ is analogous) and fix $l \in \mathbb{N}_{m_0}$.

Replacing (x, z, y, w) by $(mx, (2-m)x, ly, (2-l)y)$ in (11), we get

$$\left\|\frac{1}{2}f(mx, ly)+\frac{1}{2}f((2-m)x, (2-l)y) - f(x, y)\right\| \\ \leq \frac{cm^p(m-2)^r l^q (l-2)^s}{2} \|x\|^{p+r} \|y\|^{q+s} \tag{12}$$

for all $m \in \mathbb{N}_{m_0}$ and $x, y \in U$. Fix $m \in \mathbb{N}_{m_0}$ and write

$$\mathscr{T}_m \xi(x, y) := \frac{1}{2}\xi(mx, ly) + \frac{1}{2}\xi((2-m)x, (2-l)y),$$

$$\varepsilon_m(x, y) := \frac{cm^p(m-2)^r l^q (l-2)^s}{2} \|x\|^{p+r} \|y\|^{q+s}$$

for every $\xi \in Y^{U\times U}$ and $x, y \in U$. Then inequality (12) takes the form

$$\left\|\mathscr{T}_m f(x, y) - f(x, y)\right\| \leq \varepsilon_m(x, y), \qquad x, y \in U.$$

Let

$$\Lambda_m \delta(x, y) := \frac{1}{2}\delta(mx, ly) + \frac{1}{2}\delta((2-m)x, (2-l)y)$$

for $x, y \in U$ and $\delta \in \mathbb{R}_+{}^{U\times U}$. Then the operator Λ_m has the form described in (**H3**) with $k = 2$,

$$f_1(x, y) \equiv (mx, ly), \qquad f_2(x, y) \equiv ((2-m)x, (2-l)y),$$

$$L_1(x, y) \equiv L_2(x, y) \equiv 1/2$$

for all $x, y \in U$. Moreover, for every $\xi, \mu \in Y^{U\times U}$ and $x, y \in U$, we obtain

$$\begin{aligned}\left\|\mathscr{T}_m \xi(x, y) - \mathscr{T}_m \mu(x, y)\right\| \\ &= \left\|\frac{1}{2}\xi(mx, ly) + \frac{1}{2}\xi((2-m)x, (2-l)y) \right. \\ &\quad \left. - \frac{1}{2}\mu(mx, ly) - \frac{1}{2}\mu((2-m)x, (2-l)y)\right\| \\ &\leq \frac{1}{2}\|(\xi-\mu)(mx, ly)\| + \frac{1}{2}\|(\xi-\mu)((2-m)x, (2-l)y)\| \\ &= \sum_{i=1}^{2} L_i(x, y)\left\|(\xi-\mu)(f_i(x, y))\right\|.\end{aligned}$$

with $(\xi - \mu)(x, y) \equiv \xi(x, y) + \mu(x, y)$. So, **(H2)** is valid for $\mathscr{T}_m$. Note yet that

$$\Lambda_m \varepsilon_m(x, y) \leq a_m \varepsilon_m(x, y), \qquad m \in \mathbb{N}_{m_0},\ x, y \in U, \tag{13}$$

with

$$a_m := \frac{1}{2} m^{p+r} l^{q+s} + \frac{1}{2}(m-2)^{p+r}(l-2)^{q+s}.$$

Clearly, there is $m_1 \in \mathbb{N}_{m_0}$, such that

$$a_m < 1, \qquad m \in \mathbb{N}_{m_1}.$$

Therefore, by (13), we obtain that

$$\begin{aligned}\varepsilon_m^*(x, y) :&= \sum_{n=0}^{\infty} \Lambda_m^n \varepsilon_m(x, y) \leq \varepsilon_m(x, y) \sum_{n=0}^{\infty} (a_m)^n \\ &= \frac{\varepsilon_m(x, y)}{1 - a_m}, \qquad x, y \in U,\ \ m \in \mathbb{N}_{m_1}.\end{aligned}$$

Thus, according to Theorem 2, for each $m \in \mathbb{N}_{m_1}$ the function $J_m : U \times U \to Y$, given by $J_m(x, y) = \lim_{n \to \infty} \mathscr{T}_m^n f(x, y)$ for $x, y \in U$, is a unique fixed point of $\mathscr{T}_m$, i.e.,

$$J_m(x, y) = \frac{1}{2} J_m(mx, ly) + \frac{1}{2} J_m((2-m)x, (2-l)y)$$

for all $x, y \in U$; moreover

$$\left\| J_m(x, y) - f(x, y) \right\| \leq \frac{\varepsilon_m(x, y)}{1 - a_m}, \qquad x, y \in U.$$

We show that

$$\left\| 2\mathscr{T}_m^n f\left(\frac{x+z}{2}, \frac{y+w}{2}\right) - \mathscr{T}_m^n f(x, y) - \mathscr{T}_m^n f(z, w) \right\| \leq c a_m^n \|x\|^p \|y\|^q \|z\|^r \|w\|^s \tag{14}$$

for every $n \in \mathbb{N}_0$ and $x, y, z, w \in U$ with $\frac{x+z}{2}, \frac{y+w}{2} \in U$.

Clearly, if $n = 0$, then (14) is simply (11). So, fix $n \in \mathbb{N}_0$ and suppose that (14) holds for n and every $x, y, z, w \in U$ with $\frac{x+z}{2}, \frac{y+w}{2} \in U$. Then, for every $x, y, z, w \in U$ with $\frac{x+z}{2}, \frac{y+w}{2} \in U$,

$$\begin{aligned}
&\left\|2\mathcal{T}_m^{n+1} f\Big(\frac{x+z}{2}, \frac{y+w}{2}\Big) - \mathcal{T}_m^{n+1} f(x,y) - \mathcal{T}_m^{n+1} f(z,w)\right\| \\
&\quad = \Big\| 2\Big(\frac{1}{2}\mathcal{T}_m^n f\Big(m\frac{x+z}{2}, l\frac{y+w}{2}\Big) + \frac{1}{2}\mathcal{T}_m^n f\Big((2-m)\frac{x+z}{2}, (2-l)\frac{y+w}{2}\Big)\Big) \\
&\qquad - \frac{1}{2}\mathcal{T}_m^n f(mx, ly) - \frac{1}{2}\mathcal{T}_m^n f((2-m)x, (2-l)y) \\
&\qquad - \frac{1}{2}\mathcal{T}_m^n f(mz, lw) - \frac{1}{2}\mathcal{T}_m^n f((2-m)z, (2-l)w)\Big\| \\
&\quad \le \frac{1}{2}\Big\| 2\mathcal{T}_m^n f\Big(m\frac{x+z}{2}, l\frac{y+w}{2}\Big) - \mathcal{T}_m^n f(mx, ly) - \mathcal{T}_m^n f(mz, lw)\Big\| \\
&\qquad + \frac{1}{2}\Big\| 2\mathcal{T}_m^n f\Big((2-m)\frac{x+z}{2}, (2-l)\frac{y+w}{2}\Big) - \mathcal{T}_m^n f((2-m)x, (2-l)y) \\
&\qquad - \mathcal{T}_m^n f((2-m)z, (2-l)w)\Big\| \\
&\quad \le \frac{1}{2} c a_m^n \,\|mx\|^p\, \|ly\|^q\, \|mz\|^r\, \|lw\|^s \\
&\qquad + \frac{1}{2} c a_m^n \|(2-m)x\|^p \|(2-l)y\|^q \|(2-m)z\|^r \|(2-l)w\|^s \\
&\quad = c a_m^n \Big[\frac{1}{2} m^{p+r} l^{q+s} + \frac{1}{2}(m-2)^{p+r}(l-2)^{q+s}\Big] \|x\|^p \|y\|^q \|z\|^r \|w\|^s \\
&\quad = c (a_m)^{n+1} \|x\|^p \|y\|^q \|z\|^r \|w\|^s.
\end{aligned}$$

Thus, by induction, we have shown that (14) holds for all $x, y, z, w \in U$ such that $\frac{x+z}{2}, \frac{y+w}{2} \in U$ and for all $n \in \mathbb{N}_0$. Letting $n \to \infty$ in (14), we obtain that

$$2J_m\Big(\frac{x+z}{2}, \frac{y+w}{2}\Big) = J_m(x,y) + J_m f(z,w) \tag{15}$$

for every $x, y, z, w \in U$ with $\frac{x+z}{2}, \frac{y+w}{2} \in U$.

In this way, for each $m \in \mathbb{N}_{m_0}$, we obtain a function J_m such that (15) holds for $x, y, z, w \in U$ with $\frac{x+z}{2}, \frac{y+w}{2} \in U$ and

$$\big\| f(x,y) - J_m(x,y)\big\| \le \frac{\varepsilon_m(x,y)}{1-a_m}, \qquad x, y \in U, \ m \in \mathbb{N}_{m_1}.$$

Since

$$\lim_{m\to\infty} a_m = 0, \qquad \lim_{m\to\infty} \varepsilon_m(x,y) = 0, \qquad x, y \in U,$$

it follows, with $m \to \infty$, that f fulfills (8). □

In a similar way we can prove the following theorems.

Theorem 4 *Assume that $U \subset X_0$ is nonempty and there is $m_0 \in \mathbb{N}$, with*

$$\frac{1}{n}x, \frac{1}{2}\Big(1+\frac{1}{n}\Big)x \in U, \qquad x \in U,\ n \in \mathbb{N},\ n \geq m_0. \tag{16}$$

If $f : U \times U \to Y$ satisfies (11) *with some $c \geq 0$ and $p, q, r, s \in \mathbb{R}$ such that* "$p+r > 1$ *and* $q+s \geq 0$" *or* "$q+s > 1$ *and* $p+r \geq 0$", *then* (8) *holds.*

Proof Without loss of generality we can assume that Y is complete, because if this is not the case, then we can simply replace Y by its completion. Assume that $p+r > 1$ with $q+s \geq 0$ (the case $q+s > 1$ with $p+r \geq 0$ is analogous) and fix $l \in \mathbb{N}_{m_0}$.

Replacing (z, w) by $\left(\frac{1}{m}x, \frac{1}{l}y\right)$ in (11), we get

$$\Big\|2f\Big(\frac{m+1}{2m}x, \frac{l+1}{2l}y\Big) - f(x, y) - f\Big(\frac{x}{m}, \frac{y}{l}\Big)\Big\| \leq \frac{c}{m^r l^s}\, \|x\|^{p+r}\, \|y\|^{q+s} \tag{17}$$

for all $m \in \mathbb{N}_{m_0}$ and $x, y \in U$. Fix $m \in \mathbb{N}_{m_0}$ and we define

$$\mathscr{T}_m\xi(x, y) := 2\xi\Big(\frac{m+1}{2m}x, \frac{l+1}{2l}y\Big) - \xi\Big(\frac{x}{m}, \frac{y}{l}\Big), \quad \xi \in Y^{U\times U}$$

$$\varepsilon_m(x, y) := \frac{c}{m^r l^s}\, \|x\|^{p+r}\, \|y\|^{q+s}$$

$$\Lambda_m\delta(x, y) := 2\delta\Big(\frac{m+1}{2m}x, \frac{l+1}{2l}y\Big) + \delta\Big(\frac{x}{m}, \frac{y}{l}\Big), \quad \delta \in \mathbb{R}_+^{U\times U}$$

for every $x, y \in U$. Then inequality (17) takes the form

$$\big\|\mathscr{T}_m f(x, y) - f(x, y)\big\| \leq \varepsilon_m(x, y), \qquad x, y \in U,$$

and the operator Λ_m has the form described in (**H3**) with $k = 2$,

$$f_1(x, y) \equiv \Big(\frac{m+1}{2m}x, \frac{l+1}{2l}y\Big), \qquad f_2(x, y) \equiv \Big(\frac{x}{m}, \frac{y}{l}\Big),$$

$$L_1(x, y) \equiv 2, \qquad L_2(x, y) \equiv 1$$

for all $x, y \in U$. Moreover, for every $\xi, \mu \in Y^{U\times U}$ and $x, y \in U$, we obtain

$$\begin{aligned}\big\|\mathscr{T}_m\xi(x, y) - \mathscr{T}_m\mu(x, y)\big\| &\leq 2\big\|(\xi-\mu)(f_1(x, y))\big\| + \big\|(\xi-\mu)(f_2(x, y))\big\| \\ &= \sum_{i=1}^{2} L_i(x, y)\big\|(\xi-\mu)(f_i(x, y))\big\|.\end{aligned}$$

So, **(H2)** is valid for $\mathcal{T}_m$. Note yet that

$$\Lambda_m \varepsilon_m(x, y) \le b_m \varepsilon_m(x, y), \qquad m \in \mathbb{N}_{m_0}, \ \ x, y \in U, \tag{18}$$

with

$$b_m := 2\Big(\frac{1+m}{2m}\Big)^{p+r}\Big(\frac{1+l}{2l}\Big)^{q+s} + \frac{1}{m^{p+r} l^{q+s}}.$$

Clearly, there is $m_1 \in \mathbb{N}_{m_0}$, such that

$$b_m < 1, \qquad m \in \mathbb{N}_{m_1}.$$

Therefore, by (18), we obtain that

$$\begin{aligned}\varepsilon_m^*(x, y) :&= \sum_{n=0}^{\infty} \Lambda_m^n \varepsilon_m(x, y) \le \varepsilon_m(x, y) \sum_{n=0}^{\infty} (a_m)^n \\ &= \frac{\varepsilon_m(x, y)}{1 - b_m}, \qquad x, y \in U, \ \ m \in \mathbb{N}_{m_1}.\end{aligned}$$

Hence, according to Theorem 2, for each $m \in \mathbb{N}_{m_1}$ the function $J_m : U \times U \to Y$, given by $J_m(x, y) = \lim_{n\to\infty} \mathcal{T}_m^n f(x, y)$ for $x, y \in U$, is a unique fixed point of $\mathcal{T}_m$, i.e.,

$$J_m(x, y) = 2J_m\Big(\frac{1+m}{2m}x, \frac{1+l}{2l}y\Big) - J_m\Big(\frac{x}{m}, \frac{y}{l}\Big)$$

for all $x, y \in U$; moreover

$$\big\| J_m(x, y) - f(x, y)\big\| \le \frac{\varepsilon_m(x, y)}{1 - b_m}, \qquad x, y \in U.$$

Similarly as in the proof of Theorem 3, we show that

$$\Big\|2\mathcal{T}_m^n f\Big(\frac{x+z}{2}, \frac{y+w}{2}\Big) - \mathcal{T}_m^n f(x, y) - \mathcal{T}_m^n f(z, w)\Big\| \le c b_m^n \|x\|^p \|y\|^q \|z\|^r \|w\|^s \tag{19}$$

for every $n \in \mathbb{N}$ and $x, y, z, w \in U$ with $\frac{x+z}{2}, \frac{y+w}{2} \in U$.

Moreover, we obtain a function J_m satisfies (8) and

$$\big\| f(x, y) - J_m(x, y)\big\| \le \frac{\varepsilon_m(x, y)}{1 - b_m}, \qquad x, y \in U, \ m \in \mathbb{N}_{m_1}.$$

Since $p+r>1$, one of p, r must be positive, let $r>0$, then we obtain

$$\lim_{m\to\infty} b_m < 1, \qquad \lim_{m\to\infty} \varepsilon_m(x,y) = 0, \qquad x,y\in U,$$

it follows, with $m\to\infty$, that f fulfills (8). □

Theorem 5 *Assume that $U\subset X_0$ is nonempty and there is $m_0\in\mathbb{N}$, with*

$$\Big(2+\frac{1}{n}\Big)x, -\frac{1}{n}x, \in U, \qquad x\in U,\ n\in\mathbb{N},\ n\geq m_0. \tag{20}$$

If $f: U\times U\to Y$ satisfies (11)*, with some $c\geq 0$ and $p,q,r,s\in\mathbb{R}$ such that "$0<p+r<1$ and $q+s\leq 0$" or "$0<q+s<1$ and $p+r\leq 0$", then* (8) *holds.*

Proof Assume that Y is complete, $0<p+r<1$ and $q+s\leq 0$ (the case $0<q+s<1$ and $p+r\leq 0$ is analogous) and fix $l\in\mathbb{N}_{m_0}$. Then, one of p, r must be positive, let $p>0$.

Replacing (x,z,y,w) by $\Big(-\frac{1}{m}x, (2+\frac{1}{m})x, -\frac{1}{l}y, (2+\frac{1}{l})y\Big)$ in (11), we get

$$\begin{aligned}\Big\|\frac{1}{2}f\Big(-\frac{x}{m},-\frac{y}{l}\Big)+\frac{1}{2}f\Big(\Big(2+\frac{1}{m}\Big)x,\Big(2+\frac{1}{l}\Big)y\Big)-f(x,y)\Big\|&\\ \leq \frac{c}{2m^p l^q}\Big(2+\frac{1}{m}\Big)^r\Big(2+\frac{1}{l}\Big)^s\|x\|^{p+r}\|y\|^{q+s}&\end{aligned} \tag{21}$$

for all $m\in\mathbb{N}_{m_0}$ and $x,y\in U$. Fix $m\in\mathbb{N}_{m_0}$ and similarly as previously we define

$$\mathcal{T}_m\xi(x,y) := \frac{1}{2}\xi\Big(-\frac{x}{m},-\frac{y}{l}\Big)+\frac{1}{2}\xi\Big(\Big(2+\frac{1}{m}\Big)x,\Big(2+\frac{1}{l}\Big)y\Big),\quad \xi\in Y^{U\times U}$$

$$\varepsilon_m(x,y) := \frac{c}{2m^p l^q}\Big(2+\frac{1}{m}\Big)^r\Big(2+\frac{1}{l}\Big)^s\|x\|^{p+r}\|y\|^{q+s}$$

$$\Lambda_m\delta(x,y) := \frac{1}{2}\delta\Big(-\frac{x}{m},-\frac{y}{l}\Big)+\frac{1}{2}\delta\Big(\Big(2+\frac{1}{m}\Big)x,\Big(2+\frac{1}{l}\Big)y\Big),\quad \delta\in\mathbb{R}_+^{U\times U}$$

for every $x,y\in U$. Then inequality (21) takes the form

$$\big\|\mathcal{T}_m f(x,y)-f(x,y)\big\|\leq\varepsilon_m(x,y), \qquad x,y\in U.$$

Obviously Λ_m has the form described in (**H3**) with $k=2$,

$$f_1(x,y)\equiv\Big(-\frac{x}{m},-\frac{y}{l}\Big), \qquad f_2(x,y)\equiv\Big(\Big(2+\frac{1}{m}\Big)x,\Big(2+\frac{1}{l}\Big)y\Big),$$

$$L_1(x,y)\equiv L_2(x,y)\equiv 1/2$$

for all $x, y \in U$. It is clear that, for every $\xi, \mu \in Y^{U\times U}$ and $x, y \in U$, we obtain

$$\left\| \mathscr{T}_m\xi(x,y) - \mathscr{T}_m\mu(x,y)\right\| \le \sum_{i=1}^{2} L_i(x,y)d\left\|(\xi-\mu)(f_i(x,y))\right\|.$$

So, **(H2)** is valid for $\mathscr{T}_m$. Note yet that

$$\Lambda_m\varepsilon_m(x,y) \le d_m\varepsilon_m(x,y), \qquad m \in \mathbb{N}_{m_0},\ \ x,y \in U, \tag{22}$$

with

$$d_m := \frac{1}{2}\Big(2+\frac{1}{m}\Big)^{p+r}\Big(2+\frac{1}{l}\Big)^{q+s} + \frac{1}{2m^{p+r}l^{q+s}}.$$

Clearly, there is $m_1 \in \mathbb{N}_{m_0}$, such that

$$d_m < 1, \qquad m \in \mathbb{N}_{m_1}.$$

Therefore, by (22), we obtain that

$$\varepsilon_m^*(x,y) \le \varepsilon_m(x,y)\sum_{n=0}^{\infty}(a_m)^n = \frac{\varepsilon_m(x,y)}{1-d_m}, \qquad x,y\in U,\ \ m \in \mathbb{N}_{m_1}.$$

The remaining reasonings are analogous as in the proof of that Theorem 3. □

Remark 1 Let $c \ge 0$ and $p, q, r, s \in \mathbb{R}$ such that $p+q+r+s \in \mathbb{R}\setminus\{0,1\}$. If $U = X_0$ and $f : X \to Y$ satisfies (11) on X_0, then f satisfies (8) on X_0.

Theorem 6 *Let U be a nonempty subset of $X\setminus\{0\}$ fulfilling condition* (10) *with some $m_0 \in \mathbb{N}$. Let $c \ge 0$ and $p_1, p_2, p_3, p_4, t \in \mathbb{R}$ be such that $tp_i < 0$ for $i = 1,2,3,4$. If $f : U^2 \to Y$ satisfies the functional inequality*

$$\left\|2f\Big(\frac{x+z}{2},\frac{y+w}{2}\Big) - f(x,y) - f(z,w)\right\| \le c(\|x\|^{p_1}+\|y\|^{p_2}+\|z\|^{p_3}+\|w\|^{p_4})^t, \tag{23}$$

$$x,y,z,w \in U, \quad \frac{x+z}{2}, \frac{y+w}{2} \in U,$$

then (8) *holds.*

Proof As in the proof of Theorem 3, without loss of generality we can assume that Y is complete. Write $p(m) = m^{tp_0}$ for $m \in \mathbb{N}_3$, where

$$p_0 := \begin{cases} \max\{p_1,p_2,p_3,p_4\} & \text{if } t > 0; \\ \min\{p_1,p_2,p_3,p_4\} & \text{if } t < 0. \end{cases}$$

Clearly, if $t > 0$, then $p_i < 0$ for $i = 0, \dots, 4$ and consequently

$$\max\{m^{p_1}, m^{p_2}, m^{p_3}, m^{p_4}\} = m^{p_0}, \qquad m \in \mathbb{N}_3. \tag{24}$$

Analogously, if $t < 0$, then $p_i > 0$ for $i = 0, \dots, 4$ and

$$\min\{m^{p_1}, m^{p_2}, m^{p_3}, m^{p_4}\} = m^{p_0}, \qquad m \in \mathbb{N}_3. \tag{25}$$

Replacing (x, z, y, w) by $(mx, (2-m)x, my, (2-m)y)$ in (23), we get

$$\begin{aligned}\Big\| \frac{1}{2} f(mx, my) + \frac{1}{2} f((2-m)x, (2-m)y) - f(x, y) \Big\| \\ \leq \frac{c}{2}\big(\|mx\|^{p_1} + \|my\|^{p_2} + \|(2-m)x\|^{p_3} + \|(2-m)y\|^{p_4} \big)^t\end{aligned} \tag{26}$$

for all $x, y \in U$ and $m \in \mathbb{N}_{m_0}$. Let

$$\varepsilon_m(x, y) := \frac{c}{2}\big(\|mx\|^{p_1} + \|my\|^{p_2} + \|(2-m)x\|^{p_3} + \|(2-m)y\|^{p_4} \big)^t,$$

$$\mathcal{T}_m \xi(x) := \frac{1}{2}\xi(mx, my) + \frac{1}{2}\xi((2-m)x, (2-m)y)$$

for $x, y \in U$, $m \in \mathbb{N}_{m_0}$ and $\xi \in Y^{U \times U}$. Then, by (24) (if $t > 0$) and (25) (if $t < 0$), we get

$$\varepsilon_m(\pm mx, \pm my) \leq p(m)\varepsilon(x, y), \qquad x, y \in U, m \in \mathbb{N}_{m_0}, \tag{27}$$

and inequality (26) takes the form

$$\|\mathcal{T}_m f(x, y) - f(x, y)\| \leq \varepsilon_m(x, y), \qquad x, y \in U, m \in \mathbb{N}_{m_0}.$$

Write

$$\Lambda_m \delta(x, y) = \frac{1}{2}\delta(mx, my) + \frac{1}{2}\delta((2-m)x, (2-m)y)$$

for $x, y \in U$, $m \in \mathbb{N}_{m_0}$ and $\delta \in \mathbb{R}_+{}^{U \times U}$. Then, for each $m \in \mathbb{N}_{m_0}$, operator Λ_m has the form described in **(H3)** with $k = 3$ and

$$f_1(x, y) \equiv (mx, my), \qquad f_2(x, y) \equiv ((2-m)x, (2-m)y), \qquad L_1(x, y) \equiv L_2(x, y) \equiv 1/2.$$

Moreover, for every $\xi, \mu \in Y^{U \times U}$, $m \in \mathbb{N}_{m_0}$ and $x, y \in U$, we have

$$\|\mathcal{T}_m\xi(x,y)-\mathcal{T}_m\mu(x,y)\| \leq \sum_{i=1}^{3} L_i(x,y)\,\|(\xi-\mu)(f_i(x,y))\|.$$

So, **(H2)** is valid. Next, it is easily seen that, by induction on n, from (27) we obtain

$$\Lambda_m^n \varepsilon_m(x,y) \leq \alpha_m^n \varepsilon(x,y), \qquad n,m\in\mathbb{N}_{m_0}, x,y\in U, \tag{28}$$

where $\alpha_m := \frac{1}{2}p(m)+\frac{1}{2}p(m-2)$. Note that we can find $m_1\in\mathbb{N}_{m_0}$ with

$$\alpha_m < 1, \qquad m\in\mathbb{N}_{m_1},$$

which means that

$$\varepsilon_m^*(x,y) := \sum_{n=0}^{\infty}\Lambda_m^n\varepsilon(x,y) \leq \varepsilon_m(x,y)\sum_{n=0}^{\infty}(\alpha_m)^n = \frac{\varepsilon_m(x,y)}{1-\alpha_m}$$

for all $x,y\in U$ and $m\in\mathbb{N}_{m_1}$.

Similarly as in the proof of Theorem 3, we show that

$$\left\|2\mathcal{T}_m^n f\Big(\frac{x+z}{2},\frac{y+w}{2}\Big)-\mathcal{T}_m^n f(x,y)-\mathcal{T}_m^n f(z,w)\right\| \leq c\alpha_m^n(\|x\|^{p_1}+\|y\|^{p_2}+\|z\|^{p_3}+\|w\|^{p_4})^t \tag{29}$$

for every $n\in\mathbb{N}$ and $x,y,z,w\in U$ with $\frac{x+z}{2},\frac{y+w}{2}\in U$. Also the remaining reasonings are analogous as in the proof of that theorem. □

The next theorem shows the hyperstability of the two-variable Jensen functional equation on the set containing 0.

Theorem 7 *Assume that Y is complete and $U\subset X$ is nonempty with 0, such that $2U\subset U$ and $\frac{1}{2}U\subset U$. If $f: U\times U\to Y$ satisfies*

$$\left\|2f\Big(\frac{x+z}{2},\frac{y+w}{2}\Big)-f(x,y)-f(z,w)\right\| \leq c\|x\|^p\|y\|^q\|z\|^r\|w\|^s, \tag{30}$$

$$x,y,z,w\in U, \quad \frac{x+z}{2},\frac{y+w}{2}\in U,$$

with some $c\geq 0$ and $p,q,r,s>0$ such that $p+q+r+s\neq 1$, then (8) *holds.*

Proof Putting $z=w=0$ in (30), we obtain

$$2f\Big(\frac{x}{2},\frac{y}{2}\Big) = f(x,y)+f(0,0), \qquad x,y\in U,$$

i.e.,

$$2\Big(f\Big(\frac{x}{2},\frac{y}{2}\Big)-f(0,0)\Big)=f(x,y)-f(0,0),\qquad x,y\in U.$$

Thus g defined as $g(x,y)\equiv f(x,y)-f(0,0)$ satisfies (30) and

$$2g\Big(\frac{x}{2},\frac{y}{2}\Big)=g(x,y),\qquad x,y\in U. \tag{31}$$

Next we divide the proof into two cases.

Case 1: $p+q+r+s<1$. Using (31) to (30) we can prove by induction that for every $n\in\mathbb{N}_0$

$$\Big\|2g\Big(\frac{x+z}{2},\frac{y+w}{2}\Big)-g(x,y)-g(z,w)\Big\|\le c\Big(\frac{2^{p+q+r+s}}{2}\Big)^n\|x\|^p\|y\|^q\|z\|^r\|w\|^s, \tag{32}$$

$$x,y,z,w\in U,\quad \frac{x+z}{2},\frac{y+w}{2}\in U,$$

Indeed, if $n=0$, then (32) is simply (30). So, fix $n\in\mathbb{N}_0$ and assume that (32) holds for n. Then using (31) to (32) we have

$$\Big\|4g\Big(\frac{x+z}{4},\frac{y+w}{4}\Big)-2g\Big(\frac{x}{2},\frac{y}{2}\Big)-2g\Big(\frac{z}{2},\frac{w}{2}\Big)\Big\|\le c\Big(\frac{2^{p+q+r+s}}{2}\Big)^n\|x\|^p\|y\|^q\|z\|^r\|w\|^s, \tag{33}$$

$$x,y,z,w\in U,\quad \frac{x+z}{2},\frac{y+w}{2}\in U,$$

dividing by 2 and replacing (x,y,z,w) by $(2x,2y,2z,2w)$ in the last inequality we obtain

$$\Big\|2g\Big(\frac{x+z}{2},\frac{y+w}{2}\Big)-g(x,y)-g(z,w)\Big\|\le c\Big(\frac{2^{p+q+r+s}}{2}\Big)^{n+1}\|x\|^p\|y\|^q\|z\|^r\|w\|^s, \tag{34}$$

$$x,y,z,w\in U,\quad \frac{x+z}{2},\frac{y+w}{2}\in U,$$

so (32) holds for every $n\in\mathbb{N}_0$. As $p+q+r+s<1$, letting $n\to\infty$ in (32), we obtain that g satisfies (8) on U. Obviously f satisfies (8) on U, too.

Case 2: $p+q+r+s>1$. Replacing (x,y) by $(2x,2y)$ in (31) we get

$$2g(x,y)=g(2x,2y),\qquad x,y\in U. \tag{35}$$

Similarly as in 1) using (30), (35) and induction we obtain

$$\left\|2g\left(\frac{x+z}{2}, \frac{y+w}{2}\right) - g(x,y) - g(z,w)\right\| \le c\left(\frac{2}{2^{p+q+r+s}}\right)^n \|x\|^p\|y\|^q\|z\|^r\|w\|^s, \tag{36}$$

$$x, y, z, w \in U, \quad \frac{x+z}{2}, \frac{y+w}{2} \in U,$$

for every $n \in \mathbb{N}_0$. With $n \to \infty$ in the last inequality we have

$$2g\left(\frac{x+z}{2}, \frac{y+w}{2}\right) = g(x,y)+g(z,w), \quad x, y, z, w \in U, \quad \frac{x+z}{2}, \frac{y+w}{2} \in U$$

Thus f also satisfies (8) on U. □

4 Some Applications and Examples

The above theorems imply in particular the following corollary, which shows their simple application.

Corollary 1 *Let $U \subset X$ be nonempty and $F : U^4 \to Y$ be a function such that $F(x_0, y_0, z_0, w_0) \neq 0$ for some $x_0, y_0, z_0, w_0 \in U$ with $\frac{x_0+z_0}{2}, \frac{y_0+w_0}{2} \in U$ and*

$$\|F(x,y,z,w)\| \le c\|x\|^p\|y\|^q\|z\|^r\|z\|^s, \quad x, y, z, w \in U, \quad \frac{x+z}{2}, \frac{y+w}{2} \in U, \tag{37}$$

or

$$\|F(x,y,z,w)\| \le c\big(\|x\|^{p_1} + \|y\|^{p_2} + \|z\|^{p_3} + \|w\|^{p_4}\big)^t,$$

$$x, y, z, w \in U, \quad \frac{x+z}{2}, \frac{y+w}{2} \in U, \tag{38}$$

where $c \ge 0$ and $p, q, r, s, p_1, p_2, p_3, p_4, t \in \mathbb{R}$. Assume that one of the conditions (i)–(iv) *is valid in a case F satisfies* (37) *and* (v) *is valid in a case F satisfies* (38), *where*

(i) *$p + r < 0$ or $q + s < 0$, $0 \notin U$, and* (10) *holds with some $m_0 \in \mathbb{N}_4$,*
(ii) *$p + r > 1$ and $q + s \ge 0$ (or $q + s > 1$ and $p + r \ge 0$), $0 \notin U$ and* (16) *holds with some $m_0 \in \mathbb{N}$,*
(iii) *$0 < p + r < 1$ and $q + s \le 0$ (or $0 < q + s < 1$ and $p + r \le 0$), $0 \notin U$ and* (20) *holds with some $m_0 \in \mathbb{N}$,*
(iv) *$p, q, r, s > 0$ such that $p + q + r + s \neq 1$, $0 \in U$, $2U \subset U$ and $\frac{1}{2}U \subset U$,*
(v) *$tp_i < 0$ for $i = 1, 2, 3, 4$, $0 \notin U$ and* (10) *holds with some $m_0 \in \mathbb{N}\backslash\{1, 2\}$.*

Then the functional equation

$$2f_0\Big(\frac{x+z}{2}, \frac{y+w}{2}\Big) = f_0(x,y) + f_0(z,w) + F(x,y,z,w), \tag{39}$$

$$x, y, z, w \in U \quad \frac{x+z}{2}, \frac{y+w}{2} \in U,$$

has no solution in the class of functions $f_0 : U \to Y$.

Proof Suppose that there exists a solution $f_0 : U \to Y$ to (39). Then (11) or (23) holds, and consequently, according to the above theorems, f_0 is a solution to (8), which means that $F(x_0, y_0, z_0, w_0) = 0$ for some $x_0, y_0, z_0, w_0 \in U$ with $\frac{x_0+z_0}{2}, \frac{y_0+w_0}{2} \in U$. This is a contradiction. □

Now, we give some examples which show that in the above theorems the additional assumption on U are necessary.

Example 1 Let $X = Y = \mathbb{R}$, $U = [-1, 1]\backslash\{0\}$, $p, q, r, s < 0$, $c = 4$ and $f : U^2 \to \mathbb{R}$ be defined by $f(x, y) = |x + y|$. Then f satisfies

$$\Big|2f\Big(\frac{x+z}{2}, \frac{y+w}{2}\Big) - f(x,y) - f(z,w)\Big| \leq 4|x|^p|y|^q|z|^r|w|^s, \quad x, y, z, w \in U$$

but f is not a solution of equation (8) on U. We see that $0 \notin U$ and U does not satisfy the assumption of Theorem 3 .

Example 2 Let $X = Y = \mathbb{R}$, $U = [1, \infty)$, $p, q, r, s > 0$, $c = 4$ and $f : U^2 \to \mathbb{R}$ be defined by $f(x, y) = \frac{1}{x} + \frac{1}{y}$. Then f satisfies

$$\Big|2f\Big(\frac{x+z}{2}, \frac{y+w}{2}\Big) - f(x,y) - f(z,w)\Big| \leq 4|x|^p|y|^q|z|^r|w|^s, \quad x, y, z, w \in U$$

but f is not a solution of equation (8) on U. It is easy to check that the assumptions of Theorems 4, 5, and 7 are not satisfied.

In this example, we show that the condition $-x \in U$ for every $x \in U$ in Theorem 6 is necessary.

Example 3 Let $X = Y = \mathbb{R}$, $U = (0, \infty)$, $t = 1$, $p_i = p < 0$ for $i = 1, \ldots, 4$, and $f : U^2 \to \mathbb{R}$ be defined by $f(x, y) = x^p + y^p$. Then f satisfies

$$\Big|2f\Big(\frac{x+z}{2}, \frac{y+w}{2}\Big) - f(x,y) - f(z,w)\Big| \leq 2^{1-p}(|x|^p + |y|^p + |z|^p + |w|^p),$$

$$x, y, z, w \in U$$

but f is not a solution of equation (8) on U, which shows that in Theorem 6 the assumption that $-x \in U$ for every $x \in U$ is necessary.

We end the paper with an open problem.

Remark 2 For the cases $p + r = q + s = 0$ and $tp_i = 0$ for $i = 1, \ldots, 4$, the method used in the proofs of the above theorems cannot be applied, thus this is still an open problem.

References

1. M.R. Abdollahpour, R. Aghayari, M. Th. Rassias, Hyers-Ulam stability of associated Laguerre differential equations in a subclass of analytic functions. J. Math. Anal. Appl. **437**(1), 605–612 (2016)
2. A. Aghajani, F. Zahedi, General solution and generalized Hyers-Ulam stability of a two-variable Jensen functional equation. Math. Sci. **7**(45) (2013). https://doi.org/10.1186/2251-7456-7-45
3. T. Aoki, On the stability of the linear transformation in Banach spaces. J. Math. Soc. Jpn. **2**, 64–66 (1950)
4. J.H. Bae, W.G. Park, On the solution of a bi-Jensen functional equation and its stability. Bull. Korean Math. Soc. **43**, 499–507 (2006)
5. A. Bahyrycz, M. Piszczek, Hyperstability of the Jensen functional equation. Acta Math. Hungar. **142**, 353–365 (2014)
6. D.G. Bourgin, Approximately isometric and multiplicative transformations on continuous function rings. Duke Math. J. **16**, 385–397 (1949)
7. D.G. Bourgin, Classes of transformations and bordering transformations. Bull. Am. Math. Soc. **57**, 223–237 (1951)
8. J. Brzdęk, Hyperstability of the Cauchy equation on restricted domains. Acta Math. Hungar. **141**(1–2), 58–67 (2013)
9. J. Brzdęk, Remarks on hyperstability of the Cauchy equation. Aequations Math. **86**, 255–267 (2013)
10. J. Brzdęk, A hyperstability result for the Cauchy equation. Bull. Aust. Math. Soc. **89**, 33–40 (2014)
11. J. Brzdęk, K. Ciepliński, Hyperstability and superstability. Abstr. Appl. Anal. **2013** (2013). Article ID 401756, 13 pp.
12. J. Brzdęk, J. Chudziak, Z. Páles, A fixed point approach to stability of functional equations. Nonlinear Anal. **74**, 6728–6732 (2011)
13. J. Brzdęk, W. Fechner, M.S. Moslehian, J. Sikorska, Recent developments of the conditional stability of the homomorphism equation. Banach J. Math. Anal. **9**, 278–327 (2015)
14. J. Brzdęk, K. Ciepliński, Th.M. Rassias (eds.), *Developments in Functional Equations and Related Topics* (Springer, New York, 2017)
15. Y.J. Cho, Th.M. Rassias, R. Saadati, *Stability of Functional Equations in Random Normed Spaces* (Springer, New York, 2013)
16. Y.J. Cho, C. Park, Th.M. Rassias, R. Saadati, *Stability of Functional Equations in Banach Algebras* (Springer, New York, 2015)
17. Iz. EL-Fassi, On a new type of hyperstability for radical cubic functional equation in non-Archimedean metric spaces. Results Math. **72**, 991–1005 (2017)
18. Iz. El-Fassi, Approximate solution of radical quartic functional equation related to additive mapping in 2-Banach spaces. J. Math. Anal. Appl. **455**, 2001–2013 (2017)
19. Iz. EL-Fassi, S. Kabbaj, On the hyperstability of a Cauchy-Jensen type functional equation in Banach spaces. Proyecciones J. Math. **34**, 359–375 (2015)
20. Iz. EL-Fassi, S. Kabbaj, A. Charifi, Hyperstability of Cauchy-Jensen functional equations. Indagationes Math. **27**, 855–867 (2016)
21. Iz. El-Fassi, J. Brzdęk, A. Chahbi, S. Kabbaj, On hyperstability of the biadditive functional equation. Acta Math. Sci. **37**B(6), 1727–1739 (2017)

22. Z. Gajda, On stability of additive mappings. Int. J. Math. Math. Sci. **14**, 431–434 (1991)
23. E. Gselmann, Hyperstability of a functional equation. Acta Math. Hungar. **124**, 179–188 (2009)
24. D.H. Hyers, On the stability of the linear functional equation. Proc. Nat. Acad. Sci. U.S.A. **27**, 222–224 (1941)
25. D.H. Hyers, Th.M. Rassias, Approximate homomorphisms. Aequat. Math. **44**, 125–153 (1992)
26. G. Isac, Th.M. Rassias, On the Hyers - Ulam stability of ψ- additive mappings. J. Approx. Theory **72**, 131–137 (1993)
27. K.W. Jun, M.H. Han, Y.H. Lee, On the Hyers-Ulam-Rassias stability of the bi-Jensen functional equation. Kyungpook Math. J. **48**, 705–720 (2008)
28. K.W. Jun, Y.H. Lee, J.H. Oh, On the Rassias stability of a bi-Jensen functional equation. J. Math. Inequalities **2**, 363–375 (2008)
29. S.-M. Jung, *Hyers-Ulam-Rassias Stability of Functional Equations in Nonlinear Analysis* (Springer, New York, 2011)
30. Pl. Kannappan, *Functional Equations and Inequalities with Applications* (Springer, New York, 2009)
31. Z. Kominek, On a local stability of the Jensen functional equation. Demonstratio Math. **22**, 499–507 (1989)
32. Y.-H. Lee, On the stability of the monomial functional equation. Bull. Korean Math. Soc. **45**, 397–403 (2008)
33. G. Maksa, Z. Páles, Hyperstability of a class of linear functional equations. Acta Math. **17**, 107– 112 (2001)
34. C. Mortici, M.Th. Rassias, S.-M. Jung, On the stability of a functional equation associated with the Fibonacci numbers. Abstr. Appl. Anal. (2014). Art. ID 546046, 6 pp.
35. Th.M. Rassias, On the stability of the linear mapping in Banach spaces. Proc. Am. Math. Soc. **72**, 297–300 (1978)
36. Th.M. Rassias, On a modified Hyers-Ulam sequence. J. Math. Anal. Appl. **158**, 106–113 (1991)
37. Th.M. Rassias, J. Brzdęk (eds.), *Functional Equations in Mathematical Analysis* (Springer, New York, 2012)
38. Th.M. Rassias, P. Semrl, On the behavior of mappings which do not satisfy Hyers-Ulam stability. Proc. Am. Math. Soc. **114**, 989–993 (1992)
39. S.M. Ulam, *Problems in Modern Mathematics*, Chapter IV, Science Editions (Wiley, New York, 1960)

On the Study of Circuit Chains Associated with a Random Walk with Jumps in Fixed, Random Environments: Criteria of Recurrence and Transience

Chrysoula Ganatsiou

Abstract By considering a nonhomogeneous random walk with jumps (with steps -1 or $+1$ or in the same position having a right-elastic barrier at 0) we investigate the unique representations by directed circuits and weights of the corresponding Markov chains (circuit chains) in fixed, random environments. This will give us the possibility to find suitable criteria regarding the properties of recurrence and transience of the above-mentioned circuit chains in fixed, random environments.

2010 AMS Mathematics Subject Classification 60J10, 60G50, 60K37

1 Introduction

In recent years a systematic research has been developed (Kalpazidou [10], MacQueen [12], Qian Minping and Qian Min [13], Zemanian [16] and others) in order to investigate representations of the finite-dimensional distributions of Markov processes (with discrete or continuous parameter) having an invariant measure, as decompositions in terms of the *circuit* (or cycle) *passage functions*

$$J_c(i,j) = \begin{cases} 1, \text{ if } i, j \text{ are consecutive states of c,} \\ 0, \text{ otherwise,} \end{cases}$$

for any directed sequence $c = (i_1, i_2, \ldots, i_v, i_1)$ (or $\hat{c} = (i_1, i_2, \ldots, i_v)$) of states, called a *circuit* (or a *cycle*), $v > 1$ of the corresponding Markov process. This research has stimulated a motivation towards the representation of Markov processes through directed circuits (or cycles) and weights in terms of circuit (or cycle) passage functions in fixed or random environments as well as the study of specific problems associated with Markov processes in a different way. The

C. Ganatsiou (✉)
Department of Digital Systems, University of Thessaly, Larissa, Greece
e-mail: chganats@uth.gr; ganatsio@otenet.gr

T. M. Rassias, P. M. Pardalos (eds.), *Mathematical Analysis and Applications*, Springer Optimization and Its Applications 154,
https://doi.org/10.1007/978-3-030-31339-5_6

representations are called *circuit (or cycle) representations* while the corresponding discrete parameter Markov chains generated by directed weighted circuits are called *circuit chains* [1, 10].

In parallel, *random walks* are one of the most basic and well-studied topics in probability theory and one of the most fundamental types of stochastic processes formed by successive summation of independent, identically distributed random variables. For random walks on the integer lattice Z^d the main reference is the classic book by F. Spitzer [15]. They have a long rich history [2, 3, 8] which has been advanced according to many directions of investigation. The term "random walk" was coined by Karl Pearson [14], and the study of random walks dates back to the "Gamblers Ruin" problem analyzed by Pascal, Fermat, Huygens, Bernoulli, and others. Theoretical developments of random walks have involved mathematics (especially probability theory), computer science, statistical physics, operations research, and more. Random walk models have also been applied in various domains, ranging from locomotion and foraging of animals, the dynamics of neuronal firing and decision-making in the brain to population genetics, polymer chains, descriptions of financial markets, rankings systems, dimension reduction, and feature extraction from high-dimensional data (e.g., in the form of "diffusion maps"), sports statistics, prediction of the arrival times of diseases spreading on networks, etc.

Usually they are studied from the Markov chain point of view, where the random mechanism of spatial motion is determined by the given transition probabilities (probabilities of jumps) at each state in a *non-random (fixed) environment*. Although random walks provide a simple conventional model to describe various transport processes in many cases, the medium where the system evolves is highly irregular due to many irregularities (defects, fluctuations, etc.) known as random environments which lead to the choice of the local characteristics of the motion at random according to certain probability distribution. Such models are referred to as *random walks in random environments*. The definition of these random walks involves two special ingredients: the *environment* (randomly chosen but still fixed throughout the time evolution) and the *random walk* (whose transition probabilities are determined by the environment) [8].

It is known also that in various applications (physics, chemistry, genetics, etc.) we are led to study Markov chains obtained by restricting the motion of a "particle" which performs a random walk. This is done by introducing barriers. In this case the Markov chain defined in this way having no longer independent increments is called a *random walk with barriers* while its state space is a proper subset of Z. Furthermore except for the *homogeneous random walks* with independent and identically distributed increments there is a class of random walks formed by successive summation of independent random variables which are no longer identically distributed. This means that they still have independent increments which are no longer identically distributed. These random walks are called *nonhomogeneous* and they can be investigated from the Markov chain point of view which in general coincides with that for chains with independent increments.

Let us consider the nonhomogeneous random walk with state space S=N, right-elastic barrier at 0 [7] and transition probabilities given by $p_{ij} = 0$, if $|i - j| > 1$, $p_{i,i-1} = q_i$, $p_{i,i} = r_i$, $p_{i,i+1} = p_i$, $p_i + q_i + r_i = 1$, $i \geq 1$, $p_{00} = r_0$, $p_{01} = p_0 = 1 - r_0$, $p_i > 0$, $q_{i+1} > 0$, $r_i \geq 0$, $i \geq 0$, which expresses the movement of a particle depending on the time that the particle begins to move. It is obvious that all states form an essential class. It is known that regarding the classification of the states through the use of proper theorems ensuring a bounded solution of the system of equations

$$z_i = \sum_{j=1}^{\infty} p_{ij} z_j, \quad i \geq 1$$

we have that: the states are *positive recurrent* if and only if

$$\sum_{i=1}^{\infty} r_i = +\infty \qquad \text{and} \qquad \sum_{i=1}^{\infty} \frac{1}{r_i p_i} < +\infty$$

and *null recurrent* if and only if

$$\sum_{i=1}^{\infty} r_i = \sum_{i=1}^{\infty} \frac{1}{r_i \cdot p_i} = +\infty \qquad \text{where} \qquad r_i = \frac{q_1 \cdots q_i}{p_1 \cdots p_i}, \ i \geq 1 \text{ [9]}$$

The main purpose of this work is to bring together the two subjects—random walks and circuit chains—by discussing their interconnection. In particular following the context of the theory of Markov processes'cycle-circuit representation, the present work arises as an attempt to study the circuit and weight representation of the above-mentioned nonhomogeneous random walk with jumps in fixed, random environments as well as to investigate proper criteria regarding recurrence and transience of the corresponding "adjoint" Markov chains (circuit chains) describing uniquely the above-mentioned random walk by directed circuits and weights in fixed, random environments giving a new perspective in the whole study and especially in the classification of states.

The work is organized as follows. In Section 2, we give a brief account of certain concepts of circuit-cycle representation theory of Markov processes that we shall need throughout the paper. In Section 3, the above-mentioned nonhomogeneous random walk with jumps (having one right-elastic barrier at 0) is considered and the unique representations by directed circuits and weights of the corresponding Markov chains (circuit chains) are investigated in fixed, random environments. These representations will give us the possibility to find proper criteria regarding positive/null recurrence and transience of the above-mentioned circuit chains in fixed, random environments [4–6], as it is described in Section 4.

Throughout the paper, we shall need the following notations:

$$\mathbb{N} = \{0, 1, 2, \dots\}, \quad \mathbb{N}^* = \{1, 2, \dots\}, \quad \mathbb{Z} = \{\dots, -1, 0, 1, \dots\},$$
$$\mathbb{Z}_+^* = \{1, 2, 3, \dots\}, \quad \mathbb{Z}_-^* = \{\dots, -2, -1\}$$

2 Preliminaries

Let S be a denumerable set. The directed sequence $c = (i_1, i_2, \dots, i_v, i_1)$ modulo the cyclic permutations, where $i_1, i_2, \dots, i_v \in S, v > 1$, completely defines a directed circuit in S. The ordered sequence $\hat{c} = (i_1, i_2, \dots, i_v)$ associated with the given directed circuit c is called a *directed cycle* in S. A directed circuit may be considered as $c = (c(m), c(m+1), \dots, c(m+v-1), c(m+v))$, if there exists an $m \in \mathbb{Z}$, such that $i_1 = c(m+0), i_2 = c(m+1), \dots, i_v = c(m+v-1), i_1 = c(m+v)$, that is a periodic function from $\mathbb{Z}$ to S. The corresponding directed cycle is defined by the ordered sequence $\hat{c} = (c(m), c(m+1), \dots, c(m+v-1))$. The values $c(k)$ are the *points* of c, while the directed pairs , $(c(k), c(k+1)), k \in Z$, are the *directed edges* of c. The smallest integer $p \equiv p(c) \geq 1$ satisfying the equation $c(m+p) = c(m)$, for all $m \in \mathbb{Z}$, is the *period* of c. A directed circuit c such that $p(c) = 1$ is called a *loop*. (In the present work, we shall use directed circuits with distinct point elements.)

Let a directed circuit c (or a directed cycle $\hat{c}$) with period $p(c) > 1$. Then we may define by

$$J_c^{(n)}(i, j) = \begin{cases} 1, & \text{if there exists an } m \in \mathbb{Z} \text{ such that } i = c(m), j = c(m+n), m \in \mathbb{Z} \\ 0, & \text{otherwise} \end{cases}$$

the *n-step passage function* associated with the directed circuit c, for any $i, j \in S, n \geq 1$.

We may also define by

$$J_c(i) = \begin{cases} 1, & \text{if there exists an m } \in \mathbb{Z} \text{such that } i = c(m), \\ 0, & \text{otherwise} \end{cases}$$

the *passage function* associated with the directed circuit c, for any $i \in S$. The above definitions are due to MacQueen [12] and Kalpazidou [10].

Given a denumerable set S and an infinite denumerable class C of overlapping directed circuits (or directed cycles) with distinct points (except for the terminals) in S such that all the points of S can be reached from one another following paths of circuit-edges, that is, for each two distinct points i and j of S there exists a finite sequence $c_1, c_2 \dots, c_k, k \geq 1$, of circuits (or cycles) of C such that i lies on c_1 and j lies on c_k and any pair of consecutive circuits (c_n, c_{n+1}) have at least one point in common. We may assume also that the class C contains, among its elements, circuits (or cycles) with period greater than or equal to 2. With each directed circuit (or directed cycle) let us associate a *strictly positive weight* w_c which must be independent of the choice of the representative of c, that is, it must satisfy the consistency condition, $w_{c\circ t_k} = w_c$, $k \in \mathbb{Z}$, where t_k is the translation of length k.

For a given class C of overlapping directed circuits (or cycles) and for a given sequence $(w_c)_{c\in C}$ of weights we may define by

$$p_{ij} = \frac{\sum\limits_{c\in C} w_c \cdot J_c^{(1)}(i, j)}{\sum\limits_{c\in C} w_c \cdot J_c(i)} \tag{2.1}$$

the elements of a Markov transition matrix on S, if and only if $\sum\limits_{c\in C} w_c \cdot J_c(i) < \infty$, for any $i \in S$. This means that a given Markov transition matrix $P = (p_{ij}), i, j \in S$ can be represented by directed circuits (or cycles) and weights if and only if there exists a class of overlapping directed circuits (or cycles) C and a sequence of positive weights $(w_c)_{c\in C}$ such that the formula (2.1) holds. In this case, the representation of the distribution of Markov process (with discrete or continuous parameter) having an invariant measure as decomposition in terms of the circuit (or cycle) passage functions is called *circuit (or cycle) representation* while the corresponding discrete parameter Markov chain generated by directed circuits (or cycles) is called *circuit (or cycle) chain* with Markov transition matrix P given by (2.1) and unique stationary distribution p (a solution of $p.P = p$) defined by

$$p(i) = \sum_{c\in C} w_c \cdot J_c(i), i \in S.$$

It is known that the following classes of Markov chains may be represented uniquely by directed circuits (or cycles) and weights:

(i) *the recurrent Markov chains* [13],
(ii) *the reversible Markov chains.*

3 Circuit and Weight Representations

3.1 Fixed Environments

Let us consider the Markov chain $(X_n)_{n\in N}$ on $\mathbb{N}$ (X_n expresses the location of a particle at time $n, n \in N$) which describes the nonhomogeneous random walk with jumps having a right-elastic barrier at 0, with transitions $k \to (k+1), k \to (k-1)$ and $k \to k$, in a fixed environment, whose elements of the corresponding Markov transition matrix (transition probabilities) are defined by

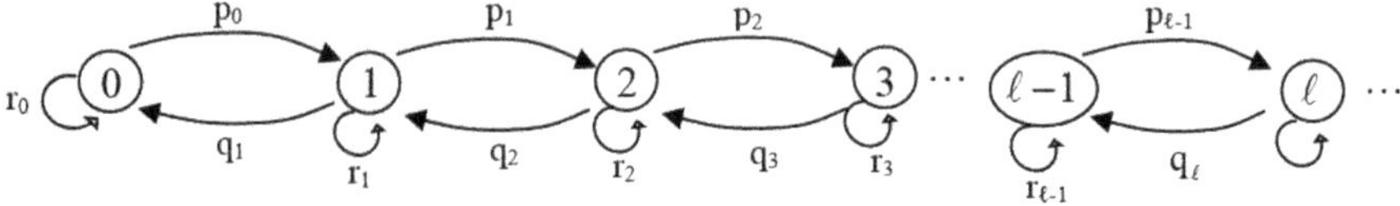

Fig. 1 The Markov chain $(X_n)_{n\in\mathbb{N}}$ (fixed environments)

$$P(X_{n+1} = 0/X_n = 0) = r_0,$$
$$P(X_{n+1} = 1/X_n = 0) = p_0,\ p_0 = 1 - r_0$$
$$P(X_{n+1} = k+1/X_n = k) = p_k,\ k \geq 1$$
$$P(X_{n+1} = k/X_n = k) = r_k,\ k \geq 1$$
$$P(X_{n+1} = k-1/X_n = k) = q_k,\ k \geq 1$$

such that $p_k + q_k + r_k = 1$, $p_k > 0$, $q_{k+1} > 0$, $r_k \geq 0$, for every $k \in \mathbb{N}$, as it is shown in Figure 1.

Assume that $(p_k)_{k\in N}$ and $(r_k)_{k\in N}$ are arbitrary fixed sequences with $0 < p_0 = 1 - r_0 \leq 1$, $p_k > 0$, $q_{k+1} > 0$, $r_k \geq 0$, for every $k \in \mathbb{N}$. If we consider the directed circuits $c_k = (k, k+1, k)$, $c'_k = (k, k)$, $k \in \mathbb{N}$ and the collections of weights $(w_{c_k})_{k\in N}$ and $(w_{c'_k})_{k\in N}$ respectively, then we may obtain the corresponding transition probabilities

$$p_k = \frac{w_{c_k}}{w_{c_{k-1}} + w_{c_k} + w_{c'_k}},$$

with

$$p_0 = \frac{w_{c_0}}{w_{c_0} + w_{c'_0}}$$

and

$$q_k = \frac{w_{c_{k-1}}}{w_{c_{k-1}} + w_{c_k} + w_{c'_k}}, \qquad r_k = \frac{w_{c'_k}}{w_{c_{k-1}} + w_{c_k} + w_{c'_k}}$$

such that $p_k + q_k + r_k = 1$, for every $k \geq 1$, with $r_0 = 1 - p_0 = \dfrac{w_{c'_0}}{w_{c_0} + w_{c'_0}}$.

Here the class $C(k)$ contains the directed circuits $c_k = (k, k+1, k)$, $c'_k = (k, k)$ and $c_{k-1} = (k-1, k, k-1)$.

Equivalently the transition matrix $P = (p_{ij})$ with

$$p_{ij} = \frac{\sum_{k \in N} w_{c_k} \cdot J_{c_k}^{(1)}(i, j)}{\sum_{k \in N} \left[w_{c_k} \cdot J_{c_k}(i) + w_{c'_k} \cdot J_{c'_k}(i) \right]}, \text{ for } i \neq j, \tag{3.1}$$

$$p_{ii} = \frac{\sum_{k \in N} w_{c'_k} \cdot J_{c'_k}^{(1)}(i, i)}{\sum_{k \in N} \left[w_{c_k} \cdot J_{c_k}(i) + w_{c'_k} \cdot J_{c'_k}(i) \right]}, \tag{3.2}$$

where $J_{c_k}^{(1)}(i, j) = 1$, if i, j are consecutive points of the circuit c_k, $J_{c_k}(i) = 1$, if i is a point of the circuit c_k, and $J_{c'_k}(i) = 1$, if i is a point of the circuit c'_k, expresses the representation of the Markov chain $(X_n)_{n \in N}$ by directed circuits and weights.

Furthermore let us consider also the "*adjoint*" Markov chain $(X'_n)_{n \in N}$ on N whose elements of the corresponding Markov transition matrix are defined by

$$\begin{aligned}
&P(X'_{n+1} = 0/X'_n = 0) = r'_0, \\
&P(X'_{n+1} = 1/X'_n = 0) = q'_0, q'_0 = 1 - r'_0, \\
&P(X'_{n+1} = k - 1/X'_n = k) = p'_k, k \geq 1, \\
&P(X'_{n+1} = k/X'_n = k) = r'_k, k \geq 1, \\
&P(X'_{n+1} = k + 1/X'_n = k) = q'_k, k \geq 1
\end{aligned}$$

such that $p'_k + q'_k + r'_k = 1$,$p'_{k+1} > 0$,$q'_k > 0$, $r'_k \geq 0$ for every $k \in \mathbb{N}$, as it is shown in Figure 2.

Assume that $(q'_k)_{k \in N}$, $(r'_k)_{k \in N}$ are arbitrary fixed sequences with $0 < q'_0 = 1 - r'_0 \leq 1$, $p'_{k+1} > 0$, $q'_k > 0$, $r'_k \geq 0$, for every $k \in \mathbb{N}$. If we consider the directed circuits $c''_k = (k + 1, k, k + 1)$, $c'''_k = (k, k)$ $k \in \mathbb{N}$, and the collections of weights $(w_{c''_k})_{k \in N}$, $(w_{c'''_k})_{k \in N}$, respectively, then we may have that

$$q'_k = \frac{w_{c''_k}}{w_{c''_{k-1}} + w_{c''_k} + w_{c'''_k}},$$

with

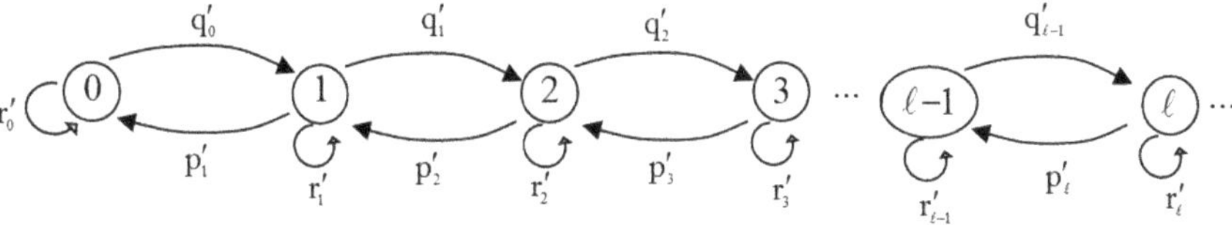

Fig. 2 The "adjoiont" Markov chain $(X'_n)_{n \in \mathbb{N}}$ (fixed environments)

$$q_0' = \frac{w_{c_0''}}{w_{c_0''} + w_{c_0'''}}$$

and

$$p_k' = \frac{w_{c_{k-1}''}}{w_{c_{k-1}''} + w_{c_k''} + w_{c_k'''}}, \qquad r_k' = \frac{w_{c_k'''}}{w_{c_{k-1}''} + w_{c_k''} + w_{c_k'''}}$$

such that $p_k' + q_k' + r_k' = 1$, for every $k \geq 1$, with $r_0' = 1 - q_0' = \dfrac{w_{c_0'''}}{w_{c_0''} + w_{c_0'''}}$.

Here the class $C'(k)$ contains the directed circuits $c_k'' = (k+1, k, k+1)$,$c_{k-1}'' = (k, k-1, k)$ and $c_k''' = (k, k)$. As a consequence, the transition matrix $P' = (p_{ij}')$ with elements equivalent to that given by the above-mentioned formulas (3.1), (3.2) expresses also the representation of the "adjoint" Markov chain $(X_n')_{n \in N}$ by directed circuits and weights.

Consequently we have the following:

Proposition 1 *The Markov chain $(X_n)_{n \in N}$ defined as above has a unique representation by directed circuits and weights.*

Proof Let us consider the set of directed circuits $c_k = (k, k+1, k)$ and $c_k' = (k, k)$, for every $k \in \mathbb{N}$, since only the transitions from k to $k+1$, k to $k-1$ and k to k are possible. There are three circuits through each point $k \geq 1 c_{k-1}, c_k, c_k'$, and two circuits through $0 : c_0, c_0'$.

The problem we have to manage is the definition of the weights. We may symbolize by w_k the weight w_{c_k} of the circuit c_k and by w_k' the weight $w_{c_k'}$ of the circuit c_k', for any $k \in \mathbb{N}$. The sequences $(w_k)_{k \in N}$,$(w_k')_{k \in N}$ must be a solution of

$$p_k = \frac{w_k}{w_{k-1} + w_k + w_k'}, k \geq 1 \qquad \text{with} \qquad p_0 = \frac{w_0}{w_0 + w_0'},$$

$$r_k = \frac{w_k'}{w_{k-1} + w_k + w_k'}, k \geq 1 \qquad \text{with} \qquad r_0 = \frac{w_0'}{w_0 + w_0'},$$

$$q_k = 1 - p_k - r_k, k \geq 1.$$

Let us take by $b_k = \dfrac{w_k}{w_{k-1}}, \gamma_k = \dfrac{w_k'}{w_{k-1}'}, k \geq 1$. As a consequence we may have

$$b_k = \frac{p_k}{q_k} = \frac{p_k}{1 - p_k - r_k}, \gamma_k = \frac{r_k}{r_{k-1}} \frac{p_{k-1}}{p_k} b_k, \text{ for every } k \geq 1.$$

Given the sequences $(p_k)_{k\in N}$ and $(r_k)_{k\in N}$ it is clear that the above sequences $(b_k)_{k\geq 1}$,$(\gamma_k)_{k\geq 1}$ exist and are unique. This means that the sequences $(w_k)_{k\in N}$,$(w'_k)_{k\in N}$ are defined uniquely, up to multiplicative constant factors, by

$$w_k = w_0 \cdot b_1 \dots b_k,$$
$$w'_k = w'_0 \cdot \gamma_1 \dots \gamma_k$$

(the unicity is understood up to the constant factors w_0, w'_0).

Proposition 2 *The "adjoint" Markov chain $(X'_n)_{n\in N}$ defined as above has a unique representation by directed circuits and weights.*

Proof Following an analogous way of that given in the proof of Proposition 1 the problem we have also to manage here is the definition of the weights. To this direction we may symbolize by w''_k the weight $w_{c''_k}$ of the circuit c''_k and by w'''_k the weight $w_{c'''_k}$ of the circuit c'''_k , for every $k \in \mathbb{N}$. The sequences $(w''_k)_{k\in N}$,$(w'''_k)_{k\in N}$ must be solutions of

$$q'_k = \frac{w''_k}{w''_{k-1} + w''_k + w'''_k}, k \geq 1 \qquad \text{with} \qquad q'_0 = \frac{w''_0}{w''_0 + w'''_0},$$

$$r'_k = \frac{w'''_k}{w''_{k-1} + w''_k + w'''_k}, k \geq 1 \qquad \text{with} \qquad r'_0 = \frac{w'''_0}{w''_0 + w'''_0},$$

$$p'_k = 1 - q'_k - r'_k, \qquad k \geq 1$$

By considering the sequences $(s_k)_k$,$(t_k)_k$ where $s_k = \frac{w''_{k-1}}{w''_k}$, $t_k = \frac{w'''_{k-1}}{w'''_k}$, $k \geq 1$ we may obtain that

$$s_k = \frac{1 - q'_k - r'_k}{q'_k}, t_k = \frac{r'_{k-1}}{r'_k} \cdot \frac{q'_k}{q'_{k-1}} \cdot s_k, \qquad \text{for every} \qquad k \geq 1.$$

For given sequences $(q'_k)_{k\in N}$, $(r'_k)_{k\in N}$ it is obvious that $(s_k)_{k\geq 1}$,$(t_k)_{k\geq 1}$ exist and are unique for those sequences, that is, the sequences $(w''_k)_{k\in N}$,$(w'''_k)_{k\in N}$ are defined uniquely, up to multiplicative constant factors, by

$$w''_k = \frac{w''_0}{s_1 \cdot s_2 \dots s_k}$$

$$w'''_k = \frac{w'''_0}{t_1 \cdot t_2 \dots t_k}$$

(the unicity is based on the constant factors w''_0, w'''_0).

3.2 Random Environments

Let us consider the random walk on $\mathbb{Z}$, with transitions $k \to (k+1)$, $k \to (k-1)$ and $k \to k$ whose transition probabilities $(p_k)_{k\in\mathbb{Z}}$, $(r_k)_{k\in\mathbb{Z}}$ constitute stationary ergodic sequences. A realization of these stationary ergodic sequences is called a *random environment* for this random walk. In order to investigate the unique circuit and weight representation of this random walk in random environments, for almost every environment, let us consider a probability space $(\Omega, \mathscr{F}, \mu)$, a measure preserving ergodic automorphism of this space $m : \Omega \mapsto \Omega$ and the measurable functions $p : \Omega \mapsto (0, 1)$, $r : \Omega \mapsto (0, 1)$ such that every $\omega \in \Omega$ generates the random environment $p_k \equiv p(m^k\omega)$, $r_k \equiv r(m^k\omega)$, $k \in \mathbb{Z}$. Since m is measure preserving and ergodic, the sequences $(p_k)_{k\in\mathbb{Z}}$, $(r_k)_{k\in\mathbb{Z}}$ are stationary ergodic sequences of random variables.

Let also $S = \mathbb{Z}^N$ be the infinite product space with coordinates $(X_n)_{n\in N}$. Then we may define a family $(\mathbb{P}^\omega)_{\omega\in\Omega}$ of probability measures such that, for every $\omega \in \Omega$, the sequence $(X_n)_{n\in N}$ forms a Markov chain on $\mathbb{Z}$ whose elements of the corresponding Markov transition matrix are defined by

$$\mathbb{P}^\omega(X_0 = 0) = 1,$$

$$\mathbb{P}^\omega(X_{n+1} = k + 1/X_n = k) = p(m^k\omega),$$

$$\mathbb{P}^\omega(X_{n+1} = k/X_n = k) = r(m^k\omega),$$

$$\mathbb{P}^\omega(X_{n+1} = k - 1/X_n = k) = 1 - p(m^k\omega) - r(m^k\omega) \equiv q(m^k\omega), k \in \mathbb{Z},$$

as it is shown in Figure 3.

We have the following:

Proposition 3 *For μ almost every environment $\omega \in \Omega$ the chain $(X_n)_{n\in N}$ has a unique circuit and weight representation.*

Proof Following an analogous way of that given in Section 3.1, let us consider the set of directed circuits $c_k = (k, k + 1, k)$ and $c'_k = (k, k)$, for every $k \in \mathbb{Z}$, since only the transitions from k to $k + 1$, k to $k - 1$ and k to k are possible. There are three circuits through each point $k \in \mathbb{Z} : c_{k-1}, c_k$ and c'_k.

The problem we have to manage is the definition of the weights of the circuits. We may symbolize by $w_k(\omega)$ the weight of the circuit c_k and by $w'_k(\omega)$ the weight

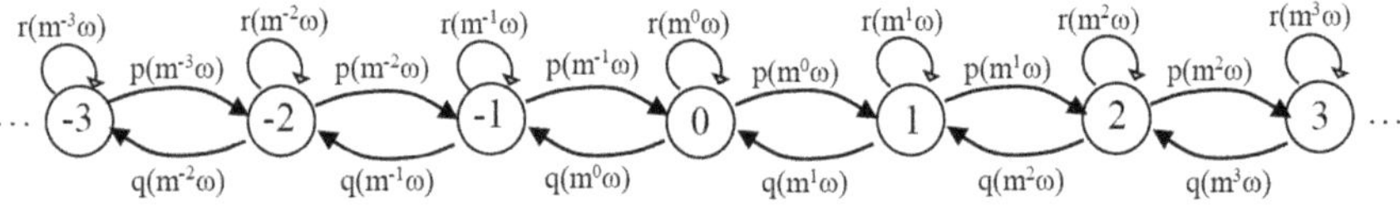

Fig. 3 The Markov chain $(X_n)_{n\in\mathbb{N}}$ (random environments)

of the circuit c'_k , for every $k \in \mathbb{Z}$. For the definition of weights let us consider the sequences $(b_k(\omega))_{k\in\mathbb{Z}}, (\gamma_k(\omega))_{k\in\mathbb{Z}}$ defined by

$$b_k(\omega) = \frac{w_k(\omega)}{w_{k-1}(\omega)}, \gamma_k(\omega) = \frac{w'_k(\omega)}{w'_{k-1}(\omega)}, k \in \mathbb{Z}.$$

As a consequence, we may have

$$b_k(\omega) = \frac{p(m^k\omega)}{1 - p(m^k\omega) - r(m^k\omega)} = \frac{p(m^k\omega)}{q(m^k\omega)} \equiv \frac{p}{q}(m^k\omega), \tag{3.3}$$

$$\gamma_k(\omega) = \frac{r(m^k\omega)}{r(m^{k-1}\omega)} \cdot \frac{p(m^{k-1}\omega)}{p(m^k\omega)} \cdot b_k(\omega), \text{ for every } k \in \mathbb{Z}. \tag{3.4}$$

Given the stationary ergodic sequences $(p_k)_{k\in\mathbb{Z}}$,$(r_k)_{k\in\mathbb{Z}}$, for which every $\omega \in \Omega$ generates the random environment $p_k \equiv p(m^k\omega), r_k \equiv r(m^k\omega), k \in \mathbb{Z}$, we have that the preceding equations (3.3), (3.4) give a unique definition of the sequences $(b_k(\omega))_{k\in\mathbb{Z}}, (\gamma_k(\omega))_{k\in\mathbb{Z}}$ for μ-almost every ω, by the ergodicity of m. Then the sequences of weights $(w_k(\omega))_{k\in\mathbb{Z}}$ and $(w'_k(\omega))_{k\in\mathbb{Z}}$ are defined uniquely by

$$w_k(\omega) = w_0(\omega)b_1(\omega) \cdot b_2(\omega) \dots b_k(\omega), k \in \mathbb{Z}^*_+,$$

$$w_k(\omega) = \frac{w_0(\omega)}{b_0(\omega) \cdot b_{-1}(\omega) \cdot b_{-2}(\omega) \cdots b_{k+1}(\omega)}, k \in \mathbb{Z}^*_-,$$

and

$$w'_k(\omega) = w'_0(\omega)\gamma_1(\omega) \cdot \gamma_2(\omega) \cdots \gamma_k(\omega), k \in \mathbb{Z}^*_+,$$

$$w'_k(\omega) = \frac{w'_0(\omega)}{\gamma_0(\omega) \cdot \gamma_{-1}(\omega) \cdot \gamma_{-2}(\omega) \cdots \gamma_{k+1}(\omega)}, k \in \mathbb{Z}^*_-.$$

(the unicity of the weight sequences $(w_k(\omega))_{k\in\mathbb{Z}}, (w'_k(\omega))_{k\in\mathbb{Z}}$ is understood up to the constant factors $w_0(\omega)$ and $w'_0(\omega)$).

Let us now introduce the "adjoint" random walk in random environment $(X'_n)_{n\in N}$. For every $\omega \in \Omega$ and for the family $(\mathbb{P}^\omega)_{\omega\in\Omega}$ of probability measures, the sequence $(X'_n)_{n\in N}$ is a Markov chain on $\mathbb{Z}$ whose elements of the corresponding Markov transition matrix are defined by

$$\mathbb{P}^\omega(X'_0 = 0) = 1,$$

$$\mathbb{P}^\omega(X'_{n+1} = k - 1/X'_n = k) = p(m^k\omega),$$

$$\mathbb{P}^\omega(X'_{n+1} = k/X'_n = k) = r(m^k\omega),$$

$$\mathbb{P}^\omega(X'_{n+1} = k + 1/X'_n = k) = 1 - p(m^k\omega) - r(m^k\omega) \equiv q(m^k\omega), k \in \mathbb{Z},$$

as it is shown in Figure 4.

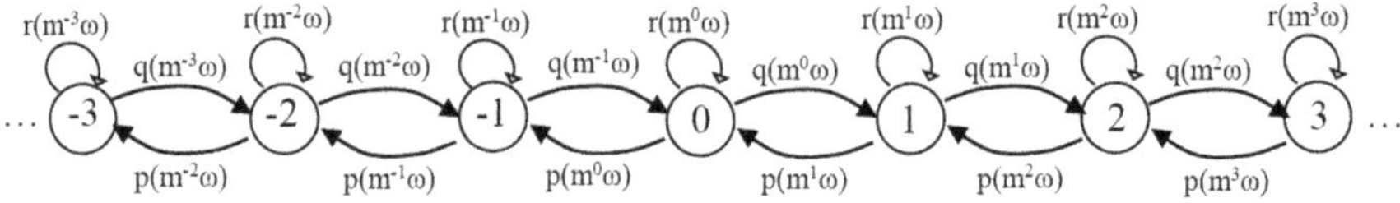

Fig. 4 The "adjoint" Markov chain $(X'_n)_{n\in\mathbb{N}}$ (random environments)

So we have the following:

Proposition 4 *For μ almost every environment $\omega \in \Omega$, the chain $(X'_n)_{n\in N}$ has a unique circuit and weight representation.*

Proof As in Proposition 3, the problem we have also to manage here is the definition of the weights of the circuits. To this direction we may denote by $w''_k(\omega)$ the weight of the circuit $c''_k = (k+1, k, k+1)$ and by $w'''_k(\omega)$ the weight of the circuit $c'''_k = (k, k)$, for every $k \in \mathbb{Z}$. By using an analogous way of that given before for the chain $(X_n)_{n\in N}$, let us consider the sequences $(\ell_k(\omega))_{k\in\mathbb{Z}}$, $(t_k(\omega))_{k\in\mathbb{Z}}$, defined by

$$\ell_k(\omega) = \frac{w''_{k-1}(\omega)}{w''_k(\omega)}, \qquad t_k(\omega) = \frac{w'''_{k-1}(\omega)}{w'''_k(\omega)},$$

such that

$$\ell_k(\omega) = \frac{p(m^k\omega)}{1 - p(m^k\omega) - r(m^k\omega)} = \frac{p(m^k\omega)}{q(m^k\omega)} \equiv \frac{p}{q}(m^k\omega), \tag{3.5}$$

$$t_k(\omega) = \frac{r(m^{k-1}\omega)}{r(m^k\omega)} \cdot \frac{1 - p(m^k\omega) - r(m^k\omega)}{1 - p(m^{k-1}\omega) - r(m^{k-1}\omega)} \cdot \ell_k(\omega), \tag{3.6}$$

for every $k \in \mathbb{Z}$.

Then the sequences of weights $(w''_k(\omega))_{k\in\mathbb{Z}}$, $(w'''_k(\omega))_{k\in\mathbb{Z}}$ are defined uniquely by

$$w''_k(\omega) = \frac{w''_0(\omega)}{\ell_1(\omega)\cdot\ell_2(\omega)\cdot\ell_3(\omega)\cdots\ell_k(\omega)}, \ k \in \mathbb{Z}^*_+,$$

$$w''_k(\omega) = w''_0(\omega)\cdot\ell_0(\omega)\cdot\ell_{-1}(\omega)\cdot\ell_{-2}(\omega)\cdots\ell_{k+3}(\omega)\cdot\ell_{k+2}(\omega)\cdot\ell_{k+1}(\omega), \quad k \in \mathbb{Z}^*_-$$

and

$$w'''_k(\omega) = \frac{w'''_0(\omega)}{t_1(\omega)\cdot t_2(\omega)\cdots t_k(\omega)}, \qquad k \in \mathbb{Z}^*_+$$

$$w'''_k(\omega) = w'''_0(\omega)t_0(\omega)\cdot t_{-1}(\omega)\cdot t_{-2}(\omega)\cdots t_{k+3}(\omega)\cdot t_{k+2}(\omega)\cdot t_{k+1}(\omega), \quad k \in \mathbb{Z}^*_-$$

(the unicity of the weight sequences $(w''_k(\omega))_{k\in\mathbb{Z}}$, $(w'''_k(\omega))_{k\in\mathbb{Z}}$ is understood up to the constant factors $w''_0(\omega)$, $w'''_0(\omega)$).

4 Recurrence and Transience

4.1 Fixed Environments

We have that for the chain $(X_n)_{n\in N}$ there is a unique invariant measure up to a multiplicative constant factor $\mu_k = w_{k-1} + w_k + w'_k$, $k \geq 1$, $\mu_0 = w_0 + w'_0$, while for the chain $(X'_n)_{n\in N}$, $\mu'_k = w''_{k-1} + w''_k + w'''_k$, $k \geq 1$ with $\mu'_0 = w''_0 + w'''_0$. In the case that an irreducible chain is recurrent there is only and only one invariant measure (finite or not), so we may obtain the following:

Proposition 5

(i) The chain $(X_n)_{n\in N}$ defined as above is positive recurrent if and only if

$$\sum_{k=1}^{\infty}(b_1 b_2 \ldots b_k) < +\infty \left(or \frac{1}{w_0} \cdot \sum_{k=1}^{\infty} w_k < +\infty \right),$$

$$\sum_{k=1}^{\infty}(\gamma_1 \cdot \gamma_2 \cdots \gamma_k) < +\infty \left(or \frac{1}{w'_0} \cdot \sum_{k=1}^{\infty} w'_k < +\infty \right).$$

(ii) The chain $(X'_n)_{n\in N}$ defined as above is positive recurrent if and only if

$$\sum_{k=1}^{\infty} \frac{1}{s_1 \cdot s_2 \cdots s_k} < +\infty \left(or \frac{1}{w''_0} \cdot \sum_{k=1}^{\infty} w''_k < +\infty \right),$$

$$\sum_{k=1}^{\infty} \frac{1}{t_1 \cdot t_2 \cdots t_k} < +\infty \left(or \frac{1}{w'''_0} \cdot \sum_{k=1}^{\infty} w'''_k < +\infty \right).$$

In order to obtain recurrence and transience criteria for the chains $(X_n)_{n\in N}$, $(X'_n)_{n\in N}$ we shall need the following proposition [11]:

Proposition 6 *Let us consider a Markov chain on $\mathcal{N}$ which is irreducible. Then if there exists a strictly increasing function that is harmonic on the complement of a finite interval and that is bounded, then the chain is transient. In the case that there exists such a function which is unbounded the chain is recurrent.*

Following this direction we shall use a well-known method-theorem based on the Foster-Kendall theorem ([11]) by considering the harmonic function $g = (g_k, k \geq 1)$. For the chain $(X_n)_{n\in N}$ this is a solution of

$$p_0 \cdot g_1 + r_0 \cdot g_0 = g_0,$$

$$p_k \cdot g_{k+1} + q_k \cdot g_{k-1} + r_k \cdot g_k = g_k, k \geq 1.$$

Since $\Delta g_k = g_k - g_{k-1}$, for every $k \geq 1$, we obtain that

$$p_k \cdot g_{k+1} + q_k \cdot g_k - q_k \cdot g_k + q_k \cdot g_{k-1} + r_k \cdot g_k = g_k$$

or

$$p_k \cdot (\Delta g_{k+1} + g_k) + q_k \cdot g_k - q_k \cdot g_k + q_k \cdot g_{k-1} + r_k \cdot g_k = g_k$$

or

$$p_k \cdot \Delta g_{k+1} + (p_k + q_k + r_k) \cdot g_k - q_k \cdot g_k + q_k \cdot g_{k-1} = g_k$$

or

$$p_k \cdot \Delta g_{k+1} - q_k \cdot (g_k - g_{k-1}) = 0$$

or

$$p_k \cdot \Delta g_{k+1} = q_k \cdot \Delta g_k.$$

If we put $\alpha_k = \dfrac{\Delta g_k}{\Delta g_{k+1}}$ we get $\alpha_k = \dfrac{p_k}{q_k}$ (with $p_k = 1 - q_k - r_k$), $k \geq 1$, which is the equation of the definition of the sequences $(s_k)_{k\geq 1}$ and $(t_k)_{k\geq 1}$ (as a multiplicative factor of the $(s_k)_{k\geq 1}$) for the chain $(X'_n)_{n\in N}$ such that $q'_k = q_k$, $r'_k = r_k$, for every $k \geq 1$. This means that the strictly increasing harmonic functions of the chain $(X_n)_{n\in N}$ are in correspondence with the weight representations of the chain $(X'_n)_{n\in N}$ such that

$$\begin{aligned}
q'_k &= P(X'_{n+1} = k+1/X'_n = k) = P(X_{n+1} = k-1/X_n = k) = q_k,\\
r'_k &= P(X'_{n+1} = k/X'_n = k) = P(X_{n+1} = k/X_n = k) = r_k, \qquad (4.1)\\
p'_k &= 1 - q'_k - r'_k = 1 - q_k - r_k = p_k, \qquad \text{for every } k \geq 1.
\end{aligned}$$

To express this kind of duality we shall call the chain $(X'_n)_{n\in N}$, the *adjoint* of the chain $(X_n)_{n\in N}$ and reciprocally in the case that the relation (4.1) is satisfied.

Equivalently for the chain $(X'_n)_{n\in N}$ the harmonic function $g' = (g'_k, k \geq 1)$ satisfies the equation

$$\begin{aligned}
&r'_0 \cdot g'_0 + q'_0 \cdot g'_1 = g'_0,\\
&q'_k \cdot g'_{k+1} + p'_k \cdot g'_{k-1} + r'_k \cdot g'_k = g'_k, k \geq 1.
\end{aligned}$$

Since $\Delta g'_k = g'_k - g'_{k-1}$, for every $k \geq 1$, we have that

$$q'_k \cdot (\Delta g'_{k+1} + g'_k) + p'_k \cdot g'_k - p'_k \cdot g'_k + p'_k \cdot g'_{k-1} + r'_k \cdot g'_k = g'_k$$

or

$$(p_k' + q_k' + r_k') \cdot g_k' + q_k' \cdot \Delta g_{k+1}' - p_k' \cdot g_k' + p_k' \cdot g_{k-1}' = g_k'$$

or

$$q_k' \cdot \Delta g_{k+1}' = p_k' \cdot (g_k' - g_{k-1}') = p_k' \cdot \Delta g_k'.$$

If we put $\beta_k = \frac{\Delta g_{k+1}'}{\Delta g_k'}$ we get $\beta_k = \frac{p_k'}{q_k'}$ (with $q_k' = 1 - p_k' - r_k'$),$k \geq 1$, which is the equation of the definition of the sequences $(b_k)_{k\geq 1}$ and $(\gamma_k)_{k\geq 1}$ (as a multiplicative factor of the $(b_k)_{k\geq 1}$) for the chain $(X_n)_{n\in N}$ such that $p_k' = p_k, r_k' = r_k$ for every $k \geq 1$. By considering a similar approximation of that given before for the chain $(X_n)_{n\in N}$ we may say that the strictly increasing harmonic functions of the chain $(X_n')_{n\in N}$ are in correspondence with the weight representations of the chain $(X_n)_{n\in N}$ such that equivalent equations of (4.1) are satisfied.

So we may have the following:

Proposition 7 *The chain $(X_n)_{n\in N}$ defined as above is transient if and only if the adjoint chain $(X_n')_{n\in N}$ is positive recurrent and reciprocal. Moreover the adjoint chains $(X_n)_{n\in N}, (X_n')_{n\in N}$ are null recurrent simultaneously.*

In particular

(i) The chain $(X_n)_{n\in N}$ defined as above is transient if and only if

$$\frac{1}{w_0''} \cdot \sum_{k=1}^{\infty} w_k'' < +\infty \text{ and } \frac{1}{w_0'''} \cdot \sum_{k=1}^{\infty} w_k''' < +\infty.$$

(ii) The chain $(X_n')_{n\in N}$ defined as above is transient if and only if

$$\frac{1}{w_0} \cdot \sum_{k=1}^{\infty} w_k < +\infty \text{ and } \frac{1}{w_0'} \cdot \sum_{k=1}^{\infty} w_k' < +\infty.$$

(iii) The adjoint chains $(X_n)_{n\in N}, (X_n')_{n\in N}$ are null recurrent if

$$\frac{1}{w_0} \cdot \sum_{k=1}^{\infty} w_k = \frac{1}{w_0'} \cdot \sum_{k=1}^{\infty} w_k' = +\infty \text{ and } \frac{1}{w_0''} \cdot \sum_{k=1}^{\infty} w_k'' = \frac{1}{w_0'''} \cdot \sum_{k=1}^{\infty} w_k''' = +\infty.$$

Proof The proof of Proposition 7 is an application mainly of Proposition 6 as well as of Proposition 5.

4.2 Random Environments

Regarding the criteria of recurrence and transience in the case of fixed environments, we have already proved that the behaviors of recurrence and transience for the "adjoint" chains $(X_n)_{n\in N}$,$(X'_n)_{n\in N}$ are tied together and depend on the convergence or not of the series

$$\sum_{k=1}^{+\infty} w_k, \sum_{k=1}^{+\infty} w'_k, \sum_{k=1}^{+\infty} w''_k \text{ and } \sum_{k=1}^{+\infty} w'''_k.$$

In the case of random environments the recurrence and transience are properties which are true for μ almost every environment $\omega \in \Omega$ or for μ almost no environment, because the system $(\Omega, \mathcal{F}, \mu, m)$ is supposed to be ergodic. This is true in general for a random walk in a random environment which is irreducible.

In order to investigate suitable criteria for the transience and recurrence of the corresponding uniquely defined circuit chains describing the above-mentioned random walk with jumps in a random environment, we may use the criteria given in the study for fixed environments for the chains $(X_n)_{n\in N}$, $(X'_n)_{n\in N}$ restricted to the half-lines $[i, +\infty)$ with reflection in i. According to the criterion in the case that

$$\sum_{k=1}^{+\infty} w_k(\omega) < +\infty \qquad \text{and} \qquad \sum_{k=1}^{+\infty} w'_k(\omega) < +\infty, \quad \mu - a.e.$$

we have that the restricted chain $(X_n)_{n\in N}$ is positive recurrent on $[i, +\infty)$, while the restricted "adjoint" chain $(X'_n)_{n\in N}$ is transient on $[i, +\infty)$, since it is known that the chain $(X_n)_{n\in N}$ defined as above is positive recurrent if and only if its "adjoint" chain $(X'_n)_{n\in N}$ is transient and reciprocal. An analogous result is obtained in the case of the half-lines $(-\propto, j]$ with reflection in j.

Therefore we have the following:

Proposition 8 *The random walk $(X_n)_{n\in N}$ in random environments defined as above is transient, for $\mu - a.e.$ environment $\omega \in \Omega$, if and only if its "adjoint" random walk $(X'_n)_{n\in N}$ is positive recurrent and reciprocal. Moreover the adjoint random walks $(X_n)_{n\in N}$ and $(X'_n)_{n\in N}$ are null recurrent simultaneously.*

Proof Taking into account the Birkoff's ergodic theorem for the sequences $(b_k(\omega))_{k\in\mathbb{Z}}$, $(\gamma_k(\omega))_{k\in\mathbb{Z}}$ for μ-almost every ω (see relations (3.3), (3.4)), we may write

$$w_k(\omega) = w_0(\omega) \prod_{d=1}^{k} b_d(\omega) \sim e^{kc}, \; k \in \mathbb{Z}^*_+,$$

$$w_k(\omega) = w_0(\omega) \left[\prod_{d=0}^{-(k+1)} b_{-d}(\omega) \right]^{-1} \sim e^{-kc}, \; k \in \mathbb{Z}^*_-,$$

$$w'_k(\omega) = w'_0(\omega) \prod_{d=1}^{k} \gamma_d(\omega) \sim e^{kc},\ k \in \mathbb{Z}_+^*,$$

$$w'_k(\omega) = w'_0(\omega) \left[\prod_{d=0}^{-(k+1)} \gamma_{-d}(\omega) \right]^{-1} \sim e^{-kc},\ k \in \mathbb{Z}_-^*$$

for the sequences of weights $(w_k(\omega))_{k\in\mathbb{Z}}$, $(w'_k(\omega))_{k\in\mathbb{Z}}$ of the chain $(X_n)_{n\in N}$. Following an analogous way for the "adjoint" chain $(X'_n)_{n\in N}$ we have

$$w''_k(\omega) = w''_0(\omega) \left[\prod_{d=1}^{k} \ell_d(\omega) \right]^{-1} \sim e^{-kc},\ k \in \mathbb{Z}_+^*,$$

$$w''_k(\omega) = w''_0(\omega) \prod_{d=0}^{-(k+1)} \ell_{-d}(\omega) \sim e^{kc},\ k \in \mathbb{Z}_-^*,$$

$$w'''_k(\omega) = w'''_0(\omega) \left[\prod_{d=1}^{k} t_d(\omega) \right]^{-1} \sim e^{-kc},\ k \in \mathbb{Z}_+^*,$$

$$w'''_k(\omega) = w'''_0(\omega) \prod_{d=0}^{-(k+1)} t_{-d}(\omega) \sim e^{kc},\ k \in \mathbb{Z}_-^*,$$

for the sequences of weights $(w''_k(\omega))_{k\in\mathbb{Z}}$, $(w'''_k(\omega))_{k\in\mathbb{Z}}$, of the chain $(X'_n)_{n\in N}$. We take into account the following cases:

(i) $c < 0$. We get

$$\sum_{k=1}^{+\infty} w_k(\omega) < +\infty,\ \sum_{k=-\infty}^{0} w_k(\omega) < +\infty, \sum_{k=1}^{+\infty} w'_k(\omega) < +\infty,\ \sum_{k=-\infty}^{0} w'_k(\omega) < +\infty,$$

$$\sum_{k=1}^{+\infty} w''_k(\omega) = +\infty,\ \sum_{k=-\infty}^{0} w''_k(\omega) = +\infty, \sum_{k=1}^{+\infty} w'''_k(\omega) = +\infty,\ \sum_{k=-\infty}^{0} w'''_k(\omega) = +\infty.$$

By using the criterion given in subsection 4.1 for the chains $(X_n)_{n\in N}$ and $(X'_n)_{n\in N}$ restricted

(a) to the half-lines $[i, +\propto)$ with reflection in i, we have that the restricted chain $(X_n)_{n\in N}$ is positive recurrent on $[i, +\propto)$, while the restricted chain $(X'_n)_{n\in N}$ is transient,
(b) to the half-lines $(-\propto, j]$ with reflection in j, we have also that the restricted chain $(X_n)_{n\in N}$ is positive recurrent on $(-\propto, j]$, while its adjoint chain $(X'_n)_{n\in N}$ is transient.

(ii) $c > 0$. We get

$$\sum_{k=1}^{+\infty} w_k(\omega) = +\infty, \sum_{k=-\infty}^{0} w_k(\omega) = +\infty, \sum_{k=1}^{+\infty} w'_k(\omega) = +\infty, \sum_{k=-\infty}^{0} w'_k(\omega) = +\infty,$$

$$\sum_{k=1}^{+\infty} w''_k(\omega) < +\infty, \sum_{k=-\infty}^{0} w''_k(\omega) < +\infty, \sum_{k=1}^{+\infty} w'''_k(\omega) < +\infty, \sum_{k=-\infty}^{0} w'''_k(\omega) < +\infty.$$

Regarding the criterion given in subsection 4.1 for the chains $(X_n)_{n\in N}$ and $(X'_n)_{n\in N}$ restricted

(a) to the half-lines $[i, +\propto)$ with reflection in i, we have that the restricted chain $(X_n)_{n\in N}$ is transient on $[i, +\propto)$, while the restricted chain $(X'_n)_{n\in N}$ is positive recurrent,
(b) to the half-lines $(-\propto, j]$ with reflection in j, we have also that the restricted chain $(X_n)_{n\in N}$ is transient on $(-\propto, j]$, while its adjoint chain $(X'_n)_{n\in N}$ is positive recurrent.

(iii) $c = 0$. Regarding the ergodic theorem, it is well-known that the averages $\frac{1}{k}\sum_{n=0}^{k-1}(\text{fom}^n)$ take infinitely many values greater than the limit and infinitely many values smaller than the limit. This means that in the sequences of weights

$$(w_k(\omega))_{k\in\mathbb{Z}}, (w'_k(\omega))_{k\in\mathbb{Z}}, (w''_k(\omega))_{k\in\mathbb{Z}}, (w'''_k(\omega))_{k\in\mathbb{Z}},$$

for a.e. $\omega \in \Omega$, infinitely many values in both directions are greater than 1. As a consequence, we may have that

$$\sum_{k=1}^{+\infty} w_k(\omega) = +\infty, \sum_{k=-\infty}^{0} w_k(\omega) = +\infty, \sum_{k=1}^{+\infty} w'_k(\omega) = +\infty, \sum_{k=-\infty}^{0} w'_k(\omega) = +\infty,$$

$$\sum_{k=1}^{+\infty} w''_k(\omega) = +\infty, \sum_{k=-\infty}^{0} w''_k(\omega) = +\infty, \sum_{k=1}^{+\infty} w'''_k(\omega) = +\infty, \sum_{k=-\infty}^{0} w'''_k(\omega) = +\infty.$$

By using the criterion of null recurrence for the chains $(X_n)_{n\in N}$, $(X'_n)_{n\in N}$ restricted to the half-lines $[i, +\infty)$ and $(-\propto, j]$ with reflection in i, j respectively, in the case of fixed environments, we may have also that both chains are null recurrent on $\mathbb{Z}$, for $\mu - a.e. \omega \in \Omega$.

References

1. Yv. Derriennic, Random walks with jumps in random environments (examples of cycle and weight representations), in *Probability Theory and Mathematical Statistics: Proceedings of the 7th Vilnius Conference 1998*, ed. by B. Grigelionis, J. Kubilius, V. Paulauskas, V.A. Statulevicius, and H. Pragarauskas (VSP, Vilnius, 1999), pp. 199–212
2. W. Feller, An Introduction to Probability Theory and Its Applications, vol. I, 3rd edn. (Wiley, Hoboken, NJ, 1968)
3. W. Feller, An Introduction to Probability Theory and Its Applications, vol. II, 2nd edn. (Wiley, Hoboken, NJ, 1971)
4. Ch. Ganatsiou, On cycle representations of random walks in fixed, random environments, in *Proceedings of the 58th World Congress of the International Statistical Institute*, Dublin, 21–26 August 2011
5. Ch. Ganatsiou, On the study of transience and recurrence of the Markov chain defined by directed weighted circuits associated with a random walk in fixed environment. J. Prob. Stat. **2013** (2013), Article ID 424601, 5 pp.
6. Ch. Ganatsiou, On the study of transience and recurrence of circuit chains associated with a random walk with jumps in random environments, in *Proceedings of the First Congress of Greek Mathematicians (FCGM-2018), Hellenic Mathematical Society (HMS)*, Special Session in Probability and Statistics, Athens, 25–30 June 2018
7. T.E. Harris, First passage and recurrence distributions. Trans. Am. Math. Soc. **73**, 471–486 (1952)
8. B.D. Hughes, *Random Walks and Random Environments, Volume 1: Random Walks* (Oxford University Press, Oxford, 1995)
9. M. Iosifescu, P. Tăutu, *Stochastic Processes and Applications in Biology and Medicine I*, 1st edn. (Springer, Berlin/Heidelberg/New York, 1973)
10. S. Kalpazidou, *Cycle Representations of Markov Processes* (Springer, New York, NY, 1995)
11. S. Karlin, H. Taylor, *A First Course in Stochastic Processes* (Academic, New York, NY, 1975)
12. J. MacQueen, Circuit processes. Ann. Probab. **9**, 604–610 (1981)
13. Q. Minping, Q. Min, Circulation for recurrent Markov chains. Z.Wahrsch. Verw. Gebiete **59**(2), 203–210 (1982)
14. K. Pearson, The problem of the random walk, Nature **72**, 294 (1905)
15. F. Spitzer, *Principles of Random Walks*, 2nd edn. (Spinger, New York, NY, 1976)
16. A.H. Zemanian, *Infinite Electrical Networks* (Cambridge University Press, Cambridge, 1991)

On Selections of Some Generalized Set-Valued Inclusions

Bahman Hayati, Hamid Khodaei, and Themistocles M. Rassias

Abstract We present some results on the existence of a unique selection of a set-valued function satisfying some generalized set-valued inclusions.

1 Introduction

For a nonempty set Y we denote by $\mathfrak{F}_0(Y)$ the family of all nonempty subsets of Y. In a linear normed space Y we define the following families of sets:

$$ccl(Y) := \{A \in \mathfrak{F}_0(Y) : \; A \text{ is closed and convex set}\},$$

$$cclz(Y) := \{A \in \mathfrak{F}_0(Y) : \; A \text{ is closed and convex set containing } 0\},$$

$$ccz(Y) := \{A \in \mathfrak{F}_0(Y) : \; A \text{ is compact and convex set containing } 0\}.$$

The diameter of a set $A \in \mathfrak{F}_0(Y)$ is defined by

$$\delta(A) := \sup\{\| a - b \| : \; a, b \in A\}.$$

Let K be a nonempty set. We say that a set-valued function $F : K \to \mathfrak{F}_0(Y)$ is with bounded diameter if the function $K \ni x \mapsto \delta(F(x)) \in \mathbb{R}$ is bounded. Finally recall that a selection of a set-valued map $F : K \to \mathfrak{F}_0(Y)$ is a single-valued map $f : K \to Y$ with the property $f(x) \in F(x)$ for all $x \in K$.

B. Hayati · H. Khodaei (✉)
Department of Mathematics, Malayer University, Malayer, Iran
e-mail: hayati@malayeru.ac.ir; hkhodaei@malayeru.ac.ir

T. M. Rassias
Department of Mathematics, Zografou Campus, National Technical University of Athens, Athens, Greece
e-mail: trassias@math.ntua.gr

T. M. Rassias, P. M. Pardalos (eds.), *Mathematical Analysis and Applications*, Springer Optimization and Its Applications 154,
https://doi.org/10.1007/978-3-030-31339-5_7

Smajdor [1] and Gajda and Ger [2] proved that if $(S, +)$ is a commutative semigroup with zero and Y is a real Banach space, then $F : S \to ccl(Y)$ is a subadditive set-valued function; i.e.,

$$F(x + y) \subset F(x) + F(y), \quad x, y \in S,$$

with bounded diameter admits a unique additive selection (i.e., a unique mapping $f : S \to Y$ such that $f(x + y) = f(x) + f(y)$ and $f(x) \in F(x)$ for all $x, y \in S$). In 2001, Popa [3] proved that if $K \neq \emptyset$ is a convex cone in a real vector space X (i.e., $sK + tK \subseteq K$ for all $s, t \geq 0$) and $F : K \to ccl(Y)$ (where Y is a real Banach space) is a set-valued function with bounded diameter fulfilling the inclusion

$$F(\alpha x + \beta y) \subset \alpha F(x) + \beta F(y), \quad x, y \in K,$$

for $\alpha, \beta > 0$, $\alpha + \beta \neq 1$, then there exists exactly one additive selection of F.

Set-valued functional equations have been investigated by a number of authors and there are many interesting results concerning this problem (see [4–14]).

We determine the conditions for which a set-valued function $F : K \to \mathfrak{F}_0(Y)$ satisfying one of the following inclusions

$$\sigma_{y,z} F(\alpha x) + 8\alpha^{-1} F(x) \subseteq 2\alpha^{-1} \left(\sigma_y F(x) + \sigma_z F(x)\right) + 4\alpha F(x),$$

$$\sigma_{y,z} F(\alpha x) + 8F(x) \subseteq 2 \left(\sigma_y F(x) + \sigma_z F(x)\right) + 4\alpha^2 F(x),$$

$$\sigma_{y,z} F(\alpha x) + 8\alpha F(x) \subseteq 2\alpha \left(\sigma_y F(x) + \sigma_z F(x)\right) + 4\alpha^3 F(x),$$

$$\begin{aligned}\sigma_{y,z} F(\alpha x) + 4\alpha^2 (2F(x) + F(y) + F(z)) \subseteq{}& 2\alpha^2 \left(\sigma_y F(x) + \sigma_z F(x)\right) \\ &+ 2\sigma_z F(y) + 4\alpha^4 F(x) \end{aligned} \tag{1}$$

for all $x, y, z \in K$ and any fixed positive integers $\alpha > 1$ admits a unique selection satisfying the corresponding functional equation. Here $\sigma_y F(x)$ denotes $\sigma_y F(x) = F(x + y) + F(x - y)$, and $\sigma_{y,z} F(x)$ denotes $\sigma_{y,z} F(x) = \sigma_z \left(\sigma_y F(x)\right) = \sigma_z F(x + y) + \sigma_z F(x - y)$.

2 Selections of Set-Valued Mappings

In what follows we give some notations and present results which will be used in the sequel.

Definition 1 Let X be a real vector space. For $A, B \in \mathfrak{F}_0(X)$, the (Minkowski) addition is defined as

$$A + B = \{a + b : a \in A, b \in B\}$$

and the scalar multiplication as

$$\lambda A = \{\lambda a : a \in A\}$$

for $\lambda \in \mathbb{R}$.

Lemma 1 (*Nikodem* [15]) *Let X be a real vector space and let λ, μ be real numbers. If $A, B \in \mathfrak{F}_0(X)$, then*

$$\lambda(A + B) = \lambda A + \lambda B,$$

$$(\lambda + \mu)A \subseteq \lambda A + \mu A.$$

In particular, if A is convex and $\lambda\mu \geq 0$, then

$$(\lambda + \mu)A = \lambda A + \mu A.$$

Lemma 2 (Rådström's Cancelation Law) *Let Y be a real normed space and $A, B, C \in \mathfrak{F}_0(Y)$. Suppose that $B \in ccl(Y)$ and C is bounded. If $A + C \subseteq B + C$, then $A \subseteq B$.*

The above law has been formulated by Rådström [16], but the proof given there is valid in topological vector spaces (see [17, 18]).

Corollary 1 *Let Y be a real normed space and $A, B, C \in \mathfrak{F}_0(Y)$. Assume that $A, B \in ccl(Y)$, C is bounded, and $A + C = B + C$. Then $A = B$.*

Nikodem and Popa in [9] and Piszczek in [12] proved the following theorem.

Theorem 1 *Let K be a convex cone in a real vector space X, Y a real Banach space and $\alpha, \beta, p, q > 0$. Consider a set-valued function $F : K \to ccl(Y)$ with bounded diameter fulfilling the inclusion*

$$F(\alpha x + \beta y) \subset pF(x) + qF(y), \quad x, y \in K.$$

If $\alpha + \beta < 1$, then there exists a unique selection $f : K \to Y$ of F satisfying the equation

$$f(\alpha x + \beta y) = pf(x) + qf(y), \quad x, y \in K.$$

If $\alpha + \beta > 1$, then F is single valued.

The case of $p+q=1$ was investigated by Popa in [14], Inoan and Popa in [5]. By means of the inclusion relation, Park et al. [7, 11] investigated the approximation of some set-valued functional equations.

We now present some examples. A constant function $F : K \to ccl(Y)$, $F(x) = M$ for $x \in K$, where $K \subseteq X$ is a cone and $M \in ccl(Y)$ is fixed, satisfies the equation

$$F(\alpha x + \beta y) = pF(x) + qF(y), \quad x, y \in K,$$

and each constant function $f : K \to Y$, $f(x) = m$ for $x \in K$, where $m \in M$ is fixed, satisfies

$$f(\alpha x + \beta y) = pf(x) + qf(y), \quad x, y \in K.$$

The set-valued function $F : \mathbb{R} \to ccl(\mathbb{R})$ given by

$$F(x) = [x-1, x+1], \quad x \in \mathbb{R},$$

satisfies the equation

$$F\left(\frac{x+y}{2}\right) = \frac{F(x)+F(y)}{2}, \quad x, y \in \mathbb{R},$$

and each function $f : \mathbb{R} \to \mathbb{R}$,

$$f(x) = x + c, \quad x \in \mathbb{R},$$

where $c \in [-1, 1]$ is fixed, is a selection of F and satisfies

$$f\left(\frac{x+y}{2}\right) = \frac{f(x)+f(y)}{2}, \quad x, y \in \mathbb{R}.$$

In the rest of this paper, unless otherwise explicitly stated, we will assume that $(K, +)$ is a commutative group, Y is a real Banach space, and k is a positive integer less than or equal to 3.

Theorem 2 *Let $F : K \to cclz(Y)$ be a set-valued function with bounded diameter.*

(1) *If*

$$\alpha^{2-k}\sigma_{y,z}F(\alpha x) + 8F(x) \subseteq 2\left(\sigma_y F(x) + \sigma_z F(x)\right) + 4\alpha^2 F(x), \qquad (2)$$

for all $x, y, z \in K$, then there exists a unique selection $f : K \to Y$ of F such that, for all $x, y \in K$, (i) $f(x+y) = f(x) + f(y)$ if $k = 1$; (ii) $\sigma_y f(x) = 2f(x) + 2f(y)$ if $k = 2$; (iii) $\sigma_y f(2x) = 2\sigma_y f(x) + 12 f(x)$ if $k = 3$.

(2) *If*

$$2\left(\sigma_y F(x) + \sigma_z F(x)\right) + 4\alpha^2 F(x) \subseteq \alpha^{2-k}\sigma_{y,z} F(\alpha x) + 8F(x) \qquad (3)$$

for all $x, y, z \in K$*, then* F *is single-valued.*

Proof

(1) Letting $y = z = 0$ in (2), we have

$$\alpha^{2-k}\left(F(\alpha x) + F(\alpha x) + F(\alpha x) + F(\alpha x)\right) + 8F(x)$$
$$\subseteq 2\left(F(x) + F(x) + F(x) + F(x)\right) + 4\alpha^2 F(x)$$

for all $x \in K$. Since the set $F(x)$ is convex, we can conclude from Lemma 1 that

$$4\alpha^{2-k}F(\alpha x) + 8F(x) \subseteq 8F(x) + 4\alpha^2 F(x)$$

for all $x \in K$. Using the Rådström's cancelation law, one obtains

$$F(\alpha x) \subseteq \alpha^k F(x)$$

for all $x \in K$. Replacing x by $\alpha^n x$, $n \in \mathbb{N}$, in the last inclusion, we see that

$$\alpha^{-k(n+1)}F(\alpha^{n+1}x) \subseteq \alpha^{-kn}F(\alpha^n x)$$

for all $x \in K$. Thus $\left(\alpha^{-kn}F(\alpha^n x)\right)_{n\in\mathbb{N}_0}$ is a decreasing sequence of closed subsets of the Banach space Y. We also get

$$\delta\left(\alpha^{-kn}F(\alpha^n x)\right) = \alpha^{-kn}\delta\left(F(\alpha^n x)\right)$$

for all $x \in K$. Now since $\sup_{x\in K}\delta\left(F(x)\right) < +\infty$, we get that

$$\lim_{n\to+\infty}\delta\left(\alpha^{-kn}F(\alpha^n x)\right) = 0$$

for all $x \in K$. Hence

$$\lim_{n\to+\infty}\alpha^{-kn}F(\alpha^n x) = \bigcap_{n\in\mathbb{N}_0}\alpha^{-kn}F(\alpha^n x) =: f(x)$$

is a singleton. Thus we obtain a function $f : K \to Y$ which is a selection of F.

We will now prove that f for $m = 1$, 2, and 3 is additive, quadratic, and cubic, respectively. We have

$$\alpha^{2-k(n+1)}\sigma_{\alpha^n y,\alpha^n z}F(\alpha^{n+1}x) + 8\alpha^{-kn}F(\alpha^n x)$$
$$\subseteq 2\alpha^{-kn}\left(\sigma_{\alpha^n y}F(\alpha^n x) + \sigma_{\alpha^n z}F(\alpha^n x)\right) + 4\alpha^{-kn+2}F(\alpha^n x)$$

for all $x, y, z \in K$ and $n \in \mathbb{N}$. By the definition of f, we get

$$\begin{aligned}
&\alpha^{2-k}\sigma_{y,z}f(\alpha x) + 8f(x)\\
&\quad = \alpha^{2-k}\sigma_{\alpha^n y,\alpha^n z}\textstyle\bigcap_{n\in\mathbb{N}_0}\alpha^{-kn}F(\alpha^{n+1}x) + 8\bigcap_{n\in\mathbb{N}_0}\alpha^{-kn}F(\alpha^n x)\\
&\quad = \textstyle\bigcap_{n\in\mathbb{N}_0}\left(\alpha^{2-k(n+1)}\sigma_{\alpha^n y,\alpha^n z}F(\alpha^{n+1}x) + 8\alpha^{-kn}F(\alpha^n x)\right)\\
&\quad \subseteq \textstyle\bigcap_{n\in\mathbb{N}_0}\left(2\alpha^{-kn}\left(\sigma_{\alpha^n y}F(\alpha^n x) + \sigma_{\alpha^n z}F(\alpha^n x)\right) + 4\alpha^{-kn+2}F(\alpha^n x)\right)
\end{aligned}$$

for all $x, y, z \in K$. Thus we obtain

$$\begin{aligned}
&\left\|\alpha^{2-k}\sigma_{y,z}f(\alpha x) + 8f(x) - 2\sigma_y f(x) - 2\sigma_z f(x) - 4\alpha^2 f(x)\right\|\\
&\quad \le \delta\left(2\alpha^{-kn}\left(\sigma_{\alpha^n y}F(\alpha^n x) + \sigma_{\alpha^n z}F(\alpha^n x)\right) + 4\alpha^{-kn+2}F(\alpha^n x)\right)\\
&\quad = 2\delta\left(\alpha^{-kn}\sigma_{\alpha^n y}F(\alpha^n x)\right) + 2\delta\left(\alpha^{-kn}\sigma_{\alpha^n z}F(\alpha^n x)\right) + 4\alpha^2\delta\left(\alpha^{-kn}F(\alpha^n x)\right)
\end{aligned}$$

which tends to zero as n tends to ∞. Thus

$$\alpha^{2-k}\sigma_{y,z}f(\alpha x) = 2\left(\sigma_y f(x) + \sigma_z f(x)\right) + 4\left(\alpha^2 - 2\right)f(x) \tag{4}$$

for all $x, y, z \in K$. Setting $x = y = z = 0$ in (4), we have $f(0) = 0$. Putting $y = 0$ in (4) and using $f(0) = 0$, one gets

$$\alpha^{2-k}\sigma_z f(\alpha x) = \sigma_z f(x) + 2\left(\alpha^2 - 1\right)f(x)$$

for all $x, z \in K$. Based on Theorem 2.1 of [19] (also see [20, 21]), we conclude that, for all $x, y \in K$, if $k = 1$, then $f(x + y) = f(x) + f(y)$, if $k = 2$, then $\sigma_y f(x) = 2f(x) + 2f(y)$ and if $k = 3$, then $\sigma_y f(2x) = 2\sigma_y f(x) + 12f(x)$.

Next, let us prove the uniqueness of f. Suppose that f and g are selections of F. We have $(kn)^k f(x) = f(knx) \in F(knx)$ and $(kn)^k g(x) = g(knx) \in F(knx)$ for all $x \in K$ and $n \in \mathbb{N}$. Then we get

$$\begin{aligned}
(kn)^k\|f(x) - g(x)\| &= \left\|(kn)^k f(x) - (kn)^k g(x)\right\|\\
&= \|f(knx) - g(knx)\|\\
&\le 2\delta\left(F(knx)\right)
\end{aligned}$$

for all $x \in K$ and $n \in \mathbb{N}$. It follows from $\sup_{x \in K} \delta(F(x)) < +\infty$ that $f(x) = g(x)$ for all $x \in K$.

(2) Letting $y = z = 0$ in (3) and using the Rådström's cancelation law, one gets $F(x) \subseteq \alpha^{-k} F(\alpha x)$ for all $x \in K$. Hence,

$$F(x) \subseteq \alpha^{-kn} F(\alpha^n x) \subseteq \alpha^{-k(n+1)} F(\alpha^{n+1} x)$$

for all $x \in K$. It follows that $\left(\alpha^{-kn} F(\alpha^n x)\right)_{n \in \mathbb{N}_0}$ is an increasing sequence of sets in the Banach space Y. It follows from $\sup_{x \in K} \delta(F(x)) < +\infty$ that

$$\lim_{n \to +\infty} \delta\left(\alpha^{-kn} F(\alpha^n x)\right) = \lim_{n \to +\infty} \alpha^{-kn} \delta\left(F(\alpha^n x)\right) = 0$$

for all $x \in K$. Then, for all $n \in \mathbb{N}_0$ and $x \in K$, $\alpha^{-kn} F(\alpha^n x)$ is single-valued and

$$\alpha^{2-k} \sigma_{y,z} F(\alpha x) = 2\left(\sigma_y F(x) + \sigma_z F(x)\right) + 4\left(\alpha^2 - 2\right) F(x)$$

for all $x, y, z \in K$. By adopting the method used in case (1), we see that, for all $x, y \in K$, if $k = 1$, then $F(x + y) = F(x) + F(y)$, if $k = 2$, then $\sigma_y F(x) = 2F(x) + 2F(y)$ and if $k = 3$, then $\sigma_y F(2x) = 2\sigma_y F(x) + 12F(x)$.

Theorem 3 *Let $F : K \to cclz(Y)$ be a set-valued function with bounded diameter.*

(1) *If F satisfies the inclusion* (1), *then there exists a unique selection $f : K \to Y$ of F such that $\sigma_y f(2x) = 4\sigma_y f(x) + 24 f(x) - 6 f(y)$ for all $x, y \in K$.*

(2) *If*

$$\begin{aligned} &2\alpha^2\left(\sigma_y F(x) + \sigma_z F(x)\right) + 2\sigma_z F(y) + 4\alpha^4 F(x) \\ &\subseteq \sigma_{y,z} F(\alpha x) + 4\alpha^2\left(2F(x) + F(y) + F(z)\right) \end{aligned} \tag{5}$$

for all $x, y, z \in K$, then F is single-valued.

Proof

(1) Letting $y = z = 0$ in (1), we have

$$\begin{aligned} &F(\alpha x) + F(\alpha x) + F(\alpha x) + F(\alpha x) + 4\alpha^2\left(2F(x) + F(0) + F(0)\right) \\ &\subseteq 2\alpha^2\left(F(x) + F(x) + F(x) + F(x)\right) + 2\left(F(0) + F(0)\right) + 4\alpha^4 F(x) \end{aligned}$$

for all $x \in K$. Hence, from the convexity of $F(x)$ and Lemma 1, we see from that

$$F(\alpha x) + 2\alpha^2 F(x) + 2\alpha^2 F(0) \subseteq 2\alpha^2 F(x) + F(0) + \alpha^4 F(x) \tag{6}$$

for all $x \in K$. Setting $x = 0$ in (6), we have

$$\left(4\alpha^2 + 1\right) F(0) \subseteq \left(\alpha^4 + 2\alpha^2 + 1\right) F(0),$$

and using the Rådström's cancelation law, one obtains

$$\{0\} \subseteq F(0). \tag{7}$$

Again applying (6) and the Rådström's cancelation law, one gets

$$F(\alpha x) + (2\alpha^2 - 1)F(0) \subseteq \alpha^4 F(x) \tag{8}$$

for all $x \in K$. It follows from (7) and (8) that

$$F(\alpha x) \subseteq F(\alpha x) + (2\alpha^2 - 1)F(0) \subseteq \alpha^4 F(x)$$

for all $x \in K$. Hence

$$\alpha^{-4(n+1)} F(\alpha^{n+1} x) \subseteq \alpha^{-4n} F(\alpha^n x)$$

for all $x \in K$. In the same way as in Theorem 2, we obtain a function $f : K \to Y$ which is a selection of F and

$$\sigma_{y,z} f(\alpha x) = 2\alpha^2 \left(\sigma_y f(x) + \sigma_z f(x) + 2\left(\alpha^2 - 2\right) f(x)\right)$$
$$+2\left(\sigma_z f(y) - 2\left(\alpha^2\right) f(y)\right) - 4\alpha^2 f(z) \tag{9}$$

for all $x, y, z \in K$. Setting $x = y = z = 0$ in (9), we have $f(0) = 0$. Putting $y = 0$ in (9) and using $f(0) = 0$, one gets

$$\sigma_z f(\alpha x) = \alpha^2 \sigma_z f(x) + 2\alpha^2(\alpha^2 - 1) f(x) + 2(1 - \alpha^2) f(z)$$

for all $x, z \in K$. Based on Theorem 2.1 of [22], we conclude that f is quartic; i.e., $\sigma_y f(2x) = 4\sigma_y f(x) + 24 f(x) - 6 f(y)$ for all $x, y \in K$.

(2) Letting $y = z = 0$ in (5) and using the convexity of $F(x)$ and the Rådström's cancelation law, we obtain

$$\alpha^4 F(x) + F(0) \subseteq F(\alpha x) + 2\alpha^2 F(0)$$

for all $x \in K$. Substituting x, y, and z by zero in (5) yields

$$F(0) \subseteq \{0\}.$$

From the last two inclusions, it follows that

$$\alpha^4 F(x) \subseteq F(\alpha x) + (2\alpha^2 - 1)F(0) \subseteq F(\alpha x)$$

for all $x \in K$. Hence,

$$F(x) \subseteq \alpha^{-4n} F(\alpha^n x) \subseteq \alpha^{-4(n+1)} F(\alpha^{n+1} x)$$

for all $x \in K$. In the same way, as in Theorem 2, we deduce that F is single-valued and $\sigma_y F(2x) = 4\sigma_y F(x) + 24F(x) - 6F(y)$ for all $x, y \in K$.

3 Set-Valued Dynamics and Applications

In this section we present a few applications of the results presented in the previous sections.

Theorem 4 *If $W \in ccz(Y)$ and $f : K \to Y$ satisfies*

$$\alpha\sigma_{y,z} f(\alpha x) - 2\sigma_y f(x) - 2\sigma_z f(x) + 4\left(2 - \alpha^2\right) f(x) \in W \tag{10}$$

for all $x, y, z \in K$, then there exists a unique function $T : K \to Y$ such that

$$\begin{cases} \alpha\sigma_{y,z} T(\alpha x) = 2\left(\sigma_y T(x) + \sigma_z T(x)\right) + 4\left(\alpha^2 - 2\right) T(x), \\ T(x) - f(x) \in \frac{1}{4(\alpha^2-\alpha)} W \end{cases}$$

for all $x, y, z \in K$.

Proof Let $F(x) := f(x) + \frac{1}{4\alpha(\alpha-1)} W$ for $x \in K$. Then

$$\begin{aligned} \alpha\sigma_{y,z} F(\alpha x) + 8F(x) &= \alpha\sigma_{y,z} f(\alpha x) + 8f(x) + \tfrac{\alpha+2}{\alpha(\alpha-1)} W \\ &\subseteq 2\sigma_y f(x) + 2\sigma_z f(x) + 4\alpha^2 f(x) + \tfrac{\alpha+2}{\alpha(\alpha-1)} W + W \\ &= 2\left(\sigma_y f(x) + \tfrac{1}{2\alpha(\alpha-1)} W\right) + 2\left(\sigma_z f(x) + \tfrac{1}{2\alpha(\alpha-1)} W\right) \\ &\quad + 4\alpha^2\left(f(x) + \tfrac{1}{4\alpha(\alpha-1)} W\right) \\ &= 2\left(\sigma_y F(x) + \sigma_z F(x)\right) + 4\alpha^2 F(x) \end{aligned}$$

for all $x, y, z \in K$. Now, according to Theorem 2 with $k = 1$, there exists a unique function $T : K \to Y$ such that

$$\alpha\sigma_{y,z}T(\alpha x) = 2\left(\sigma_y T(x) + \sigma_z T(x)\right) + 4\left(\alpha^2 - 2\right)T(x)$$

for all $x, y, z \in K$ and $T(x) \in F(x)$ for all $x \in K$.

Corollary 2 *Suppose $W \in ccz(Y)$ and $f : K \to Y$ satisfies* (10) *for all $x, y, z \in K$. Then there exists a unique additive function $T : K \to Y$ such that, for all $x \in K$,*

$$T(x) - f(x) \in \frac{1}{4(\alpha^2 - \alpha)}W.$$

We recall that a semigroup $(S, +)$ is called left (right) amenable if there exists a left (right) invariant mean on the space $B(S, \mathbb{R})$ of all real bounded functions defined on S. By a left (right) invariant mean we understand a linear functional M satisfying

$$\inf_{x \in S} f(x) \leq M(f) \leq \sup_{x \in S} f(x),$$

and

$$M({}_a f) = M(f) \quad (M(f_a) = M(f))$$

for all $f \in B(S, \mathbb{R})$ and $a \in S$, where ${}_a f$ (f_a) is the left (right) translate of f defined by ${}_a f(x) = f(a + x)$, $(f_a(x) = f(x + a))$, $x \in S$. If, on the space $B(S, \mathbb{R})$, there exists a real linear functional which is simultaneously a left and right invariant mean, then we say that S is two-sided amenable or just amenable.

One can prove that every Abelian semigroup is amenable. For the theory of amenability see, for example, Greenleaf [23]. Finally, let us see a result in [24].

Theorem 5 *Let $(S, +)$ be a left amenable semigroup and let X be a Hausdorff locally convex linear space. Let $F : S \to \mathfrak{F}_0(X)$ be set-valued function such that $F(s)$ is convex and weakly compact for all $s \in S$. Then F admits an additive selection if, and only if, there exists a function $f : S \to X$ such that*

$$f(s + t) - f(t) \in F(s) \tag{11}$$

for all $s, t \in S$.

As a consequence of the above theorem, we have the following corollaries.

Corollary 3 *Let $(S, +)$ be a left amenable semigroup and let X be a reflexive Banach space. In addition, let $\rho : S \to [0, \infty)$ and $g : S \to X$ be arbitrary functions. Then there exists an additive function $a : S \to X$ such that*

$$\| a(s) - g(s) \| \leq \rho(s) \tag{12}$$

for all $s \in S$, *if, and only if, there exists a function* $f : S \to X$ *such that*

$$\| f(s+t) - f(t) - g(s) \| \leq \rho(s) \tag{13}$$

for all $s, t \in S$.

Proof Define a set valued map $F : S \to \mathfrak{F}_0(X)$ by

$$F(s) = \{x \in X : \ \| x - g(s) \| \leq \rho(s)\}$$

for all $s \in S$. Then, due to the reflexivity of X, F has weakly compact nonempty convex values. It follows from (12) that a is a selection of F, and (13) is equivalent to (11). Now, the result follows from Theorem 5.

Corollary 4 (*Ger* [25]) *Let* $(S, +)$ *be a left amenable semigroup, let* X *be a reflexive Banach space, and let* $\rho : S \to [0, \infty)$ *be an arbitrary function. If the function* $f : S \to X$ *satisfies* $\| f(s+t) - f(s) - f(t) \| \leq \rho(s)$ *for all* s, t *in* S, *then there exists an additive function* $a : S \to X$ *such that* $\| f(s) - a(s) \| \leq \rho(s)$ *holds for all* s *in* S.

References

1. W. Smajdor, Superadditive set-valued functions. Glas. Mat. **21** (1986), 343–348
2. Z. Gajda, R. Ger, Subadditive multifunctions and Hyers-Ulam stability. Numer. Math. **80**, 281–291 (1987)
3. D. Popa, Additive selections of (α, β)-subadditive set valued maps. Glas. Mat. Ser. III **36**, 11–16 (2001)
4. J. Brzdęk, D. Popa, B. Xu, Selections of set-valued maps satisfying a linear inclusion in a single variable. Nonlin. Anal. **74**, 324–330 (2011)
5. D. Inoan, D. Popa, On selections of generalized convex set-valued maps. Aequat. Math. **88**, 267–276 (2014)
6. H. Khodaei, On the stability of additive, quadratic, cubic and quartic set-valued functional equations. Results Math. **68**, 1–10 (2015)
7. G. Lu, C. Park, Hyers-Ulam stability of additive set-valued functional equations. Appl. Math. Lett. **24**, 1312–1316 (2011)
8. K. Nikodem, On quadratic set-valued functions. Publ. Math. Debrecen **30**, 297–301 (1984)
9. K. Nikodem, D. Popa, On selections of general linear inclusions. Publ. Math. Debrecen **75**, 239–249 (2009)
10. K. Nikodem, D. Popa, On single-valuedness of set-valued maps satisfying linear inclusions. Banach J. Math. Anal. **3**, 44–51 (2009)
11. C. Park, D. O'Regan, R. Saadati, Stability of some set-valued functional equations. Appl. Math. Lett. **24**, 1910–1914 (2011)
12. M. Piszczek, On selections of set-valued inclusions in a single variable with applications to several variables. Results Math. **64**, 1–12 (2013)
13. M. Piszczek, The properties of functional inclusions and Hyers-Ulam stability. Aequat. Math. **85**, 111–118 (2013)
14. D. Popa, A property of a functional inclusion connected with Hyers-Ulam stability. J. Math. Inequal.**4**, 591–598 (2009)

15. K. Nikodem, K-Convex and K-Concave Set-Valued Functions, Zeszyty Naukowe, Politech, Krakow, 1989
16. H. Rådström, An embedding theorem for space of convex sets. Proc. Am. Math. Soc. **3**, 165–169 (1952)
17. W. Smajdor, Subadditive and subquadratic set-valued functions, Prace Nauk. Uniw. Śląsk, 889, Katowice (1987)
18. R. Urbański, A generalization of the Minkowski-Rådström-Hörmander theorem. Bull. Polish Acad. Sci. **24**, 709–715 (1976)
19. M.E. Gordji, Z. Alizadeh, H. Khodaei, C. Park, On approximate homomorphisms: a fixed point approach. Math. Sci. **6**, 59 (2012)
20. J.H. Bae, W.G. Park, A functional equation having monomials as solutions. Appl. Math. Comput. **216**, 87–94 (2010)
21. M.E. Gordji, Z. Alizadeh, Y.J. Cho, H. Khodaei, On approximate C^*-ternary m-homomorphisms: a fixed point approach. Fixed Point Theory Appl. (2011), Art. ID 454093
22. Y.S. Lee, S.Y. Chung, Stability of quartic functional equation in the spaces of generalized functions. Adv. Differ. Equ. (2009), Art. ID 838347
23. F.P. Greenleaf, *Invariant Mean on Topological Groups and their Applications*, Van Nostrand Mathematical Studies, vol. 16, New York/Toronto/London/Melbourne, 1969
24. R. Badora, R. Ger, Z. Páles, Additive selections and the stability of the Cauchy functional equation. ANZIAM J. **44**, 323–337 (2003)
25. R. Ger, The singular case in the stability behaviour of linear mappings. Grazer Math. Ber. **316**, 59–70 (1992)

Certain Fractional Integral and Differential Formulas Involving the Extended Incomplete Generalized Hypergeometric Functions

Praveen Agarwal, Themistocles M. Rassias, Gurmej Singh, and Shilpi Jain

Abstract The fractional integral and differential operators involving the family of special functions have found significant importance and applications in various fields of mathematics and engineering. The goal of this chapter is to find the fractional integral and differential formulas (also known as composition formulas) involving the extended incomplete generalized hypergeometric functions by using the generalized fractional calculus operators (the Marichev–Saigo–Maeda operators). After that, we established their image formulas by using the integral transforms like: Beta transform, Laplace transform and Whittaker transform. Moreover, the reduction formulas are also considered as special cases of our main findings associated with the well-known Saigo fractional integral and differential operators, Erdélyi-Kober fractional integral and differential operators, Riemann-Liouville fractional integral and differential operators and the Weyl fractional calculus operators.

2010 MSC Primary 33B20, 44A20, 65R10; Secondary 26A33, 33C20

P. Agarwal (✉)
International Center for Basic and Applied Sciences, Jaipur, India

Department of Mathematics, Anand International College of Engineering, Jaipur, India

Department of Mathematics, Harish-Chandra Research Institute (HRI), Allahbad, India

T. M. Rassias
Department of Mathematics, Zografou Campus, National Technical University of Athens, Athens, Greece
e-mail: trassias@math.ntua.gr

G. Singh
Department of Mathematics, Mata Sahib Kaur Girls College, Bathinda, India

S. Jain
Department of Mathematics, Poornima College of Engineering, Jaipur, India

T. M. Rassias, P. M. Pardalos (eds.), *Mathematical Analysis and Applications*, Springer Optimization and Its Applications 154,
https://doi.org/10.1007/978-3-030-31339-5_8

1 Introduction

Throughout the chapter let $\mathbb{C}$, $\mathbb{R}^+$, $\mathbb{N}$ and $\mathbb{Z}_0^-$ be the sets of complex numbers, positive real numbers, positive integers and non-positive integers, respectively and let $\mathbb{R}_0^+ := \mathbb{R}^+ \cup \{0\}$, $\mathbb{N}_0 := \mathbb{N} \cup \{0\}$ and $\Gamma(\lambda)$ is the familiar Gamma function.

The fractional calculus is nowadays one of the most rapidly growing subjects of mathematical analysis in spite of the fact that it is nearly 300 years old. Yet the giants of mathematics, G.W. Leibnitz and L. Euler, thought about the possibility to perform differentiation of non-integer order. The real birth and far-reaching development of the fractional calculus is due to numerous attempts of mathematicians during the nineteenth century to beginning of twentieth century. It is practically impossible to name all important contributions made in construction of early stages of building of the fractional calculus (see the project by the Frac. Calc. Appl. Anal. [1–3]). New era in the development of this branch of mathematical science began 40–50 years ago due to numerous applications of fractional-type models and is continued up to now. One can mention a large list of areas of applications, in particular, continuum mechanics [4, 5] (viscoelasticity [6], thermodynamics [7] and anomalous diffusion [8]), astrophysics [9], nuclear physics [10], nanophysics and cosmic physics [11, 12], statistical mechanics [13], fractional order systems and control [14–16], finance and economics [17].

Among the monographs developing the theory of fractional calculus and presenting some applications, we have to point out monographs by Diethelm [18], Kiryakova [19], Kilbas, Srivastava and Trujillo [20], Miller and Ross [21], Oldham and Spanier [22], Podlubny [23], and of course "the Bible of fractional calculus", monograph by Samko, Kilbas and Marichev [24]. Interested reader can find in these books an extended list of publications on the theory and applications of fractional calculus (see also, for example, [3, 25–30]). A rich literature is available revealing the development of fractional calculus involving various special functions (see [31–36]).

A special role of the incomplete hypergeometric functions in the fractional calculus has been discovered by many scientists from different point of view. The theory of the incomplete Gamma functions, as a part of the theory of confluent hypergeometric functions, has received its first systematic exposition by Tricomi [37] in the early 1950s. Musallam and Kalla [38, 39] considered a more generalized incomplete gamma function involving the Gauss hypergeometric function and established a number of analytic properties including recurrence relations, asymptotic expansions and computation for special values of the parameters as well as some researchers have shown that incomplete Gamma functions can be used in closed form solutions to several problems in heat conduction with time-dependent boundary conditions. Following the line of the above-mentioned works several investigations related to this function, its generalizations and related special functions have been done at the beginning of this century. Nowadays this function and its numerous generalizations are involved in the different fractional models (see monographs listed above). Special role of the incomplete hypergeometric function was pointed out by Srivastava [40–42], including it into the class of special functions for fractional calculus.

Due to this exceptional role of the incomplete hypergeometric functions any new exact result involving these functions seems very interesting. This chapter is devoted to the properties of the so-called Marichev–Saigo–Maeda generalized fractional calculus operators, i.e. integral transform of the Mellin convolution type with the Appell (or Horn) function F_3. This operator was introduced nearly 40 years ago by Marichev [43] and studied in some recent papers, including the papers by Saigo and Maeda [44] and by Saigo and Saxena [45]. The aim of the present chapter is to establish fractional integral and differential formulas (also known as composition formulas) involving the extended incomplete generalized hypergeometric function by using the generalized fractional calculus operators (the Marichev–Saigo–Maeda calculus operators). After that, we present their image formulas by using the integral transforms like: Beta transform, Laplace transform and Whittaker transform. Moreover, the reduction formulas are also considered as special cases of our main findings associated with the well-known Saigo fractional integral and differential operators, Erdélyi-Kober fractional integral and differential operators, Riemann-Liouville fractional integral and differential operators, and the Weyl fractional integral and differential operators, respectively.

2 Fractional Calculus Operators and Their Formulas

In this section we recall some known facts about the Marichev-Saigo-Maeda generalized fractional calculus operators and their special cases. Let us begin with few notions and facts related to the fractional calculus operators.

In 1974, Marichev [43] introduced fractional integral operators as Mellin type convolution operator with the Appell function F_3 in their kernel. In the middle of the 1990s, these fractional integral operators were rediscovered and studied by Saigo [46–48] and which was later on extended and studied by Saigo and Maeda [44] and by Saigo and Saxena [45] as generalizations of the celebrated Saigo fractional integral operators.

The generalized fractional calculus operators (the Marichev–Saigo–Maeda operators) involving the Appell's function or the Horn's $F_3(\cdot)$ function in the kernel are defined as follows:

Definition 1 Let $\sigma, \sigma', \nu, \nu', \eta \in \mathbb{C}$ and $x > 0$, then for $\Re(\eta) > 0$

$$\begin{aligned}&\left(I_{0,x}^{\sigma,\sigma',\nu,\nu',\eta} f\right)(x)\\&= \frac{x^{-\sigma}}{\Gamma(\eta)} \int_0^x (x-t)^{\eta-1} t^{-\sigma'} F_3\left(\sigma, \sigma', \nu, \nu'; \eta; 1-\frac{t}{x}, 1-\frac{x}{t}\right) f(t)\, dt\end{aligned} \tag{2.1}$$

and

$$\left(I_{x,\infty}^{\sigma,\sigma',\nu,\nu',\eta} f\right)(x)$$
$$= \frac{x^{-\sigma'}}{\Gamma(\eta)} \int_x^{\infty} (t-x)^{\eta-1} t^{-\sigma} F_3\left(\sigma, \sigma', \nu, \nu'; \eta; 1-\frac{x}{t}, 1-\frac{t}{x}\right) f(t)\, dt, \tag{2.2}$$

provided the function $f(t)$ is so constrained such that the integrals in Equations (2.1) and (2.2) exist.

In Equations (2.1) and (2.2), $F_3(.)$ denotes Appell's hypergeometric function [49] in two variables defined as:

$$F_3(\sigma, \sigma', \nu, \nu'; \eta; x, y)$$
$$= \sum_{m,n=0}^{\infty} \frac{(\sigma)_m(\sigma')_n(\nu)_m(\nu')_n}{(\eta)_{m+n}} \frac{x^m}{m!} \frac{x^n}{n!}, \qquad (\max\{|x|, |y|\} < 1). \tag{2.3}$$

Then the above fractional integral operators in Equations (2.1) and (2.2) can be written as follows:

$$\left(I_{0,x}^{\sigma,\sigma',\nu,\nu',\eta} f\right)(x) = \left(\frac{d}{dx}\right)^k \left(I_{0,x}^{\sigma,\sigma',\nu+k,\nu',\eta+k} f\right)(x),$$
$$(\Re(\eta) \leq 0; k = [-\Re(\eta)+1]) \tag{2.4}$$

and

$$\left(I_{x,\infty}^{\sigma,\sigma',\nu,\nu',\eta} f\right)(x) = \left(-\frac{d}{dx}\right)^k \left(I_{x,\infty}^{\sigma,\sigma',\nu,\nu'+k,\eta+k} f\right)(x),$$
$$(\Re(\eta) \leq 0; k = [-\Re(\eta)+1]). \tag{2.5}$$

Remark 1 It is interesting that the Appell function defined in Equation (2.3) reduces to the Gauss hypergeometric function ${}_2F_1$ as given in the following relations:

$$F_3(\sigma, \eta-\sigma, \nu, \eta-\nu; \eta; x, y) = {}_2F_1(\sigma, \nu; \eta; x+y-xy), \tag{2.6}$$

also we have

$$F_3(\sigma, 0, \nu, \nu', \eta; x, y) = {}_2F_1(\sigma, \nu; \eta; x) \tag{2.7}$$

and

$$F_3(0, \sigma', \nu, \nu', \eta; x, y) = {}_2F_1(\sigma', \nu'; \eta; y). \tag{2.8}$$

The corresponding Marichev–Saigo–Maeda fractional differential operators are given as follows:

Definition 2 Let $\sigma, \sigma', \nu, \nu', \eta \in \mathbb{C}$ and $x > 0$, Then

$$\begin{aligned}\left(D_{0,x}^{\sigma,\sigma',\nu,\nu',\eta} f\right)(x) &= \left(I_{0,x}^{-\sigma',-\sigma,-\nu',-\nu,-\eta} f\right)(x)\\ &= \left(\frac{d}{dx}\right)^k \left(I_{0,x}^{-\sigma',-\sigma,-\nu'+k,-\nu,-\eta+k} f\right)(x), \qquad (\Re(\eta) > 0; k = [\Re(\eta)] + 1)\\ &= \frac{1}{\Gamma(k-\eta)} \left(\frac{d}{dx}\right)^k (x)^{\sigma'} \int_0^x (x-t)^{k-\eta-1} t^{\sigma}\\ &\times F_3\left(-\sigma', -\sigma, k-\nu', -\nu; k-\eta; 1-\frac{t}{x}, 1-\frac{x}{t}\right) f(t)\,dt\end{aligned} \tag{2.9}$$

and

$$\begin{aligned}\left(D_{x,\infty}^{\sigma,\sigma',\nu,\nu',\eta} f\right)(x) &= \left(I_{x,\infty}^{-\sigma',-\sigma,-\nu',-\nu,-\eta} f\right)(x)\\ &= \left(-\frac{d}{dx}\right)^k \left(I_{x,\infty}^{-\sigma',-\sigma,-\nu',-\nu+k,-\eta+k} f\right)(x), \qquad (\Re(\eta) > 0; k = [\Re(\eta)] + 1)\\ &= \frac{1}{\Gamma(k-\eta)} \left(-\frac{d}{dx}\right)^k (x)^{\sigma} \int_x^\infty (t-x)^{k-\eta-1} t^{\sigma'}\\ &\times F_3\left(-\sigma', -\sigma, -\nu', k-\nu; k-\eta; 1-\frac{x}{t}, 1-\frac{t}{x}\right) f(t)\,dt.\end{aligned} \tag{2.10}$$

In view of the above reduction formula as given in Equation (2.7), the generalized fractional calculus operators reduce to the Saigo operators [46] defined as follows:

Definition 3 Let $x > 0, \sigma, \nu, \eta \in \mathbb{C}$ and $\Re(\sigma) > 0$, then

$$\left(I_{0,x}^{\sigma,\nu,\eta} f\right)(x) = \frac{x^{-\sigma-\nu}}{\Gamma(\sigma)} \int_0^x (x-t)^{\sigma-1} {}_2F_1\left(\sigma+\nu, -\eta; \sigma; 1-\frac{t}{x}\right) f(t)dt \tag{2.11}$$

and

$$\left(I_{x,\infty}^{\sigma,\nu,\eta} f\right)(x) = \frac{1}{\Gamma(\sigma)} \int_x^\infty (t-x)^{\sigma-1} t^{-\sigma-\nu} {}_2F_1\left(\sigma+\nu, -\eta; \sigma; 1-\frac{x}{t}\right) f(t)dt. \tag{2.12}$$

where the ${}_2F_1(.)$, a special case of the generalized hypergeometric function, is the Gauss hypergeometric function and function $f(t)$ is so constrained that the integrals in Equations (2.11) and (2.12) converge.

Remark 2 The Saigo fractional integral operators, given in Equations (2.11) and (2.12) can also be written as:

Let $x > 0, \sigma, \nu, \eta \in \mathbb{C}$, then

$$\left(I_{0,x}^{\sigma,\nu,\eta} f\right)(x) = \left(\frac{d}{dx}\right)^k \left(I_{0,x}^{\sigma+k,\nu-k,\eta-k} f\right)(x), \qquad (\Re(\sigma) \leq 0;\ k = [\Re(-\sigma)] + 1) \tag{2.13}$$

and

$$\left(I_{x,\infty}^{\sigma,\nu,\eta} f\right)(x) = \left(-\frac{d}{dx}\right)^k \left(I_{x,\infty}^{\sigma-k,\nu-k,\eta} f\right)(x), \qquad (\Re(\sigma) \leq 0;\ k = [\Re(-\sigma)] + 1). \tag{2.14}$$

And the corresponding Saigo fractional differential operators are defined as:

Definition 4 Let $\sigma, \nu, \eta \in \mathbb{C}$ and $x > 0$. Then

$$\left(D_{0,x}^{\sigma,\nu,\eta} f\right)(x) = \left(I_{0,x}^{-\sigma,-\nu,\sigma+\eta} f\right)(x) = \left(\frac{d}{dx}\right)^k \left(I_{0,x}^{-\sigma+k,-\nu-k,\sigma+\eta-k} f\right)(x), \qquad (\Re(\sigma) > 0; k = [\Re(\sigma)] + 1) \tag{2.15}$$

and

$$\left(D_{x,\infty}^{\sigma,\nu,\eta} f\right)(x) = \left(I_{x,\infty}^{-\sigma,-\nu,\sigma+\eta} f\right)(x) = \left(-\frac{d}{dx}\right)^k \left(I_{x,\infty}^{-\sigma+k,-\nu-k,\sigma+\eta} f\right)(x), \qquad (\Re(\sigma) > 0; k = [\Re(\sigma)] + 1), \tag{2.16}$$

where $[x]$ denotes the greatest integer less than or equal to the real number x.

If we take $\nu = 0$ in Equations (2.11), (2.12), (2.15) and (2.16), we get the so-called Erdélyi-Kober fractional integral and derivative operators defined as follows [9, 50]:

Definition 5 Let $x > 0, \sigma, \eta \in \mathbb{C}$ with $\Re(\sigma) > 0$, then (see [20, 24])

$$\left(I_{0,x}^{\sigma,\eta} f\right)(x) = \frac{x^{-\sigma-\eta}}{\Gamma(\sigma)} \int_0^x (x-t)^{\sigma-1} t^{\eta} f(t)\, dt \tag{2.17}$$

and

$$\left(I_{x,\infty}^{\sigma,\eta} f\right)(x) = \frac{x^{\eta}}{\Gamma(\sigma)} \int_x^\infty (t-x)^{\sigma-1} t^{-\sigma-\eta} f(t)\, dt, \tag{2.18}$$

provided that integrals in Equations (2.17) and (2.18) converge.

The corresponding derivative operators are defined as:

Definition 6 Let $x > 0$, $\sigma, \eta \in \mathbb{C}$ with $\Re(\sigma) > 0$, then (see [20, 24])

$$\begin{aligned}\left(D_{0,x}^{\sigma,\eta} f\right)(x) &= x^{-\eta} \left(\frac{d}{dx}\right)^k \frac{1}{\Gamma(k-\sigma)} \int_0^x t^{\sigma+\eta} (x-t)^{k-\sigma-1} f(t)\, dt \\ &= \left(\frac{d}{dx}\right)^k \left(I_{0,x}^{-\sigma+k,-\sigma,\sigma+\eta-k} f\right)(x), \qquad (k = [\Re(\sigma)] + 1).\end{aligned} \tag{2.19}$$

and

$$\begin{aligned}\left(D_{x,\infty}^{\sigma,\eta} f\right)(x) &= x^{\eta+\sigma} \left(\frac{d}{dx}\right)^k \frac{1}{\Gamma(k-\sigma)} \int_x^\infty t^{-\eta} (t-x)^{k-\sigma-1} f(t)\, dt \\ &= (-1)^k \left(\frac{d}{dx}\right)^k \left(I_{x,\infty}^{-\sigma+k,-\sigma,\sigma+\eta} f\right)(x), \qquad (k = [\Re(\sigma)] + 1).\end{aligned} \tag{2.20}$$

When $\nu = -\sigma$, the operators in Equations (2.11), (2.12), (2.15) and (2.16) give the Riemann-Liouville and the Weyl fractional integral operators (see [9, 22]) are defined as follows:

Definition 7 Let $x > 0$, $\sigma \in \mathbb{C}$ with $\Re(\sigma) > 0$, then

$$\left(I_{0,x}^{\sigma} f\right)(x) = \frac{1}{\Gamma(\sigma)} \int_0^x (x-t)^{\sigma-1} f(t)\, dt \tag{2.21}$$

and

$$\left(I_{x,\infty}^{\sigma} f\right)(x) = \frac{1}{\Gamma(\sigma)} \int_x^\infty (t-x)^{\sigma-1} f(t)\, dt, \tag{2.22}$$

provided both integrals converge.

The corresponding derivative operators are defined as follows:

Definition 8 Let $x > 0$, $\sigma \in \mathbb{C}$ with $\Re(\sigma) > 0$, then

$$\begin{aligned}\left(D_{0,x}^{\sigma} f\right)(x) &= \left(\frac{d}{dx}\right)^k \frac{1}{\Gamma(k-\sigma)} \int_0^x (x-t)^{k-\sigma-1} f(t)\,dt \\ &= \left(\frac{d}{dx}\right)^k \left(I_{0,x}^{k-\sigma} f\right)(x), \qquad (k = [\Re(\sigma)] + 1)\end{aligned} \tag{2.23}$$

and

$$\begin{aligned}\left(D_{x,\infty}^{\sigma} f\right)(x) &= (-1)^k \left(\frac{d}{dx}\right)^k \frac{1}{\Gamma(k-\sigma)} \int_x^{\infty} (t-x)^{k-\sigma-1} f(t)\,dt \\ &= (-1)^k \left(\frac{d}{dx}\right)^k \left(I_{x,\infty}^{k-\sigma} f\right)(x), \qquad (k = [\Re(\sigma)] + 1).\end{aligned} \tag{2.24}$$

For detail of such operators along with their properties and applications, one may refer [9, 20, 24, 51, 52].

Power functions formulas of the above discussed fractional operators required for our present study are given in the following lemmas (see [44–46]):

Lemma 1 *Let $\sigma, \sigma', \nu, \nu', \eta$ and $\rho \in \mathbb{C}$, $x > 0$ be such that $\Re(\eta) > 0$, then the following formulas hold true:*

$$\begin{aligned}&\left(I_{0,x}^{\sigma,\sigma',\nu,\nu',\eta} t^{\rho-1}\right)(x) \\ &= \frac{\Gamma(\rho)\Gamma(\rho+\eta-\sigma-\sigma'-\nu)\Gamma(\rho+\nu'-\sigma')}{\Gamma(\rho+\nu')\Gamma(\rho+\eta-\sigma-\sigma')\Gamma(\rho+\eta-\sigma'-\nu)} x^{\rho+\eta-\sigma-\sigma'-1}, \\ &\qquad (\Re(\rho) > \max\{0, \Re(\sigma+\sigma'+\nu-\eta)\Re(\sigma'-\nu')\})\end{aligned} \tag{2.25}$$

and

$$\begin{aligned}&\left(I_{x,\infty}^{\sigma,\sigma',\nu,\nu',\eta} t^{\rho-1}\right)(x) \\ &= \frac{\Gamma(1-\rho-\nu)\Gamma(1-\rho-\eta+\sigma+\sigma')\Gamma(1-\rho-\eta+\sigma+\nu')}{\Gamma(1-\rho)\Gamma(1-\rho-\eta+\sigma+\sigma'+\nu')\Gamma(1-\rho+\sigma-\nu)} x^{\rho+\eta-\sigma-\sigma'-1}, \\ &\qquad (\Re(\rho) < 1 + \min\{\Re(-\nu), \Re(\sigma+\sigma'-\eta), \Re(\sigma+\nu'-\eta)\}).\end{aligned} \tag{2.26}$$

Lemma 2 *Let $\sigma, \sigma', \nu, \nu', \eta$ and $\rho \in \mathbb{C}$, $x > 0$ be such that $\Re(\eta) > 0$, then the following formulas hold true:*

$$\left(D_{0,x}^{\sigma,\sigma',\nu,\nu',\eta}t^{\rho-1}\right)(x)$$
$$=\frac{\Gamma(\rho)\Gamma(\rho-\eta+\sigma+\sigma'+\nu')\Gamma(\rho-\nu+\sigma)}{\Gamma(\rho-\nu)\Gamma(\rho-\eta+\sigma+\sigma')\Gamma(\rho-\eta+\sigma+\nu')}x^{\rho-\eta+\sigma+\sigma'-1}, \tag{2.27}$$
$$(\Re(\rho)>\max\{0,\Re(\eta-\sigma-\sigma'-\nu'),\Re(\nu-\sigma)\})$$

and

$$\left(D_{x,\infty}^{\sigma,\sigma',\nu,\nu',\eta}t^{\rho-1}\right)(x)$$
$$=\frac{\Gamma(1-\rho+\nu')\Gamma(1-\rho+\eta-\sigma-\sigma')\Gamma(1-\rho+\eta-\sigma'-\nu)}{\Gamma(1-\rho)\Gamma(1-\rho+\eta-\sigma-\sigma'-\nu)\Gamma(1-\rho-\sigma'+\nu')}x^{\rho-\eta+\sigma+\sigma'-1},$$
$$(\Re(\rho)<1+\min\{\Re(\nu'),\Re(\eta-\sigma-\sigma'),\Re(\eta-\sigma'-\nu)\}). \tag{2.28}$$

Lemma 3 *Let $\sigma,\nu,\eta,\rho\in\mathbb{C}, x>0$ be such that $\Re(\sigma)>0$, then the following formulas hold true:*

$$\left(I_{0,x}^{\sigma,\nu,\eta}t^{\rho-1}\right)(x)=\frac{\Gamma(\rho)\Gamma(\rho+\eta-\nu)}{\Gamma(\rho-\nu)\Gamma(\rho+\eta+\sigma)}x^{\rho-\nu-1},$$
$$(\Re(\rho)>\max\{0,\Re(\nu-\eta)\}). \tag{2.29}$$

and

$$\left(I_{x,\infty}^{\sigma,\nu,\eta}t^{\rho-1}\right)(x)=\frac{\Gamma(1-\rho+\nu)\Gamma(1-\rho+\eta)}{\Gamma(1-\rho)\Gamma(1-\rho+\eta+\sigma+\nu)}x^{\rho-\nu-1},$$
$$(\Re(\rho)<1+\min\{\Re(\nu),\Re(\eta)\}). \tag{2.30}$$

Lemma 4 *Let $\sigma,\nu,\eta,\rho\in\mathbb{C}, x>0$ be such that $\Re(\sigma)>0$, then the following formulas hold true:*

$$\left(D_{0,x}^{\sigma,\nu,\eta}t^{\rho-1}\right)(x)=\frac{\Gamma(\rho)\Gamma(\rho+\eta+\sigma+\nu)}{\Gamma(\rho+\eta)\Gamma(\rho+\nu)}x^{\rho+\nu-1},$$
$$(\Re(\rho)>-\min\{0,\Re(\sigma+\nu+\eta)\}) \tag{2.31}$$

and

$$\left(D_{x,\infty}^{\sigma,\nu,\eta}t^{\rho-1}\right)(x)=\frac{\Gamma(1-\rho-\nu)\Gamma(1-\rho+\sigma+\eta)}{\Gamma(1-\rho+\eta-\nu)\Gamma(1-\rho)}x^{\rho+\nu-1},$$
$$(\Re(\rho)<1+\min\{\Re(-\nu-n),\Re(\eta+\sigma)\}\ \text{and}\ n=[\Re(\sigma)]+1). \tag{2.32}$$

Lemma 5 *Let $\sigma, \eta, \rho \in \mathbb{C}$, $x > 0$ be such that $\Re(\sigma) > 0$, then the following formulas hold true:*

$$\left(I_{0,x}^{\sigma,\eta} t^{\rho-1}\right)(x) = \frac{\Gamma(\rho+\eta)}{\Gamma(\rho+\eta+\sigma)} x^{\rho-1}, \qquad (\Re(\rho) > -\Re(\eta)) \tag{2.33}$$

and

$$\left(I_{x,\infty}^{\sigma,\eta} t^{\rho-1}\right)(x) = \frac{\Gamma(1-\rho+\eta)}{\Gamma(1-\rho+\eta+\sigma)} x^{\rho-1}, \qquad (\Re(\rho) < 1+\Re(\eta)). \tag{2.34}$$

Lemma 6 *Let $\sigma, \eta, \rho \in \mathbb{C}$, $x > 0$ be such that $\Re(\sigma) > 0$, then the following formulas hold true:*

$$\left(D_{0,x}^{\sigma,\eta} t^{\rho-1}\right)(x) = \frac{\Gamma(\rho+\eta+\sigma)}{\Gamma(\rho+\eta)} x^{\rho-1}, \qquad (\Re(\rho) > -\Re(\eta+\sigma)) \tag{2.35}$$

and

$$\left(D_{x,\infty}^{\sigma,\eta} t^{\rho-1}\right)(x) = \frac{\Gamma(1-\rho+\sigma+\eta)}{\Gamma(1-\rho+\eta)} x^{\rho-1}, \tag{2.36}$$
$$(\Re(\rho) < 1+\Re(\eta+\sigma) - n \;\; \textit{and} \;\; n = [\Re(\sigma)]+1).$$

Lemma 7 *Let $\sigma, \rho \in \mathbb{C}$, $x > 0$ be such that $\Re(\sigma) > 0$, then the following formulas hold true:*

$$\left(I_{0,x}^{\sigma} t^{\rho-1}\right)(x) = \frac{\Gamma(\rho)}{\Gamma(\rho+\sigma)} x^{\rho+\sigma-1}, \qquad (\Re(\rho) > 0) \tag{2.37}$$

and

$$\left(I_{x,\infty}^{\sigma} t^{\rho-1}\right)(x) = \frac{\Gamma(1-\rho-\sigma)}{\Gamma(1-\rho)} x^{\rho+\sigma-1}, \qquad (0 < \Re(\sigma) < 1-\Re(\rho)). \tag{2.38}$$

Lemma 8 *Let $\sigma, \rho \in \mathbb{C}$, $x > 0$ be such that $\Re(\sigma) > 0$, then the following formulas hold true:*

$$\left(D_{0,x}^{\sigma} t^{\rho-1}\right)(x) = \frac{\Gamma(\rho)}{\Gamma(\rho-\sigma)} x^{\rho-\sigma-1}, \qquad (\Re(\rho) > \Re(\sigma) > 0) \tag{2.39}$$

and

$$\left(D^{\sigma}_{x,\infty}t^{\rho-1}\right)(x)=\frac{\Gamma(1-\rho+\sigma)}{\Gamma(1-\rho)}x^{\rho-\sigma-1},$$
$$(\Re(\rho)<1+\Re(\sigma)-n \ \ and \ \ n=[\Re(\sigma)]+1). \tag{2.40}$$

3 Incomplete Gamma Functions and Their Generalizations

The incomplete gamma functions $\gamma(\lambda,x)$ and $\Gamma(\lambda,x)$, both of which are certain generalizations of the classical gamma function $\Gamma(x)$, given in Equations (3.1) and (3.2) respectively, have been investigated by many authors. The incomplete gamma functions have proved to be of great importance for physicists and engineers as well as mathematicians. For more details, one may refer to the books [20, 49, 53–60] and the recent papers [40, 41, 61–64] and [65, 66] on the subject.

The familiar incomplete gamma functions $\gamma(\lambda,x)$ and $\Gamma(\lambda,x)$ are defined by (see [63])

Definition 9 For $\Re(\lambda)>0;\ x\geq 0$

$$\gamma(\lambda,x)=\int_0^x t^{\lambda-1}e^{-t}dt. \tag{3.1}$$

Definition 10 For $x\geq 0;\ \Re(\lambda)>0$ when $x=0$

$$\Gamma(\lambda,x)=\int_x^{\infty} t^{\lambda-1}e^{-t}dt, \tag{3.2}$$

respectively, satisfying the following decomposition formula

$$\gamma(\lambda,x)+\Gamma(\lambda,x)=\Gamma(\lambda), \qquad (\Re(\lambda)>0).$$

where $\Gamma(\lambda)$ is the well-known Euler's gamma function defined by (see [67–69]).

Definition 11 Let $\lambda\in\mathbb{C};\ \Re(\lambda)>0$, then

$$\Gamma(\lambda)=\int_0^{\infty} t^{\lambda-1}e^{-t}dt. \tag{3.3}$$

We also recall the Pochhammer symbol $(\lambda)_n$ [69] defined by

Definition 12 Let $\lambda\in\mathbb{C}$, then

$$(\lambda)_n = \begin{cases} 1 & (n = 0;\ \lambda \in \mathbb{C} \setminus \{0\}) \\ \lambda(\lambda+1)\cdots(\lambda+n-1) & (n \in \mathbb{N};\ \lambda \in \mathbb{C}) \end{cases}$$
$$= \frac{\Gamma(\lambda+n)}{\Gamma(\lambda)} \qquad (\lambda \in \mathbb{C} \setminus \mathbb{Z}_0^-). \tag{3.4}$$

where $\mathbb{Z}_0^-$ denotes the set of non-positive integers (see, e.g., [70, p. 2 and p. 5]).

Very recently, Srivastava et al. [63] introduced and studied some fundamental properties and characteristics of a family of two potentially useful and generalized incomplete hypergeometric functions, defined as follows:

Definition 13 Let $z, a_1 \cdots a_r \in \mathbb{C}$: $b_1 \cdots b_s \in \mathbb{C} \setminus \mathbb{Z}_0^-$; $r, s \in \mathbb{N}_0$, then

$${}_r\gamma_s\left[\begin{matrix}(a_1,x), a_2, \cdots, a_r; \\ b_1, \cdots, b_s;\end{matrix}\ z\right] = \sum_{n=0}^{\infty} \frac{(a_1;x)_n (a_2)_n \cdots (a_r)_n}{(b_1)_n \cdots (b_s)_n} \cdot \frac{z^n}{n!} \tag{3.5}$$

and

$${}_r\Gamma_s\left[\begin{matrix}(a_1,x), a_2, \cdots, a_r; \\ b_1, \cdots, b_s;\end{matrix}\ z\right] = \sum_{n=0}^{\infty} \frac{[a_1;x]_n (a_2)_n \cdots (a_r)_n}{(b_1)_n \cdots (b_s)_n} \cdot \frac{z^n}{n!}, \tag{3.6}$$

where $(a_1; x)_n$ and $[a_1; x]_n$ are interesting generalizations of the Pochhammer symbol $(\lambda)_n$, in terms of the incomplete gamma type functions $\gamma(\lambda, x)$ and $\Gamma(\lambda, x)$ defined as follows (see [63]):

Definition 14 Let $\lambda,\ k \in \mathbb{C};\ \ x \geq 0$, then

$$(\lambda; x)_k := \frac{\gamma(\lambda+k, x)}{\Gamma(\lambda)} \tag{3.7}$$

and

$$[\lambda; x]_k := \frac{\Gamma(\lambda+k, x)}{\Gamma(\lambda)}. \tag{3.8}$$

These incomplete Pochhammer symbols $(\lambda; x)_k$ and $[\lambda; x]_k$ satisfy the following decomposition relation

$$(\lambda; x)_k + [\lambda; x]_k = (\lambda)_k, \qquad (\lambda, k \in \mathbb{C}; x \geq 0).$$

In Equations (3.1), (3.2), (3.5), (3.6), (3.7) and (3.8), the argument $x \geq 0$ is independent of the argument $z \in \mathbb{C}$ which occurs in the result (3.5) and (3.6) and also in the results presented in this chapter (see, e.g., [63, p. 675]).

We repeat the remark given by Srivastava et al. [63, Remark 7] for completeness and an easier reference.

Remark 3 Since

$$|(\lambda; x)_n| \le |(\lambda)_n| \text{ and } |[\lambda; x]_n| \le |(\lambda)_n| \qquad (n \in \mathbb{N}_0; \lambda \in \mathbb{C}; x \ge 0), \tag{3.9}$$

the precise (sufficient) conditions under which the infinite series in the definitions (3.5) and (3.6) would converge absolutely can be derived from those that are well-documented in the case of the generalized hypergeometric function ${}_rF_s$ $(r, s \in \mathbb{N}_0)$ (see, for details, [69, pp. 73–74] and [49, p. 20]; see also [57]).

The p-extension of the familiar incomplete Gamma functions $\gamma(\lambda, x)$ and $\Gamma(\lambda, x)$ presented by Chaudhary and Zubair in [71] is given as:

Definition 15 For $\Re(\lambda) > 0;\ x \ge 0;\ \Re(p) > 0;\ p = 0$:

$$\gamma(\lambda, x; p) := \int_0^x t^{\lambda-1} e^{-t-\frac{p}{t}} dt. \tag{3.10}$$

Definition 16 For $\Re(\lambda) > 0;\ x \ge 0;\ \Re(p) > 0$:

$$\Gamma(\lambda, x; p) := \int_x^\infty t^{\lambda-1} e^{-t-\frac{p}{t}} dt, \tag{3.11}$$

respectively, satisfying the decomposition formula given by:

$$\begin{aligned} &\gamma(\lambda, x; p) + \Gamma(\lambda, x; p) = \Gamma_p(\lambda) \\ &= \int_0^\infty t^{\lambda-1} e^{-t-\frac{p}{t}} dt = 2p^{\lambda/2} K_\lambda(2\sqrt{p}), \qquad (\Re(p) > 0). \end{aligned} \tag{3.12}$$

where $K_\lambda(x)$ denotes the familiar modified Bessel function [54, 66].

The generalized form of the Pochhammer symbols $(\lambda; x, p)_k$ and $[\lambda; x, p]_k$ $(\lambda, k \in \mathbb{C}; p \ge 0)$, in terms of the generalized incomplete Gamma functions $\gamma(\lambda, x; p)$ and $\Gamma(\lambda, x; p)$ given in Equations (3.10) and (3.11), are defined as [71]:

Definition 17 Let $\lambda, k \in \mathbb{C};\ x \ge 0;\ p \ge 0$, then

$$(\lambda; x, p)_k = \frac{\gamma(\lambda + k, x; p)}{\Gamma(\lambda)} \tag{3.13}$$

and

$$[\lambda; x, p]_k = \frac{\Gamma(\lambda + k, x; p)}{\Gamma(\lambda)}. \tag{3.14}$$

These forms of the Pochhammer symbols $(\lambda; x, p)_k$ and $[\lambda; x, p]_k$ satisfy the following decomposition relation:

$$(\lambda; x, p)_k + [\lambda; x, p]_k \equiv (\lambda; p)_k = \begin{cases} \dfrac{\Gamma_p(\lambda+k)}{\Gamma(\lambda)} & (\Re(p) > 0; \lambda, k \in \mathbb{C}) \\ \displaystyle\int_0^\infty t^{\lambda+k-1} e^{-t-\frac{p}{t}} dt & (\Re(p) > 0) \end{cases} \tag{3.15}$$

where $(\lambda; p)_k$ is the generalized Pochhammer symbol [66, 71].

In the view of relations (3.10), (3.13) and (3.11), (3.14), we have the integral representations of the extended forms of the Pochhammer symbols given as:

$$(\lambda; x, p)_k = \frac{1}{\Gamma(\lambda)} \int_0^x t^{\lambda+k-1} e^{-t-\frac{p}{t}} dt \tag{3.16}$$

and

$$[\lambda; x, p]_k = \frac{1}{\Gamma(\lambda)} \int_x^\infty t^{\lambda+k-1} e^{-t-\frac{p}{t}} dt. \tag{3.17}$$

In terms of the extended incomplete forms of the Pochhammer symbols $(\lambda; x, p)_k$ and $[\lambda; x, p]_k$ defined, respectively, in Equations (3.13) and (3.14), two families of the extended incomplete generalized hypergeometric functions ${}_r\gamma_s^p(z)$ and ${}_r\Gamma_s^p(z)$, respectively, involving r-numerator and s-denominator parameters are defined as [66]:

Definition 18 Let $z, \alpha_1, \cdots, \alpha_r \in \mathbb{C}: \ \beta_1, \cdots, \beta_s \in \mathbb{C} \setminus \mathbb{Z}_0^-; \ r, s \in \mathbb{N}_0$, then

$${}_r\gamma_s^p(z) = {}_r\gamma_s^p\left[\begin{matrix} (\alpha_1, x; p), \alpha_2, \ldots, \alpha_r; \\ \beta_1, \ldots, \beta_s; \end{matrix} z\right] = \sum_{n=0}^{\infty} \frac{(\alpha_1; x, p)_n (\alpha_2)_n \cdots (\alpha_r)_n}{(\beta_1)_n \cdots (\beta_s)_n} \frac{z^n}{n!} \tag{3.18}$$

and

$${}_r\Gamma_s^p(z) = {}_r\Gamma_s^p\left[\begin{matrix} (\alpha_1, x; p), \alpha_2, \ldots, \alpha_r; \\ \beta_1, \ldots, \beta_s; \end{matrix} z\right] = \sum_{n=0}^{\infty} \frac{[\alpha_1; x, p]_n (\alpha_2)_n \cdots (\alpha_r)_n}{(\beta_1)_n \cdots (\beta_s)_n} \frac{z^n}{n!}. \tag{3.19}$$

where the series on the right-hand side of Equations (3.18) and (3.19) are convergent.

These families of extended incomplete generalized hypergeometric functions satisfy the following decomposition relation:

$$
{}_r\gamma_s\left[\begin{matrix}(\alpha_1,x;\,p),\alpha_2,\ldots,\alpha_r;\\ \beta_1,\ldots,\beta_s;\end{matrix}\;z\right]+{}_r\Gamma_s\left[\begin{matrix}(\alpha_1,x;\,p),\alpha_2,\ldots,\alpha_r;\\ \beta_1,\ldots,\beta_s;\end{matrix}\;z\right]
$$

$$
=\sum_{n=0}^{\infty}\frac{(\alpha_1;\,p)_n(\alpha_2)_n\cdots(\alpha_r)_n}{(\beta_1)_n\cdots(\beta_s)_n}\frac{z^n}{n!}={}_rF_s\left[\begin{matrix}(\alpha_1,\,p),\alpha_2,\ldots,\alpha_r;\\ \beta_1,\ldots,\beta_s;\end{matrix}\;z\right]. \tag{3.20}
$$

If we take $r=2$ and $s=1$, we get extended incomplete Gauss hypergeometric function ${}_2\Gamma_1^p(z)$ given as:

$$
{}_2\Gamma_1^p(z)={}_2\Gamma_1\left[\begin{matrix}(\alpha_1,x;\,p),\alpha_2;\\ \beta;\end{matrix}\;z\right]=\sum_{n=0}^{\infty}\frac{[\alpha_1;x,p]_n(\alpha_2)_n}{(\beta)_n}\frac{z^n}{n!} \tag{3.21}
$$

and the corresponding extension of the confluent (Kummers) hypergeometric function ${}_1\Gamma_1^p(z)$ can be expressed as:

$$
{}_1\Gamma_1^p(z)={}_1\Gamma_1\left[\begin{matrix}(\alpha,x;\,p);\\ \beta;\end{matrix}\;z\right]=\sum_{n=0}^{\infty}\frac{[\alpha;x,p]_n}{(\beta)_n}\frac{z^n}{n!}. \tag{3.22}
$$

4 Fractional Integral Formulas Involving Extended Incomplete Generalized Hypergeometric Functions

In this section, we consider composition of the fractional integral operators $\left(I_{0,x}^{\sigma,\sigma',\nu,\nu',\eta}f\right)(x)$ and $\left(I_{x,\infty}^{\sigma,\sigma',\nu,\nu',\eta}f\right)(x)$ given in Equations (2.1) and (2.2), respectively with the extended incomplete generalized hypergeometric functions ${}_r\gamma_s^p(z)$ and ${}_r\Gamma_s^p(z)$ defined in Equations (3.18) and (3.19), respectively. Some special cases involving well-known fractional integral operators are also presented.

Theorem 1 *Let* $x>0, a,\sigma,\sigma',\nu,\nu',\eta,\rho\in\mathbb{C}$ *and* $\alpha_1,\ldots,\alpha_r\in\mathbb{C},\beta_1,\ldots,\beta_s\in\mathbb{C}\setminus\mathbb{Z}_0^-$ *be such that* $\Re(\eta)>0,\Re(a)>0$ *and* $\Re(\rho+n)>\max\{0,\Re(\sigma+\sigma'+\nu-\eta),\Re(\sigma'-\nu')\}$, *then the following fractional integral formula holds true:*

$$
\left(I_{0,x}^{\sigma,\sigma',\nu,\nu',\eta}t^{\rho-1}{}_r\gamma_s^p\left[\begin{matrix}(\alpha_1,x;\,p),\alpha_2,\ldots,\alpha_r;\\ \beta_1,\ldots,\beta_s;\end{matrix}\;at\right]\right)(x)
$$

$$
=x^{\rho+\eta-\sigma-\sigma'-1}\frac{\Gamma(\rho)\Gamma(\rho+\eta-\sigma-\sigma'-\nu)\Gamma(\rho+\nu'-\sigma')}{\Gamma(\rho+\nu')\Gamma(\rho+\eta-\sigma-\sigma')\Gamma(\rho+\eta-\sigma'-\nu)}
$$

$$
\times\,{}_{r+3}\gamma_{s+3}^p\left[\begin{matrix}(\alpha_1,x;\,p),\alpha_2,\ldots,\alpha_r,\rho,\rho+\eta-\sigma-\sigma'-\nu,\rho+\nu'-\sigma';\\ \beta_1,\ldots,\beta_s,\rho+\nu',\rho+\eta-\sigma-\sigma',\rho+\eta-\sigma'-\nu;\end{matrix}\;ax\right]. \tag{4.1}
$$

Proof For convenience, we denote the left-hand side of the result (4.1) by $\mathscr{I}$. Then by using Equation (3.18) and then interchanging the order of integration and summation, we have

$$\mathscr{I} = \sum_{n=0}^{\infty} \frac{(\alpha_1; x, p)_n (\alpha_2)_n \cdots (\alpha_r)_n}{(\beta_1)_n \cdots (\beta_s)_n} \frac{a^n}{n!} \left(I_{0,x}^{\sigma,\sigma',\nu,\nu',\eta} t^{\rho+n-1} \right)(x), \tag{4.2}$$

using the result (2.25), the above Equation (4.2) reduces to

$$\begin{aligned} \mathscr{I} &= \sum_{n=0}^{\infty} \frac{(\alpha_1; x, p)_n (\alpha_2)_n \cdots (\alpha_r)_n}{(\beta_1)_n \cdots (\beta_s)_n} \frac{a^n}{n!} \\ &\times \frac{\Gamma(\rho+n)\Gamma(\rho+n+\eta-\sigma-\sigma'-\nu)\Gamma(\rho+n+\nu'-\sigma')}{\Gamma(\rho+n+\nu')\Gamma(\rho+n+\eta-\sigma-\sigma')\Gamma(\rho+n+\eta-\sigma'-\nu)} \\ &\times x^{\rho+n+\eta-\sigma-\sigma'-1}, \end{aligned} \tag{4.3}$$

after simplification, the above Equation (4.3) reduces to

$$\begin{aligned} \mathscr{I} &= x^{\rho+\eta-\sigma-\sigma'-1} \frac{\Gamma(\rho)\Gamma(\rho+\eta-\sigma-\sigma'-\nu)\Gamma(\rho+\nu'-\sigma')}{\Gamma(\rho+\nu')\Gamma(\rho+\eta-\sigma-\sigma')\Gamma(\rho+\eta-\sigma'-\nu)} \\ &\times \sum_{n=0}^{\infty} \frac{(\alpha_1; x, p)_n (\alpha_2)_n \cdots (\alpha_r)_n}{(\beta_1)_n \cdots (\beta_s)_n} \\ &\times \frac{(\rho)_n (\rho+\eta-\sigma-\sigma'-\nu)_n (\rho+\nu'-\sigma')_n}{(\rho+\nu')_n (\rho+\eta-\sigma-\sigma')_n (\rho+\eta-\sigma'-\nu)_n} \frac{(ax)^n}{n!}, \end{aligned} \tag{4.4}$$

the above Equation (4.4), in view of Equation (3.18), gives the required result. □

Theorem 2 *Let* $x > 0, a, \sigma, \sigma', \nu, \nu', \eta, \rho \in \mathbb{C}$ *and* $\alpha_1, \ldots, \alpha_r \in \mathbb{C}, \beta_1, \ldots, \beta_s \in \mathbb{C} \setminus \mathbb{Z}_0^-$ *be such that* $\Re(\eta) > 0, \Re(a) > 0$ *and* $\Re(\rho+n) > \max\{0, \Re(\sigma+\sigma'+\nu-\eta), \Re(\sigma'-\nu')\}$, *then the following fractional integral formula holds true:*

$$\begin{aligned} &\left(I_{0,x}^{\sigma,\sigma',\nu,\nu',\eta} t^{\rho-1} {}_r\Gamma_s^p \left[\begin{matrix} (\alpha_1, x; p), \alpha_2, \ldots, \alpha_r; \\ \beta_1, \ldots, \beta_s; \end{matrix} at \right] \right)(x) \\ &= x^{\rho+\eta-\sigma-\sigma'-1} \frac{\Gamma(\rho)\Gamma(\rho+\eta-\sigma-\sigma'-\nu)\Gamma(\rho+\nu'-\sigma')}{\Gamma(\rho+\nu')\Gamma(\rho+\eta-\sigma-\sigma')\Gamma(\rho+\eta-\sigma'-\nu)} \\ &\times {}_{r+3}\Gamma_{s+3}^p \left[\begin{matrix} (\alpha_1, x; p), \alpha_2, \ldots, \alpha_r, \rho, \rho+\eta-\sigma-\sigma'-\nu, \rho+\nu'-\sigma'; \\ \beta_1, \ldots, \beta_s, \rho+\nu', \rho+\eta-\sigma-\sigma', \rho+\eta-\sigma'-\nu; \end{matrix} ax \right]. \end{aligned} \tag{4.5}$$

Theorem 3 *Let* $x > 0, a, \sigma, \sigma', \nu, \nu', \eta, \rho \in \mathbb{C}$ *and* $\alpha_1, \ldots, \alpha_r \in \mathbb{C}, \beta_1, \ldots, \beta_s \in \mathbb{C} \setminus \mathbb{Z}_0^-$ *be such that* $\Re(\eta) > 0, \Re(a) > 0$ *and* $\Re(\rho-n) < 1+\min\{\Re(-\nu), \Re(\sigma+\sigma'-\eta), \Re(\sigma+\nu'-\eta)\}$, *then the following fractional integral formula holds true:*

$$\left(I_{x,\infty}^{\sigma,\sigma',\nu,\nu',\eta} t^{\rho-1} {}_r\gamma_s^p \left[\begin{matrix} (\alpha_1, x; p), \alpha_2, \ldots, \alpha_r; \frac{a}{t} \\ \beta_1, \ldots, \beta_s; \end{matrix} \right]\right)(x)$$
$$= x^{\rho+\eta-\sigma-\sigma'-1} \frac{\Gamma(1-\rho-\nu)\Gamma(1-\rho-\eta+\sigma+\sigma')\Gamma(1-\rho-\eta+\sigma+\nu')}{\Gamma(1-\rho)\Gamma(1-\rho-\eta+\sigma+\sigma'+\nu')\Gamma(1-\rho+\sigma-\nu)}$$
$$\times {}_{r+3}\gamma_{s+3}^p \left[\begin{matrix} (\alpha_1, x; p), \alpha_2, \ldots, \alpha_r, \\ \beta_1, \ldots, \beta_s, \end{matrix} \right.$$
$$\left. \begin{matrix} 1-\rho-\nu, 1-\rho-\eta+\sigma+\sigma', 1-\rho-\eta+\sigma+\nu'; \frac{a}{x} \\ 1-\rho, 1-\rho-\eta+\sigma+\sigma'+\nu', 1-\rho+\sigma-\nu; \end{matrix} \right]. \tag{4.6}$$

Theorem 4 *Let $x > 0, a, \sigma, \sigma', \nu, \nu', \eta, \rho \in \mathbb{C}$ and $\alpha_1, \ldots, \alpha_r \in \mathbb{C}, \beta_1, \ldots, \beta_s \in \mathbb{C} \setminus \mathbb{Z}_0^-$ be such that $\Re(\eta) > 0, \Re(a) > 0$ and $\Re(\rho - n) < 1 + \min\{\Re(-\nu), \Re(\sigma + \sigma' - \eta), \Re(\sigma + \nu' - \eta)\}$, then the following fractional integral formula holds true:*

$$\left(I_{x,\infty}^{\sigma,\sigma',\nu,\nu',\eta} t^{\rho-1} {}_r\Gamma_s^p \left[\begin{matrix} (\alpha_1, x; p), \alpha_2, \ldots, \alpha_r; \frac{a}{t} \\ \beta_1, \ldots, \beta_s; \end{matrix} \right]\right)(x)$$
$$= x^{\rho+\eta-\sigma-\sigma'-1} \frac{\Gamma(1-\rho-\nu)\Gamma(1-\rho-\eta+\sigma+\sigma')\Gamma(1-\rho-\eta+\sigma+\nu')}{\Gamma(1-\rho)\Gamma(1-\rho-\eta+\sigma+\sigma'+\nu')\Gamma(1-\rho+\sigma-\nu)}$$
$$\times {}_{r+3}\Gamma_{s+3}^p \left[\begin{matrix} (\alpha_1, x; p), \alpha_2, \ldots, \alpha_r, \\ \beta_1, \ldots, \beta_s, \end{matrix} \right.$$
$$\left. \begin{matrix} 1-\rho-\nu, 1-\rho-\eta+\sigma+\sigma', 1-\rho-\eta+\sigma+\nu'; \frac{a}{x} \\ 1-\rho, 1-\rho-\eta+\sigma+\sigma'+\nu', 1-\rho+\sigma-\nu; \end{matrix} \right]. \tag{4.7}$$

Proof The proof of Theorems 2–4 would run parallel to Theorem 1. We, therefore, choose to skip the details involved. □

The following corollaries are easy consequences of the results involved in Theorems 1, 2, 3 and 4, respectively

4.1 Special Cases

It is interesting to mention some special cases by choosing suitable values of the parameters $\sigma, \sigma', \nu, \nu'$ and η. If we put $\sigma = \sigma + \nu, \sigma' = \nu' = 0, \nu = -\eta, \eta = \sigma$ in Theorems 1, 2, 3 and 4, we get certain interesting results concerning the Saigo fractional integral operators given by the following corollaries.

Corollary 1 *Let $x > 0, a, \sigma, \nu, \eta, \rho \in \mathbb{C}$ and $\alpha_1, \ldots, \alpha_r \in \mathbb{C}, \beta_1, \ldots, \beta_s \in \mathbb{C} \setminus \mathbb{Z}_0^-$ be such that $\Re(\sigma) > 0, \Re(a) > 0$ and $\Re(\rho + n) > \max\{0, \Re(\nu - \eta)\}$, then the following fractional integral formula holds true:*

$$\left(I_{0,x}^{\sigma,\nu,\eta} t^{\rho-1} {}_r\gamma_s^p \left[\begin{matrix} (\alpha_1, x; p), \alpha_2, \ldots, \alpha_r; \\ \beta_1, \ldots, \beta_s; \end{matrix} at \right]\right)(x)$$

$$= x^{\rho-\nu-1} \frac{\Gamma(\rho)\Gamma(\rho+\eta-\nu)}{\Gamma(\rho-\nu)\Gamma(\rho+\eta+\sigma)} \tag{4.8}$$

$$\times {}_{r+2}\gamma_{s+2}^p \left[\begin{matrix} (\alpha_1, x; p), \alpha_2, \ldots, \alpha_r, \rho, \rho+\eta-\nu; \\ \beta_1, \ldots, \beta_s, \rho-\nu, \rho+\eta+\sigma; \end{matrix} ax \right].$$

Corollary 2 *Let $x > 0, a, \sigma, \nu, \eta, \rho \in \mathbb{C}$ and $\alpha_1, \ldots, \alpha_r \in \mathbb{C}, \beta_1, \ldots, \beta_s \in \mathbb{C}\setminus\mathbb{Z}_0^-$ be such that $\Re(\sigma) > 0, \Re(a) > 0$ and $\Re(\rho+n) > \max\{0, \Re(\nu-\eta)\}$, then the following fractional integral formula holds true:*

$$\left(I_{0,x}^{\sigma,\nu,\eta} t^{\rho-1} {}_r\Gamma_s^p \left[\begin{matrix} (\alpha_1, x; p), \alpha_2, \ldots, \alpha_r; \\ \beta_1, \ldots, \beta_s; \end{matrix} at \right]\right)(x)$$

$$= x^{\rho-\nu-1} \frac{\Gamma(\rho)\Gamma(\rho+\eta-\nu)}{\Gamma(\rho-\nu)\Gamma(\rho+\eta+\sigma)} \tag{4.9}$$

$$\times {}_{r+2}\Gamma_{s+2}^p \left[\begin{matrix} (\alpha_1, x; p), \alpha_2, \ldots, \alpha_r, \rho, \rho+\eta-\nu; \\ \beta_1, \ldots, \beta_s, \rho-\nu, \rho+\eta+\sigma; \end{matrix} ax \right].$$

Corollary 3 *Let $x > 0, a, \sigma, \nu, \eta, \rho \in \mathbb{C}$ and $\alpha_1, \ldots, \alpha_r \in \mathbb{C}, \beta_1, \ldots, \beta_s \in \mathbb{C}\setminus\mathbb{Z}_0^-$ be such that $\Re(\sigma) > 0, \Re(a) > 0$ and $\Re(\rho-n) < 1+\min\{\Re(\nu), \Re(\eta)\}$, then the following fractional integral formula holds true:*

$$\left(I_{x,\infty}^{\sigma,\nu,\eta} t^{\rho-1} {}_r\gamma_s^p \left[\begin{matrix} (\alpha_1, x; p), \alpha_2, \ldots, \alpha_r; \\ \beta_1, \ldots, \beta_s; \end{matrix} \frac{a}{t} \right]\right)(x)$$

$$= x^{\rho-\nu-1} \frac{\Gamma(1-\rho+\nu)\Gamma(1-\rho+\eta)}{\Gamma(1-\rho)\Gamma(1-\rho+\sigma+\nu+\eta)} \tag{4.10}$$

$$\times {}_{r+2}\gamma_{s+2}^p \left[\begin{matrix} (\alpha_1, x; p), \alpha_2, \ldots, \alpha_r, 1-\rho+\nu, 1-\rho+\eta; \\ \beta_1, \ldots, \beta_s, 1-\rho, 1-\rho+\sigma+\nu+\eta; \end{matrix} \frac{a}{x} \right].$$

Corollary 4 *Let $x > 0, a, \sigma, \nu, \eta, \rho \in \mathbb{C}$ and $\alpha_1, \ldots, \alpha_r \in \mathbb{C}, \beta_1, \ldots, \beta_s \in \mathbb{C}\setminus\mathbb{Z}_0^-$ be such that $\Re(\sigma) > 0, \Re(a) > 0$ and $\Re(\rho-n) < 1+\min\{\Re(\nu), \Re(\eta)\}$, then the following fractional integral formula holds true:*

$$\left(I_{x,\infty}^{\sigma,\nu,\eta} t^{\rho-1} {}_r\Gamma_s^p \left[\begin{matrix} (\alpha_1, x; p), \alpha_2, \ldots, \alpha_r; \\ \beta_1, \ldots, \beta_s; \end{matrix} \frac{a}{t} \right]\right)(x)$$

$$= x^{\rho-\nu-1} \frac{\Gamma(1-\rho+\nu)\Gamma(1-\rho+\eta)}{\Gamma(1-\rho)\Gamma(1-\rho+\sigma+\nu+\eta)} \tag{4.11}$$

$$\times {}_{r+2}\Gamma_{s+2}^p \left[\begin{matrix} (\alpha_1, x; p), \alpha_2, \ldots, \alpha_r, 1-\rho+\nu, 1-\rho+\eta; \\ \beta_1, \ldots, \beta_s, 1-\rho, 1-\rho+\sigma+\nu+\eta; \end{matrix} \frac{a}{x} \right].$$

By setting $\nu = 0$ in Corollaries 1–4, the Saigo fractional integrals formulas reduce to the Erdélyi-Kober type fractional integral formulas given as follows:

Corollary 5 *Let* $x > 0, a, \sigma, \eta, \rho \in \mathbb{C}$ *and* $\alpha_1, \ldots, \alpha_r \in \mathbb{C}, \beta_1, \ldots, \beta_s \in \mathbb{C} \setminus \mathbb{Z}_0^-$ *be such that* $\Re(\sigma) > 0, \Re(a) > 0$ *and* $\Re(\rho + n) > -\Re(\eta)$*, then the following fractional integral formula holds true:*

$$\begin{aligned}&\left(I_{0,x}^{\sigma,\eta} t^{\rho-1} {}_r\gamma_s^p \left[\begin{matrix}(\alpha_1, x; p), \alpha_2, \ldots, \alpha_r; \\ \beta_1, \ldots, \beta_s;\end{matrix} at\right]\right)(x)\\ &= x^{\rho-1} \frac{\Gamma(\rho+\eta)}{\Gamma(\rho+\sigma+\eta)} {}_{r+1}\gamma_{s+1}^p \left[\begin{matrix}(\alpha_1, x; p), \alpha_2, \ldots, \alpha_r, \rho+\eta; \\ \beta_1, \ldots, \beta_s, \rho+\sigma+\eta;\end{matrix} ax\right].\end{aligned} \tag{4.12}$$

Corollary 6 *Let* $x > 0, a, \sigma, \eta, \rho \in \mathbb{C}$ *and* $\alpha_1, \ldots, \alpha_r \in \mathbb{C}, \beta_1, \ldots, \beta_s \in \mathbb{C} \setminus \mathbb{Z}_0^-$ *be such that* $\Re(\sigma) > 0, \Re(a) > 0$ *and* $\Re(\rho + n) > -\Re(\eta)$*, then the following fractional integral formula holds true:*

$$\begin{aligned}&\left(I_{0,x}^{\sigma,\eta} t^{\rho-1} {}_r\Gamma_s^p \left[\begin{matrix}(\alpha_1, x; p), \alpha_2, \ldots, \alpha_r; \\ \beta_1, \ldots, \beta_s;\end{matrix} at\right]\right)(x)\\ &= x^{\rho-1} \frac{\Gamma(\rho+\eta)}{\Gamma(\rho+\sigma+\eta)} {}_{r+1}\Gamma_{s+1}^p \left[\begin{matrix}(\alpha_1, x; p), \alpha_2, \ldots, \alpha_r, \rho+\eta; \\ \beta_1, \ldots, \beta_s, \rho+\sigma+\eta;\end{matrix} ax\right].\end{aligned} \tag{4.13}$$

Corollary 7 *Let* $x > 0, a, \sigma, \eta, \rho \in \mathbb{C}$ *and* $\alpha_1, \ldots, \alpha_r \in \mathbb{C}, \beta_1, \ldots, \beta_s \in \mathbb{C} \setminus \mathbb{Z}_0^-$ *be such that* $\Re(\sigma) > 0, \Re(a) > 0$ *and* $\Re(\rho - n) < 1 + \Re(\eta)$*, then the following fractional integral formula holds true:*

$$\begin{aligned}&\left(I_{x,\infty}^{\sigma,\eta} t^{\rho-1} {}_r\gamma_s^p \left[\begin{matrix}(\alpha_1, x; p), \alpha_2, \ldots, \alpha_r; \\ \beta_1, \ldots, \beta_s;\end{matrix} \frac{a}{t}\right]\right)(x)\\ &= x^{\rho-1} \frac{\Gamma(1-\rho+\eta)}{\Gamma(1-\rho+\sigma+\eta)} {}_{r+1}\gamma_{s+1}^p \left[\begin{matrix}(\alpha_1, x; p), \alpha_2, \ldots, \alpha_r, 1-\rho+\eta; \\ \beta_1, \ldots, \beta_s, 1-\rho+\sigma+\eta;\end{matrix} \frac{a}{x}\right].\end{aligned} \tag{4.14}$$

Corollary 8 *Let* $x > 0, a, \sigma, \eta, \rho \in \mathbb{C}$ *and* $\alpha_1, \ldots, \alpha_r \in \mathbb{C}, \beta_1, \ldots, \beta_s \in \mathbb{C} \setminus \mathbb{Z}_0^-$ *be such that* $\Re(\sigma) > 0, \Re(a) > 0$ *and* $\Re(\rho - n) < 1 + \Re(\eta)$*, then the following fractional integral formula holds true:*

$$\begin{aligned}&\left(I_{x,\infty}^{\sigma,\eta} t^{\rho-1} {}_r\Gamma_s^p \left[\begin{matrix}(\alpha_1, x; p), \alpha_2, \ldots, \alpha_r; \\ \beta_1, \ldots, \beta_s;\end{matrix} \frac{a}{t}\right]\right)(x)\\ &= x^{\rho-1} \frac{\Gamma(1-\rho+\eta)}{\Gamma(1-\rho+\sigma+\eta)} {}_{r+1}\Gamma_{s+1}^p \left[\begin{matrix}(\alpha_1, x; p), \alpha_2, \ldots, \alpha_r, 1-\rho+\eta; \\ \beta_1, \ldots, \beta_s, 1-\rho+\sigma+\eta;\end{matrix} \frac{a}{x}\right].\end{aligned} \tag{4.15}$$

Further, by setting $\nu = -\sigma$ in Corollaries 1–4, then the Saigo fractional integrals formulas reduce to the following Riemann-Liouville and the Weyl type fractional integral formulas as given below:

Corollary 9 *Let* $x > 0, a, \sigma, \rho \in \mathbb{C}$ *and* $\alpha_1, \ldots, \alpha_r \in \mathbb{C}, \beta_1, \ldots, \beta_s \in \mathbb{C} \setminus \mathbb{Z}_0^-$ *be such that* $\Re(\sigma) > 0, \Re(a) > 0$ *and* $\Re(\rho + n) > 0$*, then the following fractional integral formula holds true:*

$$\left(I_{0,x}^{\sigma} t^{\rho-1} {}_r\gamma_s^p \left[\begin{matrix} (\alpha_1, x; p), \alpha_2, \ldots, \alpha_r; \\ \beta_1, \ldots, \beta_s; \end{matrix} at\right]\right)(x) = x^{\rho+\sigma-1} \frac{\Gamma(\rho)}{\Gamma(\rho+\sigma)} {}_{r+1}\gamma_{s+1}^p \left[\begin{matrix} (\alpha_1, x; p), \alpha_2, \ldots, \alpha_r, \rho; \\ \beta_1, \ldots, \beta_s, \rho+\sigma; \end{matrix} ax\right]. \tag{4.16}$$

Corollary 10 *Let* $x > 0, a, \sigma, \rho \in \mathbb{C}$ *and* $\alpha_1, \ldots, \alpha_r \in \mathbb{C}, \beta_1, \ldots, \beta_s \in \mathbb{C} \setminus \mathbb{Z}_0^-$ *be such that* $\Re(\sigma) > 0, \Re(a) > 0$ *and* $\Re(\rho + n) > 0$*, then the following fractional integral formula holds true:*

$$\left(I_{0,x}^{\sigma} t^{\rho-1} {}_r\Gamma_s^p \left[\begin{matrix} (\alpha_1, x; p), \alpha_2, \ldots, \alpha_r; \\ \beta_1, \ldots, \beta_s; \end{matrix} at\right]\right)(x) = x^{\rho+\sigma-1} \frac{\Gamma(\rho)}{\Gamma(\rho+\sigma)} {}_{r+1}\Gamma_{s+1}^p \left[\begin{matrix} (\alpha_1, x; p), \alpha_2, \ldots, \alpha_r, \rho; \\ \beta_1, \ldots, \beta_s, \rho+\sigma; \end{matrix} ax\right]. \tag{4.17}$$

Corollary 11 *Let* $x > 0, a, \sigma, \rho \in \mathbb{C}$ *and* $\alpha_1, \ldots, \alpha_r \in \mathbb{C}, \beta_1, \ldots, \beta_s \in \mathbb{C} \setminus \mathbb{Z}_0^-$ *be such that* $\Re(\sigma) > 0, \Re(a) > 0$ *and* $0 < \Re(\sigma) < 1 - \Re(\rho - n)$*, then the following fractional integral formula holds true:*

$$\left(I_{x,\infty}^{\sigma} t^{\rho-1} {}_r\gamma_s^p \left[\begin{matrix} (\alpha_1, x; p), \alpha_2, \ldots, \alpha_r; \\ \beta_1, \ldots, \beta_s; \end{matrix} \frac{a}{t}\right]\right)(x) = x^{\rho+\sigma-1} \frac{\Gamma(1-\rho-\sigma)}{\Gamma(1-\rho)} {}_{r+1}\gamma_{s+1}^p \left[\begin{matrix} (\alpha_1, x; p), \alpha_2, \ldots, \alpha_r, 1-\rho-\sigma; \\ \beta_1, \ldots, \beta_s, 1-\rho; \end{matrix} \frac{a}{x}\right]. \tag{4.18}$$

Corollary 12 *Let* $x > 0, a, \sigma, \rho \in \mathbb{C}$ *and* $\alpha_1, \ldots, \alpha_r \in \mathbb{C}, \beta_1, \ldots, \beta_s \in \mathbb{C} \setminus \mathbb{Z}_0^-$ *be such that* $\Re(\sigma) > 0, \Re(a) > 0$ *and* $0 < \Re(\sigma) < 1 - \Re(\rho - n)$*, then the following fractional integral formula holds true:*

$$\left(I_{x,\infty}^{\sigma} t^{\rho-1} {}_r\Gamma_s^p \left[\begin{matrix} (\alpha_1, x; p), \alpha_2, \ldots, \alpha_r; \\ \beta_1, \ldots, \beta_s; \end{matrix} \frac{a}{t}\right]\right)(x) = x^{\rho+\sigma-1} \frac{\Gamma(1-\rho-\sigma)}{\Gamma(1-\rho)} {}_{r+1}\Gamma_{s+1}^p \left[\begin{matrix} (\alpha_1, x; p), \alpha_2, \ldots, \alpha_r, 1-\rho-\sigma; \\ \beta_1, \ldots, \beta_s, 1-\rho; \end{matrix} \frac{a}{x}\right]. \tag{4.19}$$

5 Fractional Differential Formulas Involving the Extended Incomplete Generalized Hypergeometric Functions

In this section, we consider composition of the fractional differential operators $\left(D_{0,x}^{\sigma,\sigma',\nu,\nu',\eta} f\right)(x)$ and $\left(D_{x,\infty}^{\sigma,\sigma',\nu,\nu',\eta} f\right)(x)$ given in Equations (2.9) and (2.10), respectively with the extended incomplete generalized hypergeometric functions ${}_r\gamma_s^p(z)$ and ${}_r\Gamma_s^p(z)$ defined in Equations (3.18) and (3.19), respectively. Some special cases involving well-known fractional differential operators are also established.

Theorem 5 *Let $x > 0, a, \sigma, \sigma', \nu, \nu', \eta, \rho \in \mathbb{C}$ and $\alpha_1, \ldots, \alpha_r \in \mathbb{C}, \beta_1, \ldots, \beta_s \in \mathbb{C} \setminus \mathbb{Z}_0^-$ be such that $\Re(\eta) > 0$, $\Re(a) > 0$ and $\Re(\rho + n) > \max\{0, \Re(\eta - \sigma - \sigma' - \nu'), \Re(\nu - \sigma)\}$, then the following fractional derivative formula holds true:*

$$\left(D_{0,x}^{\sigma,\sigma',\nu,\nu',\eta} t^{\rho-1} {}_r\gamma_s^p \left[\begin{matrix}(\alpha_1, x; p), \alpha_2, \ldots, \alpha_r; \\ \beta_1, \ldots, \beta_s;\end{matrix} at\right]\right)(x)$$
$$= x^{\rho-\eta+\sigma+\sigma'-1} \frac{\Gamma(\rho)\Gamma(\rho-\eta+\sigma+\sigma'+\nu')\Gamma(\rho-\nu+\sigma)}{\Gamma(\rho-\nu)\Gamma(\rho-\eta+\sigma+\sigma')\Gamma(\rho-\eta+\sigma+\nu')}$$
$$\times {}_{r+3}\gamma_{s+3}^p \left[\begin{matrix}(\alpha_1, x; p), \alpha_2, \ldots, \alpha_r, \rho, \rho-\eta+\sigma+\sigma'+\nu', \rho-\nu+\sigma; \\ \beta_1, \ldots, \beta_s, \rho-\nu, \rho-\eta+\sigma+\sigma', \rho-\eta+\sigma+\nu';\end{matrix} ax\right]. \tag{5.1}$$

Proof For convenience, we denote the left-hand side of the result (5.1) by $\mathscr{D}$. Then by using (3.18) and interchanging the order of differentiation and summation, we get

$$\mathscr{D} = \sum_{n=0}^{\infty} \frac{(\alpha_1; x, p)_n (\alpha_2)_n \cdots (\alpha_r)_n}{(\beta_1)_n \cdots (\beta_s)_n} \frac{a^n}{n!} \times \left(D_{0,x}^{\sigma,\sigma',\nu,\nu',\eta} t^{\rho+n-1}\right)(x), \tag{5.2}$$

applying the result (2.27), the above Equation (5.2) reduces to

$$\mathscr{D} = \sum_{n=0}^{\infty} \frac{(\alpha_1; x, p)_n (\alpha_2)_n \cdots (\alpha_r)_n}{(\beta_1)_n \cdots (\beta_s)_n} \frac{a^n}{n!} \times \frac{\Gamma(\rho+n)\Gamma(\rho+n-\eta+\sigma+\sigma'+\nu')\Gamma(\rho+n-\nu+\sigma)}{\Gamma(\rho+n-\nu)\Gamma(\rho+n-\eta+\sigma+\sigma')\Gamma(\rho+n-\eta+\sigma+\nu')} \times x^{\rho+n-\eta+\sigma+\sigma'-1}, \tag{5.3}$$

after simplification, the above Equation (5.3) reduces to

$$\mathscr{D} = x^{\rho-\eta+\sigma+\sigma'-1} \frac{\Gamma(\rho)\Gamma(\rho-\eta+\sigma+\sigma'+\nu')\Gamma(\rho-\nu+\sigma)}{\Gamma(\rho-\nu)\Gamma(\rho-\eta+\sigma+\sigma')\Gamma(\rho-\eta+\sigma+\nu')}$$
$$\times \sum_{n=0}^{\infty} \frac{(\alpha_1; x, p)_n(\alpha_2)_n \cdots (\alpha_r)_n}{(\beta_1)_n \cdots (\beta_s)_n} \tag{5.4}$$
$$\times \frac{(\rho)_n(\rho-\eta+\sigma+\sigma'+\nu')_n(\rho-\nu+\sigma)_n}{(\rho-\nu)_n(\rho-\eta+\sigma+\sigma')_n(\rho-\eta+\sigma+\nu')_n} \frac{(ax)^n}{n!},$$

the above Equation (5.4), in view of Equation (3.18), gives the required result. □

Theorem 6 *Let* $x > 0, a, \sigma, \sigma', \nu, \nu', \eta, \rho \in \mathbb{C}$ *and* $\alpha_1, \ldots, \alpha_r \in \mathbb{C}, \beta_1, \ldots, \beta_s \in \mathbb{C} \setminus \mathbb{Z}_0^-$ *be such that* $\Re(\eta) > 0, \Re(a) > 0$ *and* $\Re(\rho+n) > \max\{0, \Re(\eta-\sigma-\sigma'-\nu'), \Re(\nu-\sigma)\}$, *then the following fractional derivative formula holds true:*

$$\left(D_{0,x}^{\sigma,\sigma',\nu,\nu',\eta} t^{\rho-1} {}_r\Gamma_s^p \left[\begin{matrix} (\alpha_1, x; p), \alpha_2, \ldots, \alpha_r; \\ \beta_1, \ldots, \beta_s; \end{matrix} at\right]\right)(x)$$
$$= x^{\rho-\eta+\sigma+\sigma'-1} \frac{\Gamma(\rho)\Gamma(\rho-\eta+\sigma+\sigma'+\nu')\Gamma(\rho-\nu+\sigma)}{\Gamma(\rho-\nu)\Gamma(\rho-\eta+\sigma+\sigma')\Gamma(\rho-\eta+\sigma+\nu')}$$
$$\times {}_{r+3}\Gamma_{s+3}^p \left[\begin{matrix} (\alpha_1, x; p), \alpha_2, \ldots, \alpha_r, \rho, \rho-\eta+\sigma+\sigma'+\nu', \rho-\nu+\sigma; \\ \beta_1, \ldots, \beta_s, \rho-\nu, \rho-\eta+\sigma+\sigma', \rho-\eta+\sigma+\nu'; \end{matrix} ax\right]. \tag{5.5}$$

Theorem 7 *Let* $x > 0, a, \sigma, \sigma', \nu, \nu', \eta, \rho \in \mathbb{C}$ *and* $\alpha_1, \ldots, \alpha_r \in \mathbb{C}, \beta_1, \ldots, \beta_s \in \mathbb{C} \setminus \mathbb{Z}_0^-$ *be such that* $\Re(\eta) > 0, \Re(a) > 0$ *and* $\Re(\rho-n) < 1+\min\{\Re(\nu'), \Re(\eta-\sigma-\sigma'), \Re(\eta-\sigma'-\nu)\}$, *then the following fractional derivative formula holds true:*

$$\left(D_{x,\infty}^{\sigma,\sigma',\nu,\nu',\eta} t^{\rho-1} {}_r\gamma_s^p \left[\begin{matrix} (\alpha_1, x; p), \alpha_2, \ldots, \alpha_r; \\ \beta_1, \ldots, \beta_s; \end{matrix} \frac{a}{t}\right]\right)(x)$$
$$= x^{\rho-\eta+\sigma+\sigma'-1} \frac{\Gamma(1-\rho+\nu')\Gamma(1-\rho+\eta-\sigma-\sigma')\Gamma(1-\rho+\eta-\sigma'-\nu)}{\Gamma(1-\rho)\Gamma(1-\rho+\eta-\sigma-\sigma'-\nu)\Gamma(1-\rho-\sigma'+\nu')}$$
$$\times {}_{r+3}\gamma_{s+3}^p \left[\begin{matrix} (\alpha_1, x; p), \alpha_2, \ldots, \alpha_r, \\ \beta_1, \ldots, \beta_s, \end{matrix}\right.$$
$$\left.\begin{matrix} 1-\rho+\nu', 1-\rho+\eta-\sigma-\sigma', 1-\rho+\eta-\sigma'-\nu; \\ 1-\rho, 1-\rho+\eta-\sigma-\sigma'-\nu, 1-\rho-\sigma'+\nu'; \end{matrix} \frac{a}{x}\right]. \tag{5.6}$$

Theorem 8 *Let* $x > 0, a, \sigma, \sigma', \nu, \nu', \eta, \rho \in \mathbb{C}$ *and* $\alpha_1, \ldots, \alpha_r \in \mathbb{C}, \beta_1, \ldots, \beta_s \in \mathbb{C} \setminus \mathbb{Z}_0^-$ *be such that* $\Re(\eta) > 0, \Re(a) > 0$ *and* $\Re(\rho-n) < 1+\min\{\Re(\nu'), \Re(\eta-\sigma-\sigma'), \Re(\eta-\sigma'-\nu)\}$, *then the following fractional derivative formula holds true:*

$$\left(D_{x,\infty}^{\sigma,\sigma',\nu,\nu',\eta} t^{\rho-1} {}_r\Gamma_s^p \left[\begin{matrix} (\alpha_1, x; p), \alpha_2, \ldots, \alpha_r; \\ \beta_1, \ldots, \beta_s; \end{matrix} \frac{a}{t} \right]\right)(x)$$

$$= x^{\rho-\eta+\sigma+\sigma'-1} \frac{\Gamma(1-\rho+\nu')\Gamma(1-\rho+\eta-\sigma-\sigma')\Gamma(1-\rho+\eta-\sigma'-\nu)}{\Gamma(1-\rho)\Gamma(1-\rho+\eta-\sigma-\sigma'-\nu)\Gamma(1-\rho-\sigma'+\nu')}$$

$$\times {}_{r+3}\Gamma_{s+3}^p \left[\begin{matrix} (\alpha_1, x; p), \alpha_2, \ldots, \alpha_r, \\ \beta_1, \ldots, \beta_s, \end{matrix} \right.$$

$$\left. \begin{matrix} 1-\rho+\nu', 1-\rho+\eta-\sigma-\sigma', 1-\rho+\eta-\sigma'-\nu; \\ 1-\rho, 1-\rho+\eta-\sigma-\sigma'-\nu, 1-\rho-\sigma'+\nu'; \end{matrix} \frac{a}{x} \right]. \tag{5.7}$$

Proof The proof of Theorems 6–8 would run parallel to Theorem 5. We, therefore, choose to skip the details involved. □

5.1 Special Cases

By setting $\sigma = \sigma + \nu, \sigma' = \nu' = 0, \nu = -\eta, \eta = \sigma$ in Theorems 5–8, we get certain interesting results concerning the Saigo fractional derivative operator given in the following corollaries.

Corollary 13 *Let* $x > 0, a, \sigma, \nu, \eta, \rho \in \mathbb{C}$ *and* $\alpha_1, \ldots, \alpha_r \in \mathbb{C}, \beta_1, \ldots, \beta_s \in \mathbb{C} \setminus \mathbb{Z}_0^-$ *be such that* $\Re(\sigma) > 0, \Re(a) > 0$ *and* $\Re(\rho + n) > -\min\{0, \Re(\sigma+\nu+\eta)\}$ *then the following fractional derivative formula holds true:*

$$\left(D_{0,x}^{\sigma,\nu,\eta} t^{\rho-1} {}_r\gamma_s^p \left[\begin{matrix} (\alpha_1, x; p), \alpha_2, \ldots, \alpha_r; \\ \beta_1, \ldots, \beta_s; \end{matrix} at \right]\right)(x)$$

$$= x^{\rho+\nu-1} \frac{\Gamma(\rho)\Gamma(\rho+\eta+\sigma+\nu)}{\Gamma(\rho+\eta)\Gamma(\rho+\nu)} \tag{5.8}$$

$$\times {}_{r+2}\gamma_{s+2}^p \left[\begin{matrix} (\alpha_1, x; p), \alpha_2, \ldots, \alpha_r, \rho, \rho+\eta+\sigma+\nu; \\ \beta_1, \ldots, \beta_s, \rho+\eta, \rho+\nu; \end{matrix} ax \right].$$

Corollary 14 *Let* $x > 0, a, \sigma, \nu, \eta, \rho \in \mathbb{C}$ *and* $\alpha_1, \ldots, \alpha_r \in \mathbb{C}, \beta_1, \ldots, \beta_s \in \mathbb{C} \setminus \mathbb{Z}_0^-$ *be such that* $\Re(\sigma) > 0, \Re(a) > 0$ *and* $\Re(\rho + n) > -\min\{0, \Re(\sigma+\nu+\eta)\}$, *then the following fractional derivative formula holds true:*

$$\left(D_{0,x}^{\sigma,\nu,\eta} t^{\rho-1} {}_r\Gamma_s^p \left[\begin{matrix} (\alpha_1, x; p), \alpha_2, \ldots, \alpha_r; \\ \beta_1, \ldots, \beta_s; \end{matrix} at \right]\right)(x)$$
$$= x^{\rho+\nu-1} \frac{\Gamma(\rho)\Gamma(\rho+\eta+\sigma+\nu)}{\Gamma(\rho+\eta)\Gamma(\rho+\nu)} \tag{5.9}$$
$$\times {}_{r+2}\Gamma_{s+2}^p \left[\begin{matrix} (\alpha_1, x; p), \alpha_2, \ldots, \alpha_r, \rho, \rho+\eta+\sigma+\nu; \\ \beta_1, \ldots, \beta_s, \rho+\eta, \rho+\nu; \end{matrix} ax \right].$$

Corollary 15 *Let $x > 0, a, \sigma, \nu, \eta, \rho \in \mathbb{C}$ and $\alpha_1, \ldots, \alpha_r \in \mathbb{C}, \beta_1, \ldots, \beta_s \in \mathbb{C} \setminus \mathbb{Z}_0^-$ be such that $\Re(\sigma) > 0, \Re(a) > 0$ and $\Re(\rho - n) < 1 + \min\{\Re(-\nu - n*), \Re(\eta + \sigma)\}$ and $n* = [\Re(\sigma)] + 1$, then the following fractional derivative formula holds true:*

$$\left(D_{x,\infty}^{\sigma,\nu,\eta} t^{\rho-1} {}_r\gamma_s^p \left[\begin{matrix} (\alpha_1, x; p), \alpha_2, \ldots, \alpha_r; \\ \beta_1, \ldots, \beta_s; \end{matrix} \frac{a}{t} \right]\right)(x)$$
$$= x^{\rho+\nu-1} \frac{\Gamma(1-\rho-\nu)\Gamma(1-\rho+\sigma+\eta)}{\Gamma(1-\rho)\Gamma(1-\rho+\eta-\nu)}$$
$$\times {}_{r+2}\gamma_{s+2}^p \left[\begin{matrix} (\alpha_1, x; p), \alpha_2, \ldots, \alpha_r, 1-\rho-\nu, 1-\rho+\sigma+\eta; \\ \beta_1, \ldots, \beta_s, 1-\rho, 1-\rho+\eta-\nu; \end{matrix} \frac{a}{x} \right]. \tag{5.10}$$

Corollary 16 *Let $x > 0, a, \sigma, \nu, \eta, \rho \in \mathbb{C}$ and $\alpha_1, \ldots, \alpha_r \in \mathbb{C}, \beta_1, \ldots, \beta_s \in \mathbb{C} \setminus \mathbb{Z}_0^-$ be such that $\Re(\sigma) > 0, \Re(a) > 0$ and $\Re(\rho - n) < 1 + \min\{\Re(-\nu - n*), \Re(\eta + \sigma)\}$ and $n* = [\Re(\sigma)] + 1$, then the following fractional derivative formula holds true:*

$$\left(D_{x,\infty}^{\sigma,\nu,\eta} t^{\rho-1} {}_r\Gamma_s^p \left[\begin{matrix} (\alpha_1, x; p), \alpha_2, \ldots, \alpha_r; \\ \beta_1, \ldots, \beta_s; \end{matrix} \frac{a}{t} \right]\right)(x)$$
$$= x^{\rho+\nu-1} \frac{\Gamma(1-\rho-\nu)\Gamma(1-\rho+\sigma+\eta)}{\Gamma(1-\rho)\Gamma(1-\rho+\eta-\nu)}$$
$$\times {}_{r+2}\Gamma_{s+2}^p \left[\begin{matrix} (\alpha_1, x; p), \alpha_2, \ldots, \alpha_r, 1-\rho-\nu, 1-\rho+\sigma+\eta; \\ \beta_1, \ldots, \beta_s, 1-\rho, 1-\rho+\eta-\nu; \end{matrix} \frac{a}{x} \right]. \tag{5.11}$$

Further, on setting $\nu = 0$ in Corollaries 13–16, the Saigo fractional differential formulas reduce to the following fractional differential formulas as given below:

Corollary 17 *Let $x > 0, a, \sigma, \eta, \rho \in \mathbb{C}$ and $\alpha_1, \ldots, \alpha_r \in \mathbb{C}, \beta_1, \ldots, \beta_s \in \mathbb{C} \setminus \mathbb{Z}_0^-$ be such that $\Re(\sigma) > 0, \Re(a) > 0$ and $\Re(\rho + n) > -\Re(\eta + \sigma)$, then the following fractional derivative formula holds true:*

$$\left(D_{0,x}^{\sigma,\eta} t^{\rho-1} {}_r\gamma_s^p \left[\begin{matrix} (\alpha_1, x;\, p), \alpha_2, \ldots, \alpha_r; \\ \beta_1, \ldots, \beta_s; \end{matrix} \, at \right]\right)(x)$$

$$= x^{\rho-1} \frac{\Gamma(\rho+\eta+\sigma)}{\Gamma(\rho+\eta)} {}_{r+1}\gamma_{s+1}^p \left[\begin{matrix} (\alpha_1, x;\, p), \alpha_2, \ldots, \alpha_r, \rho+\eta+\sigma; \\ \beta_1, \ldots, \beta_s, \rho+\eta; \end{matrix} \, ax \right]. \tag{5.12}$$

Corollary 18 *Let* $x > 0, a, \sigma, \eta, \rho \in \mathbb{C}$ *and* $\alpha_1, \ldots, \alpha_r \in \mathbb{C}, \beta_1, \ldots, \beta_s \in \mathbb{C} \setminus \mathbb{Z}_0^-$ *be such that* $\Re(\sigma) > 0, \Re(a) > 0$ *and* $\Re(\rho + n) > -\Re(\eta + \sigma)$*, then the following fractional derivative formula holds true:*

$$\left(D_{0,x}^{\sigma,\eta} t^{\rho-1} {}_r\Gamma_s^p \left[\begin{matrix} (\alpha_1, x;\, p), \alpha_2, \ldots, \alpha_r; \\ \beta_1, \ldots, \beta_s; \end{matrix} \, at \right]\right)(x)$$

$$= x^{\rho-1} \frac{\Gamma(\rho+\eta+\sigma)}{\Gamma(\rho+\eta)} {}_{r+1}\Gamma_{s+1}^p \left[\begin{matrix} (\alpha_1, x;\, p), \alpha_2, \ldots, \alpha_r, \rho+\eta+\sigma; \\ \beta_1, \ldots, \beta_s, \rho+\eta; \end{matrix} \, ax \right]. \tag{5.13}$$

Corollary 19 *Let* $x > 0, a, \sigma, \eta, \rho \in \mathbb{C}$ *and* $\alpha_1, \ldots, \alpha_r \in \mathbb{C}, \beta_1, \ldots, \beta_s \in \mathbb{C} \setminus \mathbb{Z}_0^-$ *be such that* $\Re(\sigma) > 0, \Re(a) > 0$ *and* $\Re(\rho - n) < 1 + \Re(\eta + \sigma) - n*$ *and* $n* = [\Re(\sigma)] + 1$*, then the following fractional derivative formula holds true:*

$$\left(D_{x,\infty}^{\sigma,\eta} t^{\rho-1} {}_r\gamma_s^p \left[\begin{matrix} (\alpha_1, x;\, p), \alpha_2, \ldots, \alpha_r; \\ \beta_1, \ldots, \beta_s; \end{matrix} \, \frac{a}{t} \right]\right)(x)$$

$$= x^{\rho-1} \frac{\Gamma(1-\rho+\sigma+\eta)}{\Gamma(1-\rho+\eta)} {}_{r+1}\gamma_{s+1}^p \left[\begin{matrix} (\alpha_1, x;\, p), \alpha_2, \ldots, \alpha_r, 1-\rho+\sigma+\eta; \\ \beta_1, \ldots, \beta_s, 1-\rho+\eta; \end{matrix} \, \frac{a}{x} \right]. \tag{5.14}$$

Corollary 20 *Let* $x > 0, a, \sigma, \eta, \rho \in \mathbb{C}$ *and* $\alpha_1, \ldots, \alpha_r \in \mathbb{C}, \beta_1, \ldots, \beta_s \in \mathbb{C} \setminus \mathbb{Z}_0^-$ *be such that* $\Re(\sigma) > 0, \Re(a) > 0$ *and* $\Re(\rho - n) < 1 + \Re(\eta + \sigma) - n*$ *and* $n* = [\Re(\sigma)] + 1$*, then the following fractional derivative formula holds true:*

$$\left(D_{x,\infty}^{\sigma,\eta} t^{\rho-1} {}_r\Gamma_s^p \left[\begin{matrix} (\alpha_1, x;\, p), \alpha_2, \ldots, \alpha_r; \\ \beta_1, \ldots, \beta_s; \end{matrix} \, \frac{a}{t} \right]\right)(x)$$

$$= x^{\rho-1} \frac{\Gamma(1-\rho+\sigma+\eta)}{\Gamma(1-\rho+\eta)} {}_{r+1}\Gamma_{s+1}^p \left[\begin{matrix} (\alpha_1, x;\, p), \alpha_2, \ldots, \alpha_r, 1-\rho+\sigma+\eta; \\ \beta_1, \ldots, \beta_s, 1-\rho+\eta; \end{matrix} \, \frac{a}{x} \right]. \tag{5.15}$$

Further, if we set $\nu = -\sigma$ in Corollaries 13–16, then these Saigo fractional derivatives reduce to the following Riemann-Liouville and the Weyl type fractional derivative operators as given below:

Corollary 21 *Let $x > 0, a, \sigma, \rho \in \mathbb{C}$ and $\alpha_1, \ldots, \alpha_r \in \mathbb{C}, \beta_1, \ldots, \beta_s \in \mathbb{C} \setminus \mathbb{Z}_0^-$ be such that $\Re(\sigma) > 0, \Re(a) > 0$ and $\Re(\rho + n) > \Re(\sigma) > 0$, then the following fractional derivative formula holds true:*

$$\begin{aligned} &\left(D_{0,x}^{\sigma} t^{\rho-1} {}_r\gamma_s^p \left[\begin{matrix} (\alpha_1, x; p), \alpha_2, \ldots, \alpha_r; \\ \beta_1, \ldots, \beta_s; \end{matrix} at \right]\right)(x) \\ &= x^{\rho-\sigma-1} \frac{\Gamma(\rho)}{\Gamma(\rho-\sigma)} {}_{r+1}\gamma_{s+1}^p \left[\begin{matrix} (\alpha_1, x; p), \alpha_2, \ldots, \alpha_r, \rho; \\ \beta_1, \ldots, \beta_s, \rho-\sigma; \end{matrix} ax \right]. \end{aligned} \tag{5.16}$$

Corollary 22 *Let $x > 0, a, \sigma, \rho \in \mathbb{C}$ and $\alpha_1, \ldots, \alpha_r \in \mathbb{C}, \beta_1, \ldots, \beta_s \in \mathbb{C} \setminus \mathbb{Z}_0^-$ be such that $\Re(\sigma) > 0, \Re(a) > 0$ and $\Re(\rho + n) > \Re(\sigma) > 0$, then the following fractional derivative formula holds true:*

$$\begin{aligned} &\left(D_{0,x}^{\sigma} t^{\rho-1} {}_r\Gamma_s^p \left[\begin{matrix} (\alpha_1, x; p), \alpha_2, \ldots, \alpha_r; \\ \beta_1, \ldots, \beta_s; \end{matrix} at \right]\right)(x) \\ &= x^{\rho-\sigma-1} \frac{\Gamma(\rho)}{\Gamma(\rho-\sigma)} {}_{r+1}\Gamma_{s+1}^p \left[\begin{matrix} (\alpha_1, x; p), \alpha_2, \ldots, \alpha_r, \rho; \\ \beta_1, \ldots, \beta_s, \rho-\sigma; \end{matrix} ax \right]. \end{aligned} \tag{5.17}$$

Corollary 23 *Let $x > 0, a, \sigma, \rho \in \mathbb{C}$ and $\alpha_1, \ldots, \alpha_r \in \mathbb{C}, \beta_1, \ldots, \beta_s \in \mathbb{C} \setminus \mathbb{Z}_0^-$ be such that $\Re(\sigma) > 0, \Re(a) > 0$ and $\Re(\rho - n) < 1 + \Re(\sigma) - n*$ and $n* = [\Re(\sigma)] + 1$, then the following fractional derivative formula holds true:*

$$\begin{aligned} &\left(D_{x,\infty}^{\sigma} t^{\rho-1} {}_r\gamma_s^p \left[\begin{matrix} (\alpha_1, x; p), \alpha_2, \ldots, \alpha_r; \\ \beta_1, \ldots, \beta_s; \end{matrix} \frac{a}{t} \right]\right)(x) \\ &= x^{\rho-\sigma-1} \frac{\Gamma(1-\rho+\sigma)}{\Gamma(1-\rho)} {}_{r+1}\gamma_{s+1}^p \left[\begin{matrix} (\alpha_1, x; p), \alpha_2, \ldots, \alpha_r, 1-\rho+\sigma; \\ \beta_1, \ldots, \beta_s, 1-\rho; \end{matrix} \frac{a}{x} \right]. \end{aligned} \tag{5.18}$$

Corollary 24 *Let $x > 0, a, \sigma, \rho \in \mathbb{C}$ and $\alpha_1, \ldots, \alpha_r \in \mathbb{C}, \beta_1, \ldots, \beta_s \in \mathbb{C} \setminus \mathbb{Z}_0^-$ be such that $\Re(\sigma) > 0, \Re(a) > 0$ and $\Re(\rho - n) < 1 + \Re(\sigma) - n*$ and $n* = [\Re(\sigma)] + 1$, then the following fractional derivative formula holds true:*

$$\begin{aligned} &\left(D_{x,\infty}^{\sigma} t^{\rho-1} {}_r\Gamma_s^p \left[\begin{matrix} (\alpha_1, x; p), \alpha_2, \ldots, \alpha_r; \\ \beta_1, \ldots, \beta_s; \end{matrix} \frac{a}{t} \right]\right)(x) \\ &= x^{\rho-\sigma-1} \frac{\Gamma(1-\rho+\sigma)}{\Gamma(1-\rho)} {}_{r+1}\Gamma_{s+1}^p \left[\begin{matrix} (\alpha_1, x; p), \alpha_2, \ldots, \alpha_r, 1-\rho+\sigma; \\ \beta_1, \ldots, \beta_s, 1-\rho; \end{matrix} \frac{a}{x} \right]. \end{aligned} \tag{5.19}$$

6 Integral Transform Formulas of the Extended Incomplete Generalized Hypergeometric Functions

In this section, we established certain theorems involving the results obtained in the previous sections associated with the integral transforms like Beta transform, Laplace transform and Whittaker transform.

6.1 Beta Transform

Definition 19 The Beta transform of function $f(z)$ is defined as [72]:

$$B\{f(z):l,m\}=\int_0^1 z^{l-1}(1-z)^{m-1}f(z)dz \tag{6.1}$$

Theorem 9 *Let* $x>0, a, \sigma, \sigma', \nu, \nu', \eta, \rho \in \mathbb{C}$ *and* $\alpha_1, \ldots, \alpha_r \in \mathbb{C}, \beta_1, \ldots, \beta_s \in \mathbb{C}\setminus\mathbb{Z}_0^-$ *be such that* $\Re(\eta)>0, \Re(a)>0$ *and* $\Re(\rho+n)>\max\{0, \Re(\sigma+\sigma'+\nu-\eta), \Re(\sigma'-\nu')\}$, *then the following formula holds:*

$$\begin{aligned}
&B\left\{\left(I_{0,x}^{\sigma,\sigma',\nu,\nu',\eta}t^{\rho-1}{}_r\gamma_s^p\left[\begin{matrix}(\alpha_1,x;p),\alpha_2,\ldots,\alpha_r;\\ \beta_1,\ldots,\beta_s;\end{matrix}\,atz\right]\right)(x):l,m\right\}\\
&=x^{\rho+\eta-\sigma-\sigma'-1}\frac{B(l,m)\Gamma(\rho)\Gamma(\rho+\eta-\sigma-\sigma'-\nu)\Gamma(\rho+\nu'-\sigma')}{\Gamma(\rho+\nu')\Gamma(\rho+\eta-\sigma-\sigma')\Gamma(\rho+\eta-\sigma'-\nu)}\\
&\times{}_{r+4}\gamma_{s+4}^p\left[\begin{matrix}(\alpha_1,x;p),\alpha_2,\ldots,\alpha_r,\rho,\rho+\eta-\sigma-\sigma'-\nu,\rho+\nu'-\sigma',l;\\ \beta_1,\ldots,\beta_s,\rho+\nu',\rho+\eta-\sigma-\sigma',\rho+\eta-\sigma'-\nu,l+m;\end{matrix}\,ax\right].
\end{aligned}\tag{6.2}$$

Proof Denote the left-hand side of the result (6.2) by $\mathscr{B}$. Using definition of beta transform as given in Equation (6.1), we get

$$\begin{aligned}
\mathscr{B}=&\int_0^1 z^{l-1}(1-z)^{m-1}\\
&\times\left\{\left(I_{0,x}^{\sigma,\sigma',\nu,\nu',\eta}t^{\rho-1}{}_r\gamma_s^p\left[\begin{matrix}(\alpha_1,x;p),\alpha_2,\ldots,\alpha_r;\\ \beta_1,\ldots,\beta_s;\end{matrix}\,atz\right]\right)(x)\right\}dz,
\end{aligned}\tag{6.3}$$

using the result (4.4), the above Equation (6.3) reduces to

$$\mathscr{B}=\int_0^1 z^{l+n-1}(1-z)^{m-1}x^{\rho+\eta-\sigma-\sigma'-1}\frac{\Gamma(\rho)\Gamma(\rho+\eta-\sigma-\sigma'-\nu)\Gamma(\rho+\nu'-\sigma')}{\Gamma(\rho+\nu')\Gamma(\rho+\eta-\sigma-\sigma')\Gamma(\rho+\eta-\sigma'-\nu)}$$
$$\times\sum_{n=0}^{\infty}\frac{(\alpha_1;x,p)_n(\alpha_2)_n\cdots(\alpha_r)_n}{(\beta_1)_n\cdots(\beta_s)_n}\frac{(\rho)_n(\rho+\eta-\sigma-\sigma'-\nu)_n(\rho+\nu'-\sigma')_n}{(\rho+\nu')_n(\rho+\eta-\sigma-\sigma')_n(\rho+\eta-\sigma'-\nu)_n}\frac{(ax)^n}{n!}dz, \tag{6.4}$$

Interchanging the order of integration and summation, we have

$$\mathscr{B}=x^{\rho+\eta-\sigma-\sigma'-1}\frac{\Gamma(\rho)\Gamma(\rho+\eta-\sigma-\sigma'-\nu)\Gamma(\rho+\nu'-\sigma')}{\Gamma(\rho+\nu')\Gamma(\rho+\eta-\sigma-\sigma')\Gamma(\rho+\eta-\sigma'-\nu)}$$
$$\times\sum_{n=0}^{\infty}\frac{(\alpha_1;x,p)_n(\alpha_2)_n\cdots(\alpha_r)_n}{(\beta_1)_n\cdots(\beta_s)_n}\frac{(\rho)_n(\rho+\eta-\sigma-\sigma'-\nu)_n}{(\rho+\nu')_n(\rho+\eta-\sigma-\sigma')_n} \tag{6.5}$$
$$\times\frac{(\rho+\nu'-\sigma')_n}{(\rho+\eta-\sigma'-\nu)_n}\frac{(ax)^n}{n!}\int_0^1 z^{l+n-1}(1-z)^{m-1}dz,$$

after simplification, we have

$$\mathscr{B}=x^{\rho+\eta-\sigma-\sigma'-1}\frac{B(l,m)\Gamma(\rho)\Gamma(\rho+\eta-\sigma-\sigma'-\nu)\Gamma(\rho+\nu'-\sigma')}{\Gamma(\rho+\nu')\Gamma(\rho+\eta-\sigma-\sigma')\Gamma(\rho+\eta-\sigma'-\nu)}$$
$$\times\sum_{n=0}^{\infty}\frac{(\alpha_1;x,p)_n(\alpha_2)_n\cdots(\alpha_r)_n}{(\beta_1)_n\cdots(\beta_s)_n} \tag{6.6}$$
$$\times\frac{(\rho)_n(\rho+\eta-\sigma-\sigma'-\nu)_n(\rho+\nu'-\sigma')_n(l)_n}{(\rho+\nu')_n(\rho+\eta-\sigma-\sigma')_n(\rho+\eta-\sigma'-\nu)_n(l+m)_n}\frac{(ax)^n}{n!},$$

interpreting the above equation, in view of (3.18), we have the required result. □

Theorem 10 *Let $x>0, a, \sigma, \sigma', \nu, \nu', \eta, \rho\in\mathbb{C}$ and $\alpha_1,\ldots,\alpha_r\in\mathbb{C}, \beta_1,\ldots,\beta_s\in\mathbb{C}\setminus\mathbb{Z}_0^-$ be such that $\Re(\eta)>0$, $\Re(a)>0$ and $\Re(\rho+n)>\max\{0,\Re(\sigma+\sigma'+\nu-\eta),\Re(\sigma'-\nu')\}$, then the following formula holds:*

$$B\left\{\left(I_{0,x}^{\sigma,\sigma',\nu,\nu',\eta}t^{\rho-1}{}_r\Gamma_s^p\left[\begin{matrix}(\alpha_1,x;p),\alpha_2,\ldots,\alpha_r;\\ \beta_1,\ldots,\beta_s;\end{matrix}\,atz\right]\right)(x):l,m\right\}$$
$$=x^{\rho+\eta-\sigma-\sigma'-1}\frac{B(l,m)\Gamma(\rho)\Gamma(\rho+\eta-\sigma-\sigma'-\nu)\Gamma(\rho+\nu'-\sigma')}{\Gamma(\rho+\nu')\Gamma(\rho+\eta-\sigma-\sigma')\Gamma(\rho+\eta-\sigma'-\nu)}$$
$$\times{}_{r+4}\Gamma_{s+4}^p\left[\begin{matrix}(\alpha_1,x;p),\alpha_2,\ldots,\alpha_r,\rho,\rho+\eta-\sigma-\sigma'-\nu,\rho+\nu'-\sigma',l;\\ \beta_1,\ldots,\beta_s,\rho+\nu',\rho+\eta-\sigma-\sigma',\rho+\eta-\sigma'-\nu,l+m;\end{matrix}\,ax\right]. \tag{6.7}$$

Theorem 11 *Let $x>0, a, \sigma, \sigma', \nu, \nu', \eta, \rho\in\mathbb{C}$ and $\alpha_1,\ldots,\alpha_r\in\mathbb{C}, \beta_1,\ldots,\beta_s\in\mathbb{C}\setminus\mathbb{Z}_0^-$ be such that $\Re(\eta)>0$, $\Re(a)>0$ and $\Re(\rho-n)<1+\min\{\Re(-\nu),\Re(\sigma+$*

$\sigma' - \eta), \Re(\sigma + \nu' - \eta)\}$, *then the following formula holds:*

$$B\left\{\left(I_{x,\infty}^{\sigma,\sigma',\nu,\nu',\eta} t^{\rho-1} {}_r\gamma_s^p\left[\begin{matrix}(\alpha_1, x; p), \alpha_2, \ldots, \alpha_r; \\ \beta_1, \ldots, \beta_s;\end{matrix} \frac{az}{t}\right]\right)(x) : l, m\right\}$$
$$= x^{\rho+\eta-\sigma-\sigma'-1} \frac{B(l,m)\Gamma(1-\rho-\nu)\Gamma(1-\rho-\eta+\sigma+\sigma')\Gamma(1-\rho-\eta+\sigma+\nu')}{\Gamma(1-\rho)\Gamma(1-\rho-\eta+\sigma+\sigma'+\nu')\Gamma(1-\rho+\sigma-\nu)}$$
$$\times {}_{r+4}\gamma_{s+4}^p\left[\begin{matrix}(\alpha_1, x; p), \alpha_2, \ldots, \alpha_r, \\ \beta_1, \ldots, \beta_s,\end{matrix}\right.$$
$$\left.\begin{matrix}1-\rho-\nu, 1-\rho-\eta+\sigma+\sigma', 1-\rho-\eta+\sigma+\nu', l; \\ 1-\rho, 1-\rho-\eta+\sigma+\sigma'+\nu', 1-\rho+\sigma-\nu, l+m;\end{matrix} \frac{a}{x}\right]. \tag{6.8}$$

Theorem 12 *Let* $x > 0, a, \sigma, \sigma', \nu, \nu', \eta, \rho \in \mathbb{C}$ *and* $\alpha_1, \ldots, \alpha_r \in \mathbb{C}, \beta_1, \ldots, \beta_s \in \mathbb{C} \setminus \mathbb{Z}_0^-$ *be such that* $\Re(\eta) > 0, \Re(a) > 0$ *and* $\Re(\rho - n) < 1 + \min\{\Re(-\nu), \Re(\sigma + \sigma' - \eta), \Re(\sigma + \nu' - \eta)\}$, *then the following formula holds:*

$$B\left\{\left(I_{x,\infty}^{\sigma,\sigma',\nu,\nu',\eta} t^{\rho-1} {}_r\Gamma_s^p\left[\begin{matrix}(\alpha_1, x; p), \alpha_2, \ldots, \alpha_r; \\ \beta_1, \ldots, \beta_s;\end{matrix} \frac{az}{t}\right]\right)(x) : l, m\right\}$$
$$= x^{\rho+\eta-\sigma-\sigma'-1} \frac{B(l,m)\Gamma(1-\rho-\nu)\Gamma(1-\rho-\eta+\sigma+\sigma')\Gamma(1-\rho-\eta+\sigma+\nu')}{\Gamma(1-\rho)\Gamma(1-\rho-\eta+\sigma+\sigma'+\nu')\Gamma(1-\rho+\sigma-\nu)}$$
$$\times {}_{r+4}\Gamma_{s+4}^p\left[\begin{matrix}(\alpha_1, x; p), \alpha_2, \ldots, \alpha_r, \\ \beta_1, \ldots, \beta_s,\end{matrix}\right.$$
$$\left.\begin{matrix}1-\rho-\nu, 1-\rho-\eta+\sigma+\sigma', 1-\rho-\eta+\sigma+\nu', l; \\ 1-\rho, 1-\rho-\eta+\sigma+\sigma'+\nu', 1-\rho+\sigma-\nu, l+m;\end{matrix} \frac{a}{x}\right]. \tag{6.9}$$

Proof The proof of Theorems 10–12 would run parallel to Theorem 9, so we omit the details involved. □

Theorem 13 *Let* $x > 0, a, \sigma, \sigma', \nu, \nu', \eta, \rho \in \mathbb{C}$ *and* $\alpha_1, \ldots, \alpha_r \in \mathbb{C}, \beta_1, \ldots, \beta_s \in \mathbb{C} \setminus \mathbb{Z}_0^-$ *be such that* $\Re(\eta) > 0, \Re(a) > 0$ *and* $\Re(\rho + n) > \max\{0, \Re(\eta - \sigma - \sigma' - \nu'), \Re(\nu - \sigma)\}$, *then the following formula holds:*

$$B\left\{\left(D_{0,x}^{\sigma,\sigma',\nu,\nu',\eta} t^{\rho-1} {}_r\gamma_s^p\left[\begin{matrix}(\alpha_1, x; p), \alpha_2, \ldots, \alpha_r; \\ \beta_1, \ldots, \beta_s;\end{matrix} atz\right]\right)(x) : l, m\right\}$$
$$= x^{\rho-\eta+\sigma+\sigma'-1} \frac{B(l,m)\Gamma(\rho)\Gamma(\rho-\eta+\sigma+\sigma'+\nu')\Gamma(\rho-\nu+\sigma)}{\Gamma(\rho-\nu)\Gamma(\rho-\eta+\sigma+\sigma')\Gamma(\rho-\eta+\sigma+\nu')}$$
$$\times {}_{r+4}\gamma_{s+4}^p\left[\begin{matrix}(\alpha_1, x; p), \alpha_2, \ldots, \alpha_r, \rho, \rho-\eta+\sigma+\sigma'+\nu', \rho-\nu+\sigma, l; \\ \beta_1, \ldots, \beta_s, \rho-\nu, \rho-\eta+\sigma+\sigma', \rho-\eta+\sigma+\nu', l+m;\end{matrix} ax\right]. \tag{6.10}$$

Proof Denote the left-hand side of the result (6.10) by $\mathscr{B}$, then using definition of beta transform as given in Equation (6.1), we have

$$\mathscr{B} = \int_0^1 z^{l-1}(1-z)^{m-1} \times \left\{ \left(D_{0,x}^{\sigma,\sigma',\nu,\nu',\eta} t^{\rho-1} {}_r\gamma_s^p \left[\begin{matrix} (\alpha_1, x; p), \alpha_2, \ldots, \alpha_r; \\ \beta_1, \ldots, \beta_s; \end{matrix} atz \right] \right)(x) \right\} dz, \tag{6.11}$$

using the result (5.4), the above Equation (6.11) reduces to

$$\mathscr{B} = \int_0^1 z^{l+n-1}(1-z)^{m-1} \left\{ x^{\rho-\eta+\sigma+\sigma'-1} \frac{\Gamma(\rho)\Gamma(\rho-\eta+\sigma+\sigma'+\nu')}{\Gamma(\rho-\nu)\Gamma(\rho-\eta+\sigma+\sigma')} \right.$$
$$\times \frac{\Gamma(\rho-\nu+\sigma)}{\Gamma(\rho-\eta+\sigma+\nu')} \sum_{n=0}^{\infty} \frac{(\alpha_1; x, p)_n (\alpha_2)_n \cdots (\alpha_r)_n}{(\beta_1)_n \cdots (\beta_s)_n}$$
$$\left. \times \frac{(\rho)_n (\rho-\eta+\sigma+\sigma'+\nu')_n (\rho-\nu+\sigma)_n}{(\rho-\nu)_n (\rho-\eta+\sigma+\sigma')_n (\rho-\eta+\sigma+\nu')_n} \frac{(ax)^n}{n!} \right\} dz, \tag{6.12}$$

interchanging the order of integration and summation, we have

$$\mathscr{B} = x^{\rho-\eta+\sigma+\sigma'-1} \frac{\Gamma(\rho)\Gamma(\rho-\eta+\sigma+\sigma'+\nu')\Gamma(\rho-\nu+\sigma)}{\Gamma(\rho-\nu)\Gamma(\rho-\eta+\sigma+\sigma')\Gamma(\rho-\eta+\sigma+\nu')}$$
$$\times \sum_{n=0}^{\infty} \frac{(\alpha_1; x, p)_n (\alpha_2)_n \cdots (\alpha_r)_n}{(\beta_1)_n \cdots (\beta_s)_n} \frac{(\rho)_n (\rho-\eta+\sigma+\sigma'+\nu')_n}{(\rho-\nu)_n (\rho-\eta+\sigma+\sigma')_n} \tag{6.13}$$
$$\times \frac{(\rho-\nu+\sigma)_n}{(\rho-\eta+\sigma+\nu')_n} \frac{(ax)^n}{n!} \int_0^1 z^{l+n-1}(1-z)^{m-1} dz,$$

after simplification, we have

$$\mathscr{B} = x^{\rho-\eta+\sigma+\sigma'-1} \frac{B(l,m)\Gamma(\rho)\Gamma(\rho-\eta+\sigma+\sigma'+\nu')\Gamma(\rho-\nu+\sigma)}{\Gamma(\rho-\nu)\Gamma(\rho-\eta+\sigma+\sigma')\Gamma(\rho-\eta+\sigma+\nu')}$$
$$\times \sum_{n=0}^{\infty} \frac{(\alpha_1; x, p)_n (\alpha_2)_n \cdots (\alpha_r)_n}{(\beta_1)_n \cdots (\beta_s)_n}$$
$$\times \frac{(\rho)_n (\rho-\eta+\sigma+\sigma'+\nu')_n (\rho-\nu+\sigma)_n (l)_n}{(\rho-\nu)_n (\rho-\eta+\sigma+\sigma')_n (\rho-\eta+\sigma+\nu')_n (l+m)_n} \frac{(ax)^n}{n!}, \tag{6.14}$$

interpreting the above equation, in view of (3.18), we have the required result. □

Theorem 14 *Let* $x > 0, a, \sigma, \sigma', \nu, \nu', \eta, \rho \in \mathbb{C}$ *and* $\alpha_1, \ldots, \alpha_r \in \mathbb{C}, \beta_1, \ldots, \beta_s \in \mathbb{C} \setminus \mathbb{Z}_0^-$ *be such that* $\Re(\eta) > 0, \Re(a) > 0$ *and* $\Re(\rho + n) > \max\{0, \Re(\eta - \sigma - \sigma' - \nu'), \Re(\nu - \sigma)\}$, *then the following formula holds:*

$$B\left\{\left(D_{0,x}^{\sigma,\sigma',\nu,\nu',\eta} t^{\rho-1} {}_r\Gamma_s^p\left[\begin{matrix}(\alpha_1, x; p), \alpha_2, \ldots, \alpha_r; \\ \beta_1, \ldots, \beta_s;\end{matrix} atz\right]\right)(x) : l, m\right\}$$
$$= x^{\rho-\eta+\sigma+\sigma'-1} \frac{B(l,m)\Gamma(\rho)\Gamma(\rho-\eta+\sigma+\sigma'+\nu')\Gamma(\rho-\nu+\sigma)}{\Gamma(\rho-\nu)\Gamma(\rho-\eta+\sigma+\sigma')\Gamma(\rho-\eta+\sigma+\nu')}$$
$$\times {}_{r+4}\Gamma_{s+4}^p\left[\begin{matrix}(\alpha_1, x; p), \alpha_2, \ldots, \alpha_r, \rho, \rho-\eta+\sigma+\sigma'+\nu', \rho-\nu+\sigma, l; \\ \beta_1, \ldots, \beta_s, \rho-\nu, \rho-\eta+\sigma+\sigma', \rho-\eta+\sigma+\nu', l+m;\end{matrix} ax\right]. \tag{6.15}$$

Theorem 15 *Let* $x > 0, a, \sigma, \sigma', \nu, \nu', \eta, \rho \in \mathbb{C}$ *and* $\alpha_1, \ldots, \alpha_r \in \mathbb{C}, \beta_1, \ldots, \beta_s \in \mathbb{C} \setminus \mathbb{Z}_0^-$ *be such that* $\Re(\eta) > 0, \Re(a) > 0$ *and* $\Re(\rho - n) < 1 + \min\{\Re(\nu'), \Re(\eta - \sigma - \sigma'), \Re(\eta - \sigma' - \nu)\}$, *then the following formula holds:*

$$B\left\{\left(D_{x,\infty}^{\sigma,\sigma',\nu,\nu',\eta} t^{\rho-1} {}_r\gamma_s^p\left[\begin{matrix}(\alpha_1, x; p), \alpha_2, \ldots, \alpha_r; \\ \beta_1, \ldots, \beta_s;\end{matrix} \frac{az}{t}\right]\right)(x) : l, m\right\}$$
$$= x^{\rho-\eta+\sigma+\sigma'-1} \frac{B(l,m)\Gamma(1-\rho+\nu')\Gamma(1-\rho+\eta-\sigma-\sigma')}{\Gamma(1-\rho)\Gamma(1-\rho+\eta-\sigma-\sigma'-\nu)}$$
$$\times \frac{\Gamma(1-\rho+\eta-\sigma'-\nu)}{\Gamma(1-\rho-\sigma'+\nu')} {}_{r+4}\gamma_{s+4}^p\left[\begin{matrix}(\alpha_1, x; p), \alpha_2, \ldots, \alpha_r, \\ \beta_1, \ldots, \beta_s,\end{matrix}\right.$$
$$\left.\begin{matrix}1-\rho+\nu', 1-\rho+\eta-\sigma-\sigma', 1-\rho+\eta-\sigma'-\nu, l; \\ 1-\rho, 1-\rho+\eta-\sigma-\sigma'-\nu, 1-\rho-\sigma'+\nu', l+m;\end{matrix} \frac{a}{x}\right]. \tag{6.16}$$

Theorem 16 *Let* $x > 0, a, \sigma, \sigma', \nu, \nu', \eta, \rho \in \mathbb{C}$ *and* $\alpha_1, \ldots, \alpha_r \in \mathbb{C}, \beta_1, \ldots, \beta_s \in \mathbb{C} \setminus \mathbb{Z}_0^-$ *be such that* $\Re(\eta) > 0, \Re(a) > 0$ *and* $\Re(\rho - n) < 1 + \min\{\Re(\nu'), \Re(\eta - \sigma - \sigma'), \Re(\eta - \sigma' - \nu)\}$, *then the following formula holds:*

$$B\left\{\left(D_{x,\infty}^{\sigma,\sigma',\nu,\nu',\eta} t^{\rho-1} {}_r\Gamma_s^p\left[\begin{matrix}(\alpha_1, x; p), \alpha_2, \ldots, \alpha_r; \\ \beta_1, \ldots, \beta_s;\end{matrix} \frac{az}{t}\right]\right)(x) : l, m\right\}$$
$$= x^{\rho-\eta+\sigma+\sigma'-1} \frac{B(l,m)\Gamma(1-\rho+\nu')\Gamma(1-\rho+\eta-\sigma-\sigma')}{\Gamma(1-\rho)\Gamma(1-\rho+\eta-\sigma-\sigma'-\nu)}$$
$$\times \frac{\Gamma(1-\rho+\eta-\sigma'-\nu)}{\Gamma(1-\rho-\sigma'+\nu')} {}_{r+4}\Gamma_{s+4}^p\left[\begin{matrix}(\alpha_1, x; p), \alpha_2, \ldots, \alpha_r, \\ \beta_1, \ldots, \beta_s,\end{matrix}\right.$$
$$\left.\begin{matrix}1-\rho+\nu', 1-\rho+\eta-\sigma-\sigma', 1-\rho+\eta-\sigma'-\nu, l; \\ 1-\rho, 1-\rho+\eta-\sigma-\sigma'-\nu, 1-\rho-\sigma'+\nu', l+m;\end{matrix} \frac{a}{x}\right]. \tag{6.17}$$

Proof The proof of Theorems 14–16 would run parallel to Theorem 13, so details are omitted here. □

6.2 Laplace Transform

Definition 20 The Laplace transform of $f(z)$ is defined as [72, 73]:

$$L\{f(z)\} = \int_0^\infty e^{-\omega z} f(z)dz \tag{6.18}$$

Theorem 17 *Let* $x > 0, a, \sigma, \sigma', \nu, \nu', \eta, \rho \in \mathbb{C}$ *and* $\alpha_1, \ldots, \alpha_r \in \mathbb{C}, \beta_1, \ldots, \beta_s \in \mathbb{C} \setminus \mathbb{Z}_0^-$ *be such that* $\Re(\eta) > 0, \Re(a) > 0$ *and* $\Re(\rho + n) > \max\{0, \Re(\sigma + \sigma' + \nu - \eta), \Re(\sigma' - \nu')\}$, *then the following formula holds:*

$$L\left\{z^{l-1}\left(I_{0,x}^{\sigma,\sigma',\nu,\nu',\eta} t^{\rho-1} {}_r\gamma_s^p\left[\begin{matrix}(\alpha_1, x; p), \alpha_2, \ldots, \alpha_r; \\ \beta_1, \ldots, \beta_s;\end{matrix} atz\right]\right)(x)\right\}$$
$$= \frac{x^{\rho+\eta-\sigma-\sigma'-1}}{\omega^l} \frac{\Gamma(l)\Gamma(\rho)\Gamma(\rho+\eta-\sigma-\sigma'-\nu)\Gamma(\rho+\nu'-\sigma')}{\Gamma(\rho+\nu')\Gamma(\rho+\eta-\sigma-\sigma')\Gamma(\rho+\eta-\sigma'-\nu)}$$
$$\times {}_{r+4}\gamma_{s+3}^p\left[\begin{matrix}(\alpha_1, x; p), \alpha_2, \ldots, \alpha_r, l, \rho, \rho+\eta-\sigma-\sigma'-\nu, \rho+\nu'-\sigma'; \\ \beta_1, \ldots, \beta_s, \rho+\nu', \rho+\eta-\sigma-\sigma', \rho+\eta-\sigma'-\nu;\end{matrix} \frac{ax}{\omega}\right]. \tag{6.19}$$

Proof Denote the left-hand side of the result (6.19) by $\mathscr{L}$. Using definition of Laplace transform as given in Equation (6.18), we have

$$\mathscr{L} = \int_0^\infty e^{-\omega z} z^{l-1}$$
$$\times \left\{\left(I_{0,x}^{\sigma,\sigma',\nu,\nu',\eta} t^{\rho-1} {}_r\gamma_s^p\left[\begin{matrix}(\alpha_1, x; p), \alpha_2, \ldots, \alpha_r; \\ \beta_1, \ldots, \beta_s;\end{matrix} atz\right]\right)(x)\right\} dz, \tag{6.20}$$

using the result (4.4) and interchanging the order of integration and summation, the above Equation (6.20) reduces to

$$\mathscr{L} = x^{\rho+\eta-\sigma-\sigma'-1} \frac{\Gamma(\rho)\Gamma(\rho+\eta-\sigma-\sigma'-\nu)\Gamma(\rho+\nu'-\sigma')}{\Gamma(\rho+\nu')\Gamma(\rho+\eta-\sigma-\sigma')\Gamma(\rho+\eta-\sigma'-\nu)}$$
$$\times \sum_{n=0}^{\infty} \frac{(\alpha_1; x, p)_n (\alpha_2)_n \cdots (\alpha_r)_n}{(\beta_1)_n \cdots (\beta_s)_n}$$
$$\times \frac{(\rho)_n (\rho+\eta-\sigma-\sigma'-\nu)_n (\rho+\nu'-\sigma')_n}{(\rho+\nu')_n (\rho+\eta-\sigma-\sigma')_n (\rho+\eta-\sigma'-\nu)_n} \frac{(ax)^n}{n!} \int_0^\infty z^{l+n-1} e^{-\omega z} dz, \tag{6.21}$$

after simplification, we have

$$\mathscr{L} = \frac{x^{\rho+\eta-\sigma-\sigma'-1}}{\omega^l} \frac{\Gamma(l)\Gamma(\rho)\Gamma(\rho+\eta-\sigma-\sigma'-\nu)\Gamma(\rho+\nu'-\sigma')}{\Gamma(\rho+\nu')\Gamma(\rho+\eta-\sigma-\sigma')\Gamma(\rho+\eta-\sigma'-\nu)}$$
$$\times \sum_{n=0}^{\infty} \frac{(\alpha_1; x, p)_n(\alpha_2)_n \cdots (\alpha_r)_n}{(\beta_1)_n \cdots (\beta_s)_n} \tag{6.22}$$
$$\times \frac{(l)_n(\rho)_n(\rho+\eta-\sigma-\sigma'-\nu)_n(\rho+\nu'-\sigma')_n}{(\rho+\nu')_n(\rho+\eta-\sigma-\sigma')_n(\rho+\eta-\sigma'-\nu)_n} \left(\frac{ax}{\omega}\right)^n \frac{1}{n!},$$

interpreting the above equation, in view of Equation (3.18), we have the required result. □

Theorem 18 *Let* $x > 0, a, \sigma, \sigma', \nu, \nu', \eta, \rho \in \mathbb{C}$ *and* $\alpha_1, \ldots, \alpha_r \in \mathbb{C}, \beta_1, \ldots, \beta_s \in \mathbb{C} \setminus \mathbb{Z}_0^-$ *be such that* $\Re(\eta) > 0, \Re(a) > 0$ *and* $\Re(\rho+n) > \max\{0, \Re(\sigma+\sigma'+\nu-\eta), \Re(\sigma'-\nu')\}$*, then the following formula holds:*

$$L\left\{z^{l-1}\left(I_{0,x}^{\sigma,\sigma',\nu,\nu',\eta} t^{\rho-1} {}_r\Gamma_s^p \left[\begin{matrix}(\alpha_1, x; p), \alpha_2, \ldots, \alpha_r; \\ \beta_1, \ldots, \beta_s;\end{matrix}\, atz\right]\right)(x)\right\}$$
$$= \frac{x^{\rho+\eta-\sigma-\sigma'-1}}{\omega^l} \frac{\Gamma(l)\Gamma(\rho)\Gamma(\rho+\eta-\sigma-\sigma'-\nu)\Gamma(\rho+\nu'-\sigma')}{\Gamma(\rho+\nu')\Gamma(\rho+\eta-\sigma-\sigma')\Gamma(\rho+\eta-\sigma'-\nu)}$$
$$\times {}_{r+4}\Gamma_{s+3}^p \left[\begin{matrix}(\alpha_1, x; p), \alpha_2, \ldots, \alpha_r, l, \rho, \rho+\eta-\sigma-\sigma'-\nu, \rho+\nu'-\sigma'; \\ \beta_1, \ldots, \beta_s, \rho+\nu', \rho+\eta-\sigma-\sigma', \rho+\eta-\sigma'-\nu;\end{matrix}\, \frac{ax}{\omega}\right]. \tag{6.23}$$

Theorem 19 *Let* $x > 0, a, \sigma, \sigma', \nu, \nu', \eta, \rho \in \mathbb{C}$ *and* $\alpha_1, \ldots, \alpha_r \in \mathbb{C}, \beta_1, \ldots, \beta_s \in \mathbb{C} \setminus \mathbb{Z}_0^-$ *be such that* $\Re(\eta) > 0, \Re(a) > 0$ *and* $\Re(\rho-n) < 1+\min\{\Re(-\nu), \Re(\sigma+\sigma'-\eta), \Re(\sigma+\nu'-\eta)\}$*, then the following formula holds:*

$$L\left\{z^{l-1}\left(I_{x,\infty}^{\sigma,\sigma',\nu,\nu',\eta} t^{\rho-1} {}_r\gamma_s^p \left[\begin{matrix}(\alpha_1, x; p), \alpha_2, \ldots, \alpha_r; \\ \beta_1, \ldots, \beta_s;\end{matrix}\, \frac{az}{t}\right]\right)(x)\right\}$$
$$= \frac{x^{\rho+\eta-\sigma-\sigma'-1}}{\omega^l} \frac{\Gamma(l)\Gamma(1-\rho-\nu)\Gamma(1-\rho-\eta+\sigma+\sigma')}{\Gamma(1-\rho)\Gamma(1-\rho-\eta+\sigma+\sigma'+\nu')}$$
$$\times \frac{\Gamma(1-\rho-\eta+\sigma+\nu')}{\Gamma(1-\rho+\sigma-\nu)} {}_{r+4}\gamma_{s+3}^p \left[\begin{matrix}(\alpha_1, x; p), \alpha_2, \ldots, \alpha_r, \\ \beta_1, \ldots, \beta_s,\end{matrix}\right.$$
$$\left.\begin{matrix}l, 1-\rho-\nu, 1-\rho-\eta+\sigma+\sigma', 1-\rho-\eta+\sigma+\nu'; \\ 1-\rho, 1-\rho-\eta+\sigma+\sigma'+\nu', 1-\rho+\sigma-\nu;\end{matrix}\, \frac{a}{\omega x}\right]. \tag{6.24}$$

Theorem 20 *Let* $x > 0, a, \sigma, \sigma', \nu, \nu', \eta, \rho \in \mathbb{C}$ *and* $\alpha_1, \ldots, \alpha_r \in \mathbb{C}, \beta_1, \ldots, \beta_s \in \mathbb{C} \setminus \mathbb{Z}_0^-$ *be such that* $\Re(\eta) > 0, \Re(a) > 0$ *and* $\Re(\rho-n) < 1+min\{\Re(-\nu), \Re(\sigma+\sigma'-\eta), \Re(\sigma+\nu'-\eta)\}$*, then the following formula holds:*

$$L\left\{z^{l-1}\left(I_{x,\infty}^{\sigma,\sigma',\nu,\nu',\eta}t^{\rho-1}{}_r\Gamma_s^p\left[\begin{matrix}(\alpha_1,x;p),\alpha_2,\ldots,\alpha_r; & \dfrac{az}{t}\\ \beta_1,\ldots,\beta_s; & \end{matrix}\right]\right)(x)\right\}$$
$$=\frac{x^{\rho+\eta-\sigma-\sigma'-1}}{\omega^l}\frac{\Gamma(l)\Gamma(1-\rho-\nu)\Gamma(1-\rho-\eta+\sigma+\sigma')}{\Gamma(1-\rho)\Gamma(1-\rho-\eta+\sigma+\sigma'+\nu')}$$
$$\times\frac{\Gamma(1-\rho-\eta+\sigma+\nu')}{\Gamma(1-\rho+\sigma-\nu)}\cdot{}_{r+4}\Gamma_{s+3}^p\left[\begin{matrix}(\alpha_1,x;p),\alpha_2,\ldots,\alpha_r,\\ \beta_1,\ldots,\beta_s,\end{matrix}\right.$$
$$\left.\begin{matrix}l,1-\rho-\nu,1-\rho-\eta+\sigma+\sigma',1-\rho-\eta+\sigma+\nu'; & \dfrac{a}{\omega x}\\ 1-\rho,1-\rho-\eta+\sigma+\sigma'+\nu',1-\rho+\sigma-\nu; & \end{matrix}\right]. \tag{6.25}$$

Proof The proof of Theorems 18–20 would run parallel to Theorem 17, so we omit the details involved. □

Theorem 21 *Let $x>0,a,\sigma,\sigma',\nu,\nu',\eta,\rho\in\mathbb{C}$ and $\alpha_1,\ldots,\alpha_r\in\mathbb{C},\beta_1,\ldots,\beta_s\in\mathbb{C}\setminus\mathbb{Z}_0^-$ be such that $\Re(\eta)>0,\Re(a)>0$ and $\Re(\rho+n)>\max\{0,\Re(\eta-\sigma-\sigma'-\nu'),\Re(\nu-\sigma)\}$, then the following formula holds:*

$$L\left\{z^{l-1}\left(D_{0,x}^{\sigma,\sigma',\nu,\nu',\eta}t^{\rho-1}{}_r\gamma_s^p\left[\begin{matrix}(\alpha_1,x;p),\alpha_2,\ldots,\alpha_r; & atz\\ \beta_1,\ldots,\beta_s; & \end{matrix}\right]\right)(x)\right\}$$
$$=\frac{x^{\rho-\eta+\sigma+\sigma'-1}}{\omega^l}\frac{\Gamma(l)\Gamma(\rho)\Gamma(\rho-\eta+\sigma+\sigma'+\nu')\Gamma(\rho-\nu+\sigma)}{\Gamma(\rho-\nu)\Gamma(\rho-\eta+\sigma+\sigma')\Gamma(\rho-\eta+\sigma+\nu')}$$
$$\times{}_{r+4}\gamma_{s+3}^p\left[\begin{matrix}(\alpha_1,x;p),\alpha_2,\ldots,\alpha_r,l,\rho,\rho-\eta+\sigma+\sigma'+\nu',\rho-\nu+\sigma; & \dfrac{ax}{\omega}\\ \beta_1,\ldots,\beta_s,\rho-\nu,\rho-\eta+\sigma+\sigma',\rho-\eta+\sigma+\nu'; & \end{matrix}\right]. \tag{6.26}$$

Proof Denote the left-hand side of the result (6.26) by $\mathscr{L}$. Using definition of Laplace transform as given in Equation (6.18), we have

$$\mathscr{L}=\int_0^\infty e^{-\omega z}\left\{z^{l-1}\left(D_{0,x}^{\sigma,\sigma',\nu,\nu',\eta}t^{\rho-1}{}_r\gamma_s^p\left[\begin{matrix}(\alpha_1,x;p),\alpha_2,\ldots,\alpha_r; & atz\\ \beta_1,\ldots,\beta_s; & \end{matrix}\right]\right)(x)\right\}dz, \tag{6.27}$$

using the result (5.4), the above Equation (6.27) reduces to

$$\mathscr{L}=x^{\rho-\eta+\sigma+\sigma'-1}\frac{\Gamma(\rho)\Gamma(\rho-\eta+\sigma+\sigma'+\nu')\Gamma(\rho-\nu+\sigma)}{\Gamma(\rho-\nu)\Gamma(\rho-\eta+\sigma+\sigma')\Gamma(\rho-\eta+\sigma+\nu')}$$
$$\times\sum_{n=0}^{\infty}\frac{(\alpha_1;x,p)_n(\alpha_2)_n\cdots(\alpha_r)_n}{(\beta_1)_n\cdots(\beta_s)_n}$$
$$\times\frac{(\rho)_n(\rho-\eta+\sigma+\sigma'+\nu')_n(\rho-\nu+\sigma)_n}{(\rho-\nu)_n(\rho-\eta+\sigma+\sigma')_n(\rho-\eta+\sigma+\nu')_n}\frac{(ax)^n}{n!}\int_0^\infty z^{l+n-1}e^{-\omega z}dz, \tag{6.28}$$

after simplification, we have

$$\mathscr{L} = \frac{x^{\rho-\eta+\sigma+\sigma'-1}}{\omega^l} \frac{\Gamma(l)\Gamma(\rho)\Gamma(\rho-\eta+\sigma+\sigma'+\nu')\Gamma(\rho-\nu+\sigma)}{\Gamma(\rho-\nu)\Gamma(\rho-\eta+\sigma+\sigma')\Gamma(\rho-\eta+\sigma+\nu')} \times \sum_{n=0}^{\infty} \frac{(\alpha_1; x, p)_n (\alpha_2)_n \cdots (\alpha_r)_n}{(\beta_1)_n \cdots (\beta_s)_n} \times \frac{(l)_n(\rho)_n(\rho-\eta+\sigma+\sigma'+\nu')_n(\rho-\nu+\sigma)_n}{(\rho-\nu)_n(\rho-\eta+\sigma+\sigma')_n(\rho-\eta+\sigma+\nu')_n} \left(\frac{ax}{\omega}\right)^n \frac{1}{n!}, \tag{6.29}$$

interpreting the above equation, in view of Equation (3.18), we have the required result. □

Theorem 22 *Let* $x > 0, a, \sigma, \sigma', \nu, \nu', \eta, \rho \in \mathbb{C}$ *and* $\alpha_1, \ldots, \alpha_r \in \mathbb{C}, \beta_1, \ldots, \beta_s \in \mathbb{C} \setminus \mathbb{Z}_0^-$ *be such that* $\Re(\eta) > 0, \Re(a) > 0$ *and* $\Re(\rho+n) > \max\{0, \Re(\eta-\sigma-\sigma'-\nu'), \Re(\nu-\sigma)\}$, *then the following formula holds:*

$$L\left\{z^{l-1}\left(D_{0,x}^{\sigma,\sigma',\nu,\nu',\eta} t^{\rho-1} {}_r\Gamma_s^p \left[\begin{matrix}(\alpha_1, x; p), \alpha_2, \ldots, \alpha_r; \\ \beta_1, \ldots, \beta_s;\end{matrix}\ atz\right]\right)(x)\right\} = \frac{x^{\rho-\eta+\sigma+\sigma'-1}}{\omega^l} \frac{\Gamma(l)\Gamma(\rho)\Gamma(\rho-\eta+\sigma+\sigma'+\nu')\Gamma(\rho-\nu+\sigma)}{\Gamma(\rho-\nu)\Gamma(\rho-\eta+\sigma+\sigma')\Gamma(\rho-\eta+\sigma+\nu')} \times {}_{r+4}\Gamma_{s+3}^p \left[\begin{matrix}(\alpha_1, x; p), \alpha_2, \ldots, \alpha_r, l, \rho, \rho-\eta+\sigma+\sigma'+\nu', \rho-\nu+\sigma; \\ \beta_1, \ldots, \beta_s, \rho-\nu, \rho-\eta+\sigma+\sigma', \rho-\eta+\sigma+\nu';\end{matrix}\ \frac{ax}{\omega}\right]. \tag{6.30}$$

Theorem 23 *Let* $x > 0, a, \sigma, \sigma', \nu, \nu', \eta, \rho \in \mathbb{C}$ *and* $\alpha_1, \ldots, \alpha_r \in \mathbb{C}, \beta_1, \ldots, \beta_s \in \mathbb{C} \setminus \mathbb{Z}_0^-$ *be such that* $\Re(\eta) > 0, \Re(a) > 0$ *and* $\Re(\rho-n) < 1 + \min\{\Re(\nu'), \Re(\eta-\sigma-\sigma'), \Re(\eta-\sigma'-\nu)\}$, *then the following formula holds:*

$$L\left\{z^{l-1}\left(D_{x,\infty}^{\sigma,\sigma',\nu,\nu',\eta} t^{\rho-1} {}_r\gamma_s^p \left[\begin{matrix}(\alpha_1, x; p), \alpha_2, \ldots, \alpha_r; \\ \beta_1, \ldots, \beta_s;\end{matrix}\ \frac{az}{t}\right]\right)(x)\right\} = \frac{x^{\rho-\eta+\sigma+\sigma'-1}}{\omega^l} \frac{\Gamma(l)\Gamma(1-\rho+\nu')\Gamma(1-\rho+\eta-\sigma-\sigma')}{\Gamma(1-\rho)\Gamma(1-\rho+\eta-\sigma-\sigma'-\nu)} \times \frac{\Gamma(1-\rho+\eta-\sigma'-\nu)}{\Gamma(1-\rho-\sigma'+\nu')} {}_{r+4}\gamma_{s+3}^p \left[\begin{matrix}(\alpha_1, x; p), \alpha_2, \ldots, \alpha_r, l, 1-\rho+\nu', 1-\rho+\eta-\sigma-\sigma', 1-\rho+\eta-\sigma'-\nu; \\ \beta_1, \ldots, \beta_s, 1-\rho, 1-\rho+\eta-\sigma-\sigma'-\nu, 1-\rho-\sigma'+\nu';\end{matrix}\ \frac{a}{\omega x}\right]. \tag{6.31}$$

Theorem 24 *Let* $x > 0, a, \sigma, \sigma', \nu, \nu', \eta, \rho \in \mathbb{C}$ *and* $\alpha_1, \ldots, \alpha_r \in \mathbb{C}, \beta_1, \ldots, \beta_s \in \mathbb{C} \setminus \mathbb{Z}_0^-$ *be such that* $\Re(\eta) > 0, \Re(a) > 0$ *and* $\Re(\rho-n) < 1 + \min\{\Re(\nu'), \Re(\eta-\sigma-\sigma'), \Re(\eta-\sigma'-\nu)\}$, *then the following formula holds:*

$$L\left\{z^{l-1}\left(D_{x,\infty}^{\sigma,\sigma',\nu,\nu',\eta}t^{\rho-1}{}_r\Gamma_s^p\left[\begin{matrix}(\alpha_1,x;p),\alpha_2,\ldots,\alpha_r; & \frac{az}{t}\\ \beta_1,\ldots,\beta_s; & \end{matrix}\right]\right)(x)\right\}$$

$$=\frac{x^{\rho-\eta+\sigma+\sigma'-1}}{\omega^l}\frac{\Gamma(l)\Gamma(1-\rho+\nu')\Gamma(1-\rho+\eta-\sigma-\sigma')}{\Gamma(1-\rho)\Gamma(1-\rho+\eta-\sigma-\sigma'-\nu)}$$

$$\times\frac{\Gamma(1-\rho+\eta-\sigma'-\nu)}{\Gamma(1-\rho-\sigma'+\nu')}{}_{r+4}\Gamma_{s+3}^p\left[\begin{matrix}(\alpha_1,x;p),\alpha_2,\ldots,\alpha_r,\\ \beta_1,\ldots,\beta_s,\end{matrix}\right.$$

$$\left.\begin{matrix}l,1-\rho+\nu',1-\rho+\eta-\sigma-\sigma',1-\rho+\eta-\sigma'-\nu; & \frac{a}{\omega x}\\ 1-\rho,1-\rho+\eta-\sigma-\sigma'-\nu,1-\rho-\sigma'+\nu'; & \end{matrix}\right]. \tag{6.32}$$

Proof The proof of Theorems 22–24 would run parallel to Theorem 21, so details are omitted here. □

6.3 Whittaker Transform

Definition 21 Whittaker Transform is defined as [72]:

$$\int_0^\infty t^{l-1}e^{-t/2}W_{\lambda,\mu}(t)dt=\frac{\Gamma(1/2+\mu+l)\Gamma(1/2-\mu+l)}{\Gamma(1/2-\lambda+l)}. \tag{6.33}$$

Theorem 25 *Let* $x>0, a,\sigma,\sigma',\nu,\nu',\eta,\rho\in\mathbb{C}$ *and* $\alpha_1,\ldots,\alpha_r\in\mathbb{C},\beta_1,\ldots,\beta_s\in\mathbb{C}\setminus\mathbb{Z}_0^-$ *be such that* $\Re(\eta)>0$, $\Re(a)>0$ *and* $\Re(\rho+n)>\max\{0,\Re(\sigma+\sigma'+\nu-\eta),\Re(\sigma'-\nu')\}$, *then the following formula holds:*

$$\int_0^\infty z^{l-1}e^{-\delta z/2}W_{\lambda,\mu}(\delta z)\left\{\left(I_{0,x}^{\sigma,\sigma',\nu,\nu',\eta}t^{\rho-1}{}_r\gamma_s^p\left[\begin{matrix}(\alpha_1,x;p),\alpha_2,\ldots,\alpha_r; & atz\\ \beta_1,\ldots,\beta_s; & \end{matrix}\right]\right)(x)\right\}dz$$

$$=\frac{x^{\rho+\eta-\sigma-\sigma'-1}}{\delta^l}\frac{\Gamma(1/2+\mu+l)\Gamma(1/2-\mu+l)\Gamma(\rho)}{\Gamma(1/2-\lambda+l)\Gamma(\rho+\nu')}$$

$$\times\frac{\Gamma(\rho+\eta-\sigma-\sigma'-\nu)\Gamma(\rho+\nu'-\sigma')}{\Gamma(\rho+\eta-\sigma-\sigma')\Gamma(\rho+\eta-\sigma'-\nu)}{}_{r+5}\gamma_{s+4}^p\left[\begin{matrix}(\alpha_1,x;p),\alpha_2,\ldots,\alpha_r,\\ \beta_1,\ldots,\beta_s,\end{matrix}\right.$$

$$\left.\begin{matrix}1/2+\mu+l,1/2-\mu+l,\rho,\rho+\eta-\sigma-\sigma'-\nu,\rho+\nu'-\sigma'; & \frac{ax}{\delta}\\ 1/2-\lambda+l,\rho+\nu',\rho+\eta-\sigma-\sigma',\rho+\eta-\sigma'-\nu; & \end{matrix}\right]. \tag{6.34}$$

Proof Denote the left-hand side of the result (6.34) by $\mathscr{W}$. Using definition of Whittaker transform as given in Equation (6.33), we have

$$\mathscr{W} = \int_0^\infty z^{l-1} e^{-\delta z/2} W_{\lambda,\mu}(\delta z) \times \left\{ \left(I_{0,x}^{\sigma,\sigma',\nu,\nu',\eta} t^{\rho-1} {}_r\gamma_s^p \left[\begin{matrix} (\alpha_1, x; p), \alpha_2, \ldots, \alpha_r; \\ \beta_1, \ldots, \beta_s; \end{matrix} atz \right] \right)(x) \right\} dz, \tag{6.35}$$

using the result (4.4), the above Equation (6.35) reduces to

$$\begin{aligned} \mathscr{W} &= x^{\rho+\eta-\sigma-\sigma'-1} \frac{\Gamma(\rho)\Gamma(\rho+\eta-\sigma-\sigma'-\nu)\Gamma(\rho+\nu'-\sigma')}{\Gamma(\rho+\nu')\Gamma(\rho+\eta-\sigma-\sigma')\Gamma(\rho+\eta-\sigma'-\nu)} \\ &\times \sum_{n=0}^\infty \frac{(\alpha_1; x, p)_n (\alpha_2)_n \cdots (\alpha_r)_n}{(\beta_1)_n \cdots (\beta_s)_n} \frac{(\rho)_n (\rho+\eta-\sigma-\sigma'-\nu)_n}{(\rho+\nu')_n (\rho+\eta-\sigma-\sigma')_n} \\ &\times \frac{(\rho+\nu'-\sigma')_n}{(\rho+\eta-\sigma'-\nu)_n} \frac{(ax)^n}{n!} \int_0^\infty z^{l+n-1} e^{-\delta z/2} W_{\lambda,\mu}(\delta z) dz, \end{aligned} \tag{6.36}$$

by substituting $\delta z = y$ and after little simplification, we have

$$\begin{aligned} \mathscr{W} &= \frac{x^{\rho+\eta-\sigma-\sigma'-1}}{\delta^l} \frac{\Gamma(\rho)\Gamma(\rho+\eta-\sigma-\sigma'-\nu)\Gamma(\rho+\nu'-\sigma')}{\Gamma(\rho+\nu')\Gamma(\rho+\eta-\sigma-\sigma')\Gamma(\rho+\eta-\sigma'-\nu)} \\ &\times \sum_{n=0}^\infty \frac{(\alpha_1; x, p)_n (\alpha_2)_n \cdots (\alpha_r)_n}{(\beta_1)_n \cdots (\beta_s)_n} \frac{(\rho)_n (\rho+\eta-\sigma-\sigma'-\nu)_n}{(\rho+\nu')_n (\rho+\eta-\sigma-\sigma')_n} \\ &\times \frac{(\rho+\nu'-\sigma')_n}{(\rho+\eta-\sigma'-\nu)_n} \left(\frac{ax}{\delta}\right)^n \frac{1}{n!} \int_0^\infty y^{l+n-1} e^{-y/2} W_{\lambda,\mu}(y) dy, \end{aligned} \tag{6.37}$$

by using the integral formula involving Whittaker function, we have

$$\begin{aligned} \mathscr{W} &= \frac{x^{\rho+\eta-\sigma-\sigma'-1}}{\delta^l} \frac{\Gamma(1/2+\mu+l)\Gamma(1/2-\mu+l)}{\Gamma(1/2-\lambda+l)} \\ &\times \frac{\Gamma(\rho)\Gamma(\rho+\eta-\sigma-\sigma'-\nu)\Gamma(\rho+\nu'-\sigma')}{\Gamma(\rho+\nu')\Gamma(\rho+\eta-\sigma-\sigma')\Gamma(\rho+\eta-\sigma'-\nu)} \\ &\times \sum_{n=0}^\infty \frac{(\alpha_1; x, p)_n (\alpha_2)_n \cdots (\alpha_r)_n}{(\beta_1)_n \cdots (\beta_s)_n} \frac{(1/2+\mu+l)_n (1/2-\mu+l)_n}{(1/2-\lambda+l)_n} \\ &\times \frac{(\rho)_n (\rho+\eta-\sigma-\sigma'-\nu)_n (\rho+\nu'-\sigma')_n}{(\rho+\nu')_n (\rho+\eta-\sigma-\sigma')_n (\rho+\eta-\sigma'-\nu)_n} \left(\frac{ax}{\delta}\right)^n \frac{1}{n!}, \end{aligned} \tag{6.38}$$

interpreting the above Equation (6.38), in view of Equation (3.18), we have the required result. □

Theorem 26 *Let* $x > 0, a, \sigma, \sigma', \nu, \nu', \eta, \rho \in \mathbb{C}$ *and* $\alpha_1, \ldots, \alpha_r \in \mathbb{C}, \beta_1, \ldots, \beta_s \in \mathbb{C} \setminus \mathbb{Z}_0^-$ *be such that* $\Re(\eta) > 0, \Re(a) > 0$ *and* $\Re(\rho + n) > \max\{0, \Re(\sigma + \sigma' + \nu - \eta), \Re(\sigma' - \nu')\}$, *then the following formula holds:*

$$\int_0^\infty z^{l-1} e^{-\delta z/2} W_{\lambda,\mu}(\delta z) \left\{ \left(I_{0,x}^{\sigma,\sigma',\nu,\nu',\eta} t^{\rho-1} {}_r\Gamma_s^p \left[\begin{matrix} (\alpha_1, x;\, p), \alpha_2, \ldots, \alpha_r; \\ \beta_1, \ldots, \beta_s; \end{matrix}\; atz \right] \right)(x) \right\} dz$$
$$= \frac{x^{\rho+\eta-\sigma-\sigma'-1}}{\delta^l} \frac{\Gamma(1/2+\mu+l)\Gamma(1/2-\mu+l)}{\Gamma(1/2-\lambda+l)} \frac{\Gamma(\rho)}{\Gamma(\rho+\nu')}$$
$$\times \frac{\Gamma(\rho+\eta-\sigma-\sigma'-\nu)\Gamma(\rho+\nu'-\sigma')}{\Gamma(\rho+\eta-\sigma-\sigma')\Gamma(\rho+\eta-\sigma'-\nu)} {}_{r+5}\Gamma_{s+4}^p \left[\begin{matrix} (\alpha_1, x;\, p), \alpha_2, \ldots, \alpha_r, \\ \beta_1, \ldots, \beta_s, \end{matrix} \right.$$
$$\left. \begin{matrix} 1/2+\mu+l, 1/2-\mu+l, \rho, \rho+\eta-\sigma-\sigma'-\nu, \rho+\nu'-\sigma'; \\ 1/2-\lambda+l, \rho+\nu', \rho+\eta-\sigma-\sigma', \rho+\eta-\sigma'-\nu; \end{matrix}\; \frac{ax}{\delta} \right]. \tag{6.39}$$

Theorem 27 *Let* $x > 0, a, \sigma, \sigma', \nu, \nu', \eta, \rho \in \mathbb{C}$ *and* $\alpha_1, \ldots, \alpha_r \in \mathbb{C}, \beta_1, \ldots, \beta_s \in \mathbb{C} \setminus \mathbb{Z}_0^-$ *be such that* $\Re(\eta) > 0, \Re(a) > 0$ *and* $\Re(\rho - n) < 1 + \min\{\Re(-\nu), \Re(\sigma + \sigma' - \eta), \Re(\sigma + \nu' - \eta)\}$, *then the following formula holds:*

$$\int_0^\infty z^{l-1} e^{-\delta z/2} W_{\lambda,\mu}(\delta z)$$
$$\times \left\{ \left(I_{x,\infty}^{\sigma,\sigma',\nu,\nu',\eta} t^{\rho-1} {}_r\gamma_s^p \left[\begin{matrix} (\alpha_1, x;\, p), \alpha_2, \ldots, \alpha_r; \\ \beta_1, \ldots, \beta_s; \end{matrix}\; \frac{az}{t} \right] \right)(x) \right\} dz$$
$$= \frac{x^{\rho+\eta-\sigma-\sigma'-1}}{\delta^l} \frac{\Gamma(1/2+\mu+l)\Gamma(1/2-\mu+l)}{\Gamma(1/2-\lambda+l)}$$
$$\times \frac{\Gamma(1-\rho-\nu)\Gamma(1-\rho-\eta+\sigma+\sigma')\Gamma(1-\rho-\eta+\sigma+\nu')}{\Gamma(1-\rho)\Gamma(1-\rho-\eta+\sigma+\sigma'+\nu')\Gamma(1-\rho+\sigma-\nu)}$$
$$\times {}_{r+5}\gamma_{s+4}^p \left[\begin{matrix} (\alpha_1, x;\, p), \alpha_2, \ldots, \alpha_r, 1/2+\mu+l, 1/2-\mu+l, \\ \beta_1, \ldots, \beta_s, 1/2-\lambda+l, \end{matrix} \right.$$
$$\left. \begin{matrix} 1-\rho-\nu, 1-\rho-\eta+\sigma+\sigma', 1-\rho-\eta+\sigma+\nu'; \\ 1-\rho, 1-\rho-\eta+\sigma+\sigma'+\nu', 1-\rho+\sigma-\nu; \end{matrix}\; \frac{a}{\delta x} \right]. \tag{6.40}$$

Theorem 28 *Let* $x > 0, a, \sigma, \sigma', \nu, \nu', \eta, \rho \in \mathbb{C}$ *and* $\alpha_1, \ldots, \alpha_r \in \mathbb{C}, \beta_1, \ldots, \beta_s \in \mathbb{C} \setminus \mathbb{Z}_0^-$ *be such that* $\Re(\eta) > 0, \Re(a) > 0$ *and* $\Re(\rho - n) < 1 + \min\{\Re(-\nu), \Re(\sigma + \sigma' - \eta), \Re(\sigma + \nu' - \eta)\}$, *then the following formula holds:*

$$\int_0^\infty z^{l-1}e^{-\delta z/2}W_{\lambda,\mu}(\delta z)$$
$$\times\left\{\left(I_{x,\infty}^{\sigma,\sigma',\nu,\nu',\eta}t^{\rho-1}{}_r\Gamma_s^p\left[\begin{matrix}(\alpha_1,x;p),\alpha_2,\ldots,\alpha_r; & \dfrac{az}{t}\\ \beta_1,\ldots,\beta_s; & \end{matrix}\right]\right)(x)\right\}dz$$
$$=\frac{x^{\rho+\eta-\sigma-\sigma'-1}}{\delta^l}\frac{\Gamma(1/2+\mu+l)\Gamma(1/2-\mu+l)}{\Gamma(1/2-\lambda+l)}$$
$$\times\frac{\Gamma(1-\rho-\nu)\Gamma(1-\rho-\eta+\sigma+\sigma')\Gamma(1-\rho-\eta+\sigma+\nu')}{\Gamma(1-\rho)\Gamma(1-\rho-\eta+\sigma+\sigma'+\nu')\Gamma(1-\rho+\sigma-\nu)}$$
$$\times{}_{r+5}\Gamma_{s+4}^p\left[\begin{matrix}(\alpha_1,x;p),\alpha_2,\ldots,\alpha_r,1/2+\mu+l,1/2-\mu+l,\\ \beta_1,\ldots,\beta_s,1/2-\lambda+l,\end{matrix}\right.$$
$$\left.\begin{matrix}1-\rho-\nu,1-\rho-\eta+\sigma+\sigma',1-\rho-\eta+\sigma+\nu'; & \dfrac{a}{\delta x}\\ 1-\rho,1-\rho-\eta+\sigma+\sigma'+\nu',1-\rho+\sigma-\nu; & \end{matrix}\right]. \tag{6.41}$$

Proof The proof of Theorems 26–28 would run parallel to Theorem 25, so details are omitted here. □

Theorem 29 *Let* $x>0,a,\sigma,\sigma',\nu,\nu',\eta,\rho\in\mathbb{C}$ *and* $\alpha_1,\ldots,\alpha_r\in\mathbb{C},\beta_1,\ldots,\beta_s\in\mathbb{C}\setminus\mathbb{Z}_0^-$ *be such that* $\Re(\eta)>0,\Re(a)>0$ *and* $\Re(\rho+n)>\max\{0,\Re(\eta-\sigma-\sigma'-\nu'),\Re(\nu-\sigma)\}$*, then the following formula holds:*

$$\int_0^\infty z^{l-1}e^{-\delta z/2}W_{\lambda,\mu}(\delta z)$$
$$\times\left\{\left(D_{0,x}^{\sigma,\sigma',\nu,\nu',\eta}t^{\rho-1}{}_r\gamma_s^p\left[\begin{matrix}(\alpha_1,x;p),\alpha_2,\ldots,\alpha_r; & atz\\ \beta_1,\ldots,\beta_s; & \end{matrix}\right]\right)(x)\right\}dz$$
$$=\frac{x^{\rho-\eta+\sigma+\sigma'-1}}{\delta^l}\frac{\Gamma(1/2+\mu+l)\Gamma(1/2-\mu+l)}{\Gamma(1/2-\lambda+l)}$$
$$\times\frac{\Gamma(\rho)\Gamma(\rho-\eta+\sigma+\sigma'+\nu')\Gamma(\rho-\nu+\sigma)}{\Gamma(\rho-\nu)\Gamma(\rho-\eta+\sigma+\sigma')\Gamma(\rho-\eta+\sigma+\nu')}$$
$$\times{}_{r+5}\gamma_{s+4}^p\left[\begin{matrix}(\alpha_1,x;p),\alpha_2,\ldots,\alpha_r,1/2+\mu+l,1/2-\mu+l,\\ \beta_1,\ldots,\beta_s,1/2-\lambda+l,\end{matrix}\right.$$
$$\left.\begin{matrix}\rho,\rho-\eta+\sigma+\sigma'+\nu',\rho-\nu+\sigma; & \dfrac{ax}{\delta}\\ \rho-\nu,\rho-\eta+\sigma+\sigma',\rho-\eta+\sigma+\nu'; & \end{matrix}\right]. \tag{6.42}$$

Proof Denote the left-hand side of the result (6.42) by $\mathscr{W}$. Using definition of Whittaker transform as given in Equation (6.33), we have

$$\mathscr{W} = \int_0^\infty z^{l-1} e^{-\delta z/2} W_{\lambda,\mu}(\delta z) \times \left\{ \left(D_{0,x}^{\sigma,\sigma',\nu,\nu',\eta} t^{\rho-1} {}_r\gamma_s^p \left[\begin{matrix} (\alpha_1, x; p), \alpha_2, \ldots, \alpha_r; \\ \beta_1, \ldots, \beta_s; \end{matrix} atz \right] \right)(x) \right\} dz, \tag{6.43}$$

using the result (5.4), the above Equation (6.43) reduces to

$$\mathscr{W} = x^{\rho-\eta+\sigma+\sigma'-1} \frac{\Gamma(\rho)\Gamma(\rho-\eta+\sigma+\sigma'+\nu')\Gamma(\rho-\nu+\sigma)}{\Gamma(\rho-\nu)\Gamma(\rho-\eta+\sigma+\sigma')\Gamma(\rho-\eta+\sigma+\nu')} \times \sum_{n=0}^{\infty} \frac{(\alpha_1; x, p)_n (\alpha_2)_n \cdots (\alpha_r)_n}{(\beta_1)_n \cdots (\beta_s)_n} \frac{(\rho)_n (\rho-\eta+\sigma+\sigma'+\nu')_n}{(\rho-\nu)_n (\rho-\eta+\sigma+\sigma')_n} \times \frac{(\rho-\nu+\sigma)_n}{(\rho-\eta+\sigma+\nu')_n} \frac{(ax)^n}{n!} \int_0^\infty z^{l+n-1} e^{-\delta z/2} W_{\lambda,\mu}(\delta z) dz, \tag{6.44}$$

by putting $\delta z = y$ and after little simplification, we have

$$\mathscr{W} = \frac{x^{\rho-\eta+\sigma+\sigma'-1}}{\delta^l} \frac{\Gamma(\rho)\Gamma(\rho-\eta+\sigma+\sigma'+\nu')\Gamma(\rho-\nu+\sigma)}{\Gamma(\rho-\nu)\Gamma(\rho-\eta+\sigma+\sigma')\Gamma(\rho-\eta+\sigma+\nu')} \times \sum_{n=0}^{\infty} \frac{(\alpha_1; x, p)_n (\alpha_2)_n \cdots (\alpha_r)_n}{(\beta_1)_n \cdots (\beta_s)_n} \frac{(\rho)_n (\rho-\eta+\sigma+\sigma'+\nu')_n}{(\rho-\nu)_n (\rho-\eta+\sigma+\sigma')_n} \times \frac{(\rho-\nu+\sigma)_n}{(\rho-\eta+\sigma+\nu')_n} \left(\frac{ax}{\delta}\right)^n \frac{1}{n!} \int_0^\infty y^{l+n-1} e^{-y/2} W_{\lambda,\mu}(y) dy, \tag{6.45}$$

by using the definition of Whittaker transform, we have

$$\mathscr{W} = \frac{x^{\rho-\eta+\sigma+\sigma'-1}}{\delta^l} \frac{\Gamma(1/2+\mu+l)\Gamma(1/2-\mu+l)}{\Gamma(1/2-\lambda+l)} \times \frac{\Gamma(\rho)\Gamma(\rho-\eta+\sigma+\sigma'+\nu')\Gamma(\rho-\nu+\sigma)}{\Gamma(\rho-\nu)\Gamma(\rho-\eta+\sigma+\sigma')\Gamma(\rho-\eta+\sigma+\nu')} \times \sum_{n=0}^{\infty} \frac{(\alpha_1; x, p)_n (\alpha_2)_n \cdots (\alpha_r)_n}{(\beta_1)_n \cdots (\beta_s)_n} \frac{(1/2+\mu+l)_n (1/2-\mu+l)_n}{(1/2-\lambda+l)_n} \times \frac{(\rho)_n (\rho-\eta+\sigma+\sigma'+\nu')_n (\rho-\nu+\sigma)_n}{(\rho-\nu)_n (\rho-\eta+\sigma+\sigma')_n (\rho-\eta+\sigma+\nu')_n} \left(\frac{ax}{\delta}\right)^n \frac{1}{n!}, \tag{6.46}$$

the above Equation (6.46) in the view of Equation (3.18) gives the required result. □

Theorem 30 *Let* $x > 0, a, \sigma, \sigma', \nu, \nu', \eta, \rho \in \mathbb{C}$ *and* $\alpha_1, \ldots, \alpha_r \in \mathbb{C}, \beta_1, \ldots, \beta_s \in \mathbb{C} \setminus \mathbb{Z}_0^-$ *be such that* $\Re(\eta) > 0, \Re(a) > 0$ *and* $\Re(\rho + n) > \max\{0, \Re(\eta - \sigma - \sigma' - \nu'), \Re(\nu - \sigma)\}$*, then the following formula holds:*

$$\begin{aligned}
&\int_0^\infty z^{l-1} e^{-\delta z/2} W_{\lambda,\mu}(\delta z) \\
&\quad \times \left\{ \left(D_{0,x}^{\sigma,\sigma',\nu,\nu',\eta} t^{\rho-1} {}_r\Gamma_s^p \left[\begin{matrix} (\alpha_1, x; p), \alpha_2, \ldots, \alpha_r; \\ \beta_1, \ldots, \beta_s; \end{matrix} atz \right] \right) (x) \right\} dz \\
&= \frac{x^{\rho-\eta+\sigma+\sigma'-1}}{\delta^l} \frac{\Gamma(1/2+\mu+l)\Gamma(1/2-\mu+l)}{\Gamma(1/2-\lambda+l)} \\
&\quad \times \frac{\Gamma(\rho)\Gamma(\rho-\eta+\sigma+\sigma'+\nu')\Gamma(\rho-\nu+\sigma)}{\Gamma(\rho-\nu)\Gamma(\rho-\eta+\sigma+\sigma')\Gamma(\rho-\eta+\sigma+\nu')} \\
&\quad \times {}_{r+5}\Gamma_{s+4}^p \left[\begin{matrix} (\alpha_1, x; p), \alpha_2, \ldots, \alpha_r, 1/2+\mu+l, 1/2-\mu+l, \\ \beta_1, \ldots, \beta_s, 1/2-\lambda+l, \end{matrix} \right. \\
&\qquad \left. \begin{matrix} \rho, \rho-\eta+\sigma+\sigma'+\nu', \rho-\nu+\sigma; \\ \rho-\nu, \rho-\eta+\sigma+\sigma', \rho-\eta+\sigma+\nu'; \end{matrix} \frac{ax}{\delta} \right].
\end{aligned} \tag{6.47}$$

Theorem 31 *Let* $x > 0, a, \sigma, \sigma', \nu, \nu', \eta, \rho \in \mathbb{C}$ *and* $\alpha_1, \ldots, \alpha_r \in \mathbb{C}, \beta_1, \ldots, \beta_s \in \mathbb{C} \setminus \mathbb{Z}_0^-$ *be such that* $\Re(\eta) > 0, \Re(a) > 0$ *and* $\Re(\rho - n) < 1 + \min\{\Re(\nu'), \Re(\eta - \sigma - \sigma'), \Re(\eta - \sigma' - \nu)\}$*, then the following formula holds:*

$$\begin{aligned}
&\int_0^\infty z^{l-1} e^{-\delta z/2} W_{\lambda,\mu}(\delta z) \\
&\quad \times \left\{ z^{l-1} \left(D_{x,\infty}^{\sigma,\sigma',\nu,\nu',\eta} t^{\rho-1} {}_r\gamma_s^p \left[\begin{matrix} (\alpha_1, x; p), \alpha_2, \ldots, \alpha_r; \\ \beta_1, \ldots, \beta_s; \end{matrix} \frac{az}{t} \right] \right) (x) \right\} dz \\
&= \frac{x^{\rho-\eta+\sigma+\sigma'-1}}{\delta^l} \frac{\Gamma(1/2+\mu+l)\Gamma(1/2-\mu+l)}{\Gamma(1/2-\lambda+l)} \\
&\quad \times \frac{\Gamma(1-\rho+\nu')\Gamma(1-\rho+\eta-\sigma-\sigma')\Gamma(1-\rho+\eta-\sigma'-\nu)}{\Gamma(1-\rho)\Gamma(1-\rho+\eta-\sigma-\sigma'-\nu)\Gamma(1-\rho-\sigma'+\nu')} \\
&\quad \times {}_{r+5}\gamma_{s+4}^p \left[\begin{matrix} (\alpha_1, x; p), \alpha_2, \ldots, \alpha_r, 1/2+\mu+l, 1/2-\mu+l, \\ \beta_1, \ldots, \beta_s, 1/2-\lambda+l, \end{matrix} \right. \\
&\qquad \left. \begin{matrix} 1-\rho+\nu', 1-\rho+\eta-\sigma-\sigma', 1-\rho+\eta-\sigma'-\nu; \\ 1-\rho, 1-\rho+\eta-\sigma-\sigma'-\nu, 1-\rho-\sigma'+\nu'; \end{matrix} \frac{a}{\delta x} \right].
\end{aligned} \tag{6.48}$$

Theorem 32 *Let* $x > 0, a, \sigma, \sigma', \nu, \nu', \eta, \rho \in \mathbb{C}$ *and* $\alpha_1, \ldots, \alpha_r \in \mathbb{C}, \beta_1, \ldots, \beta_s \in \mathbb{C} \setminus \mathbb{Z}_0^-$ *be such that* $\Re(\eta) > 0, \Re(a) > 0$ *and* $\Re(\rho - n) < 1 + \min\{\Re(\nu'), \Re(\eta - \sigma - \sigma'), \Re(\eta - \sigma' - \nu)\}$, *then the following formula holds:*

$$\int_0^\infty z^{l-1} e^{-\delta z/2} W_{\lambda,\mu}(\delta z)$$

$$\left\{z^{l-1}\left(D_{x,\infty}^{\sigma,\sigma',\nu,\nu',\eta} t^{\rho-1} {}_r\Gamma_s^p\left[\begin{matrix}(\alpha_1, x; p), \alpha_2, \ldots, \alpha_r; \frac{az}{t} \\ \beta_1, \ldots, \beta_s;\end{matrix}\right]\right)(x)\right\} dz$$

$$= \frac{x^{\rho-\eta+\sigma+\sigma'-1}}{\delta^l} \frac{\Gamma(1/2+\mu+l)\Gamma(1/2-\mu+l)}{\Gamma(1/2-\lambda+l)}$$

$$\times \frac{\Gamma(1-\rho+\nu')\Gamma(1-\rho+\eta-\sigma-\sigma')\Gamma(1-\rho+\eta-\sigma'-\nu)}{\Gamma(1-\rho)\Gamma(1-\rho+\eta-\sigma-\sigma'-\nu)\Gamma(1-\rho-\sigma'+\nu')}$$

$$\times {}_{r+5}\Gamma_{s+4}^p\left[\begin{matrix}(\alpha_1, x; p), \alpha_2, \ldots, \alpha_r, 1/2+\mu+l, 1/2-\mu+l, \\ \beta_1, \ldots, \beta_s, 1/2-\lambda+l,\end{matrix}\right.$$

$$\left.\begin{matrix}1-\rho+\nu', 1-\rho+\eta-\sigma-\sigma', 1-\rho+\eta-\sigma'-\nu; \frac{a}{\delta x} \\ 1-\rho, 1-\rho+\eta-\sigma-\sigma'-\nu, 1-\rho-\sigma'+\nu';\end{matrix}\right]. \tag{6.49}$$

Proof The proof of Theorems 30–32 would run parallel to Theorem 29, so details are omitted here. □

7 Fractional Kinetic Equations

Fractional kinetic equations gained remarkable interest due to their applications not only in mathematics but also in physics, dynamical systems, control systems and engineering and to create the mathematical model of many physical phenomena. Especially, the kinetic equations describe the continuity of motion of substance. For the extensions and generalizations of fractional kinetic equations involving many fractional operators, see the available literature [74–92].

In view of the effectiveness and a great importance of the fractional kinetic equations in certain astrophysical problems, the authors develop a further generalized form of the fractional kinetic equation involving extended incomplete generalized hypergeometric functions.

If an arbitrary reaction is described by a time dependent quantity $N = N(t)$, then the fractional differential equation between rate of change of the reaction, the destruction rate and the production rate of N was established by Haubold and Mathai in [85] is given as follows:

$$\frac{dN}{dt} = -d(N_t) + p(N_t), \tag{7.1}$$

where $N = N(t)$ denotes the rate of reaction, $d = d(N)$ is the rate of destruction, $p = p(N)$ denotes the rate of production and N_t denotes the function defined by

$$N_t(t^*) = N(t - t^*), \quad t^* > 0. \tag{7.2}$$

The following differential equation gives the special case of Equation (7.1), when spatial fluctuations and inhomogeneities in the quantity $N(t)$ are neglected:

$$\frac{dN}{dt} = -c_i N_i(t), \tag{7.3}$$

such that $N_i(t = 0) = N_0$ is the number density of the species i at time $t = 0$ and $c_i > 0$ constant.

The solution of equation (7.3) is given as [93]:

$$N_i(t) = N_0 e^{c_i t}. \tag{7.4}$$

If we remove the index i and integrate the standard kinetic equation (7.3), we have

$$N(t) - N_0 = -c\,{}_0D_t^{-1} N(t), \tag{7.5}$$

where c is a constant and ${}_0D_t^{-1}$ is the special case of the Riemann-Liouville fractional integral operator ${}_0D_t^{-\nu}$ defined as:

$${}_0D_t^{-\nu} f(t) = \frac{1}{\Gamma(\nu)} \int_0^t (t-u)^{\nu-1} f(u)du, \qquad (t > 0,\ \Re(\nu) > 0). \tag{7.6}$$

The fractional generalization of the standard kinetic equation (7.5) is given by Haubold and Mathai [85] as follows:

$$N(t) - N_0 = -c^{\nu}\,{}_0D_t^{-\nu} N(t). \tag{7.7}$$

They obtained the solution of equation (7.7) given as follows:

$$N(t) = N_0 \sum_{k=0}^{\infty} \frac{(-1)^k}{\Gamma(\nu k + 1)} (ct)^{\nu k}. \tag{7.8}$$

The exponential solution given in Equation (7.4) of standard kinetic equation (7.3) can be obtained by taking $\nu = 1$ in Equation (7.8).

Further, Saxena and Kalla [90] considered the following fractional kinetic equation:

$$N(t) - N_0 f(t) = -c^{\nu}{}_0D_t^{-\nu} N(t), \qquad (\Re(\nu) > 0, c > 0), \tag{7.9}$$

where $N(t)$ denotes the number density of a given species at time t, $N_0 = N(0)$ is the number density of that species at time $t = 0$, c is a constant and $f \in \mathcal{L}(0, \infty)$.

In available literature, there are several integral transforms which are extensively used to solve fractional kinetic differential equations. One of them is Laplace transform.

Let $f(t)$ be a real or complex valued function of variable t and p is a real or complex parameter, then Laplace transform of $f(t)$ is defined as (see [94]):

$$F(p) = L\{f(t); p\} = \int_0^{\infty} e^{-pt} f(t) dt, \qquad (\Re(p) > 0). \tag{7.10}$$

The Laplace transform of Riemann-Liouville fractional integral operator is given as (Erdelyi et al. [95], Srivastava and Saxena [52]):

$$L\left\{{}_0D_t^{-\nu} f(t); p\right\} = p^{-\nu} F(p), \tag{7.11}$$

where $F(p)$ is defined in Equation (7.10).

By applying the Laplace transform to Equation (7.9), we have (see [86]):

$$\begin{aligned} L\{N(t); p\} &= N_0 \frac{F(p)}{1 + c^{\nu} p^{-\nu}} \\ &= N_0 \left(\sum_{n=0}^{\infty} (-c^{\nu})^n p^{-\nu n} \right) F(p), \qquad \left(n \in N_0, \left| \frac{c}{p} \right| < 1 \right). \end{aligned} \tag{7.12}$$

8 Solution of Generalized Fractional Kinetic Equations Involving Extended Incomplete Generalized Hypergeometric Functions

In this section, we investigated the solutions of the generalized fractional kinetic equations involving the extended incomplete generalized hypergeometric functions.

Remark 4 The solutions of the fractional kinetic equations in this section are obtained in terms of the generalized Mittag-Leffler function $E_{\alpha,\beta}(x)$ (Mittag-Leffler[96]), which is defined as:

$$E_{\alpha,\beta}(z) = \sum_{n=0}^{\infty} \frac{z^n}{\Gamma(\alpha n + \beta)}, \qquad \Re(\alpha) > 0, \Re(\beta) > 0. \tag{8.1}$$

Theorem 33 *If* $a > 0, d > 0, \sigma > 0, \alpha_1, \cdots, \alpha_r \in \mathbb{C}$ *and* $\beta_1, \cdots, \beta_s \in \mathbb{C} \setminus \mathbb{Z}_0^-$, *then the solution of the equation*

$$N(t) = N_0\, {}_r\gamma_s^p \left[\begin{matrix} (\alpha_1, x; p), \alpha_2, \ldots, \alpha_r; \\ \beta_1, \ldots, \beta_s; \end{matrix} d^\sigma t^\sigma \right] - a^\sigma {}_0D_t^{-\sigma} N(t) \tag{8.2}$$

is given by

$$N(t) = N_0 \sum_{n=0}^{\infty} \frac{(\alpha_1; x, p)_n (\alpha_2)_n \cdots (\alpha_r)_n \Gamma(\sigma n + 1)}{(\beta_1)_n \cdots (\beta_s)_n} \frac{(d^\sigma t^\sigma)^n}{n!} E_{\sigma, \sigma n+1}(-a^\sigma t^\sigma). \tag{8.3}$$

where $E_{\alpha,\beta}(x)$ *is a generalized Mittag-Leffler function [96] defined in Equation* (8.1).

Proof Applying Laplace transform to both sides of Equation (8.2), we have

$$L\{N(t); u\}$$
$$= N_0 L \left\{ {}_r\gamma_s^p \left[\begin{matrix} (\alpha_1, x; p), \alpha_2, \ldots, \alpha_r; \\ \beta_1, \ldots, \beta_s; \end{matrix} d^\sigma t^\sigma \right]; u \right\} - a^\sigma L\left\{ {}_0D_t^{-\sigma} N(t); u \right\}, \tag{8.4}$$

using the result as given in Equation (7.11), we have

$$N(u) = N_0 \int_0^\infty e^{-ut}\, {}_r\gamma_s^p \left[\begin{matrix} (\alpha_1, x; p), \alpha_2, \ldots, \alpha_r; \\ \beta_1, \ldots, \beta_s; \end{matrix} d^\sigma t^\sigma \right] dt - a^\sigma u^{-\sigma} N(u), \tag{8.5}$$

using Equation (3.18), we have

$$N(u) = N_0 \left(\int_0^\infty e^{-ut} \sum_{n=0}^{\infty} \frac{(\alpha_1; x, p)_n (\alpha_2)_n \cdots (\alpha_r)_n}{(\beta_1)_n \cdots (\beta_s)_n} \frac{(d^\sigma t^\sigma)^n}{n!} dt \right) - a^\sigma u^{-\sigma} N(u), \tag{8.6}$$

after little simplification, we have

$$N(u) + a^\sigma u^{-\sigma} N(u) = N_0 \sum_{n=0}^{\infty} \frac{(\alpha_1; x, p)_n (\alpha_2)_n \cdots (\alpha_r)_n}{(\beta_1)_n \cdots (\beta_s)_n} \frac{(d^\sigma)^n}{n!} \int_0^\infty e^{-ut} t^{\sigma n} dt, \tag{8.7}$$

using the result (6.18), we have

$$\left(1+\left(\frac{u}{a}\right)^{-\sigma}\right)N(u)=N_0\sum_{n=0}^{\infty}\frac{(\alpha_1;x,p)_n(\alpha_2)_n\cdots(\alpha_r)_n}{(\beta_1)_n\cdots(\beta_s)_n}\frac{(d^\sigma)^n}{n!}\frac{\Gamma(\sigma n+1)}{u^{\sigma n+1}},\tag{8.8}$$

after simplification, we have

$$\begin{aligned}N(u)=N_0\sum_{n=0}^{\infty}&\frac{(\alpha_1;x,p)_n(\alpha_2)_n\cdots(\alpha_r)_n}{(\beta_1)_n\cdots(\beta_s)_n}\frac{(d^\sigma)^n}{n!}\\&\times\Gamma(\sigma n+1)\left\{u^{-(\sigma n+1)}\sum_{l=0}^{\infty}\left[-\left(\frac{u}{a}\right)^{-\sigma}\right]^l\right\},\end{aligned}\tag{8.9}$$

taking Laplace inverse of Equation (8.9) and by using

$$L^{-1}\left\{u^{-\sigma};t\right\}=\frac{t^{\sigma-1}}{\Gamma(\sigma)},\qquad(\Re(\sigma)>0),\tag{8.10}$$

we have

$$\begin{aligned}L^{-1}\left\{N(u)\right\}=N_0\sum_{n=0}^{\infty}&\frac{(\alpha_1;x,p)_n(\alpha_2)_n\cdots(\alpha_r)_n}{(\beta_1)_n\cdots(\beta_s)_n}\frac{(d^\sigma)^n}{n!}\\&\times\Gamma(\sigma n+1)L^{-1}\left\{\sum_{l=0}^{\infty}(-1)^l a^{\sigma l}u^{-[\sigma(n+l)+1]}\right\},\end{aligned}\tag{8.11}$$

after simplification, we have

$$\begin{aligned}\text{i.e.}\quad N(t)=N_0\sum_{n=0}^{\infty}&\frac{(\alpha_1;x,p)_n(\alpha_2)_n\cdots(\alpha_r)_n}{(\beta_1)_n\cdots(\beta_s)_n}\frac{(d^\sigma)^n}{n!}\\&\times\Gamma(\sigma n+1)\left\{\sum_{l=0}^{\infty}(-1)^l a^{\sigma l}\frac{t^{\sigma(n+l)}}{\Gamma(\sigma(n+l)+1)}\right\},\end{aligned}\tag{8.12}$$

that implies

$$\begin{aligned}N(t)=N_0\sum_{n=0}^{\infty}&\frac{(\alpha_1;x,p)_n(\alpha_2)_n\cdots(\alpha_r)_n}{(\beta_1)_n\cdots(\beta_s)_n}\frac{(d^\sigma t^\sigma)^n}{n!}\\&\times\Gamma(\sigma n+1)\left\{\sum_{l=0}^{\infty}(-1)^l\frac{(a^\sigma t^\sigma)^l}{\Gamma(\sigma n+\sigma l+1)}\right\},\end{aligned}\tag{8.13}$$

the Equation (8.13) can be written as

$$N(t) = N_0 \sum_{n=0}^{\infty} \frac{(\alpha_1; x, p)_n (\alpha_2)_n \cdots (\alpha_r)_n \Gamma(\sigma n + 1)}{(\beta_1)_n \cdots (\beta_s)_n} \frac{(d^\sigma t^\sigma)^n}{n!} E_{\sigma, \sigma n+1}(-a^\sigma t^\sigma). \tag{8.14}$$

□

Theorem 34 *If $a > 0, d > 0, \sigma > 0, \alpha_1, \cdots, \alpha_r \in \mathbb{C}$ and $\beta_1, \cdots, \beta_s \in \mathbb{C} \setminus \mathbb{Z}_0^-$, then the solution of the equation*

$$N(t) = N_0 \, {}_r\Gamma_s^p \left[\begin{matrix} (\alpha_1, x; p), \alpha_2, \ldots, \alpha_r; \\ \beta_1, \ldots, \beta_s; \end{matrix} d^\sigma t^\sigma \right] - a^\sigma {}_0D_t^{-\sigma} N(t) \tag{8.15}$$

is given by

$$N(t) = N_0 \sum_{n=0}^{\infty} \frac{[\alpha_1; x, p]_n (\alpha_2)_n \cdots (\alpha_r)_n \Gamma(\sigma n + 1)}{(\beta_1)_n \cdots (\beta_s)_n} \frac{(d^\sigma t^\sigma)^n}{n!} E_{\sigma, \sigma n+1}(-a^\sigma t^\sigma). \tag{8.16}$$

Proof The proof of Theorem 34 would run parallel to Theorem 33, so we omit the details involved. □

Theorem 35 *If $d > 0, \sigma > 0, \alpha_1, \cdots, \alpha_r \in \mathbb{C}$ and $\beta_1, \cdots, \beta_s \in \mathbb{C} \setminus \mathbb{Z}_0^-$, then the solution of the equation*

$$N(t) = N_0 \, {}_r\gamma_s^p \left[\begin{matrix} (\alpha_1, x; p), \alpha_2, \ldots, \alpha_r; \\ \beta_1, \ldots, \beta_s; \end{matrix} d^\sigma t^\sigma \right] - d^\sigma {}_0D_t^{-\sigma} N(t) \tag{8.17}$$

is given by

$$N(t) = N_0 \sum_{n=0}^{\infty} \frac{(\alpha_1; x, p)_n (\alpha_2)_n \cdots (\alpha_r)_n \Gamma(\sigma n + 1)}{(\beta_1)_n \cdots (\beta_s)_n} \frac{(d^\sigma t^\sigma)^n}{n!} E_{\sigma, \sigma n+1}(-d^\sigma t^\sigma). \tag{8.18}$$

Proof Applying Laplace transform to both sides of Equation (8.17), we have

$$L\{N(t); u\} = N_0 L \left\{ {}_r\gamma_s^p \left[\begin{matrix} (\alpha_1, x; p), \alpha_2, \ldots, \alpha_r; \\ \beta_1, \ldots, \beta_s; \end{matrix} d^\sigma t^\sigma \right]; u \right\} - d^\sigma L \left\{ {}_0D_t^{-\sigma} N(t); u \right\}, \tag{8.19}$$

using the result as given in Equation (7.11), we have

$$N(u) = N_0 \int_0^\infty e^{-ut} {}_r\gamma_s^p \left[\begin{matrix} (\alpha_1, x; p), \alpha_2, \ldots, \alpha_r; \\ \beta_1, \ldots, \beta_s; \end{matrix} d^\sigma t^\sigma \right] dt - d^\sigma u^{-\sigma} N(u), \tag{8.20}$$

using Equation (3.18), we have

$$N(u) = N_0 \left(\int_0^\infty e^{-ut} \sum_{n=0}^\infty \frac{(\alpha_1; x, p)_n (\alpha_2)_n \cdots (\alpha_r)_n}{(\beta_1)_n \cdots (\beta_s)_n} \frac{(d^\sigma t^\sigma)^n}{n!} dt \right) - d^\sigma u^{-\sigma} N(u), \tag{8.21}$$

interchanging the order of integration and summation in Equation (8.21) and after little simplification, we have

$$N(u) + d^\sigma u^{-\sigma} N(u) = N_0 \sum_{n=0}^\infty \frac{(\alpha_1; x, p)_n (\alpha_2)_n \cdots (\alpha_r)_n}{(\beta_1)_n \cdots (\beta_s)_n} \frac{(d^\sigma)^n}{n!} \int_0^\infty e^{-ut} t^{\sigma n} dt, \tag{8.22}$$

using the definition of Laplace transform as given in Equation (6.18), we have

$$\left(1 + \left(\frac{u}{d}\right)^{-\sigma}\right) N(u) = N_0 \sum_{n=0}^\infty \frac{(\alpha_1; x, p)_n (\alpha_2)_n \cdots (\alpha_r)_n}{(\beta_1)_n \cdots (\beta_s)_n} \frac{(d^\sigma)^n}{n!} \frac{\Gamma(\sigma n + 1)}{u^{\sigma n+1}}, \tag{8.23}$$

after simplification, we have

$$\begin{aligned} N(u) = N_0 &\sum_{n=0}^\infty \frac{(\alpha_1; x, p)_n (\alpha_2)_n \cdots (\alpha_r)_n}{(\beta_1)_n \cdots (\beta_s)_n} \frac{(d^\sigma)^n}{n!} \\ &\times \Gamma(\sigma n + 1) \left\{ u^{-(\sigma n+1)} \sum_{l=0}^\infty \left[-\left(\frac{u}{d}\right)^{-\sigma} \right]^l \right\}, \end{aligned} \tag{8.24}$$

taking Laplace inverse of (8.24) and by using

$$L^{-1}\left\{u^{-\sigma}; t\right\} = \frac{t^{\sigma-1}}{\Gamma(\sigma)}, \qquad (R(\sigma) > 0), \tag{8.25}$$

we have

$$\begin{aligned} L^{-1}\{N(u)\} = N_0 &\sum_{n=0}^\infty \frac{(\alpha_1; x, p)_n (\alpha_2)_n \cdots (\alpha_r)_n}{(\beta_1)_n \cdots (\beta_s)_n} \frac{(d^\sigma)^n}{n!} \\ &\times \Gamma(\sigma n + 1) L^{-1} \left\{ \sum_{l=0}^\infty (-1)^l d^{\sigma l} u^{-[\sigma(n+l)+1]} \right\}, \end{aligned} \tag{8.26}$$

after simplification, we have

$$\text{i.e.} \quad N(t) = N_0 \sum_{n=0}^{\infty} \frac{(\alpha_1; x, p)_n (\alpha_2)_n \cdots (\alpha_r)_n}{(\beta_1)_n \cdots (\beta_s)_n} \frac{(d^\sigma)^n}{n!} \times \Gamma(\sigma n + 1) \left\{ \sum_{l=0}^{\infty} (-1)^l d^{\sigma l} \frac{t^{\sigma(n+l)}}{\Gamma(\sigma(n+l)+1)} \right\}, \tag{8.27}$$

that implies

$$N(t) = N_0 \sum_{n=0}^{\infty} \frac{(\alpha_1; x, p)_n (\alpha_2)_n \cdots (\alpha_r)_n}{(\beta_1)_n \cdots (\beta_s)_n} \frac{(d^\sigma t^\sigma)^n}{n!} \times \Gamma(\sigma n + 1) \left\{ \sum_{l=0}^{\infty} (-1)^l \frac{(d^\sigma t^\sigma)^l}{\Gamma(\sigma n + \sigma l + 1)} \right\}, \tag{8.28}$$

the Equation (8.28) can be written as

$$N(t) = N_0 \sum_{n=0}^{\infty} \frac{(\alpha_1; x, p)_n (\alpha_2)_n \cdots (\alpha_r)_n \Gamma(\sigma n + 1)}{(\beta_1)_n \cdots (\beta_s)_n} \frac{(d^\sigma t^\sigma)^n}{n!} E_{\sigma, \sigma n+1}(-d^\sigma t^\sigma). \tag{8.29}$$

□

Theorem 36 *If* $d > 0, \sigma > 0, \alpha_1, \cdots, \alpha_r \in \mathbb{C}$ *and* $\beta_1, \cdots, \beta_s \in \mathbb{C} \setminus \mathbb{Z}_0^-$, *then the solution of the equation*

$$N(t) = N_0 \, {}_r\Gamma_s^p \left[\begin{matrix} (\alpha_1, x; p), \alpha_2, \ldots, \alpha_r; \\ \beta_1, \ldots, \beta_s; \end{matrix} \, d^\sigma t^\sigma \right] - d^\sigma \, {}_0D_t^{-\sigma} N(t) \tag{8.30}$$

is given by

$$N(t) = N_0 \sum_{n=0}^{\infty} \frac{[\alpha_1; x, p]_n (\alpha_2)_n \cdots (\alpha_r)_n \Gamma(\sigma n + 1)}{(\beta_1)_n \cdots (\beta_s)_n} \frac{(d^\sigma t^\sigma)^n}{n!} E_{\sigma, \sigma n+1}(-d^\sigma t^\sigma). \tag{8.31}$$

Proof The proof of Theorem 36 would run parallel to Theorem 35, so we omit the details involved. □

Theorem 37 *If* $d > 0, \sigma > 0, \alpha_1, \cdots, \alpha_r \in \mathbb{C}$ *and* $\beta_1, \cdots, \beta_s \in \mathbb{C} \setminus \mathbb{Z}_0^-$, *then the solution of the equation*

$$N(t) = N_0\, {}_r\gamma_s^p \left[\begin{matrix} (\alpha_1, x; p), \alpha_2, \ldots, \alpha_r; \\ \beta_1, \ldots, \beta_s; \end{matrix} t \right] - d^\sigma\, {}_0D_t^{-\sigma} N(t) \tag{8.32}$$

is given by

$$N(t) = N_0 \sum_{n=0}^{\infty} \frac{(\alpha_1; x, p)_n (\alpha_2)_n \cdots (\alpha_r)_n \ t^n}{(\beta_1)_n \cdots (\beta_s)_n} E_{\sigma, n+1}(-d^\sigma t^\sigma). \tag{8.33}$$

Proof Applying Laplace transform to both sides of Equation (8.32), we have

$$L\left\{N(t); u\right\} = N_0 L \left\{ {}_r\gamma_s^p \left[\begin{matrix} (\alpha_1, x; p), \alpha_2, \ldots, \alpha_r; \\ \beta_1, \ldots, \beta_s; \end{matrix} t \right]; u \right\} - d^\sigma L\left\{ {}_0D_t^{-\sigma} N(t); u \right\}, \tag{8.34}$$

using the result as given in Equation (7.11), we have

$$N(u) = N_0 \int_0^\infty e^{-ut}\, {}_r\gamma_s^p \left[\begin{matrix} (\alpha_1, x; p), \alpha_2, \ldots, \alpha_r; \\ \beta_1, \ldots, \beta_s; \end{matrix} t \right] dt - d^\sigma u^{-\sigma} N(u), \tag{8.35}$$

using Equation (3.18), we have

$$N(u) = N_0 \left(\int_0^\infty e^{-ut} \sum_{n=0}^{\infty} \frac{(\alpha_1; x, p)_n (\alpha_2)_n \cdots (\alpha_r)_n}{(\beta_1)_n \cdots (\beta_s)_n} \frac{t^n}{n!} dt \right) - d^\sigma u^{-\sigma} N(u), \tag{8.36}$$

interchanging the order of integration and summation in Equation (8.36) and after little simplification, we have

$$N(u) + d^\sigma u^{-\sigma} N(u) = N_0 \sum_{n=0}^{\infty} \frac{(\alpha_1; x, p)_n (\alpha_2)_n \cdots (\alpha_r)_n}{(\beta_1)_n \cdots (\beta_s)_n} \frac{1}{n!} \int_0^\infty e^{-ut} t^n dt, \tag{8.37}$$

using the definition of Laplace transform as given in Equation (6.18), we have

$$\left(1 + \left(\frac{u}{d}\right)^{-\sigma}\right) N(u) = N_0 \sum_{n=0}^{\infty} \frac{(\alpha_1; x, p)_n (\alpha_2)_n \cdots (\alpha_r)_n}{(\beta_1)_n \cdots (\beta_s)_n} \frac{1}{n!} \frac{\Gamma(n+1)}{u^{n+1}}, \tag{8.38}$$

after simplification, we have

$$N(u) = N_0 \sum_{n=0}^{\infty} \frac{(\alpha_1; x, p)_n (\alpha_2)_n \cdots (\alpha_r)_n}{(\beta_1)_n \cdots (\beta_s)_n} \left\{ u^{-(n+1)} \sum_{l=0}^{\infty} \left[-\left(\frac{u}{d}\right)^{-\sigma} \right]^l \right\}, \tag{8.39}$$

taking Laplace inverse of (8.39) and by using

$$L^{-1}\left\{u^{-\sigma}; t\right\} = \frac{t^{\sigma-1}}{\Gamma(\sigma)}, \qquad (R(\sigma) > 0), \tag{8.40}$$

we have

$$L^{-1}\{N(u)\} = N_0 \sum_{n=0}^{\infty} \frac{(\alpha_1; x, p)_n (\alpha_2)_n \cdots (\alpha_r)_n}{(\beta_1)_n \cdots (\beta_s)_n} L^{-1} \left\{ \sum_{l=0}^{\infty} (-1)^l d^{\sigma l} u^{-[\sigma l + n + 1]} \right\}, \tag{8.41}$$

after simplification, we have

$$N(t) = N_0 \sum_{n=0}^{\infty} \frac{(\alpha_1; x, p)_n (\alpha_2)_n \cdots (\alpha_r)_n}{(\beta_1)_n \cdots (\beta_s)_n} \left\{ \sum_{l=0}^{\infty} (-1)^l d^{\sigma l} \frac{t^{\sigma l + n}}{\Gamma(\sigma l + n + 1)} \right\}, \tag{8.42}$$

that implies

$$N(t) = N_0 \sum_{n=0}^{\infty} \frac{(\alpha_1; x, p)_n (\alpha_2)_n \cdots (\alpha_r)_n t^n}{(\beta_1)_n \cdots (\beta_s)_n} \left\{ \sum_{l=0}^{\infty} (-1)^l \frac{(d^{\sigma} t^{\sigma})^l}{\Gamma(\sigma l + n + 1)} \right\}, \tag{8.43}$$

the Equation (8.43) can be written as

$$N(t) = N_0 \sum_{n=0}^{\infty} \frac{(\alpha_1; x, p)_n (\alpha_2)_n \cdots (\alpha_r)_n t^n}{(\beta_1)_n \cdots (\beta_s)_n} E_{\sigma, n+1}(-d^{\sigma} t^{\sigma}). \tag{8.44}$$

□

Theorem 38 *If $d > 0, \sigma > 0, \alpha_1, \cdots, \alpha_r \in \mathbb{C}$ and $\beta_1, \cdots, \beta_s \in \mathbb{C} \setminus \mathbb{Z}_0^-$, then the solution of the equation*

$$N(t) = N_0 \, {}_r\Gamma_s^p \left[\begin{matrix} (\alpha_1, x; p), \alpha_2, \ldots, \alpha_r; \\ \beta_1, \ldots, \beta_s; \end{matrix} \, t \right] - d^{\sigma} \, {}_0D_t^{-\sigma} N(t) \tag{8.45}$$

is given by

$$N(t) = N_0 \sum_{n=0}^{\infty} \frac{[\alpha_1; x, p]_n (\alpha_2)_n \cdots (\alpha_r)_n \; t^n}{(\beta_1)_n \cdots (\beta_s)_n} E_{\sigma, n+1}(-d^\sigma t^\sigma). \tag{8.46}$$

Proof The proof of Theorem 38 would run parallel to Theorem 37, so we omit the details involved. □

9 Conclusion

Recent developments in the theory of fractional calculus show its importance (see, for example, [45, 97–105]). Therefore, the fractional integral and differential formulas (of Marichev-Saigo-Maeda type) involving the extended generalized incomplete hypergeometric functions established in this chapter will be useful for investigators in various disciplines of applied sciences and engineering. We are also trying to find certain possible applications of these results presented here to some other research areas due to presence of the extended incomplete generalized hypergeometric functions ${}_r\gamma_s^p(z)$ and ${}_r\Gamma_s^p(z)$ defined by (3.18) and (3.19), respectively, possess the advantage that a number of incomplete gamma functions and hypergeometric function happen to be the particular cases of these functions. Further, applications of these functions in communication theory, probability theory and groundwater pumping modeling are shown by many authors. Therefore, we conclude this investigation by noting that the results deduced above are significant and can lead to yield numerous other fractional integral and derivative formulas and integral transforms involving various special functions by suitable specializations of arbitrary parameters in the main findings. More importantly, they are expected to find some applications in probability theory and to the solutions of differential equations.

Acknowledgements This work was supported by a research grant from SERB Project No. TAR/2018/000001 to the first author.

References

1. J.A.T. Machado, V. Kiryakova, F. Mainardi, A poster about the old history of fractional calculus. Fract. Calc. Appl. Anal. **13**(4), 447–454 (2010)
2. J.A.T. Machado, V. Kiryakova, F. Mainardi, A poster about the recent history of fractional calculus. Fract. Calc. Appl. Anal. **13**(3), 329–334 (2010)
3. J.A.T. Machado, V. Kiryakova, F. Mainardi, Recent history of fractional calculus. Commun. Nonlinear Sci. Numer. Simul. **16**(3), 1140–1153 (2011). https://doi.org/10.1016/j.cnsns.2010.05.027
4. M. Caputo, F. Mainardi, Linear models of dissipation in anelastic solids. Riv. Nuovo Cimento (Ser. II) **1**, 161–198 (1971)
5. Y.N. Rabotnov, *Elements of Hereditary Solid Mechanics* (MIR, Moscow, 1980)

6. F. Mainardi, *Fractional Calculus and Waves in Linear Viscoelasticity* (Imperial College Press, London, 2010)
7. R. Hilfer, *Applications of Fractional Calculus in Physics* (World Scientific Publishing Company, Singapore, 2000)
8. R.B.L.S. Prakasa, *Statistical Inference for Fractional Diffusion Processes* (Wiley, Chichester, 2010)
9. A. Mathai, R. Saxena, H. Haubold, *The H-Functions: Theory and Applications* (Springer, New York, NY, 2010)
10. V.E. Tarasov, Fractional dynamics: application of fractional calculus to dynamics of particles, in *Fields and Media* (Springer, Heidelberg /Higher Education Press, Beijing, 2010)
11. V.V. Uchaikin, *Fractional Derivatives for Physicists and Engineers. Vol. I. Background and Theory* (Springer, Berlin/Higher Education Press, Beijing, 2013)
12. V.V. Uchaikin, *Fractional Derivatives for Physicists and Engineers, Vol. II. Applications* (Springer, Berlin /Higher Education Press, Beijing, 2013)
13. G.M. Zaslavsky, *Hamiltonian Chaos and Fractional Dynamics* (Oxford University Press, Oxford, 2005)
14. R. Caponetto, G. Dongola, L. Fortuna, I. Petráš, *Fractional Order Systems: Modeling and Control Applications* (World Scientific Publishing Co Inc., Singapore, 2010)
15. M. Axtell, M.E. Bise, Fractional calculus applications in control systems, in *Proceedings of the 1990 National Aerospace and Electronics Conference*, Dayton, OH, 1990
16. A. Azar, S.Vaidyanathan, A. Ouannas, *Fractional Order Control and Synchronization of Chaotic Systems*. Studies in Computational Intelligence, vol. 688 (Springer, Heidelberg, 2017)
17. D. Baleanu, K. Diethelm, E. Scalas, J.J. Trujillo, *Fractional Calculus: Models and Numerical Methods* (World Scientific, London, Singapore, Berlin, 2012)
18. K. Diethelm, *The Analysis of Fractional Differential Equations. An Application-Oriented Exposition Using Differential Operators of Caputo Type*. Springer Lecture Notes in Mathematics, vol. 2004 (Springer, Berlin, 2010)
19. V. Kiryakova, *Generalized Fractional Calculus and Applications*. Pitman Research Notes in Mathematics Series (Longman Scientific & Technical, Harlow, Longman, 1994)
20. A.A. Kilbas, H.M. Srivastava, J.J. Trujillo, *Theory and Applications of Fractional Differential Equations*. North-Holland Mathematics Studies, vol. 204 (Elsevier (North-Holland) Science Publishers, Amsterdam, London, New York, 2006)
21. K.S. Miller, B. Ross, *An Introduction to the Fractional Calculus and Fractional Differential Equations* (Wiley, New York, 1993)
22. K. Oldham, J. Spanier, *Fractional Calculus: Theory and Applications of Differentiation and Integration of Arbitrary Order* (Academic, New York, London, 1974)
23. I. Podlubny, *Fractional Differential Equations: An Introduction to Fractional Derivatives, Fractional Differential Equations, to Methods of Their Solution and Some of Their Applications*, vol. 198 (Academic, New York, London, Sydney, Tokyo, Toronto, 1999)
24. S.G. Samko, A.A. Kilbas, O.I. Marichev, *Fractional Integrals and Derivatives: Theory and Applications* (Gordon and Breach Science Publishers, Yverdon, New York, London, 1993)
25. Y.A. Brychkov, *Handbook of Special Functions, Derivatives, Integrals, Series and Other Formulas* (CRC Press, Taylor & Francis Group, Boca Raton, London, New York, 2008)
26. J. Choi, P. Agarwal, Certain integral transform and fractional integral formulas for the generalized Gauss hypergeometric functions. Abstr. Appl. Anal. **2014** (2014). https://doi.org/10.1155/2014/735946
27. R. Gorenflo, F. Mainardi, Fractional calculus: integral and differential equations of fractional order, in *Fractals and Fractional Calculus in Continuum Mechanics*, ed. by A. Carpinteri, F. Mainardi (Springer, Wien, 1997)
28. V. Kiryakova, A brief story about the operators of the generalized fractional calculus. Fract. Calc. Appl. Anal. **11**(2), 203220 (2008)
29. A.A. Kilbas, Fractional calculus of the generalized Wright function. Fract. Calc. Appl. Anal. **8**(2), 113–126 (2005)

30. V. Kiryakova, Some special functions related to fractional calculus and fractional (non-integer) order control systems and equations, in *Facta Universitatis (Sci. J. of University of Nis). Series: Automatic Control and Robotics*, vol. 7(1) (2008), pp. 79–98
31. P. Agarwal, F. Qi, M. Chand, G. Singh, Some fractional differential equations involving generalized hypergeometric functions. J. Appl. Anal. **25**(1), 37–44 (2019)
32. G. Singh, P. Agarwal, S. Araci, M. Ackigoz, Certain fractional calculus formulas involving extended generalized Mathieu series, in *Advances in Difference Equations*, vol. 2018 (2018). https://doi.org/10.1186/s13662-018-1596-9
33. J. Choi, P. Agarwal, S. Mathur, S. Purohit, Certain new integral formulas involving the generalized Bessel functions. Bull. Korean Math. Soc. **51**(4), 995–1003 (2014)
34. P. Agarwal, M. Chand, J. Choi, G. Singh, Certain fractional integrals and image formulas of generalized k-Bessel function. Commun. Korean Math. Soc. **33**(2), 423–436 (2018)
35. P. Agarwal, Q. Ai-Mdallal, Y. Cho, S. Jain, Fractional differential equations for the generalized Mittag-Leffler function. Adv. Differ. Equ. **2018**(1) (2018), 58
36. I. KIyamz, A. Cetinkaya, P. Agarwal, An extension of Caputo fractional derivative operator and its applications. J. Nonlinear Sci. Appl. **9**, 3611–3621 (2016)
37. F.G. Tricomi, Sulla funzione gamma incompleta. Ann. Mat. Pura Appl. **31**(4), 263–279 (1950)
38. F.A. Musallam, S.L. Kalla, Asymptotic expansions for generalized gamma and incomplete gamma functions. Appl. Anal. **66**, 173–187 (1997)
39. F.A. Musallam, S.L. Kalla, Further results on a generalized gamma function occurring in diffraction theory. Integr. Transf. Spec. Funct. **7**(3–4), 175–190 (1998)
40. R. Srivastava, Some properties of a family of incomplete hypergeometric functions. Russian J. Math. Phys. **20**, 121–128 (2013)
41. R. Srivastava, Some generalizations of Pochhammer symbol and their associated families of hypergeometric functions and hypergeometric polynomials. Appl. Math. Inf. Sci. **7**(6), 2195–2206 (2013)
42. R. Srivasrava, R. Agawal, S. Jain, A family of the incomplete hypergeometric functions and associated integral transform and fractional derivative formulas. Filomat **31**(1), 125–140 (2017)
43. O.I. Marichev, Volterra equation of Mellin convolution type with a Horn function in the kernel. Izv. ANBSSR Ser. Fiz.-Mat. Nauk. **1**, 128–129 (1974)
44. M. Saigo, N. Maeda, More generalization of fractional calculus, in *Transform Methods and Special Functions* (Bulgarian Academy of Sciences, Sofia, Varna, 1996)
45. R. Saxena, M. Saigo, Generalized fractional calculus of the H-function associated with the Appell function. J. Frac. Calc. **19**, 89104 (2001)
46. M. Saigo, A remark on integral operators involving the Gauss hypergeometric functions. Math. Rep. Kyushu Univ. **11**(2), 135143 (1978)
47. M. Saigo, A certain boundary value problem for the Euler-Darboux equation I. Math. Japon. **24**(4), 377385 (1979)
48. M. Saigo, A certain boundary value problem for the Euler-Darboux equation II. Math. Japon. **25**(2), 211220 (1980)
49. H.M. Srivastava, P.W. Karlsson, *Multiple Gaussian Hypergeometric Series* (Halsted Press (Ellis Horwood Limited, Chichester), Wiley, New York, Chichester, Brisbane, Toronto, 1985)
50. H. Kober, On fractional integrals and derivatives. Quart. J. Math. Oxford **11**, 193212 (1940)
51. V. Kiryakova, *Generalized Fractional Calculus and Applications*. Pitman Research Notes in Mathematics Series (Longman Scientific and Technical, Harlow/Wiley, New York, NY, 1993)
52. H. Srivastava, R. Saxena, Operators of fractional integration and their applications. Appl. Math. Comput. **118**, 1–52 (2001)
53. L.C. Andrews, *Special Functions for Engineers and Applied Mathematicians* (Macmillan Company, New York, 1984)
54. F. Olver, D.W. Lozier, R.F. Boisvert, C.W. Clark, *Handbook of Mathematical Functions* (US Department of Commerce, National Institute of Standards and Technology, Washington, DC; Cambridge University Press, Cambridge, London, New York, 2010)

55. M.A. Chaudhry, S.M. Zubair, *On a Class of Incomplete Gamma Functions with Applications* (Chapman and Hall (CRC Press), Boca Raton, London, New York, Washington, DC, 2001)
56. A. Erdélyi, W. Magnus, F. Oberhettinger, F.G. Tricomi, *Higher Transcendental Functions*, vol. II (McGraw-Hill Book Company, New York, Toronto, London, 1953)
57. Y.L. Luke, *Mathematical Functions and Their Approximations* (Academic, New York, San Francisco, London, 1975)
58. A.M. Mathai, R.K. Saxena, *Generalized Hypergeometric Functions with Applications in Statistics and Physical Sciences*. Lecture Note Series, vol. 348 (Springer, New York, 1973)
59. A.M. Mathai, H.J. Haubold, *Special Functions for Applied Scientists* (Springer, New York, 2008)
60. N.M. Temme, *Special Functions: An Introduction to Classical Functions of Mathematical Physics* (Wiley, New York, Chichester, Brisbane, Toronto, 1996)
61. J. Choi, P. Agarwal, Certain class of generating functions for the incomplete hypergeometric functions. Abstr. Appl. Anal. **2014**, 5 pp. (2014). https://doi.org/10.1155/2014/714560
62. H. Srivastava, P. Agarwal, Certain fractional integral operators and the generalized incomplete Hypergeometric functions. Appl. Appl. Math. **8**(2), 333–345 (2013)
63. H.M. Srivastava, M.A. Chaudhry, R.P. Agarwal, The incomplete Pochhammer symbols and their applications to hypergeometric and related functions. Integr. Transf. Spec. Funct. **23**, 659–683 (2012)
64. R. Srivastava, N.E. Cho, Generating functions for a certain class of incomplete hypergeometric polynomials. Appl. Math. Comput. **219**, 3219–3225 (2012)
65. H. Srivastava, A. Çetinkaya, I. Kymaz, A certain generalized Pochhammer symbol and its applications to hypergeometric functions. Appl. Math. Comput. **226**, 484–491 (2014) https://doi.org/10.1016/j.amc.2013.10.032
66. R. Parmar, R. Raina, On the Extended Incomplete Pochhammer Symbols and Hypergeometric Functions (2017). arXiv:1701.04159v1[math.CA]
67. A. Erdélyi, W. Magnus, F. Oberhettinger, F.G. Tricomi, **Higher Transcendental Functions**, vols. I–III (Krieger Pub., Melbourne, FL, 1981)
68. G. E. Andrews, R. Askey, R. Roy, *Special Functions* (Cambridge University Press, Cambridge, 1999)
69. E.D. Rainville, *Special Functions* (The Macmillan Company, New York, 1960)
70. H.M. Srivastava, J. Choi, *Zeta and q-Zeta Functions and Associated Series and Integrals* (Elsevier Science Publishers, Amsterdam, London, New York, 2012)
71. M. Chaudhry, S.M. Zubair, Generalized incomplete gamma functions with applications. J. Comput. Appl. Math. **55**, 99124 (1994)
72. I. Sneddon, *The Use of Integral Transforms* (Tata McGraw-Hill, New Delhi, 1979)
73. J.L. Schiff, *The Laplace Transform, Theory and Applications* (Springer, New York, 1999)
74. G. Singh, P. Agarwal, M. Chand, S. Jain, Certain fractional kinetic equations involving generalized k-Bessel function. Trans. A. Razmadze Math. Inst. **172**(3), 559–570 (2018). https://doi.org/10.1186/s13662-018-1596-9
75. P. Agarwal, M. Chand, G. Singh, Certain fractional kinetic equations involving the product of generalized k-Bessel function. Alex. Eng. J. **55**(4), 3053–3059 (2016)
76. P. Agarwal, S.K. Ntouyas, S. Jain, M. Chand, G. Singh, Fractional kinetic equations involving generalized k-Bessel function via Sumudu transform. Alex. Eng. J. **57**, 1937–1942 (2018)
77. J. Choi, D. Kumar, Solutions of generalized fractional kinetic equations involving Aleph functions. Math. Commun. **20**, 113–123 (2015)
78. J. Choi, P. Agarwal, Certain unified integrals associated with Bessel functions. Bound. Value Probl. **95**(1) (2013). https://doi.org/10.1186/1687-2770-2013-95
79. V. Chaurasia, S. Pandey, On the new computable solution of the generalized fractional kinetic equations involving the generalized function for the fractional calculus and related functions. Astrophys. Space Sci. **317**, 213–219 (2008)
80. M. Chand, J.C. Prajapati, E. Bonyah, Fractional integrals and solution of fractional kinetic equations involving k-Mittag-Leffler function. Trans. A. Razmadze Math. Inst. **171**(2), 144–166 (2017). https://doi.org/101016/j.trmi.2017.03.003

81. A. Chouhan, S.D. Purohit, S. Saraswat, On solution of generalized kinetic equation of fractional order. Int. J. Math. Sci. Appl. **2**, 813–818 (2012)
82. A. Chouhan, S.D. Purohit, S. Saraswat, An alternative method for solving generalized differential equations of fractional order. Kragujevac J. Math. **37**(2), 299–306 (2013)
83. V.G. Gupta, B. Sharma, On the solutions of generalized fractional kinetic equations. Appl. Math. Sci. **5**(19), 899–910 (2011)
84. A. Gupta, C.L. Parihar, On solutions of generalized kinetic equations of fractional order. Bol. Soc. Paran. Mat. **32**(1), 181–189 (2014)
85. H. Haubold, A.M. Mathai, The fractional kinetic equation and thermonuclear functions. Astrophys. Space Sci. **327**, 53–63 (2000)
86. D. Kumar, S.D. Purohit, A. Secer, A. Atangana, On generalized fractional kinetic equations involving generalized Bessel function of the first kind. Math. Probl. Eng. **7** (2015). https://doi.org/10.1155/2015/289387
87. R. Saxena, A.M. Mathai, H.J. Haubold, On fractional kinetic equations. Astrophys. Space Sci. **282**, 281–287 (2002)
88. R. Saxena, A.M. Mathai, H.J. Haubold, On generalized fractional kinetic equations. Physica A **344**, 657–664 (2004)
89. R. Saxena, A.M. Mathai, H.J. Haubold, Solution of generalized fractional reaction-diffusion equations. Astrophys. Space Sci. **305**, 305–313 (2006)
90. R.K. Saxena, S.L. Kalla, On the solutions of certain fractional kinetic equations. Appl. Math. Comput. **199**, 504–511 (2008)
91. G. Zaslavsky, Fractional kinetic equation for Hamiltonian chaos. Physica D **76**, 110–122 (1994)
92. A. Saichev, M. Zaslavsky, Fractional kinetic equations: solutions and applications. Chaos **7**, 753–764 (1997)
93. V. Kourganoff, *Introduction to the Physics of Stellar Interiors* (D. Reidel Publishing Company, Dordrecht, 1973)
94. M. Spiegel, *Theory and Problems of Laplace Transforms*. Schaums Outline Series (McGraw-Hill, New York, 1965)
95. A. Erdélyi, W. Magnus, F. Oberhettinger, F.G. Tricomi, *Tables of Integral Transforms*, vol. 1 (McGraw-Hill, New York, Toronto, London, 1954)
96. G.M. Mittag-Leffler, Sur la representation analytiqie d'une fonction monogene cinquieme note. Acta Math. **29**, 101–181 (1905)
97. P. Agarwal, Further results on fractional calculus of Saigo operators. Appl. Appl. Math. **7**(2), 585–594 (2012)
98. P. Agarwal, J. Choi, Fractional calculus operators and their images formulas. J. Korean Math. Soc. **53**(5), 1183–1210 (2016)
99. P. Agarwal, S. Jain, Further results on fractional calculus of Srivastava polynomials. Bull. Math. Anal. Appl. **3**(2), 167–174 (2011)
100. A. McBride, Fractional Calculus and Integral Transforms of Generalized Functions. Research Notes in Mathematics (Pitman Publishing Limited, London, 1979)
101. A. McBride, Fractional powers of a class of ordinary differential operators. Proc. Lond. Math. Soc. (III) **45**, 519546 (1982)
102. M. Saigo, On generalized fractional calculus operators. Recent Advances in Applied Mathematics. Proc. Internat. Workshop held at Kuwait Univ., Kuwait (1996), pp. 441–450
103. R. Saxena, K. Nishimoto, Fractional calculus of generalized Mittag-Leffler functions. J. Fract. Calc. **37**, 43–52 (2010)
104. R. Saxena, J. Ram, M. Vishnoi, Fractional integration and fractional differentiation of generalized Mittag-Leffler functions. J. Indian Acad. Math. **32**(1), 153–161 (2010)
105. H. Srivastava, Ž. Tomovski, Fractional calculus with an integral operator containing generalized Mittag-Leffler function in the kernel. Appl. Math. Comput. **211**(1), 198–210 (2009). https://doi.org/10.1016/j.amc.2009.01.055

On the Stability of the Triangular Equilibrium Points in the Elliptic Restricted Three-Body Problem with Radiation and Oblateness

Vassilis S. Kalantonis, Angela E. Perdiou, and Efstathios A. Perdios

Abstract The elliptic restricted three-body problem when the primary is a source of radiation and the secondary is an oblate spheroid is considered and the stability of the triangular equilibrium points is studied. The transition curves separating stable from unstable regions are determined in the parametric space both analytically and numerically. Our results show that the oblateness and radiation parameters do not cause significant changes on the topology of the stability regions in the parametric plane defined by the mass parameter and eccentricity. However, in the remaining parametric planes, we observe that by increasing the values of the parameters which are kept fixed stability gives place to instability.

MSC 70F07; 70F15; 70K20; 70K42

1 Introduction

The restricted three-body problem is the most widespread problem in Celestial Mechanics and during the past of the last ages too many works have been devoted for its study either from mathematical or from practical point of view ([8, 11, 17, 20, 31, 35, 46], among many others). This problem consists of two bodies, known as primaries, which rotate around their common center of mass and a massless body which moves in the plane of motion of the primaries under their gravitational attraction and does not affect their motion. If the two primaries track circular orbits around their center of mass, it is called circular restricted three-body problem (CR3BP), while if they describe elliptic orbits, it is called elliptic

V. S. Kalantonis (✉) · E. A. Perdios
Department of Electrical and Computer Engineering, University of Patras, Patras, Greece
e-mail: kalantonis@upatras.gr; eperdios@upatras.gr

A. E. Perdiou
Department of Civil Engineering, University of Patras, Patras, Greece
e-mail: aperdiou@upatras.gr

T. M. Rassias, P. M. Pardalos (eds.), *Mathematical Analysis and Applications*, Springer Optimization and Its Applications 154,
https://doi.org/10.1007/978-3-030-31339-5_9

(ER3BP). A major difference that exists for the study of these two problems is that although the CR3BP possesses the well-known Jacobi integral, the ER3BP does not and this makes the investigation of the latter more complicated. However, both the CR3BP and the ER3BP admit five equilibrium points, three of which are located on the Ox-axis joining the primaries and the remaining two form equilateral triangles with them. Due to their immediate connection with real applications, e.g., in space mission design, the dynamics around them is of special importance (see, for example, [14, 19], and the references therein).

In the framework of the CR3BP, various modifications have been proposed in which additional forces are incorporated so as to make the models of our Solar system more accurate. These variants take into account the oblateness of the primary bodies and/or the radiation pressure when the primaries are radiation sources ([1, 3, 4, 10, 15, 32, 37, 38, 42, 45], among others). For example, Oberti and Vienne [30] have shown that the inclusion of the oblateness effect of the primary bodies provides significant improvements in the theory of motion of the Lagrangian satellites Telesto, Calypso, and Helene in the Saturn-Tethys or Saturn-Dione and a satellite three-body systems. An interesting variant of CR3BP has been proposed recently by Bosanac et al. [6] in which an additional coupled three-body interaction effect is considered. In addition, Zeng et al. [44] dealt with a rotating mass dipole with oblateness of one primary in order to approximate accurately the potential distribution of nearly axisymmetrical elongated celestial bodies.

On the other hand, in relation to the ER3BP certain amount of work has been done during the past. Some of these works deal with its periodic orbits such as the work by Broucke [7] where he systematically studied these orbits together with their stability, the work by Markellos [21] where, through periodic solutions, he extensively studied the stability regions of retrograde satellites or that by Perdios [33] in which doubly asymptotic orbits at the unstable equilibrium points were determined both analytically and numerically. Also, Haghighipour et al. [13] studied 1:2 resonant periodic orbits, Voyatzis et al. [43] considered the Hill limiting case of the ER3BP and determined a large set of families of periodic orbits together with their stability and more recently, Antoniadou and Libert [2] computed periodic orbits of the ER3BP by originating from periodic solutions of the CR3BP. Additionally, several studies in the ER3BP discuss about the dynamics around the triangular equilibrium points. The main results can be found in the works by Danby [9], Bennett [5], Nayfeh [29], and Meire [25] where the linear stability of the triangular equilibrium points was investigated focusing especially on the transition curves separating regions of stability and instability in the (μ, e) parametric plane. Also, Valente et al. [41] considered the non-linear stability zones around the triangular points, Érdi et al. [12] investigated the size distribution of their stability region while Kovács [18] presented the stability chart of the triangular points using the energy-rate method.

With regard to certain modifications of the ER3BP we may refer here that for the photogravitational version of the ER3BP Markellos et al. [22–24] studied the linear stability of the Lagrangian points as well as that of the inner equilibrium point by determining the stability regions in the space of the parameters of the problem. Also, in the case where the first primary is an oblate luminous and the second one an oblate non-luminous body Narayan and Kumar [27, 28] investigated the effects of radiation and oblateness to the location and stability of the triangular equilibrium points while Singh and Umar [39] provided the critical mass ratio for which the triangular points change their stability properties in the framework of the ER3BP where the primary is an oblate and the secondary a luminous body. Recently, Ruth and Sharma [36] considered the ER3BP when the larger primary is a source of radiation and the smaller one an oblate spheroid and studied the positions and stability of all equilibria as well as tadpole orbits. They used the mean anomaly to average the distance between the primaries and they approximated the mean motion. By making use of those approximations, semi-analytical formulae for the positions of the triangular equilibrium points were derived and based on them they obtained the transition curves in the (μ, e) parametric plane with an accuracy of order two.

So, in view of our previous discussion and based on the work by Ruth and Sharma [36] we consider in our study the latter version of the ER3BP and extend their results by presenting analytical expressions for the positions of the triangular equilibrium points and by determining accurately the stability boundaries which separate the stability regions in all parametric planes. An analytical solution is given for one of the transition curves (they are three in total) which depends only on the parameters of the problem. The remaining two transition curves which delimit the stability of motion around the triangular equilibrium points, in any parametric plane, are computed numerically by applying Floquet theory for a system with periodic coefficients. The nature of instability is also presented in all the considered cases. Our work is organized as follows. In Section 2 the equations of motion of the considered problem are recalled and the positions of the triangular equilibrium points are presented. In Section 3 the stability of motion of the test particle around the respective equilibria is studied analytically and numerically while in Section 4 our numerical results are shown by illustrating the determined transition curves in several parametric planes. Finally, our paper ends in Section 5 with some concluding remarks.

2 Equations of Motion and Triangular Equilibrium Points

We consider a rotating pulsating non-dimensional system x, y, where two primary bodies with masses $m_1 = 1 - \mu$ and $m_2 = \mu$, with $\mu = m_2/(m_1 + m_2) \leqslant 0.5$ being the mass parameter, have fixed positions at the Ox−axis and move in elliptic Keplerian orbit. Also, the more massive primary m_1 is considered to be a source of radiation and the secondary m_2 an oblate spheroid. The equations describing the motion of a third body of negligible mass which always moves in the same plane with the two primaries and does not affect their motion are [27]:

$$\ddot{x} - 2\dot{y} = \sigma(\theta)\frac{\partial \Omega}{\partial x}, \qquad \ddot{y} + 2\dot{x} = \sigma(\theta)\frac{\partial \Omega}{\partial y}, \tag{1}$$

where Ω is the potential function given by:

$$\Omega = \frac{x^2 + y^2}{2} + \frac{1}{n^2}\left[\frac{(1-\mu)q_1}{r_1} + \frac{\mu}{r_2} + \frac{\mu A_2}{2r_2^3}\right], \tag{2}$$

while

$$r_1 = \sqrt{(x+\mu)^2 + y^2}, \quad r_2 = \sqrt{(x+\mu-1)^2 + y^2}, \tag{3}$$

are the distances of the third body from the two primaries, respectively, and:

$$n^2 = 1 + \frac{3A_2}{2}, \quad \sigma(\theta) = \frac{1}{1 + e\cos\theta}. \tag{4}$$

The true anomaly θ is the independent variable and the dots in system (1) denote differentiation with respect to θ, i.e. $d/d\theta$, n is the perturbed mean motion and e is the eccentricity of the elliptical orbits of the primaries. Also, A_2 is the oblateness coefficient of the second primary defined by the formula $A_2 = (R_E^2 - R_P^2)/5R^2$ where R_E and R_P are the equatorial and polar radii of the said primary body, respectively, and R is the distance between the primaries while $q_1 = 1 - F_p/F_g$ stands for the mass reduction factor with F_g and F_p being the gravitational and radiation pressure forces, correspondingly. Note that for $q_1 = 1$ and $A_2 = 0$ the problem reduces to the purely gravitational elliptic three-body problem and by also considering $e = 0$ it becomes the classical circular restricted three-body problem.

The partial derivatives of the potential function Ω involved in (1) are:

$$\frac{\partial\Omega}{\partial x} = x - \frac{1}{n^2}\left[\frac{q_1(1-\mu)(x+\mu)}{r_1^3} + \frac{\mu(x+\mu-1)}{r_2^3} + \frac{3\mu A_2(x+\mu-1)}{2r_2^5}\right],$$
$$\frac{\partial\Omega}{\partial y} = y - \frac{1}{n^2}\left[\frac{q_1(1-\mu)y}{r_1^3} + \frac{\mu y}{r_2^3} + \frac{3\mu A_2 y}{2r_2^5}\right], \tag{5}$$

or equivalently

$$\Omega_x = B_1 x + B_2\mu(1-\mu), \qquad \Omega_y = B_1 y, \tag{6}$$

where we have set for abbreviation:

$$B_1 = 1 - \frac{1}{n^2}\left[\frac{q_1(1-\mu)}{r_1^3} + \frac{\mu}{r_2^3} + \frac{3A_2\mu}{2r_2^5}\right],$$
$$B_2 = -\frac{1}{n^2}\left[\frac{q_1}{r_1^3} - \frac{1}{r_2^3} - \frac{3A_2}{2r_2^5}\right]. \tag{7}$$

Thus, the equations of motion (1) are finally written in the form:

$$\begin{aligned} \ddot{x} - 2\dot{y} &= \sigma(\theta)\,[B_1 x + B_2 \mu(1-\mu)]\,, \\ \ddot{y} + 2\dot{x} &= \sigma(\theta) B_1 y. \end{aligned} \tag{8}$$

The positions of the classical equilateral triangle Lagrangian equilibrium points $L_{4,5}$, for which $x, y \neq 0$, are determined by the conditions:

$$B_1 x + B_2 \mu(1-\mu) = 0, \qquad B_1 = 0, \tag{9}$$

which also give that $B_2 = 0$. With the help of the latter equation, the second equation of (9) gives the following relation for the distance of the Lagrangian equilibrium points $L_{4,5}$ from the first primary:

$$r_1 = \left(\frac{q_1}{n^2}\right)^{1/3}, \tag{10}$$

from which, evidently, we obtain that the triangular equilibrium points exist only for $q_1 \in [0, 1]$. The second equation of (9) also gives that:

$$n^2 = \frac{1}{r_2^3} + \frac{2A_2}{2r_2^5}, \tag{11}$$

and after some simple calculations leads to:

$$r_2 = 1, \tag{12}$$

which means that the distance between $L_{4,5}$ and the second primary is always fixed to one, independently of the parameters of the problem. It is now easy from (3) to obtain analytically the exact coordinates of the positions of the triangular equilibrium points $L_{4,5}$ in the form:

$$x_0 = \frac{1}{2}\left(\frac{q_1}{n^2}\right)^{2/3} - \mu, \qquad y_0 = \pm\frac{1}{2}\left(\frac{q_1}{n^2}\right)^{1/3}\sqrt{4 - \left(\frac{q_1}{n^2}\right)^{2/3}}. \tag{13}$$

Note here that for $A_2 = 0$ in (13) we get the positions of the triangular equilibrium points of the photogravitational problem:

$$x_0 = \frac{1}{2}q_1^{2/3} - \mu, \;\; y_0 = \pm\frac{1}{2}q_1^{1/3}\sqrt{4 - q_1^{2/3}}, \tag{14}$$

which conform with those given by Markellos et al. [22] with non-luminous secondary. If we also set $q_1 = 1$ the corresponding positions $(x_0, y_0) = (1/2 - \mu, \pm\sqrt{3}/2)$ for the classical problem are obtained.

3 Stability of the Triangular Equilibrium Points

By setting $x = x_0 + \xi$ and $y = y_0 + \eta$ we transfer the origin of the coordinate system at the equilateral equilibrium point $L_{4,5}$ introducing thus the new variables ξ and η in the equations of motion (8):

$$\begin{aligned} \ddot{\xi} - 2\dot{\eta} &= \sigma(\theta)\,[B_1(x_0+\xi) + B_2\mu(1-\mu)]\,, \\ \ddot{\eta} + 2\dot{\xi} &= \sigma(\theta) B_1(y_0+\eta). \end{aligned} \tag{15}$$

By linearizing system (15) with respect to the new variables we obtain:

$$\dot{\Xi} = \mathrm{A}(\theta)\Xi, \tag{16}$$

where $\Xi = (\xi, \eta, \dot{\xi}, \dot{\eta})^{\mathrm{T}}$ and:

$$\mathrm{A}(\theta) = \begin{bmatrix} \mathrm{O} & \mathrm{I} \\ \sigma(\theta)\mathrm{C} & 2\mathrm{D} \end{bmatrix}, \tag{17}$$

with the corresponding blocks:

$$\mathrm{O} = \begin{bmatrix} 0\,0 \\ 0\,0 \end{bmatrix}, \quad \mathrm{I} = \begin{bmatrix} 1\,0 \\ 0\,1 \end{bmatrix}, \quad \mathrm{C} = \begin{bmatrix} \Omega^{(0)}_{\xi\xi} & \Omega^{(0)}_{\xi\eta} \\ \Omega^{(0)}_{\eta\xi} & \Omega^{(0)}_{\eta\eta} \end{bmatrix}, \quad \mathrm{D} = \begin{bmatrix} 0 & 1 \\ -1 & 0 \end{bmatrix}. \tag{18}$$

The superscript in the partial derivatives of the potential function involved in C block above denotes that the corresponding derivatives have been evaluated at the triangular equilibrium points (x_0, y_0) and are given by the following analytical formulae depending only on the parameters of the problem:

$$\begin{aligned} \Omega^{(0)}_{\xi\xi} &= \frac{(5n^2)(4n^{8/3} + q_1^{4/3})\mu}{4n^{14/3}} - \frac{[(23\mu-3)n^2 - 8\mu]q_1^{2/3}}{4n^{10/3}}, \\ \Omega^{(0)}_{\xi\eta} &= \left[\frac{3-13\mu}{2} + \frac{2\mu}{n^2} + \frac{\mu(5n^2-2)q_1^{2/3}}{2n^{10/3}}\right] y_0, \\ \Omega^{(0)}_{\eta\eta} &= 3(1-\mu) + \frac{[(23\mu-3)n^2 - 8\mu]}{4n^{10/3}} q_1^{2/3} - \frac{(5n^2-2)\mu}{4n^{14/3}} q_1^{4/3}, \end{aligned} \tag{19}$$

while we note that $\Omega^{(0)}_{\eta\xi} = \Omega^{(0)}_{\xi\eta}$.

The determination of the stability of motion around the triangular equilibrium points can be accomplished through the computation of the characteristic roots of the variational equations (16). The study of the nature of these roots allows us to configure the regions of stability and instability in any parametric plane, e.g. (μ, e), and the corresponding transition curves which separate these regions can be computed numerically using the classical Floquet theory. Following [5], [34] as well as [16], we will briefly describe here this technique. The four sought characteristic roots $\lambda_k,\ k = 1, 2, 3, 4$, are the solutions of the characteristic equation:

$$\det(\mathrm{B} - \lambda \mathrm{I}) = 0, \tag{20}$$

with $\mathrm{I} = \mathrm{diag}[1, 1, 1, 1]$ being the 4×4 identity matrix and $\mathrm{B} = \mathtt{X}^{-1}(\theta)\mathtt{X}(\theta + T)$, where $\mathtt{X}(\theta)$ is a fundamental solution of equations (16) while T is the period of the coefficients of (16) which is equal to 2π. Without loss of generality, we can set in the latter equation $\mathtt{X}(0) = \mathrm{I}$ which means that for matrix B we obtain the form $\mathrm{B} = \mathtt{X}(T)$. If the roots of the characteristic equation (20) are distinct, there are four independent solutions ξ_k satisfying the property $\xi_k(t + T) = \lambda_k \xi_k(t)$, $k = 1, 2, 3, 4$, and obviously a solution is periodic if $\lambda_k = 1$, while the case $|\lambda_k| < 1$ means that the motion is bounded and the case $|\lambda_k| > 1$ denotes unbounded motion. The corresponding characteristic equation is quartic and can be written as the product of two quadratic factors:

$$(\lambda^2 + a_1\lambda + 1)\,(\lambda^2 + a_2\lambda + 1) = 0, \tag{21}$$

with

$$a_1 = \frac{1}{2}(p_1 + \sqrt{D}), \qquad a_2 = \frac{1}{2}(p_1 - \sqrt{D}), \qquad D = p_1^2 - 4(p_2 - 2), \tag{22}$$

where we have abbreviated:

$$p_1 = -\mathrm{Tr}\, B, \qquad p_2 = \sum_{j=i+1}^{4} \sum_{i=1}^{4} (b_{ii}b_{jj} - b_{ij}b_{ji}), \tag{23}$$

and $b_{ij},\ i, j = 1, 2, 3, 4$ are the elements of matrix B. So, for stability we have the following conditions:

$$D > 0, \qquad |a_1| < 2, \qquad |a_2| < 2, \tag{24}$$

while the cases $D = 0$, $|a_1| = 2$ and $|a_2| = 2$ form the required transition curves which define the stability regions of motion around the triangular equilibrium points in a specific parametric plane. We recall here that our studied problem admits the four parameters μ, e, q_1, and A_2 which means that in order to determine the corresponding transition curves on a particular parametric plane we have to fix the values of two of them.

It is worth to mention here that the transition curve which corresponds to the case $D = 0$ may be determined analytically as follows [26]. Using the transformation:

$$\mathrm{X} = \begin{bmatrix} \mathrm{R} & 0 \\ 0 & \mathrm{R}^{-1} \end{bmatrix} \Xi, \qquad \mathrm{R} = \begin{bmatrix} \cos\phi & -\sin\phi \\ \sin\phi & \cos\phi \end{bmatrix}, \tag{25}$$

where R is the two-dimensional rotation matrix, we can compute the angle ϕ:

$$\phi = \frac{1}{2}\arctan\left[\frac{2\Omega_{\xi\eta}^{(0)}}{\Omega_{\xi\xi}^{(0)} - \Omega_{\eta\eta}^{(0)}}\right], \tag{26}$$

for which the matrix C is diagonalized in the form:

$$\mathrm{C}^* = \begin{bmatrix} C_1 & 0 \\ 0 & C_2 \end{bmatrix}, \tag{27}$$

with

$$C_{1,2} = \frac{1}{2}\left[\Omega_{\xi\xi}^{(0)} + \Omega_{\eta\eta}^{(0)} \mp \sqrt{[\Omega_{\xi\xi}^{(0)} - \Omega_{\eta\eta}^{(0)}]^2 + 4[\Omega_{\xi\eta}^{(0)}]^2}\right]. \tag{28}$$

By substituting the positions of the triangular equilibrium points (13) in the latter coefficients, which correspond to the elements of matrix C^*, these obtain, after some calculations, the following special form:

$$C_{1,2} = \frac{3}{2}(1 \mp \sqrt{1 - g^{**}}), \tag{29}$$

where

$$g^{**} = \frac{\mu q_1^{2/3}(1-\mu)(2-5n^2)}{3n^{10/3}} - \frac{4\mu}{9n^4}[2\mu+n^2(3-10\mu)-n^4(12-17\mu)+\sqrt{G}(1-n^2)], \tag{30}$$

with

$$G = [n^2(8\mu - 3) - 2\mu]^2 + 3n^{2/3}\mu q_1^{2/3}(1 - \mu)(5n^2 - 2). \tag{31}$$

Note here that, in case where we consider that the second primary is not an oblate spheroid, i.e. $A_2 = 0$ which also means that $n = 1$, we obtain g^{**} in the following simplified form:

$$g^* = \mu(1 - \mu)(4 - q_1^{2/3}), \tag{32}$$

which is in agreement with the corresponding results given by Markellos et al. [22] for the photogravitational case where only the first primary is luminous. In addition, if we also consider that $q_1 = 1$, i.e. we have the classical gravitational case, the above quantity g^* obviously reduces to:

$$g = 3\mu(1 - \mu), \tag{33}$$

as this was given by Tschauner [40].

We consider now System (16) where its block C has been diagonalized as it is given by the relation (27). By setting $\Upsilon = [\xi, \eta]^{\mathrm{T}}$ and introducing the transformation provided by Meire [26]:

$$\begin{bmatrix} \Upsilon \\ \Upsilon' \end{bmatrix} = \begin{bmatrix} \mathrm{I} & \mathrm{I} \\ S_1 & S_2 \end{bmatrix} \begin{bmatrix} \Omega_1 \\ \Omega_2 \end{bmatrix}, \tag{34}$$

the equations of motion around the triangular equilibrium points become:

$$\begin{bmatrix} \Omega_1 \\ \Omega_2 \end{bmatrix}' = \begin{bmatrix} S_1 & 0 \\ 0 & S_2 \end{bmatrix} \begin{bmatrix} \Omega_1 \\ \Omega_2 \end{bmatrix}, \tag{35}$$

where S_1 and S_2 are the solutions of a Riccati matrix differential equation (see [26]) which produces the following analytical expression:

$$1 - 9g^{**} + 2e^2 + \frac{e^4}{1 - g^{**}} = 0, \tag{36}$$

for the transition curve which corresponds to the case $D = 0$ and depends only on the four parameters of the problem. The analytical form of this curve is equivalent with that given by Meire [26], for the classical case, as well as with that by Markellos et al. [22], for the photogravitational problem, with the difference that g and g^* in those papers are replaced now by the coefficient (30), respectively. Equation (36) may provide the transition curve $D = 0$ in any parametric plane if the remaining two parameters are kept fixed.

4 Numerical Results

For the numerical determination of the transition curves in any parametric plane, say the (μ, e) plane, we use the stability conditions (24) and look either for the stability parameters a_i, $i = 1$ or 2, to obtain the critical values ± 2 or for the discriminant D to be equal to zero. Since these, in general, do not hold we seek corrections $\delta\mu$ and δe such that:

$$a_i(\mu + \delta\mu, e + \delta e) = \pm 2 \qquad \text{or} \qquad D(\mu + \delta\mu, e + \delta e) = 0, \tag{37}$$

and by linearizing we obtain:

$$\frac{\partial a_i}{\partial \mu}\delta\mu + \frac{\partial a_i}{\partial e}\delta e = \pm 2 - a_i \qquad \text{or} \qquad \frac{\partial D}{\partial \mu}\delta\mu + \frac{\partial D}{\partial e}\delta e = -D, \tag{38}$$

$i = 1$ or $i = 2$, from which we can construct the relevant corrector-predictor algorithms. For example, by keeping constant the value of the mass parameter, i.e. $\delta\mu = 0$, we obtain from (38) the corresponding corrections for the eccentricity:

$$\delta e = \frac{\pm 2 - a_i}{\partial a_i/\partial e} \qquad \text{or} \qquad \delta e = \frac{-D}{\partial D/\partial e}, \tag{39}$$

while in case where the conditions (37) have been satisfied with the desired accuracy, by slightly varying the value of the mass parameter, i.e. $\delta\mu = \epsilon$, the corresponding predictions for the eccentricity:

$$\Delta e = -\frac{\partial a_i/\partial \mu}{\partial a_i/\partial e}\,\epsilon \qquad \text{or} \qquad \Delta e = -\frac{\partial D/\partial \mu}{\partial D/\partial e}\,\epsilon, \tag{40}$$

are obtained from (38), respectively. This procedure has to be applied successively for all the admissible values of the corresponding parameters in order to obtain the requested transition curves in the (μ, e) parametric plane. Note that, a similar procedure can be implemented for any other selected plane of the parameters of the problem.

In our study, the transition curve corresponding to the critical value $D = 0$ has been determined, for all the considered cases, by using the analytical expression (36), and not from the previous described numerical technique, while the remaining two transition curves have been computed by applying the aforementioned predictor-corrector algorithms. The stability regions, as they separated by the transition curves, are presented in several parametric planes of the initial four-dimensional parametric space (μ, e, q_1, A_2) of the considered problem.

In Figure 1 we present the transition curves in the (μ, e) parametric plane for certain values of the remaining parameters q_1 and A_2 of the problem which are given in the figure's caption. The isolated point A in this figure is of special importance since at this point simultaneously holds $|a_1| = 2$ and $|a_2| = 2$ which also means that the discriminant of the characteristic equation is zero, i.e. $D = 0$. The numbers 2, 3, and 4 in this figure represent the nature of instability and correspond to the cases $|a_2| > 2$, $|a_1| > 2$ and $D < 0$, respectively, while S denotes the regions where

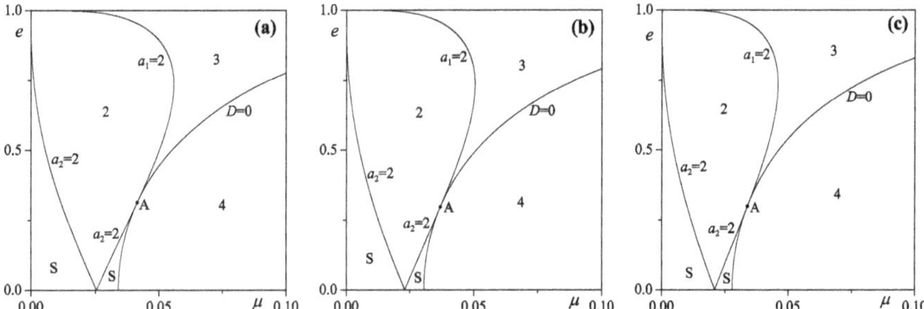

Fig. 1 The transition curves in the (μ, e) parametric plane for (**a**) $q_1 = 0.5$, $A_2 = 0$, (**b**) $q_1 = 1$, $A_2 = 0.2$, and (**c**) $q_1 = 0.5$ and $A_2 = 0.2$. In all frames, the numbers 2, 3, and 4 represent the instability type corresponding to $|a_2| > 2$, $|a_1| > 2$ and $D < 0$, respectively. Point A represents the case for which simultaneously holds $|a_1| = 2$, $|a_2| = 2$ and $D = 0$. Letter S stands for stability

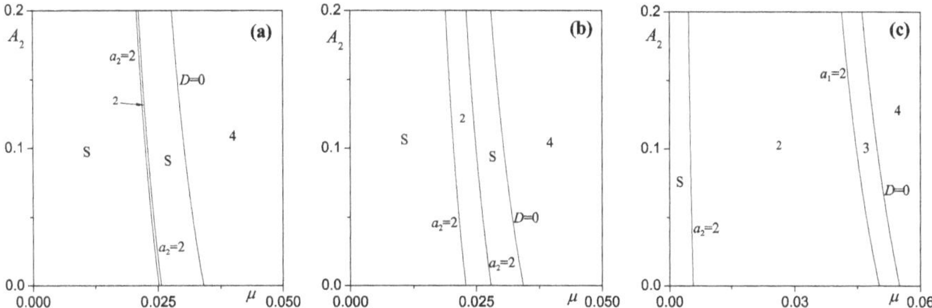

Fig. 2 The transition curves in the (μ, A_2) parametric plane for (**a**) $q_1 = 0.5$, $e = 0.005$, (**b**) $q_1 = 0.5$, $e = 0.05$, and (**c**) $q_1 = 0.5$, $e = 0.5$. The numbers 2, 3, and 4 represent the instability type corresponding to $|a_2| > 2$, $|a_1| > 2$ and $D < 0$, respectively. Letter S stands for stability

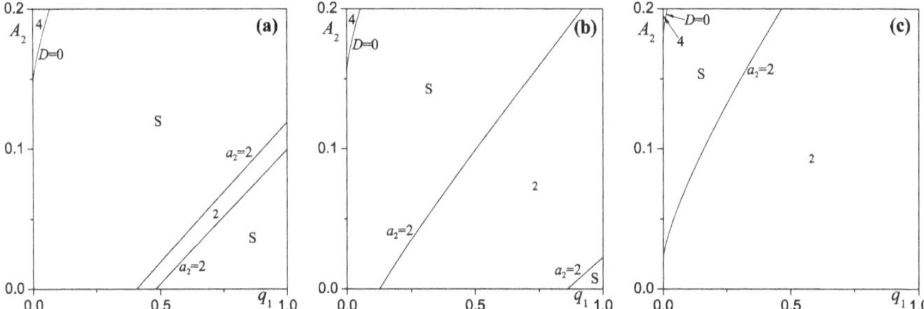

Fig. 3 The transition curves in the (q_1, A_2) parametric plane for (**a**) $\mu = 0.025$, $e = 0.005$, (**b**) $\mu = 0.025$, $e = 0.05$, and (**c**) $\mu = 0.025$, $e = 0.1$. In all frames, the numbers 2 and 4 represent the instability type corresponding to $|a_2| > 2$ and $D < 0$, respectively. The instability type arising from the condition $|a_1| > 2$ does not appear in these cases. Letter S stands for stability

the motion of the test particle in the vicinity of the triangular equilibrium points is stable. As we may observe, the transition curves in this plane are not evidently affected by the radiation factor q_1 and the oblateness parameter A_2 and the occupied stability regions preserve their form.

Figure 2 shows the respective stability regions in the (μ, A_2) parametric plane for the specific values of the parameters q_1 and e which are provided in its caption. The notations in this figure are the same as they were described previously. We see now that by changing the values of the eccentricity from lower to higher values, the leftmost transition curve corresponding to the critical value $|a_2| = 2$ shifts left while the corresponding rightmost curve shifts right; the stability regions shrink giving space to instability of type 2.

A similar behavior with the latter one is observed in Figure 3 where the parametric plane (q_1, A_2) has been chosen to be presented. It is clearly shown, in this figure, that increasing the values of the eccentricity parameter results in the reduction of the stability region and the instability of type 2 occupies the larger part of the parametric plane. In addition, the stability region S, frame (a) of this figure, existing approximately for values of the radiation factor $q_1 \geqslant 0.5$ and for values of the oblateness parameter $A_1 \leqslant 0.1$ for $e = 0.005$ has been lost for the eccentricity value $e = 0.1$ as shown in frame (c).

5 Conclusions

We considered the elliptic restricted three-body problem when the primary is a source of radiation and the secondary is an oblate spheroid and studied the motion of the massless body around the triangular equilibrium points. This problem may be considered more appropriate for solar system applications than that of the classical one, since our Sun radiates and many planets are sufficiently oblate. We showed that the positions of the triangular equilibrium points are given by analytical formulae in which only the parameters of the problem are involved. Their linear stability was determined both analytically and numerically. With regard to the analytical solution of the respective boundary curve which separate the regions of stability it was shown that it depends only on the four parameters of the problem. The numerical determination of the remaining criticality conditions was accomplished by certain predictor-corrector algorithms. For each studied case the nature of instability was also investigated.

A natural extension of the present work would be to study the case where both primaries are oblate luminous bodies in order to obtain the corresponding results for the motion of dust particles around two companion stars of a binary system. Furthermore, a suitable extension of our current results is to investigate the periodic orbits around the triangular equilibrium points. These solutions together with the present results will provide the basic dynamical features of the considered problem and we intend to do it so in a future correspondence.

References

1. E.I. Abouelmagd, J.L.G. Guirao, A. Mostafa, Numerical integration of the restricted three-body problem with Lie series. Astrophys. Space Sci. **354**, 369–378 (2014)
2. K.I. Antoniadou, A.S. Libert, Origin and continuation of 3/2, 5/2, 3/1, 4/1 and 5/1 resonant periodic orbits in the circular and elliptic restricted three-body problem. Celest. Mech. Dyn. Astron. **130**(41), 30 (2006)
3. H.X. Baoyin, C.R. McInnes, Solar sail halo orbits at the sun–earth artificial L1 point. Celest. Mech. Dyn. Astron. **94**, 155–171 (2006)
4. A.S. Beevi, R.K. Sharma, Oblateness effect of Saturn on periodic orbits in the Saturn-Titan restricted three-body problem. Astrophys. Space Sci. **340**, 245–261 (2012)
5. A. Bennett, Characteristic exponents of the five equilibrium solutions in the elliptically restricted problem. Icarus **4**, 177–187 (1965)
6. N. Bosanac, K.C. Howell, E. Fischbach, Exploring the impact of a three-body interaction added to the gravitational potential function in the restricted three-body problem, in *23rd AAS/AIAA Space Flight Mechanics Meeting* (Hawaii, AAS, 2013), pp. 13-490

7. R. Broucke, Stability of periodic orbits in the elliptic, restricted three-body problem. AIAA J. **7**, 1003–1009 (1969)
8. C.C. Conley, Low energy transit orbits in the restricted three-body problem. SIAM J. Appl. Math. **16**, 732–746 (1968)
9. J.M.A. Danby, Stability of the triangular points in the elliptic restricted problem of three bodies. Astron. J. **69**, 165–172 (1964)
10. M.K. Das, P. Narang, S. Mahajan, M. Yuasa, Effect of radiation on the stability of a retrograde particle orbit in different stellar systems. Planet. Space Sci. **57**, 836–845 (2009)
11. Á. Dena, M. Rodríguez, S. Serrano, R. Barrio, High-precision continuation of periodic orbits. Abstr. Appl. Anal. **2012**, Article ID 716024, 12 (2012)
12. B. Érdi, I. Nagy, Zs. Sándor, Á. Süli, G. Fröhlich, Secondary resonances of co-orbital motions. Mon. Not. R. Astron. Soc. **381**, 33–40 (2007)
13. N. Haghighipour, J. Couetdic, F. Varadi, W.B. Moore, Stable 1:2 resonant periodic orbits in elliptic three-body systems. Astrophys. J., **596**, 1332–1340 (2003)
14. K.C. Howell, D.C. Davis, A.F. Haapala, Application of periapse maps for the design of trajectories near the smaller primary in multi–body regimes. Math. Probl. Eng. **2012**, Article ID 351759, 22 (2012)
15. V.S. Kalantonis, C.N. Douskos, E.A. Perdios, Numerical determination of homoclinic and heteroclinic orbits at collinear equilibria in the restricted three-body problem with oblateness. Celest. Mech. Dyn. Astr. **94**, 135–153 (2006)
16. V.S. Kalantonis, E.A. Perdios, A.E. Perdiou, The Sitnikov family and the associated families of 3D periodic orbits in the photogravitational RTBP with oblateness. Astrophys. Space Sci. **315**, 323–334 (2008)
17. T.A. Kotoulas, G. Voyatzis, Three dimensional periodic orbits in exterior mean motion resonances with Neptune. Astron. Astrophys. **441**, 807–814 (2005)
18. T. Kovács, Stability chart of the triangular points in the elliptic-restricted problem of three bodies. Mon. Not. R. Astron. Soc. **430**, 2755–2760 (2013)
19. H. Lei, B. Xu, Families of impulsive transfers between libration points in the restricted three-body problem. Mon. Not. R. Astron. Soc. **461**, 1786–1803 (2016)
20. J. Llibre, R. Martínez, C. Simó, Transversality of the invariant manifolds associated to the Lyapunov family of the periodic orbits near L_2 in the restricted three-body problem. J. Diff. Eqs **58**, 104–156 (1985)
21. V.V. Markellos, Numerical investigation of the planar restricted three-body problem III: closed branches of family f and related periodic orbits of the elliptic problem. Celes. Mech. **12**, 215–224 (1975)
22. V.V. Markellos, E.A. Perdios, P. Labropoulou, Linear stability of the triangular equilibrium points in the photogravitational elliptic restricted problem, I. Astrophys. Space Sci. **194**, 207–213 (1992)
23. V.V. Markellos, E.A. Perdios, C. Georgiou, Linear stability of the triangular equilibrium points in the photogravitational elliptic restricted problem, II. Astrophys. Space Sci. **199**, 23–33 (1993)
24. V.V. Markellos, E.A. Perdios, K. Papadakis, The stability of inner collinear equilibrium point in the photogravitational elliptic restricted problem. Astrophys. Space Sci. **199**, 139–146 (1993)
25. R. Meire, A contribution to the stability of the triangular points in the elliptic restricted three-body problem. Bull. Astron. Inst. Czechosl. **31**, 312–316 (1980)
26. R. Meire, The stability of the triangular points in the elliptic restricted problem. Celest. Mech. **23**, 89–95 (1981)
27. A. Narayan, C.R. Kumar, Effects of photogravitational and oblateness on the triangular Lagrangian points in the elliptical restricted three body problem. Int. J. Pure Appl. Math. **68**, 201–224 (2011)
28. A. Narayan, C.R. Kumar, Stability of triangular equilibrium points in elliptical restricted three body problem under the effects of photogravitational and oblateness of primaries. Int. J. Pure Appl. Math. **70**, 735–754 (2011)

29. A.H. Nayfeh, Characteristic exponents for the triangular points in the elliptic restricted problem of three bodies. AIAA J. **8**, 1916–1917 (1970)
30. P. Oberti, A. Vienne, An upgraded theory for Helene, Telesto, and Calypso. Astron. Astrophys. **397**, 353–359 (2003)
31. M. Ollé, J.R. Pacha, The 3D elliptic restricted three-body problem: periodic orbits which bifurcate from limiting restricted problems. Astron. Astrophys. **351**, 1149–1164 (1999)
32. N. Pathak, V.O. Thomas, Evolution of the f family orbits in the photo gravitational Sun-Saturn system with oblateness. Int. J. Astron. Astrophys. **6**, 254–271 (2016)
33. E. Perdios, Doubly asymptotic orbits at the unstable equilibrium in the elliptic restricted problem, in ed. by V.V. Markellos, Y. Kozai. *Dynamical Trapping and Evolution in the Solar System*. Astrophysics and Space Science Library (A Series of Books on the Recent Developments of Space Science and of General Geophysics and Astrophysics Published in Connection with the Journal Space Science Reviews), vol. 106, (Springer, Dordrecht, 1983)
34. E.A. Perdios, V.V. Markellos, Stability and bifurcations of Sitnikov motions. Celest. Mech. **42**, 187–200 (1988)
35. I.A. Robin, V.V. Markellos, Numerical determination of three-dimensional periodic orbits generated from vertical self-resonant satellite orbits. Celest. Mech. **21**, 395–434 (1980)
36. Y.S. Ruth, R.K. Sharma, Periodic orbits in the photogravitational elliptic restricted three-body problem. Advanc. Astrophys. **3**, 154–170 (2018)
37. R.K. Sharma, P.V. Subba Rao, Collinear equilibria and their characteristic exponents in the restricted three-body problem when the primaries are oblate spheroids. Celest. Mech. **12**, 189–201 (1975)
38. J.F.L. Simmons, A.J.C. McDonald, J.C. Brown, The restricted 3-body problem with radiation pressure. Celes. Mech. **35**, 145–187 (1985)
39. J. Singh, A. Umar, Motion in the photogravitational elliptic restricted three-body problem under an oblate body. Astron. J. **143**(109), 22 (2012)
40. J. Tschauner, Die Bewegung in der Nahe der Dreieckspunkte des elliptischen eingeschrankten Dreikorperproblems. Celes. Mech. **3**, 189–196 (1971)
41. C. Valente, L. Marino, G. Orecchini, Stability of the motion near L_4 equilateral Lagrangian point. Acta Astronaut. **46**, 501–506 (2000)
42. P. Verrier, T. Waters, J. Sieber, Evolution of the L_1 halo family in the radial solar sail circular restricted three-body problem. Celest. Mech. Dyn. Astr. **120**, 373–400 (2014)
43. G. Voyatzis, I. Gkolias, H. Varvoglis, The dynamics of the elliptic Hill problem: periodic orbits and stability regions. Celest. Mech. Dyn. Astr. **113**, 125–139 (2012)
44. X.Y. Zeng, H.X. Baoyin, J.F. Li, Updated rotating mass dipole with oblateness of one primary (I): equilibria in the equator and their stability. Astrophys. Space Sci. **361**(14), 12 (2015)
45. E.E. Zotos, J. Nagler, On the classification of orbits in the three-dimensional Copenhagen problem with oblate primaries. Int. J. Non Linear Mech. **108**, 55–71 (2019)
46. E.E. Zotos, Y. Qi, Near-optimal capture in the planar circular restricted Pluto-Charon system. Planet. Space Sci. **165**, 85–98 (2019)

Some Different Type Integral Inequalities and Their Applications

Artion Kashuri and Rozana Liko

Abstract In this article, we first present some integral inequalities for Gauss-Jacobi type quadrature formula involving generalized-$\mathbf{m}$-$((h_1^p, h_2^q); (\eta_1, \eta_2))$-convex mappings. Secondly, an identity pertaining twice differentiable mappings defined on $\mathbf{m}$-invex set is used. By using the notion of generalized-$\mathbf{m}$-$((h_1^p, h_2^q); (\eta_1, \eta_2))$-convexity and the obtained identity as an auxiliary result, some new estimates with respect to Hermite-Hadamard, Ostrowski, and Simpson type inequalities via fractional integrals are established. It is pointed out that some new special cases can be deduced from main results. At the end, some applications to special means for different positive real numbers are provided as well.

1 Introduction

The following notations are used throughout this paper. We use I to denote an interval on the real line $\mathbb{R} = (-\infty, +\infty)$. For any subset $K \subseteq \mathbb{R}^n$, K° is the interior of K. The set of integrable functions on the interval $[a, b]$ is denoted by $L[a, b]$.

The following inequality, named Hermite-Hadamard inequality, is one of the most famous inequalities in the literature for convex functions.

Theorem 1 *Let $f : I \subseteq \mathbb{R} \longrightarrow \mathbb{R}$ be a convex function on I and $a, b \in I$ with $a < b$. Then the following inequality holds:*

$$f\left(\frac{a+b}{2}\right) \leq \frac{1}{b-a} \int_a^b f(x)dx \leq \frac{f(a)+f(b)}{2}. \tag{1}$$

This inequality (1) is also known as trapezium inequality.

A. Kashuri (✉) · R. Liko
Department of Mathematics, Faculty of Technical Science, University Ismail Qemali of Vlora, Vlorë, Albania

T. M. Rassias, P. M. Pardalos (eds.), *Mathematical Analysis and Applications*, Springer Optimization and Its Applications 154,
https://doi.org/10.1007/978-3-030-31339-5_10

The trapezium type inequality has remained an area of great interest due to its wide applications in the field of mathematical analysis. For other recent results which generalize, improve, and extend the inequality (1) through various classes of convex functions interested readers are referred to [1–47].

Also the following result is known in the literature as the Ostrowski inequality [33], which gives an upper bound for the approximation of the integral average $\frac{1}{b-a}\int_a^b f(t)dt$ by the value $f(x)$ at point $x \in [a, b]$.

Theorem 2 *Let $f : I \longrightarrow \mathbb{R}$, where $I \subseteq \mathbb{R}$ is an interval, be a mapping differentiable in the interior I° of I, and let $a, b \in I^\circ$ with $a < b$. If $|f'(x)| \leq M$ for all $x \in [a, b]$, then*

$$\left|f(x) - \frac{1}{b-a}\int_a^b f(t)dt\right| \leq M(b-a)\left[\frac{1}{4} + \frac{\left(x - \frac{a+b}{2}\right)^2}{(b-a)^2}\right], \quad \forall x \in [a, b]. \tag{2}$$

The following inequality is well known in the literature as Simpson's inequality:

Theorem 3 *Let $f : [a, b] \longrightarrow \mathbb{R}$ be four time differentiable on the interval (a, b) and having the fourth derivative bounded on (a, b), that is $\|f^{(4)}\|_\infty = \sup_{x\in(a,b)} |f^{(4)}|$ $< \infty$. Then, we have*

$$\left|\int_a^b f(t)dt - \frac{b-a}{3}\left[\frac{f(a)+f(b)}{2} + 2f\left(\frac{a+b}{2}\right)\right]\right| \leq \frac{1}{2880}\|f^{(4)}\|_\infty (b-a)^5. \tag{3}$$

Inequality (3) gives an error bound for the classical Simpson quadrature formula, which is one of the most used quadrature formulae in practical applications.

In recent years, various generalizations, extensions, and variants of such inequalities have been obtained. For other recent results concerning Ostrowski type inequalities, see [21, 33]. For other recent results concerning Simpson type inequalities, see [32, 38].

Gauss-Jacobi type quadrature formula [40] is defined as follows:

$$\int_a^b (x-a)^p (b-x)^q f(x)dx = \sum_{k=0}^{+\infty} B_{m,k} f(\gamma_k) + R_m^\star |f|, \tag{4}$$

for certain $B_{m,k}$, γ_k and rest $R_m^\star |f|$. In [31], Liu obtained integral inequalities for P-function related to the left-hand side of (4), and in [48], Özdemir et al. also presented several integral inequalities concerning the left-hand side of (4) via some kinds of convexity.

Let us recall some special functions and evoke some basic definitions as follows:

Definition 1 The Euler beta function is defined for $a, b > 0$ as

$$\beta(a,b) = \int_0^1 t^{a-1}(1-t)^{b-1}dt. \tag{5}$$

Definition 2 ([34]) Let $f \in L[a,b]$. The Riemann-Liouville integrals $J_{a+}^{\alpha} f$ and $J_{b-}^{\alpha} f$ of order $\alpha > 0$ with $a \geq 0$ are defined by

$$J_{a+}^{\alpha} f(x) = \frac{1}{\Gamma(\alpha)} \int_a^x (x-t)^{\alpha-1} f(t)dt, \quad x > a$$

and

$$J_{b-}^{\alpha} f(x) = \frac{1}{\Gamma(\alpha)} \int_x^b (t-x)^{\alpha-1} f(t)dt, \quad b > x,$$

where $\Gamma(\alpha) = \int_0^{+\infty} e^{-u} u^{\alpha-1} du$. Here $J_{a+}^{0} f(x) = J_{b-}^{0} f(x) = f(x)$.
Note that $\alpha = 1$, the fractional integral reduces to the classical integral.

Definition 3 ([49]) A set $S \subseteq \mathbb{R}^n$ is said to be invex set with respect to the mapping $\eta : S \times S \longrightarrow \mathbb{R}^n$, if $x + t\eta(y,x) \in S$ for every $x, y \in S$ and $t \in [0,1]$.

The invex set S is also termed an η-connected set.

Definition 4 ([50]) Let $h : [0,1] \longrightarrow \mathbb{R}$ be a non-negative function and $h \neq 0$. The function f on the invex set K is said to be h-preinvex with respect to η, if

$$f\big(x + t\eta(y,x)\big) \leq h(1-t)f(x) + h(t)f(y) \tag{6}$$

for each $x, y \in K$ and $t \in [0,1]$ where $f(\cdot) > 0$.

Clearly, when putting $h(t) = t$ in Definition 4, f becomes a preinvex function [51]. If the mapping $\eta(y,x) = y - x$ in Definition 4, then the non-negative function f reduces to h-convex mappings [52].

Definition 5 ([53]) Let $S \subseteq \mathbb{R}^n$ be an invex set with respect to $\eta : S \times S \longrightarrow \mathbb{R}^n$. A function $f : S \longrightarrow [0, +\infty)$ is said to be s-preinvex (or s-Breckner-preinvex) with respect to η and $s \in (0,1]$, if for every $x, y \in S$ and $t \in [0,1]$,

$$f\big(x + t\eta(y,x)\big) \leq (1-t)^s f(x) + t^s f(y). \tag{7}$$

Definition 6 ([54]) A function $f : K \longrightarrow \mathbb{R}$ is said to be s-Godunova-Levin-Dragomir-preinvex of second kind, if

$$f\big(x + t\eta(y,x)\big) \leq (1-t)^{-s} f(x) + t^{-s} f(y), \tag{8}$$

for each $x, y \in K$, $t \in (0,1)$ and $s \in (0,1]$.

Definition 7 ([55]) A non-negative function $f : K \subseteq \mathbb{R} \longrightarrow \mathbb{R}$ is said to be *tgs*-convex on K if the inequality

$$f\big((1-t)x+ty\big) \leq t(1-t)[f(x)+f(y)] \tag{9}$$

grips for all $x, y \in K$ and $t \in (0, 1)$.

Definition 8 ([33]) A function $f : I \subseteq \mathbb{R} \longrightarrow \mathbb{R}$ is said to MT-convex functions, if it is non-negative and $\forall\, x, y \in I$ and $t \in (0, 1)$ satisfies the subsequent inequality

$$f(tx+(1-t)y) \leq \frac{\sqrt{t}}{2\sqrt{1-t}}f(x)+\frac{\sqrt{1-t}}{2\sqrt{t}}f(y). \tag{10}$$

Definition 9 ([38]) Let $K \subseteq \mathbb{R}$ be an open m-invex set respecting $\eta : K \times K \longrightarrow \mathbb{R}$ and $h_1, h_2 : [0, 1] \longrightarrow [0, +\infty)$. A function $f : K \longrightarrow \mathbb{R}$ is said to be generalized (m, h_1, h_2)-preinvex, if

$$f\big(mx+t\eta(y, mx)\big) \leq mh_1(t)f(x)+h_2(t)f(y) \tag{11}$$

is valid for all $x, y \in K$ and $t \in [0, 1]$, for some fixed $m \in (0, 1]$.

The concept of η-convex functions (at the beginning was named by φ-convex functions), considered in [14], has been introduced as the following.

Definition 10 Consider a convex set $I \subseteq \mathbb{R}$ and a bifunction $\eta : f(I) \times f(I) \longrightarrow \mathbb{R}$. A function $f : I \longrightarrow \mathbb{R}$ is called convex with respect to η (briefly η-convex), if

$$f\big(\lambda x+(1-\lambda)y\big) \leq f(y)+\lambda\eta(f(x), f(y)), \tag{12}$$

is valid for all $x, y \in I$ and $\lambda \in [0, 1]$.

Geometrically it says that if a function is η-convex on I, then for any $x, y \in I$, its graph is on or under the path starting from $(y, f(y))$ and ending at $(x, f(y)+\eta(f(x), f(y)))$. If $f(x)$ should be the end point of the path for every $x, y \in I$, then we have $\eta(x, y) = x - y$ and the function reduces to a convex one. For more results about η-convex functions, see [7, 8, 13, 14].

Definition 11 ([1]) Let $I \subseteq \mathbb{R}$ be an invex set with respect to $\eta_1 : I \times I \longrightarrow \mathbb{R}$. Consider $f : I \longrightarrow \mathbb{R}$ and $\eta_2 : f(I) \times f(I) \longrightarrow \mathbb{R}$. The function f is said to be (η_1, η_2)-convex if

$$f\big(x+\lambda\eta_1(y, x)\big) \leq f(x)+\lambda\eta_2(f(y), f(x)), \tag{13}$$

is valid for all $x, y \in I$ and $\lambda \in [0, 1]$.

Motivated by the above literatures, the main objective of this article is to establish in Section 2 integral inequalities using two lemmas as auxiliary results for the left-hand side of Gauss-Jacobi type quadrature formula and some new estimates on

Hermite-Hadamard, Ostrowski, and Simpson type inequalities via fractional integrals associated with generalized-**m**-$((h_1^p, h_2^q); (\eta_1, \eta_2))$-convex mappings. Also, some new special cases will be deduced. In Section 3, some applications to special means for different positive real numbers will be given as well. In Section 4, some conclusion and future research are given.

2 Main Results

The following definitions will be used in this section.

Definition 12 Let $\mathbf{m} : [0, 1] \longrightarrow (0, 1]$ be a function. A set $K \subseteq \mathbb{R}^n$ is named as **m**-invex with respect to the mapping $\eta : K \times K \longrightarrow \mathbb{R}^n$, if $\mathbf{m}(t)x + \xi\eta(y, \mathbf{m}(t)x) \in K$ holds for each $x, y \in K$ and any $t, \xi \in [0, 1]$.

Remark 1 In Definition 12, under certain conditions, the mapping $\eta(y, \mathbf{m}(t)x)$ for any $t, \xi \in [0, 1]$ could reduce to $\eta(y, mx)$. For example, when $\mathbf{m}(t) = m$ for all $t \in [0, 1]$, then the **m**-invex set degenerates an m-invex set on K.

We next introduce the concept of generalized-**m**-$((h_1^p, h_2^q); (\eta_1, \eta_2))$-convex mappings.

Definition 13 Let $K \subseteq \mathbb{R}$ be an open **m**-invex set with respect to the mapping $\eta_1 : K \times K \longrightarrow \mathbb{R}$ and $\mathbf{m} : [0, 1] \longrightarrow (0, 1]$. Suppose $h_1, h_2 : [0, 1] \longrightarrow [0, +\infty)$ and $\varphi : I \longrightarrow K$ are continuous. Consider $f : K \longrightarrow (0, +\infty)$ and $\eta_2 : f(K) \times f(K) \longrightarrow \mathbb{R}$. The mapping f is said to be generalized-**m**-$((h_1^p, h_2^q); (\eta_1, \eta_2))$-convex if

$$f\big(\mathbf{m}(t)\varphi(x) + \xi\eta_1(\varphi(y), \mathbf{m}(t)\varphi(x))\big)$$

$$\leq \left[\mathbf{m}(\xi)h_1^p(\xi)f^r(x) + h_2^q(\xi)\eta_2(f^r(y), f^r(x))\right]^{\frac{1}{r}}, \tag{14}$$

holds for all $x, y \in I$, $r \neq 0$, $t, \xi \in [0, 1]$ and any fixed $p, q > -1$.

Remark 2 In Definition 13, if we choose $\mathbf{m} = p = q = r = 1$ and $\varphi(x) = x$, then we get Definition 11.

Remark 3 In Definition 13, if we choose $\mathbf{m} = p = q = r = 1$, $h_1(t) = 1$, $h_2(t) = t$, $\eta_1(\varphi(y), \mathbf{m}(t)\varphi(x)) = \varphi(y) - \mathbf{m}(t)\varphi(x)$, $\eta_2(f^r(y), f^r(x)) = \eta(f^r(y), f^r(x))$ and $\varphi(x) = x$, $\forall x \in I$, then we get Definition 10. Also, in Definition 13, if we choose $\mathbf{m} = p = q = r = 1$, $h_1(t) = 1$, $h_2(t) = t$ and $\varphi(x) = x$, $\forall x \in I$, then we get Definition 11. Under some suitable choices as we have done above, we can get also the Definitions 5 and 6.

Remark 4 Let us discuss some special cases in Definition 13 as follows:

(I) If taking $h_1(t) = h(1-t)$ and $h_2(t) = h(t)$, then we get generalized-**m**-$((h^p(1-t), h^q(t)); (\eta_1, \eta_2))$-convex mappings.
(II) If taking $h_1(t) = (1-t)^s$ and $h_2(t) = t^s$ for $s \in (0, 1]$, then we get generalized-**m**-$(((1-t)^{sp}, t^{sq}); (\eta_1, \eta_2))$-Breckner-convex mappings.
(III) If taking $h_1(t) = (1-t)^{-s}$ and $h_2(t) = t^{-s}$ for $s \in (0, 1]$, then we get generalized-**m**-$(((1-t)^{-sp}, t^{-sq}); (\eta_1, \eta_2))$-Godunova-Levin-Dragomir-convex mappings.
(IV) If taking $h_1(t) = h_2(t) = t(1-t)$, then we get generalized-**m**-$((t(1-t))^{sp}, (t(1-t))^{sq}); (\eta_1, \eta_2))$-convex mappings.
(V) If taking $h_1(t) = \dfrac{\sqrt{1-t}}{2\sqrt{t}}$ and $h_2(t) = \dfrac{\sqrt{t}}{2\sqrt{1-t}}$, then we get generalized-**m**-$\left(\left(\left(\frac{\sqrt{1-t}}{2\sqrt{t}}\right)^p, \left(\frac{\sqrt{t}}{2\sqrt{1-t}}\right)^q\right); (\eta_1, \eta_2)\right)$-convex mappings.

It is worth to mention here that to the best of our knowledge all the special cases discussed above are new in the literature.
Let us see the following example of a generalized-**m**-$((h_1^p, h_2^q); (\eta_1, \eta_2))$-convex mapping which is not convex.

Example 1 Let us take $\mathbf{m} = r = \frac{1}{2}$, $h_1(t) = t^l$, $h_2(t) = t^s$ for all $l, s \in [0, 1]$, any fixed $p, q \geq 1$ and φ an identity function. Consider the function $f : [0, +\infty) \longrightarrow [0, +\infty)$ by

$$f(x) = \begin{cases} x, & 0 \leq x \leq 1; \\ 2, & x > 1. \end{cases}$$

Define two bifunctions $\eta_1 : [0, +\infty) \times [0, +\infty) \longrightarrow \mathbb{R}$ and $\eta_2 : [0, +\infty) \times [0, +\infty) \longrightarrow [0, +\infty)$ by

$$\eta_1(x, y) = \begin{cases} -y, & 0 \leq y \leq 1; \\ x + y, & y > 1, \end{cases}$$

and

$$\eta_2(x, y) = \begin{cases} x + y, & x \leq y; \\ 2(x + y), & x > y. \end{cases}$$

Then f is generalized $\frac{1}{2}$-$((t^{lp}, t^{sq}); (\eta_1, \eta_2))$-convex mapping. But f is not preinvex with respect to η_1 and also it is not convex (consider $x = 0, y = 2$ and $t \in (0, 1]$).

We claim the following integral identity.

Lemma 1 *Let* $\varphi : I \longrightarrow K$ *be a continuous function and* $\boldsymbol{m} : [0, 1] \longrightarrow (0, 1]$. *Assume that* $f : K = [\boldsymbol{m}(t)\varphi(a), \boldsymbol{m}(t)\varphi(a) + \eta(\varphi(b), \boldsymbol{m}(t)\varphi(a))] \longrightarrow$

$\mathbb{R}$ *is a continuous function on* K° *with respect to* $\eta : K \times K \longrightarrow \mathbb{R}$ *for* $\eta(\varphi(b), \boldsymbol{m}(t)\varphi(a)) > 0$ *and* $\forall\, t \in [0, 1]$. *Then for any fixed* $p, q > 0$, *we have*

$$\int_{\boldsymbol{m}(t)\varphi(a)}^{\boldsymbol{m}(t)\varphi(a)+\eta(\varphi(b),\boldsymbol{m}(t)\varphi(a))} (x - \boldsymbol{m}(t)\varphi(a))^p \times (\boldsymbol{m}(t)\varphi(a) + \eta(\varphi(b), \boldsymbol{m}(t)\varphi(a)) - x)^q f(x)dx$$
$$= \eta^{p+q+1}(\varphi(b), \boldsymbol{m}(t)\varphi(a)) \int_0^1 \xi^p(1-\xi)^q f(\boldsymbol{m}(t)\varphi(a)+\xi\eta(\varphi(b), \boldsymbol{m}(t)\varphi(a)))d\xi.$$

We denote

$$T_f^{p,q}(\eta, \varphi, \boldsymbol{m}; a, b) := \int_{\boldsymbol{m}(t)\varphi(a)}^{\boldsymbol{m}(t)\varphi(a)+\eta(\varphi(b),\boldsymbol{m}(t)\varphi(a))} (x - \boldsymbol{m}(t)\varphi(a))^p \times(\boldsymbol{m}(t)\varphi(a) + \eta(\varphi(b), \boldsymbol{m}(t)\varphi(a)) - x)^q f(x)dx. \tag{15}$$

Proof We observe that

$$T_f^{p,q}(\eta, \varphi, \mathbf{m}; a, b)$$
$$= \eta(\varphi(b), \mathbf{m}(t)\varphi(a)) \int_0^1 (\mathbf{m}(t)\varphi(a) + \xi\eta(\varphi(b), \mathbf{m}(t)\varphi(a)) - \mathbf{m}(t)\varphi(a))^p$$
$$\times (\mathbf{m}(t)\varphi(a) + \eta(\varphi(b), \mathbf{m}(t)\varphi(a)) - \mathbf{m}(t)\varphi(a) - \xi\eta(\varphi(b), \mathbf{m}(t)\varphi(a)))^q$$
$$\times f(\mathbf{m}(t)\theta(a) + \xi\Lambda(\theta(b), \mathbf{m}(t)\theta(a)))d\xi$$
$$= \eta^{p+q+1}(\varphi(b), \mathbf{m}(t)\varphi(a)) \int_0^1 \xi^p(1-\xi)^q f(\mathbf{m}(t)\varphi(a)+\xi\eta(\varphi(b), \mathbf{m}(t)\varphi(a)))d\xi.$$

This completes the proof of the lemma.

Remark 5 In Lemma 1, if we choose $\mathbf{m}(t) \equiv 1$ for any $t \in [0, 1]$, $\eta(\varphi(b), \mathbf{m}(t)\varphi(a)) = \varphi(b) - \mathbf{m}(t)\varphi(a)$ and $\varphi(x) = x$ for all $x \in I$, then we get the left-hand side of (4).

With the help of Lemma 1, we have the following results.

Theorem 4 *Let* $k > 1$, $0 < r \le 1$ *and* $p_1, p_2 > -1$. *Suppose* $h_1, h_2 : [0, 1] \longrightarrow [0, +\infty)$, $\varphi : I \longrightarrow K$ *are continuous functions and* $\boldsymbol{m} : [0, 1] \longrightarrow (0, 1]$. *Assume that* $f : K = [\boldsymbol{m}(t)\varphi(a), \boldsymbol{m}(t)\varphi(a) + \eta_1(\varphi(b), \boldsymbol{m}(t)\varphi(a))] \longrightarrow (0, +\infty)$ *is a continuous mapping on* K° *with respect to* $\eta_1 : K \times K \longrightarrow \mathbb{R}$ *for* $\eta_1(\varphi(b), \boldsymbol{m}(t)\varphi(a)) > 0$ *for all* $t \in [0, 1]$ *and* $\eta_2 : f(K) \times f(K) \longrightarrow \mathbb{R}$. *If* $f^{\frac{k}{k-1}}$ *is generalized-*$\boldsymbol{m}$*-*$((h_1^{p_1}, h_2^{p_2}); (\eta_1, \eta_2))$*-convex mapping on an open* $\boldsymbol{m}$*-invex set* K, *then for any fixed* $p, q > 0$, *we have*

$$\left|T_f^{p,q}(\eta_1,\varphi,\boldsymbol{m};a,b)\right| \leq \eta_1^{p+q+1}(\varphi(b),\boldsymbol{m}(t)\varphi(a))\sqrt[k]{\beta(kp+1,kq+1)} \tag{16}$$

$$\times\left[f^{\frac{rk}{k-1}}(a)I^r(h_1(\xi),\boldsymbol{m}(\xi);p_1,r)+\eta_2\left(f^{\frac{rk}{k-1}}(b),f^{\frac{rk}{k-1}}(a)\right)I^r(h_2(\xi);p_2,r)\right]^{\frac{k-1}{rk}},$$

where

$$I(h_1(\xi),\boldsymbol{m}(\xi);p_1,r):=\int_0^1 \boldsymbol{m}^{\frac{1}{r}}(\xi)h_1^{\frac{p_1}{r}}(\xi)d\xi,\quad I(h_2(\xi);p_2,r):=\int_0^1 h_2^{\frac{p_2}{r}}(\xi)d\xi.$$

Proof Since $f^{\frac{k}{k-1}}$ is generalized-$\mathbf{m}$-$((h_1^{p_1},h_2^{p_2});(\eta_1,\eta_2))$-convex mapping on K, combining with Lemma 1, Hölder inequality, Minkowski inequality, and properties of the modulus, we get

$$\left|T_f^{p,q}(\eta_1,\varphi,\mathbf{m};a,b)\right| \leq |\eta_1(\varphi(b),\mathbf{m}(t)\varphi(a))|^{p+q+1}\left[\int_0^1 \xi^{kp}(1-\xi)^{kq}d\xi\right]^{\frac{1}{k}}$$

$$\times\left[\int_0^1 \left|f(\mathbf{m}(t)\varphi(a)+\xi\eta_1(\varphi(b),\mathbf{m}(t)\varphi(a)))\right|^{\frac{k}{k-1}}d\xi\right]^{\frac{k-1}{k}}$$

$$\leq \eta_1^{p+q+1}(\varphi(b),\mathbf{m}(t)\varphi(a))\sqrt[k]{\beta(kp+1,kq+1)}$$

$$\times\left[\int_0^1\left[\mathbf{m}(\xi)h_1^{p_1}(\xi)f^{\frac{rk}{k-1}}(a)+h_2^{p_2}(\xi)\eta_2\left(f^{\frac{rk}{k-1}}(b),f^{\frac{rk}{k-1}}(a)\right)\right]^{\frac{1}{r}}d\xi\right]^{\frac{k-1}{k}}$$

$$\leq \eta_1^{p+q+1}(\varphi(b),\mathbf{m}(t)\varphi(a))\sqrt[k]{\beta(kp+1,kq+1)}$$

$$\times\left\{\left(\int_0^1 \mathbf{m}^{\frac{1}{r}}(\xi)h_1^{\frac{p_1}{r}}(\xi)f^{\frac{k}{k-1}}(a)d\xi\right)^r\right.$$

$$\left.+\left(\int_0^1 \eta_2^{\frac{1}{r}}\left(f^{\frac{rk}{k-1}}(b),f^{\frac{rk}{k-1}}(a)\right)h_2^{\frac{p_2}{r}}(\xi)d\xi\right)^r\right\}^{\frac{k-1}{rk}}$$

$$= \eta_1^{p+q+1}(\varphi(b),\mathbf{m}(t)\varphi(a))\sqrt[k]{\beta(kp+1,kq+1)}$$

$$\times\left[f^{\frac{rk}{k-1}}(a)I^r(h_1(\xi),\mathbf{m}(\xi);p_1,r)+\eta_2\left(f^{\frac{rk}{k-1}}(b),f^{\frac{rk}{k-1}}(a)\right)I^r(h_2(\xi);p_2,r)\right]^{\frac{k-1}{rk}}.$$

So, the proof of this theorem is completed.

We point out some special cases of Theorem 4.

Corollary 1 *In Theorem 4 for $k = 2$, we get*

$$\left|T_f^{p,q}(\eta_1, \varphi, \boldsymbol{m}; a, b)\right| \leq \eta_1^{p+q+1}(\varphi(b), \boldsymbol{m}(t)\varphi(a))\sqrt{\beta(2p+1, 2q+1)} \tag{17}$$

$$\times \sqrt[2r]{f^{2r}(a)I^r(h_1(\xi), \boldsymbol{m}(\xi); p_1, r) + \eta_2\left(f^{2r}(b), f^{2r}(a)\right) I^r(h_2(\xi); p_2, r)}.$$

Corollary 2 *In Theorem 4 for $h_1(t) = h(1-t)$, $h_2(t) = h(t)$ and $\boldsymbol{m}(t) = m \in (0, 1]$ for all $t \in [0, 1]$, we get*

$$\left|T_f^{p,q}(\eta_1, \varphi, m; a, b)\right| \leq \eta_1^{p+q+1}(\varphi(b), m\varphi(a))\sqrt[k]{\beta(kp+1, kq+1)} \tag{18}$$

$$\times\left[mf^{\frac{rk}{k-1}}(a)I^r(h(1-\xi); p_1, r) + \eta_2\left(f^{\frac{rk}{k-1}}(b), f^{\frac{rk}{k-1}}(a)\right) I^r(h(\xi); p_2, r)\right]^{\frac{k-1}{rk}}.$$

Corollary 3 *In Corollary 2 for $h_1(t) = (1-t)^s$ and $h_2(t) = t^s$, we get*

$$\left|T_f^{p,q}(\eta_1, \varphi, m; a, b)\right| \leq \eta_1^{p+q+1}(\varphi(b), m\varphi(a))\sqrt[k]{\beta(kp+1, kq+1)} \tag{19}$$

$$\times\left[mf^{\frac{rk}{k-1}}(a)\left(\frac{r}{r+sp_1}\right)^r + \eta_2\left(f^{\frac{rk}{k-1}}(b), f^{\frac{rk}{k-1}}(a)\right)\left(\frac{r}{r+sp_2}\right)^r\right]^{\frac{k-1}{rk}}.$$

Corollary 4 *In Corollary 2 for $h_1(t) = (1-t)^{-s}$, $h_2(t) = t^{-s}$ and $r > s \cdot \max\{p_1, p_2\}$, we get*

$$\left|T_f^{p,q}(\eta_1, \varphi, m; a, b)\right| \leq \eta_1^{p+q+1}(\varphi(b), m\varphi(a))\sqrt[k]{\beta(kp+1, kq+1)} \tag{20}$$

$$\times\left[mf^{\frac{rk}{k-1}}(a)\left(\frac{r}{r-sp_1}\right)^r + \eta_2\left(f^{\frac{rk}{k-1}}(b), f^{\frac{rk}{k-1}}(a)\right)\left(\frac{r}{r-sp_2}\right)^r\right]^{\frac{k-1}{rk}}.$$

Corollary 5 *In Theorem 4 for $h_1(t) = h_2(t) = t(1-t)$ and $\boldsymbol{m}(t) = m \in (0, 1]$ for all $t \in [0, 1]$, we get*

$$\left|T_f^{p,q}(\eta_1, \varphi, m; a, b)\right| \leq \eta_1^{p+q+1}(\varphi(b), m\varphi(a))\sqrt[k]{\beta(kp+1, kq+1)} \tag{21}$$

$$\times\left[mf^{\frac{rk}{k-1}}(a)\beta^r\left(1+\frac{p_1}{r}, 1+\frac{p_1}{r}\right)+\eta_2\left(f^{\frac{rk}{k-1}}(b), f^{\frac{rk}{k-1}}(a)\right)\beta^r\left(1+\frac{p_2}{r}, 1+\frac{p_2}{r}\right)\right]^{\frac{k-1}{rk}}.$$

Corollary 6 *In Corollary 2 for $h_1(t) = \dfrac{\sqrt{1-t}}{2\sqrt{t}}$, $h_2(t) = \dfrac{\sqrt{t}}{2\sqrt{1-t}}$ and $r > \frac{1}{2} \cdot \max\{p_1, p_2\}$, we get*

$$\left|T_f^{p,q}(\eta_1, \varphi, m; a, b)\right| \leq \eta_1^{p+q+1}(\varphi(b), m\varphi(a))\sqrt[k]{\beta(kp+1, kq+1)} \tag{22}$$

$$\times\left[mf^{\frac{rk}{k-1}}(a)\beta^r\left(1-\frac{p_1}{2r}, 1+\frac{p_1}{2r}\right)+\eta_2\left(f^{\frac{rk}{k-1}}(b), f^{\frac{rk}{k-1}}(a)\right)\beta^r\left(1-\frac{p_2}{2r}, 1+\frac{p_2}{2r}\right)\right]^{\frac{k-1}{rk}}.$$

Theorem 5 *Let $l \geq 1$, $0 < r \leq 1$ and $p_1, p_2 > -1$. Suppose $h_1, h_2 : [0, 1] \longrightarrow [0, +\infty)$, $\varphi : I \longrightarrow K$ are continuous functions and $\boldsymbol{m} : [0, 1] \longrightarrow (0, 1]$. Assume that $f : K = [\boldsymbol{m}(t)\varphi(a), \boldsymbol{m}(t)\varphi(a) + \eta_1(\varphi(b), \boldsymbol{m}(t)\varphi(a))] \longrightarrow (0, +\infty)$ is a continuous mapping on K° with respect to $\eta_1 : K \times K \longrightarrow \mathbb{R}$ for $\eta_1(\varphi(b), \boldsymbol{m}(t)\varphi(a)) > 0$ for all $t \in [0, 1]$ and $\eta_2 : f(K) \times f(K) \longrightarrow \mathbb{R}$. If f^l is generalized-$\boldsymbol{m}$-$((h_1^{p_1}, h_2^{p_2}); (\eta_1, \eta_2))$-convex mapping on an open $\boldsymbol{m}$-invex set K, then for any fixed $p, q > 0$, we have*

$$\left|T_f^{p,q}(\eta_1, \varphi, \boldsymbol{m}; a, b)\right| \leq \eta_1^{p+q+1}(\varphi(b), \boldsymbol{m}(t)\varphi(a))\beta^{\frac{l-1}{l}}(p+1, q+1) \tag{23}$$

$$\times\sqrt[rl]{f^{rl}(a)J^r(h_1(\xi), \boldsymbol{m}(\xi); p, q, p_1, r) + \eta_2\left(f^{rl}(b), f^{rl}(a)\right)J^r(h_2(\xi); p, q, p_2, r)},$$

where

$$J(h_1(\xi), \boldsymbol{m}(\xi); p, q, p_1, r) := \int_0^1 \boldsymbol{m}^{\frac{1}{r}}(\xi)\xi^p(1-\xi)^q h_1^{\frac{p_1}{r}}(\xi)d\xi;$$

$$J(h_2(\xi); p, q, p_2, r) := \int_0^1 \xi^p(1-\xi)^q h_2^{\frac{p_2}{r}}(\xi)d\xi.$$

Proof Since f^l is generalized-$\mathbf{m}$-$((h_1^{p_1}, h_2^{p_2}); (\eta_1, \eta_2))$-convex mapping on K, combining with Lemma 1, the well-known power mean inequality, Minkowski inequality, and properties of the modulus, we get

$$\left|T_f^{p,q}(\eta_1, \varphi, \mathbf{m}; a, b)\right| = \left|\eta_1^{p+q+1}(\varphi(b), \mathbf{m}(t)\varphi(a))\right|$$

$$\times\left|\int_0^1 \left[\xi^p(1-\xi)^q\right]^{\frac{l-1}{l}}\left[\xi^p(1-\xi)^q\right]^{\frac{1}{l}} f(\mathbf{m}(t)\varphi(a) + \xi\eta_1(\varphi(b), \mathbf{m}(t)\varphi(a)))d\xi\right|$$

$$\leq |\eta_1(\varphi(b), \mathbf{m}(t)\varphi(a))|^{p+q+1}\left[\int_0^1 \xi^p(1-\xi)^q d\xi\right]^{\frac{l-1}{l}}$$

$$\times\left[\int_0^1 \xi^p(1-\xi)^q\left|f(\mathbf{m}(t)\varphi(a) + \xi\eta_1(\varphi(b), \mathbf{m}(t)\varphi(a)))\right|^l d\xi\right]^{\frac{1}{l}}$$

$$\leq \eta_1^{p+q+1}(\varphi(b), \mathbf{m}(t)\varphi(a))\beta^{\frac{l-1}{l}}(p+1, q+1)$$

$$\times\left[\int_0^1 \xi^p(1-\xi)^q\left[\mathbf{m}(\xi)h_1^{p_1}(\xi)f^{rl}(a)+h_2^{p_2}(\xi)\eta_2\left(f^{rl}(b),f^{rl}(a)\right)\right]^{\frac{1}{r}}d\xi\right]^{\frac{1}{l}}$$

$$\leq \eta_1^{p+q+1}(\varphi(b),\mathbf{m}(t)\varphi(a))\beta^{\frac{l-1}{l}}(p+1,q+1)$$

$$\times\left\{\left(\int_0^1 \mathbf{m}^{\frac{1}{r}}(\xi)\xi^p(1-\xi)^q h_1^{\frac{p_1}{r}}(\xi)f^l(a)d\xi\right)^r\right.$$

$$\left.+\left(\int_0^1 \xi^p(1-\xi)^q h_2^{\frac{p_2}{r}}(\xi)\eta_2^{\frac{1}{r}}\left(f^{rl}(b),f^{rl}(a)\right)d\xi\right)^r\right\}^{\frac{1}{rl}}$$

$$=\eta_1^{p+q+1}(\varphi(b),\mathbf{m}(t)\varphi(a))\beta^{\frac{l-1}{l}}(p+1,q+1)$$

$$\times\sqrt[rl]{f^{rl}(a)J^r(h_1(\xi),\mathbf{m}(\xi);p,q,p_1,r)+\eta_2\left(f^{rl}(b),f^{rl}(a)\right)J^r(h_2(\xi);p,q,p_2,r)}.$$

So, the proof of this theorem is completed.

Let us discuss some special cases of Theorem 5.

Corollary 7 *In Theorem 5 for $l=1$, we get*

$$\left|T_f^{p,q}(\eta_1,\varphi,\boldsymbol{m};a,b)\right|\leq \eta_1^{p+q+1}(\varphi(b),\boldsymbol{m}(t)\varphi(a)) \tag{24}$$

$$\times\sqrt[r]{f^r(a)J^r(h_1(\xi),\boldsymbol{m}(\xi);p,q,p_1,r)+\eta_2\left(f^r(b),f^r(a)\right)J^r(h_2(\xi);p,q,p_2,r)}.$$

Corollary 8 *In Theorem 5 for $h_1(t)=h(1-t)$, $h_2(t)=h(t)$ and $\mathbf{m}(t)=m\in(0,1]$ for all $t\in[0,1]$, we get*

$$\left|T_f^{p,q}(\eta_1,\varphi,m;a,b)\right|\leq \eta_1^{p+q+1}(\varphi(b),m\varphi(a))\beta^{\frac{l-1}{l}}(p+1,q+1) \tag{25}$$

$$\times\sqrt[rl]{mf^{rl}(a)J^r(h(1-\xi);p,q,p_1,r)+\eta_2\left(f^{rl}(b),f^{rl}(a)\right)J^r(h(\xi);p,q,p_2,r)}.$$

Corollary 9 *In Corollary 8 for $h_1(t)=(1-t)^s$, $h_2(t)=t^s$ and $0<s\leq r$, we get*

$$\left|T_f^{p,q}(\eta_1,\varphi,m;a,b)\right|\leq \eta_1^{p+q+1}(\varphi(b),m\varphi(a))\beta^{\frac{l-1}{l}}(p+1,q+1) \tag{26}$$

$$\times\sqrt[rl]{mf^{rl}(a)\beta^r\left(p+1,q+\frac{sp_1}{r}+1\right)+\eta_2\left(f^{rl}(b),f^{rl}(a)\right)\beta^r\left(q+1,p+\frac{sp_2}{r}+1\right)}.$$

Corollary 10 *In Corollary 8 for $h_1(t)=(1-t)^{-s}$ and $h_2(t)=t^{-s}$, we get*

$$\left|T_f^{p,q}(\eta_1,\varphi,m;a,b)\right|\leq \eta_1^{p+q+1}(\varphi(b),m\varphi(a))\beta^{\frac{l-1}{l}}(p+1,q+1) \tag{27}$$

$$\times \sqrt[rl]{mf^{rl}(a)\beta^{r}\left(p+1, q-\frac{sp_1}{r}+1\right)+\eta_2\left(f^{rl}(b), f^{rl}(a)\right)\beta^{r}\left(q+1, p-\frac{sp_2}{r}+1\right)}.$$

Corollary 11 *In Theorem 5 for* $h_1(t) = h_2(t) = t(1-t)$ *and* $\boldsymbol{m}(t) = m \in (0, 1]$ *for all* $t \in [0, 1]$, *we get*

$$\left|T_f^{p,q}(\eta_1, \varphi, m; a, b)\right| \leq \eta_1^{p+q+1}(\varphi(b), m\varphi(a))\beta^{\frac{l-1}{l}}(p+1, q+1) \tag{28}$$

$$\times\Big[mf^{rl}(a)\beta^{r}\left(p+\frac{p_1}{r}+1, q+\frac{p_1}{r}+1\right)$$

$$+\eta_2\left(f^{rl}(b), f^{rl}(a)\right)\beta^{r}\left(p+\frac{p_2}{r}+1, q+\frac{p_2}{r}+1\right)\Big]^{\frac{1}{rq}}.$$

Corollary 12 *In Corollary 8 for* $h_1(t) = \dfrac{\sqrt{1-t}}{2\sqrt{t}}$ *and* $h_2(t) = \dfrac{\sqrt{t}}{2\sqrt{1-t}}$, *we get*

$$\left|T_f^{p,q}(\eta_1, \varphi, m; a, b)\right| \leq \eta_1^{p+q+1}(\varphi(b), m\varphi(a))\beta^{\frac{l-1}{l}}(p+1, q+1) \tag{29}$$

$$\times\Big[mf^{rl}(a)\beta^{r}\left(p-\frac{p_1}{2r}+1, q+\frac{p_1}{2r}+1\right)$$

$$+\eta_2\left(f^{rl}(b), f^{rl}(a)\right)\beta^{r}\left(p+\frac{p_2}{2r}+1, q-\frac{p_2}{2r}+1\right)\Big]^{\frac{1}{rq}}.$$

For establishing our second main results regarding generalizations of Hermite-Hadamard, Ostrowski, and Simpson type inequalities associated with generalized-$\boldsymbol{m}$-$((h_1^{p_1}, h_2^{p_2}); (\eta_1, \eta_2))$-convexity via fractional integrals, we need the following lemma.

Lemma 2 *Let* $\varphi : I \longrightarrow K$ *be a continuous function and* $\boldsymbol{m} : [0, 1] \longrightarrow (0, 1]$. *Suppose* $K = [\boldsymbol{m}(t)\varphi(a), \boldsymbol{m}(t)\varphi(a) + \eta(\varphi(b), \boldsymbol{m}(t)\varphi(a))] \subseteq \mathbb{R}$ *be an open* $\boldsymbol{m}$*-invex subset with respect to* $\eta : K \times K \longrightarrow \mathbb{R}$ *and let* $\eta(\varphi(b), \boldsymbol{m}(t)\varphi(a)) > 0$ *for all* $t \in [0, 1]$. *Assume that* $f : K \longrightarrow \mathbb{R}$ *be a twice differentiable mapping on* K° *and* $f'' \in L(K)$. *Then for any* $\lambda \in [0, 1]$ *and* $\alpha > 0$, *the following identity holds:*

$$\frac{\lambda-1}{\eta(\varphi(b), \boldsymbol{m}(t)\varphi(a))}\Big\{\eta^{\alpha+1}(\varphi(x), \boldsymbol{m}(t)\varphi(a))f'(\boldsymbol{m}(t)\varphi(a)+\eta(\varphi(x), \boldsymbol{m}(t)\varphi(a)))$$

$$+\eta^{\alpha+1}(\varphi(x), \boldsymbol{m}(t)\varphi(b))f'(\boldsymbol{m}(t)\varphi(b)+\eta(\varphi(x), \boldsymbol{m}(t)\varphi(b)))\Big\}$$

$$+\frac{1+\alpha-\lambda}{\eta(\varphi(b), \boldsymbol{m}(t)\varphi(a))}\Big\{\eta^{\alpha}(\varphi(x), \boldsymbol{m}(t)\varphi(a))f(\boldsymbol{m}(t)\varphi(a)+\eta(\varphi(x), \boldsymbol{m}(t)\varphi(a)))$$

$$+\eta^{\alpha}(\varphi(x), \boldsymbol{m}(t)\varphi(b))f(\boldsymbol{m}(t)\varphi(b)+\eta(\varphi(x), \boldsymbol{m}(t)\varphi(b)))\Big\}$$

$$+\frac{\lambda}{\eta(\varphi(b),\boldsymbol{m}(t)\varphi(a))}$$

$$\times\Big\{\eta^{\alpha}(\varphi(x),\boldsymbol{m}(t)\varphi(a))f(\boldsymbol{m}(t)\varphi(a))+\eta^{\alpha}(\varphi(x),\boldsymbol{m}(t)\varphi(b))f(\boldsymbol{m}(t)\varphi(b))\Big\}$$

$$-\frac{\Gamma(\alpha+2)}{\eta(\varphi(b),\boldsymbol{m}(t)\varphi(a))}\times\Big[J^{\alpha}_{(\boldsymbol{m}(t)\varphi(a)+\eta(\varphi(x),\boldsymbol{m}(t)\varphi(a)))^{-}}f(\boldsymbol{m}(t)\varphi(a))$$

$$+J^{\alpha}_{(\boldsymbol{m}(t)\varphi(b)+\eta(\varphi(x),\boldsymbol{m}(t)\varphi(b)))^{-}}f(\boldsymbol{m}(t)\varphi(b))\Big]$$

$$=\frac{\eta^{\alpha+2}(\varphi(x),\boldsymbol{m}(t)\varphi(a))}{\eta(\varphi(b),\boldsymbol{m}(t)\varphi(a))} \tag{30}$$

$$\times\int_0^1\xi(\lambda-\xi^{\alpha})f''(\boldsymbol{m}(t)\varphi(a)+\xi\eta(\varphi(x),\boldsymbol{m}(t)\varphi(a)))d\xi$$

$$+\frac{\eta^{\alpha+2}(\varphi(x),\boldsymbol{m}(t)\varphi(b))}{\eta(\varphi(b),\boldsymbol{m}(t)\varphi(a))}$$

$$\times\int_0^1\xi(\lambda-\xi^{\alpha})f''(\boldsymbol{m}(t)\varphi(b)+\xi\eta(\varphi(x),\boldsymbol{m}(t)\varphi(b)))d\xi.$$

We denote

$$\Delta^{\alpha}_{f}(\eta,\varphi,\boldsymbol{m};\lambda,x,a,b):=\frac{\eta^{\alpha+2}(\varphi(x),\boldsymbol{m}(t)\varphi(a))}{\eta(\varphi(b),\boldsymbol{m}(t)\varphi(a))} \tag{31}$$

$$\times\int_0^1\xi(\lambda-\xi^{\alpha})f''(\boldsymbol{m}(t)\varphi(a)+\xi\eta(\varphi(x),\boldsymbol{m}(t)\varphi(a)))d\xi$$

$$+\frac{\eta^{\alpha+2}(\varphi(x),\boldsymbol{m}(t)\varphi(b))}{\eta(\varphi(b),\boldsymbol{m}(t)\varphi(a))}\int_0^1\xi(\lambda-\xi^{\alpha})f''(\boldsymbol{m}(t)\varphi(b)+\xi\eta(\varphi(x),\boldsymbol{m}(t)\varphi(b)))d\xi.$$

Proof A simple proof of the equality (30) can be done by performing two integration by parts in the integrals above and changing the variables. The details are left to the interested reader. This completes the proof of our lemma.

Using Lemma 2, we now state the following theorems for the corresponding version for power of second derivative.

Theorem 6 *Let* $0<r\le 1$ *and* $p_1,p_2>-1$. *Suppose* $h_1,h_2:[0,1]\longrightarrow[0,+\infty)$, $\varphi:I\longrightarrow K$ *are continuous functions and* $\boldsymbol{m}:[0,1]\longrightarrow(0,1]$. *Let* $K=[\boldsymbol{m}(t)\varphi(a),\boldsymbol{m}(t)\varphi(a)+\eta_1(\varphi(b),\boldsymbol{m}(t)\varphi(a))]\subseteq\mathbb{R}$ *be an open* $\boldsymbol{m}$*-invex subset with respect to* $\eta_1:K\times K\longrightarrow\mathbb{R}$ *and let* $\eta_1(\varphi(b),\boldsymbol{m}(t)\varphi(a))>0$ *for all* $t\in[0,1]$ *and* $\eta_2:f(K)\times f(K)\longrightarrow\mathbb{R}$. *Assume that* $f:K\longrightarrow(0,+\infty)$ *be a* *twice differentiable mapping on* K°. *If* f''^{q} *is generalized-*$\boldsymbol{m}$*-*$((h_1^{p_1}, h_2^{p_2}); (\eta_1, \eta_2))$*-convex mapping on* K, $q > 1$, $p^{-1} + q^{-1} = 1$, *then for any* $\lambda \in [0, 1]$ *and* $\alpha > 0$, *the following inequality for fractional integrals holds:*

$$\left|\Delta_f^{\alpha}(\eta_1, \varphi, \boldsymbol{m}; \lambda, x, a, b)\right| \leq \frac{\sqrt[p]{D(\alpha, \lambda, p)}}{\eta_1(\varphi(b), \boldsymbol{m}(t)\varphi(a))}$$

$$\times \left\{|\eta_1(\varphi(x), \boldsymbol{m}(t)\varphi(a))|^{\alpha+2}\right.$$

$$\times \sqrt[rq]{(f''(a))^{rq} I^r(h_1(\xi), \boldsymbol{m}(\xi); p_1, r) + \eta_2\left((f''(x))^{rq}, (f''(a))^{rq}\right) I^r(h_2(\xi); p_2, r)}$$

$$+ |\eta_1(\varphi(x), \boldsymbol{m}(t)\varphi(b))|^{\alpha+2}$$

$$\left.\times \sqrt[rq]{(f''(b))^{rq} I^r(h_1(\xi), \boldsymbol{m}(\xi); p_1, r) + \eta_2\left((f''(x))^{rq}, (f''(b))^{rq}\right) I^r(h_2(\xi); p_2, r)}\right\}, \tag{32}$$

where

$$D(\alpha, \lambda, p) := \int_0^1 |\xi(\lambda - \xi^{\alpha})|^p d\xi$$

and $I(h_1(\xi), \boldsymbol{m}(\xi); p_1, r)$, $I(h_2(\xi); p_2, r)$ *are defined as in Theorem 4.*

Proof Using relation (31), generalized-$\mathbf{m}$-$((h_1^{p_1}, h_2^{p_2}); (\eta_1, \eta_2))$-convexity of f''^{q}, Hölder inequality, Minkowski inequality, and properties of the modulus, we have

$$\left|\Delta_f^{\alpha}(\eta_1, \varphi, \mathbf{m}; \lambda, x, a, b)\right| \leq \frac{|\eta_1(\varphi(x), \mathbf{m}(t)\varphi(a))|^{\alpha+2}}{|\eta_1(\varphi(b), \mathbf{m}(t)\varphi(a))|}$$

$$\times \int_0^1 |\xi(\lambda - \xi^{\alpha})||f''(\mathbf{m}(t)\varphi(a) + \xi\eta_1(\varphi(x), \mathbf{m}(t)\varphi(a)))|d\xi$$

$$+ \frac{|\eta_1(\varphi(x), \mathbf{m}(t)\varphi(b))|^{\alpha+2}}{|\eta_1(\varphi(b), \mathbf{m}(t)\varphi(a))|}$$

$$\times \int_0^1 |\xi(\lambda - \xi^{\alpha})||f''(\mathbf{m}(t)\varphi(b) + \xi\eta_1(\varphi(x), \mathbf{m}(t)\varphi(b)))|d\xi$$

$$\leq \frac{|\eta_1(\varphi(x), \mathbf{m}(t)\varphi(a))|^{\alpha+2}}{\eta_1(\varphi(b), \mathbf{m}(t)\varphi(a))} \left(\int_0^1 |\xi(\lambda - \xi^{\alpha})|^p d\xi\right)^{\frac{1}{p}}$$

$$\times \left(\int_0^1 \left(f''(\mathbf{m}(t)\varphi(a) + \xi\eta_1(\varphi(x), \mathbf{m}(t)\varphi(a)))\right)^q d\xi\right)^{\frac{1}{q}}$$

$$+ \frac{|\eta_1(\varphi(x), \mathbf{m}(t)\varphi(b))|^{\alpha+2}}{\eta_1(\varphi(b), \mathbf{m}(t)\varphi(a))} \left(\int_0^1 |\xi(\lambda - \xi^{\alpha})|^p d\xi\right)^{\frac{1}{p}}$$

$$\times\left(\int_0^1 \left(f''(\mathbf{m}(t)\varphi(b)+\xi\eta_1(\varphi(x),\mathbf{m}(t)\varphi(b)))\right)^q d\xi\right)^{\frac{1}{q}}$$

$$\leq \frac{|\eta_1(\varphi(x),\mathbf{m}(t)\varphi(a))|^{\alpha+2}}{\eta_1(\varphi(b),\mathbf{m}(t)\varphi(a))}\sqrt[p]{D(\alpha,\lambda,p)}$$

$$\times\left(\int_0^1 \left[\mathbf{m}(\xi)h_1^{p_1}(\xi)(f''(a))^{rq}+h_2^{p_2}(\xi)\eta_2\left((f''(x))^{rq},(f''(a))^{rq}\right)\right]^{\frac{1}{r}} d\xi\right)^{\frac{1}{q}}$$

$$+\frac{|\eta_1(\varphi(x),\mathbf{m}(t)\varphi(b))|^{\alpha+2}}{\eta_1(\varphi(b),\mathbf{m}(t)\varphi(a))}\sqrt[p]{D(\alpha,\lambda,p)}$$

$$\times\left(\int_0^1 \left[\mathbf{m}(\xi)h_1^{p_1}(\xi)(f''(b))^{rq}+h_2^{p_2}(\xi)\eta_2\left((f''(x))^{rq},(f''(b))^{rq}\right)\right]^{\frac{1}{r}} d\xi\right)^{\frac{1}{q}}$$

$$\leq \frac{|\eta_1(\varphi(x),\mathbf{m}(t)\varphi(a))|^{\alpha+2}}{\eta_1(\varphi(b),\mathbf{m}(t)\varphi(a))}\sqrt[p]{D(\alpha,\lambda,p)}$$

$$\times\left\{\left(\int_0^1 \mathbf{m}^{\frac{1}{r}}(\xi)h_1^{\frac{p_1}{r}}(\xi)(f''(a))^q d\xi\right)^r\right.$$

$$\left.+\left(\int_0^1 h_2^{\frac{p_2}{r}}(\xi)\eta_2^{\frac{1}{r}}\left((f''(x))^{rq},(f''(a))^{rq}\right)d\xi\right)^r\right\}^{\frac{1}{rq}}$$

$$+\frac{|\eta_1(\varphi(x),\mathbf{m}(t)\varphi(b))|^{\alpha+2}}{\eta_1(\varphi(b),\mathbf{m}(t)\varphi(a))}\sqrt[p]{D(\alpha,\lambda,p)}$$

$$\times\left\{\left(\int_0^1 \mathbf{m}^{\frac{1}{r}}(\xi)h_1^{\frac{p_1}{r}}(\xi)(f''(b))^q d\xi\right)^r\right.$$

$$\left.+\left(\int_0^1 h_2^{\frac{p_2}{r}}(\xi)\eta_2^{\frac{1}{r}}\left((f''(x))^{rq},(f''(b))^{rq}\right)d\xi\right)^r\right\}^{\frac{1}{rq}}$$

$$=\frac{\sqrt[p]{D(\alpha,\lambda,p)}}{\eta_1(\varphi(b),\mathbf{m}(t)\varphi(a))}$$

$$\times\left\{|\eta_1(\varphi(x),\mathbf{m}(t)\varphi(a))|^{\alpha+2}\right.$$

$$\times\sqrt[rq]{(f''(a))^{rq}I^r(h_1(\xi),\mathbf{m}(\xi);p_1,r)+\eta_2\left((f''(x))^{rq},(f''(a))^{rq}\right)I^r(h_2(\xi);p_2,r)}$$

$$+|\eta_1(\varphi(x),\mathbf{m}(t)\varphi(b))|^{\alpha+2}$$

$$\left.\times\sqrt[rq]{(f''(b))^{rq}I^r(h_1(\xi),\mathbf{m}(\xi);p_1,r)+\eta_2\left((f''(x))^{rq},(f''(b))^{rq}\right)I^r(h_2(\xi);p_2,r)}\right\}.$$

So, the proof of this theorem is completed.

Let us discuss some special cases of Theorem 6.

Corollary 13 *In Theorem 6 for $p = q = 2$, we get*

$$\left|\Delta_f^\alpha(\eta_1, \varphi, \boldsymbol{m}; \lambda, x, a, b)\right| \le \frac{1}{\eta_1(\varphi(b), \boldsymbol{m}(t)\varphi(a))}\sqrt{\frac{\lambda^2}{3} + \frac{1}{2\alpha+3} - \frac{2\lambda}{\alpha+3}}$$
$$\times \Big\{ |\eta_1(\varphi(x), \boldsymbol{m}(t)\varphi(a))|^{\alpha+2}$$
$$\times \sqrt[2r]{(f''(a))^{2r} I^r(h_1(\xi), \boldsymbol{m}(\xi); p_1, r) + \eta_2\left((f''(x))^{2r}, (f''(a))^{2r}\right) I^r(h_2(\xi); p_2, r)}$$
$$+ |\eta_1(\varphi(x), \boldsymbol{m}(t)\varphi(b))|^{\alpha+2}$$
$$\times \sqrt[2r]{(f''(b))^{2r} I^r(h_1(\xi), \boldsymbol{m}(\xi); p_1, r) + \eta_2\left((f''(x))^{2r}, (f''(b))^{2r}\right) I^r(h_2(\xi); p_2, r)}\Big\}. \tag{33}$$

Corollary 14 *In Theorem 6, if we choose $\eta_1(\varphi(y), \boldsymbol{m}(t)\varphi(x)) = \varphi(y) - \boldsymbol{m}(t)\varphi(x)$ and $\lambda = \boldsymbol{m}(t) \equiv 1, \forall t \in [0, 1]$, we get the following generalized Hermite-Hadamard type inequality for fractional integrals:*

$$\left|\Delta_f^\alpha(\varphi; 1, x, a, b)\right| = \Big|\frac{\alpha}{(\varphi(b) - \varphi(a))}\Big\{\left((\varphi(x) - \varphi(a))^\alpha + (\varphi(b) - \varphi(x))^\alpha\right) f(\varphi(x))\Big\}$$
$$+ \frac{1}{(\varphi(b) - \varphi(a))}\Big\{(\varphi(x) - \varphi(a))^\alpha f(\varphi(a)) + (\varphi(b) - \varphi(x))^\alpha f(\varphi(b))\Big\}$$
$$- \frac{\Gamma(\alpha+2)}{(\varphi(b) - \varphi(a))} \times \left[J^\alpha_{(\varphi(x))^-} f(\varphi(a)) + J^\alpha_{(\varphi(x))^-} f(\varphi(b))\right]\Big|$$

$$\le \frac{1}{(\varphi(b) - \varphi(a))}\sqrt[p]{\frac{\alpha}{2(\alpha+2)}}$$
$$\times \Big\{ (\varphi(x) - \varphi(a))^{\alpha+2}$$
$$\times \sqrt[rq]{(f''(a))^{rq} I^r(h_1(\xi); p_1, r) + \eta_2\left((f''(x))^{rq}, (f''(a))^{rq}\right) I^r(h_2(\xi); p_2, r)}$$
$$+ (\varphi(b) - \varphi(x))^{\alpha+2}$$
$$\times \sqrt[rq]{(f''(b))^{rq} I^r(h_1(\xi); p_1, r) + \eta_2\left((f''(x))^{rq}, (f''(b))^{rq}\right) I^r(h_2(\xi); p_2, r)}\Big\}. \tag{34}$$

Corollary 15 *In Theorem 6, if we choose $\eta_1(\varphi(y), \boldsymbol{m}(t)\varphi(x)) = \varphi(y) - \boldsymbol{m}(t)\varphi(x)$, $\lambda = 0$ and $\boldsymbol{m}(t) \equiv 1, \forall t \in [0, 1]$, we get the following generalized Ostrowski type inequality for fractional integrals:*

$$\left|\Delta_f^\alpha(\varphi; 0, x, a, b)\right|$$

$$= \left| \frac{1}{(\varphi(a) - \varphi(b))} \left\{ \left((\varphi(x) - \varphi(a))^{\alpha+1} + (\varphi(b) - \varphi(x))^{\alpha+1} \right) f'(\varphi(x)) \right\} \right.$$

$$+ \frac{1+\alpha}{(\varphi(b) - \varphi(a))} \left\{ \left((\varphi(x) - \varphi(a))^{\alpha} + (\varphi(b) - \varphi(x))^{\alpha} \right) f(\varphi(x)) \right\}$$

$$\left. - \frac{\Gamma(\alpha+2)}{(\varphi(b) - \varphi(a))} \times \left[J^{\alpha}_{(\varphi(x))^-} f(\varphi(a)) + J^{\alpha}_{(\varphi(x))^-} f(\varphi(b)) \right] \right|$$

$$\leq \frac{1}{(\varphi(b) - \varphi(a))} \sqrt[p]{\frac{1}{p(\alpha+1)+1}}$$

$$\times \left\{ (\varphi(x) - \varphi(a))^{\alpha+2} \right.$$

$$\times \sqrt[rq]{(f''(a))^{rq} I^r(h_1(\xi); p_1, r) + \eta_2\left((f''(x))^{rq}, (f''(a))^{rq}\right) I^r(h_2(\xi); p_2, r)}$$

$$+ (\varphi(b) - \varphi(x))^{\alpha+2}$$

$$\left. \times \sqrt[rq]{(f''(b))^{rq} I^r(h_1(\xi); p_1, r) + \eta_2\left((f''(x))^{rq}, (f''(b))^{rq}\right) I^r(h_2(\xi); p_2, r)} \right\}. \tag{35}$$

Corollary 16 *In Theorem 6, if we choose* $\eta_1(\varphi(y), \boldsymbol{m}(t)\varphi(x)) = \varphi(y) - \boldsymbol{m}(t)\varphi(x)$, $x = \frac{a+b}{2}$ *and* $\boldsymbol{m}(t) \equiv 1, \forall t \in [0,1]$, *we get the following generalized Simpson type inequality for fractional integrals:*

$$\left| \Delta^{\alpha}_{f}\left(\varphi; \lambda, \frac{a+b}{2}, a, b\right) \right| = \left| \frac{\lambda - 1}{(\varphi(b) - \varphi(a))} \right.$$

$$\times \left\{ \left(\left(\varphi\left(\frac{a+b}{2}\right) - \varphi(a) \right)^{\alpha+1} + \left(\varphi(b) - \varphi\left(\frac{a+b}{2}\right) \right)^{\alpha+1} \right) f'\left(\varphi\left(\frac{a+b}{2}\right)\right) \right\}$$

$$+ \frac{1+\alpha-\lambda}{(\varphi(b) - \varphi(a))}$$

$$\times \left\{ \left(\left(\varphi\left(\frac{a+b}{2}\right) - \varphi(a) \right)^{\alpha} + \left(\varphi(b) - \varphi\left(\frac{a+b}{2}\right) \right)^{\alpha} \right) f\left(\varphi\left(\frac{a+b}{2}\right)\right) \right\}$$

$$+ \frac{\lambda}{(\varphi(b) - \varphi(a))}$$

$$\times\left\{\left(\varphi\left(\frac{a+b}{2}\right)-\varphi(a)\right)^{\alpha}f(\varphi(a))+\left(\varphi(b)-\varphi\left(\frac{a+b}{2}\right)\right)^{\alpha}f(\varphi(b))\right\}$$

$$-\frac{\Gamma(\alpha+2)}{(\varphi(b)-\varphi(a))}\times\left[J^{\alpha}_{\left(\varphi\left(\frac{a+b}{2}\right)\right)^{-}}f(\varphi(a))+J^{\alpha}_{\left(\varphi\left(\frac{a+b}{2}\right)\right)^{-}}f(\varphi(b))\right]\Bigg|$$

$$\leq\frac{\sqrt[p]{D(\alpha,\lambda,p)}}{(\varphi(b)-\varphi(a))}$$

$$\times\left\{\left(\varphi\left(\frac{a+b}{2}\right)-\varphi(a)\right)^{\alpha+2}\left[(f''(a))^{rq}I^{r}(h_1(\xi);p_1,r)\right.\right.$$

$$\left.+\eta_2\left(\left(f''\left(\frac{a+b}{2}\right)\right)^{rq},(f''(a))^{rq}\right)I^{r}(h_2(\xi);p_2,r)\right]^{\frac{1}{rq}}\tag{36}$$

$$+\left(\varphi(b)-\varphi\left(\frac{a+b}{2}\right)\right)^{\alpha+2}\left[(f''(b))^{rq}I^{r}(h_1(\xi);p_1,r)\right.$$

$$\left.\left.+\eta_2\left(\left(f''\left(\frac{a+b}{2}\right)\right)^{rq},(f''(b))^{rq}\right)I^{r}(h_2(\xi);p_2,r)\right]^{\frac{1}{rq}}\right\}.$$

Corollary 17 *In Theorem 6 for* $h_1(t)=h(1-t)$, $h_2(t)=h(t)$ *and* $\mathbf{m}(t)=m\in(0,1]$ *for all* $t\in[0,1]$, *we get the following inequality for generalized-m-*$((h^{p_1}(1-t),h^{p_2}(t));(\eta_1,\eta_2))$*-convex mappings:*

$$\left|\Delta^{\alpha}_{f}(\eta_1,\varphi,m;\lambda,x,a,b)\right|\leq\frac{\sqrt[p]{D(\alpha,\lambda,p)}}{\eta_1(\varphi(b),m\varphi(a))}$$

$$\times\left\{|\eta_1(\varphi(x),m\varphi(a))|^{\alpha+2}\right.$$

$$\times\sqrt[rq]{m(f''(a))^{rq}I^{r}(h(1-\xi);p_1,r)+\eta_2\left((f''(x))^{rq},(f''(a))^{rq}\right)I^{r}(h(\xi);p_2,r)}$$

$$+|\eta_1(\varphi(x),m\varphi(b))|^{\alpha+2}$$

$$\left.\times\sqrt[rq]{m(f''(b))^{rq}I^{r}(h(1-\xi);p_1,r)+\eta_2\left((f''(x))^{rq},(f''(b))^{rq}\right)I^{r}(h(\xi);p_2,r)}\right\}.\tag{37}$$

Corollary 18 *In Corollary 17 for* $h_1(t)=(1-t)^s$ *and* $h_2(t)=t^s$, *we get the following inequality for generalized-m-*$(((1-t)^{sp_1},t^{sp_2});(\eta_1,\eta_2))$*-Breckner-convex mappings:*

$$\left|\Delta_f^{\alpha}(\eta_1,\varphi,m;\lambda,x,a,b)\right| \leq \frac{\sqrt[p]{D(\alpha,\lambda,p)}}{\eta_1(\varphi(b),m\varphi(a))}$$
$$\times\Bigg\{|\eta_1(\varphi(x),m\varphi(a))|^{\alpha+2}$$
$$\times\sqrt[rq]{m(f''(a))^{rq}\left(\frac{r}{r+sp_1}\right)^r+\eta_2\left((f''(x))^{rq},(f''(a))^{rq}\right)\left(\frac{r}{r+sp_2}\right)^r} \tag{38}$$
$$+|\eta_1(\varphi(x),m\varphi(b))|^{\alpha+2}$$
$$\times\sqrt[rq]{m(f''(b))^{rq}\left(\frac{r}{r+sp_1}\right)^r+\eta_2\left((f''(x))^{rq},(f''(b))^{rq}\right)\left(\frac{r}{r+sp_2}\right)^r}\Bigg\}.$$

Corollary 19 *In Corollary 17 for* $h_1(t) = (1-t)^{-s}$, $h_2(t) = t^{-s}$ *and* $r > s\cdot\max\{p_1,p_2\}$, *we get the following inequality for generalized-m-*$(((1-t)^{-sp_1}, t^{-sp_2}); (\eta_1,\eta_2))$*-Godunova-Levin-Dragomir-convex mappings:*

$$\left|\Delta_f^{\alpha}(\eta_1,\varphi,m;\lambda,x,a,b)\right| \leq \frac{\sqrt[p]{D(\alpha,\lambda,p)}}{\eta_1(\varphi(b),m\varphi(a))}$$
$$\times\Bigg\{|\eta_1(\varphi(x),m\varphi(a))|^{\alpha+2}$$
$$\times\sqrt[rq]{m(f''(a))^{rq}\left(\frac{r}{r-sp_1}\right)^r+\eta_2\left((f''(x))^{rq},(f''(a))^{rq}\right)\left(\frac{r}{r-sp_2}\right)^r} \tag{39}$$
$$+|\eta_1(\varphi(x),m\varphi(b))|^{\alpha+2}$$
$$\times\sqrt[rq]{m(f''(b))^{rq}\left(\frac{r}{r-sp_1}\right)^r+\eta_2\left((f''(x))^{rq},(f''(b))^{rq}\right)\left(\frac{r}{r-sp_2}\right)^r}\Bigg\}.$$

Corollary 20 *In Theorem 6 for* $h_1(t) = h_2(t) = t(1-t)$ *and* $\mathbf{m}(t) = m \in (0,1]$ *for all* $t \in [0,1]$, *we get the following inequality for generalized-m-*$((t(1-t))^{sp_1}, (t(1-t))^{sp_2}); (\eta_1,\eta_2))$*-convex mappings:*

$$\left|\Delta_f^{\alpha}(\eta_1,\varphi,m;\lambda,x,a,b)\right| \leq \frac{\sqrt[p]{D(\alpha,\lambda,p)}}{\eta_1(\varphi(b),m\varphi(a))}$$
$$\times\Bigg\{|\eta_1(\varphi(x),m\varphi(a))|^{\alpha+2}\Big[m(f''(a))^{rq}\beta^r\left(1+\frac{p_1}{r},1+\frac{p_1}{r}\right) \tag{40}$$
$$+\eta_2\left((f''(x))^{rq},(f''(a))^{rq}\right)\beta^r\left(1+\frac{p_2}{r},1+\frac{p_2}{r}\right)\Big]^{\frac{1}{rq}}$$

$$+ |\eta_1(\varphi(x), m\varphi(b))|^{\alpha+2}\Big[m(f''(b))^{rq}\beta^r\left(1+\frac{p_1}{r}, 1+\frac{p_1}{r}\right)$$

$$+ \eta_2\left((f''(x))^{rq}, (f''(b))^{rq}\right)\beta^r\left(1+\frac{p_2}{r}, 1+\frac{p_2}{r}\right)\Big]^{\frac{1}{rq}}\Bigg\}.$$

Corollary 21 *In Corollary 17 for* $h_1(t) = \dfrac{\sqrt{1-t}}{2\sqrt{t}}$, $h_2(t) = \dfrac{\sqrt{t}}{2\sqrt{1-t}}$ *and* $r > \frac{1}{2}\cdot\max\{p_1, p_2\}$, *we get the following inequality for generalized-m* -$\left(\left(\left(\frac{\sqrt{1-t}}{2\sqrt{t}}\right)^p, \left(\frac{\sqrt{t}}{2\sqrt{1-t}}\right)^q\right); (\eta_1, \eta_2)\right)$*-convex mappings:*

$$\left|\Delta_f^{\alpha}(\eta_1, \varphi, m; \lambda, x, a, b)\right| \le \frac{\sqrt[p]{D(\alpha, \lambda, p)}}{\eta_1(\varphi(b), m\varphi(a))}$$

$$\times\Bigg\{|\eta_1(\varphi(x), m\varphi(a))|^{\alpha+2}\Big[m(f''(a))^{rq}\beta^r\left(1-\frac{p_1}{2r}, 1+\frac{p_1}{2r}\right)$$

$$+ \eta_2\left((f''(x))^{rq}, (f''(a))^{rq}\right)\beta^r\left(1-\frac{p_2}{2r}, 1+\frac{p_2}{2r}\right)\Big]^{\frac{1}{rq}} \tag{41}$$

$$+ |\eta_1(\varphi(x), m\varphi(b))|^{\alpha+2}\Big[m(f''(b))^{rq}\beta^r\left(1-\frac{p_1}{2r}, 1+\frac{p_1}{2r}\right)$$

$$+ +\eta_2\left((f''(x))^{rq}, (f''(b))^{rq}\right)\beta^r\left(1-\frac{p_2}{2r}, 1+\frac{p_2}{2r}\right)\Big]^{\frac{1}{rq}}\Bigg\}.$$

Theorem 7 *Let* $0 < r \le 1$ *and* $p_1, p_2 > -1$. *Suppose* $h_1, h_2 : [0, 1] \longrightarrow [0, +\infty)$, $\varphi : I \longrightarrow K$ *are continuous functions and* $\boldsymbol{m} : [0, 1] \longrightarrow (0, 1]$. *Let* $K = [\boldsymbol{m}(t)\varphi(a), \boldsymbol{m}(t)\varphi(a) + \eta_1(\varphi(b), \boldsymbol{m}(t)\varphi(a))] \subseteq \mathbb{R}$ *be an open* $\boldsymbol{m}$*-invex subset with respect to* $\eta_1 : K \times K \longrightarrow \mathbb{R}$ *and let* $\eta_1(\varphi(b), \boldsymbol{m}(t)\varphi(a)) > 0$ *for all* $t \in [0, 1]$ *and* $\eta_2 : f(K) \times f(K) \longrightarrow \mathbb{R}$. *Assume that* $f : K \longrightarrow (0, +\infty)$ *be a twice differentiable mapping on* K°. *If* f''^q *is generalized-*$\boldsymbol{m}$-$((h_1^{p_1}, h_2^{p_2}); (\eta_1, \eta_2))$*-convex mapping on* K *and* $q \ge 1$, *then for any* $\lambda \in [0, 1]$ *and* $\alpha > 0$, *the following inequality for fractional integrals holds:*

$$\left|\Delta_f^{\alpha}(\eta_1, \varphi, \boldsymbol{m}; \lambda, x, a, b)\right| \le \frac{C^{1-\frac{1}{q}}(\alpha, \lambda)}{\eta_1(\varphi(b), \boldsymbol{m}(t)\varphi(a))}$$

$$\times\Bigg\{|\eta_1(\varphi(x), \boldsymbol{m}(t)\varphi(a))|^{\alpha+2}\Big[(f''(a))^{rq}F^r(h_1(\xi), \boldsymbol{m}(\xi); \lambda, \alpha, p_1, r)$$

$$+ \eta_2\left((f''(x))^{rq}, (f''(a))^{rq}\right)F^r(h_2(\xi); \lambda, \alpha, p_2, r)\Big]^{\frac{1}{rq}} \tag{42}$$

$$+ |\eta_1(\varphi(x), \boldsymbol{m}(t)\varphi(b))|^{\alpha+2}\Big[(f''(b))^{rq}F^r(h_1(\xi), \boldsymbol{m}(\xi); \lambda, \alpha, p_1, r)$$

$$+ \eta_2\left((f''(x))^{rq}, (f''(b))^{rq}\right)F^r(h_2(\xi); \lambda, \alpha, p_2, r)\Big]^{\frac{1}{rq}}\Bigg\},$$

where

$$C(\alpha,\lambda):=\frac{\alpha\lambda^{1+\frac{2}{\alpha}}+1}{\alpha+2}-\frac{\lambda}{2};$$

$$F(h_1(\xi),\boldsymbol{m}(\xi);\lambda,\alpha,p_1,r):=\int_0^1 \boldsymbol{m}^{\frac{1}{r}}(\xi)|\xi(\lambda-\xi^{\alpha})|h_1^{\frac{p_1}{r}}(\xi)d\xi;$$

$$F(h_2(\xi);\lambda,\alpha,p_2,r):=\int_0^1 |\xi(\lambda-\xi^{\alpha})|h_2^{\frac{p_2}{r}}(\xi)d\xi.$$

Proof Using relation (31), generalized-$\mathbf{m}$-$((h_1^{p_1},h_2^{p_2});(\eta_1,\eta_2))$-convexity of f''^q, the well-known power mean inequality, Minkowski inequality, and properties of the modulus, we have

$$\left|\Delta_f^{\alpha}(\eta_1,\varphi,\mathbf{m};\lambda,x,a,b)\right|\leq\frac{|\eta_1(\varphi(x),\mathbf{m}(t)\varphi(a))|^{\alpha+2}}{|\eta_1(\varphi(b),\mathbf{m}(t)\varphi(a))|}$$

$$\times\int_0^1|\xi(\lambda-\xi^{\alpha})||f''(\mathbf{m}(t)\varphi(a)+\xi\eta_1(\varphi(x),\mathbf{m}(t)\varphi(a)))|d\xi$$

$$+\frac{|\eta_1(\varphi(x),\mathbf{m}(t)\varphi(b))|^{\alpha+2}}{|\eta_1(\varphi(b),\mathbf{m}(t)\varphi(a))|}$$

$$\times\int_0^1|\xi(\lambda-\xi^{\alpha})||f''(\mathbf{m}(t)\varphi(b)+\xi\eta_1(\varphi(x),\mathbf{m}(t)\varphi(b)))|d\xi$$

$$\leq\frac{|\eta_1(\varphi(x),\mathbf{m}(t)\varphi(a))|^{\alpha+2}}{\eta_1(\varphi(b),\mathbf{m}(t)\varphi(a))}\left(\int_0^1|\xi(\lambda-\xi^{\alpha})|d\xi\right)^{1-\frac{1}{q}}$$

$$\times\left(\int_0^1|\xi(\lambda-\xi^{\alpha})|\left(f''(\mathbf{m}(t)\varphi(a)+\xi\eta_1(\varphi(x),\mathbf{m}(t)\varphi(a)))\right)^q d\xi\right)^{\frac{1}{q}}$$

$$+\frac{|\eta_1(\varphi(x),\mathbf{m}(t)\varphi(b))|^{\alpha+2}}{\eta_1(\varphi(b),\mathbf{m}(t)\varphi(a))}\left(\int_0^1|\xi(\lambda-\xi^{\alpha})|d\xi\right)^{1-\frac{1}{q}}$$

$$\times\left(\int_0^1|\xi(\lambda-\xi^{\alpha})|\left(f''(\mathbf{m}(t)\varphi(b)+\xi\eta_1(\varphi(x),\mathbf{m}(t)\varphi(b)))\right)^q d\xi\right)^{\frac{1}{q}}$$

$$\leq\frac{|\eta_1(\varphi(x),\mathbf{m}(t)\varphi(a))|^{\alpha+2}}{\eta_1(\varphi(b),\mathbf{m}(t)\varphi(a))}C^{1-\frac{1}{q}}(\alpha,\lambda)$$

$$\times\left(\int_0^1|\xi(\lambda-\xi^{\alpha})|\Big[\mathbf{m}(\xi)h_1^{p_1}(\xi)(f''(a))^{rq}+h_2^{p_2}(\xi)\eta_2\left((f''(x))^{rq},(f''(a))^{rq}\right)\Big]^{\frac{1}{r}}d\xi\right)^{\frac{1}{q}}$$

$$+\frac{|\eta_1(\varphi(x),\mathbf{m}(t)\varphi(b))|^{\alpha+2}}{\eta_1(\varphi(b),\mathbf{m}(t)\varphi(a))}C^{1-\frac{1}{q}}(\alpha,\lambda)$$

$$\times\left(\int_0^1|\xi(\lambda-\xi^\alpha)|\Big[\mathbf{m}(\xi)h_1^{p_1}(\xi)(f''(b))^{rq}+h_2^{p_2}(\xi)\eta_2\left((f''(x))^{rq},(f''(b))^{rq}\right)\Big]^{\frac{1}{r}}d\xi\right)^{\frac{1}{q}}$$

$$\leq\frac{|\eta_1(\varphi(x),\mathbf{m}(t)\varphi(a))|^{\alpha+2}}{\eta_1(\varphi(b),\mathbf{m}(t)\varphi(a))}C^{1-\frac{1}{q}}(\alpha,\lambda)$$

$$\times\left\{\left(\int_0^1\mathbf{m}^{\frac{1}{r}}(\xi)|\xi(\lambda-\xi^\alpha)|h_1^{\frac{p_1}{r}}(\xi)(f''(a))^q d\xi\right)^r\right.$$

$$\left.+\left(\int_0^1|\xi(\lambda-\xi^\alpha)|h_2^{\frac{p_2}{r}}(\xi)\eta_2^{\frac{1}{r}}\left((f''(x))^{rq},(f''(a))^{rq}\right)d\xi\right)^r\right\}^{\frac{1}{rq}}$$

$$+\frac{|\eta_1(\varphi(x),\mathbf{m}(t)\varphi(b))|^{\alpha+2}}{\eta_1(\varphi(b),\mathbf{m}(t)\varphi(a))}C^{1-\frac{1}{q}}(\alpha,\lambda)$$

$$\times\left\{\left(\int_0^1\mathbf{m}^{\frac{1}{r}}(\xi)|\xi(\lambda-\xi^\alpha)|h_1^{\frac{p_1}{r}}(\xi)(f''(b))^q d\xi\right)^r\right.$$

$$\left.+\left(\int_0^1|\xi(\lambda-\xi^\alpha)|h_2^{\frac{p_2}{r}}(\xi)\eta_2^{\frac{1}{r}}\left((f''(x))^{rq},(f''(b))^{rq}\right)d\xi\right)^r\right\}^{\frac{1}{rq}}$$

$$=\frac{C^{1-\frac{1}{q}}(\alpha,\lambda)}{\eta_1(\varphi(b),\mathbf{m}(t)\varphi(a))}$$

$$\times\left\{|\eta_1(\varphi(x),\mathbf{m}(t)\varphi(a))|^{\alpha+2}\Big[(f''(a))^{rq}F^r(h_1(\xi),\mathbf{m}(\xi);\lambda,\alpha,p_1,r)\right.$$

$$+\eta_2\left((f''(x))^{rq},(f''(a))^{rq}\right)F^r(h_2(\xi);\lambda,\alpha,p_2,r)\Big]^{\frac{1}{rq}}$$

$$+|\eta_1(\varphi(x),\mathbf{m}(t)\varphi(b))|^{\alpha+2}\Big[(f''(b))^{rq}F^r(h_1(\xi),\mathbf{m}(\xi);\lambda,\alpha,p_1,r)$$

$$\left.+\eta_2\left((f''(x))^{rq},(f''(b))^{rq}\right)F^r(h_2(\xi);\lambda,\alpha,p_2,r)\Big]^{\frac{1}{rq}}\right\}.$$

So, the proof of this theorem is completed.

Let us discuss some special cases of Theorem 7.

Corollary 22 *In Theorem 7 for $q = 1$, we get*

$$\begin{aligned}
&\left|\Delta_f^{\alpha}(\eta_1, \varphi, \boldsymbol{m}; \lambda, x, a, b)\right| \leq \frac{1}{\eta_1(\varphi(b), \boldsymbol{m}(t)\varphi(a))} \\
&\times \Bigg\{ |\eta_1(\varphi(x), \boldsymbol{m}(t)\varphi(a))|^{\alpha+2} \Big[(f''(a))^r F^r(h_1(\xi), \boldsymbol{m}(\xi); \lambda, \alpha, p_1, r) \\
&+ \eta_2\left((f''(x))^r, (f''(a))^r\right) F^r(h_2(\xi); \lambda, \alpha, p_2, r) \Big]^{\frac{1}{r}} \\
&+ |\eta_1(\varphi(x), \boldsymbol{m}(t)\varphi(b))|^{\alpha+2} \Big[(f''(b))^r F^r(h_1(\xi), \boldsymbol{m}(\xi); \lambda, \alpha, p_1, r) \\
&+ \eta_2\left((f''(x))^r, (f''(b))^r\right) F^r(h_2(\xi); \lambda, \alpha, p_2, r) \Big]^{\frac{1}{r}} \Bigg\}.
\end{aligned} \tag{43}$$

Corollary 23 *In Theorem 7, if we choose $\eta_1(\varphi(y), \boldsymbol{m}(t)\varphi(x)) = \varphi(y) - \boldsymbol{m}(t)\varphi(x)$ and $\lambda = \boldsymbol{m}(t) \equiv 1, \forall t \in [0, 1]$, we get the following generalized Hermite-Hadamard type inequality for fractional integrals:*

$$\begin{aligned}
&\left|\Delta_f^{\alpha}(\varphi; 1, x, a, b)\right| \leq \left(\frac{\alpha}{2(\alpha+2)}\right)^{1-\frac{1}{q}} \frac{1}{(\varphi(b) - \varphi(a))} \\
&\times \Bigg\{ (\varphi(x) - \varphi(a))^{\alpha+2} \Big[(f''(a))^{rq} F^r(h_1(\xi); 1, \alpha, p_1, r) \\
&+ \eta_2\left((f''(x))^{rq}, (f''(a))^{rq}\right) F^r(h_2(\xi); 1, \alpha, p_2, r) \Big]^{\frac{1}{rq}} \\
&+ (\varphi(b) - \varphi(x))^{\alpha+2} \Big[(f''(b))^{rq} F^r(h_1(\xi); 1, \alpha, p_1, r) \\
&+ \eta_2\left((f''(x))^{rq}, (f''(b))^{rq}\right) F^r(h_2(\xi); 1, \alpha, p_2, r) \Big]^{\frac{1}{rq}} \Bigg\}.
\end{aligned} \tag{44}$$

Corollary 24 *In Theorem 7, if we choose $\eta_1(\varphi(y), \boldsymbol{m}(t)\varphi(x)) = \varphi(y) - \boldsymbol{m}(t)\varphi(x)$, $\lambda = 0$ and $\boldsymbol{m}(t) \equiv 1, \forall t \in [0, 1]$, we get the following generalized Ostrowski type inequality for fractional integrals:*

$$\begin{aligned}
&\left|\Delta_f^{\alpha}(\varphi; 0, x, a, b)\right| \leq \left(\frac{1}{\alpha+2}\right)^{1-\frac{1}{q}} \frac{1}{(\varphi(b) - \varphi(a))} \\
&\times \Bigg\{ (\varphi(x) - \varphi(a))^{\alpha+2} \Big[(f''(a))^{rq} F^r(h_1(\xi); 0, \alpha, p_1, r) \\
&+ \eta_2\left((f''(x))^{rq}, (f''(a))^{rq}\right) F^r(h_2(\xi); 0, \alpha, p_2, r) \Big]^{\frac{1}{rq}}
\end{aligned} \tag{45}$$

$$+ (\varphi(b) - \varphi(x))^{\alpha+2}\Big[(f''(b))^{rq} F^r(h_1(\xi); 0, \alpha, p_1, r)$$

$$+ \eta_2\left((f''(x))^{rq}, (f''(b))^{rq}\right) F^r(h_2(\xi); 0, \alpha, p_2, r)\Big]^{\frac{1}{rq}}\Bigg\}.$$

Corollary 25 *In Theorem 7, if we choose* $\eta_1(\varphi(y), \boldsymbol{m}(t)\varphi(x)) = \varphi(y) - \boldsymbol{m}(t)\varphi(x)$, $x = \dfrac{a+b}{2}$ *and* $\boldsymbol{m}(t) \equiv 1, \forall t \in [0, 1]$, *we get the following generalized Simpson type inequality for fractional integrals:*

$$\left|\Delta_f^{\alpha}\left(\varphi; \lambda, \frac{a+b}{2}, a, b\right)\right| \leq \frac{C^{1-\frac{1}{q}}(\alpha, \lambda)}{(\varphi(b) - \varphi(a))}$$

$$\times \Bigg\{\left(\varphi\left(\frac{a+b}{2}\right) - \varphi(a)\right)^{\alpha+2}\Bigg[(f''(a))^{rq} F^r(h_1(\xi); \lambda, \alpha, p_1, r)$$

$$+ \eta_2\left(\left(f''\left(\frac{a+b}{2}\right)\right)^{rq}, (f''(a))^{rq}\right) F^r(h_1(\xi); \lambda, \alpha, p_2, r)\Bigg]^{\frac{1}{rq}} \tag{46}$$

$$+ \left(\varphi(b) - \varphi\left(\frac{a+b}{2}\right)\right)^{\alpha+2}\Bigg[(f''(b))^{rq} F^r(h_1(\xi); \lambda, \alpha, p_1, r)$$

$$+ \eta_2\left(\left(f''\left(\frac{a+b}{2}\right)\right)^{rq}, (f''(b))^{rq}\right) F^r(h_1(\xi); \lambda, \alpha, p_2, r)\Bigg]^{\frac{1}{rq}}\Bigg\}.$$

Corollary 26 *In Theorem 7 for* $h_1(t) = h(1-t)$, $h_2(t) = h(t)$ *and* $\boldsymbol{m}(t) = m \in (0, 1]$ *for all* $t \in [0, 1]$, *we get the following inequality for generalized-m-*$((h^{p_1}(1-t), h^{p_2}(t)); (\eta_1, \eta_2))$*-convex mappings:*

$$\left|\Delta_f^{\alpha}(\eta_1, \varphi, m; \lambda, x, a, b)\right| \leq \frac{C^{1-\frac{1}{q}}(\alpha, \lambda)}{\eta_1(\varphi(b), m\varphi(a))}$$

$$\times \Bigg\{|\eta_1(\varphi(x), m\varphi(a))|^{\alpha+2}\Big[m(f''(a))^{rq} F^r(h(1-\xi); \lambda, \alpha, p_1, r)$$

$$+ \eta_2\left((f''(x))^{rq}, (f''(a))^{rq}\right) F^r(h(\xi); \lambda, \alpha, p_2, r)\Big]^{\frac{1}{rq}} \tag{47}$$

$$+ |\eta_1(\varphi(x), m\varphi(b))|^{\alpha+2}\Big[m(f''(b))^{rq} F^r(h(1-\xi); \lambda, \alpha, p_1, r)$$

$$+ \eta_2\left((f''(x))^{rq}, (f''(b))^{rq}\right) F^r(h(\xi); \lambda, \alpha, p_2, r)\Big]^{\frac{1}{rq}}\Bigg\}.$$

Corollary 27 *In Corollary 26 for $h_1(t) = (1-t)^s$ and $h_2(t) = t^s$, we get the following inequality for generalized-m-$(((1-t)^{sp_1}, t^{sp_2}); (\eta_1, \eta_2))$-Breckner-convex mappings:*

$$\begin{aligned}
\left|\Delta_f^{\alpha}(\eta_1, \varphi, m; \lambda, x, a, b)\right| &\leq \frac{C^{1-\frac{1}{q}}(\alpha, \lambda)}{\eta_1(\varphi(b), m\varphi(a))} \\
&\times \Bigg\{ |\eta_1(\varphi(x), m\varphi(a))|^{\alpha+2} \Big[m(f''(a))^{rq} F^r((1-\xi)^s; \lambda, \alpha, p_1, r) \\
&+ \eta_2\left((f''(x))^{rq}, (f''(a))^{rq}\right) F^r(\xi^s; \lambda, \alpha, p_2, r) \Big]^{\frac{1}{rq}} \\
&+ |\eta_1(\varphi(x), m\varphi(b))|^{\alpha+2} \Big[m(f''(b))^{rq} F^r((1-\xi)^s; \lambda, \alpha, p_1, r) \\
&+ \eta_2\left((f''(x))^{rq}, (f''(b))^{rq}\right) F^r(\xi^s; \lambda, \alpha, p_2, r) \Big]^{\frac{1}{rq}} \Bigg\}.
\end{aligned} \tag{48}$$

Corollary 28 *In Corollary 26 for $h_1(t) = (1-t)^{-s}$ and $h_2(t) = t^{-s}$, we get the following inequality for generalized-m-$(((1-t)^{-sp_1}, t^{-sp_2}); (\eta_1, \eta_2))$-Godunova-Levin-Dragomir-convex mappings:*

$$\begin{aligned}
\left|\Delta_f^{\alpha}(\eta_1, \varphi, m; \lambda, x, a, b)\right| &\leq \frac{C^{1-\frac{1}{q}}(\alpha, \lambda)}{\eta_1(\varphi(b), m\varphi(a))} \\
&\times \Bigg\{ |\eta_1(\varphi(x), m\varphi(a))|^{\alpha+2} \Big[m(f''(a))^{rq} F^r((1-\xi)^{-s}; \lambda, \alpha, p_1, r) \\
&+ \eta_2\left((f''(x))^{rq}, (f''(a))^{rq}\right) F^r(\xi^{-s}; \lambda, \alpha, p_2, r) \Big]^{\frac{1}{rq}} \\
&+ |\eta_1(\varphi(x), m\varphi(b))|^{\alpha+2} \Big[m(f''(b))^{rq} F^r((1-\xi)^{-s}; \lambda, \alpha, p_1, r) \\
&+ \eta_2\left((f''(x))^{rq}, (f''(b))^{rq}\right) F^r(\xi^{-s}; \lambda, \alpha, p_2, r) \Big]^{\frac{1}{rq}} \Bigg\}.
\end{aligned} \tag{49}$$

Corollary 29 *In Theorem 7 for $h_1(t) = h_2(t) = t(1-t)$ and $\mathbf{m}(t) = m \in (0, 1]$ for all $t \in [0, 1]$, we get the following inequality for generalized-m-$((t(1-t))^{sp_1}, (t(1-t))^{sp_2}); (\eta_1, \eta_2))$-convex mappings:*

$$\begin{aligned}
\left|\Delta_f^{\alpha}(\eta_1, \varphi, m; \lambda, x, a, b)\right| &\leq \frac{C^{1-\frac{1}{q}}(\alpha, \lambda)}{\eta_1(\varphi(b), m\varphi(a))} \\
&\times \Bigg\{ |\eta_1(\varphi(x), m\varphi(a))|^{\alpha+2} \Big[m(f''(a))^{rq} F^r(\xi(1-\xi); \lambda, \alpha, p_1, r)
\end{aligned}$$

$$+\eta_2\left((f''(x))^{rq},(f''(a))^{rq}\right)F^r(\xi(1-\xi);\lambda,\alpha,p_2,r)\Big]^{\frac{1}{rq}}$$

$$+|\eta_1(\varphi(x),m\varphi(b))|^{\alpha+2}\Big[m(f''(b))^{rq}F^r(\xi(1-\xi);\lambda,\alpha,p_1,r) \tag{50}$$

$$+\eta_2\left((f''(x))^{rq},(f''(b))^{rq}\right)F^r(\xi(1-\xi);\lambda,\alpha,p_2,r)\Big]^{\frac{1}{rq}}\Bigg\}.$$

Corollary 30 *In Corollary 26 for* $h_1(t)=\dfrac{\sqrt{1-t}}{2\sqrt{t}}$ *and* $h_2(t)=\dfrac{\sqrt{t}}{2\sqrt{1-t}}$, *we get the following inequality for generalized-m-*$\left(\left(\left(\frac{\sqrt{1-t}}{2\sqrt{t}}\right)^p,\left(\frac{\sqrt{t}}{2\sqrt{1-t}}\right)^q\right);(\eta_1,\eta_2)\right)$*-convex mappings:*

$$\left|\Delta_f^{\alpha}(\eta_1,\varphi,m;\lambda,x,a,b)\right|\le\frac{C^{1-\frac{1}{q}}(\alpha,\lambda)}{\eta_1(\varphi(b),m\varphi(a))}$$

$$\times\Bigg\{|\eta_1(\varphi(x),m\varphi(a))|^{\alpha+2}\Big[m(f''(a))^{rq}F^r\left(\left(\frac{\sqrt{1-\xi}}{2\sqrt{\xi}}\right);\lambda,\alpha,p_1,r\right)$$

$$+\eta_2\left((f''(x))^{rq},(f''(a))^{rq}\right)F^r\left(\left(\frac{\sqrt{\xi}}{2\sqrt{1-\xi}}\right);\lambda,\alpha,p_2,r\right)\Big]^{\frac{1}{rq}} \tag{51}$$

$$+|\eta_1(\varphi(x),m\varphi(b))|^{\alpha+2}\Big[m(f''(b))^{rq}F^r\left(\left(\frac{\sqrt{1-\xi}}{2\sqrt{\xi}}\right);\lambda,\alpha,p_1,r\right)$$

$$+\eta_2\left((f''(x))^{rq},(f''(b))^{rq}\right)F^r\left(\left(\frac{\sqrt{\xi}}{2\sqrt{1-\xi}}\right);\lambda,\alpha,p_2,r\right)\Big]^{\frac{1}{rq}}\Bigg\}.$$

Remark 6 For $\alpha=1$, by our Theorems 6 and 7, we can get some new special Hermite-Hadamard, Ostrowski, and Simpson type inequalities for classical integrals associated with generalized-**m**-$((h_1^{p_1},h_2^{p_2});(\eta_1,\eta_2))$-convex mappings.

Remark 7 Also, applying our Theorems 6 and 7, for different values of $\lambda\in(0,1)$, for different values of $p_1,p_2>-1$, for different choices of function $\mathbf{m}(t)$ and if $0<f''(x)\le L$ for all $x\in I$, we can get some new special Hermite-Hadamard, Ostrowski, and Simpson type inequalities for fractional integrals associated with generalized-**m**-$((h_1^{p},h_2^{p_2});(\eta_1,\eta_2))$-convex mappings.

3 Applications to Special Means

Definition 14 A function $M:\mathbb{R}_+^2\longrightarrow\mathbb{R}_+$ is called a Mean function if it has the following properties:

1. Homogeneity: $M(ax, ay) = aM(x, y)$, for all $a > 0$,
2. Symmetry: $M(x, y) = M(y, x)$,
3. Reflexivity: $M(x, x) = x$,
4. Monotonicity: If $x \le x'$ and $y \le y'$, then $M(x, y) \le M(x', y')$,
5. Internality: $\min\{x, y\} \le M(x, y) \le \max\{x, y\}$.

We consider some means for different positive real numbers α, β.

1. The arithmetic mean:

$$A := A(\alpha, \beta) = \frac{\alpha + \beta}{2}.$$

2. The geometric mean:

$$G := G(\alpha, \beta) = \sqrt{\alpha\beta}.$$

3. The harmonic mean:

$$H := H(\alpha, \beta) = \frac{2}{\frac{1}{\alpha} + \frac{1}{\beta}}.$$

4. The power mean:

$$P_r := P_r(\alpha, \beta) = \left(\frac{\alpha^r + \beta^r}{2}\right)^{\frac{1}{r}}, \quad r \ge 1.$$

5. The identric mean:

$$I := I(\alpha, \beta) = \begin{cases} \frac{1}{e}\left(\frac{\beta^\beta}{\alpha^\alpha}\right), & \alpha \neq \beta; \\ \alpha, & \alpha = \beta. \end{cases}$$

6. The logarithmic mean:

$$L := L(\alpha, \beta) = \frac{\beta - \alpha}{\ln\beta - \ln\alpha}.$$

7. The generalized log-mean:

$$L_p := L_p(\alpha, \beta) = \left[\frac{\beta^{p+1} - \alpha^{p+1}}{(p+1)(\beta - \alpha)}\right]^{\frac{1}{p}}; \quad p \in \mathbb{R} \setminus \{-1, 0\}.$$

It is well known that L_p is monotonic nondecreasing over $p \in \mathbb{R}$ with $L_{-1} := L$ and $L_0 := I$. In particular, we have the following inequality $H \le G \le L \le I \le A$. Now, let a and b be positive real numbers such that $a < b$. Let us consider continuous functions $\varphi : I \longrightarrow K$, $\eta_1 : K \times K \longrightarrow \mathbb{R}$, $\eta_2 : f(K) \times f(K) \longrightarrow \mathbb{R}$ and $\overline{M} := M(\varphi(a), \varphi(b)) : [\varphi(a), \varphi(a) + \eta_1(\varphi(b), \varphi(a))] \times [\varphi(a), \varphi(a) + \eta_1(\varphi(b), \varphi(a))] \longrightarrow \mathbb{R}_+$, which is one of the above-mentioned means. Therefore one can obtain various inequalities using the results of Section 2 for these means as follows. If we take $\mathbf{m}(t) \equiv 1$, $\forall t \in [0, 1]$ and replace $\eta_1(\varphi(x), \mathbf{m}(t)\varphi(y)) = M(\varphi(x), \varphi(y))$ for all $x, y \in I$ for value $m = 1$, in (32) and (42), one can obtain the following interesting inequalities involving means:

$$\begin{aligned} \left|\Delta_f^{\alpha}(\overline{M}, \varphi; \lambda, x, a, b)\right| &\leq \frac{\sqrt[p]{D(\alpha, \lambda, p)}}{\overline{M}} \\ &\times \Big\{ M^{\alpha+2}(\varphi(x), \varphi(a)) \\ &\times \sqrt[rq]{(f''(a))^{rq} I^r(h_1(\xi); p_1, r) + \eta_2\left((f''(x))^{rq}, (f''(a))^{rq}\right) I^r(h_2(\xi); p_2, r)} \\ &+ M^{\alpha+2}(\varphi(x), \varphi(b)) \\ &\times \sqrt[rq]{(f''(b))^{rq} I^r(h_1(\xi); p_1, r) + \eta_2\left((f''(x))^{rq}, (f''(b))^{rq}\right) I^r(h_2(\xi); p_2, r)} \Big\}, \end{aligned} \tag{52}$$

$$\begin{aligned} \left|\Delta_f^{\alpha}(\overline{M}, \varphi; \lambda, x, a, b)\right| &\leq \frac{C^{1-\frac{1}{q}}(\alpha, \lambda)}{\overline{M}} \\ &\times \Big\{ M^{\alpha+2}(\varphi(x), \varphi(a)) \Big[(f''(a))^{rq} F^r(h_1(\xi); \lambda, \alpha, p_1, r) \\ &+ \eta_2\left((f''(x))^{rq}, (f''(a))^{rq}\right) F^r(h_2(\xi); \lambda, \alpha, p_2, r)\Big]^{\frac{1}{rq}} \\ &+ M^{\alpha+2}(\varphi(x), \varphi(b)) \Big[(f''(b))^{rq} F^r(h_1(\xi); \lambda, \alpha, p_1, r) \\ &+ \eta_2\left((f''(x))^{rq}, (f''(b))^{rq}\right) F^r(h_2(\xi); \lambda, \alpha, p_2, r)\Big]^{\frac{1}{rq}} \Big\}. \end{aligned} \tag{53}$$

Letting $\overline{M} := A, G, H, P_r, I, L, L_p$ in (52) and (53), we get the inequalities involving means for a particular choices of f''^q that are generalized-1-$((h_1^{p_1}, h_2^{p_2}); (\eta_1, \eta_2))$-convex mappings.

Remark 8 Also, applying our Theorems 6 and 7 for appropriate choices of functions h_1 and h_2 (see Remark 4) such that f''^q to be generalized-1-$((h_1^{p_1}, h_2^{p_2}); (\eta_1, \eta_2))$-convex mappings (for example $f(x) = x^{\alpha}$, where $\alpha > 1, \forall x > 0$; $f(x) = \frac{1}{x}$, $\forall x > 0$ etc.), we can deduce some new inequalities using above special means. The details are left to the interested reader.

4 Conclusion

In this article, we first presented some integral inequalities for Gauss-Jacobi type quadrature formula involving generalized-$\mathbf{m}$-$((h_1^p, h_2^q); (\eta_1, \eta_2))$-convex mappings. Secondly, an identity pertaining twice differentiable mappings defined on $\mathbf{m}$-invex set is used for derived some new estimates with respect to Hermite-Hadamard, Ostrowski, and Simpson type inequalities via fractional integrals associated with generalized-$\mathbf{m}$-$((h_1^{p_1}, h_2^{p_2}); (\eta_1, \eta_2))$-convex mappings. Also, some new special cases are given. At the end, some applications to special means for different positive real numbers are provided as well. Motivated by this interesting class we can indeed see to be vital for fellow researchers and scientists working in the same domain. We conclude that our methods considered here may be a stimulant for further investigations concerning Hermite-Hadamard, Ostrowski, and Simpson type integral inequalities for various kinds of convex and preinvex functions involving local fractional integrals, fractional integral operators, Caputo k-fractional derivatives, q-calculus, (p, q)-calculus, time scale calculus, and conformable fractional integrals.

References

1. S.M. Aslani, M.R. Delavar, S.M. Vaezpour, Inequalities of Fejér type related to generalized convex functions with applications. Int. J. Anal. Appl. **16**(1), 38–49 (2018)
2. F. Chen, A note on Hermite-Hadamard inequalities for products of convex functions via Riemann-Liouville fractional integrals. Ital. J. Pure Appl. Math. **33**, 299–306 (2014)
3. Y.-M. Chu, M.A. Khan, T. Ali, S.S. Dragomir, Inequalities for α-fractional differentiable functions. J. Inequal. Appl. **2017**(93), 12 (2017)
4. Y.-M. Chu, G.D. Wang, X.H. Zhang, Schur convexity and Hadamard's inequality. Math. Inequal. Appl. **13**(4), 725–731 (2010)
5. Y.-M. Chu, M.A. Khan, T.U. Khan, T. Ali, Generalizations of Hermite-Hadamard type inequalities for MT-convex functions. J. Nonlinear Sci. Appl. **9**(5), 4305–4316 (2016)
6. Z. Dahmani, On Minkowski and Hermite-Hadamard integral inequalities via fractional integration. Ann. Funct. Anal. **1**(1), 51–58 (2010)
7. M.R. Delavar, S.S. Dragomir, On η-convexity. Math. Inequal. Appl. **20**, 203–216 (2017)
8. M.R. Delavar, M. De La Sen, Some generalizations of Hermite-Hadamard type inequalities. SpringerPlus **5**, 1661 (2016)
9. S.S. Dragomir, J. Pečarić, L.E. Persson, Some inequalities of Hadamard type. Soochow J. Math. **21**, 335–341 (1995)
10. T.S. Du, J.G. Liao, Y.J. Li, Properties and integral inequalities of Hadamard-Simpson type for the generalized (s, m)-preinvex functions. J. Nonlinear Sci. Appl. **9**, 3112–3126 (2016)
11. G. Farid, A.U. Rehman, Generalizations of some integral inequalities for fractional integrals. Ann. Math. Sil. **31**, 14 (2017)
12. G. Farid, A. Javed, A.U. Rehman, On Hadamard inequalities for n-times differentiable functions which are relative convex via Caputo k-fractional derivatives. Nonlinear Anal. Forum **22**(2), 17–28 (2017)
13. M.E. Gordji, S.S. Dragomir, M.R. Delavar, An inequality related to η-convex functions (II). Int. J. Nonlinear Anal. Appl. **6**(2), 26–32 (2016)
14. M.E. Gordji, M.R. Delavar, M. De La Sen, Onφ-convex functions. J. Math. Inequal. Wiss **10**(1), 173–183 (2016)

15. A. Iqbal, M.A. Khan, S. Ullah, Y.-M. Chu, A. Kashuri, Hermite-Hadamard type inequalities pertaining conformable fractional integrals and their applications. AIP Adv. **8**(7), 18 (2018)
16. A. Kashuri, R. Liko, Some different type integral inequalities pertaining generalized relative semi-**m**-$(r; h_1, h_2)$-preinvex mappings and their applications. Electron. J. Math. Analysis Appl. **7**(1), 351–373 (2019)
17. A. Kashuri, R. Liko, On Hermite-Hadamard type inequalities for generalized (s, m, φ)-preinvex functions via k-fractional integrals. Adv. Inequal. Appl. **6**, 1–12 (2017)
18. A. Kashuri, R. Liko, Generalizations of Hermite-Hadamard and Ostrowski type inequalities for MT_m-preinvex functions. Proyecciones **36**(1), 45–80 (2017)
19. A. Kashuri, R. Liko, Hermite-Hadamard type fractional integral inequalities for generalized $(r; s, m, \varphi)$-preinvex functions. Eur. J. Pure Appl. Math. **10**(3), 495–505 (2017)
20. A. Kashuri, R. Liko, Hermite-Hadamard type fractional integral inequalities for twice differentiable generalized (s, m, φ)-preinvex functions. Konuralp J. Math. **5**(2), 228–238 (2017)
21. A. Kashuri, R. Liko, Some new Ostrowski type fractional integral inequalities for generalized $(r; g, s, m, \varphi)$-preinvex functions via Caputo k-fractional derivatives. Int. J. Nonlinear Anal. Appl. **8**(2), 109–124 (2017)
22. A. Kashuri, R. Liko, Hermite-Hadamard type inequalities for generalized (s, m, φ)-preinvex functions via k-fractional integrals. Tbil. Math. J. **10**(4), 73–82 (2017)
23. A. Kashuri, R. Liko, Hermite-Hadamard type fractional integral inequalities for $MT_{(m,\varphi)}$-preinvex functions. Stud. Univ. Babeş-Bolyai Math. **62**(4), 439–450 (2017)
24. A. Kashuri, R. Liko, Hermite-Hadamard type fractional integral inequalities for twice differentiable generalized beta-preinvex functions. J. Fract. Calc. Appl. **9**(1), 241–252 (2018)
25. M.A. Khan, Y.-M. Chu, A. Kashuri, R. Liko, Hermite-Hadamard type fractional integral inequalities for $MT_{(r;g,m,\varphi)}$-preinvex functions. J. Comput. Anal. Appl. **26**(8), 1487–1503 (2019)
26. M.A. Khan, T. Ali, S.S. Dragomir, M.Z. Sarikaya, Hermite-Hadamard type inequalities for conformable fractional integrals. Rev. Real Acad. Cienc. Exact. Fís. Natur. Ser. A. Math. (2017) https://doi.org/10.1007/s13398-017-0408-5
27. M.A. Khan, Y.-M. Chu, A. Kashuri, R. Liko, G. Ali, New Hermite-Hadamard inequalities for conformable fractional integrals. J. Funct. Spaces **2018**, 9. Article ID 6928130
28. M.A. Khan, Y.-M. Chu, T.U. Khan, J. Khan, Some new inequalities of Hermite-Hadamard type for s-convex functions with applications. Open Math. **15**, 1414–1430 (2017)
29. M.A. Khan, Y. Khurshid, T. Ali, N. Rehman, Inequalities for three times differentiable functions. J. Math. Punjab Univ. **48**(2), 35–48 (2016)
30. M.A. Khan, Y. Khurshid, T. Ali, Hermite-Hadamard inequality for fractional integrals via η-convex functions. Acta Math. Univ. Comenian. **79**(1), 153–164 (2017)
31. W. Liu, New integral inequalities involving η-function via P-convexity. Miskolc Math. Notes **15**(2), 585–591 (2014)
32. W.J. Liu, Some Simpson type inequalities for h-convex and (α, m)-convex functions. J. Comput. Anal. Appl. **16**(5), 1005–1012 (2014)
33. W. Liu, W. Wen, J. Park, Ostrowski type fractional integral inequalities for MT-convex functions. Miskolc Math. Notes **16**(1), 249–256 (2015)
34. W. Liu, W. Wen, J. Park, Hermite-Hadamard type inequalities for MT-convex functions via classical integrals and fractional integrals. J. Nonlinear Sci. Appl. **9**, 766–777 (2016)
35. C. Luo, T.S. Du, M.A. Khan, A. Kashuri, Y. Shen, Some k-fractional integrals inequalities through generalized $\lambda_{\phi m}$-MT-preinvexity. J. Comput. Anal. Appl. **27**(4), 690–705 (2019)
36. S. Mubeen, G.M. Habibullah, k-Fractional integrals and applications. Int. J. Contemp. Math. Sci. **7**, 89–94 (2012)
37. O. Omotoyinbo, A. Mogbodemu, Some new Hermite-Hadamard integral inequalities for convex functions. Int. J. Sci. Innov. Tech. **1**(1), 1–12 (2014)
38. C. Peng, C. Zhou, T.S. Du, Riemann-Liouville fractional Simpson's inequalities through generalized (m, h_1, h_2)-preinvexity. Ital. J. Pure Appl. Math., **38**, 345–367 (2017)
39. E. Set, Some new generalized Hermite-Hadamard type inequalities for twice differentiable functions (2018) https://www.researchgate.net/publication/327601181

40. D.D. Stancu, G. Coman, P. Blaga, Analiză numerică şi teoria aproximării. Cluj-Napoca: Presa Universitară Clujeană **2** (2002)
41. E. Set, A. Gözpinar, J. Choi, Hermite-Hadamard type inequalities for twice differentiable m-convex functions via conformable fractional integrals. Far East J. Math. Sci. **101**(4), 873–891 (2017)
42. E. Set, S.S. Karataş, M.A. Khan, Hermite-Hadamard type inequalities obtained via fractional integral for differentiable m-convex and (α, m)-convex functions. Int. J. Anal. **2016**, 8. Article ID 4765691
43. E. Set, M.Z. Sarikaya, A. Gözpinar, Some Hermite-Hadamard type inequalities for convex functions via conformable fractional integrals and related inequalities. Creat. Math. Inform. **26**(2), 221–229 (2017)
44. H.N. Shi, Two Schur-convex functions related to Hadamard-type integral inequalities. Publ. Math. Debrecen **78**(2), 393–403 (2011)
45. H. Wang, T.S. Du, Y. Zhang, k-fractional integral trapezium-like inequalities through (h, m)-convex and (α, m)-convex mappings. J. Inequal. Appl. **2017**(311), 20 (2017)
46. X.M. Zhang, Y.-M. Chu, X.H. Zhang, The Hermite-Hadamard type inequality of GA-convex functions and its applications. J. Inequal. Appl. (2010), 11. Article ID 507560
47. Y. Zhang, T.S. Du, H. Wang, Y.J. Shen, A. Kashuri, Extensions of different type parameterized inequalities for generalized (m, h)-preinvex mappings via k-fractional integrals. J. Inequal. Appl. **2018**(49), 30 (2018)
48. M.E. Özdemir, E. Set, M. Alomari, Integral inequalities via several kinds of convexity. Creat. Math. Inform. **20**(1), 62–73 (2011)
49. T. Weir, B. Mond, Preinvex functions in multiple objective optimization. J. Math. Anal. Appl. **136**, 29–38 (1988)
50. M. Matłoka, Inequalities for h-preinvex functions. Appl. Math. Comput. **234**, 52–57 (2014)
51. R. Pini, Invexity and generalized convexity. Optimization **22**, 513–525 (1991)
52. S. Varošanec, On h-convexity. J. Math. Anal. Appl. **326**(1), 303–311 (2007)
53. Y. Wang, S.H. Wang, F. Qi, Simpson type integral inequalities in which the power of the absolute value of the first derivative of the integrand is s-preinvex. Facta Univ. Ser. Math. Inform. **28**(2), 151–159 (2013)
54. M.A. Noor, K.I. Noor, M.U. Awan, S. Khan, Hermite-Hadamard inequalities for s-Godunova-Levin preinvex functions. J. Adv. Math. Stud. **7**(2), 12–19 (2014)
55. M. Tunç, E. Göv, Ü. Şanal, On tgs-convex function and their inequalities. Facta Univ. Ser. Math. Inform. **30**(5), 679–691 (2015)

Extensions of Kannappan's and Van Vleck's Functional Equations on Semigroups

Keltouma Belfakih, Elhoucien Elqorachi, and Ahmed Redouani

Abstract This paper treats two functional equations, the Kannappan-Van Vleck functional equation

$$\mu(y)f(x\tau(y)z_0) \pm f(xyz_0) = 2f(x)f(y), \ \ x, y \in S$$

and the following variant of it

$$\mu(y)f(\tau(y)xz_0) \pm f(xyz_0) = 2f(x)f(y), \ \ x, y \in S,$$

in the setting of semigroups S that need not be abelian or unital, τ is an involutive morphism of S, $\mu : S \longrightarrow C$ is a multiplicative function such that $\mu(x\tau(x)) = 1$ for all $x \in S$ and z_0 is a fixed element in the center of S.

We find the complex-valued solutions of these equations in terms of multiplicative functions and solutions of d'Alembert's functional equation.

1 Introduction

Van Vleck [1, 2] studied the continuous solutions $f : R \longrightarrow R$, $f \neq 0$, of the following functional equation

$$f(x - y + z_0) - f(x + y + z_0) = 2f(x)f(y), \ \ x, y \in R, \tag{1}$$

where $z_0 > 0$ is fixed. He showed that any continuous solution of (1) with minimal period $4z_0$ is $f(x) = \cos(\frac{\pi}{2z_0}(x - z_0))$, $x \in R$.

K. Belfakih · E. Elqorachi (✉) · A. Redouani
Department of Mathematics, Faculty of Sciences, University Ibn Zohr, Agadir, Morocco

T. M. Rassias, P. M. Pardalos (eds.), *Mathematical Analysis and Applications*, Springer Optimization and Its Applications 154,
https://doi.org/10.1007/978-3-030-31339-5_11

Stetkær [3, Exercise 9.18] found the complex-valued solutions of equation

$$f(xy^{-1}z_0) - f(xyz_0) = 2f(x)f(y),\ x, y \in G, \tag{2}$$

on groups that need not be abelian and z_0 is a fixed element in the center of G.

Perkins and Sahoo [4] replaced the group inversion by an involution $\tau\colon G \longrightarrow G$ and obtained the abelian, complex-valued solutions of the equation

$$f(x\tau(y)z_0) - f(xyz_0) = 2f(x)f(y),\ x, y \in G, \tag{3}$$

by means of d'Alembert's functional equation

$$g(xy) + g(x\tau(y)) = 2g(x)g(y),\ x, y \in G. \tag{4}$$

Stetkær [5] extended the results of [4] about equation (3) to semigroups and derived an explicit formula for the solutions in terms of multiplicative maps. In particular, Stetkær proved that all solutions of (3) are abelian. So, the restriction to abelian solutions in [4] is not needed.

D'Alembert's classic functional equation

$$g(x + y) + g(x - y) = 2g(x)g(y),\ x, y \in R \tag{5}$$

has solutions $g\colon R \longrightarrow C$ that are periodic, for instance $g(x) = \cos(x)$, and solutions that are not, for instance $g(x) = \cosh(x)$.

Kannappan [6] proved that any solution of the extension of (5)

$$f(x - y + z_0) + f(x + y + z_0) = 2f(x)f(y),\ x, y \in R, \tag{6}$$

where $z_0 \neq 0$ is a real constant has the form $f(x) = g(x - z_0)$, where $g\colon R \longrightarrow C$ is a periodic solution of (5) with period $2z_0$.

Perkins and Sahoo [4] considered the following version of Kannappan's functional equation

$$f(xyz_0) + f(xy^{-1}z_0) = 2f(x)f(y),\ x, y \in G \tag{7}$$

on groups and they found the form of any abelian solution f of (7).

Stetkær [7] took z_0 in the center and expressed the complex-valued solutions of Kannappan's functional equation

$$f(xyz_0) + f(x\tau(y)z_0) = 2f(x)f(y),\ x, y \in S \tag{8}$$

on semigroups with involution τ in terms of solutions of d'Alembert's functional equation (4).

In the very special case of z_0 being the neutral element of a monoid S equation (8) becomes (4) which has been solved by Davison [8].

Here we shall consider the following functional equations

$$f(xyz_0) + \mu(y)f(x\tau(y)z_0) = 2f(x)f(y),\ x, y \in S, \tag{9}$$

$$f(xyz_0) + \mu(y)f(\tau(y)xz_0) = 2f(x)f(y),\ x, y \in S, \tag{10}$$

$$\mu(y)f(x\tau(y)z_0) - f(xyz_0) = 2f(x)f(y),\ x, y \in S \tag{11}$$

and

$$\mu(y)f(\tau(y)xz_0) - f(xyz_0) = 2f(x)f(y),\ x, y \in S, \tag{12}$$

where S is a semigroup, τ is an involutive morphism of S. That is, τ is an involutive automorphism: $\tau(xy) = \tau(x)\tau(y)$ and $\tau(\tau(x)) = x$ for all $x, y \in S$ or τ is an involutive anti-automorphism: $\tau(xy) = \tau(y)\tau(x)$ and $\tau(\tau(x)) = x$ for all $x, y \in S$. The map $\mu : S \longrightarrow C$ is a multiplicative function such that $\mu(x\tau(x)) = 1$ for all $x \in S$ and z_0 is a fixed element in the center of S. By algebraic methods:

(1) We find all solutions of (11) and (12). Only multiplicative functions occur in the solution formulas.
(2) We find the solutions of (10) for the particular case of τ being an involutive automorphism and
(3) We express the solutions of (9) and (10) in terms of solutions of d'Alembert's μ-functional equation

$$g(xy) + \mu(y)g(x\tau(y)) = 2g(x)g(y),\ x, y \in S. \tag{13}$$

Of course we are not the first to consider trigonometric functional equations having a multiplicative function μ in front of terms like $f(x\tau(y))$ or $f(\tau(y)x)$. The μ-d'Alembert's functional equation (13) which is an extension of d'Alembert's functional equation (4) has been treated systematically by Stetkær [3, 9] on groups with involution. The non-zero solutions of (13) on groups with involution are the normalized traces of certain representation of S on C^2.

Stetkær [10] obtained the complex-valued solution of the following variant of d'ĄAlembert's functional equation

$$f(xy) + f(\tau(y)x)) = 2f(x)f(y),\ x, y \in S, \tag{14}$$

where τ is an involutive automorphism of S.

Elqorachi and Redouani [11] proved that the solutions of the variant of d'Alembert's functional equation

$$f(xy) + \mu(y)f(\tau(y)x)) = 2f(x)f(y),\ x, y \in S \tag{15}$$

are of the form $f(x) = \frac{\chi(x)+\mu(x)\chi(\tau(x))}{2}$, $x \in S$, where τ is an involutive automorphism of S and $\chi\colon S \longrightarrow S$ is a multiplicative function.

Bouikhalene and Elqorachi [12] obtained the solutions of (11) for involutive anti-automorphism τ. In the same paper they also found the solutions of (11) for involutive automorphism τ, but on monoids only.

Throughout this paper S denotes a semigroup with an involutive morphism τ: $S \longrightarrow S$, μ: $S \longrightarrow C$ denotes a multiplicative function such that $\mu(x\tau(x)) = 1$ for all $x \in S$ and z_0 a fixed element in the center of S.

In all proofs of the results of this paper we use without explicit mentioning the assumption that z_0 is contained in the center of S and its consequence $\tau(z_0)$ is contained in the center of S.

2 Solutions of Equation (9) on Semigroups

In this section we express the solutions of (9) in terms of solutions of d'Alembert's functional equation (13). The following lemma will be used later.

Lemma 1 *If f: $S \longrightarrow C$ is a solution of (9), then for all $x \in S$*

$$f(x) = \mu(x)f(\tau(x)), \tag{16}$$

$$f(x\tau(z_0)z_0) = \mu(\tau(z_0))f(z_0)f(x), \tag{17}$$

$$f(xz_0^2) = f(x)f(z_0), \tag{18}$$

$$f(z_0) \neq 0 \Longleftrightarrow f \neq 0. \tag{19}$$

Proof Equation (16): By replacing y by $\tau(y)$ in (9) and multiplying the result obtained by $\mu(y)$ and using $\mu(y\tau(y)) = 1$ we get by computation that

$$\mu(y)2f(x)f(\tau(y)) = \mu(y)f(x\tau(y)z_0) + \mu(y\tau(y))f(xyz_0)$$

$$= \mu(y)f(x\tau(y)z_0) + f(xyz_0) = 2f(x)f(y),$$

which implies (16).

Equation (17): Replacing x by $\tau(z_0)$ in (9) and using (16) two times we get by a computation that

$$f(\tau(z_0)yz_0) + \mu(y)f(\tau(z_0)\tau(y)z_0) = 2f(\tau(z_0))f(y) = 2\mu(\tau(z_0))f(z_0)f(y)$$

and

$$f(\tau(z_0)yz_0) + \mu(y)f(\tau(z_0)\tau(y)z_0) = 2f(\tau(z_0))f(y)$$

$$= f(\tau(z_0)yz_0) + \mu(y)\mu(\tau(z_0)\tau(y)z_0)f(\tau(z_0)yz_0) = 2f(\tau(z_0)yz_0).$$

This proves (17).

Equation (18): Putting $y = z_0$ in (9) and using (17) we obtain (18).

Equation (19): Assume that $f(z_0) = 0$. By replacing x by xz_0 and y by yz_0 in (9) and using (17) and (18) we get by a computation that

$$2f(xz_0)f(yz_0) = f(xz_0yz_0^2) + \mu(yz_0)f(xz_0\tau(y)\tau(z_0)z_0)$$

$$= f(z_0)f(xyz_0) + \mu(y)f(z_0)f(x\tau(y)z_0) = 0 \text{ for all } x, y \in S,$$

which implies that $f(xz_0) = 0$ for all $x \in S$. So, from equation (9) we get $2f(x)f(y) = 0$ for all $x, y \in S$, and then $f(x) = 0$ for all $x \in S$. Conversely, it's clear that $f(x) = 0$ for all $x \in S$ implies that $f(z_0) = 0$.

For the rest of this section we use the following notations [7].

- $\mathscr{A}$ consists of the solutions of $g : S \longrightarrow C$ of d'Alembert's functional equation (13) with $g(z_0) \neq 0$ and satisfying the condition

$$g(xz_0) = g(z_0)g(x) \text{ for all } x \in S. \tag{20}$$

- To any $g \in \mathscr{A}$ we associate the function $Tg = g(z_0)g : S \longrightarrow C$.
- $\mathscr{K}$ consists of the non-zero solutions $f : S \longrightarrow C$ of Kannappan's functional equation (9).

In the following main result of the present section, the complex solutions of equation (9) are expressed by means of solutions of d'Alembert's functional equation (13).

Theorem 1

(1) T is a bijection of $\mathscr{A}$ onto $\mathscr{K}$. The inverse $T^{-1}: \mathscr{K} \longrightarrow \mathscr{A}$ is given by the formula

$$(T^{-1}f)(x) = \frac{f(xz_0)}{f(z_0)}$$

for all $f \in \mathscr{K}$ and $x \in S$.

(2) Any non-zero solution $f: S \longrightarrow C$ of the Kannappan's functional equation (9) is of the form $f = T(g) = g(z_0)g$, where $g \in \mathscr{A}$. Furthermore,

$$f(x) = g(xz_0) = \mu(z_0)g(x\tau(z_0)) = g(z_0)g(x)$$

for all $x \in S$.

(3) f is central, i.e. $f(xy) = f(yx)$ for all $x, y \in S$ if and only if g is central.

(4) f is abelian [3, Definition B.3] if and only if g is abelian.

(5) If S is equipped with a topology, then f is continuous if and only if g is continuous.

Proof For any $g \in \mathscr{A}$ and for all $x, y \in S$ we have

$$T(g)(xyz_0) + \mu(y)T(g)(x\tau(y)z_0) = g(z_0)[g(xyz_0) + \mu(y)g(x\tau(y)z_0)]$$

$$= g(z_0)^2[g(xy) + \mu(y)g(x\tau(y))] = 2g(z_0)g(x)g(z_0)g(y) = 2T(g)(x)T(g)(y).$$

On the other hand, $T(g)(z_0) = g(z_0)^2 \neq 0$, so we get $T(\mathscr{A}) \subseteq \mathscr{K}$.

By adapting the proof of [7, Lemma 3] T is injective. Now, we will show that T is surjective. Let $f \in \mathscr{K}$. Then from (19) we have $f(z_0) \neq 0$ and we can define the function $g(x) = \frac{f(xz_0)}{f(z_0)}$. In the following we will show that $g \in \mathscr{A}$ and $T(g) = f$. By using the definition of g and (17)–(18) we have

$$f(z_0)^2[g(xy) + \mu(y)g(x\tau(y))] = f(z_0)f(xyz_0) + \mu(y)f(z_0)f(x\tau(y)z_0)$$

$$= f(xyz_0^3) + \mu(y)\mu(z_0)f(x\tau(y)z_0{}^2\tau(z_0))$$

$$= f(xz_0yz_0z_0) + \mu(yz_0)f(xz_0\tau(yz_0)z_0) = 2f(xz_0)f(yz_0) = 2f(z_0)^2g(x)g(y)$$

for all $x, y \in S$. This shows that g is a solution of d'Alembert's functional equation (13).

By replacing x by xz_0^2 and y by z_0 in (9) we get

$$f(xz_0^4) + \mu(z_0)f(xz_0^3\tau(z_0)) = 2f(xz_0^2)f(z_0). \tag{21}$$

By replacing x by xz_0 and y by z_0^2 in (9) we have

$$f(xz_0^4) + \mu(z_0^2)f(xz_0^2\tau(z_0^2)) = 2f(z_0^2)f(xz_0). \tag{22}$$

From (17) and (18) we have

$$f(xz_0^3\tau(z_0)) = \mu(\tau(z_0))f(x)(f(z_0))^2$$

and

$$f(xz_0^2\tau(z_0^2)) = (\mu(\tau(z_0)))^2 f(x)(f(z_0))^2.$$

In view of (21) and (22) we deduce that $f(z_0^2)f(xz_0) = f(xz_0^2)f(z_0)$. So, by using the definition of g we obtain $g(xz_0) = g(x)g(z_0)$ for all $x \in S$. In particular, $g(z_0^2) = g(z_0)^2 = \frac{f(z_0^2z_0)}{f(z_0)} = \frac{f(z_0)f(z_0)}{f(z_0)} = f(z_0) \neq 0$. Furthermore, $T(g)(x) = g(z_0)g(x) = g(xz_0) = \frac{f(xz_0^2)}{f(z_0)} = \frac{f(x)f(z_0)}{f(z_0)} = f(x)$.

The statements (2)–(5) are obvious. This completes the proof.

Now, we extend Stetkær's result [7] from anti-automorphisms to the more general case of morphism as follows.

Corollary 1 *Let z_0 be a fixed element in the center of a semigroup S and let τ be an involutive morphism of S. Then, any non-zero solution $f\colon S \longrightarrow C$ of the functional equation (8) is of the form $f = g(z_0)g$, where g is a solution of d'Alembert's functional equation (4) with $g(z_0) \neq 0$ and satisfying the condition $g(xz_0) = g(z_0)g(x)$ for all $x \in S$.*

We will in the following propositions determine all abelian (resp. central) solutions f of Kannappan's functional equation (9).

Proposition 1 *Let z_0 be a fixed element in the center of a semigroup S. Let $\tau\colon S \longrightarrow S$ be an involutive anti-automorphism of S and let $\mu\colon S \longrightarrow C$ be a multiplicative function such that $\mu(x\tau(x)) = 1$ for all $x \in S$. The non-zero abelian solutions of Kannappan's functional equation (9) are the functions of the form*

$$f(x) = \frac{\chi(x) + \mu(x)\chi(\tau(x))}{2}\chi(z_0), \quad x \in S,$$

where $\chi : S \longrightarrow C$ is a multiplicative function such that $\chi(z_0) \neq 0$ and $\mu(z_0)\chi(\tau(z_0)) = \chi(z_0)$.

Proof Verifying that the function f defined in Proposition 1 is an abelian solution of (9) consists of simple computations that we omit.

Let $f\colon S \longrightarrow C$ be a non-zero solution of (9). From Theorem 1(2) and (4) the function f has the form $f = g(z_0)g$ where $g \in \mathscr{A}$ and g is abelian. From [3, Proposition 9.31] there exists a non-zero multiplicative function $\chi\colon S \longrightarrow C$ such that $g = \frac{\chi + \mu\chi\circ\tau}{2}$. Since $g \in \mathscr{A}$, it satisfies (20). If we replace x by z_0 in (20) we get $g(z_0^2) = g(z_0)^2$, which via computation gives that $\chi(z_0) = \mu(z_0)\chi(\tau(z_0))$. This implies that f has the desired form. This completes the proof.

By using [11, Lemma 3.2] and the proof of the preceding proposition we get

Proposition 2 *Let z_0 be a fixed element in the center of a semigroup S. Let $\tau\colon S \longrightarrow S$ be an involutive automorphism of S and let $\mu\colon S \longrightarrow C$ be a multiplicative function such that $\mu(x\tau(x)) = 1$ for all $x \in S$. The non-zero central solutions of the Kannappan's functional equation (9) are the functions of the form*

$$f(x) = \frac{\chi(x) + \mu(x)\chi(\tau(x))}{2}\chi(z_0), \quad x \in S,$$

where $\chi : S \longrightarrow C$ is a multiplicative function such that $\chi(z_0) \neq 0$ and $\mu(z_0)\chi(\tau(z_0)) = \chi(z_0)$.

3 Solutions of Equation (10) on Semigroups

In this section we determine the complex-valued solutions of (10) for any involutive morphism $\tau\colon S \longrightarrow S$. By help of Theorem 1 we express them in terms of solutions of d'Alembert's functional equation (13). We first prove the following two useful lemmas.

Lemma 2 *If $f: S \longrightarrow C$ is a solution of (10), then for all $x \in S$*

$$f(x) = \mu(x)f(\tau(x)), \tag{23}$$

$$f(x\tau(z_0)z_0) = \mu(\tau(z_0))f(z_0)f(x), \tag{24}$$

$$f(xz_0^2) = f(x)f(z_0), \tag{25}$$

$$f(z_0) \neq 0 \Longleftrightarrow f \neq 0. \tag{26}$$

Proof Equation (23): Interchanging x and y in (10) and multiplying the two members of the equation by $\mu(\tau(y))$ we get

$$\mu(x)\mu(\tau(y))f(\tau(x)yz_0) + \mu(\tau(y))f(yxz_0) = 2f(x)\mu(\tau(y))f(y), \ x, y \in S. \tag{27}$$

Replacing y by $\tau(y)$ in (10) we obtain

$$\mu(\tau(y))f(yxz_0) + f(x\tau(y)z_0) = 2f(x)f(\tau(y)), \ x, y \in S. \tag{28}$$

By subtracting (28) from (27) we get

$$\mu(x\tau(y))f(\tau(x)yz_0) - f(x\tau(y)z_0) = 2f(x)[\mu(\tau(y))f(y) - f(\tau(y)], \ x, y \in S. \tag{29}$$

By replacing x by $\tau(x)$ in (29) we have

$$\mu(\tau(x)\tau(y))f(xyz_0) - f(\tau(x)\tau(y)z_0) = 2f(\tau(x))[\mu(\tau(y))f(y) - f(\tau(y)], \ x, y \in S. \tag{30}$$

Replacing y by $\tau(y)$ in (29) and multiplying the two members of the equation by $\mu(\tau(y)\tau(x))$ we obtain

$$f(\tau(x)\tau(y)z_0) - \mu(\tau(x)\tau(y))f(xyz_0) = 2f(x)\mu(\tau(x))[f(\tau(y)) - \mu(\tau(y))f(y)], \ x, y \in S. \tag{31}$$

Now, by adding (30) and (31) we get $[f(\tau(x)) - \mu(\tau(x))f(x)][f(\tau(y)) - \mu(\tau(y))f(y)] = 0$ for all $x, y \in S$. This proves (23).

Equation (24): Taking $x = \tau(z_0)$ in (10) and using (23) we get

$$\mu(y)f(\tau(y)\tau(z_0)z_0) + f(\tau(z_0)yz_0) = 2\mu(\tau(z_0))f(z_0)f(y)$$

$$= f(\tau(z_0)yz_0) + \mu(y)\mu(\tau(y)\tau(z_0)z_0)f(\tau(z_0)yz_0) = 2f(\tau(z_0)yz_0),$$

which implies (23).

Equation (25): By replacing y by z_0 in (10) and using (24) we obtain

$$\mu(z_0)f(\tau(z_0)xz_0) + f(xz_0{}^2) = 2f(z_0)f(x)$$

$$= \mu(z_0)\mu(\tau(z_0))f(z_0)f(x) + f(xz_0{}^2).$$

So, we deduce (24).

Equation (25): The proof is similar to the proof of (19).

Lemma 3 *Let $\mathcal{M}$ consist of the solutions $g\colon S \longrightarrow C$ of the variant d'Alembert's functional equation (15) with $g(z_0) \neq 0$ and satisfying the condition (20). Let $\mathcal{N}$ consist of the non-zero solutions $f\colon S \longrightarrow C$ of the variant Kannappan's functional equation (10); Then*

(1) The map $J\colon \mathcal{M} \longrightarrow \mathcal{N}$ defined by $Jh := h(z_0)h\colon S \longrightarrow C$ is a bijection. The inverse J^{-1}; $\mathcal{N} \longrightarrow \mathcal{M}$ is given by the formula $(J^{-1}f)(x) = \frac{f(xz_0)}{f(z_0)} = g(x)$ for all $x \in S$ and for all $f \in \mathcal{N}$. Furthermore,
(2) If $\tau\colon S \longrightarrow S$ is an involutive automorphism, the function g has the form $g = \frac{\chi+\mu\chi\circ\tau}{2}$, where $\chi : S \longrightarrow C$, $\chi \neq 0$, is a multiplicative function.
(3) If $\tau\colon S \longrightarrow S$ is an involutive anti-automorphism, the function g satisfies the d'Alembert's functional equation (13).

Proof For all $h \in \mathcal{M}$ we have

$$Jh(xyz_0) + \mu(y)Jh(\tau(y)xz_0) = h(z_0)h(xyz_0) + \mu(y)h(z_0)h(\tau(y)xz_0)$$

$$= h(z_0)^2[h(xy) + \mu(y)h(\tau(y)x)] = 2h(z_0)h(x)h(z_0)h(y) = 2Jh(x)Jh(y).$$

Furthermore, $Jh(z_0) = h(z_0)^2 \neq 0$. So, $Jh \in \mathcal{N}$. By adapting the proof of [7, Lemma 3] J is injective. Now, let $f \in \mathcal{N}$ and let $g(x) := \frac{f(xz_0)}{f(z_0)}$ for $x \in S$. By using the definition of g, equations (10), (24), and (25) we get

$$f(z_0)^2[g(xy) + \mu(y)g(\tau(y)x) - 2g(x)g(y)]$$

$$= f(z_0)f(xyz_0) + \mu(y)f(z_0)f(\tau(y)xz_0) - 2f(xz_0)f(yz_0)$$

$$= f(xyz_0z_0^2) + \mu(y)\mu(z_0)f(\tau(y)xz_0\tau(z_0)z_0) - 2f(xz_0)f(yz_0)$$

$$f(xz_0yz_0z_0) + \mu(yz_0)f(\tau(yz_0)xz_0z_0) - 2f(xz_0)f(yz_0) = 0.$$

Since $f(z_0) \neq 0$ then g satisfies (15). By using similar computations as in the proof of Theorem 1 we get that $g(xz_0) = g(z_0)g(x)$ for all $x \in S$.

(2) If $\tau\colon S \longrightarrow S$ is an involutive automorphism then from [11, Lemma 3.2] g has the form $g = \frac{\chi+\mu\chi\circ\tau}{2}$, where $\chi : S \longrightarrow C$, $\chi \neq 0$, is a multiplicative function.

(3) If $\tau \colon S \longrightarrow S$ is an involutive anti-automorphism then by adapting the proof of [11, Theorem 2.1(1)(i)] for $\delta = 0$ we get that g satisfies the d'Alembert's functional equation (13).

Theorem 2

(1) Let $\tau \colon S \longrightarrow S$ be an involutive automorphism. The non-zero solutions $f : S \longrightarrow C$ of the functional equation (10) are the functions of the form

$$f = \frac{\chi + \mu\chi \circ \tau}{2}\chi(z_0), \tag{32}$$

where $\chi : S \longrightarrow C$ is a multiplicative function such that $\chi(z_0) \neq 0$ and $\mu(z_0)\chi(\tau(z_0)) = \chi(z_0)$.

(2) Let $\tau \colon S \longrightarrow S$ be an involutive anti-automorphism. The non-zero solutions $f : S \longrightarrow C$ of the functional equation (10) are the functions of the form $f = g(z_0)g$, where g is a solution of d'Alembert's functional equation (13) with $g(z_0) \neq 0$ and satisfying the condition $g(xz_0) = g(z_0)g(x)$ for all $x \in S$.

Proof Let $f \colon S \longrightarrow S$ be a non-zero solution of equation (10). From Theorem 1(2) $f = g(z_0)g(x) = g(xz_0)$, where g is a solution of d'Alembert's functional equation (4). We will discuss two possibilities.

(1) τ is an involutive automorphism of S. From Lemma 3, there exists $\chi : S \longrightarrow C$ a multiplicative function such that $g = \frac{\chi+\mu\chi\circ\tau}{2}$. So,

$$f = g(z_0) = \frac{\chi + \mu\chi \circ \tau}{2}g(z_0) = \frac{\chi(z_0) + \mu(z_0)\chi \circ \tau(z_0)}{2}\frac{\chi + \mu\chi \circ \tau}{2}. \tag{33}$$

By using $g(z_0^2) = g(z_0)^2$ we get after simple computation that $\chi(z_0) = \mu(z_0)\chi(\tau(z_0))$. This proves (1).

(2) τ is an involutive anti-automorphism of S. Combining Theorem 1 and Lemma 3(2) we find (2). This completes the proof.

4 Solutions of Equation (11)

The solutions of the functional equation (11) with τ an involutive anti-automorphism are explicitly obtained by Bouikhalene and Elqorachi [12] on semigroups not necessarily abelian in terms of multiplicative functions. In this section we obtain a similar formula for the solutions of the functional equation (11) when τ was an involutive automorphism. The following lemma is obtained in [12] for the case where τ is an involutive anti-automorphism. It still holds for the case where τ is an involutive automorphism.

Lemma 4 *Let $f \neq 0$ be a solution of (11). Then for all $x \in S$ we have*

$$f(x) = -\mu(x)f(\tau(x)), \tag{34}$$

$$f(z_0) \neq 0, \tag{35}$$

$$f(z_0^2) = 0, \tag{36}$$

$$f(x\tau(z_0)z_0) = \mu(\tau(z_0))f(x)f(z_0), \tag{37}$$

$$f(xz_0^2) = -f(z_0)f(x), \tag{38}$$

$$\mu(x)f(\tau(x)z_0) = f(xz_0). \tag{39}$$

The function $g(x) = \frac{f(xz_0)}{f(z_0)}$ is a non-zero solution of d'Alembert's functional equation (13).

Now, we are ready to prove the main result of this section.

In [12] we used [3, Proposition 8.14] to prove that the function g defined in Lemma 4 is an abelian solution of (13), where τ is an involutive anti-automorphism of S. This reasoning no longer works for the present situation. We will use another approach.

Theorem 3 *The non-zero solutions $f : S \longrightarrow C$ of the functional equation (11), where τ is an involutive morphism of S are the functions of the form*

$$f = \chi(z_0)\frac{\mu\chi \circ \tau - \chi}{2}, \tag{40}$$

where $\chi : S \longrightarrow C$ is a multiplicative function such that $\chi(z_0) \neq 0$ and $\mu(z_0)\chi(\tau(z_0)) = -\chi(z_0)$.

If S is a topological semigroup and that $\tau : S \longrightarrow S$, $\mu : S \longrightarrow C$ are continuous, then the non-zero solution f of equation (11) is continuous if and only if χ is continuous.

Proof Let f be a non-zero solution of (11). Replacing x by xz_0 in (11) and using (38) we get

$$-\mu(y)f(x\tau(y)) + f(xy) = 2f(y)g(x), \; x, y \in S, \tag{41}$$

where g is the function defined in Lemma 4.

If we replace y by yz_0 in (11) and use (37) and (38) we get

$$\mu(yz_0)\mu(\tau(z_0))f(x\tau(y)) + f(xy) = 2f(x)g(y) = \mu(y)f(x\tau(y)) + f(xy), \; x, y \in S. \tag{42}$$

By adding (41) and (42) we get that the pair f, g satisfies the sine addition law

$$f(xy) = f(x)g(y) + f(y)g(x) \text{ for all } x, y \in S.$$

Now, in view of [13, Lemma 3.4], [3, Theorem 4.1] g is abelian. Since g is a non-zero solution of d'Alembert's functional equation (13), then from [3, Proposition 9.31] there exists a non-zero multiplicative function $\chi\colon S \longrightarrow C$ such that $g = \frac{\chi+\mu\chi\circ\tau}{2}$. The rest of the proof is similar to the one used in [12].

5 Solutions of Equation (12)

The solutions of (12) were obtained in [12] on monoids for τ an involutive automorphism. In this section we determine the solutions of (12) for the general case where S is assumed to be a semigroup and τ an involutive morphism of S.

The following useful lemmas will be used later.

Lemma 5 *Let $f\colon S \longrightarrow C$ be a solution of equation (12). Then for all $x, y \in S$ we have*

$$f(x) = -\mu(x)f(\tau(x)), \tag{43}$$

$$f \neq 0 \Longleftrightarrow f(z_0) \neq 0, \tag{44}$$

$$\mu(y)f(\tau(y)x) = -\mu(x)f(\tau(x)y), \tag{45}$$

$$f(x\tau(z_0)z_0) = \mu(\tau(z_0))f(z_0)f(x), \tag{46}$$

$$f(xz_0^2) = -f(z_0)f(x), \tag{47}$$

$$\mu(x)f(\tau(x)z_0) = f(xz_0), \tag{48}$$

$$f(x\tau(z_0)) = \mu(x)f(\tau(x)\tau(z_0)), \tag{49}$$

$$f(z_0^2) = f(z_0\tau(z_0)) = 0. \tag{50}$$

Proof Equation (44): Let $f \neq 0$ be a non-zero solution of equation (12). We will derive (44) by contradiction. Assume that $f(z_0) = 0$. Putting $y = z_0$ in equation (12) we get

$$\mu(z_0)f(\tau(z_0)xz_0) - f(xz_0z_0) = 2f(x)f(z_0) = 0 \tag{51}$$

Replacing y by yz_0 in (12) and using (51) and (12) we get

$$\begin{aligned}\mu(yz_0)f(\tau(y)xz_0\tau(z_0)) - f(xyz_0z_0) &= 2f(x)f(yz_0)\\ &= \mu(y)f(\tau(y)xz_0z_0) - f(xz_0yz_0)\\ &= 2f(y)f(xz_0).\end{aligned}$$

So, we deduce that $f(y)f(xz_0) = f(x)f(yz_0)$ for all $x, y \in S$. Since $f \neq 0$, then there exists $\alpha \in C$ such that $f(xz_0) = \alpha f(x)$ for all $x \in S$. Furthermore, $\alpha \neq 0$, because if $\alpha = 0$ we get $f(xz_0) = 0$ for all $x \in S$ and equation (12) implies that $f = 0$. This contradicts the assumption that $f \neq 0$.

Now, by substituting $f(xz_0) = \alpha f(x)$ into (12) we get

$$\mu(y)f(\tau(y)x) - f(xy) = \frac{2}{\alpha}f(x)f(y) \text{ for all } x, y \in S. \tag{52}$$

Switching x and y in (52) we get

$$-f(yx) + \mu(x)f(\tau(x)y) = \frac{2}{\alpha}f(x)f(y), \; x, y \in S. \tag{53}$$

If we replace y by $\tau(y)$ in (52) and multiplying the result obtained by $\mu(y)$ we get

$$-\mu(y)f(x\tau(y)) + f(yx) = \frac{2}{\alpha}f(x)\mu(y)f(\tau(y)), \; x, y \in S. \tag{54}$$

By adding (54) and (53) we obtain

$$-\mu(y)f(x\tau(y)) + \mu(x)f(\tau(x)y) = \frac{2}{\alpha}f(x)[\mu(y)f(\tau(y)) + f(y)], \; x, y \in S. \tag{55}$$

By replacing x by $\tau(x)$ in (55) and multiplying the result obtained by $\mu(x)$ we get

$$f(xy) - \mu(xy)f(\tau(x)\tau(y)) = \frac{2}{\alpha}\mu(x)f(\tau(x))[\mu(y)f(\tau(y)) + f(y)]. \tag{56}$$

By replacing y by $\tau(y)$ in (55) and multiplying the result obtained by $\mu(y)$ we get

$$\mu(xy)f(\tau(x)\tau(y)) - f(xy) = \frac{2}{\alpha}f(x)[f(y) + \mu(y)f(\tau(y))]. \tag{57}$$

By adding (56) and (57) we obtain

$$[f(x) + \mu(x)f(\tau(x))][\mu(y)f(\tau(y)) + f(y)] = 0, \; x, y \in S. \tag{58}$$

So, $\mu(x)f(\tau(x)) = -f(x)$ for all $x \in S$. Now, we will discuss the following two cases.

(1) τ is an involutive anti-automorphism. By using $\mu(x)f(\tau(x)) = -f(x)$ for all $x \in S$ we get $f(\tau(y)x) = -\mu(\tau(y)x)f(\tau(x)y)$ for all $x, y \in S$. Substituting this in equation (52) we obtain

$$f(xy) + \mu(x)f(\tau(x)y) = 2\frac{-f(x)}{\alpha}f(y), \; x, y \in S. \tag{59}$$

By replacing x by $\tau(x)$ in (59) and multiplying the result obtained by $\mu(x)$ we deduce that $f(x) = \mu(x)f(\tau(x))$ for all $x \in S$. So, we have $f(x) = -\mu(x)f(\tau(x)) = -f(x)$, which implies that $f = 0$. This contradicts the assumption that $f \neq 0$.

(2) τ is an involutive automorphism. Then from $\mu(x)f(\tau(x)) = -f(x)$ for all $x \in S$ we get $f(\tau(y)x) = -\mu(\tau(y)x)f(y\tau(x))$ for all $x, y \in S$. Substituting this in equation (52) we obtain

$$f(xy) + \mu(x)f(y\tau(x)) = 2\frac{-f(x)}{\alpha}f(y) \text{ for all } x, y \in S. \tag{60}$$

By replacing x by $\tau(x)$ in (60) and multiplying the result obtained by $\mu(x)$ and using $\mu(x)f(\tau(x)) = -f(x)$ we get

$$h(yx) + \mu(x)h(\tau(x)y) = 2h(x)h(y) \text{ for all } x, y \in S.$$

where $h = \frac{f}{\alpha}$. So, from [11] $\mu(x)f(\tau(x)) = f(x)$ for all $x \in S$. Consequently, $\mu(x)f(\tau(x)) = f(x) = -f(x)$ for all $x \in S$, which implies that $f = 0$. This contradicts the assumption that $f \neq 0$ and this proves (44).

Equation (45): By replacing y by yz_0 in (12) we get

$$\mu(yz_0)f(\tau(y)xz_0\tau(z_0)) - f(xyz_0z_0) = 2f(x)f(yz_0). \tag{61}$$

Replacing x by xz_0 in (12) we get

$$\mu(y)f(\tau(y)xz_0z_0) - f(xyz_0z_0) = 2f(y)f(xz_0). \tag{62}$$

Subtracting these equations results in

$$\begin{aligned}\mu(yz_0)f(\tau(y)xz_0\tau(z_0)) - \mu(y)f(\tau(y)xz_0z_0) \\ = 2f(x)f(yz_0) - 2f(y)f(xz_0).\end{aligned} \tag{63}$$

On the other hand, from (12) we have

$$\begin{aligned}&\mu(yz_0)f(\tau(y)xz_0\tau(z_0)) - \mu(y)f(\tau(y)xz_0z_0) \\ &= \mu(y)[\mu(z_0)f(\tau(z_0)\tau(y)xz_0) - f(\tau(y)xz_0z_0)] \\ &= 2\mu(y)f(z_0)f(\tau(y)x).\end{aligned}$$

This implies that

$$f(x)f(yz_0) - f(y)f(xz_0) = \mu(y)f(\tau(y)x)f(z_0) \tag{64}$$

for all $x, y \in S$. Since $f(x)f(yz_0)-f(y)f(xz_0) = -[f(y)f(xz_0)-f(x)f(yz_0)]$, then we deduce $\mu(y)f(\tau(y)x)f(z_0) = -\mu(x)f(\tau(x)y)f(z_0)$. Now, by using (44) we deduce (45).

Equation (49): By replacing x by $x\tau(z_0)$ in (12) we get

$$\mu(y)f(\tau(y)x\tau(z_0)z_0) - f(xy\tau(z_0)z_0) \tag{65}$$

$$= 2f(y)f(x\tau(z_0)).$$

From (45) we have $\mu(\tau(x))f(xy\tau(z_0)z_0) = \mu(\tau(x))f(\tau(\tau(x))(y\tau(z_0)z_0)) = -\mu(y)f(\tau(y)\tau(x)\tau(z_0)z_0)$ and then equation (65) can be written as follows:

$$f(\tau(y)x\tau(z_0)z_0) + \mu(x)f(\tau(y)\tau(x)\tau(z_0)z_0) \tag{66}$$

$$= 2f(y)\mu(\tau(y))f(x\tau(z_0)).$$

By replacing x by $\tau(x)$ in (66) and multiplying the result obtained by $\mu(x)$ and using $f \neq 0$ we get (49).

From equations (45) and (49) we have

$$\mu(\tau(x))f(xz_0) = -\mu(z_0)f(\tau(z_0)\tau(x))$$

$$= -\mu(z_0)\mu(\tau(x))f(x\tau(z_0)) = f(\tau(x)z_0).$$

This proves (48).

Equation (43): Replacing x by $\tau(x)$ in (12) we get

$$\mu(y)f(\tau(y)\tau(x)z_0) - f(\tau(x)yz_0) = 2f(\tau(x))f(y), \ x, y \in S. \tag{67}$$

We will discuss the following two possibilities.

(1) τ is an involutive automorphism. From (48) we have

$$f(\tau(y)\tau(x)z_0) = f(\tau(yx)z_0) = \mu(\tau(yx))f(yxz_0)$$

and in view of (67) we obtain

$$\mu(\tau(x))f(yxz_0) - f(\tau(x)yz_0) = 2f(\tau(x))f(y), \ x, y \in S.$$

Since

$$\begin{aligned}\mu(\tau(x))f(yxz_0) - f(\tau(x)yz_0) &= -\mu(\tau(x))[\mu(x)f(\tau(x)yz_0) - f(yxz_0)] \\ &= -\mu(\tau(x))2f(y)f(x),\end{aligned}$$

then we deduce that

$$-2\mu(\tau(x))f(x)f(y) = 2f(\tau(x))f(y)$$

for all $x, y \in S$. Since $f \neq 0$ then we have (43).

(2) τ is an involutive anti-automorphism. Using (48) we have

$$f(\tau(y)\tau(x)z_0) = f(\tau(xy)z_0) = \mu(\tau(yx))f(xyz_0)$$

and $f(\tau(x)yz_0) = \mu(\tau(x)y)f(\tau(y)xz_0)$. Now, equation (67) can be written as follows:

$$\mu(\tau(x))f(xyz_0) - \mu(\tau(x)y)f(\tau(y)xz_0) = 2f(\tau(x))f(y)$$

$$= -\mu(\tau(x))[\mu(y)f(\tau(y)xz_0) - f(xyz_0)]$$

$$= -\mu(\tau(x))2f(x)f(y).$$

Since $f \neq 0$ then we obtain again (43).

Equation (46): Putting $x = \tau(z_0)$ in (12), using (43) we get

$$\mu(y)f(\tau(y)\tau(z_0)z_0) - f(\tau(z_0)yz_0) = 2f(y)f(\tau(z_0))$$

$$= -2f(y)\mu(\tau(z_0))f(z_0).$$

Since

$$\mu(y)f(\tau(y)\tau(z_0)z_0) = -\mu(\tau(z_0)z_0)f(yz_0\tau(z_0)) = -f(yz_0\tau(z_0))$$

then we obtain

$$f(\tau(z_0)yz_0) = \mu(\tau(z_0))f(y)f(z_0)$$

for all $y \in S$. We see that we deal with (46).

Equation (47): Replacing y by z_0 in (12) and using (46) we get

$$\mu(z_0)f(\tau(z_0)xz_0) - f(xz_0z_0)$$

$$= 2f(x)f(z_0) = f(x)f(z_0) - f(xz_0z_0),$$

which proves (46).

Equation (50): By replacing x by z_0 in (48) we get $\mu(z_0)f(\tau(z_0)z_0) = f(z_0^2)$. From (43) we have $f(\tau(z_0)z_0) = -f(\tau(z_0)z_0)$, then we conclude that

$$f(\tau(z_0)z_0) = f(z_0^2) = 0,$$

which proves (50). This completes the proof.

Lemma 6 *Let $f: S \longrightarrow C$ be a non-zero solution of equation (12). Then*

(1) *The function defined by*

$$g(x) := \frac{f(xz_0)}{f(z_0)} \text{ for } x \in S$$

is a non-zero solution of the variant of d'Alembert's functional equation (15).

(2) *The function g from (1) has the form $g = \frac{\chi + \mu\chi\circ\tau}{2}$, where $\chi : S \longrightarrow C$, $\chi \neq 0$, is a multiplicative function.*

Proof

(1) From (46), (47), (12) and the definition of g we have

$$(f(z_0))^2[g(xy) + \mu(y)g(\tau(y)x)] = f(z_0)\mu(y)f(\tau(y)xz_0) + f(z_0)f(xyz_0)$$

$$= \mu(y)\mu(z_0)f(\tau(y)xz_0\tau(z_0)z_0) - f(xyz_0z_0^2)$$

$$= \mu(yz_0)f(\tau(yz_0)(xz_0)z_0) - f((xz_0)(yz_0)z_0)$$

$$= 2f(xz_0)f(yz_0).$$

Dividing the last equation by $(f(z_0))^2$ we get g satisfies the variant of d'Alembert's functional equation (15). In view of (47) and the definition of g we get

$$g(z_0^2) = \frac{f(z_0z_0^2)}{f(z_0)}$$

$$= \frac{-f(z_0)f(z_0)}{f(z_0)} = -f(z_0) \neq 0.$$

Then g is a non-zero solution of equation (15).

(2) By replacing x by xz_0 in (12) we get

$$\mu(y)f(\tau(y)xz_0^2) - f(xyz_0^2) = 2f(y)f(xz_0). \tag{68}$$

By using (47), equation (68) can be written as follows:

$$-\mu(y)f(\tau(y)x) + f(xy) = 2f(y)g(x), \; x, y \in S, \tag{69}$$

where g is the function defined above. If we replace y by yz_0 in (12) we get

$$\mu(yz_0)f(\tau(y)x\tau(z_0)z_0) - f(xyz_0z_0) = 2f(x)f(yz_0). \tag{70}$$

By using (46), (47) we obtain

$$\mu(y)f(\tau(y)x) + f(xy) = 2f(x)g(y),\ x, y \in S. \tag{71}$$

By adding (71) and (69) we get that the pair f, g satisfies the sine addition law

$$f(xy) = f(x)g(y) + f(y)g(x) \text{ for all } x, y \in S.$$

Now, in view of [13, Lemma 3.4.] g is abelian. Since g is a non-zero solution of d'Alembert's functional equation (15) then from [3, Proposition 9.31] there exists a non-zero multiplicative function $\chi\colon S \longrightarrow C$ such that $g = \frac{\chi+\mu\chi\circ\tau}{2}$. This completes the proof.

The following theorem is the main result of this section.

Theorem 4 *The non-zero solutions $f : S \longrightarrow C$ of the functional equation (12) are the functions of the form*

$$f = \frac{\mu\chi\circ\tau - \chi}{2}\chi(z_0), \tag{72}$$

where $\chi : S \longrightarrow C$ is a multiplicative function such that $\chi(z_0) \neq 0$ and $\mu(z_0)\chi(\tau(z_0)) = -\chi(z_0)$.

If S is a topological semigroup and that $\tau : S \longrightarrow S$ and $\mu\colon S \longrightarrow C$ are continuous, then the non-zero solution f of equation (12) is continuous if and only if χ is continuous.

Proof Simple computations show that f defined by (72) is a solution of (12). Conversely, let $f : S \longrightarrow C$ be a non-zero solution of the functional equation (12). By putting $y = z_0$ in (12) we get

$$f(x) = \frac{\mu(z_0)f(\tau(z_0)xz_0) - f(xz_0z_0)}{2f(z_0)} \tag{73}$$

$$= \frac{1}{2}(\mu(z_0)g(\tau(z_0)x) - g(xz_0)),$$

where g is the function defined by $g(x) = \frac{f(xz_0)}{f(z_0)}$ and that from Lemma 6 has the form $g = \frac{\chi+\mu\chi\circ\tau}{2}$, where $\chi : S \longrightarrow C$, $\chi \neq 0$ is a multiplicative function. Substituting this into (73) we find that f has the form

$$f = \frac{\chi(z_0) - \mu(z_0)\chi(\tau(z_0))}{2}\frac{\mu\chi\circ\tau - \chi}{2}. \tag{74}$$

Furthermore, from (48) f satisfies $\mu(x)f(\tau(x)z_0) = f(xz_0)$ for all $x \in S$. By applying the last expression of f in (48) we get after computations that

$$[\mu(z_0)\chi(\tau(z_0)) + \chi(z_0)][\chi - \mu\chi \circ \tau] = 0.$$

Since $\chi \neq \mu\chi \circ \tau$, we obtain $\mu(z_0)\chi(\tau(z_0)) + \chi(z_0) = 0$ and then from (74) we have

$$f = \frac{\mu\chi \circ \tau - \chi}{2}\chi(z_0).$$

For the topological statement we use [3, Theorem 3.18(d)]. This completes the proof.

References

1. E.B. Vleck Van, A functional equation for the sine. Ann. Math. Second Ser. **11**(4), 161–165 (1910)
2. E.B. Vleck Van, A functional equation for the sine, Additional note. Ann. Math. Second Ser. **13**(1/4), 154 (1911–1912)
3. H. Stetkær, *Functional Equations on Groups* (World Scientific, Singapore, 2013)
4. A.M. Perkins, P.K. Sahoo, On two functional equations with involution on groups related to sine and cosine functions. Aequationes Math. **89**(5), 1251–1263 (2015)
5. H. Stetkær, Van Vleck's functional equation for the sine. Aequationes Math. **90**(1), 25–34 (2016)
6. P.L. Kannappan, A functional equation for the cosine. Canad. Math. Bull. **2**, 495–498 (1968)
7. H. Stetkær, Kannappan's functional equation on semigroups with involution. Semigroup Forum, 1–14 (2016) https://doi.org/10.1007/s00233-015-9756-7
8. T.M.K. Davison, D'Alembert's functional equation on topological monoids. Publ. Math. Debr. **75**(1/2), 41–66 (2009)
9. H. Stetkær, d'Alembert's functional equation on groups, in *Recent Developments in Functional Equations and Inequalities. Banach Center Publications*, vol. 99. (Polish Academy of Sciences, Institute of Mathematics, Warszawa, 2013), pp. 173–191
10. H. Stetkær, A variant of d'Alembert's functional equation. Aequationes Math. **89**(3), 657–662 (2015)
11. E. Elqorachi, A. Redouani, Solutions and stability of a variant of Wilson's functional equation (Preprint), arXiv:1505.06512v1 [math.CA] (2015)
12. B. Bouikhalene, E. Elqorachi, An extension of Van Vleck's functional equation for the sine. Acta Math. Hungar. **150**(1), 258–267 (2016)
13. B.R. Ebanks, H. Stetkær, D'Alembert's other functional equation on monoids with involution. Aequationes Math. **89**(1), 187–206 (2015)

Recent Advances of Convexity Theory and Its Inequalities

Jichang C. Kuang

Abstract In this chapter, we introduce some new notions of generalized convex functionals in normed linear spaces. It unifies and generalizes the many known and new classes of convex functions. The corresponding Schur, Jensen, and Hermite-Hadamard type inequalities are also established.

Mathematics Subject Classification 26A51, 39B62

1 Introduction

Definition 1 A function $f : [a, b] \to \mathbb{R}$ is called convex if

$$f(\lambda x_1 + (1 - \lambda)x_2) \leq \lambda f(x_1) + (1 - \lambda) f(x_2), \tag{1}$$

$\forall x_1, x_2 \in [a, b], \forall \lambda \in [0, 1]$.

This classical inequality (1) plays an important role in analysis, optimization and in the theory of inequalities, and it has a huge literature dealing with its applications, various generalizations and refinements. Further, convexity is one of the most fundamental and important notions in mathematics. Convexity theory and its inequalities are fields of interest of numerous mathematicians and there are many paper, books, and monographs devoted to these fields and various applications (see, e.g., [1, 4, 6–14, 16, 18–22] and the references therein).

In this chapter, we introduce some new notions of generalized convex functionals in normed linear spaces in Section 2. It unifies and generalizes the many known and new classes of convex functions. Some new basic inequalities are presented in Section 3. New generalized Hermite-Hadamard type inequalities are presented in

J. C. Kuang (✉)
Department of Mathematics, Hunan Normal University, Changsha, Hunan, People's Republic of China

T. M. Rassias, P. M. Pardalos (eds.), *Mathematical Analysis and Applications*, Springer Optimization and Its Applications 154,
https://doi.org/10.1007/978-3-030-31339-5_12

Section 4. In Sections 5 and 6, strongly convex functional and the corresponding inequalities in normed linear spaces are also given.

2 Generalized Convex Functionals in Normed Linear Spaces

In what follows, $(X, \|\cdot\|)$ denotes the real normed linear spaces, D be a convex subset of X, $h : (0, 1) \to (0, \infty)$ is a given function, whose h is not identical to 0.

In this section, we introduce and study a new class of generalized convex functionals, that is, $(\alpha, \beta, \lambda, \lambda_0, t, \xi, h)$ convex functionals.

Definition 2 A functional $f : D \to (0, \infty)$ is called $(\alpha, \beta, \lambda, \lambda_0, t, \xi, h)$ convex if

$$f((\lambda\|x_1\|^\alpha + \lambda_0(1-\lambda)\|x_2\|^\alpha)^{1/\alpha}) \leq \{h(t^\xi)f^\beta(\|x_1\|) + \lambda_0 h(1-t^\xi)f^\beta(\|x_2\|)\}^{1/\beta}, \tag{2}$$

$\forall x_1, x_2 \in D$, $\forall \lambda, \lambda_0, t, \xi \in [0, 1]$, α, β are real numbers, and $\alpha, \beta \neq 0$.

If $\lambda_0 = 1$ in (2), that is,

$$f((\lambda\|x_1\|^\alpha + (1-\lambda)\|x_2\|^\alpha)^{1/\alpha}) \leq \{h(t^\xi)f^\beta(\|x_1\|) + h(1-t^\xi)f^\beta(\|x_2\|)\}^{1/\beta}, \tag{3}$$

we say that f is a $(\alpha, \beta, \lambda, t, \xi, h)$ convex functional.

If $\xi = 1$ in (3), that is,

$$f((\lambda\|x_1\|^\alpha + (1-\lambda)\|x_2\|^\alpha)^{1/\alpha}) \leq \{h(t)f^\beta(\|x_1\|) + h(1-t)f^\beta(\|x_2\|)\}^{1/\beta},$$

we say that f is a $(\alpha, \beta, \lambda, t, h)$ convex functional.

For $t = \lambda$ in (2), that is,

$$f((\lambda\|x_1\|^\alpha + \lambda_0(1-\lambda)\|x_2\|^\alpha)^{1/\alpha}) \leq \{h(\lambda^\xi)f^\beta(\|x_1\|) + \lambda_0 h(1-\lambda^\xi)f^\beta(\|x_2\|)\}^{1/\beta}, \tag{4}$$

we say that f is a $(\alpha, \beta, \lambda, \lambda_0, \xi, h)$ convex functional.

If $\xi = 1$ in (4), that is,

$$f((\lambda\|x_1\|^\alpha + \lambda_0(1-\lambda)\|x_2\|)^{1/\alpha}) \leq \{h(\lambda)f^\beta(\|x_1\|) + \lambda_0 h(1-\lambda)f^\beta(\|x_2\|)\}^{1/\beta}, \tag{5}$$

we say that f is a $(\alpha, \beta, \lambda, \lambda_0, h)$ convex functional.

For $\lambda_0 = 1$ in (5), that is,

$$f((\lambda\|x_1\|^\alpha + (1-\lambda)\|x_2\|^\alpha)^{1/\alpha}) \leq \{h(\lambda)f^\beta(\|x_1\|) + h(1-\lambda)f^\beta(\|x_2\|)\}^{1/\beta}, \tag{6}$$

we say that f is a $(\alpha, \beta, \lambda, h)$ convex functional.

In particular, if $h(\lambda) = \lambda^s$, $0 < |s| \leq 1$ in (6), that is,

$$f((\lambda\|x_1\|^\alpha + (1-\lambda)\|x_2\|^\alpha)^{1/\alpha}) \leq \{\lambda^s f^\beta(\|x_1\|) + (1-\lambda)^s f^\beta(\|x_2\|)\}^{1/\beta}, \tag{7}$$

we say that f is a $(\alpha, \beta, \lambda, s)$ convex functional. If $s = 1$, then (7) reduces to (α, β, λ) convex functional.

For $\alpha = \beta = 1$ in (2), that is,

$$f(\lambda\|x_1\| + \lambda_0(1-\lambda)\|x_2\|) \leq h(t^{\xi})f(\|x_1\|) + \lambda_0 h(1-t^{\xi})f(\|x_2\|), \tag{8}$$

we say that f is a $(\lambda, \lambda_0, t, \xi, h)$ convex functional.

If $\lambda_0 = 1$ in (8), that is,

$$f(\lambda\|x_1\| + (1-\lambda)\|x_2\|) \leq h(t^{\xi})f(\|x_1\|) + h(1-t^{\xi})f(\|x_2\|), \tag{9}$$

we say that f is a (λ, t, ξ, h) convex functional.

In particular, if $t = \lambda$ in (8), that is,

$$f(\lambda\|x_1\| + \lambda_0(1-\lambda)\|x_2\|) \leq h(\lambda^{\xi})f(\|x_1\|) + \lambda_0 h(1-\lambda^{\xi})f(\|x_2\|), \tag{10}$$

we say that f is a $(\lambda, \lambda_0, \xi, h)$ convex functional.

If $\xi = 1$ in (10), that is,

$$f(\lambda\|x_1\| + \lambda_0(1-\lambda)\|x_2\|) \leq h(\lambda)f(\|x_1\|) + \lambda_0 h(1-\lambda)f(\|x_2\|), \tag{11}$$

we say that f is a (λ, λ_0, h) convex functional.

If $\lambda_0 = 1$ in (11), that is,

$$f(\lambda\|x_1\| + (1-\lambda)\|x_2\|) \leq h(\lambda)f(\|x_1\|) + h(1-\lambda)f(\|x_2\|), \tag{12}$$

we say that f is an h-convex functional.

In the following Examples 1–6, we make appointment that

$$X = [0, \infty), D \subset [0, \infty), f : D \to [0, \infty).$$

Then (2) reduces to

$$f((\lambda x_1^{\alpha} + \lambda_0(1-\lambda)x_2^{\alpha})^{1/\alpha}) \leq \{h(t^{\xi})f^{\beta}(x_1) + \lambda_0 h(1-t^{\xi})f^{\beta}(x_2)\}^{1/\beta}, \tag{13}$$

$\forall x_1, x_2 \in D, \forall \lambda, \lambda_0, t, \xi \in [0, 1], \alpha, \beta$ are real numbers, and $\alpha, \beta \neq 0$.

If $\xi = 1$ in (13), that is,

$$f((\lambda x_1^{\alpha} + \lambda_0(1-\lambda)x_2^{\alpha})^{1/\alpha}) \leq \{h(t)f^{\beta}(x_1) + \lambda_0 h(1-t)f^{\beta}(x_2)\}^{1/\beta}, \tag{14}$$

$\forall x_1, x_2 \in D, \forall \lambda, t \in [0, 1]$, we say that f is a $(\alpha, \beta, \lambda, \lambda_0, t, h)$ convex function.

If $t = \lambda$ in (14), that is,

$$f((\lambda x_1^{\alpha} + \lambda_0(1-\lambda)x_2^{\alpha})^{1/\alpha}) \leq \{h(\lambda)f^{\beta}(x_1) + \lambda_0 h(1-\lambda)f^{\beta}(x_2)\}^{1/\beta}, \tag{15}$$

$\forall x_1, x_2 \in D, \forall \lambda \in [0, 1]$, we say that f is a $(\alpha, \beta, \lambda, \lambda_0, h)$ convex function.

If $\lambda_0 = 1$ in (14), that is,

$$f((\lambda x_1^\alpha + (1-\lambda)x_2^\alpha)^{1/\alpha}) \le \{h(t)f^\beta(x_1) + h(1-t)f^\beta(x_2)\}^{1/\beta}, \tag{16}$$

$\forall x_1, x_2 \in D, \forall \lambda, t \in [0,1]$, we say that f is a $(\alpha, \beta, \lambda, t, h)$ convex function. If $t = \lambda$ in (16), that is,

$$f((\lambda x_1^\alpha + (1-\lambda)x_2^\alpha)^{1/\alpha}) \le \{h(\lambda)f^\beta(x_1) + h(1-\lambda)f^\beta(x_2)\}^{1/\beta}, \tag{17}$$

$\forall x_1, x_2 \in D, \forall \lambda \in [0,1]$, we say that f is a $(\alpha, \beta, \lambda, h)$ convex function.

If $h(\lambda) = \lambda^s, 0 < |s| \le 1$ in (16), (17), that is,

$$f((\lambda x_1^\alpha + (1-\lambda)x_2^\alpha)^{1/\alpha}) \le \{t^s f^\beta(x_1) + (1-t)^s f^\beta(x_2)\}^{1/\beta}, \tag{18}$$

$$f((\lambda x_1^\alpha + (1-\lambda)x_2^\alpha)^{1/\alpha}) \le \{\lambda^s f^\beta(x_1) + (1-\lambda)^s f^\beta(x_2)\}^{1/\beta}, \tag{19}$$

we say that f is a $(\alpha, \beta, \lambda, t, s)$, $(\alpha, \beta, \lambda, s)$ convex function, respectively.

In particular, if $s = 1$, then (18), (19) reduce to $(\alpha, \beta, \lambda, t)$, (α, β, λ) convex function, respectively.

Example 1 If $\alpha = \beta = 1$ in (13), then

$$f(\lambda x_1 + \lambda_0(1-\lambda)x_2) \le h(t^\xi)f(x_1) + \lambda_0 h(1-t^\xi)f(x_2), \tag{20}$$

$\forall x_1, x_2 \in D, \forall \lambda, \lambda_0, t, \xi \in [0,1]$, we say that f is a $(\lambda, \lambda_0, t, \xi, h)$ convex function.

In particular, if $\lambda_0 = 1$ in (20), that is,

$$f(\lambda x_1 + (1-\lambda)x_2) \le h(t^\xi)f(x_1) + h(1-t^\xi)f(x_2), \tag{21}$$

$\forall x_1, x_2 \in D, \forall \lambda, t, \xi \in [0,1]$, we say that f is a (λ, t, ξ, h) convex function.

For $h(t) = t^s, 0 < |s| \le 1, \xi = 1$ in (21), that is,

$$f(\lambda x_1 + (1-\lambda)x_2) \le t^s f(x_1) + (1-t)^s f(x_2), \tag{22}$$

$\forall x_1, x_2 \in D, \forall \lambda, t \in [0,1]$, we say that f is a (λ, t, s) convex function.

In particular, when $s = 1$ in (22), that is,

$$f(\lambda x_1 + (1-\lambda)x_2) \le tf(x_1) + (1-t)f(x_2), \tag{23}$$

$\forall x_1, x_2 \in D, \forall \lambda, t \in [0,1]$, we say that f is a (λ, t) convex function (see, e.g., [8]).

For $t = \lambda$ in (20), that is,

$$f(\lambda x_1 + \lambda_0(1-\lambda)x_2) \le h(\lambda^\xi)f(x_1) + \lambda_0 h(1-\lambda^\xi)f(x_2), \tag{24}$$

$\forall x_1, x_2 \in D, \forall \lambda, \lambda_0, \xi \in [0,1]$, we say that f is a $(\lambda, \lambda_0, \xi, h)$ convex function.

If $\xi = 1$ in (24), that is,

$$f(\lambda x_1 + \lambda_0(1-\lambda)x_2) \le h(\lambda)f(x_1) + \lambda_0 h(1-\lambda)f(x_2), \tag{25}$$

$\forall x_1, x_2 \in D, \forall \lambda \in [0, 1]$, we say that f is a (λ, λ_0, h) convex function. In particular, when $\lambda_0 = 1$,(25) reduces to h-convex function (see [4, 19]), that is,

$$f(\lambda x_1 + (1-\lambda)x_2) \le h(\lambda)f(x_1) + h(1-\lambda)f(x_2). \tag{26}$$

If $h(\lambda) = \lambda$, then (26) reduces to (1).

If $h(\lambda) = \lambda^s, 0 < s \le 1$ in (26), that is,

$$f(\lambda x_1 + (1-\lambda)x_2) \le \lambda^s f(x_1) + (1-\lambda)^s f(x_2), \tag{27}$$

$\forall x_1, x_2 \in D, \forall \lambda \in [0, 1]$, we say that f is a s-Breckner convex function (see, e.g., [4, 5, 8]).

If $h(\lambda) = \lambda^{-s}, 0 < s \le 1$ in (26), that is,

$$f(\lambda x_1 + (1-\lambda)x_2) \le \lambda^{-s} f(x_1) + (1-\lambda)^{-s} f(x_2), \tag{28}$$

$\forall x_1, x_2 \in D, \forall \lambda \in [0, 1]$, we say that f is a s-Godunova-Levin function (see [4]). In particular, when $s = 1$, (28) reduces to Godunova-Levin function (see, e.g., [5, 8])

If $h(\lambda) = 1$ in (26), that is,

$$f(\lambda x_1 + (1-\lambda)x_2) \le f(x_1) + f(x_2), \tag{29}$$

$\forall x_1, x_2 \in D, \forall \lambda \in [0, 1]$, we say that f is a P-function (see, e.g., [5]).

If $h(\lambda) = \lambda^s, 0 < |s| \le 1$, in (24), that is,

$$f(\lambda x_1 + \lambda_0(1-\lambda)x_2) \le \lambda^{s\xi} f(x_1) + \lambda_0(1-\lambda^{\xi})^s f(x_2), \tag{30}$$

$\forall x_1.x_2 \in D, \forall \lambda, \lambda_0, \xi \in [0, 1]$, we say that f is a $(\lambda, \lambda_0, \xi, s)$ convex function. In particular, if $\xi = s = 1$ in (30), that is,

$$f(\lambda x_1 + \lambda_0(1-\lambda)x_2) \le \lambda f(x_1) + \lambda_0(1-\lambda)f(x_2), \tag{31}$$

we say that f is a λ_0-convex function (that is, m-convex function in [2]).

If $\lambda_0 = 0$ in (20), then

$$f(\lambda x) \le h(t^{\xi})f(x), \ \ x \in D. \tag{32}$$

When $t = \lambda, \xi = 1, h(t) = t$ in (32), that is,

$$f(\lambda x) \le \lambda f(x), \tag{33}$$

we say that f is a starshaped function (see [2])

Example 2 If $\beta = 1$ in (13), then

$$f((\lambda x_1^\alpha + \lambda_0(1-\lambda)x_2^\alpha)^{1/\alpha}) \le h(t^\xi)f(x_1) + \lambda_0 h(1-t^\xi)f(x_2), \tag{34}$$

$\forall x_1, x_2 \in D, \forall \lambda, \lambda_0, t, \xi \in [0,1], \alpha \ne 0$, we say that f is a $(\alpha, \lambda, \lambda_0, t, \xi, h)$ convex function.

For $t = \lambda$ in (34), that is,

$$f((\lambda x_1^\alpha + \lambda_0(1-\lambda)x_2^\alpha)^{1/\alpha}) \le h(\lambda^\xi)f(x_1) + \lambda_0 h(1-\lambda^\xi)f(x_2), \tag{35}$$

$\forall x_1, x_2 \in D, \forall \lambda, \lambda_0, t, \xi \in [0,1], \alpha \ne 0$, we say that f is a $(\alpha, \lambda, \lambda_0, \xi, h)$ convex function. When $\lambda_0 = 1, \xi = 1$, (35) reduces to (α, h) convex function (that is, (p, h) convex function in [5]). In particular, if $\lambda_0 = 1, \xi = 1, h(t) = t$, (35) reduces to α-convex function (that is, p-convex function in [5, 22])

Example 3 If $\alpha = 1, \beta = q$ in (13), then

$$f(\lambda x_1 + \lambda_0(1-\lambda)x_2) \le \{h(t^\xi)f^q(x_1) + \lambda_0 h(1-t^\xi)f^q(x_2)\}^{1/q}, \tag{36}$$

$\forall x_1, x_2 \in D, \forall \lambda, \lambda_0, t, \xi \in [0,1], q \ne 0$, we say that f is a $(q, \lambda, \lambda_0, t, \xi, h)$ convex function.

For $t = \lambda$ in (36), that is,

$$f(\lambda x_1 + \lambda_0(1-\lambda)x_2) \le \{h(\lambda^\xi)f^q(x_1) + \lambda_0 h(1-\lambda^\xi)f^q(x_2)\}^{1/q}, \tag{37}$$

$\forall x_1, x_2 \in D, \forall \lambda, \lambda_0, t, \xi \in [0,1], q \ne 0$, we say that f is a $(q, \lambda, \lambda_0, \xi, h)$ convex function. When $\lambda_0 = 1, \xi = 1, h(t) = t$, (37) reduces to q-convex function (see, e.g., [8]).

Example 4 If $\alpha = 1, \beta = -1$ in (13), then

$$f(\lambda x_1 + \lambda_0(1-\lambda)x_2) \le \{h(t^\xi)f^{-1}(x_1) + \lambda_0 h(1-t^\xi)f^{-1}(x_2)\}^{-1}, \tag{38}$$

$\forall x_1, x_2 \in D, \forall \lambda, \lambda_0, t, \xi \in [0,1]$, we say that f is a $(AH, \lambda, \lambda_0, t, \xi, h)$ convex function, where AH means the arithmetic-harmonic means.

For $t = \lambda$ in (38), that is,

$$f(\lambda x_1 + \lambda_0(1-\lambda)x_2) \le \{h(\lambda^\xi)f^{-1}(x_1) + \lambda_0 h(1-\lambda^\xi)f^{-1}(x_2)\}^{-1}, \tag{39}$$

$\forall x_1, x_2 \in D, \forall \lambda, \lambda_0, \xi \in [0,1]$, we say that f is a $(AH, \lambda, \lambda_0, \xi, h)$ convex function.

For $h(\lambda) = \lambda^s, 0 < |s| \le 1, \lambda_0 = \xi = 1$ in (39), that is,

$$f(\lambda x_1 + (1-\lambda)x_2) \le \{\lambda^s f^{-1}(x_1) + (1-\lambda)^s f^{-1}(x_2)\}^{-1}, \tag{40}$$

$\forall x_1, x_2 \in D, \forall\lambda \in [0, 1]$, we say f is a (AH, λ, s) convex function. In particular, if $s = 1$, then (40) reduces to AH convex function.

Example 5 If $\alpha = -1, \lambda_0 = 1, h(\lambda) = \lambda$ in (15), then

$$f(\frac{x_1x_2}{\lambda x_2 + (1-\lambda)x_1}) \leq \{\lambda f^{\beta}(x_1) + (1-\lambda)f^{\beta}(x_2)\}^{1/\beta},$$

$\forall x_1, x_2 \in D, \forall\lambda \in [0, 1], \beta \neq 0$, we say that f is a harmonically β-convex functions, see [15].

Example 6 If $\alpha = -1, \beta = 1$ in (13), then

$$f((\lambda x_1^{-1} + \lambda_0(1-\lambda)x_2^{-1})^{-1}) \leq h(t^{\xi})f(x_1) + \lambda_0 h(1-t^{\xi})f(x_2), \tag{41}$$

$\forall x_1, x_2 \in D, \forall\lambda, \lambda_0, t, \xi \in [0, 1]$, we say that f is a $(HA, \lambda, \lambda_0, t, \xi, h)$ convex function.

For $t = \lambda$ in (41), that is,

$$f((\lambda x_1^{-1} + \lambda_0(1-\lambda)x_2^{-1})^{-1}) \leq h(\lambda^{\xi})f(x_1) + \lambda_0 h(1-\lambda^{\xi})f(x_2), \tag{42}$$

$\forall x_1, x_2 \in D, \forall\lambda, \lambda_0, \xi \in [0, 1]$, we say that f is a $(HA, \lambda, \lambda_0, \xi, h)$ convex function.

For $h(\lambda) = \lambda^s, 0 < |s| \leq 1, \lambda_0 = \xi = 1$ in (42), that is,

$$f((\lambda x_1^{-1} + (1-\lambda)x_2^{-1})^{-1}) \leq \lambda^s f(x_1) + (1-\lambda)^s f(x_2), \tag{43}$$

$\forall x_1, x_2 \in D, \forall\lambda \in [0, 1]$, we say f is a (HA, λ, s) convex function. In particular, if $s = 1$, then (43) reduces to HA convex function.

Example 7 If $\alpha = \beta = -2$ in (13), then

$$f((\lambda x_1^{-2} + \lambda_0(1-\lambda)x_2^{-2})^{-(1/2)}) \leq \{h(t^{\xi})f^{-2}(x_1) + \lambda_0 h(1-t^{\xi})f^{-2}(x_2)\}^{-(1/2)}, \tag{44}$$

$\forall x_1, x_2 \in D, \forall\lambda, \lambda_0, t, \xi \in [0, 1]$, we say that f is a $(HS, \lambda, \lambda_0, t, \xi, h)$ convex function.

For $t = \lambda$ in (44), that is,

$$f((\lambda x_1^{-2} + \lambda_0(1-\lambda)x_2^{-2})^{-(1/2)}) \leq \{h(\lambda^{\xi})f^{-2}(x_1) + \lambda_0 h(1-\lambda^{\xi})f^{-2}(x_2)\}^{-(1/2)}, \tag{45}$$

$\forall x_1, x_2 \in D, \forall\lambda, \lambda_0, \xi \in [0, 1]$, we say that f is a $(HS, \lambda, \lambda_0, \xi, h)$ convex function.

For $h(\lambda) = \lambda^s, 0 < |s| \leq 1, \lambda_0 = \xi = 1$ in (45), that is,

$$f((\lambda x_1^{-2} + (1-\lambda)x_2^{-2})^{-(1/2)}) \leq \{\lambda^s f^{-2}(x_1) + (1-\lambda)^s f^{-2}(x_2)\}^{-(1/2)}, \tag{46}$$

$\forall x_1, x_2 \in D, \forall \lambda \in [0, 1]$, we say that f is a (HS, λ, s) convex function, that is, f is the harmonic square s-convex function. In particular, if $s = 1$, then (46) reduces to HS convex function.

Example 8 Let X be a real normed linear space, and D be a convex subset of X, $h : (0, 1) \to (0, \infty)$ is a given function. If

$$\lambda = \frac{t_1}{t_1 + t_2}, h(\lambda) = \frac{\lambda(t_1)}{\lambda(t_1 + t_2)}, \quad 0 < t_1, t_2 < \infty,$$

then

$$1 - \lambda = \frac{t_2}{t_1 + t_2}, h(1 - \lambda) = \frac{\lambda(t_2)}{\lambda(t_1 + t_2)},$$

and by (25), we get

$$f(\frac{t_1 x_1 + \lambda_0 t_2 x_2}{t_1 + t_2}) \le \frac{\lambda(t_1) f(x_1) + \lambda_0 \lambda(t_2) f(x_2)}{\lambda(t_1 + t_2)}, \tag{47}$$

$\forall x_1, x_2 \in D$, $\forall \lambda, \lambda_0 \in [0, 1]$, we say that f is a (λ, λ_0) convex function. When $\lambda_0 = 1$,(47) reduces to λ-convex function (see, e.g., [3, 4]).

Hence, Definition 2 unifies and generalizes the many known and new classes of convex functions.

3 Some New Basic Inequalities

The classical Schur, Jensen, and Hermite-Hadamard inequalities play an important role in analysis, optimization and in the theory of inequalities, and it has a huge literature dealing with its applications, various generalizations and refinements (see, e.g., [2, 6–9, 11, 12, 18–22], and the references therein). In this and next section, we present the corresponding inequalities for $(\alpha, \beta, \lambda, \lambda_0, t, \xi, h)$ convex functionals.

Definition 3 ([19]) A function $h : (0, 1) \to (0, \infty)$ is called a super-multiplicative function if

$$h(tu) \ge h(t)h(u), \tag{48}$$

for all $t, u \in (0, 1)$.

Lemma 1 *Let $g(\|x\|) = f^{\beta}(\|x\|^{1/\alpha})$, $x \in D$, α, β are real numbers, and $\alpha, \beta \ne 0$. Then a functional $f : D \to (0, \infty)$ is $(\alpha, \beta, \lambda, \lambda_0, t, \xi, h)$ convex if and only if the functional $g : D \to (0, \infty)$ is $(\lambda, \lambda_0, t, \xi, h)$ convex. In particular, a functional $f : D \to (0, \infty)$ is $(\alpha, \beta, \lambda, \lambda_0, \xi, h)$ convex if and only if the functional $g : D \to$*

$(0, \infty)$ *is* $(\lambda, \lambda_0, \xi, h)$ *convex, and a functional* $f : D \to (0, \infty)$ *is* $(\alpha, \beta, \lambda, h)$ *convex if and only if the functional* $g : D \to (0, \infty)$ *is* h*-convex.*

Proof Setting $\|u\| = \|x\|^{1/\alpha}, x \in D$. Assume that f is $(\alpha, \beta, \lambda, \lambda_0, t, \xi, h)$ convex, then for all $x_1, x_2 \in D$, we get

$$\begin{aligned}
&g(\lambda\|x_1\| + \lambda_0(1-\lambda)\|x_2\|) = g(\lambda\|u_1\|^\alpha + \lambda_0(1-\lambda)\|u_2\|^\alpha) \\
&= f^\beta((\lambda\|u_1\|^\alpha + \lambda_0(1-\lambda)\|u_2\|^\alpha)^{1/\alpha}) \le h(t^\xi) f^\beta(\|u_1\|) + \lambda_0 h(1-t^\xi) f^\beta(\|u_2\|) \\
&= h(t^\xi) g(\|u_1\|^\alpha) + \lambda_0 h(1-t^\xi) g(\|u_2\|^\alpha) = h(t^\xi) g(\|x_1\|) + \lambda_0 h(1-t^\xi) g(\|x_2\|),
\end{aligned}$$

which proves that g is $(\lambda, \lambda_0, t, \xi, h)$ convex.

Conversely, if g is $(\lambda, \lambda_0, t, \xi, h)$ convex, then

$$\begin{aligned}
f^\beta((\lambda\|u_1\|^\alpha + \lambda_0(1-\lambda)\|u_2\|^\alpha)^{1/\alpha}) &= g(\lambda\|u_1\|^\alpha + \lambda_0(1-\lambda)\|u_2\|^\alpha) \\
&= g(\lambda\|x_1\| + \lambda_0(1-\lambda)\|x_2\|) \le h(t^\xi) g(\|x_1\|) + \lambda_0 h(1-t^\xi) g(\|x_2\|) \\
&= h(t^\xi) f^\beta(\|x_1\|^{1/\alpha}) + \lambda_0 h(1-t^\xi) f^\beta(\|x_2\|^{1/\alpha}) \\
&= h(t^\xi) f^\beta(\|u_1\|) + \lambda_0 h(1-t^\xi) f^\beta(\|u_2\|),
\end{aligned}$$

which proves that f is $(\alpha, \beta, \lambda, \lambda_0, t, \xi, h)$ convex.

First of all, we establish Schur type inequalities of $(\alpha, \beta, \lambda, h)$ convex functionals.

Theorem 1 *Let* $f : D \to (0, \infty)$ *be a* h*-convex functional and* $h : (0, 1) \to (0, \infty)$ *is a super-multiplicative function, then for all* $x_1, x_2, x_3 \in D$*, such that* $\|x_1\| < \|x_2\| < \|x_3\|$*, and* $0 < \|x_3\| - \|x_1\| < 1$*, the following generalized Schur inequality holds:*

$$f(\|x_2\|) \le \frac{h(\|x_3\| - \|x_2\|)}{h(\|x_3\| - \|x_1\|)} f(\|x_1\|) + \frac{h(\|x_2\| - \|x_1\|)}{h(\|x_3\| - \|x_1\|)} f(\|x_3\|). \tag{49}$$

Proof Setting

$$\lambda = \frac{\|x_3\| - \|x_2\|}{\|x_3\| - \|x_1\|},$$

we have $0 < \lambda < 1$,

$$1 - \lambda = \frac{\|x_2\| - \|x_1\|}{\|x_3\| - \|x_1\|},$$

and $\|x_2\| = \lambda\|x_1\| + (1-\lambda)\|x_3\|$. By (12), we get

$$\begin{aligned}
f(\|x_2\|) &= f(\lambda\|x_1\| + (1-\lambda)\|x_3\|) \\
&\le h(\lambda) f(\|x_1\|) + h(1-\lambda) f(\|x_3\|).
\end{aligned} \tag{50}$$

By (48), we get

$$h(\|x_3\| - \|x_2\|) = h(\lambda(\|x_3\| - \|x_1\|)) \geq h(\lambda)h(\|x_3\| - \|x_1\|).$$

Hence,

$$h(\lambda) \leq \frac{h(\|x_3\| - \|x_2\|)}{h(\|x_3\| - \|x_1\|)}. \tag{51}$$

Similarly, we get

$$h(1-\lambda) \leq \frac{h(\|x_2\| - \|x_1\|)}{h(\|x_3\| - \|x_1\|)}. \tag{52}$$

Therefore, (49) follows from (50), (51), and (52). The proof is complete.

Using Lemma 1, we get

Corollary 1 *Let $f : D \to (0, \infty)$ be a $(\alpha, \beta, \lambda, h)$ convex functional, and $h : (0, 1) \to (0, \infty)$ is a super-multiplicative function, then for all $x_1, x_2, x_3 \in D$, such that $\|x_1\|^\alpha < \|x_2\|^\alpha < \|x_3\|^\alpha$, and $0 < \|x_3\|^\alpha - \|x_1\|^\alpha < 1$, the following Schur-type inequalities holds:*

$$f^\beta(\|x_2\|) \leq \frac{h(\|x_3\|^\alpha - \|x_2\|^\alpha)}{h(\|x_3\|^\alpha - \|x_1\|^\alpha)} f^\beta(\|x_1\|) + \frac{h(\|x_2\|^\alpha - \|x_1\|^\alpha)}{h(\|x_3\|^\alpha - \|x_1\|^\alpha)} f^\beta(\|x_3\|). \tag{53}$$

Corollary 2 *Let $f : (0, \infty) \to (0, \infty)$ be a $(\alpha, \beta, \lambda, h)$ convex function and $h : (0, 1) \to (0, \infty)$ is a super-multiplicative function, then for all $x_1, x_2, x_3 \in (0, \infty)$, such that $x_1^\alpha < x_2^\alpha < x_3^\alpha$, and $0 < x_3^\alpha - x_1^\alpha < 1$, the following generalized Schur inequality holds:*

$$f^\beta(x_2) \leq \frac{h(x_3^\alpha - x_2^\alpha)}{h(x_3^\alpha - x_1^\alpha)} f^\beta(x_1) + \frac{h(x_2^\alpha - x_1^\alpha)}{h(x_3^\alpha - x_1^\alpha)} f^\beta(x_3). \tag{54}$$

Next by using the definition of (λ, t, ξ, h) convex functional and induction, one obtains the following new generalized Jensen inequality:

Theorem 2 *Let $f : D \to (0, \infty)$ be a (λ, t, ξ, h) convex functional and $h : (0, 1) \to (0, \infty)$ is a super-multiplicative function, then*

$$f(\sum_{k=1}^{n} \lambda_k \|x_k\|) \leq \sum_{k=1}^{n} h(t_k^\xi) f(\|x_k\|), \tag{55}$$

for any $x_k \in D, \lambda_k, t_k, \xi \in [0, 1], 1 \leq k \leq n$, with $\sum_{k=1}^{n} \lambda_k = 1$ and $\sum_{k=1}^{n} t_k^\xi = 1$.

Proof For $n = 2$, this is just the definition of (λ, t, ξ, h) convex functional, and for $n > 2$ it follows by induction. Assume that (55) is true for some positive integer $n > 2$, we shall prove that

$$f(\sum_{k=1}^{n+1}\lambda_k\|x_k\|) \le \sum_{k=1}^{n+1} h(t_k^{\xi}) f(\|x_k\|), \tag{56}$$

for any $x_k \in D$, $\lambda_k, t_k, \xi \in [0,1]$, $1 \le k \le n+1$, with $\sum_{k=1}^{n+1}\lambda_k = 1$ and $\sum_{k=1}^{n+1} t_k^{\xi} = 1$.

To show that (56) is true, we note that

$$\begin{aligned}
f(\sum_{k=1}^{n+1}\lambda_k\|x_k\|) &= f(\sum_{k=1}^{n-1}\lambda_k\|x_k\| + \lambda_n\|x_n\| + \lambda_{n+1}\|x_{n+1}\|)\\
&= f\{\sum_{k=1}^{n-1}\lambda_k\|x_k\| + (\lambda_n+\lambda_{n+1})(\frac{\lambda_n}{\lambda_n+\lambda_{n+1}}\|x_n\| + \frac{\lambda_{n+1}}{\lambda_n+\lambda_{n+1}}\|x_{n+1}\|)\}\\
&\le \sum_{k=1}^{n-1} h(t_k^{\xi}) f(\|x_k\|) + h(t_n^{\xi}+t_{n+1}^{\xi}) f(\frac{\lambda_n}{\lambda_n+\lambda_{n+1}}\|x_n\| + \frac{\lambda_{n+1}}{\lambda_n+\lambda_{n+1}}\|x_{n+1}\|).
\end{aligned} \tag{57}$$

By (48), we get

$$\begin{aligned}
h(t_n^{\xi}) &= h(\frac{t_n^{\xi}}{t_n^{\xi}+t_{n+1}^{\xi}} \times (t_n^{\xi}+t_{n+1}^{\xi}))\\
&\ge h(\frac{t_n^{\xi}}{t_n^{\xi}+t_{n+1}^{\xi}}) h(t_n^{\xi}+t_{n+1}^{\xi}),
\end{aligned}$$

that is,

$$h(\frac{t_n^{\xi}}{t_n^{\xi}+t_{n+1}^{\xi}}) \le \frac{h(t_n^{\xi})}{h(t_n^{\xi}+t_{n+1}^{\xi})}. \tag{58}$$

Similarly, we get

$$h(\frac{t_{n+1}^{\xi}}{t_n^{\xi}+t_{n+1}^{\xi}}) \le \frac{h(t_{n+1}^{\xi})}{h(t_n^{\xi}+t_{n+1}^{\xi})}. \tag{59}$$

Using (9), (58), and (59), we have

$$f(\frac{\lambda_n}{\lambda_n+\lambda_{n+1}}\|x_n\| + \frac{\lambda_{n+1}}{\lambda_n+\lambda_{n+1}}\|x_{n+1}\|)$$

$$\leq h(\frac{t_n^\xi}{t_n^\xi + t_{n+1}^\xi})f(\|x_n\|) + h(\frac{t_{n+1}^\xi}{t_n^\xi + t_{n+1}^\xi})f(\|x_{n+1}\|)$$

$$\leq \frac{1}{h(t_n^\xi + t_{n+1}^\xi)}\{h(t_n^\xi)f(\|x_n\|) + h(t_{n+1}^\xi)f(\|x_{n+1}\|)\}. \tag{60}$$

Hence, (56) follows from (57) and (60). The proof is complete.

Corollary 3 *Let $f : D \to (0, \infty)$ be a $(\alpha, \beta, \lambda, t, \xi, h)$ convex functional and $h : (0, 1) \to (0, \infty)$ is a super-multiplicative function, then*

$$f((\sum_{k=1}^n \lambda_k \|x_k\|^\alpha)^{1/\alpha}) \leq \{\sum_{k=1}^n h(t_k^\xi) f^\beta(\|x_k\|)\}^{1/\beta},$$

for any $x_k \in D, \lambda_k, t_k, \xi \in [0, 1], 1 \leq k \leq n$, with $\sum_{k=1}^n \lambda_k = 1$ and $\sum_{k=1}^n t_k^\xi = 1$.

Corollary 4 *Let $f : (0, \infty) \to (0, \infty)$ be a $(\alpha, \beta, \lambda, t, \xi, h)$ convex function and $h : (0, 1) \to (0, \infty)$ is a super-multiplicative function, then*

$$f((\sum_{k=1}^n \lambda_k x_k^\alpha)^{1/\alpha}) \leq \{\sum_{k=1}^n h(t_k^\xi) f^\beta(x_k)\}^{1/\beta},$$

for any $x_k \in (0, \infty), \lambda_k, t_k, \xi \in [0, 1], 1 \leq k \leq n$, with $\sum_{k=1}^n \lambda_k = 1$ and $\sum_{k=1}^n t_k^\xi = 1$.

Corollary 5 *Let $f : (0, \infty) \to (0, \infty)$ be a $(\alpha, \beta, \lambda, t, s)$ convex function, then*

$$f((\sum_{k=1}^n \lambda_k x_k^\alpha)^{1/\alpha}) \leq \{\sum_{k=1}^n t_k^s f^\beta(x_k)\}^{1/\beta},$$

for any $x_k \in (0, \infty), \lambda_k, t_k, s \in [0, 1], 1 \leq k \leq n$, with $\sum_{k=1}^n \lambda_k = 1$, and $\sum_{k=1}^n t_k^s = 1$.

Corollary 6 *Let $f : (0, \infty) \to (0, \infty)$ be a (λ, t, s) convex function, then*

$$f(\sum_{k=1}^n \lambda_k x_k) \leq \sum_{k=1}^n t_k^s f(x_k),$$

for any $x_k \in (0, \infty), \lambda_k, t_k, s \in [0, 1], 1 \leq k \leq n$, with $\sum_{k=1}^n \lambda_k = 1$, and $\sum_{k=1}^n t_k^s = 1$.

4 New Generalized Hermite-Hadamard Type Inequalities

In this section, we present a counterpart of the Hermite-Hadamard type inequality for $(\alpha, \beta, \lambda, \lambda_0, h)$ convex functional. In what follows, we write

$$E_n(p)=\{x=(x_1, x_2, \cdots, x_n): x_k\in\mathbb{R}^1, 1\le k\le n, \|x\|_p=(\sum_{k=1}^{n}|x_k|^p)^{1/p}, 1\le p<\infty\},$$

$$B(0, r) = \{x \in E_n(p) : \|x\|_p \le r\}.$$

In particular, $E_n(2)$ is an n-dimensional Euclidean space $\mathbb{R}^n$.

Theorem 3 *Let $B(0, r_1)$ be an n-ball of radius r_1 in $E_n(p)$, $E = B(0, r_2) - B(0, r_1)$, $0 < r_1 < r_2 < \infty$. Let $f : E \to (0, \infty)$ be a $(\alpha, \beta, \lambda, \lambda_0, h)$ convex functional. If $\int_E \|x\|_p^{\alpha-n} f^\beta(\|x\|_p)dx < \infty$, and $h \in L(0, 1)$, then*

$$\begin{aligned}
&\frac{1}{2h(1/2)} f^\beta((\frac{r_1^{\alpha/p}+\lambda_0 r_2^{\alpha/p}}{2})^{1/\alpha})\\
&\le \frac{\alpha p^{n-1}\Gamma(n/p)}{(\lambda_0 r_2^{\alpha/p}-r_1^{\alpha/p})(\Gamma(1/p))^n}\int_E \|x\|_p^{\alpha-n} f^\beta(\|x\|_p)dx\\
&\le \{f^\beta(r_1^{1/p})+\lambda_0 f^\beta(r_2^{1/p})\}\int_0^1 h(u)du, \qquad (61)
\end{aligned}$$

where $\Gamma(\alpha)$ is the Gamma function:

$$\Gamma(\alpha) = \int_0^\infty x^{\alpha-1}e^{-x}dx \ (\alpha > 0).$$

Proof By transforming the integral to polar coordinates (see [9]), we have

$$\int_E \|x\|_p^{\alpha-n} f^\beta(\|x\|_p)dx = \frac{(\Gamma(1/p))^n}{p^n\Gamma(n/p)}\int_{r_1}^{r_2} r^{(\alpha/p)-1} f^\beta(r^{1/p})dr. \qquad (62)$$

Setting $r = \left(\frac{r_2-u}{r_2-r_1}r_1^{\alpha/p} + \lambda_0\frac{u-r_1}{r_2-r_1}r_2^{\alpha/p}\right)^{p/\alpha}$, we have

$$\begin{aligned}
\int_{r_1}^{r_2} r^{(\alpha/p)-1} f^\beta(r^{1/p})dr &= \frac{p(\lambda_0 r_2^{\alpha/p}-r_1^{\alpha/p})}{\alpha(r_2-r_1)}\\
&\quad\times\int_{r_1}^{r_2} f^\beta((\frac{r_2-u}{r_2-r_1}r_1^{\alpha/p}+\lambda_0\frac{u-r_1}{r_2-r_1}r_2^{\alpha/p})^{1/\alpha})du.
\end{aligned} \qquad (63)$$

By (5) in Definition 2, we get

$$f^{\beta}((\frac{r_2-u}{r_2-r_1}r_1^{\alpha/p}+\lambda_0\frac{u-r_1}{r_2-r_1}r_2^{\alpha/p})^{1/\alpha})$$
$$\leq h(\frac{r_2-u}{r_2-r_1})f^{\beta}(r_1^{1/p})+\lambda_0 h(\frac{u-r_1}{r_2-r_1})f^{\beta}(r_2^{1/p}). \tag{64}$$

Thus, by (62), (63), and (64), we obtain

$$\int_E \|x\|_p^{\alpha-n} f^{\beta}(\|x\|_p)dx$$
$$\leq \frac{(\Gamma(1/p))^n(\lambda_0 r_2^{\alpha/p}-r_1^{\alpha/p})}{\alpha p^{n-1}(r_2-r_1)\Gamma(n/p)}$$
$$\times\{f^{\beta}(r_1^{1/p})\int_{r_1}^{r_2} h(\frac{r_2-u}{r_2-r_1})du+\lambda_0 f^{\beta}(r_2^{1/p})\int_{r_1}^{r_2} h(\frac{u-r_1}{r_2-r_1})du\}$$
$$=\frac{(\Gamma(1/p))^n(\lambda_0 r_2^{\alpha/p}-r_1^{\alpha/p})}{\alpha p^{n-1}\Gamma(n/p)}\{f^{\beta}(r_1^{1/p})+\lambda_0 f^{\beta}(r_2^{1/p})\}\int_0^1 h(u)du$$

which gives the right-hand inequality in (61).

To show the left-hand inequality in (61), setting $u=\frac{1}{2}(r_1+r_2)+t$, then

$$r^{\alpha/p}=\frac{r_2-u}{r_2-r_1}r_1^{\alpha/p}+\lambda_0\frac{u-r_1}{r_2-r_1}r_2^{\alpha/p}$$
$$=\frac{1}{2}(r_1^{\alpha/p}+\lambda_0 r_2^{\alpha/p})+\frac{\lambda_0 r_2^{\alpha/p}-r_1^{\alpha/p}}{r_2-r_1}t. \tag{65}$$

Setting

$$\|x_1\|_p=\{\frac{1}{2}(r_1^{\alpha/p}+\lambda_0 r_2^{\alpha/p})-\frac{\lambda_0 r_2^{\alpha/p}-r_1^{\alpha/p}}{r_2-r_1}t\}^{p/\alpha},$$
$$\|x_2\|_p=\{\frac{1}{2}(r_1^{\alpha/p}+\lambda_0 r_2^{\alpha/p})+\frac{\lambda_0 r_2^{\alpha/p}-r_1^{\alpha/p}}{r_2-r_1}t\}^{p/\alpha}$$

we get

$$\|x_1\|_p^{\alpha/p}+\|x_2\|_p^{\alpha/p}=r_1^{\alpha/p}+\lambda_0 r_2^{\alpha/p}.$$

Thus, by the definition of (α,β,λ,h) convex functional, we have

$$f^{\beta}((\frac{r_1^{\alpha/p}+\lambda_0 r_2^{\alpha/p}}{2})^{1/\alpha})=f^{\beta}((\frac{1}{2}\|x_1\|_p^{\alpha/p}+\frac{1}{2}\|x_2\|_p^{\alpha/p})^{1/\alpha})$$
$$\leq h(1/2)\{f^{\beta}(\|x_1\|_p^{1/p})+f^{\beta}(\|x_2\|_p^{1/p})\}. \tag{66}$$

Hence, by (63), (65), and (66), we get

$$\int_{r_1}^{r_2} r^{\frac{\alpha}{p}-1} f^{\beta}(r^{1/p})dr = \frac{p(\lambda_0 r_2^{\alpha/p} - r_1^{\alpha/p})}{\alpha(r_2 - r_1)}$$

$$\times \int_{-(r_2-r_1)/2}^{(r_2-r_1)/2} f^{\beta}((\frac{1}{2}(r_1^{\alpha/p} + \lambda_0 r_2^{\alpha/p}) + \frac{\lambda_0 r_2^{\alpha/p} - r_1^{\alpha/p}}{r_2 - r_1} t)^{1/\alpha})dt$$

$$= \frac{p(\lambda_0 r_2^{\alpha/p} - r_1^{\alpha/p})}{\alpha(r_2 - r_1)} \int_0^{(r_2-r_1)/2} (f^{\beta}(\|x_1\|_p^{1/p}) + f^{\beta}(\|x_2\|_p^{1/p}))dt$$

$$\geq \frac{p(\lambda_0 r_2^{\alpha/p} - r_1^{\alpha/p})}{\alpha(r_2 - r_1)h(1/2)} \int_0^{(r_2-r_1)/2} f^{\beta}((\frac{r_1^{\alpha/p} + \lambda_0 r_2^{\alpha/p}}{2})^{1/\alpha})dt$$

$$= \frac{p(\lambda_0 r_2^{\alpha/p} - r_1^{\alpha/p})}{2\alpha h(1/2)} f^{\beta}((\frac{r_1^{\alpha/p} + \lambda_0 r_2^{\alpha/p}}{2})^{1/\alpha}). \quad (67)$$

By (62) and (67), we get

$$\int_E \|x\|_p^{\alpha-n} f^{\beta}(\|x\|_p)dx = \frac{(\Gamma(1/p))^n}{p^n \Gamma(n/p)} \int_{r_1}^{r_2} r^{(\alpha/p)-1} f^{\beta}(r^{1/p})dr$$

$$\geq \frac{(\Gamma(1/p))^n (\lambda_0 r_2^{\alpha/p} - r_1^{\alpha/p})}{2\alpha p^{n-1} h(1/2)\Gamma(n/p)} f^{\beta}((\frac{r_1^{\alpha/p} + \lambda_0 r_2^{\alpha/p}}{2})^{1/\alpha}),$$

which finishes the proof.

Corollary 7 *Let $f : E_n(p) \to (0, \infty)$ be a $(\alpha, \beta, \lambda, \lambda_0, h)$ convex functional. If $h \in L(0, 1)$ and $\int_{B(0,r)} \|x\|_p^{\alpha-n} f^{\beta}(\|x\|_p)dx < \infty$, then*

$$\frac{1}{2h(1/2)} f^{\beta}((\frac{\lambda_0}{2})^{1/\alpha} r^{1/p}) \leq \frac{\alpha p^{n-1}\Gamma(n/p)}{\lambda_0 r^{\alpha/p}(\Gamma(1/p))^n} \int_{B(0,r)} \|x\|_p^{\alpha-n} f^{\beta}(\|x\|_p)dx$$

$$\leq \{f^{\beta}(0) + \lambda_0 f^{\beta}(r^{1/p})\} \int_0^1 h(u)du. \quad (68)$$

Corollary 8 *Let $f : (0, \infty) \to (0, \infty)$ be a $(\alpha, \beta, \lambda, \lambda_0, h)$ convex function. If $h \in L(0, 1)$, and $\int_a^b x^{\alpha-1} f^{\beta}(x)dx < \infty$, $0 < a < b < \infty$, then*

$$\frac{1}{2h(1/2)} f^{\beta}((\frac{a^{\alpha} + \lambda_0 b^{\alpha}}{2})^{1/\alpha}) \leq \frac{\alpha}{\lambda_0 b^{\alpha} - a^{\alpha}} \int_a^b x^{\alpha-1} f^{\beta}(x)dx$$

$$\leq \{f^{\beta}(a) + \lambda_0 f^{\beta}(b)\} \int_0^1 h(u)du. \quad (69)$$

Corollary 9 ([4]) *Let $f : (0,\infty) \to (0,\infty)$ be an h-convex function. If $h \in L(0,1)$, $f \in L[a,b]$, $[a,b] \subset (0,\infty)$, then*

$$\frac{1}{2h(1/2)} f(\frac{a+b}{2}) \leq \frac{1}{b-a}\int_a^b f(x)dx \leq \{f(a)+f(b)\}\int_0^1 h(u)du. \tag{70}$$

In particular, if $h(t) = t^s, 0 < s \leq 1$, then (70) reduces to (25) in [4]; if $h(t) = t^{-s}, 0 < s \leq 1$, then (70) reduces to (26) in [4].

Remark 1 If $\beta = 1$ and $\lambda_0 = 1$, then (69) reduces to Theorem 5 in [5]. For $h(t) = t, \lambda_0 = 1$ in (69), we get

$$f^\beta((\frac{a^\alpha + b^\alpha}{2})^{1/\alpha}) \leq \frac{\alpha}{b^\alpha - a^\alpha}\int_a^b x^{\alpha-1} f^\beta(x)dx \leq \frac{1}{2}\{f^\beta(a) + f^\beta(b)\}. \tag{71}$$

For $\alpha = 1$ in (71), we get

$$f^\beta(\frac{a+b}{2}) \leq \frac{1}{b-a}\int_a^b f^\beta(x)dx \leq \frac{1}{2}\{f^\beta(a) + f^\beta(b)\}. \tag{72}$$

If $\beta = 1$, then (72) reduces to the classical Hermite-Hadamard inequality:

$$f(\frac{a+b}{2}) \leq \frac{1}{b-a}\int_a^b f(x)dx \leq \frac{1}{2}\{f(a)+f(b)\}. \tag{73}$$

Remark 2 Inequality (72) is proved by Yang Zhen-hang, but he adds the conditions: $\beta \geq 1$ and $f(x), f^{''}(x) > 0$. (see [11, P. 12]).

Theorem 4 *Let $f : (0,\infty) \to (0,\infty)$ be a (α,β,λ,h) convex function and $h : (0,1) \to (0,\infty)$ is a super-multiplicative function, if $[a,b] \subset (0,\infty)$ and $a^\alpha = x_1^\alpha < x_2^\alpha < \cdots < x_n^\alpha = b^\alpha$ be equidistant points, then*

$$\frac{1}{h(1/n)} f^\beta((\frac{x_1^\alpha + x_n^\alpha}{2})^{1/\alpha}) \leq \sum_{k=1}^n f^\beta(x_k) \leq f^\beta(x_1)\sum_{k=1}^n h(1-\lambda_k) + f^\beta(x_n)\sum_{k=1}^n h(\lambda_k), \tag{74}$$

where $\lambda_k = \frac{k-1}{n-1}, k = 1, 2, \cdots, n, n > 1$.

Proof Since the points $x_1^\alpha, \cdots, x_n^\alpha$ are equidistant, putting $t = \frac{x_n^\alpha - x_1^\alpha}{n-1}$, we have $x_k^\alpha = x_1^\alpha + (k-1)t, k = 1, 2, \cdots, n$ and $\frac{1}{n}\sum_{k=1}^n x_k^\alpha = \frac{1}{2}(x_1^\alpha + x_n^\alpha)$. By Corollary 4, we get

$$f^\beta((\frac{x_1^\alpha + x_n^\alpha}{2})^{1/\alpha}) = f^\beta((\frac{1}{n}\sum_{k=1}^n x_k^\alpha)^{1/\alpha})$$
$$\le \sum_{k=1}^n h(\frac{1}{n}) f^\beta(x_k) = h(\frac{1}{n})\sum_{k=1}^n f^\beta(x_k),$$

which gives the left-hand inequality in (74).

To show the right-hand inequality in (74), we note that $x_k^\alpha = x_1^\alpha + (k-1)t$ can be written as $x_k^\alpha = (1-\lambda_k)x_1^\alpha + \lambda_k x_n^\alpha$, where $\lambda_k = \frac{k-1}{n-1}, k = 1, 2, \cdots, n$. By the definition of $(\alpha, \beta, \lambda, h)$ convex function, we get

$$f^\beta(x_k) = f^\beta(((1-\lambda_k)x_1^\alpha + \lambda_k x_n^\alpha)^{1/\alpha})$$
$$\le h(1-\lambda_k) f^\beta(x_1) + h(\lambda_k) f^\beta(x_n).$$

Summing up the above inequalities, we get

$$\sum_{k=1}^n f^\beta(x_k) \le f^\beta(x_1)\sum_{k=1}^n h(1-\lambda_k) + f^\beta(x_n)\sum_{k=1}^n h(\lambda_k),$$

which finishes the proof.

Corollary 10 *Let $f : (0, \infty) \to (0, \infty)$ be an h-convex function and $h : (0, 1) \to (0, \infty)$ is a super-multiplicative function. If $[a, b] \subset (0, \infty)$ and $a = x_1 < x_2 < \cdots < x_n = b$ be equidistant points, then*

$$\frac{1}{h(1/n)} f(\frac{x_1 + x_n}{2}) \le \sum_{k=1}^n f(x_k)$$
$$\le f(x_1)\sum_{k=1}^n h(1-\lambda_k) + f(x_n)\sum_{k=1}^n h(\lambda_k), \tag{75}$$

where $\lambda_k = \frac{k-1}{n-1}, k = 1, 2, \cdots, n, n > 1$.

Remark 3 Using Lemma 1, we also obtain Theorem 4 from Corollary 10. For $h(t) = t^s, 0 < |s| \le 1$ in (74), we get

$$f^\beta((\frac{x_1^\alpha + x_n^\alpha}{2})^{1/\alpha}) \le \frac{1}{n^s}\sum_{k=1}^n f^\beta(x_n) \le \frac{1}{(n(n-1))^s}(\sum_{k=1}^{n-1} k^s)\{f^\beta(x_1) + f^\beta(x_n)\}.$$

In particular, when $s = 1$, we get

$$f^{\beta}((\frac{x_1^{\alpha}+x_n^{\alpha}}{2})^{1/\alpha}) \leq \frac{1}{n}\sum_{k=1}^{n} f^{\beta}(x_k) \leq \frac{1}{2}\{f^{\beta}(x_1)+f^{\beta}(x_n)\}. \quad (76)$$

If $\alpha = \beta = 1$, then (76) reduces to the discrete analogous of the classical Hermite-Hadamard inequality (73) (see [13]):

$$f(\frac{x_1+x_n}{2}) \leq \frac{1}{n}\sum_{k=1}^{n} f(x_k) \leq \frac{1}{2}\{f(x_1)+f(x_n)\}.$$

5 Strongly Convex Functionals in Normed Linear Spaces

Strongly convex functions have been introduced by Polyak [17] and they play an important role in optimization theory, mathematical economics, and other branches of pure and applied mathematics. Many properties and applications of them can be found in the literature (see, for instance, [1, 4, 10, 13, 14, 16, 17], and the references therein).

In what follows, $(X, \|\cdot\|)$ denotes the real normed linear spaces, D be a convex subset of X, $h : (0, 1) \to (0, \infty)$ is a given function and c be a positive constant.

Definition 4 (See [13]) A function $f : D \to \mathbb{R}$ is called strongly convex with modulus c, if

$$f(\lambda x_1 + (1-\lambda)x_2) \leq \lambda f(x_1) + (1-\lambda)f(x_2) - c\lambda(1-\lambda)\|x_1 - x_2\|^2, \quad (77)$$

$\forall x_1, x_2 \in D, \forall \lambda \in [0, 1]$.

In this section, we introduce a new class of strongly convex functional with modulus c in real normed linear spaces, that is, $(\alpha, \beta, \lambda, t, h)$ strongly convex functional with modulus c in real normed linear spaces, and present the new Schur, Jensen, and Hermite-Hadamard type inequalities for these strongly convex functional with modulus c. They are significant generalizations of the corresponding inequalities for the classical convex functions.

Definition 5 A functional $f : D \to (0, \infty)$ is said to be a $(\alpha, \beta, \lambda, t, h)$ strongly convex with modulus c, if

$$\begin{aligned} f^{\beta}((\lambda\|x_1\|^{\alpha} + (1-\lambda)\|x_2\|^{\alpha})^{1/\alpha}) &\leq h(t)f^{\beta}(\|\|x_1\|) + h(1-t)f^{\beta}(\|x_2\|) \\ &\quad - ch(t)h(1-t)|\|x_1\|^{\alpha} - \|x_2\|^{\alpha}|^2, \quad (78) \end{aligned}$$

$\forall x_1, x_2 \in D, \forall \lambda, t \in [0, 1], \alpha, \beta$ are real numbers, and $\alpha, \beta \neq 0$.

For $c = 0$ in (78), we get

$$f((\lambda\|x_1\|^\alpha+(1-\lambda)\|x_2\|^\alpha)^{1/\alpha}) \leq \{h(t)f^\beta(\|x_1\|)+h(1-t)f^\beta(\|x_2\|)\}^{1/\beta}, \quad (79)$$

that is, f reduces to $(\alpha, \beta, \lambda, t, h)$ convex functional in Section 2.

For $t = \lambda$ in (78), that is,

$$f^\beta((\lambda\|x_1\|^\alpha + (1-\lambda)\|x_2\|^\alpha)^{1/\alpha}) \leq h(\lambda)f^\beta(\|x_1\|) + h(1-\lambda)f^\beta(\|x_2\|) - ch(\lambda)h(1-\lambda)|\|x_1\|^\alpha - \|x_2\|^\alpha|^2, \quad (80)$$

then f is said to be a $(\alpha, \beta, \lambda, h)$ strongly convex functional with modulus c. If $c = 0$ in (80), then f reduces to $(\alpha, \beta, \lambda, h)$ convex functional in Section 2.

If $h(\lambda) = \lambda^s, 0 < |s| \leq 1$, then $(\alpha, \beta, \lambda, t, h)$, $(\alpha, \beta, \lambda, h)$ strongly convex functional with modulus c reduce to $(\alpha, \beta, \lambda, t, s)$, $(\alpha, \beta, \lambda, s)$ strongly convex functional with modulus c, respectively. In particular, if $s = 1$, then f is said to be a $(\alpha, \beta, \lambda, t)$, (α, β, λ) strongly convex functional with modulus c, respectively.

If $D = (0, \infty)$ in (78), that is, if a function $f : (0, \infty) \to (0, \infty)$ satisfies

$$f^\beta((\lambda x_1^\alpha + (1-\lambda)x_2^\alpha)^{1/\alpha}) \leq h(t)f^\beta(x_1) + h(1-t)f^\beta(x_2) - ch(t)h(1-t)|x_2^\alpha - x_1^\alpha|^2, \quad (81)$$

$\forall x_1, x_2 \in (0, \infty)$, $\forall\lambda, t \in [0, 1]$, α, β are real numbers, and $\alpha, \beta \neq 0$, then f is said to be a $(\alpha, \beta, \lambda, t, h)$ strongly convex function with modulus c.

For $c = 0$ in (81), we get

$$f((\lambda x_1^\alpha + (1-\lambda)x_2^\alpha)^{1/\alpha}) \leq \{h(t)f^\beta(x_1) + h(1-t)f^\beta(x_2)\}^{1/\beta}, \quad (82)$$

that is, f reduces to $(\alpha, \beta, \lambda, t, h)$ convex function in Section 2.

If $t = \lambda$ in (81), that is,

$$f^\beta((\lambda x_1^\alpha + (1-\lambda)x_2^\alpha)^{1/\alpha}) \leq h(\lambda)f^\beta(x_1) + h(1-\lambda)f^\beta(x_2) - ch(\lambda)h(1-\lambda)|x_2^\alpha - x_1^\alpha|^2, \quad (83)$$

$\forall x_1, x_2 \in (0, \infty)$, $\forall\lambda \in [0, 1]$, α, β are real numbers, and $\alpha, \beta \neq 0$, then f is said to be a $(\alpha, \beta, \lambda, h)$ strongly convex function with modulus c.

For $c = 0$ in (83), we get

$$f((\lambda x_1^\alpha + (1-\lambda)x_2^\alpha)^{1/\alpha}) \leq \{h(\lambda)f^\beta(x_1) + h(1-\lambda)f^\beta(x_2)\}^{1/\beta}, \quad (84)$$

that is, f reduces to $(\alpha, \beta, \lambda, h)$ convex function in Section 2.

If $h(\lambda) = \lambda^s, 0 < |s| \leq 1$ in (83), we get

$$f^\beta((\lambda x_1^\alpha + (1-\lambda)x_2^\alpha)^{1/\alpha}) \leq \lambda^s f^\beta(x_1) + (1-\lambda)^s f^\beta(x_2) - c\lambda^s(1-\lambda)^s|x_2^\alpha - x_1^\alpha|^2, \tag{85}$$

then f is said to be a $(\alpha, \beta, \lambda, s)$ strongly convex function with modulus c.

For $c = 0$ in (85), that is,

$$f((\lambda x_1^\alpha + (1-\lambda)x_2^\alpha)^{1/\alpha}) \leq \{\lambda^s f^\beta(x_1) + (1-\lambda)^s f^\beta(x_2)\}^{1/\beta}, \tag{86}$$

that is, f reduces to $(\alpha, \beta, \lambda, s)$ convex function in Section 2.

Remark 4 If $\alpha = \beta = 1, s = 1$, then (86) reduces to the classical convex function. In fact, the notion of $(\alpha, \beta, \lambda, t, h)$ strongly convex functional with modulus c unifies and generalizes the many known and new classes of convex functions, see, e.g., [1, 10, 13, 14, 16, 17], and the references therein.

6 New Schur, Jensen, Hermite-Hadamard Type Inequalities

In this section, we present the Schur, Jensen, and Hermite-Hadamard type inequalities for (α, β, λ) strongly convex functional with modulus c.

Lemma 2 *Let $g = \{f^\beta - c\|\cdot\|^{2\alpha}\}^{1/\beta}$ with $f^\beta(\|x\|) \geq c\|x\|^{2\alpha}, x \in D$, then a functional $f : D \to (0, \infty)$ is (α, β, λ) strongly convex with modulus c if and only if the functional $g : D \to (0, \infty)$ is (α, β, λ) convex.*

Proof Assume that f is (α, β, λ) strongly convex with modulus c, then

$$\begin{aligned}
&g^\beta((\lambda\|x_1\|^\alpha + (1-\lambda)\|x_2\|^\alpha)^{1/\alpha}) \\
&\quad = f^\beta((\lambda\|x_1\|^\alpha + (1-\lambda)\|x_2\|^\alpha)^{1/\alpha}) - c|(\lambda\|x_1\|^\alpha + (1-\lambda)\|x_2\|^\alpha)^{1/\alpha}|^{2\alpha} \\
&\quad \leq \lambda f^\beta(\|x_1\|) + (1-\lambda) f^\beta(\|x_2\|) - c\lambda(1-\lambda)|\|x_1\|^\alpha - \|x_2\|^\alpha|^2 \\
&\quad -c|\lambda\|x_1\|^\alpha + (1-\lambda)\|x_2\|^\alpha|^2 \\
&\quad = \lambda f^\beta(\|x_1\|) + (1-\lambda) f^\beta(\|x_2\|) - c\lambda\|x_1\|^{2\alpha} - c(1-\lambda)\|x_2\|^{2\alpha} \\
&\quad = \lambda g^\beta(\|x_1\|) + (1-\lambda) g^\beta(\|x_2\|),
\end{aligned}$$

which proves that g is (α, β, λ) convex.

Conversely, if g is (α, β, λ) convex, then

$$\begin{aligned}
&f^\beta((\lambda\|x_1\|^\alpha + (1-\lambda)\|x_2\|^\alpha)^{1/\alpha}) \\
&= g^\beta((\lambda\|x_1\|^\alpha + (1-\lambda)\|x_2\|^\alpha)^{1/\alpha}) + c\|(\lambda\|x_1\|^\alpha + (1-\lambda)\|x_2\|^\alpha)^{1/\alpha}\|^{2\alpha} \\
&\leq \lambda g^\beta(\|x_1\|) + (1-\lambda) g^\beta(\|x_2\|) + c|\lambda\|x_1\|^\alpha + (1-\lambda)\|x_2\|^\alpha|^2 \\
&= \lambda f^\beta(\|x_1\|) + (1-\lambda) f^\beta(\|x_2\|) - c\lambda(1-\lambda)|\|x_1\|^\alpha - \|x_2\|^\alpha|^2,
\end{aligned}$$

which proves that f is (α, β, λ) strongly convex with modulus c.

Using Corollary 1 (with $h(t) = t$) and Lemma 2, and the definition of (α, β, λ) strongly convex with modulus c, we get

Theorem 5 *Let a functional $f : D \to (0, \infty)$ be (α, β, λ) strongly convex with modulus c, and $h : (0, 1) \to (0, \infty)$ is a super-multiplicative function, then for all $x_1, x_2, x_3 \in D$, such that $\|x_1\|^\alpha < \|x_2\|^\alpha < \|x_3\|^\alpha$, and $0 < \|x_3\|^\alpha - \|x_1\|^\alpha < 1$, the following Schur-type inequalities holds:*

$$f^\beta(\|x_2\|) \leq \frac{\|x_3\|^\alpha - \|x_2\|^\alpha}{\|x_3\|^\alpha - \|x_1\|^\alpha} f^\beta(\|x_1\|) + \frac{\|x_2\|^\alpha - \|x_1\|^\alpha}{\|x_3\|^\alpha - \|x_1\|^\alpha} f^\beta(\|x_3\|)$$
$$- c\{\|x_2\|^{2\alpha} + (\|x_1\|^\alpha - \|x_2\|^\alpha)\|x_3\|^\alpha - \|x_1\|^\alpha\|x_2\|^\alpha\}. \tag{87}$$

Using Corollary 3 (with $h(\lambda_k = \lambda_k, t_k = \lambda_k, \xi_k = 1))$ and Lemma 2, and the definition of (α, β, λ) strongly convex with modulus c, one obtains the following new Jensen-type inequality:

Theorem 6 *Let a functional $f : D \to (0, \infty)$ be (α, β, λ) strongly convex with modulus c, and $f^\beta(\|x\|) \geq c\|x\|^{2\alpha}, x \in D$, and $h : (0, 1) \to (0, \infty)$ is a super-multiplicative function, then*

$$f^\beta((\sum_{k=1}^{n} \lambda_k\|x_k\|^\alpha)^{1/\alpha}) \leq \sum_{k=1}^{n} \lambda_k f^\beta(\|x_k\|)$$
$$-c\{\sum_{k=1}^{n} \lambda_k\|x_k\|^{2\alpha} - (\sum_{k=1}^{n} \lambda_k\|x_k\|^\alpha)^2\}, \tag{88}$$

for any $x_k \in D, \lambda_k \in [0, 1], 1 \leq k \leq n$, with $\sum_{k=1}^{n} \lambda_k = 1$.

We present a counterpart of the Hermite-Hadamard inequality for $(\alpha, \beta, \lambda, h)$ strongly convex functional with modulus c. In what follows, we use the notations in Section 4.

Theorem 7 *Let $B(0, r_1)$ be an n-ball of radius r_1 in $E_n(p)$, $E = B(0, r_2) - B(0, r_1), 0 < r_1 < r_2 < \infty$. Let $f : E \to (0, \infty)$ be a $(\alpha, \beta, \lambda, h)$ strongly convex functional with modulus c. If $\int_E \|x\|_p^{\alpha-n} f^\beta(\|x\|_p)dx < \infty$, and $h \in L(0, 1)$, then*

$$\frac{1}{2h(1/2)} f^\beta((\frac{r_2^{\alpha/p} + r_1^{\alpha/p}}{2})^{1/\alpha}) + \frac{c}{6}(r_2^{\alpha/p} - r_1^{\alpha/p})^2$$
$$\leq \frac{\alpha p^{n-1}\Gamma(n/p)}{(r_2^{\alpha/p} - r_1^{\alpha/p})\Gamma^n(1/p)} \int_E \|x\|_p^{\alpha-n} f^\beta(\|x\|_p)dx$$
$$\leq \{f^\beta(r_1^{1/p}) + f^\beta(r_2^{1/p})\} \int_0^1 h(u)du$$
$$-c|r_2^{\alpha/p} - r_1^{\alpha/p}|^2 \int_0^1 h(t)h(1-t)dt. \tag{89}$$

Proof By transforming the integral to polar coordinates (see [9]), we have

$$\int_E \|x\|_p^{\alpha-n} f^\beta(\|x\|_p)dx = \frac{(\Gamma(1/p))^n}{p^n\Gamma(n/p)} \int_{r_1}^{r_2} r^{(\alpha/p)-1} f^\beta(r^{1/p})dr. \tag{90}$$

Setting $r = \left(\frac{r_2-u}{r_2-r_1} r_1^{\alpha/p} + \frac{u-r_1}{r_2-r_1} r_2^{\alpha/p}\right)^{p/\alpha}$, we have

$$\int_{r_1}^{r_2} r^{(\alpha/p)-1} f^\beta(r^{1/p})dr = \frac{p(r_2^{\alpha/p} - r_1^{\alpha/p})}{\alpha(r_2 - r_1)} \\ \times \int_{r_1}^{r_2} f^\beta((\frac{r_2-u}{r_2-r_1} r_1^{\alpha/p} + \frac{u-r_1}{r_2-r_1} r_2^{\alpha/p})^{1/\alpha})du. \tag{91}$$

By the definition of $(\alpha, \beta, \lambda, h)$ strongly convex with modulus c, we get

$$f^\beta((\frac{r_2-u}{r_2-r_1} r_1^{\alpha/p} + \frac{u-r_1}{r_2-r_1} r_2^{\alpha/p})^{1/\alpha}) \\ \leq h(\frac{r_2-u}{r_2-r_1}) f^\beta(r_1^{1/p}) + h(\frac{u-r_1}{r_2-r_1}) f^\beta(r_2^{1/p}) \\ -ch(\frac{r_2-u}{r_2-r_1})h(\frac{u-r_1}{r_2-r_1})|r_2^{\alpha/p} - r_1^{\alpha/p}|^2. \tag{92}$$

Thus, by (90), (91) and (92), we obtain

$$\int_E \|x\|_p^{\alpha-n} f^\beta(\|x\|_p)dx \leq \frac{(\Gamma(1/p))^n (r_2^\alpha - r_1^\alpha)}{\alpha p^{n-1}(r_2 - r_1)\Gamma(n/p)} \\ \times\{f^\beta(r_1^{1/p}) \int_{r_1}^{r_2} h(\frac{r_2-u}{r_2-r_1})du + f^\beta(r_2^{1/p}) \int_{r_1}^{r_2} h(\frac{u-r_1}{r_2-r_1})du \\ -c|r_2^{\alpha/p} - r_1^{\alpha/p}|^2 \int_{r_1}^{r_2} h(\frac{r_2-u}{r_2-r_1})h(\frac{u-r_1}{r_2-r_1})du\} \\ = \frac{(\Gamma(1/\alpha))^n (r_2^{\alpha/p} - r_1^{\alpha/p})}{\alpha p^{n-1}\Gamma(n/p)}\{(f^\beta(r_1^{1/p}) + f^\beta(r_2^{1/p})) \\ \times \int_0^1 h(t)dt - c|r_2^{\alpha/p} - r_1^{\alpha/p}|^2 \int_0^1 h(t)h(1-t)dt\},$$

which gives the right-hand inequality in (89).

To show the left-hand inequality in (89), setting $u = \frac{1}{2}(r_1 + r_2) + t$, then

$$r^{\alpha/p} = \frac{r_2-u}{r_2-r_1} r_1^{\alpha/p} + \frac{u-r_1}{r_2-r_1} r_2^{\alpha/p} \\ = \frac{1}{2}(r_1^{\alpha/p} + r_2^{\alpha/p}) + \frac{r_2^{\alpha/p} - r_1^{\alpha/p}}{r_2 - r_1} t. \tag{93}$$

Setting

$$\|x_1\|_p = \{\frac{1}{2}(r_1^{\alpha/p} + r_2^{\alpha/p}) - \frac{r_2^{\alpha/p} - r_1^{\alpha/p}}{r_2 - r_1}t\}^{p/\alpha}$$

$$\|x_2\|_p = \{\frac{1}{2}(r_1^{\alpha/p} + r_2^{\alpha/p}) + \frac{r_2^{\alpha/p} - r_1^{\alpha/p}}{r_2 - r_1}t\}^{p/\alpha}$$

we get

$$\|x_1\|_p^{\alpha/p} + \|x_2\|_p^{\alpha/p} = r_1^{\alpha/p} + r_2^{\alpha/p}$$

$$\|x_2\|_p^{\alpha/p} - \|x_1\|_p^{\alpha/p} = \frac{2(r_2^{\alpha} - r_1^{\alpha})}{r_2 - r_1}t.$$

Thus, by (80), we have

$$\begin{aligned} f^{\beta}((\frac{r_1^{\alpha/p} + r_2^{\alpha/p}}{2})^{1/\alpha}) &= f^{\beta}((\frac{1}{2}\|x_1\|_p^{\alpha/p} + \frac{1}{2}\|x_2\|_p^{\alpha/p})^{1/\alpha}) \\ &\leq h(\frac{1}{2})f^{\beta}(\|x_1\|_p^{1/p}) + h(\frac{1}{2})f^{\beta}(\|x_2\|_p^{1/p}) - c(h(\frac{1}{2}))^2|\|x_2\|_p^{\alpha/p} - \|x_1\|_p^{\alpha/p}|^2 \\ &= h(\frac{1}{2})\{f^{\beta}(\|x_1\|_p^{1/p}) + f^{\beta}(\|x_2\|_p^{1/p})\} - 4c(h(\frac{1}{2}))^2(\frac{r_2^{\alpha/p} - r_1^{\alpha/p}}{r_2 - r_1})^2t^2. \quad (94) \end{aligned}$$

Hence, by (91), (93), and (94), we get

$$\begin{aligned} \int_{r_1}^{r_2} r^{(\alpha/p)-1} f^{\beta}(r^{1/p})dr &= \frac{p(r_2^{\alpha/p} - r_1^{\alpha/p})}{\alpha(r_2 - r_1)} \\ &\times \int_{-(r_2-r_1)/2}^{(r_2-r_1)/2} f^{\beta}((\frac{1}{2}(r_1^{\alpha/p} + r_2^{\alpha/p}) + \frac{r_2^{\alpha/p} - r_1^{\alpha/p}}{r_2 - r_1}t)^{1/\alpha})dt \\ &= \frac{p(r_2^{\alpha/p} - r_1^{\alpha/p})}{\alpha(r_2 - r_1)} \int_0^{(r_2-r_1)/2} \{f^{\beta}(\|x_1\|_p^{1/p}) + f^{\beta}(\|x_2\|_p^{1/p})\}dt \\ &\geq \frac{p(r_2^{\alpha/p} - r_1^{\alpha/p})}{\alpha(r_2 - r_1)h(1/2)} \int_0^{(r_2-r_1)/2} \{f^{\beta}((\frac{r_1^{\alpha/p} + r_2^{\alpha/p}}{2})^{1/\alpha}) + 4ch(\frac{1}{2})(\frac{r_2^{\alpha/p} - r_1^{\alpha/p}}{r_2 - r_1})^2t^2\}dt \\ &= \frac{p(r_2^{\alpha/p} - r_1^{\alpha/p})}{2\alpha h(1/2)} f^{\beta}((\frac{r_1^{\alpha/p} + r_2^{\alpha/p}}{2})^{1/\alpha}) + c\frac{p(r_2^{\alpha/p} - r_1^{\alpha/p})^3}{6\alpha}. \quad (95) \end{aligned}$$

By (90) and (95), we get

$$\int_E \|x\|_p^{\alpha-n} f^\beta(\|x\|_p)dx = \frac{(\Gamma(1/p))^n}{p^n\Gamma(n/p)}\int_{r_1}^{r_2} r^{(\alpha/p)-1} f^\beta(r^{1/p})dr$$
$$\geq \frac{(\Gamma(1/p))^n(r_2^{\alpha/p}-r_1^{\alpha/p})}{\alpha p^{n-1}\Gamma(n/p)}$$
$$\times \{\frac{1}{2h(1/2)} f^\beta((\frac{r_1^{\alpha/p}+r_2^{\alpha/p}}{2})^{1/\alpha}) + \frac{c}{6}(r_2^{\alpha/p}-r_1^{\alpha/p})^2\},$$

which finishes the proof.

Corollary 11 *Let $X = \mathbb{R}^n$, $B(0, r_k)$ be an n-ball of radius r_k in $\mathbb{R}^n$, $E = B(0, r_2) - B(0, r_1)$, $0 < r_1 < r_2 < \infty$. Let a functional $f : E \to (0, \infty)$ be $(\alpha, \beta, \lambda, h)$ strongly convex with modulus c, $\int_E \|x\|_2^{\alpha-n} f^\beta(\|x\|_2)dx < \infty$, and $h \in L(0, 1)$, then*

$$\frac{1}{2h(1/2)} f^\beta((\frac{r_2^{\alpha/2}+r_1^{\alpha/2}}{2})^{1/\alpha}) + \frac{c}{6}(r_2^{\alpha/2}-r_1^{\alpha/2})^2$$
$$\leq \frac{\alpha 2^{n-1}\Gamma(n/2)}{\pi^{n/2}(r_2^{\alpha/2}-r_1^{\alpha/2})}\int_E \|x\|_2^{\alpha-n} f^\beta(\|x\|_2)dx$$
$$\leq \{f^\beta(r_1^{1/2}) + f^\beta(r_2^{1/2})\}\int_0^1 h(t)dt - c|r_2^{\alpha/2}-r_1^{\alpha/2}|^2\int_0^1 h(t)h(1-t)dt. \tag{96}$$

Corollary 12 *Let a function $f : (0, \infty) \to (0, \infty)$ be $(\alpha, \beta, \lambda, h)$ strongly convex with modulus c. If $\int_a^b x^{\alpha-1} f^\beta(x)dx < \infty$, $0 < a < b < \infty$, and $h \in L(0, 1)$, then*

$$\frac{1}{2h(1/2)} f^\beta((\frac{a^\alpha+b^\alpha}{2})^{1/\alpha}) + \frac{c}{6}(b^\alpha - a^\alpha)^2$$
$$\leq \frac{\alpha}{b^\alpha - a^\alpha}\int_a^b x^{\alpha-1} f^\beta(x)dx$$
$$\leq \{f^\beta(a) + f^\beta(b)\}\int_0^1 h(t)dt - c(b^\alpha - a^\alpha)^2\int_0^1 h(t)h(1-t)dt. \tag{97}$$

If $\alpha = -1$, then (97) reduces to

$$\frac{1}{2h(1/2)} f^\beta(\frac{2ab}{a+b}) + \frac{c}{6}(\frac{b-a}{ab})^2$$
$$\leq \frac{ab}{b-a}\int_a^b \frac{1}{x^2} f^\beta(x)dx$$
$$\leq \{f^\beta(a) + f^\beta(b)\}\int_0^1 h(t)dt - c(\frac{b-a}{ab})^2\int_0^1 h(t)h(1-t)dt.$$

In particular, if $c = 0$, $h(t) = t$, then the above inequality reduces to the mail result of [15]:

$$f^{\beta}(\frac{2ab}{a+b}) \le \frac{ab}{b-a}\int_a^b \frac{1}{x^2} f^{\beta}(x)dx \le \frac{1}{2}\{f^{\beta}(a) + f^{\beta}(b)\}.$$

If $\alpha = \beta = 1$, then (97) reduces to the Hermite-Hadamard inequality for strongly h-convex functions:

$$\frac{1}{2h(1/2)} f(\frac{a+b}{2}) + \frac{c}{6}(b-a)^2 \le \frac{1}{b-a}\int_a^b f(x)dx$$
$$\le \{f(a) + f(b)\}\int_0^1 h(t)dt - c(b-a)^2 \int_0^1 h(t)h(1-t)dt. \quad (98)$$

If $h(t) = t^s$, $0 < s \le 1$, then (98) reduces to:

$$2^{s-1} f(\frac{a+b}{2}) + \frac{c}{6}(b-a)^2 \le \frac{1}{b-a}\int_a^b f(x)dx$$
$$\le \frac{f(a) + f(b)}{s+1} - c(b-a)^2 \frac{(\Gamma(1+s))^2}{\Gamma(2(1+s))}. \quad (99)$$

If $h(t) = t^{-s}$, $0 < s < 1$, then (98) reduces to:

$$2^{-(s+1)} f(\frac{a+b}{2}) + \frac{c}{6}(b-a)^2 \le \frac{1}{b-a}\int_a^b f(x)dx$$
$$\le \frac{f(a) + f(b)}{1-s} - c(b-a)^2 \frac{(\Gamma(1-s))^2}{\Gamma(2(1-s))} \quad (100)$$

If $h(t) = t$, $s = 1$, then (98) reduces to:

$$f(\frac{a+b}{2}) + \frac{c}{6}(b-a)^2 \le \frac{1}{b-a}\int_a^b f(x)dx$$
$$\le \frac{f(a) + f(b)}{2} - \frac{c}{6}(b-a)^2. \quad (101)$$

If $c = 0$, then (101) reduces to the classical Hermite-Hadamard inequality (73).

Hence, the above results are some substantial refinements and generalizations of the corresponding results obtained by Nikodem [13] and Merentes and Nikodem [10].

References

1. A. Azócar, J. Giménez, K. Nikodem, J.L. Sánchez, On strongly midconvex functions. Opusc. Math. **31**(1), 15–26 (2011)
2. S.S. Dragomir, On some new inequalities of Hermite-Hadamard type for m-convex functions. Tamkang J. Math. **33**(1), 55–65 (2002)
3. S.S. Dragomir, Inequalities of Hermite-Hadamard type for λ-convex functions on linear spaces. RGMIA Rep. Coll.17 Art 13 (2014, preprint)
4. S.S. Dragomir, Bounding the čebyšev functional for a differentiable function whose derivative is h or λ-convex in absolute value and applications. Le Matematiche **71**(1), 173–202 (2016)
5. Z.B. Fang, R.J. Shi, On the (p, h)-convex function and some integral inequalities. J. Inequal. Appl. (2014, Paper No. 45)
6. G.H. Hardy, J.E. Littlewood, G. Polya, *Inequalities*, 2nd edn. (Cambridge University Press, London, 1952)
7. J.-B. Hiriart, C. Lemaréchal, *Fundamentals of Convex Analysis* (Springer, Berlin, 2001)
8. J.C. Kuang, *Applied Inequalities*, 4th edn. (Shangdong Science and Technology Press, Jinan, 2010)
9. J.C. Kuang, *Real and Functional Analysis* (continuation), vol. 2 (Higher Education Press, Beijing, 2015)
10. N. Merentes, K. Nikodem, Remarks on strongly convex functions. Aequ. Math. **80**, 193–199 (2010)
11. D.S. Mitrinović, J.E. Pečarić, A.M. Fink, *Classical and New Inequalities in Analysis* (Kluwer Academic Publishers, Dordrecht, 1993)
12. C.P. Niculescu, L.-E. Persson, *Convex Functions and Their Applications. A Contemporary Approach*. CMS Books in Mathematics, vol. 23 (Springer, New York, 2006)
13. K. Nikodem, On strongly convex functions and related classes of functions, in *Handbook of Functional Equations-Functional Inequalities*. Springer Optimization and Its Applications, ed. by T.M. Rassias, vol. 95 (Springer, Berlin, 2014)
14. K. Nikodem, Z. Páles, Characterizations of inner product spaces by strongly convex functions. Banach J. Math. Anal. **5**(1), 83–87 (2011)
15. M.A. Noor, K.I. Noor, M.U. Awan, Some new estimates of Hermite-Hadamard inequalities via harmonically r-convex functions. Le Matematiche **71**(2), 117–127 (2016)
16. E.S. Polovinkin, Strongly convex analysis. Sb. Math. **187**(2), 259–286 (1996)
17. B.T. Polyak, Existence theorems and convergence of minimizing sequence in extremum problems with restrictions. Sov. Math. Dokl. **7**, 72–75 (1966)
18. H.N. Shi, *Theory of Majorization and Analytic Inequalities* (Harbin Institute of Technology Press, Harbin, 2012)
19. S. Varšanec, On h-convexity. J. Math. Anal. Appl. **326**, 303–311 (2007)
20. W.L. Wang, *Approaches to Prove Inequalities* (Harbin Institute of Technology Press, Harbin, 2011)
21. X.M. Zhang, Y.M. Chu, *New Discussion to Analytic Inequalities* (Harbin Institute of Technology Press, Harbin, 2009)
22. K.S. Zhang, J.P. Wang, p-convex functions and their properties. Pure Appl. Math. **23**(1), 130–133 (2007)

Additive Functional Inequalities and Partial Multipliers in Complex Banach Algebras

Jung Rye Lee, Choonkil Park, and Themistocles M. Rassias

Abstract In this paper, we solve the additive functional inequalities

$$\|f(x+y+z)-f(x+y)-f(z)\| \leq \|s(f(x-y)+f(y-z)-f(x-z))\| \quad (1)$$

and

$$\|f(x-y)+f(y-z)-f(x-z)\| \leq \|s(f(x+y-z)+f(x-y+z)-2f(x))\|, \quad (2)$$

where s is a fixed nonzero complex number with $|s|<1$. Using the direct method, we prove the Hyers-Ulam stability of the additive functional inequalities (1) and (2) in complex Banach spaces. This is applied to investigate partial multipliers in Banach $*$-algebras, unital C^*-algebras, Lie C^*-algebras, JC^*-algebras, and C^*-ternary algebras, associated with the additive functional inequalities (1) and (2).

1 Introduction and Preliminaries

The stability problem of functional equations originated from a question of Ulam [27] concerning the stability of group homomorphisms.

J. R. Lee
Department of Mathematics, Daejin University, Pocheon, South Korea
e-mail: jrlee@daejin.ac.kr

C. Park (✉)
Research Institute for Natural Sciences, Hanyang University, Seoul, South Korea
e-mail: baak@hanyang.ac.kr

T. M. Rassias
Department of Mathematics, Zografou Campus, National Technical University of Athens, Athens, Greece
e-mail: trassias@math.ntua.gr

T. M. Rassias, P. M. Pardalos (eds.), *Mathematical Analysis and Applications*, Springer Optimization and Its Applications 154,
https://doi.org/10.1007/978-3-030-31339-5_13

The functional equation $f(x+y) = f(x) + f(y)$ is called the *Cauchy equation*. In particular, every solution of the Cauchy equation is said to be an *additive mapping*. Hyers [10] gave a first affirmative partial answer to the question of Ulam for Banach spaces. Hyers' Theorem was generalized by Aoki [1] for additive mappings and by Rassias [22] for linear mappings by considering an unbounded Cauchy difference. A generalization of the Rassias theorem was obtained by Găvruta [7] by replacing the unbounded Cauchy difference by a general control function in the spirit of Rassias' approach.

Gilányi [8] showed that if f satisfies the functional inequality

$$\|2f(x) + 2f(y) - f(x-y)\| \leq \|f(x+y)\| \tag{3}$$

then f satisfies the Jordan-von Neumann functional equation

$$2f(x) + 2f(y) = f(x+y) + f(x-y).$$

See also [23]. Fechner [6] and Gilányi [9] proved the Hyers-Ulam stability of the functional inequality (3).

Park [16, 17] defined additive ρ-functional inequalities and proved the Hyers-Ulam stability of the additive ρ-functional inequalities in Banach spaces and non-Archimedean Banach spaces. The stability problems of various functional equations and functional inequalities have been extensively investigated by a number of authors (see [2–5, 21, 24, 25]).

Jordan observed that $\mathscr{L}(\mathscr{H})$ is an algebra via the *anticommutator product* $x \circ y := \frac{xy+yx}{2}$. A commutative algebra X with product $x \circ y$ is called a *Jordan algebra*. A Jordan C^*-subalgebra of a C^*-algebra, endowed with the anticommutator product, is called a *JC^*-algebra*. A C^*-algebra $\mathscr{C}$, endowed with the Lie product $[x, y] = \frac{xy-yx}{2}$ on $\mathscr{C}$, is called a *Lie C^*-algebra*. (see [12, 13, 20]).

In [26], Taghavi introduced partial multipliers in complex Banach $*$-algebras as follows.

Definition 1 Let A be a complex Banach $*$-algebra. A $\mathbf{C}$-linear mapping $P : A \to A$ is called a *partial multiplier* if P satisfies

$$P \circ P(xy) = P(x)P(y)$$
$$P(x^*) = P(x)^*$$

for all $x, y \in A$.

This paper is organized as follows: In Section 2, we solve the additive functional inequality (1) and prove the Hyers-Ulam stability of the additive functional inequality (1) in complex Banach spaces.

In Section 3, we solve the additive functional inequality (2) and prove the Hyers-Ulam stability of the additive functional inequality (2) in complex Banach spaces.

In Section 4, we investigate partial multipliers in C^*-algebras associated with the additive functional inequalities (1) and (2).

In Section 5, we investigate partial multipliers in Lie C^*-algebras associated with the additive functional inequalities (1) and (2).

In Section 6, we investigate partial multipliers in JC^*-algebras associated with the additive functional inequalities (1) and (2).

In Section 7, we investigate partial multipliers in C^*-ternary algebras associated with the additive functional inequalities (1) and (2).

Throughout this paper, let X be a complex normed space with norm $\|\cdot\|$, Y a complex Banach space with norm $\|\cdot\|$, and A a complex Banach $*$-algebra with norm $\|\cdot\|$. Assume that s is a fixed nonzero complex number with $|s| < 1$.

2 Additive Functional Inequality (1)

We solve and investigate the additive functional inequality (1) in complex normed spaces.

Lemma 1 ([18, Lemma 2.1]) *If a mapping $f : X \to Y$ satisfies*

$$\|f(x+y+z) - f(x+y) - f(z)\| \le \|s(f(x-y) + f(y-z) - f(x-z))\| \quad (4)$$

for all $x, y, z \in X$, then $f : X \to Y$ is additive.

Using the direct method, we prove the Hyers-Ulam stability of the additive functional equation (4) in complex Banach spaces.

Theorem 1 *Let $\varphi : X^3 \to [0, \infty)$ be a function such that*

$$\Phi(x, y, z) := \sum_{j=1}^{\infty} 2^j \varphi\left(\frac{x}{2^j}, \frac{y}{2^j}, \frac{z}{2^j}\right) < \infty \quad (5)$$

for all $x, y, z \in X$. Let $f : X \to Y$ be a mapping satisfying $f(0) = 0$ and

$$\|f(x+y+z) - f(x+y) - f(z)\| \le \|s(f(x-y) + f(y-z) - f(x-z))\| + \varphi(x, y, z) \quad (6)$$

for all $x, y, z \in X$. Then there exists a unique additive mapping $A : X \to Y$ such that

$$\|f(x) - A(x)\| \le \frac{1}{2}\Phi(0, x, x) \quad (7)$$

for all $x \in X$.

Proof Letting $x = 0$ and replacing y and z by x and x in (6), we get

$$\|f(2x) - 2f(x)\| \le \varphi(0, x, x) \tag{8}$$

for all $x \in X$. So

$$\left\|f(x) - 2f\left(\frac{x}{2}\right)\right\| \le \varphi\left(0, \frac{x}{2}, \frac{x}{2}\right)$$

for all $x \in X$. Hence

$$\begin{aligned}\left\|2^l f\left(\frac{x}{2^l}\right) - 2^m f\left(\frac{x}{2^m}\right)\right\| &\le \sum_{j=l}^{m-1} \left\|2^j f\left(\frac{x}{2^j}\right) - 2^{j+1} f\left(\frac{x}{2^{j+1}}\right)\right\| \\ &\le \frac{1}{2} \sum_{j=l+1}^{m} 2^j \varphi\left(0, \frac{x}{2^j}, \frac{x}{2^j}\right)\end{aligned} \tag{9}$$

for all nonnegative integers m and l with $m > l$ and all $x \in X$. It follows from (9) that the sequence $\{2^k f(\frac{x}{2^k})\}$ is Cauchy for all $x \in X$. Since Y is a Banach space, the sequence $\{2^k f(\frac{x}{2^k})\}$ converges. So one can define the mapping $A : X \to Y$ by

$$A(x) := \lim_{k\to\infty} 2^k f\left(\frac{x}{2^k}\right)$$

for all $x \in X$. Moreover, letting $l = 0$ and passing the limit $m \to \infty$ in (9), we get (7).

It follows from (5) and (6) that

$$\begin{aligned}&\|A(x + y + z) - A(x + y) - A(z)\| \\ &= \lim_{n\to\infty} \left\|2^n \left(f\left(\frac{x + y + z}{2^n}\right) - f\left(\frac{x + y}{2^n}\right) - f\left(\frac{z}{2^n}\right)\right)\right\| \\ &\le \lim_{n\to\infty} \left\|2^n s\left(f\left(\frac{x + y - z}{2^n}\right) + f\left(\frac{x - y + z}{2^n}\right) - 2f\left(\frac{x}{2^n}\right)\right)\right\| \\ &\quad + \lim_{n\to\infty} 2^n \varphi\left(\frac{x}{2^n}, \frac{y}{2^n}, \frac{z}{2^n}\right) \\ &\le \|s(A(x + y - z) + A(x - y + z) - 2A(x))\|\end{aligned}$$

for all $x, y, z \in X$. So

$$\|A(x + y + z) - A(x + y) - A(z)\| \le \|s(A(x + y - z) + A(x - y + z) - 2A(x))\|$$

for all $x, y, z \in X$. By Lemma 1, the mapping $A : X \to Y$ is additive.

Now, let $T : X \to Y$ be another additive mapping satisfying (7). Then we have

$$\begin{aligned}\|A(x) - T(x)\| &= \left\|2^q A\left(\frac{x}{2^q}\right) - 2^q T\left(\frac{x}{2^q}\right)\right\| \\ &\le \left\|2^q A\left(\frac{x}{2^q}\right) - 2^q f\left(\frac{x}{2^q}\right)\right\| + \left\|2^q T\left(\frac{x}{2^q}\right) - 2^q f\left(\frac{x}{2^q}\right)\right\| \\ &\le \frac{2^{q+1}}{2}\Phi\left(0, \frac{x}{2^q}, \frac{x}{2^q}\right),\end{aligned}$$

which tends to zero as $q \to \infty$ for all $x \in X$. So we can conclude that $A(x) = T(x)$ for all $x \in X$. This proves the uniqueness of A, as desired.

Corollary 1 *Let $r > 1$ and θ be nonnegative real numbers and let $f : X \to Y$ be a mapping satisfying*

$$\begin{aligned}\|f(x+y+z) - f(x+y) - f(z)\| &\le \|s(f(x-y) + f(y-z) - f(x-z))\| \\ &\quad + \theta(\|x\|^r + \|y\|^r + \|z\|^r) \end{aligned} \tag{10}$$

for all $x, y, z \in X$. Then there exists a unique additive mapping $A : X \to Y$ such that

$$\|f(x) - A(x)\| \le \frac{2\theta}{2^r - 2}\|x\|^r$$

for all $x \in X$.

Proof The proof follows from Theorem 1 by taking $\varphi(x, y, z) = \theta(\|x\|^r + \|y\|^r + \|z\|^r)$ for all $x, y, z \in X$.

Theorem 2 *Let $\varphi : X^3 \to [0, \infty)$ be a function such that*

$$\Phi(x, y, z) := \sum_{j=0}^{\infty} \frac{1}{2^j}\varphi\left(2^j x, 2^j y, 2^j z\right) < \infty \tag{11}$$

for all $x, y, z \in X$. Let $f : X \to Y$ be a mapping satisfying $f(0) = 0$ and (6). Then there exists a unique additive mapping $A : X \to Y$ such that

$$\|f(x) - A(x)\| \le \frac{1}{2}\Phi(0, x, x)$$

for all $x \in X$.

Proof It follows from (7) that

$$\left\|f(x) - \frac{1}{2}f(2x)\right\| \le \frac{1}{2}\varphi(0, x, x)$$

for all $x \in X$.

The rest of the proof is similar to the proof of Theorem 1.

Corollary 2 *Let $r < 1$ and θ be nonnegative real numbers and let $f : X \to Y$ be a mapping satisfying* (10). *Then there exists a unique additive mapping $A : X \to Y$ such that*

$$\|f(x) - A(x)\| \leq \tfrac{2\theta}{2-2^r}\|x\|^r$$

for all $x \in X$.

Proof The proof follows from Theorem 2 by taking $\varphi(x, y, z) = \theta(\|x\|^r + \|y\|^r + \|z\|^r)$ for all $x, y, z \in X$.

3 Additive Functional Inequality (2)

We solve and investigate the additive functional inequality (2) in complex normed spaces.

Lemma 2 ([18, Lemma 3.1]) *If a mapping $f : X \to Y$ satisfies*

$$\|f(x-y)+f(y-z)-f(x-z)\| \leq \|s(f(x+y+Sz) - f(x+y) - f(z))\| \quad (12)$$

for all $x, y, z \in X$, then $f : X \to Y$ is additive.

Using the direct method, we prove the Hyers-Ulam stability of the additive functional inequality (12) in complex Banach spaces.

Theorem 3 *Let $\varphi : X^3 \to [0, \infty)$ be a function satisfying* (5). *Let $f : X \to Y$ be an odd mapping satisfying*

$$\begin{aligned}\|f(x-y) + f(y-z) - f(x-z)\| &\leq \|s(f(x+y+z) - f(x+y) - f(z))\| \\ &\quad + \varphi(x, y, z) \qquad (13)\end{aligned}$$

for all $x, y, z \in X$. Then there exists a unique additive mapping $A : X \to Y$ such that

$$\|f(x) - A(x)\| \leq \frac{1}{2}\Phi(x, 0, -x) \quad (14)$$

for all $x \in X$, where Φ is given in Theorem 1.

Proof Letting $y = 0$ and $z = -x$ in (13), we get

$$\|f(2x) - 2f(x)\| \leq \varphi(x, 0, -x) \quad (15)$$

and so

$$\left\| f(x) - 2f\left(\frac{x}{2}\right)\right\| \le \varphi\left(\frac{x}{2}, 0, -\frac{x}{2}\right)$$

for all $x \in X$. So

$$\begin{aligned}\left\|2^l f\left(\frac{x}{2^l}\right) - 2^m f\left(\frac{x}{2^m}\right)\right\| &\le \sum_{j=l}^{m-1} \left\|2^j f\left(\frac{x}{2^j}\right) - 2^{j+1} f\left(\frac{x}{2^{j+1}}\right)\right\| \\ &\le \frac{1}{2} \sum_{j=l+1}^{m} 2^j \varphi\left(\frac{x}{2^j}, 0, -\frac{x}{2^j}\right) \end{aligned} \tag{16}$$

for all nonnegative integers m and l with $m > l$ and all $x \in X$. It follows from (16) that the sequence $\{2^k f(\frac{x}{2^k})\}$ is Cauchy for all $x \in X$. Since Y is a Banach space, the sequence $\{2^k f(\frac{x}{2^k})\}$ converges. So one can define the mapping $A : X \to Y$ by

$$A(x) := \lim_{k\to\infty} 2^k f\left(\frac{x}{2^k}\right)$$

for all $x \in X$. Moreover, letting $l = 0$ and passing the limit $m \to \infty$ in (16), we get (17).

The rest of the proof is similar to the proof of Theorem 1.

Corollary 3 *Let $r > 1$ and θ be nonnegative real numbers and let $f : X \to Y$ be an odd mapping satisfying*

$$\begin{aligned}\|f(x-y) + f(y-z) - f(x-z)\| &\le \|s(f(x+y+z) - f(x+y) - f(z))\| \\ &\quad + \theta(\|x\|^r + \|y\|^r + \|z\|^r) \end{aligned} \tag{17}$$

for all $x, y, z \in X$. Then there exists a unique additive mapping $A : X \to Y$ such that

$$\|f(x) - A(x)\| \le \frac{2\theta}{2^r - 2}\|x\|^r$$

for all $x \in X$.

Proof The proof follows from Theorem 3 by taking $\varphi(x, y, z) = \theta(\|x\|^r + \|y\|^r + \|z\|^r)$ for all $x, y, z \in X$.

Theorem 4 *Let $\varphi : X^3 \to [0, \infty)$ be a function satisfying* (11). *Let $f : X \to Y$ be an odd mapping satisfying* (13). *Then there exists a unique additive mapping $A : X \to Y$ such that*

$$\|f(x) - A(x)\| \le \frac{1}{2}\Phi(x, 0, -x)$$

for all $x \in X$, where Φ is given in Theorem 2.

Proof It follows from (15) that

$$\left\| f(x) - \frac{1}{2} f(2x) \right\| \leq \frac{1}{2} \varphi(x, 0, -x)$$

for all $x \in X$.

The rest of the proof is similar to the proofs of Theorems 1 and 3.

Corollary 4 *Let $r < 1$ and θ be nonnegative real numbers and let $f : X \to Y$ be an odd mapping satisfying* (17)*. Then there exists a unique additive mapping $A : X \to Y$ such that*

$$\|f(x) - A(x)\| \leq \frac{2\theta}{2 - 2^r} \|x\|^r$$

for all $x \in X$.

Proof The proof follows from Theorem 4 by taking $\varphi(x, y, z) = \theta(\|x\|^r + \|y\|^r + \|z\|^r)$ for all $x, y, z \in X$.

4 Partial Multipliers in C^*-Algebras

In this section, we investigate partial multipliers in complex Banach $*$-algebras and unital C^*-algebras associated with the additive functional inequalities (4) and (12).

Theorem 5 *Let $\varphi : A^3 \to [0, \infty)$ be a function such that*

$$\sum_{j=1}^{\infty} 4^j \varphi\left(\frac{x}{2^j}, \frac{y}{2^j}, \frac{z}{2^j}\right) < \infty \tag{18}$$

for all $x, y, z \in A$. Let $f : A \to A$ be a mapping satisfying $f(0) = 0$ and

$$\|f(\mu(x+y+z)) - \mu(f(x+y) + f(z))\| \leq \|s(f(x-y) + f(y-z) - f(x-z))\| + \varphi(x, y, z) \tag{19}$$

for all $\mu \in \mathbf{T}^1 := \{\lambda \in \mathbf{C} \mid |\lambda| = 1\}$ and all $x, y, z \in A$. Then there exists a unique $\mathbf{C}$-linear mapping $P : A \to A$ such that

$$\|f(x) - P(x)\| \leq \frac{1}{2} \Phi(0, x, x) \tag{20}$$

for all $x \in A$, where

$$\Phi(x, y, z) := \sum_{j=1}^{\infty} 2^j \varphi\left(\frac{x}{2^j}, \frac{y}{2^j}, \frac{z}{2^j}\right)$$

for all $x, y, z \in A$.

If, in addition, the mapping $f : A \to A$ satisfies $f(2x) = 2f(x)$ and

$$\|f \circ f(xy) - f(x)f(y)\| \leq \varphi(x, y, 0), \tag{21}$$

$$\|f(x^*) - f(x)^*\| \leq \varphi(x, 0, 0) \tag{22}$$

for all $x, y \in A$, then the mapping f is a partial multiplier.

Proof Let $\mu = 1$ in (19). By Theorem 1, there is a unique additive mapping $P : A \to A$ satisfying (20) defined by

$$P(x) := \lim_{n\to\infty} 2^n f(\frac{x}{2^n})$$

for all $x \in A$.

Letting $y = z = 0$ in (19), we get

$$\|f(\mu x) - \mu f(x)\| \leq \varphi(x, 0, 0)$$

for all $x \in A$ and all $\mu \in \mathbf{T}^1$. So

$$\begin{aligned}\|P(\mu x) - \mu P(x)\| &= \lim_{n\to\infty} 2^n \left\|f\left(\mu\frac{x}{2^n}\right) - f\left(\mu\frac{x}{2^n}\right)\right\| \\ &\leq \lim_{n\to\infty} 2^n \varphi\left(\frac{x}{2^n}x, 0, 0\right) \leq \lim_{n\to\infty} 4^n \varphi\left(\frac{x}{2^n}x, 0, 0\right) = 0\end{aligned}$$

for all $x \in A$ and all $\mu \in \mathbf{T}^1$. Hence $P(\mu x) = \mu P(x)$ for all $x \in A$ and all $\mu \in \mathbf{T}^1$. By the same reasoning as in the proof of [14, Theorem 2.1], the mapping $P : A \to A$ is $\mathbf{C}$-linear.

If $f(2x) = 2f(x)$ for all $x \in A$, then we can easily show that $P(x) = f(x)$ for all $x \in A$.

It follows from (18) and (21) that

$$\begin{aligned}\|f \circ f(xy) - f(x)f(y)\| &= \|P \circ P(xy) - P(x)P(y)\| \\ &= \lim_{n\to\infty} 4^n \left\|f \circ f\left(\frac{xy}{2^n \cdot 2^n}\right) - f\left(\frac{x}{2^n}\right) f\left(\frac{y}{2^n}\right)\right\| \\ &\leq \lim_{n\to\infty} 4^n \varphi\left(\frac{x}{2^n}, \frac{y}{2^n}, 0\right) = 0\end{aligned}$$

for all $x, y \in A$. Thus

$$f \circ f(xy) = f(x)f(y)$$

for all $x, y \in A$.

It follows from (18) and (22) that

$$\|f(x^*) - f(x)^*\| = \|P(x^*) - P(x)^*\| = \lim_{n\to\infty} 2^n \left\| f\left(\frac{x^*}{2^n}\right) - f\left(\frac{x}{2^n}\right)^* \right\|$$
$$\leq \lim_{n\to\infty} 2^n \varphi\left(\frac{x}{2^n}, 0, 0\right) \leq \lim_{n\to\infty} 4^n \varphi\left(\frac{x}{2^n}, 0, 0\right) = 0$$

for all $x \in A$. Thus

$$f(x^*) = f(x)^*$$

for all $x \in A$. Hence the mapping $f : A \to A$ is a partial multiplier.

Corollary 5 *Let $r > 2$ and θ be nonnegative real numbers, and let $f : A \to A$ be a mapping satisfying $f(0) = 0$ and*

$$\|f(\mu(x+y+z)) - \mu(f(x+y) + f(z))\| \leq \|s(f(x+y-z) + f(x-y+z) - 2f(x))\| + \theta(\|x\|^r + \|y\|^r + \|z\|^r) \tag{23}$$

for all $\mu \in \mathbf{T}^1$ and all $x, y, z \in A$. Then there exists a unique **C***-linear mapping $P : A \to A$ such that*

$$\|f(x) - P(x)\| \leq \frac{2\theta}{2^r - 2}\|x\|^r \tag{24}$$

for all $x \in A$.

If, in addition, the mapping $f : A \to A$ satisfies $f(2x) = 2f(x)$ and

$$\|f \circ f(xy) - f(x)f(y)\| \leq \theta(\|x\|^r + \|y\|^r), \tag{25}$$

$$\|f(x^*) - f(x)^*\| \leq \theta\|x\|^r \tag{26}$$

for all $x, y \in A$, then the mapping f is a partial multiplier.

Proof The proof follows from Theorem 5 by taking $\varphi(x, y, z) = \theta(\|x\|^r + \|y\|^r + \|z\|^r)$ for all $x, y, z \in A$.

Theorem 6 *Let $\varphi : A^3 \to [0, \infty)$ be a function such that*

$$\Phi(x, y, z) := \sum_{j=0}^{\infty} \frac{1}{2^j}\varphi\left(2^j x, 2^j y, 2^j z\right) < \infty \tag{27}$$

for all $x, y, z \in A$. Let $f : A \to A$ be a mapping satisfying $f(0) = 0$ and (19). Then there exists a unique **C***-linear mapping $P : A \to A$ such that*

$$\|f(x) - P(x)\| \le \frac{1}{2}\Phi(0, x, x) \tag{28}$$

for all $x \in A$.

If, in addition, the mapping $f : A \to A$ satisfies $f(2x) = 2f(x)$ for all $x \in A$, (21) and (22), then the mapping f is a partial multiplier.

Proof The proof is similar to the proof of Theorem 18.

Corollary 6 *Let $r < 1$ and θ be nonnegative real numbers, and let $f : A \to A$ be a mapping satisfying $f(0) = 0$ and (23). Then there exists a unique* **C***-linear mapping $P : A \to A$ such that*

$$\|f(x) - P(x)\| \le \frac{2\theta}{2 - 2^r}\|x\|^r \tag{29}$$

for all $x \in A$.

If, in addition, the mapping $f : A \to A$ satisfies $f(2x) = 2f(x)$ for all $x \in A$, (25) and (26), then the mapping f is a partial multiplier.

Proof The proof follows from Theorem 6 by taking $\varphi(x, y, z) = \theta(\|x\|^r + \|y\|^r + \|z\|^r)$ for all $x, y, z \in A$.

Similarly, we can obtain the following results.

Theorem 7 *Let $\varphi : A^3 \to [0, \infty)$ be a function satisfying (18). Let $f : A \to A$ be an odd mapping satisfying*

$$\begin{aligned} \|f(\mu(x - y)) + f(\mu(y - z)) - \mu f(x - z)\| &\le \|s(f(x + y + z) - f(x + y) - f(z))\| \\ &\quad + \varphi(x, y, z) \end{aligned} \tag{30}$$

for all $\mu \in \mathbf{T}^1$ and all $x, y, z \in A$. Then there exists a unique **C***-linear mapping $P : A \to A$ such that*

$$\|f(x) - P(x)\| \le \frac{1}{2}\varphi(x, 0, -x) \tag{31}$$

for all $x \in A$, where Φ is given in Theorem 5.

If, in addition, the mapping $f : A \to A$ satisfies $f(2x) = 2f(x)$ for all $x \in A$, (21) and (22), then the mapping f is a partial multiplier.

Proof The proof is similar to the proof of Theorem 18.

Corollary 7 *Let $r > 2$ and θ be nonnegative real numbers, and let $f : A \to A$ be an odd mapping satisfying*

$$\|f(\mu(x+y-z)) + f(\mu(x-y+z)) - 2\mu f(x)\| \leq \|s(f(x+y+z) - f(x+y) - f(z))\| + \theta(\|x\|^r + \|y\|^r + \|z\|^r) \quad (32)$$

for all $\mu \in \mathbf{T}^1$ *and all* $x, y, z \in A$. *Then there exists a unique* **C**-*linear mapping* $P : A \to A$ *such that*

$$\|f(x) - P(x)\| \leq \frac{2\theta}{2^r - 2}\|x\|^r \quad (33)$$

for all $x \in A$.

If, in addition, the mapping $f : A \to A$ *satisfies* $f(2x) = 2f(x)$ *for all* $x \in A$, (25) *and* (26), *then the mapping* f *is a partial multiplier.*

Proof The proof follows from Theorem 7 by taking $\varphi(x, y, z) = \theta(\|x\|^r + \|y\|^r + \|z\|^r)$ for all $x, y, z \in A$.

Theorem 8 *Let* $\varphi : A^3 \to [0, \infty)$ *be a function satisfying* (27). *Let* $f : A \to A$ *be an odd mapping satisfying* (30). *Then there exists a unique* **C**-*linear mapping* $P : A \to A$ *such that*

$$\|f(x) - P(x)\| \leq \frac{1}{2}\Phi(x, 0, -x) \quad (34)$$

for all $x \in A$, *where* Φ *is given in Theorem* 6.

If, in addition, the mapping $f : A \to A$ *satisfies* $f(2x) = 2f(x)$ *for all* $x \in A$, (21) *and* (22), *then the mapping* f *is a partial multiplier.*

Corollary 8 *Let* $r < 1$ *and* θ *be nonnegative real numbers, and let* $f : A \to A$ *be an odd mapping satisfying* (32). *Then there exists a unique* **C**-*linear mapping* $P : A \to A$ *such that*

$$\|f(x) - P(x)\| \leq \frac{2\theta}{2 - 2^r}\|x\|^r \quad (35)$$

for all $x \in A$.

If, in addition, the mapping $f : A \to A$ *satisfies* $f(2x) = 2f(x)$ *for all* $x \in A$, (25) *and* (26), *then the mapping* f *is a partial multiplier.*

From now on, assume that A is a unital C^*-algebra with norm $\|\cdot\|$ and unitary group $U(A)$.

Theorem 9 *Let* $\varphi : A^3 \to [0, \infty)$ *be a function satisfying* (18). *Let* $f : A \to A$ *be a mapping satisfying* $f(0) = 0$ *and* (19). *Then there exists a unique* **C**-*linear mapping* $P : A \to A$ *satisfying* (20).

If, in addition, the mapping $f : A \to A$ *satisfies* $f(2x) = 2f(x)$ *for all* $x \in A$ *and*

$$\|f \circ f(uv) - f(u)f(v)\| \leq \varphi(u, v, 0) \quad (36)$$

$$\|f(u^*) - f(u)^*\| \leq \varphi(u, 0, 0) \tag{37}$$

for all $u, v \in U(A)$, then the mapping f is a partial multiplier.

Proof By the same reasoning as in the proof of Theorem 5, there is a unique $\mathbf{C}$-linear mapping $P : A \to A$ satisfying (20) defined by

$$P(x) := \lim_{n\to\infty} 2^n f(\frac{x}{2^n})$$

for all $x \in A$.

If $f(2x) = 2f(x)$ for all $x \in A$, then we can easily show that $P(x) = f(x)$ for all $x \in A$.

By the same reasoning as in the proof of Theorem 5, $f \circ f(uv) = f(u)f(v)$ and $f(u^*) = f(u)^*$ for all $u, v \in U(A)$.

Since f is $\mathbf{C}$-linear and each $x \in A$ is a finite linear combination of unitary elements (see [11]), i.e., $x = \sum_{j=1}^m \lambda_j u_j$ $(\lambda_j \in \mathbf{C}, u_j \in U(A))$,

$$\begin{aligned} f(x^*) &= f(\sum_{j=1}^m \overline{\lambda_j} u_j^*) = \sum_{j=1}^m \overline{\lambda_j} f(u_j^*) = \sum_{j=1}^m \overline{\lambda_j} f(u_j)^* = (\sum_{j=1}^m \lambda_j f(u_j))^* \\ &= f(\sum_{j=1}^m \lambda_j u_j)^* = f(x)^* \end{aligned}$$

for all $x \in A$.

Since f and $f \circ f$ are $\mathbf{C}$-linear and $x, y \in A$ are finite linear combinations of unitary elements, i.e., $x = \sum_{j=1}^m \lambda_j u_j$ $(\lambda_j \in \mathbf{C}, u_j \in U(A))$ and $y = \sum_{k=1}^n \beta_k v_k$ $(\beta_k \in \mathbf{C}, v_k \in U(A))$,

$$\begin{aligned} f \circ f(xy) &= f \circ f(\sum_{j=1}^m \sum_{k=1}^n \lambda_j \beta_k u_j v_k) = \sum_{j=1}^m \sum_{k=1}^n \lambda_j \beta_k f \circ f(u_j v_k) \\ &= \sum_{j=1}^m \sum_{k=1}^n \lambda_j \beta_k f(u_j) f(v_k) = f(\sum_{j=1}^m \lambda_j u_j) f(\sum_{k=1}^n \beta_k v_k) = f(x)f(y) \end{aligned}$$

for all $x, y \in A$.

Therefore, the mapping $f : A \to A$ is a partial multiplier.

Corollary 9 *Let $r > 2$ and θ be nonnegative real numbers, and let $f : A \to A$ be a mapping satisfying $f(0) = 0$ and* (23)*. Then there exists a unique $\mathbf{C}$-linear mapping $P : A \to A$ satisfying* (24).

If, in addition, the mapping $f : A \to A$ satisfies $f(2x) = 2f(x)$ for all $x \in A$ and

$$\|f \circ f(uv) - f(u)f(v)\| \leq 2\theta, \tag{38}$$

$$\|f(u^*) - f(u)^*\| \leq \theta \tag{39}$$

for all $u, v \in U(A)$, then the mapping f is a partial multiplier.

Proof The proof follows from Theorem 9 by taking $\varphi(x, y, z) = \theta(\|x\|^r + \|y\|^r + \|z\|^r)$ for all $x, y, z \in A$.

Theorem 10 *Let $\varphi : A^3 \to [0, \infty)$ be a function satisfying* (27). *Let $f : A \to A$ be a mapping satisfying $f(0) = 0$ and* (19). *Then there exists a unique* **C**-*linear mapping $P : A \to A$ satisfying* (28).

If, in addition, the mapping $f : A \to A$ satisfies $f(2x) = 2f(x)$ for all $x \in A$, (36) *and* (37), *then the mapping f is a partial multiplier.*

Proof The proof is similar to the proof of Theorem 9.

Corollary 10 *Let $r < 1$ and θ be nonnegative real numbers, and let $f : A \to A$ be a mapping satisfying $f(0) = 0$ and* (23). *Then there exists a unique* **C**-*linear mapping $P : A \to A$ satisfying* (29).

If, in addition, the mapping $f : A \to A$ satisfies $f(2x) = 2f(x)$ for all $x \in A$, (38) *and* (39), *then the mapping f is a partial multiplier.*

Proof The proof follows from Theorem 10 by taking $\varphi(x, y, z) = \theta(\|x\|^r + \|y\|^r + \|z\|^r)$ for all $x, y, z \in A$.

Similarly, we can obtain the following results. We will omit the proofs.

Theorem 11 *Let $\varphi : A^3 \to [0, \infty)$ be a function satisfying* (18). *Let $f : A \to A$ be an odd mapping satisfying* (30). *Then there exists a unique* **C**-*linear mapping $P : A \to A$ satisfying* (31).

If, in addition, the mapping $f : A \to A$ satisfies $f(2x) = 2f(x)$ for all $x \in A$, (36) *and* (37), *then the mapping f is a partial multiplier..*

Corollary 11 *Let $r > 2$ and θ be nonnegative real numbers, and let $f : A \to A$ be an odd mapping satisfying* (32). *Then there exists a unique* **C**-*linear mapping $P : A \to A$ satisfying* (33).

If, in addition, the mapping $f : A \to A$ satisfies $f(2x) = 2f(x)$ for all $x \in A$, (38) *and* (39), *then the mapping f is a partial multiplier.*

Proof The proof follows from Theorem 11 by taking $\varphi(x, y, z) = \theta(\|x\|^r + \|y\|^r + \|z\|^r)$ for all $x, y, z \in A$.

Theorem 12 *Let $\varphi : A^3 \to [0, \infty)$ be a function satisfying* (27). *Let $f : A \to A$ be an odd mapping satisfying* (30). *Then there exists a unique* **C**-*linear mapping $P : A \to A$ satisfying* (34).

If, in addition, the mapping $f : A \to A$ satisfies $f(2x) = 2f(x)$ for all $x \in A$, (36) *and* (37), *then the mapping f is a partial multiplier.*

Corollary 12 *Let $r < 1$ and θ be nonnegative real numbers, and let $f : A \to A$ be an odd mapping satisfying* (32). *Then there exists a unique* **C***-linear mapping $P : A \to A$ satisfying* (35).

If, in addition, the mapping $f : A \to A$ satisfies $f(2x) = 2f(x)$ for all $x \in A$, (38) *and* (39)*, then the mapping f is a partial multiplier.*

Proof The proof follows from Theorem 12 by taking $\varphi(x, y, z) = \theta(\|x\|^r + \|y\|^r + \|z\|^r)$ for all $x, y, z \in A$.

5 Partial Multipliers in Lie C^*-Algebras

Throughout this section, assume that A is a Lie C^*-algebra with norm $\| \cdot \|$.

Definition 2 ([19]) A **C**-linear mapping $P : A \to A$ is called a *partial multiplier* if $P : A \to A$ satisfies

$$P \circ P([x, y]) = [P(x), P(y)],$$
$$P(x^*) = P(x)^*$$

for all $x, y \in A$.

In this section, we investigate partial multipliers in Lie C^*-algebras associated with the additive functional inequalities (4) and (12).

Theorem 13 *Let $\varphi : A^3 \to [0, \infty)$ be a function satisfying* (18). *Let $f : A \to A$ be a mapping satisfying $f(0) = 0$ and* (19). *Then there exists a unique* **C***-linear mapping $P : A \to A$ satisfying* (20).

If, in addition, the mapping $f : A \to A$ satisfies $f(2x) = 2f(x)$ and

$$\|f \circ f([x, y]) - [f(x), f(y)]\| \leq \varphi(x, y, 0), \tag{40}$$

$$\|f(x^*) - f(x)^*\| \leq \varphi(x, 0, 0) \tag{41}$$

for all $x, y \in A$, then the mapping f is a partial multiplier.

Proof Let $\mu = 1$ in (19). By Theorem 5, there is a unique **C**-linear mapping $P : A \to A$ satisfying (20) defined by

$$P(x) := \lim_{n\to\infty} 2^n f(\frac{x}{2^n})$$

for all $x \in A$.

If $f(2x) = 2f(x)$ for all $x \in A$, then we can easily show that $P(x) = f(x)$ for all $x \in A$.

It follows from (40) that

$$\begin{aligned}\|f \circ f([x, y]) - [f(x), f(y)]\| &= \|P \circ P([x, y]) - [P(x), P(y)]\| \\ &= \lim_{n\to\infty} 4^n \left\| f \circ f\left(\frac{[x, y]}{2^n \cdot 2^n}\right) - \left[f\left(\frac{x}{2^n}\right), f\left(\frac{y}{2^n}\right)\right]\right\| \\ &\leq \lim_{n\to\infty} 4^n \varphi\left(\frac{x}{2^n}, \frac{y}{2^n}, 0\right) = 0\end{aligned}$$

for all $x, y \in A$. Thus

$$f \circ f([x, y]) = [f(x), f(y)]$$

for all $x, y \in A$.

It follows from (41) that

$$\begin{aligned}\|f(x^*) - f(x)^*\| = \|P(x^*) - P(x)^*\| &= \lim_{n\to\infty} 2^n \left\| f\left(\frac{x^*}{2^n}\right) - f\left(\frac{x}{2^n}\right)^*\right\| \\ &\leq \lim_{n\to\infty} 2^n \varphi\left(\frac{x}{2^n}, 0, 0\right) \leq \lim_{n\to\infty} 4^n \varphi\left(\frac{x}{2^n}, 0, 0\right) = 0\end{aligned}$$

for all $x \in A$. Thus

$$f(x^*) = f(x)^*$$

for all $x \in A$. Hence the mapping $f : A \to A$ is a partial multiplier.

Corollary 13 *Let $r > 2$ and θ be nonnegative real numbers, and let $f : A \to A$ be a mapping satisfying $f(0) = 0$ and (23). Then there exists a unique $\mathbf{C}$-linear mapping $P : A \to A$ satisfying (24).*

If, in addition, the mapping $f : A \to A$ satisfies $f(2x) = 2f(x)$ and

$$\|f \circ f([x, y]) - [f(x), f(y)]\| \leq \theta(\|x\|^r + \|y\|^r), \tag{42}$$

$$\|f(x^*) - f(x)^*\| \leq \theta\|x\|^r \tag{43}$$

for all $x, y \in A$, then the mapping f is a partial multiplier.

Proof The proof follows from Theorem 13 by taking $\varphi(x, y, z) = \theta(\|x\|^r + \|y\|^r + \|z\|^r)$ for all $x, y, z \in A$.

Theorem 14 *Let $\varphi : A^3 \to [0, \infty)$ be a function satisfying (27). Let $f : A \to A$ be a mapping satisfying $f(0) = 0$ and (19). Then there exists a unique $\mathbf{C}$-linear mapping $P : A \to A$ satisfying (28).*

If, in addition, the mapping $f : A \to A$ satisfies $f(2x) = 2f(x)$ for all $x \in A$, (40) and (41), then the mapping f is a partial multiplier.

Proof The proof is similar to the proof of Theorem 13.

Corollary 14 *Let $r < 1$ and θ be nonnegative real numbers, and let $f : A \to A$ be a mapping satisfying* (23). *Then there exists a unique* **C**-*linear mapping satisfying* $f(0) = 0$ *and* (29).

If, in addition, the mapping $f : A \to A$ satisfies $f(2x) = 2f(x)$ for all $x \in A$, (42) *and* (43), *then the mapping f is a partial multiplier.*

Proof The proof follows from Theorem 14 by taking $\varphi(x, y, z) = \theta(\|x\|^r + \|y\|^r + \|z\|^r)$ for all $x, y, z \in A$.

Similarly, we can obtain the following results. We will omit the proofs.

Theorem 15 *Let $\varphi : A^3 \to [0, \infty)$ be a function satisfying* (18). *Let $f : A \to A$ be an odd mapping satisfying* (30). *Then there exists a unique* **C**-*linear mapping $P : A \to A$ satisfying* (31).

If, in addition, the mapping $f : A \to A$ satisfies $f(2x) = 2f(x)$ for all $x \in A$, (40) *and* (41), *then the mapping f is a partial multiplier.*

Corollary 15 *Let $r > 2$ and θ be nonnegative real numbers, and let $f : A \to A$ be an odd mapping satisfying* (32). *Then there exists a unique* **C**-*linear mapping $P : A \to A$ satisfying* (33).

If, in addition, the mapping $f : A \to A$ satisfies $f(2x) = 2f(x)$ for all $x \in A$, (42) *and* (43), *then the mapping f is a partial multiplier.*

Proof The proof follows from Theorem 15 by taking $\varphi(x, y, z) = \theta(\|x\|^r + \|y\|^r + \|z\|^r)$ for all $x, y, z \in A$.

Theorem 16 *Let $\varphi : A^3 \to [0, \infty)$ be a function satisfying* (27). *Let $f : A \to A$ be an odd mapping satisfying* (30). *Then there exists a unique* **C**-*linear mapping $P : A \to A$ satisfying* (34).

If, in addition, the mapping $f : A \to A$ satisfies $f(2x) = 2f(x)$ for all $x \in A$, (40) *and* (41), *then the mapping f is a partial multiplier.*

Corollary 16 *Let $r < 1$ and θ be nonnegative real numbers, and let $f : A \to A$ be an odd mapping satisfying* (32). *Then there exists a unique* **C**-*linear mapping $P : A \to A$ satisfying* (35).

If, in addition, the mapping $f : A \to A$ satisfies $f(2x) = 2f(x)$ for all $x \in A$, (42) *and* (43), *then the mapping f is a partial multiplier.*

Proof The proof follows from Theorem 16 by taking $\varphi(x, y, z) = \theta(\|x\|^r + \|y\|^r + \|z\|^r)$ for all $x, y, z \in A$.

6 Partial Multipliers in JC^*-Algebras

Throughout this section, assume that A is a JC^*-algebra with norm $\|\cdot\|$.

Definition 3 ([19]) A **C**-linear mapping $P : A \to A$ is called a *partial multiplier* if $P : A \to A$ satisfies

$$P \circ P(x \circ y) = P(x) \circ P(y),$$
$$P(x^*) = P(x)^*$$

for all $x, y \in A$.

In this section, we investigate partial multipliers in JC^*-algebras associated with the additive functional inequalities (4) and (12).

Theorem 17 *Let $\varphi : A^3 \to [0, \infty)$ be a function satisfying* (18). *Let $f : A \to A$ be a mapping satisfying $f(0) = 0$ and* (19). *Then there exists a unique* **C**-*linear mapping $P : A \to A$ satisfying* (20).

If, in addition, the mapping $f : A \to A$ satisfies $f(2x) = 2f(x)$ and

$$\|f \circ f(x \circ y) - f(x) \circ f(y)\| \leq \varphi(x, y, 0), \tag{44}$$

$$\|f(x^*) - f(x)^*\| \leq \varphi(x, 0, 0) \tag{45}$$

for all $x, y \in A$, then the mapping f is a partial multiplier.

Proof Let $\mu = 1$ in (19). By Theorem 5, there is a unique **C**-linear mapping $P : A \to A$ satisfying (20) defined by

$$P(x) := \lim_{n\to\infty} 2^n f(\frac{x}{2^n})$$

for all $x \in A$.

If $f(2x) = 2f(x)$ for all $x \in A$, then we can easily show that $P(x) = f(x)$ for all $x \in A$.

It follows from (44) that

$$\begin{aligned}\|f \circ f(x \circ y) - f(x) \circ f(y)\| &= \|P \circ P(x \circ y) - P(x) \circ P(y)\| \\ &= \lim_{n\to\infty} 4^n \left\|f \circ f\left(\frac{x \circ y}{2^n \cdot 2^n}\right) - f\left(\frac{x}{2^n}\right) \circ f\left(\frac{y}{2^n}\right)\right\| \\ &\leq \lim_{n\to\infty} 4^n \varphi\left(\frac{x}{2^n}, \frac{y}{2^n}, 0\right) = 0\end{aligned}$$

for all $x, y \in A$. Thus

$$f \circ f(x \circ y) = f(x) \circ f(y)$$

for all $x, y \in A$.

It follows from (45) that

$$\|f(x^*) - f(x)^*\| = \|P(x^*) - P(x)^*\| = \lim_{n\to\infty} 2^n \left\| f\left(\frac{x^*}{2^n}\right) - f\left(\frac{x}{2^n}\right)^* \right\|$$
$$\leq \lim_{n\to\infty} 2^n \varphi\left(\frac{x}{2^n}, 0, 0\right) \leq \lim_{n\to\infty} 4^n \varphi\left(\frac{x}{2^n}, 0, 0\right) = 0$$

for all $x \in A$. Thus

$$f(x^*) = f(x)^*$$

for all $x \in A$. Hence the mapping $f : A \to A$ is a partial multiplier.

Corollary 17 *Let $r > 2$ and θ be nonnegative real numbers, and let $f : A \to A$ be a mapping satisfying $f(0) = 0$ and* (23)*. Then there exists a unique* **C***-linear mapping $P : A \to A$ satisfying* (24)*.*

If, in addition, the mapping $f : A \to A$ satisfies $f(2x) = 2f(x)$ and

$$\|f \circ f(x \circ y) - f(x) \circ f(y)\| \leq \theta(\|x\|^r + \|y\|^r), \tag{46}$$

$$\|f(x^*) - f(x)^*\| \leq \theta\|x\|^r \tag{47}$$

for all $x, y \in A$, then the mapping f is a partial multiplier.

Proof The proof follows from Theorem 17 by taking $\varphi(x, y, z) = \theta(\|x\|^r + \|y\|^r + \|z\|^r)$ for all $x, y, z \in A$.

Theorem 18 *Let $\varphi : A^3 \to [0, \infty)$ be a function satisfying* (27)*. Let $f : A \to A$ be a mapping satisfying $f(0) = 0$ and* (19)*. Then there exists a unique* **C***-linear mapping $P : A \to A$ satisfying* (28)*.*

If, in addition, the mapping $f : A \to A$ satisfies $f(2x) = 2f(x)$ for all $x \in A$, (44) *and* (45)*, then the mapping f is a partial multiplier.*

Proof The proof is similar to the proof of Theorem 17.

Corollary 18 *Let $r < 1$ and θ be nonnegative real numbers, and let $f : A \to A$ be a mapping satisfying $f(0) = 0$ and* (23)*. Then there exists a unique* **C***-linear mapping satisfying* (29)*.*

If, in addition, the mapping $f : A \to A$ satisfies $f(2x) = 2f(x)$ for all $x \in A$, (46) *and* (47)*, then the mapping f is a partial multiplier.*

Proof The proof follows from Theorem 18 by taking $\varphi(x, y, z) = \theta(\|x\|^r + \|y\|^r + \|z\|^r)$ for all $x, y, z \in A$.

Similarly, we can obtain the following results. We will omit the proofs.

Theorem 19 *Let $\varphi : A^3 \to [0, \infty)$ be a function satisfying* (18). *Let $f : A \to A$ be an odd mapping satisfying* (30). *Then there exists a unique* **C**-*linear mapping $P : A \to A$ satisfying* (31).

If, in addition, the mapping $f : A \to A$ satisfies $f(2x) = 2f(x)$ for all $x \in A$, (44) *and* (45), *then the mapping f is a partial multiplier.*

Corollary 19 *Let $r > 2$ and θ be nonnegative real numbers, and let $f : A \to A$ be an odd mapping satisfying* (32). *Then there exists a unique* **C**-*linear mapping $P : A \to A$ satisfying* (33).

If, in addition, the mapping $f : A \to A$ satisfies $f(2x) = 2f(x)$ for all $x \in A$, (46) *and* (47), *then the mapping f is a partial multiplier.*

Proof The proof follows from Theorem 19 by taking $\varphi(x, y, z) = \theta(\|x\|^r + \|y\|^r + \|z\|^r)$ for all $x, y, z \in A$.

Theorem 20 *Let $\varphi : A^3 \to [0, \infty)$ be a function satisfying* (27). *Let $f : A \to A$ be an odd mapping satisfying* (30). *Then there exists a unique* **C**-*linear mapping $P : A \to A$ satisfying* (34).

If, in addition, the mapping $f : A \to A$ satisfies $f(2x) = 2f(x)$ for all $x \in A$, (44) *and* (45), *then the mapping f is a partial multiplier.*

Corollary 20 *Let $r < 1$ and θ be nonnegative real numbers, and let $f : A \to A$ be an odd mapping satisfying* (32). *Then there exists a unique* **C**-*linear mapping $P : A \to A$ satisfying* (35).

If, in addition, the mapping $f : A \to A$ satisfies $f(2x) = 2f(x)$ for all $x \in A$, (46) *and* (47), *then the mapping f is a partial multiplier.*

Proof The proof follows from Theorem 20 by taking $\varphi(x, y, z) = \theta(\|x\|^r + \|y\|^r + \|z\|^r)$ for all $x, y, z \in A$.

7 Partial Multipliers in C^*-Ternary Algebras

A C^*-ternary algebra is a complex Banach space A, equipped with a ternary product $(x, y, z) \mapsto [x, y, z]$ of A^3 into A, which is **C**-linear in the outer variables, conjugate **C**-linear in the middle variable, and associative in the sense that $[x, y, [z, w, v]] = [x, [w, z, y], v] = [[x, y, z], w, v]$, and satisfies $\|[x, y, z]\| \le \|x\| \cdot \|y\| \cdot \|z\|$ and $\|[x, x, x]\| = \|x\|^3$ (see [15, 28]).

If a C^*-ternary algebra $(A, [\cdot, \cdot, \cdot])$ has an identity, i.e., an element $e \in A$ such that $x = [x, e, e] = [e, e, x]$ for all $x \in A$, then it is routine to verify that A, endowed with $x \circ y := [x, e, y]$ and $x^* := [e, x, e]$, is a unital C^*-algebra. Conversely, if $(A, \circ)$ is a unital C^*-algebra, then $[x, y, z] := x \circ y^* \circ z$ makes A into a C^*-ternary algebra.

Throughout this section, assume that A is a C^*-ternary algebra with norm $\| \cdot \|$.

Definition 4 ([19]) A **C**-linear mapping $P : A \to A$ is called a *partial multiplier* if $P : A \to A$ satisfies

$$P \circ P([x, y, z]) = [P(x), P(y), P(z)],$$
$$P(x^*) = P(x)^*$$

for all $x, y, z \in A$.

In this section, we investigate partial multipliers in C^*-ternary algebras associated with the additive functional inequalities (4) and (12).

Theorem 21 *Let $\varphi : A^3 \to [0, \infty)$ be a function satisfying*

$$\sum_{j=1}^{\infty} 8^j \varphi\left(\frac{x}{2^j}, \frac{y}{2^j}, \frac{z}{2^j}\right) < \infty \tag{48}$$

for all $x, y, z \in A$. Let $f : A \to A$ be a mapping satisfying $f(0) = 0$ and (19). Then there exists a unique **C***-linear mapping $P : A \to A$ satisfying (20), where*

$$\Phi(x, y, z) := \sum_{j=1}^{\infty} 2^j \varphi\left(\frac{x}{2^j}, \frac{y}{2^j}, \frac{z}{2^j}\right)$$

for all $x, y, z \in A$.

If, in addition, the mapping $f : A \to A$ satisfies $f(2x) = 2f(x)$ and

$$\|f \circ f([x, y, z]) - [f(x), f(y), f(z)]\| \leq \varphi(x, y, z), \tag{49}$$

$$\|f(x^*) - f(x)^*\| \leq \varphi(x, 0, 0) \tag{50}$$

for all $x, y, z \in A$, then the mapping f is a partial multiplier.

Proof Let $\mu = 1$ in (19). By Theorem 5, there is a unique **C**-linear mapping $P : A \to A$ satisfying (20) defined by

$$P(x) := \lim_{n\to\infty} 2^n f(\frac{x}{2^n})$$

for all $x \in A$.

If $f(2x) = 2f(x)$ for all $x \in A$, then we can easily show that $P(x) = f(x)$ for all $x \in A$.

It follows from (49) that

$$\begin{aligned}
&\|f \circ f([x, y, z]) - [f(x), f(y), f(z)]\| \\
&\quad = \|P \circ P([x, y, z]) - [P(x), P(y), P(z)]\|
\end{aligned}$$

$$= \lim_{n\to\infty} 8^n \left\| f \circ f\left(\frac{[x, y, z]}{8^n}\right) - \left[f\left(\frac{x}{2^n}\right), f\left(\frac{y}{2^n}\right), f\left(\frac{z}{2^n}\right)\right]\right\|$$
$$\leq \lim_{n\to\infty} 8^n \varphi\left(\frac{x}{2^n}, \frac{y}{2^n}, \frac{z}{2^n}\right) = 0$$

for all $x, y, z \in A$. Thus

$$f \circ f([x, y, z]) = [f(x), f(y), f(z)]$$

for all $x, y, z \in A$.

It follows from (50) that

$$\|f(x^*) - f(x)^*\| = \|P(x^*) - P(x)^*\| = \lim_{n\to\infty} 2^n \left\| f\left(\frac{x^*}{2^n}\right) - f\left(\frac{x}{2^n}\right)^*\right\|$$
$$\leq \lim_{n\to\infty} 2^n \varphi\left(\frac{x}{2^n}, 0, 0\right) \leq \lim_{n\to\infty} 8^n \varphi\left(\frac{x}{2^n}, 0, 0\right) = 0$$

for all $x \in A$. Thus

$$f(x^*) = f(x)^*$$

for all $x \in A$. Hence the mapping $f : A \to A$ is a partial multiplier.

Corollary 21 *Let $r > 3$ and θ be nonnegative real numbers, and let $f : A \to A$ be a mapping satisfying* (23)*. Then there exists a unique* **C***-linear mapping $P : A \to A$ satisfying* (24)*.*

If, in addition, the mapping $f : A \to A$ satisfies $f(2x) = 2f(x)$ and

$$\|f \circ f([x, y, z]) - [f(x), f(y), f(z)]\| \leq \theta(\|x\|^r + \|y\|^r + \|z\|^r), \tag{51}$$

$$\|f(x^*) - f(x)^*\| \leq \theta\|x\|^r \tag{52}$$

for all $x, y, z \in A$, then the mapping f is a partial multiplier.

Proof The proof follows from Theorem 21 by taking $\varphi(x, y, z) = \theta(\|x\|^r + \|y\|^r + \|z\|^r)$ for all $x, y, z \in A$.

Theorem 22 *Let $\varphi : A^3 \to [0, \infty)$ be a function satisfying* (27)*. Let $f : A \to A$ be a mapping satisfying $f(0) = 0$ and* (19)*. Then there exists a unique* **C***-linear mapping $P : A \to A$ satisfying* (28)*.*

If, in addition, the mapping $f : A \to A$ satisfies $f(2x) = 2f(x)$ for all $x \in A$, (49) *and* (50)*, then the mapping f is a partial multiplier.*

Proof The proof is similar to the proof of Theorem 21.

Corollary 22 *Let $r < 1$ and θ be nonnegative real numbers, and let $f : A \to A$ be a mapping satisfying* (23). *Then there exists a unique* **C**-*linear mapping satisfying* (29).

If, in addition, the mapping $f : A \to A$ satisfies $f(2x) = 2f(x)$ for all $x \in A$, (51) *and* (52), *then the mapping f is a partial multiplier.*

Proof The proof follows from Theorem 22 by taking $\varphi(x, y, z) = \theta(\|x\|^r + \|y\|^r + \|z\|^r)$ for all $x, y, z \in A$.

Similarly, we can obtain the following results. We will omit the proofs.

Theorem 23 *Let $\varphi : A^3 \to [0, \infty)$ be a function satisfying* (48). *Let $f : A \to A$ be an odd mapping satisfying* (30). *Then there exists a unique* **C**-*linear mapping $P : A \to A$ satisfying* (31).

If, in addition, the mapping $f : A \to A$ satisfies $f(2x) = 2f(x)$ for all $x \in A$, (49) *and* (50), *then the mapping f is a partial multiplier.*

Corollary 23 *Let $r > 3$ and θ be nonnegative real numbers, and let $f : A \to A$ be an odd mapping satisfying* (32). *Then there exists a unique* **C**-*linear mapping $P : A \to A$ satisfying* (33).

If, in addition, the mapping $f : A \to A$ satisfies $f(2x) = 2f(x)$, (51) *and* (52), *then the mapping f is a partial multiplier.*

Proof The proof follows from Theorem 23 by taking $\varphi(x, y, z) = \theta(\|x\|^r + \|y\|^r + \|z\|^r)$ for all $x, y, z \in A$.

Theorem 24 *Let $\varphi : A^3 \to [0, \infty)$ be a function satisfying* (27). *Let $f : A \to A$ be an odd mapping satisfying* (30). *Then there exists a unique* **C**-*linear mapping $P : A \to A$ satisfying* (34).

If, in addition, the mapping $f : A \to A$ satisfies $f(2x) = 2f(x)$ for all $x \in A$, (49) *and* (50), *then the mapping f is a partial multiplier.*

Corollary 24 *Let $r < 1$ and θ be nonnegative real numbers, and let $f : A \to A$ be an odd mapping satisfying* (32). *Then there exists a unique* **C**-*linear mapping $P : A \to A$ satisfying* (35).

If, in addition, the mapping $f : A \to A$ satisfies $f(2x) = 2f(x)$, (51) *and* (52), *then the mapping f is a partial multiplier.*

Proof The proof follows from Theorem 24 by taking $\varphi(x, y, z) = \theta(\|x\|^r + \|y\|^r + \|z\|^r)$ for all $x, y, z \in A$.

Acknowledgements C. Park was supported by Basic Science Research Program through the National Research Foundation of Korea funded by the Ministry of Education, Science and Technology (NRF-2017R1D1A1B04032937).

References

1. T. Aoki, On the stability of the linear transformation in Banach spaces. J. Math. Soc. Jpn. **2**, 64–66 (1950)
2. M. Amyari, C. Baak, M. Moslehian, Nearly ternary derivations. Taiwan. J. Math. **11**, 1417–1424 (2007)
3. M. Eshaghi Gordji, N. Ghobadipour, Stability of (α, β, γ)-derivations on Lie C^*-algebras. Int. J. Geom. Methods Mod. Phys. **7**, 1097–1102 (2010)
4. M. Eshaghi Gordji, M.B. Ghaemi, B. Alizadeh, A fixed point method for perturbation of higher ring derivations in non-Archimedean Banach algebras. Int. J. Geom. Methods Mod. Phys. **8**(7), 1611–1625 (2011)
5. M. Eshaghi Gordji, A. Fazeli, C. Park, 3-Lie multipliers on Banach 3-Lie algebras. Int. J. Geom. Meth. Mod. Phys. **9**(7), 1250052, 15pp. (2012)
6. W. Fechner, Stability of a functional inequalities associated with the Jordan-von Neumann functional equation. Aequationes Math. **71**, 149–161 (2006)
7. P. Găvruta, A generalization of the Hyers-Ulam-Rassias stability of approximately additive mappings. J. Math. Anal. Appl. **184**, 431–436 (1994)
8. A. Gilányi, Eine zur Parallelogrammgleichung äquivalente Ungleichung. Aequationes Math. **62**, 303–309 (2001)
9. A. Gilányi, On a problem by K. Nikodem. Math. Inequal. Appl. **5**, 707–710 (2002)
10. D.H. Hyers, On the stability of the linear functional equation. Proc. Nat. Acad. Sci. U.S.A. **27**, 222–224 (1941)
11. R.V. Kadison, J.R. Ringrose, *Fundamentals of the Theory of Operator Algebras: Elementary Theory* (Academic, New York, 1983)
12. C. Park, Lie $*$-homomorphisms between Lie C^*-algebras and Lie $*$-derivations on Lie C^*-algebras. J. Math. Anal. Appl. **293**, 419–434 (2004)
13. C. Park, Homomorphisms between Lie JC^*-algebras and Cauchy-Rassias stability of Lie JC^*-algebra derivations. J. Lie Theory **15**, 393–414 (2005)
14. C. Park, Homomorphisms between Poisson JC^*-algebras. Bull. Braz. Math. Soc. **36**, 79–97 (2005)
15. C. Park, Isomorphisms between C^*-ternary algebras. J. Math. Anal. Appl. **327**, 101–115 (2007)
16. C. Park, Additive ρ-functional inequalities and equations. J. Math. Inequal. **9**, 17–26 (2015)
17. C. Park, Additive ρ-functional inequalities in non-Archimedean normed spaces. J. Math. Inequal. **9**, 397–407 (2015)
18. C. Park, Functional inequalities and partial multipliers in Banach algebras (preprint)
19. C. Park, Additive s-functional inequalities and partial multipliers in Banach algebras, J. Math. Inequal. **13**, 867–877 (2019)
20. C. Park, J. Hou, S. Oh, Homomorphisms between JC^*-algebras and between Lie C^*-algebras. Acta Math. Sin. **21**, 1391–1398 (2005)
21. C. Park, K. Ghasemi, S.G. Ghaleh, S. Jang, Approximate n-Jordan $*$-homomorphisms in C^*-algebras. J. Comput. Anal. Appl. **15**, 365–368 (2013)
22. T.M. Rassias, On the stability of the linear mapping in Banach spaces. Proc. Am. Math. Soc. **72**, 297–300 (1978)
23. J. Rätz, On inequalities associated with the Jordan-von Neumann functional equation. Aequationes Math. **66**, 191–200 (2003)
24. D. Shin, C. Park, S. Farhadabadi, On the superstability of ternary Jordan C^*-homomorphisms. J. Comput. Anal. Appl. **16**, 964–973 (2014)
25. D. Shin, C. Park, S. Farhadabadi, Stability and superstability of J^*-homomorphisms and J^*-derivations for a generalized Cauchy-Jensen equation. J. Comput. Anal. Appl. **17**, 125–134 (2014)

26. A. Taghavi, On a functional equation for symmetric linear operators on C^*-algebras. Bull. Iranian Math. Soc. **42**, 1169–1177 (2016)
27. S.M. Ulam, *A Collection of the Mathematical Problems* (Interscience Publishers, New York, 1960)
28. H. Zettl, A characterization of ternary rings of operators. Adv. Math. **48**, 117–143 (1983)

Additive ρ-Functional Inequalities and Their Applications

Jung Rye Lee, Choonkil Park, Themistocles M. Rassias, and Xiaohong Zhang

Abstract In this paper, we solve the additive ρ-functional inequalities

$$\begin{aligned} &\|f(x+y+z)-f(x)-f(y)-f(z)\| \\ &\quad \le \left\|\rho\left(2f\left(\frac{x+y}{2}+z\right)-f(x)-f(y)-2f(z)\right)\right\|, \end{aligned} \tag{1}$$

where ρ is a fixed complex number with $|\rho| < 1$, and

$$\begin{aligned} &\left\|2f\left(\frac{x+y}{2}+z\right)-f(x)-f(y)-2f(z)\right\| \\ &\quad \le \|\rho(f(x+y+z)-f(x)-f(y)-f(z))\|, \end{aligned} \tag{2}$$

where ρ is a fixed complex number with $|\rho| < 1$. Furthermore, we prove the Hyers-Ulam stability of the additive ρ-functional inequalities (1) and (2) in complex Banach spaces. This is applied to investigate homomorphisms in C^*-algebras, Lie C^*-algebras and JC^*-algebras, and derivations on C^*-algebras, Lie C^*-algebras and JC^*-algebras associated with the additive ρ-functional inequalities (1) and (2).

J. R. Lee
Department of Mathematics, Daejin University, Pocheon, South Korea
e-mail: jrlee@daejin.ac.kr

C. Park (✉)
Research Institute for Natural Sciences, Hanyang University, Seoul, South Korea
e-mail: baak@hanyang.ac.kr

T. M. Rassias
Department of Mathematics, Zografou Campus, National Technical University of Athens, Athens, Greece
e-mail: trassias@math.ntua.gr

X. Zhang
Shanghai Maritime University, Shanghai, People's Republic of China
e-mail: zhangxh@shmtu.edu.cn

T. M. Rassias, P. M. Pardalos (eds.), *Mathematical Analysis and Applications*, Springer Optimization and Its Applications 154,
https://doi.org/10.1007/978-3-030-31339-5_14

1 Introduction and Preliminaries

Ulam [40] gave a talk before the Mathematics Club of the University of Wisconsin in which he discussed a number of unsolved problems. Among these was the following question concerning the stability of homomorphisms.

We are given a group G and a metric group G' with metric $\rho(\cdot, \cdot)$. Given $\epsilon > 0$, does there exist a $\delta > 0$ such that if $f : G \to G'$ satisfies $\rho(f(xy), f(x)f(y)) < \delta$ for all $x, y \in G$, then a homomorphism $h : G \to G'$ exists with $\rho(f(x), h(x)) < \epsilon$ for all $x \in G$?

Hyers [13] considered the case of approximately additive mappings $f : E \to E'$, where E and E' are Banach spaces and f satisfies *Hyers inequality*

$$\|f(x+y) - f(x) - f(y)\| \leq \epsilon$$

for all $x, y \in E$. It was shown that the limit

$$L(x) = \lim_{n\to\infty} \frac{f(2^n x)}{2^n}$$

exists for all $x \in E$ and that $L : E \to E'$ is the unique additive mapping satisfying

$$\|f(x) - L(x)\| \leq \epsilon.$$

Hyers' Theorem was generalized by Aoki [2] for additive mappings and by Rassias [32] for linear mappings by considering an unbounded Cauchy difference.

Theorem 1 (Rassias) *Let $f : E \to E'$ be a mapping from a normed vector space E into a Banach space E' subject to the inequality*

$$\|f(x+y) - f(x) - f(y)\| \leq \epsilon(\|x\|^p + \|y\|^p) \tag{3}$$

for all $x, y \in E$, where ϵ and p are constants with $\epsilon > 0$ and $p < 1$. Then the limit

$$L(x) = \lim_{n\to\infty} \frac{f(2^n x)}{2^n}$$

exists for all $x \in E$ and $L : E \to E'$ is the unique additive mapping which satisfies

$$\|f(x) - L(x)\| \leq \frac{2\epsilon}{2 - 2^p}\|x\|^p \tag{4}$$

for all $x \in E$. If $p < 0$, then inequality (3) holds for $x, y \neq 0$ and (4) for $x \neq 0$. Also, if for each $x \in E$ the function $f(tx)$ is continuous in $t \in \mathbf{R}$, then L is linear.

The inequality (3) that was introduced for the first time by Rassias [32] provided a lot of influence in the development of a generalization of the Hyers-Ulam stability concept. This new concept is known as *Hyers-Ulam-Rassias stability* of functional equations (cf. the books of Czerwik [4], Hyers et al. [14]).

A generalization of the Rassias theorem was obtained by Găvruta [10] by replacing the unbounded Cauchy difference by a general control function in the spirit of Rassias' approach. The stability of quadratic functional equation was proved by Skof [39] for mappings $f : E_1 \to E_2$, where E_1 is a normed space and E_2 is a Banach space. Cholewa [3] noticed that the theorem of Skof is still true if the relevant domain E_1 is replaced by an Abelian group. During the several papers have been published on various generalizations and applications of Hyers-Ulam stability to a number of functional equations and mappings, for example: quadratic functional equation, invariant means, multiplicative mappings—superstability, bounded nth differences, convex functions, generalized orthogonality functional equation, Euler-Lagrange functional equation, Navier-Stokes equations. Several mathematicians have contributed works on these subjects (see [1, 9, 15–26, 29, 33]).

Gilányi [11] showed that if f satisfies the functional inequality

$$\|2f(x) + 2f(y) - f(x - y)\| \leq \|f(x + y)\| \tag{5}$$

then f satisfies the Jordan-von Neumann functional equation

$$2f(x) + 2f(y) = f(x + y) + f(x - y).$$

See also [34]. Fechner [8] and Gilányi [12] proved the Hyers-Ulam stability of the functional inequality (5).

Park [27, 28] defined additive ρ-functional inequalities and proved the Hyers-Ulam stability of the additive ρ-functional inequalities in Banach spaces and non-Archimedean Banach spaces. The stability problems of various functional equations have been extensively investigated by a number of authors (see [5–7, 31, 35–38]).

Jordan observed that $\mathscr{L}(\mathscr{H})$ is an algebra via the *anticommutator product* $x \circ y := \frac{xy+yx}{2}$. A commutative algebra X with product $x \circ y$ is called a *Jordan algebra*. A Jordan C^*-subalgebra of a C^*-algebra, endowed with the anticommutator product is called a *JC^*-algebra*. A C^*-algebra $\mathscr{C}$, endowed with the Lie product $[x, y] = \frac{xy-yx}{2}$ on $\mathscr{C}$, is called a Lie *C^*-algebra* (see [20, 21, 30]).

This paper is organized as follows: In Section 2, we solve the additive ρ-functional inequality (1) and prove the Hyers-Ulam stability of the additive ρ-functional inequality (1) in complex Banach spaces.

In Section 3, we solve the additive ρ-functional inequality (2) and prove the Hyers-Ulam stability of the additive ρ-functional inequality (2) in complex Banach spaces.

In Section 4, we investigate homomorphisms and derivations in C^*-algebras associated with the additive ρ-functional inequalities (1) and (2).

In Section 5, we investigate homomorphisms and derivations in Lie C^*-algebras associated with the additive ρ-functional inequalities (1) and (2).

In Section 6, we investigate homomorphisms and derivations in JC^*-algebras associated with the additive ρ-functional inequalities (1) and (2).

Throughout this paper, let X be a complex normed space with norm $\|\cdot\|$ and Y a complex Banach space with norm $\|\cdot\|$. Assume that ρ is a fixed complex number with $|\rho| < 1$.

2 Additive ρ-Functional Inequality (1)

In this section, we solve and investigate the additive ρ-functional inequality (1) in complex Banach spaces.

Lemma 1 *If a mapping $f : X \to Y$ satisfies*

$$\begin{aligned} &\|f(x+y+z) - f(x) - f(y) - f(z)\| \\ &\quad \le \left\| \rho \left(2f\left(\frac{x+y}{2} + z\right) - f(x) - f(y) - 2f(z) \right) \right\|, \end{aligned} \tag{6}$$

for all $x, y, z \in X$, then $f : X \to Y$ is additive.

Proof Assume that $f : X \to Y$ satisfies (6).

Letting $x = y = z = 0$ in (6), we get $\|2f(0)\| \le \|2\rho f(0)\|$ and so $f(0) = 0$.

Letting $y = x$ and $z = 0$ in (6), we get $\|f(2x) - 2f(x)\| \le 0$ and so $f(2x) = 2f(x)$ for all $x \in G$. Thus

$$f\left(\frac{x}{2}\right) = \frac{1}{2} f(x) \tag{7}$$

for all $x \in X$.

Let $z = 0$ in (6). It follows from (6) and (7) that

$$\begin{aligned} \|f(x+y) - f(x) - f(y)\| &\le \left\| \rho \left(2f\left(\frac{x+y}{2}\right) - f(x) - f(y) \right) \right\| \\ &= |\rho| \|f(x+y) - f(x) - f(y)\| \end{aligned}$$

and so $f(x+y) = f(x) + f(y)$ for all $x, y \in X$, since $|\rho| < 1$.

We prove the Hyers-Ulam stability of the additive ρ-functional inequality (6) in complex Banach spaces.

Theorem 2 *Let $r > 1$ and θ be nonnegative real numbers, and let $f : X \to Y$ be a mapping satisfying*

$$\begin{aligned} &\|f(x+y+z) - f(x) - f(y) - f(z)\| \\ &\le \left\| \rho \left(2f\left(\frac{x+y}{2} + z\right) - f(x) - f(y) - 2f(z) \right) \right\| + \theta(\|x\|^r + \|y\|^r + \|z\|^r) \end{aligned} \tag{8}$$

for all $x, y, z \in X$. Then there exists a unique additive mapping $h : X \to Y$ such that

$$\|f(x) - h(x)\| \leq \frac{2\theta}{2^r - 2}\|x\|^r \tag{9}$$

for all $x \in X$.

Proof Letting $x = y = z = 0$ in (8), we get $\|2f(0)\| \leq \|2\rho f(0)\|$ and so $f(0) = 0$.
Letting $y = x$ and $z = 0$ in (8), we get

$$\|f(2x) - 2f(x)\| \leq 2\theta\|x\|^r \tag{10}$$

for all $x \in X$. So

$$\left\|f(x) - 2f\left(\frac{x}{2}\right)\right\| \leq \frac{2}{2^r}\theta\|x\|^r$$

for all $x \in X$. Hence

$$\begin{aligned}\left\|2^l f\left(\frac{x}{2^l}\right) - 2^m f\left(\frac{x}{2^m}\right)\right\| &\leq \sum_{j=l}^{m-1}\left\|2^j f\left(\frac{x}{2^j}\right) - 2^{j+1} f\left(\frac{x}{2^{j+1}}\right)\right\| \\ &\leq \frac{2}{2^r}\sum_{j=l}^{m-1}\frac{2^j}{2^{rj}}\theta\|x\|^r \end{aligned}\tag{11}$$

for all nonnegative integers m and l with $m > l$ and all $x \in X$. It follows from (11) that the sequence $\{2^n f(\frac{x}{2^n})\}$ is a Cauchy sequence for all $x \in X$. Since Y is complete, the sequence $\{2^n f(\frac{x}{2^n})\}$ converges. So one can define the mapping $h : X \to Y$ by

$$h(x) := \lim_{n\to\infty} 2^n f(\frac{x}{2^n})$$

for all $x \in X$. Moreover, letting $l = 0$ and passing the limit $m \to \infty$ in (11), we get (9).

It follows from (8) that

$$\begin{aligned}&\|h(x + y + z) - h(x) - h(y) - h(z)\| \\ &= \lim_{n\to\infty} 2^n \left\|f\left(\frac{x + y + z}{2^n}\right) - f\left(\frac{x}{2^n}\right) - f\left(\frac{y}{2^n}\right) - f\left(\frac{z}{2^n}\right)\right\| \\ &\leq \lim_{n\to\infty} 2^n|\rho| \left\|2f\left(\frac{x + y + 2z}{2^{n+1}}\right) - f\left(\frac{x}{2^n}\right) - f\left(\frac{y}{2^n}\right) - 2f\left(\frac{z}{2^n}\right)\right\|\end{aligned}$$

$$+ \lim_{n\to\infty} \frac{2^n\theta}{2^{nr}}(\|x\|^r + \|y\|^r + \|z\|^r)$$

$$= |\rho| \left\| \left(2h\left(\frac{x+y}{2}+z\right) - h(x) - h(y) - 2h(z)\right) \right\|$$

for all $x, y, z \in X$. So

$$\|h(x+y+z) - h(x) - h(y) - h(z)\|$$
$$\leq \left\| \rho \left(2h\left(\frac{x+y}{2}+z\right) - h(x) - h(y) - 2h(z)\right) \right\|$$

for all $x, y, z \in X$. By Lemma 1, the mapping $h : X \to Y$ is additive.

Now, let $T : X \to Y$ be another additive mapping satisfying (9). Then we have

$$\begin{aligned} \|h(x) - T(x)\| &= 2^n \left\| h\left(\frac{x}{2^n}\right) - T\left(\frac{x}{2^n}\right) \right\| \\ &\leq 2^n \left(\left\| h\left(\frac{x}{2^n}\right) - f\left(\frac{x}{2^n}\right) \right\| + \left\| T\left(\frac{x}{2^n}\right) - f\left(\frac{x}{2^n}\right) \right\| \right) \\ &\leq \frac{4 \cdot 2^n}{(2^r - 2)2^{nr}} \theta \|x\|^r, \end{aligned}$$

which tends to zero as $n \to \infty$ for all $x \in X$. So we can conclude that $h(x) = T(x)$ for all $x \in X$. This proves the uniqueness of h. Thus the mapping $h : X \to Y$ is a unique additive mapping satisfying (9).

Theorem 3 *Let $r < 1$ and θ be positive real numbers, and let $f : X \to Y$ be a mapping satisfying* (8)*. Then there exists a unique additive mapping $h : X \to Y$ such that*

$$\|f(x) - h(x)\| \leq \frac{2\theta}{2 - 2^r} \|x\|^r \tag{12}$$

for all $x \in X$.

Proof It follows from (10) that

$$\left\| f(x) - \frac{1}{2} f(2x) \right\| \leq \theta \|x\|^r$$

for all $x \in X$. Hence

$$\begin{aligned} \left\| \frac{1}{2^l} f(2^l x) - \frac{1}{2^m} f(2^m x) \right\| &\leq \sum_{j=l}^{m-1} \left\| \frac{1}{2^j} f(2^j x) - \frac{1}{2^{j+1}} f(2^{j+1} x) \right\| \\ &\leq \sum_{j=l}^{m-1} \frac{2^{rj}}{2^j} \theta \|x\|^r \end{aligned} \tag{13}$$

for all nonnegative integers m and l with $m > l$ and all $x \in X$. It follows from (13) that the sequence $\{\frac{1}{2^n} f(2^n x)\}$ is a Cauchy sequence for all $x \in X$. Since Y is complete, the sequence $\{\frac{1}{2^n} f(2^n x)\}$ converges. So one can define the mapping $h : X \to Y$ by

$$h(x) := \lim_{n\to\infty} \frac{1}{2^n} f(2^n x)$$

for all $x \in X$. Moreover, letting $l = 0$ and passing the limit $m \to \infty$ in (13), we get (12).

The rest of the proof is similar to the proof of Theorem 2.

Remark 1 If ρ is a real number such that $-1 < \rho < 1$ and Y is a real Banach space, then all the assertions in this section remain valid.

3 Additive ρ-Functional Inequality (2)

In this section, we solve and investigate the additive ρ-functional inequality (2) in complex Banach spaces.

Lemma 2 *If a mapping $f : X \to Y$ satisfies*

$$\begin{aligned} \left\| 2f\left(\frac{x+y}{2}+z\right) - f(x) - f(y) - 2f(z) \right\| & \\ \leq \|\rho(f(x+y+z) - f(x) - f(y) - f(z))\| & \end{aligned} \tag{14}$$

for all $x, y, z \in X$, then $f : X \to Y$ is additive.

Proof Assume that $f : X \to Y$ satisfies (14).

Letting $y = z = 0$ in (14), we get $\left\|2f\left(\frac{x}{2}\right) - f(x)\right\| \leq 0$ and so

$$2f\left(\frac{x}{2}\right) = f(x) \tag{15}$$

for all $x \in X$.

Let $z = 0$ in (14). It follows from (14) and (15) that

$$\begin{aligned} \|f(x+y) - f(x) - f(y)\| &= \left\| 2f\left(\frac{x+y}{2}\right) - f(x) - f(y) \right\| \\ &\leq |\rho| \|f(x+y) - f(x) - f(y)\| \end{aligned}$$

and so $f(x+y) = f(x) + f(y)$ for all $x, y \in X$, since $|\rho| < 1$.

We prove the Hyers-Ulam stability of the additive ρ-functional inequality (14) in complex Banach spaces.

Theorem 4 *Let $r > 1$ and θ be nonnegative real numbers, and let $f : X \to Y$ be a mapping such that*

$$\left\|2f\left(\frac{x+y}{2}+z\right)-f(x)-f(y)-2f(z)\right\| \quad (16)$$
$$\leq \|\rho(f(x+y+z)-f(x)-f(y)-f(z))\|+\theta(\|x\|^r+\|y\|^r+\|z\|^r)$$

for all $x, y, z \in X$. Then there exists a unique additive mapping $h : X \to Y$ such that

$$\|f(x)-h(x)\| \leq \frac{2^r\theta}{2^r-2}\|x\|^r \quad (17)$$

for all $x \in X$.

Proof Letting $x = y = z = 0$ in (16), we get $\|2f(0)\| \leq \|2\rho f(0)\|$ and so $f(0) = 0$.

Letting $y = z = 0$ in (16), we get

$$\left\|2f\left(\frac{x}{2}\right)-f(x)\right\| \leq \theta\|x\|^r \quad (18)$$

for all $x \in X$. So

$$\left\|2^l f\left(\frac{x}{2^l}\right)-2^m f\left(\frac{x}{2^m}\right)\right\| \leq \sum_{j=l}^{m-1}\left\|2^j f\left(\frac{x}{2^j}\right)-2^{j+1} f\left(\frac{x}{2^{j+1}}\right)\right\|$$
$$\leq \sum_{j=l}^{m-1}\frac{2^j}{2^{rj}}\theta\|x\|^r \quad (19)$$

for all nonnegative integers m and l with $m > l$ and all $x \in X$. It follows from (19) that the sequence $\{2^n f(\frac{x}{2^n})\}$ is a Cauchy sequence for all $x \in X$. Since Y is complete, the sequence $\{2^n f(\frac{x}{2^n})\}$ converges. So one can define the mapping $h : X \to Y$ by

$$h(x) := \lim_{n\to\infty} 2^n f(\frac{x}{2^n})$$

for all $x \in X$. Moreover, letting $l = 0$ and passing the limit $m \to \infty$ in (19), we get (17).

It follows from (16) that

$$\left\|2h\left(\frac{x+y}{2}+z\right)-h(x)-h(y)-2h(z)\right\|$$
$$= \lim_{n\to\infty} 2^n\left\|2f\left(\frac{x+y+2z}{2^{n+1}}\right)-f\left(\frac{x}{2^n}\right)-f\left(\frac{y}{2^n}\right)-2f\left(\frac{z}{2^n}\right)\right\|$$

$$\leq \lim_{n\to\infty} 2^n \left\| \rho \left(f\left(\frac{x+y+z}{2^n}\right) - f\left(\frac{x}{2^n}\right) - f\left(\frac{y}{2^n}\right) - f\left(\frac{z}{2^n}\right) \right) \right\|$$
$$+ \lim_{n\to\infty} \frac{2^n \theta}{2^{nr}} (\|x\|^r + \|y\|^r + \|z\|^r)$$
$$= \|\rho(h(x+y+z) - h(x) - h(y) - h(z))\|$$

for all $x, y, z \in X$. So

$$\left\| 2h\left(\frac{x+y}{2} + z\right) - h(x) - h(y) - 2h(z) \right\| \leq \|\rho(h(x+y+z)-h(x)-h(y)-h(z))\|$$

for all $x, y, z \in X$. By Lemma 2, the mapping $h : X \to Y$ is additive.

Now, let $T : X \to Y$ be another additive mapping satisfying (17). Then we have

$$\begin{aligned}
\|h(x) - T(x)\| &= 2^n \left\| h\left(\frac{x}{2^n}\right) - T\left(\frac{x}{2^n}\right) \right\| \\
&\leq 2^n \left(\left\| h\left(\frac{x}{2^n}\right) - f\left(\frac{x}{2^n}\right) \right\| + \left\| T\left(\frac{x}{2^n}\right) - f\left(\frac{x}{2^n}\right) \right\| \right) \\
&\leq \frac{2 \cdot 2^n \cdot 2^r}{(2^r - 2)2^{nr}} \theta \|x\|^r,
\end{aligned}$$

which tends to zero as $n \to \infty$ for all $x \in X$. So we can conclude that $h(x) = T(x)$ for all $x \in X$. This proves the uniqueness of h. Thus the mapping $h : X \to Y$ is a unique additive mapping satisfying (17).

Theorem 5 *Let $r < 1$ and θ be positive real numbers, and let $f : X \to Y$ be a mapping satisfying* (16)*. Then there exists a unique additive mapping $h : X \to Y$ such that*

$$\|f(x) - h(x)\| \leq \frac{2^r \theta}{2 - 2^r} \|x\|^r \tag{20}$$

for all $x \in X$.

Proof It follows from (18) that

$$\left\| f(x) - \frac{1}{2} f(2x) \right\| \leq \frac{2^r \theta}{2} \|x\|^r$$

for all $x \in X$. Hence

$$\begin{aligned}
\left\| \frac{1}{2^l} f(2^l x) - \frac{1}{2^m} f(2^m x) \right\| &\leq \sum_{j=l}^{m-1} \left\| \frac{1}{2^j} f(2^j x) - \frac{1}{2^{j+1}} f(2^{j+1} x) \right\| \\
&\leq \frac{2^r \theta}{2} \sum_{j=l}^{m-1} \frac{2^{rj}}{2^j} \|x\|^r
\end{aligned} \tag{21}$$

for all nonnegative integers m and l with $m > l$ and all $x \in X$. It follows from (21) that the sequence $\{\frac{1}{2^n} f(2^n x)\}$ is a Cauchy sequence for all $x \in X$. Since Y is complete, the sequence $\{\frac{1}{2^n} f(2^n x)\}$ converges. So one can define the mapping $h : X \to Y$ by

$$h(x) := \lim_{n\to\infty} \frac{1}{2^n} f(2^n x)$$

for all $x \in X$. Moreover, letting $l = 0$ and passing the limit $m \to \infty$ in (21), we get (20).

The rest of the proof is similar to the proof of Theorem 4.

Remark 2 If ρ is a real number such that $-1 < \rho < 1$ and Y is a real Banach space, then all the assertions in this section remain valid.

4 Homomorphisms and Derivations in C^*-Algebras

Throughout this section, assume that A is a C^*-algebra with norm $\|\cdot\|$, and that B is a C^*-algebra with norm $\|\cdot\|$.

In this section, we investigate C^*-algebra homomorphisms in C^*-algebras and derivations on C^*-algebras associated with the additive ρ-functional inequalities (6) and (14).

Theorem 6 *Let $r > 2$ and θ be nonnegative real numbers, and let $f : A \to B$ be a mapping such that*

$$\begin{aligned} &\|f(x+y+\mu z) - f(x) - f(y) - \mu f(z)\| \\ &\leq \left\|\rho\left(2f\left(\frac{x+y}{2} + \mu z\right) - f(x) - f(y) - 2f(\mu z)\right)\right\| \\ &\quad +\theta(\|x\|^r + \|y\|^r + \|z\|^r), \end{aligned} \tag{22}$$

$$\|f(xy) - f(x)f(y)\| \leq \theta(\|x\|^r + \|y\|^r), \tag{23}$$

$$\|f(x^*) - f(x)^*\| \leq \theta(\|x\|^r + \|x\|^r) \tag{24}$$

for all $\mu \in \mathbf{T}^1 := \{\lambda \in \mathbf{C} \mid |\lambda| = 1\}$ and all $x, y, z \in A$. Then there exists a unique C^*-algebra homomorphism $h : A \to B$ such that

$$\|f(x) - h(x)\| \leq \frac{2\theta}{2^r - 2}\|x\|^r \tag{25}$$

for all $x \in A$.

Proof Let $\mu = 1$ in (22). By Theorem 2, there is a unique additive mapping $h : A \to B$ satisfying (25) defined by

$$h(x) := \lim_{n\to\infty} 2^n f(\frac{x}{2^n})$$

for all $x \in A$.

Letting $x = y = 0$ and replacing z by x in (22), we get

$$\| f(\mu x) - \mu f(x)\| \le \theta \|x\|^r$$

for all $x \in A$ and all $\mu \in \mathbf{T}^1$. So

$$\|h(\mu x) - \mu h(x)\| = \lim_{n\to\infty} 2^n \left\| f\left(\mu \frac{x}{2^n}\right) - f\left(\mu \frac{x}{2^n}\right)\right\| \le \lim_{n\to\infty} \frac{2^n}{2^{rn}} \theta \|x\|^r = 0$$

for all $x \in A$ and all $\mu \in \mathbf{T}^1$. Hence $h(\mu x) = \mu h(x)$ for all $x \in A$ and all $\mu \in \mathbf{T}^1$. By the same reasoning as in the proof of [23, Theorem 2.1], the mapping $h : A \to B$ is $\mathbf{C}$-linear.

It follows from (23) that

$$\begin{aligned}\|h(xy) - h(x)h(y)\| &= \lim_{n\to\infty} 4^n \left\| f\left(\frac{xy}{2^n \cdot 2^n}\right) - f\left(\frac{x}{2^n}\right) f\left(\frac{y}{2^n}\right)\right\| \\ &\le \lim_{n\to\infty} \frac{4^n \theta}{2^{rn}} (\|x\|^r + \|y\|^r) = 0\end{aligned}$$

for all $x, y \in A$. Thus

$$h(xy) = h(x)h(y)$$

for all $x, y \in A$.

It follows from (24) that

$$\|h(x^*) - h(x)^*\| = \lim_{n\to\infty} 2^n \left\| f\left(\frac{x^*}{2^n}\right) - f\left(\frac{x}{2^n}\right)^*\right\| \le \lim_{n\to\infty} \frac{2^n \theta}{2^{nr}} (\|x\|^r + \|x\|^r) = 0$$

for all $x \in A$. Thus

$$h(x^*) = h(x)^*$$

for all $x \in A$. Hence the additive mapping $h : A \to B$ is a C^*-algebra homomorphism.

Theorem 7 *Let $r < 1$ and θ be positive real numbers, and let $f : A \to B$ be a mapping satisfying (22)–(24). Then there exists a unique C^*-algebra homomorphism $h : A \to B$ such that*

$$\|f(x) - h(x)\| \leq \frac{2\theta}{2 - 2^r}\|x\|^r \tag{26}$$

for all $x \in A$.

Proof The proof is similar to the proof of Theorem 6.

Theorem 8 *Let* $r > 2$ *and* θ *be nonnegative real numbers, and let* $f : A \to A$ *be a mapping satisfying* (22) *such that*

$$\|f(xy) - f(x)y - xf(y)\| \leq \theta(\|x\|^r + \|y\|^r) \tag{27}$$

for all $x, y \in A$. *Then there exists a unique derivation* $D : A \to A$ *such that*

$$\|f(x) - D(x)\| \leq \frac{2\theta}{2^r - 2}\|x\|^r \tag{28}$$

for all $x \in A$.

Proof By the same reasoning as in the proof of Theorem 6, there is a unique **C**-linear mapping $D : A \to A$ satisfying (28) defined by

$$D(x) := \lim_{n\to\infty} 2^n f(\frac{x}{2^n})$$

for all $x \in A$.

It follows from (27) that

$$\begin{aligned}\|D(xy) - D(x)y - xD(y)\| &= \lim_{n\to\infty} 4^n \left\| f\left(\frac{xy}{4^n}\right) - f\left(\frac{x}{2^n}\right)\frac{y}{2^n} - \frac{x}{2^n} f\left(\frac{y}{2^n}\right)\right\| \\ &\leq \lim_{n\to\infty} \frac{4^n\theta}{2^{nr}}(\|x\|^r + \|y\|^r) = 0\end{aligned}$$

for all $x, y \in A$. So

$$D(xy) = D(x)y + xD(y)$$

for all $x, y \in A$. Thus the **C**-linear mapping $D : A \to A$ is a derivation.

Theorem 9 *Let* $r < 1$ *and* θ *be positive real numbers, and let* $f : A \to A$ *be a mapping satisfying* (22) *and* (27). *Then there exists a unique derivation* $D : A \to A$ *such that*

$$\|f(x) - D(x)\| \leq \frac{2\theta}{2 - 2^r}\|x\|^r \tag{29}$$

for all $x \in A$.

Proof The proof is similar to the proofs of Theorems 6 and 8.

Theorem 10 *Let $r > 2$ and θ be nonnegative real numbers, and let $f : A \to B$ be a mapping satisfying* (23), (24) *and*

$$\left\| 2f\left(\frac{x+y}{2} + \mu z\right) - f(x) - f(y) - 2\mu f(z) \right\| \tag{30}$$
$$\leq \|\rho(f(x+y+\mu z) - f(x) - f(y) - f(\mu z))\| + \theta(\|x\|^r + \|y\|^r + \|z\|^r)$$

for all $\mu \in \mathbf{T}^1$ and all $x, y, z \in A$. Then there exists a unique C^*-algebra homomorphism $h : A \to B$ such that

$$\|f(x) - h(x)\| \leq \frac{2^r\theta}{2^r - 2}\|x\|^r \tag{31}$$

for all $x \in A$.

Proof Let $\mu = 1$ in (30). By Theorem 4, there is a unique additive mapping $h : A \to B$ satisfying (31) defined by

$$h(x) := \lim_{n\to\infty} 2^n f(\frac{x}{2^n})$$

for all $x \in A$.

Letting $x = y = 0$ and replacing z by x in (30), we get

$$\|2f(\mu x) - 2\mu f(x)\| \leq \theta\|x\|^r$$

for all $x \in A$ and all $\mu \in \mathbf{T}^1$. So

$$2\|h(\mu x) - \mu h(x)\| = 2\lim_{n\to\infty} 2^n \left\| f\left(\mu\frac{x}{2^n}\right) - f\left(\mu\frac{x}{2^n}\right)\right\| \leq \lim_{n\to\infty}\frac{2^n}{2^{rn}}\theta\|x\|^r = 0$$

for all $x \in A$ and all $\mu \in \mathbf{T}^1$. Hence $h(\mu x) = \mu h(x)$ for all $x \in A$ and all $\mu \in \mathbf{T}^1$. By the same reasoning as in the proof of [23, Theorem 2.1], the mapping $h : A \to B$ is $\mathbf{C}$-linear.

The rest of the proof is similar to the proof of Theorem 6.

Theorem 11 *Let $r < 1$ and θ be positive real numbers, and let $f : A \to B$ be a mapping satisfying* (23), (24) *and* (30). *Then there exists a unique C^*-algebra homomorphism $h : A \to B$ such that*

$$\|f(x) - h(x)\| \leq \frac{2^r\theta}{2 - 2^r}\|x\|^r$$

for all $x \in A$.

Proof The proof is similar to the proof of Theorem 10.

Theorem 12 *Let $r > 2$ and θ be nonnegative real numbers, and let $f : A \to A$ be a mapping satisfying* (27) *and* (30). *Then there exists a unique derivation $D : A \to A$ such that*

$$\|f(x) - D(x)\| \leq \frac{2^r\theta}{2^r - 2}\|x\|^r \tag{32}$$

for all $x \in A$.

Proof By the same reasoning as in the proof of Theorem 10, there is a unique **C**-linear mapping $D : A \to A$ satisfying (32) defined by

$$D(x) := \lim_{n\to\infty} 2^n f(\frac{x}{2^n})$$

for all $x \in A$.

The rest of the proof is similar to the proofs of Theorem 8.

Theorem 13 *Let $r < 1$ and θ be positive real numbers, and let $f : A \to A$ be a mapping satisfying* (27) *and* (30). *Then there exists a unique derivation $D : A \to A$ such that*

$$\|f(x) - D(x)\| \leq \frac{2^r\theta}{2 - 2^r}\|x\|^r \tag{33}$$

for all $x \in A$.

Proof The proof is similar to the proofs of Theorems 8 and 10.

5 Homomorphisms and Derivations in Lie C^*-Algebras

Throughout this section, assume that A is a Lie C^*-algebra with norm $\|\cdot\|$, and that B is a Lie C^*-algebra with norm $\|\cdot\|$.

Definition 1 ([20, 21, 30]) A **C**-linear mapping $H : A \to B$ is called a *Lie C^*-algebra homomorphism* if $H : A \to B$ satisfies

$$H([x, y]) = [H(x), H(y)]$$

for all $x, y \in A$.

Definition 2 ([20, 21, 30]) A **C**-linear mapping $D : A \to A$ is called a *Lie derivation* if $D : A \to A$ satisfies

$$D([x, y]) = [Dx, y] + [x, Dy]$$

for all $x, y \in A$.

In this section, we investigate Lie C^*-algebra homomorphisms in Lie C^*-algebras and Lie derivations on Lie C^*-algebras associated with the additive ρ-functional inequalities (6) and (14).

Theorem 14 *Let $r > 2$ and θ be nonnegative real numbers, and let $f : A \to B$ be a mapping satisfying* (22) *and*

$$\|f([x, y]) - [f(x), f(y)]\| \leq \theta(\|x\|^r + \|y\|^r) \tag{34}$$

for all $x, y \in A$. Then there exists a unique Lie C^*-algebra homomorphism h : $A \to B$ satisfying (25).

Proof By the same reasoning as in the proof of Theorem 6, there is a unique **C**-linear mapping $h : A \to B$ satisfying (25) defined by

$$h(x) := \lim_{n\to\infty} 2^n f(\frac{x}{2^n})$$

for all $x \in A$.

It follows from (34) that

$$\begin{aligned}\|h([x, y]) - [h(x), h(y)]\| &= \lim_{n\to\infty} 4^n \left\| f\left(\frac{[x, y]}{2^n \cdot 2^n}\right) - \left[f\left(\frac{x}{2^n}\right), f\left(\frac{y}{2^n}\right)\right]\right\| \\ &\leq \lim_{n\to\infty} \frac{4^n\theta}{2^{nr}}(\|x\|^r + \|y\|^r) = 0\end{aligned}$$

for all $x, y \in A$. Thus

$$h([x, y]) = [h(x), h(y)]$$

for all $x, y \in A$. Hence the **C**-linear mapping $h : A \to B$ is a Lie C^*-algebra homomorphism.

Theorem 15 *Let $r < 1$ and θ be positive real numbers, and let $f : A \to B$ be a mapping satisfying* (22) *and* (34). *Then there exists a unique Lie C^*-algebra homomorphism $h : A \to B$ satisfying (26).*

Proof The proof is similar to the proofs of Theorems 6 and 14.

Theorem 16 *Let $r > 2$ and θ be nonnegative real numbers, and let $f : A \to A$ be a mapping satisfying* (22) *such that*

$$\|f([x, y]) - [f(x), y] - [x, f(y)]\| \leq \theta(\|x\|^r + \|y\|^r) \tag{35}$$

for all $x, y \in A$. Then there exists a unique Lie derivation $D : A \to A$ satisfying (28).

Proof By the same reasoning as in the proof of Theorem 6, there is a unique **C**-linear mapping $D : A \to A$ satisfying (28) defined by

$$D(x) := \lim_{n\to\infty} 2^n f(\frac{x}{2^n})$$

for all $x \in A$.

It follows from (35) that

$$\begin{aligned}
&\|D([x, y]) - [D(x), y] - [x, D(y)]\| \\
&= \lim_{n\to\infty} 4^n \left\| f\left(\frac{[x, y]}{4^n}\right) - \left[f\left(\frac{x}{2^n}\right), \frac{y}{2^n}\right] - \left[\frac{x}{2^n}, f\left(\frac{y}{2^n}\right)\right]\right\| \\
&\leq \lim_{n\to\infty} \frac{4^n\theta}{2^{nr}}(\|x\|^r + \|y\|^r) = 0
\end{aligned}$$

for all $x, y \in A$. So

$$D([x, y]) = [D(x), y] + [x, D(y)]$$

for all $x, y \in A$. Thus the **C**-linear mapping $D : A \to A$ is a Lie derivation.

Theorem 17 *Let* $r < 1$ *and* θ *be positive real numbers, and let* $f : A \to A$ *be a mapping satisfying* (22) *and* (35)*. Then there exists a unique Lie derivation* $D : A \to A$ *satisfying* (29).

Proof The proof is similar to the proofs of Theorems 6 and 16.

Theorem 18 *Let* $r > 2$ *and* θ *be nonnegative real numbers, and let* $f : A \to B$ *be a mapping satisfying* (30) *and* (34)*. Then there exists a unique Lie* C^**-algebra homomorphism* $h : A \to B$ *satisfying* (31).

Proof The proof is similar to the proofs of Theorems 10 and 14.

Theorem 19 *Let* $r < 1$ *and* θ *be positive real numbers, and let* $f : A \to B$ *be a mapping satisfying* (30) *and* (34)*. Then there exists a unique Lie* C^**-algebra homomorphism* $h : A \to B$ *satisfying* (32).

Proof The proof is similar to the proofs of Theorems 10 and 14.

Theorem 20 *Let* $r > 2$ *and* θ *be nonnegative real numbers, and let* $f : A \to A$ *be a mapping satisfying* (30) *and* (35)*. Then there exists a unique Lie derivation* $D : A \to A$ *satisfying* (32).

Proof The proof is similar to the proof of Theorem 16.

Theorem 21 *Let* $r < 1$ *and* θ *be positive real numbers, and let* $f : A \to A$ *be a mapping satisfying* (30) *and* (35)*. Then there exists a unique Lie derivation* $D : A \to A$ *satisfying* (33).

Proof The proof is similar to the proof of Theorem 16.

6 Homomorphisms and Derivations in JC^*-Algebras

Throughout this section, assume that A is a JC^*-algebra with norm $\|\cdot\|$, and that B is a JC^*-algebra with norm $\|\cdot\|$.

Definition 3 ([21, 30]) A **C**-linear mapping $H : A \to B$ is called a *JC^*-algebra homomorphism* if $H : A \to B$ satisfies

$$H(x \circ y) = H(x) \circ H(y)$$

for all $x, y \in A$.

Definition 4 ([21, 30]) A **C**-linear mapping $D : A \to A$ is called a *Jordan derivation* if $D : A \to A$ satisfies

$$D(x \circ y) = Dx \circ y + x \circ Dy$$

for all $x, y \in A$.

In this section, we investigate JC^*-algebra homomorphisms in JC^*-algebras and Jordan derivations on JC^*-algebras associated with the additive ρ-functional inequalities (6) and (14).

Theorem 22 *Let $r > 2$ and θ be nonnegative real numbers, and let $f : A \to B$ be a mapping satisfying* (22) *and*

$$\|f(x \circ y) - f(x) \circ f(y)\| \le \theta(\|x\|^r + \|y\|^r) \tag{36}$$

for all $x, y \in A$. Then there exists a unique JC^*-algebra homomorphism $h : A \to B$ satisfying (25).

Proof By the same reasoning as in the proof of Theorem 6, there is a unique **C**-linear mapping $h : A \to B$ satisfying (25) defined by

$$h(x) := \lim_{n\to\infty} 2^n f(\frac{x}{2^n})$$

for all $x \in A$.

It follows from (36) that

$$\begin{aligned}\|h(x \circ y) - h(x) \circ h(y)\| &= \lim_{n\to\infty} 4^n \left\| f\left(\frac{x \circ y}{2^n \cdot 2^n}\right) - f\left(\frac{x}{2^n}\right) \circ f\left(\frac{y}{2^n}\right)\right\| \\ &\le \lim_{n\to\infty} \frac{4^n\theta}{2^{nr}}(\|x\|^r + \|y\|^r) = 0\end{aligned}$$

for all $x, y \in A$. Thus

$$h(x \circ y) = h(x) \circ h(y)$$

for all $x, y \in A$. Hence the **C**-linear mapping $h : A \to B$ is a JC^*-algebra homomorphism.

Theorem 23 *Let $r < 1$ and θ be positive real numbers, and let $f : A \to B$ be a mapping satisfying* (22) *and* (36). *Then there exists a unique JC^*-algebra homomorphism $h : A \to B$ satisfying (26).*

Proof The proof is similar to the proofs of Theorems 6 and 22.

Theorem 24 *Let $r > 2$ and θ be nonnegative real numbers, and let $f : A \to A$ be a mapping satisfying* (22) *such that*

$$\|f(x \circ y) - f(x) \circ y - x \circ f(y)]\| \leq \theta(\|x\|^r + \|y\|^r) \tag{37}$$

for all $x, y \in A$. Then there exists a unique Jordan derivation $D : A \to A$ satisfying (28).

Proof By the same reasoning as in the proof of Theorem 6, there is a unique **C**-linear mapping $D : A \to A$ satisfying (28) defined by

$$D(x) := \lim_{n\to\infty} 2^n f(\frac{x}{2^n})$$

for all $x \in A$.

It follows from (37) that

$$\begin{aligned}
&\|D(x \circ y) - D(x) \circ y - x \circ D(y)\| \\
&= \lim_{n\to\infty} 4^n \left\| f\left(\frac{x \circ y}{4^n}\right) - f\left(\frac{x}{2^n}\right) \circ \frac{y}{2^n} - \frac{x}{2^n} \circ f\left(\frac{y}{2^n}\right) \right\| \\
&\leq \lim_{n\to\infty} \frac{4^n\theta}{2^{nr}}(\|x\|^r + \|y\|^r) = 0
\end{aligned}$$

for all $x, y \in A$. So

$$D(x \circ y) = D(x) \circ y + x \circ D(y)$$

for all $x, y \in A$. Thus the **C**-linear mapping $D : A \to A$ is a Jordan derivation.

Theorem 25 *Let $r < 1$ and θ be positive real numbers, and let $f : A \to A$ be a mapping satisfying* (22) *and* (37). *Then there exists a unique Lie derivation $D : A \to A$ satisfying* (29).

Proof The proof is similar to the proofs of Theorems 6 and 24.

Theorem 26 *Let $r > 2$ and θ be nonnegative real numbers, and let $f : A \to B$ be a mapping satisfying* (30) *and* (36). *Then there exists a unique JC^*-algebra homomorphism $h : A \to B$ satisfying* (31).

Proof The proof is similar to the proofs of Theorems 10 and 22.

Theorem 27 *Let $r < 1$ and θ be positive real numbers, and let $f : A \to B$ be a mapping satisfying* (30) *and* (36). *Then there exists a unique JC^*-algebra homomorphism $h : A \to B$ satisfying* (32).

Proof The proof is similar to the proofs of Theorems 10 and 22.

Theorem 28 *Let $r > 2$ and θ be nonnegative real numbers, and let $f : A \to A$ be a mapping satisfying* (30) *and* (37). *Then there exists a unique Jordan derivation $D : A \to A$ satisfying* (32).

Proof The proof is similar to the proof of Theorem 24.

Theorem 29 *Let $r < 1$ and θ be positive real numbers, and let $f : A \to A$ be a mapping satisfying* (30) *and* (37). *Then there exists a unique Jordan derivation $D : A \to A$ satisfying* (33).

Proof The proof is similar to the proof of Theorem 24.

References

1. M. Amyari, C. Baak, M. Moslehian, Nearly ternary derivations. Taiwan. J. Math. **11**, 1417–1424 (2007)
2. T. Aoki, On the stability of the linear transformation in Banach spaces. J. Math. Soc. Japan **2**, 64–66 (1950)
3. P.W. Cholewa, Remarks on the stability of functional equations. Aequationes Math. **27**, 76–86 (1984)
4. P. Czerwik, *Functional Equations and Inequalities in Several Variables* (World Scientific, Singapore, 2002)
5. M. Eshaghi Gordji, N. Ghobadipour, Stability of (α, β, γ)-derivations on Lie C^*-algebras. Int. J. Geom. Meth. Mod. Phys. **7**, 1097–1102 (2010)
6. M. Eshaghi Gordji, M.B. Ghaemi, B. Alizadeh, A fixed point method for perturbation of higher ring derivations in non-Archimedean Banach algebras. Int. J. Geom. Meth. Mod. Phys. **8**(7), 1611–1625 (2011)
7. M. Eshaghi Gordji, A. Fazeli, C. Park, 3-Lie multipliers on Banach 3-Lie algebras. Int. J. Geom. Meth. Mod. Phys. **9**(7), Article ID 1250052, 15 (2012)
8. W. Fechner, Stability of a functional inequalities associated with the Jordan-von Neumann functional equation. Aequationes Math. **71**, 149–161 (2006)
9. Z. Gajda, On stability of additive mappings. Internat. J. Math. Math. Sci. **14**, 431–434 (1991)
10. P. Găvruta, A generalization of the Hyers-Ulam-Rassias stability of approximately additive mappings. J. Math. Anal. Appl. **184**, 431–436 (1994)
11. A. Gilányi, Eine zur Parallelogrammgleichung äquivalente Ungleichung. Aequationes Math. **62**, 303–309 (2001)
12. A. Gilányi, On a problem by K. Nikodem. Math. Inequal. Appl. **5**, 707–710 (2002)
13. D.H. Hyers, On the stability of the linear functional equation. Proc. Nat. Acad. Sci. U.S.A. **27**, 222–224 (1941)
14. D.H. Hyers, G. Isac, T.M. Rassias, *Stability of Functional Equations in Several Variables* (Birkhäuser, Basel, 1998)
15. K. Jun, H. Kim, On the stability of Appolonius' equation. Bull. Belg. Math. Soc. Simon Stevin **11**, 615–624 (2004)
16. S. Jung, On the Hyers-Ulam-Rassias stability of approximately additive mappings. J. Math. Anal. Appl. **204**, 221–226 (1996)
17. S. Oh, C. Park, Linear functional equations in a Hilbert module. Taiwan. J. Math. **7**, 441–448 (2003)
18. C. Park, On the stability of the linear mapping in Banach modules. J. Math. Anal. Appl. **275**, 711–720 (2002)

19. C. Park, Modified Trif's functional equations in Banach modules over a C^*-algebra and approximate algebra homomorphisms. J. Math. Anal. Appl. **278**, 93–108 (2003)
20. C. Park, Lie $*$-homomorphisms between Lie C^*-algebras and Lie $*$-derivations on Lie C^*-algebras. J. Math. Anal. Appl. **293**, 419–434 (2004)
21. C. Park, Homomorphisms between Lie JC^*-algebras and Cauchy-Rassias stability of Lie JC^*-algebra derivations, J. Lie Theory **15**, 393–414 (2005)
22. C. Park, Approximate homomorphisms on JB^*-triples. J. Math. Anal. Appl. **306**, 375–381 (2005)
23. C. Park, Homomorphisms between Poisson JC^*-algebras. Bull. Braz. Math. Soc. **36**, 79–97 (2005)
24. C. Park, Isomorphisms between unital C^*-algebras. J. Math. Anal. Appl. **307**, 753–762 (2005)
25. C. Park, Isomorphisms between C^*-ternary algebras. J. Math. Anal. Appl. **327**, 101–115 (2007)
26. C. Park, Homomorphisms in C^*-ternary algebras and JB^*-triples. J. Math. Anal. Appl. **337**, 13–20 (2008)
27. C. Park, Additive ρ-functional inequalities and equations. J. Math. Inequal. **9**, 17–26 (2015)
28. C. Park, Additive ρ-functional inequalities in non-Archimedean normed spaces. J. Math. Inequal. **9**, 397–407 (2015)
29. C. Park, W. Park, On the Jensen's equation in Banach modules. Taiwan. J. Math. **6**, 523–531 (2002)
30. C. Park, J. Hou, S. Oh, Homomorphisms between JC^*-algebras and between Lie C^*-algebras. Acta Math. Sinica **21**, 1391–1398 (2005)
31. C. Park, K. Ghasemi, S.G. Ghaleh, S. Jang, Approximate n-Jordan $*$-homomorphisms in C^*-algebras. J. Comput. Anal. Appl. **15**, 365–368 (2013)
32. T.M. Rassias, On the stability of the linear mapping in Banach spaces. Proc. Amer. Math. Soc. **72**, 297–300 (1978)
33. T.M. Rassias, *Functional Equations, Inequalities and Applications* (Kluwer Academic, Dordrecht/Boston/London, 2003)
34. J. Rätz, On inequalities associated with the Jordan-von Neumann functional equation. Aequationes Math. **66**, 191–200 (2003)
35. S. Shagholi, M. Bavand Savadkouhi, M. Eshaghi Gordji, Nearly ternary cubic homomorphism in ternary Fréchet algebras. J. Comput. Anal. Appl. **13**, 1106–1114 (2011)
36. S. Shagholi, M. Eshaghi Gordji, M. Bavand Savadkouhi, Stability of ternary quadratic derivation on ternary Banach algebras. J. Comput. Anal. Appl. **13**, 1097–1105 (2011)
37. D. Shin, C. Park, Sh. Farhadabadi, On the superstability of ternary Jordan C^*-homomorphisms. J. Comput. Anal. Appl. **16**, 964–973 (2014)
38. D. Shin, C. Park, Sh. Farhadabadi, Stability and superstability of J^*-homomorphisms and J^*-derivations for a generalized Cauchy-Jensen equation. J. Comput. Anal. Appl. **17**, 125–134 (2014)
39. F. Skof, Proprietà locali e approssimazione di operatori. Rend. Sem. Mat. Fis. Milano **53**, 113–129 (1983)
40. S.M. Ulam, *A Collection of the Mathematical Problems* (Interscience Publishers, New York, 1960)

Graphic Contraction Principle and Applications

A. Petruşel and I. A. Rus

Abstract The purpose of this paper is to emphasize the role of the graphic contractions in metric fixed point theory. Two general results about the fixed points of graphic contractions and several related examples are given. The case of non-self graphic contractions will be also considered. Existence, uniqueness, data dependence, well-posedness, Ulam-Hyers stability, and the Ostrowski property for the fixed point equation will be discussed. Some fixed point results in metric spaces endowed with a partial ordering will be also proved. Finally, open questions and research directions are presented.

2010 Mathematics Subject Classification 47H10, 54H25, 47H09, 34D10, 45GXX, 45M10

1 Introduction

Let X be a nonempty set and $f : X \to X$ be an operator. We consider the fixed point equation

$$x = f(x), \ x \in X. \tag{1}$$

We denote by F_f the fixed point set of f, i.e., $F_f := \{x \in X \mid f(x) = x\}$.

In the same context, if $f(X_\lambda) \subset X_\lambda$ for all $\lambda \in \Lambda$ and $X = \bigcup_{\lambda \in \Lambda} X_\lambda$ is a partition of the space, then we say that $\bigcup_{\lambda \in \Lambda} X_\lambda$ is an invariant partition of X with respect to the operator f.

A. Petruşel (✉) · I. A. Rus
Department of Mathematics, Babeş-Bolyai University of Cluj-Napoca, Cluj-Napoca, Romania
e-mail: petrusel@math.ubbcluj.ro; iarus@math.ubbcluj.ro

T. M. Rassias, P. M. Pardalos (eds.), *Mathematical Analysis and Applications*, Springer Optimization and Its Applications 154,
https://doi.org/10.1007/978-3-030-31339-5_15

If (X, d) is a metric space, then, by definition, f is a weakly Picard operator if

$$f^n(x) \to x^*(x) \in F_f \text{ as } n \to \infty, \text{ for all } x \in X.$$

Actually, the above definition generates the set retraction $f^\infty : X \to F_f$ given by $f^\infty(x) = \lim_{n\to\infty} f^n(x)$.

If we write $F_f = \{x^*\}$, then f has a unique fixed point and we denote it by x^*. A weakly Picard operator with a unique fixed point is, by definition, a Picard operator. We denote the attraction basin of a fixed point x^* of f by

$$(AB)_f(x^*) := \{x \in X|\ f^n(x) \to x^* \text{ as } n \to \infty\}.$$

We will also use the notation X_{x^*} for the above set.

A weakly Picard operator $f : X \to X$ for which there exists a function $\psi : \mathbb{R}_+ \to \mathbb{R}_+$ increasing, continuous in 0 and satisfying $\psi(0) = 0$, such that

$$d(x, f^\infty(x)) \le \psi(d(x, f(x)), \text{ for all } x \in X,$$

is called a weakly ψ-Picard operator.

Moreover, a Picard operator for which there exists a function $\psi : \mathbb{R}_+ \to \mathbb{R}_+$ increasing, continuous in 0 and satisfying $\psi(0) = 0$, such that

$$d(x, x^*) \le \psi(d(x, f(x)), \text{ for all } x \in X,$$

is called a ψ-Picard operator.

For more considerations on weakly Picard operator theory, see [4, 36, 44, 46–48].

Definition 1 Let (X, d) be a metric space and $f : X \to X$ be an operator. Then, f is called:

(i) a k-contraction if $k \in]0, 1[$ and

$$d(f(x), f(y)) \le kd(x, y), \text{ for every } x, y \in X.$$

(ii) a graphic k-contraction if $k \in]0, 1[$ and

$$d(f(x), f^2(x)) \le kd(x, f(x)), \text{ for every } x \in X.$$

To our best knowledge, the first fixed point theorem for graphic contractions on a closed subset of $\mathbb{R}^n$ was given by Rheinboldt [31]. In his paper [31], Rheinboldt named a mapping satisfying the above condition (ii) by the term of iterated contraction. See also [21] for related results and applications in $\mathbb{R}^n$.

In 1972, Rus proved in [33] (see also [48, p. 29]) that a graphic k-contraction on a complete metric space has at least one fixed point.

In 1974 a fixed point result for graphic k-contractions in the framework of a Banach space was given by Subrahmanyam in [49]. In his paper, Subrahmanyam used (see Definition 4 in [49]) the name of Banach operator of type k for the above concept. His result (see Corollary 2 in [49]) says that if $f : S \to S$ is a continuous Banach operator of type k (where S is a closed subset of a Banach space), then it has a fixed point.

Five years later, Hicks and Rhoades proved the following fixed point result in complete metric spaces.

Theorem 1 (See [11]) *Let (X, d) be a complete metric space and $k \in [0, 1[$. Suppose there exists $x \in X$ such that*

$$d(f(y), f^2(y)) \le kd(y, f(y)),\ \text{for every } y \in O(x, \infty)$$
$$:= \{x, f(x), f^2(x), \cdots, f^n(x), \cdots\}.$$

Then:

(A) $\lim\limits_{n\to\infty} f^n(x) = x^*$ *exists;*
(B) $d(f^n(x), x^*) \le \frac{k^n}{1-k} d(x, f(x))$;
(C) x^* *is a fixed point of f if and only if the functional $G : X \to \mathbb{R}_+$ given by $G(x) := d(x, f(x))$ is f-orbitally lower semi-continuous at x^*, i.e., if $(x_n)_{n\in\mathbb{N}}$ is a sequence in $O(x, \infty)$ and $x_n \to x^*$, then $G(x^*) \le \liminf\limits_{n\to\infty} G(x_n)$.*

The purpose of this paper is to emphasize the role of the graphic contractions in metric fixed point theory. Two general results about the fixed points of graphic contractions and several related examples are given. The case of non-self graphic contractions will be also considered. Existence, uniqueness, data dependence, well-posedness, Ulam-Hyers stability, and the Ostrowski property for the fixed point equation will be discussed. Some fixed point results in metric spaces endowed with a partial ordering will be also proved. Finally, some open questions and research directions are presented.

More precisely, the structure of our paper is the following:

1. Introduction and preliminaries
2. Two general results on self graphic contractions
3. Relevant examples of graphic contractions
4. Some general results on non-self graphic contractions
5. Data dependence and stability results
6. Ran-Reurings type results for graphic contractions
7. Open questions and some new research directions

2 Two General Results on Graphic Contractions

We recall first a result concerning some equivalent statements in the theory of weakly Picard operators.

Theorem 2 ([34]) *Let X be a nonempty set and $f : X \to X$ be an operator. The following statements are equivalent:*

(1) $F_f = F_{f^n} \neq \emptyset$, for all $n \in \mathbb{N}^*$;
(2) There exists a metric d on X such that $f : (X, d) \to (X, d)$ is a weakly Picard operator;
(3) For each $k \in]0, 1[$ there exists a complete metric d on X such that:

(a) $d(f(x), f^2(x)) \leq kd(x, f(x))$, for every $x \in X$.;
(b) f is orbitally continuous on X;

(4) For each $k \in]0, 1[$ there exist a complete metric d on X and a partition of the space $X = \bigcup_{\lambda \in \Lambda} X_\lambda$, such that:

(a) $f(X_\lambda) \subset X_\lambda$, for all $\lambda \in \Lambda$;
(b) $f : X_\lambda \to X_\lambda$ is a k-contraction, for all $\lambda \in \Lambda$;
(c) $F_f \cap X_\lambda = \{x_\lambda^*\}$, for all $\lambda \in \Lambda$.

The following concepts will be important in our main results.

Definition 2 Let (X, d) be a metric space and $f : X \to X$ be an operator. Then:

(i) the fixed point equation (1) is called well-posed if $F_f = \{x^*\}$ and for any sequence $(x_n)_{n\in\mathbb{N}}$ in X for which

$$d(x_n, f(x_n)) \to 0 \text{ as } n \to \infty,$$

we have that

$$x_n \to x^* \text{ as } n \to \infty.$$

(ii) the operator f has the Ostrowski property if $F_f = \{x^*\}$ and for any sequence $(x_n)_{n\in\mathbb{N}}$ for which

$$d(x_{n+1}, f(x_n) \to 0 \text{ as } n \to \infty,$$

we have that

$$x_n \to x^* \text{ as } n \to \infty.$$

(iii) the fixed point equation (1) has the data dependence property if $F_f = \{x^*\}$ and for any operator $g : X \to X$ for which there exists $\eta > 0$ with

$$d(f(x), g(x)) \leq \eta, \quad \text{for all } x \in X,$$

the following implication holds:

$$y^* \in F_g \text{ implies } d(x^*, y^*) \leq \psi(\eta),$$

where the function $\psi : \mathbb{R}_+ \to \mathbb{R}_+$ is increasing, continuous in zero and satisfies $\psi(0) = 0$.

In the case of $F_f \neq \emptyset$ and $r : X \to F_f$ a retraction, the above notions take the following form:

Definition 3

(iv) the fixed point equation (1) is called well-posed if for each $x^* in F_f$ and any sequence $(x_n)_{n\in\mathbb{N}}$ in $r^{-1}(x^*)$ for which

$$d(x_n, f(x_n)) \to 0 \text{ as } n \to \infty,$$

we have that

$$x_n \to x^* \text{ as } n \to \infty.$$

(v) the operator f has the Ostrowski property if for each $x^* in F_f$ and any sequence $(x_n)_{n\in\mathbb{N}}$ in $r^{-1}(x^*)$ for which

$$d(x_{n+1}, f(x_n) \to 0 \text{ as } n \to \infty,$$

we have that

$$x_n \to x^* \text{ as } n \to \infty.$$

(vi) the fixed point equation (1) has the data dependence property if for any operator $g : X \to X$ for which there exists $\eta > 0$ with

$$d(f(x), g(x)) \leq \eta, \quad \text{for all } x \in X,$$

there exists a function $\psi : \mathbb{R}_+ \to \mathbb{R}_+$ increasing, continuous in zero and satisfying $\psi(0) = 0$, such that

$$H_d(F_f, F_g) \leq \psi(\eta).$$

For the above notions, see [3, 4, 41, 44, 45, 47].

The following result is well-known.

Lemma 1 (Cauchy-Toeplitz Lemma, See [21, 47, 48]) *Let $(a_n)_{n\in\mathbb{N}}$ be a sequence in $\mathbb{R}_+$, such that the series $\sum_{n\geq 0} a_n$ is convergent and $(b_n)_{n\in\mathbb{N}} \in \mathbb{R}_+$ be a sequence such that $\lim_{n\to\infty} b_n = 0$. Then*

$$\lim_{n\to\infty} (\sum_{k=0}^{n} a_{n-k}b_k) = 0.$$

We will present now the saturated principle of graphic contractions.

Theorem 3 ([48]) *Let (X, d) be a complete metric space and $f : X \to X$ be a graphic k-contraction. Then:*

(1) the sequence $(f^n(x))_{n\in\mathbb{N}}$ converges in (X, d) and $\sum_{n\in\mathbb{N}} d(f^n(x), f^{n+1}(x)) < \infty$, for each $x \in X$;

If, in addition, $\lim_{n\to\infty} f(f^n(x)) = f(\lim_{n\to\infty} f^n(x))$ for each $x \in X$, then we have the following conclusions:

(2) $F_f = F_{f^n} \neq \emptyset$, for all $n \in \mathbb{N}^*$;

(3) f is a weakly Picard operator and $X = \bigcup_{x^*\in F_f} X_{x^*}$ is an invariant partition of X;

(4) $d(x, f^\infty(x)) \leq \frac{1}{1-k} d(x, f(x))$, for every $x \in X$;

(5) the fixed point equation (1) is well-posed;

(6) if $k < \frac{1}{3}$, then $d(f(x), f^\infty(x)) \leq \frac{k}{1-2k} d(x, f^\infty(x))$, for every $x \in X$;

(7) if $k < \frac{1}{3}$, then f has the Ostrowski property;

3 Relevant Examples of Graphic Contractions

We will present now several examples of graphic contractions.

Example 1 Let $f \in C([a, b] \times \mathbb{R}^m, \mathbb{R}^m)$. Consider the following system of first order differential equations

$$x'(t) = f(t, x(t)). \tag{2}$$

We are looking for solutions $x \in C^1([a, b], \mathbb{R}^m)$ of the above system.

In the above conditions, it is easy to see that (2) is equivalent with the following system of functional-integral equations

$$x(t) = x(a) + \int_a^t f(s, x(s))ds, \tag{3}$$

where $x \in C([a, b], \mathbb{R}^m)$.

We denote

$$X_\lambda := \{x \in C([a,b],\mathbb{R}^m) \mid x(a) = \lambda\}$$

and define

$$T : C([a,b],\mathbb{R}^m) \to C([a,b],\mathbb{R}^m) \text{ by } Tx(t) := x(a) + \int_a^t f(s,x(s))ds.$$

Then the following conclusions hold:

(i) $C([a,b],\mathbb{R}^m) = \bigcup_{\lambda\in\mathbb{R}^m} X_\lambda$ is a partition;
(ii) $T(X_\lambda) \subset X_\lambda$, for every $\lambda \in \mathbb{R}^m$;
(iii) if $f(t,\cdot) : \mathbb{R}^m \to \mathbb{R}^m$ is L-Lipschitz, then $T\Big|_{X_\lambda}$ is a contraction and we have that $X_\lambda \cap F_T = \{x_\lambda^*\}$;
(iv) the operator $T : C([a,b],\mathbb{R}^m) \to C([a,b],\mathbb{R}^m)$ is a graphic contraction with respect to a suitable Bielecki type norm;
(v) the operator $T : C([a,b],\mathbb{R}^m) \to C([a,b],\mathbb{R}^m)$ has no isolated fixed points.

Example 2 Let $f \in C([a,b] \times \mathbb{R}^m \times \mathbb{R}^m, \mathbb{R}^m)$. Consider the following system of second order differential equations

$$-x''(t) = f(t,x(t),x'(t)). \tag{4}$$

We are looking for solutions $x \in C^2([a,b],\mathbb{R}^m)$ of the above system.

In the above conditions, it is easy to see that (4) is equivalent with each of the following system of functional-integral equations

$$x(t) = x(a) + \frac{x'(a)}{2}(t-a) + \int_a^t (t-s)f(s,x(s)x'(s))ds,\ t \in [a,b], \tag{5}$$

where we are looking for solutions $x \in C^1([a,b],\mathbb{R}^m)$,

and

$$x(t) = \frac{t-a}{b-a}x(b) + \frac{b-t}{b-a}x(a) + \int_a^b G(t,s)f(s,x(s)x'(s))ds,\ t \in [a,b], \tag{6}$$

where G denotes the usual Green function corresponding to (4) and we are looking for solutions $x \in C^1([a,b],\mathbb{R}^m)$.

We denote

$$X_{u,v} := \{x \in C^1([a,b],\mathbb{R}^m) \mid x(a) = u,\ x'(a) = v\}$$

and define

$$T : C^1([a,b],\mathbb{R}^m) \to C^1([a,b],\mathbb{R}^m),$$

by

$$Tx(t) := x(a) + \frac{x'(a)}{2}(t-a) + \int_a^t (t-s)f(s, x(s)x'(s))ds,$$

respectively

$$\tilde{X}_{u,v} := \{x \in C^1([a,b], \mathbb{R}^m) | \; x(a) = u, \; x(b) = v\}$$

and

$$S : C^1([a,b], \mathbb{R}^m) \to C^1([a,b], \mathbb{R}^m),$$

by

$$Sx(t) := \frac{t-a}{b-a}x(b) + \frac{b-t}{b-a}x(a) + \int_a^b G(t,s)f(s, x(s)x'(s))ds.$$

Then the following conclusions hold:

(i) $C^1([a,b], \mathbb{R}^m) = \bigcup_{u,v \in \mathbb{R}^m} X_{u,v} = \bigcup_{u,v \in \mathbb{R}^m} \tilde{X}_{u,v}$ are partitions;
(ii) $T(X_{u,v}) \subset X_{u,v}$ and $S(\tilde{X}_{u,v}) \subset \tilde{X}_{u,v}$, for every $u, v \in \mathbb{R}^m$;
(iii) if $f(t, \cdot, \cdot) : \mathbb{R}^m \times \mathbb{R}^m \to \mathbb{R}^m$ is L-Lipschitz, then:

    (a) T is a graphic contraction with respect to a suitable Bielecki norm;
    (b) if $(b-a)$ is sufficiently small, then S is graphic contraction with respect to the $\|\cdot\|_\infty$-norm;

Example 3 In a metric space (X, d) several generalized contractions, such as:

(a) Kannan mappings
(b) Ćirić-Reich-Rus mappings
(c) Chatterjea mappings
(d) Zamfirescu mappings
(e) Hardy-Rogers mappings
(f) Berinde mappings
(g) Suzuki mappings, etc.

are graphic contractions, see [1, 10, 12, 14, 15, 32, 35, 48].

If additionally the space is complete, then, by the theorems proved by the above authors, we obtain various conclusions about the fixed point set, such as $F_f = \{x^*\}$ or $F_f \neq \emptyset$;

Example 4 Let $(X, +, \mathbb{R}, \|\cdot\|)$ be a normed space and $\Phi : X \to \mathbb{R}$ be a nontrivial linear functional. For $\lambda \in \mathbb{R}$ we consider

$$X_\lambda := \{x \in X \mid \Phi(x) = \lambda\}.$$

Then $X = \bigcup_{\lambda \in \mathbb{R}} X_\lambda$ is a partition of the space X, which will be called the partition of X corresponding to Φ. Moreover, if Φ is continuous, then X_λ is a closed set in X.

Let X be a Banach space, $\Phi : X \to \mathbb{R}$ be a nontrivial linear continuous functional and $A : X \to X$ be a bounded linear operator. We say that Φ is an invariant functional of A if

$$\Phi(A(x)) = \Phi(x), \text{ for all } x \in X.$$

In the above conditions, if we suppose:

(i) Φ is an invariant functional of A;
(ii) $k := \|A|_{X_0}\| < 1$,

then A is a graphic k-contraction.

Example 5 Consider on the Banach space $(C[0,1], \|\cdot\|_\infty)$ the Berstein operator

$$B_n(x)(t) := \sum_{k=0}^{n} f(\frac{k}{n}) \binom{k}{n} x^k (1-x)^{n-k}, \ n \in \mathbb{N}^*$$

Then, B_n is a graphic $(1 - \frac{1}{2^{n-1}})$-contraction.

Example 6 In a metric space (X, d) the identity $f = 1_{|X}$ is a graphic contraction. In this case $F_f = X$.

Example 7 Let $f : [-1, 1] \to [-1, 1]$ be defined by

$$f(x) := \begin{cases} \frac{x}{2}, \ x \neq 0 \\ \frac{1}{2}, \ x = 0, \end{cases}$$

Then f is a discontinuous graphic $\frac{1}{2}$-contraction, $f^n(x) \to 0$ as $n \to \infty$, for every $x \in [-1, 1]$ and $F_f = \emptyset$.

Example 8 Let $f : [0, 1] \to [0, 1]$ be defined by

$$f(x) := \begin{cases} 0, \ x \in [0, \frac{1}{2}[\\ 1, \ x \in [\frac{1}{2}, 1], \end{cases}$$

Then f is a discontinuous graphic k-contraction (with any $k \in]0, 1[$) and $F_f = \{0, 1\}$. Moreover, $f^n(x) \to 0$ as $n \to \infty$, for every $x \in [0, \frac{1}{2}[$ and $f^n(x) \to 1$ as $n \to \infty$, for every $x \in [\frac{1}{2}, 1]$.

For other examples of graphic contractions, see [1, 21, 23, 24, 38, 40, 44, 46, 48].

4 Some General Results on Non-self Graphic Contractions

We will present now a general result concerning the fixed point equation (1) for the case of a non-self graphic contraction.

Theorem 4 *Let (X, d) be a complete metric space, $k \in]0, 1[$, $x_0 \in X$, $R > 0$ and $f : \tilde{B}(x_0, R) \to X$ be an operator. We suppose that the following assumptions take place:*

(i) $d(x_0, f(x_0)) \leq (1-k)R$;
(ii) *if* $x, f(x) \in \tilde{B}(x_0, R)$, *then* $d(f(x), f^2(x)) \leq kd(x, f(x))$;
(iii) *if* $x \in \tilde{B}(x_0, R)$ *has the property that* $f^n(x) \in \tilde{B}(x_0, R)$ *for every* $n \in \mathbb{N}^*$ *and the sequence* $(f^n(x))_{n\in\mathbb{N}}$ *is convergent, then*

$$\lim_{n\to\infty} f(f^n(x)) = f(\lim_{n\to\infty} f^n(x)).$$

In the above conditions, we have the following conclusions:

(a) $f^n(x_0) \in \tilde{B}(x_0, R)$ *for every* $n \in \mathbb{N}$ *and* $f^n(x_0) \to f^\infty(x_0) \in F_f$ *as* $n \to \infty$.
(b) *if for some* $y \in \tilde{B}(x_0, R)$ *we have that* $f^n(y) \in \tilde{B}(x_0, R)$ *for every* $n \in \mathbb{N}^*$, *then* $f^n(y) \to f^\infty(y) \in F_f$ *as* $n \to \infty$.
(c) *if* $y_0 \in \tilde{B}(x_0, R)$ *is such that* $d(x_0, y_0) \leq \eta_1 R$ *and* $d(x_0, f(y_0)) \leq \eta_2 R$ *with* $\eta_1 + \eta_2 \leq 1 - k$, *then* $f^n(y_0) \in \tilde{B}(x_0, R)$ *for every* $n \in \mathbb{N}^*$ *and* $f^n(y_0) \to f^\infty(y_0) \in F_f$ *as* $n \to \infty$.
(d) *if* $x^* \in F_f$ *and* $x \in (AB)_f(x^*)$, *then* $d(x, x^*) \leq \frac{1}{1-k} d(x, f(x))$.
(e) *if* $x^* \in F_f$, $y_n \in (AB)_f(x^*)$ *for every* $n \in \mathbb{N}$ *is such that* $d(y_n, f(y_n)) \to 0$ *as* $n \to \infty$, *then* $y_n \to x^*$ *as* $n \to \infty$.
(f) *if* $k < \frac{1}{3}$ *and* $x^* \in F_f$, *then* $d(f(x), x^*) \leq \frac{k}{1-2k} d(x, x^*)$, *for every* $x \in (AB)_f(x^*)$.
(g) *if* $k < \frac{1}{3}$, $x^* \in F_f$ *and* $y_n \in (AB)_f(x^*)$ *for every* $n \in \mathbb{N}$ *is such that* $d(y_{n+1}, f(y_n)) \to 0$ *as* $n \to \infty$, *then* $y_n \to x^*$ *as* $n \to \infty$.

Proof

(a) Let $x_0 \in X$ such that $d(x_0, f(x_0)) \leq (1-k)R$. Then

$$d(x_0, f^2(x_0)) \leq d(x_0, f(x_0)) + d(f(x_0), f^2(x_0)) \leq$$

$$d(x_0, f(x_0)) + kd(x_0, f(x_0)) = (1+k)d(x_0, f(x_0)) \leq (1-k^2)R.$$

Hence $f^2(x_0) \in \tilde{B}(x_0, R)$. By mathematical induction, we obtain that

$$d(x_0, f^n(x_0)) \leq (1-k^n)R, \text{ for every } n \in \mathbb{N}, n \geq 2.$$

Thus, $f^n(x_0) \in \tilde{B}(x_0, R)$, for every $n \in \mathbb{N}^*$. Denote $x_n := f^n(x_0)$ and observe that, for every $n \in \mathbb{N}$, we have

$$d(x_n, x_{n+1}) \leq kd(x_{n-1}, x_n) \leq \cdots \leq k^n d(x_0, f(x_0)).$$

By a standard argument, we get that the sequence $(x_n)_{n\in\mathbb{N}}$ is Cauchy and converges, in (X, d), to an element of $\tilde{B}(x_0, R)$. By (iii) this element is a fixed point of f and we denote it by $f^\infty(x_0)$.

(b) Let $y \in \tilde{B}(x_0, R)$ such that $f^n(y) \in \tilde{B}(x_0, R)$, for every $n \in \mathbb{N}^*$. Then, in a similar way to the above proof, by (ii) and (iii), we get that the sequence $(f^n(x))_{n\in\mathbb{N}}$ is Cauchy and it converges to a fixed point $f^\infty(y) \in \tilde{B}(x_0, R)$ of f.

(c) Let $y_0 \in \tilde{B}(x_0, R)$ be such that $d(x_0, y_0) \leq \eta_1 R$ and $d(x_0.f(y_0)) \leq \eta_2 R$, where $\eta_1 + \eta_2 \leq 1 - k$. Then, since $f(y_0) \in \tilde{B}(x_0, R)$, we have

$$d(x_0, f^2(y_0)) \leq d(x_0, f(y_0)) + d(f(y_0), f^2(y_0)) \leq \eta_2 R + kd(y_0, f(y_0)) \leq$$

$$\eta_2 R + k\,(d(y_0, x_0) + d(x_0, f(y_0))) \leq \left(k(\eta_1 + \eta_2) + \eta_2\right) R \leq (1 - k^2)R.$$

Moreover,

$$d(x_0, f^3(y_0)) \leq d(x_0, f^2(y_0)) + d(f^2(y_0), f^3(y_0)) \leq$$

$$\left(k(\eta_1 + \eta_2) + \eta_2\right) R + k^2 d(y_0, f(y_0)) \leq \left[(k + k^2)(\eta_1 + \eta_2) + \eta_2\right] R \leq (1 - k^3)R.$$

By mathematical induction, we obtain

$$d(x_0, f^n(y_0)) \leq (1 - k^n)R, \text{ for every } n \in \mathbb{N}^*,$$

showing that $f^n(y_0) \in \tilde{B}(x_0, R)$ for every $n \in \mathbb{N}^*$. Then, in a similar way to the proof of (a) and (b), using (ii) and (iii), we get that the sequence $(f^n(y_0))_{n\in\mathbb{N}}$ is Cauchy and it converges to a fixed point $f^\infty(y_0) \in \tilde{B}(x_0, R)$ of f.

(d) Let $x^* \in F_f$ and $x \in (AB)_f(x^*)$. Then

$$d(x, x^*) \leq d(x, f(x)) + d(f(x), f^2(x)) + \cdots + d(f^n(x), x^*) \leq$$

$$(1 + k + \ldots k^{n-1})d(x, f(x)) + d(f^n(x), x^*) \leq \frac{1}{1-k} d(x, f(x)) + d(f^n(x), x^*).$$

Letting $n \to \infty$ we obtain the conclusion.

(e) Let $x^* \in F_f$ and $y_n \in (AB)_f(x^*)$ for every $n \in \mathbb{N}$ be such that $d(y_n, f(y_n)) \to 0$ as $n \to \infty$. Then

$$d(y_n, x^*) \leq d(y_n, f(y_n)) + \cdots + d(f^{n-1}(y_n), f^n(y_n)) + d(f^n(y_n), x^*) \leq$$

$$(1 + k + \cdots + k^{n-1})d(y_n, f(y_n)) + d(f^n(y_n), x^*) \leq$$

$$\frac{1}{1-k} d(y_n, f(y_n)) + d(f^n(y_n), x^*).$$

Letting $n \to \infty$ we obtain the conclusion.

(f) Suppose that $k < \frac{1}{3}$. Let $x^* \in F_f$ and $x \in (AB)_f(x^*)$. Then

$$d(f(x), x^*) \le d(f(x), f^2(x)) + \cdots + d(f^{n-1}(x), f^n(x)) + d(f^n(x), x^*) \le$$

$$(k+k^2+\cdots k^{n-1})d(x, f(x))+d(f^n(x), x^*) \le \frac{k}{1-k}d(x, f(x))+d(f^n(x), x^*).$$

Using the triangle inequality, we obtain that

$$d(f(x), x^*) \le \frac{k}{1-2k}d(x, x^*) + \frac{1-k}{1-2k}d(f^n(x), x^*).$$

The conclusion follows letting $n \to \infty$.

(g) Suppose that $k < \frac{1}{3}$. Let $x^* \in F_f$ and $y_n \in (AB)_f(x^*)$ for every $n \in \mathbb{N}$ be such that $d(y_{n+1}, f(y_n)) \to 0$ as $n \to \infty$. Then, we have

$$d(y_{n+1}, x^*) \le d(y_{n+1}, f(y_n)) + d(f(y_n), f^2(y_n)) + \cdots + d(f^n(y_n), x^*) \le$$

$$d(y_{n+1}, f(y_n)) + kd(y_n, f(y_n)) + \cdots + k^{n-1}d(y_n, f(y_n)) + d(f^n(y_n), x^*) \le$$

$$d(y_{n+1}, f(y_n)) + \frac{k}{1-k}d(y_n, f(y_n)) + d(f^n(y_n), x^*) \le$$

$$d(y_{n+1}, f(y_n)) + \frac{k}{1-k}\left[d(y_n, x^*) + d(x^*, y_{n+1}) + d(y_{n+1}, f(y_n))\right] +$$

$$d(f^n(y_n), x^*) = \frac{1}{1-k}d(y_{n+1}, f(y_n)) + \frac{k}{1-k}d(y_n, x^*) +$$

$$\frac{k}{1-k}d(x^*, y_{n+1}) + d(f^n(y_n), x^*).$$

Thus

$$d(y_{n+1}, x^*) \le \frac{1}{1-2k}d(y_{n+1}, f(y_n)) + \frac{k}{1-2k}d(y_n, x^*) + \frac{1-k}{1-2k}d(f^n(y_n), x^*).$$

We denote $\alpha := \frac{1}{1-2k}$, $\beta := \frac{k}{1-2k}$ and $\gamma := \frac{1-k}{1-2k}$. Notice that $\alpha, \gamma > 1$ and $\beta \in]0, 1[$.

Then we get

$$d(y_{n+1}, x^*) \le \alpha d(y_{n+1}, f(y_n)) + \beta d(y_n, x^*) + \gamma d(f^n(y_n), x^*) \le$$

$$\alpha\left[d(y_{n+1}, f(y_n)) + \beta d(y_n, f(y_{n-1}))\right] +$$

$$\gamma\left[d(f^n(y_n), x^*) + \beta d(f^{n-1}(y_{n-1}), x^*)\right] +$$

$$\beta^2 d(y_{n-1}, x^*) \leq \cdots \leq$$

$$\alpha\left[d(y_{n+1}, f(y_n)) + \beta d(y_n, f(y_{n-1})) + \cdots \beta^n d(y_1, f(y_0))\right] +$$

$$\gamma\left[d(f^n(y_n), x^*) + \beta d(f^{n-1}(y_{n-1}), x^*) + \cdots + \beta^n d(y_0, x^*)\right] + \beta^n d(y_0, x^*).$$

Letting $n \to \infty$ and using Cauchy-Toeplitz Lemma (see Lemma 1), we obtain the desired conclusion.

A second general result for non-self contractions is the following.

Theorem 5 *Let (X, d) be a complete metric space, $k \in]0, 1[$, $x_0 \in X$, $R > 0$ and $f : \tilde{B}(x_0, R) \to X$ be an operator. We suppose that the following assumptions take place:*

(i) $d(x_0, f(x_0)) \leq (1-k)R$;
(ii) *if* $x, f(x) \in \tilde{B}(x_0, R)$, *then* $d(f(x), f^2(x)) \leq kd(x, f(x))$;
(iii) *if* $x \in \tilde{B}(x_0, R)$ *has the property that* $f^n(x) \in \tilde{B}(x_0, R)$ *for every* $n \in \mathbb{N}^*$ *and the sequence* $(f^n(x))_{n\in\mathbb{N}}$ *is convergent, then*

$$\lim_{n\to\infty} f(f^n(x)) = f(\lim_{n\to\infty} f^n(x)).$$

(iv) $card(F_f) \leq 1$.

In the above conditions, we have the following conclusions:

(a) $F_f = \{x^*\}$ *and* $f^n(x_0) \to x^*$ *as* $n \to \infty$.
(b) *if for some* $y \in \tilde{B}(x_0, R)$ *we have that* $f^n(y) \in \tilde{B}(x_0, R)$ *for every* $n \in \mathbb{N}^*$, *then* $(f^n(y)) \to x^*$ *as* $n \to \infty$.
(c) *if* $y_0 \in \tilde{B}(x_0, R)$ *is such that* $d(x_0, y_0) \leq \eta_1 R$ *and* $d(x_0, f(y_0)) \leq \eta_2 R$ *with* $\eta_1 + \eta_2 \leq 1 - k$, *then* $f^n(y_0) \in \tilde{B}(x_0, R)$ *for every* $n \in \mathbb{N}^*$ *and* $(f^n(y_0)) \to x^*$ *as* $n \to \infty$.
(d) $d(x, x^*) \leq \frac{1}{1-k} d(x, f(x))$, *for every* $x \in (AB)_f(x^*)$.
(e) *if* $y_n \in (AB)_f(x^*)$ *for every* $n \in \mathbb{N}$ *is such that* $d(y_n, f(y_n)) \to 0$ *as* $n \to \infty$, *then* $y_n \to x^*$ *as* $n \to \infty$.
(f) *if* $k < \frac{1}{3}$, *then* $d(f(x), x^*) \leq \frac{k}{1-2k} d(x, x^*)$, *for every* $x \in (AB)_f(x^*)$.
(g) *if* $k < \frac{1}{3}$ *and* $y_n \in (AB)_f(x^*)$ *for every* $n \in \mathbb{N}$ *is such that* $d(y_{n+1}, f(y_n)) \to 0$ *as* $n \to \infty$, *then* $y_n \to x^*$ *as* $n \to \infty$.

Proof The conclusions follow by Theorem 4 and the hypothesis (iv).

For some general considerations on fixed point theory for nonself contractions, see [1–3, 6, 9, 17, 19, 21, 48].

5 Data Dependence and Stability Results for Graphic Contractions

In this section we will present some continuous data dependence and stability results for the fixed point equation (1) governed by a graphic contraction.

We will start with the data dependence problem. More precisely, let (X, d) be a metric space, $x_0 \in X$, $R > 0$ and $f, g : \tilde{B}(x_0, R) \to X$ are two operators. We suppose that the following assumptions take place:

(i) $F_f \neq \emptyset$ and, for $x^* \in F_f$ we have that $(AB)_f(x^*) \neq \{x^*\}$.
(ii) there exists $\eta > 0$ such that $d(f(x), g(x)) \leq \eta$, for each $x \in \tilde{B}(x_0, R)$.

The problem is in which conditions there exists a function $\psi : \mathbb{R}_+ \to \mathbb{R}_+$ increasing, continuous in zero and satisfying $\psi(0) = 0$, such that

$$d(y^*, x^*) \leq \psi(\eta), \text{ for every } x^* \in F_f \text{ and every } y^* \in F_g \cap (AB)_f(x^*).$$

In this context, we have the following result.

Theorem 6 *Let (X, d) be a complete metric space, $x_0 \in X$, $R > 0$ and $f, g : \tilde{B}(x_0, R) \to X$ be two operators. We suppose that the following assumptions take place:*

(i) f satisfies all the assumptions in Theorem 4.
(ii) there exists $\eta > 0$ such that $d(f(x), g(x)) \leq \eta$, for each $x \in \tilde{B}(x_0, R)$.

Then

$$d(x^*, y^*) \leq \frac{\eta}{1-k}, \text{ for every } x^* \in F_f \cap \tilde{B}(x_0, R) \text{ and every } y^* \in F_g \cap (AB)_f(x^*).$$

Proof By Theorem 4 there exists $x^* \in F_f \cap \tilde{B}(x_0, R)$. Suppose $F_g \cap (AB)_f(x^*)$ is nonempty (otherwise, we have nothing to prove). Then, we have

$$d(x^*, y^*) = d(f(x^*), g(y^*)) \leq d(f(x^*), f(y^*)) +$$

$$d(f(y^*), g(y^*)) \leq d(x^*, f(y^*)) + \eta \leq$$

$$d(x^*, f^n(y^*)) + d(f^n(y^*), f^{n-1}(y^*)) + \cdots + d(f^2(y^*), f(y^*)) + \eta \leq$$

$$d(x^*, f^n(y^*)) + k^{n-1} d(f(y^*), y^*) + \cdots + kd(f(y^*), y^*) + \eta \leq$$

$$d(x^*, f^n(y^*)) + (k^{n-1} + \cdots + k + 1)\eta \leq d(x^*, f^n(y^*)) + \frac{\eta}{1-k}.$$

The conclusion follows letting $n \to \infty$.

For some general considerations on data dependence problem, see [4, 6–9, 13, 18–20, 23, 26, 30, 35, 39, 43, 47, 48].

We will consider now the Ulam-Hyers stability of the fixed point equation (1).

Definition 4 Let (X, d) be a complete metric space, $x_0 \in X$, $R > 0$ and $f : \tilde{B}(x_0, R) \to X$ be an operator. Then, the fixed point equation (1) is Ulam-Hyers stable if there exists $c > 0$ such that for every $\varepsilon > 0$ and every ε-solution y^* of the fixed point equation (1), i.e.,

$$d(y^*, f(y^*)) \leq \varepsilon,$$

there exists a solution $x^* \in \tilde{B}(x_0, R)$ of the fixed point equation (1) such that

$$d(x^*, y^*) \leq c\varepsilon.$$

For the above problem, we have the following result.

Theorem 7 *Let (X, d) be a complete metric space, $x_0 \in X$, $R > 0$ and $f : \tilde{B}(x_0, R) \to X$ be an operator. We suppose that the following assumptions take place:*

(i) f satisfies all the assumptions in Theorem 4.
(ii) let $\varepsilon > 0$ and let $y^* \in \tilde{B}(x_0, R)$ be such that

$$d(y^*, f(y^*)) \leq \varepsilon \text{ and } y^* \in (AB)_f(x^*).$$

Then

$$d(x^*, y^*) \leq \frac{\varepsilon}{1-k}.$$

Proof By Theorem 4 there exists $x^* \in F_f \cap \tilde{B}(x_0, R)$. Let $\varepsilon > 0$ and let $y^* \in \tilde{B}(x_0, R)$ be such that

$$d(y^*, f(y^*)) \leq \varepsilon \text{ and } y^* \in (AB)_f(x^*).$$

Then, we have

$$d(x^*, y^*) \leq d(x^*, f^n(y^*)) + d(f^n(y^*), f^{n-1}(y^*)) + \cdots + d(f(y^*), y^*) \leq$$

$$d(x^*, f^n(y^*)) + k^{n-1}d(f(y^*), y^*) + \cdots + kd(f(y^*), y^*) + d(f(y^*), y^*) \leq$$

$$d(x^*, f^n(y^*)) + (k^{n-1} + \cdots + k + 1)d(f(y^*), y^*) \leq d(x^*, f^n(y^*)) + \frac{\varepsilon}{1-k}.$$

The conclusion follows letting $n \to \infty$.

For other considerations and results on Ulam-Hyers stability, see [5, 29, 38, 41, 42, 44].

6 Operators on a Complete Metric Space Which Are Graphic Contractions on an Invariant Subset

In this section, we will prove a fixed point theorem for an operator $f : X \to X$ satisfying the graphic contraction condition on an invariant (not necessary closed) subset of a complete metric space.

Let (X, d) be a complete metric space and $f : X \to X$ be an operator. Let us suppose that there exists a nonempty subset Y of X such that $f(Y) \subset Y$ and $f : Y \to Y$ is a graphic contraction. Then, for every $x \in Y$ the sequence of successive approximations $(f^n(x))_{n\in\mathbb{N}}$ is convergent in (X, d). Let us denote $f^\infty(x) := \lim_{n\to\infty} f^n(x), x \in Y$. Notice that if $f \circ f^\infty = f^\infty \circ f$, then $f^\infty(x) \in F_f$. Moreover, the operator

$$f : Y \cup f^\infty(Y) \to Y \cup f^\infty(Y)$$

is weakly Picard and a graphic contraction too. For this operator, we are in the conditions of Theorem 3, since $Y \cup f^\infty(Y)$ is complete with respect to the sequences $(f^n(x))_{n\in\mathbb{N}}$ with $x \in Y \cup f^\infty(Y)$, see [37]. As a consequence, we have the following general result.

Theorem 8 *Let (X, d) be a complete metric space and $f : X \to X$ be an operator. We suppose that there exists a nonempty subset Y of X such that $f(Y) \subset Y$ and $f : Y \to Y$ is a graphic k-contraction. Then the following conclusions take place:*

(a) for every $x \in Y$ the sequence of successive approximations $(f^n(x))_{n\in\mathbb{N}}$ is convergent in (X, d) and

$$\sum_{n\in\mathbb{N}} d(f^n(x), f^{n+1}(x)) < \infty;$$

If, in addition,

$$\lim_{n\to\infty} f(f^n(x)) = f(\lim_{n\to\infty} f^n(x)), \text{ for each } x \in Y,$$

then we also have the following conclusions:

(b) $F_f \cap Y = F_{f^n} \cap Y \neq \emptyset$, for every $n \in \mathbb{N}^*$;

(c) $f : Y \cup f^\infty(Y) \to Y \cup f^\infty(Y)$ is a weakly Picard and

$$Z := Y \cup f^\infty(Y) = \bigcup_{x^* \in F_f \cap Z} Z_{x^*}$$

is an invariant partition of Z;

(d) $d(x, f^\infty(x)) \leq \frac{1}{1-k} d(x, f(x))$, for every $x \in Y$;

(e) *if* $y_n \in Z_{x^*}$ *for every* $n \in \mathbb{N}$ *is such that* $d(y_n, f(y_n)) \to 0$ *as* $n \to \infty$, *then* $y_n \to x^*$ *as* $n \to \infty$, *for every* $x^* \in F_f \cap Z$;

(f) *if* $k < \frac{1}{3}$, *then* $d(f(x), f^\infty(x)) \leq \frac{k}{1-2k} d(x, f^\infty(x))$, *for every* $x \in Y$;

(g) *if* $k < \frac{1}{3}$, *then* $f : Z \to Z$ *has the Ostrowski property.*

Actually, following the above ideas, the problem is in which conditions on the metric space (X, d) and on the operator f there exists such a subset Y with the above properties. For example, we can do the following construction.

Let X be a nonempty set endowed with a partial order "$\preceq$" and $d : X \times X \to \mathbb{R}_+$ be a complete metric. Let $f : X \to X$ be an operator which is increasing with respect to "$\preceq$". Then the following subsets are invariant with respect to f:

(1) $Y := \{x \in X : x \preceq f(x)\}$;
(2) $Y := \{x \in X : f(x) \preceq x\}$;
(3) $Y := \{x \in X : x \preceq f(x) \text{ or } f(x) \preceq x\}$.

Hence, in order to apply Theorem 8 we need to impose the following conditions:

(i) $Y \neq \emptyset$;
(ii) $f : Y \to Y$ is a graphic contraction.

For example, we have the following results.

Theorem 9 *Let* X *be a nonempty set endowed with a partial order "*$\preceq$*" and* $d : X \times X \to \mathbb{R}_+$ *be a complete metric. Let* $f : X \to X$ *be an operator which is increasing with respect to "*$\preceq$*". Suppose that there exist a constant* $k \in]0, 1[$ *and an element* $x_0 \in X$ *such that:*

(i) $x \in X$ *with* $x \preceq f(x)$ *(or reversely) implies* $d(f(x), f^2(x)) \leq kd(x, f(x))$;
(ii) $x_0 \preceq f(x_0)$ *(or reversely);*
(iii) $\lim_{n\to\infty} f(f^n(x_0)) = f(\lim_{n\to\infty} f^n(x_0))$.

Then $F_f \neq \emptyset$ *and the sequence of successive approximations* $(f^n(x_0))_{n\in\mathbb{N}}$ *converges to a fixed point of* f.

Proof Let $x_0 \in X$ such that $x_0 \preceq f(x_0)$. Then, the sequence $(f^n(x_0))_{n\in\mathbb{N}}$ is increasing with respect to $\preceq$. Hence, we can apply the graphic contraction condition (i) and we obtain that f is asymptotically regular at x_0, i.e.,

$$d(f^n(x_0), f^{n+1}(x_0)) \leq k^n d(x_0, f(x_0)) \to 0 \text{ as } n \to \infty.$$

Now, in order to show that $(f^n(x_0))_{n\in\mathbb{N}}$ is a Cauchy sequence, we observe that, for $n \in \mathbb{N}$ and $p \in \mathbb{N}^*$, we have

$$d(f^n(x_0), f^{n+p}(x_0)) \leq$$

$$d\left(f^n(x_0), f^{n+1}(x_0)\right) + \ldots + d\left(f^{n+p-1}(x_0), f^{n+p}(x_0)\right) \leq$$

$$k^n(1+k+\cdots+k^{p-1})d(x_0, f(x_0))$$

Since $k < 1$, we get that

$$d(f^n(x_0), f^{n+p}(x_0)) \leq \frac{k^n}{1-k} d(x_0, f(x_0)) \to 0 \text{ as } n, p \to \infty.$$

Hence $(f^n(x_0))_{n\in\mathbb{N}}$ is a Cauchy sequence and by the completeness of the metric d, there exists $x^* \in X$ such that $(f^n(x_0))_{n\in\mathbb{N}} \to x^*$ as $n \to \infty$. By (iii) it follows that $x^* \in F_f$.

A more general result is the following one.

Theorem 10 *Let X be a nonempty set endowed with a partial order "$\preceq$" and $d : X \times X \to \mathbb{R}_+$ be a complete metric. Let $f : X \to X$ be an operator such that the following assumptions are satisfied:*

(i) the set $X_{\preceq} := \{x \in X\ : x \preceq f(x) \text{ or } f(x) \preceq x\}$ is nonempty;
(ii) $X_{\preceq}$ is invariant with respect to f, i.e., $f(X_{\preceq}) \subseteq X_{\preceq}$;
(iii) there exists a constant $k \in]0, 1[$ such that

$$d(f(x), f^2(x)) \leq kd(x, f(x)), \text{ for all } x \in X_{\preceq};$$

(iv) $\lim_{n\to\infty} f(f^n(x)) = f(\lim_{n\to\infty} f^n(x))$, for each $x \in X_{\preceq}$.

Then $F_f \neq \emptyset$ and, for all $x \in X_{\preceq}$, the sequence of successive approximations $(f^n(x))_{n\in\mathbb{N}}$ converges to a fixed point of f.

Proof For $x_0 \in X_{\preceq}$ we consider the sequence $(f^n(x_0))_{n\in\mathbb{N}}$. Then $f^n(x_0) \in X_{\preceq}$, for each $n \in \mathbb{N}$. Hence, by the graphic contraction condition (i), we obtain that

$$d(f^n(x_0), f^{n+1}(x_0)) \leq k^n d(x_0, f(x_0)) \to 0 \text{ as } n \to \infty.$$

As before, we can show that $(f^n(x_0))_{n\in\mathbb{N}}$ is a Cauchy sequence. Thus, by the completeness of the metric d, there exists $x^* \in X$ such that $(f^n(x_0))_{n\in\mathbb{N}} \to x^*$ as $n \to \infty$. By (iv) it follows that $x^* \in F_f$.

Remark 1 Theorem 9 is a generalization of the well-known fixed point theorem for contraction mappings given by Ran and Reurings in [27]. Theorem 10 extends to the case of graphic contractions, one of the main result of the paper [22]. The results of this section are in connection and generalize some other known results, see, for example, [16, 28, 35] and the references therein.

7 Open Questions and New Research Directions

We will formulate now some open questions and related research directions.

7.1 *Generalized Contractions as Graphic Contractions*

Which are those generalized contractions (self or non-self) which are graphic contractions?

In the case of various applications of these results, the problem is to improve it from the saturated principle of graphic contractions point of view.

References: [1, 20, 32, 35, 48].

7.2 *Nonlinear Graphic Contractions*

The problem is to prove similar results for the case of nonlinear graphic contractions (also called graphic φ-contractions), i.e., in the case when f satisfies the following assumption

$$d(f(x), f^2(x)) \leq \varphi(d(x, f(x)), \text{ for every } x \in X(\text{ or } x \in \tilde{B}(x_0, R),$$

where $\varphi : \mathbb{R}_+ \to \mathbb{R}_+$ is a comparison function, see [35].

References: [1, 4, 35, 48].

7.3 *Generalized Metric Spaces*

Another open question and research direction is to study the above problems in various generalized metric spaces, such as:

(1) $\mathbb{R}^m_+$-metric spaces;
(2) $s(\mathbb{R}_+)$-metric spaces;
(3) b-metric (quasi-metric) spaces;
(4) Banach spaces with f a differentiable operator.

References: [21, 35, 46, 48].

7.4 Coupled Fixed Point Problems Via Graphic Contraction Conditions

Using Theorems 9 and 10 and the approach presented in [25], the problem is to give existence results for the following coupled fixed point problem:

$$\begin{cases} x = T(x, y) \\ y = T(y, x). \end{cases}$$

7.5 Y-Contractions

The problems studied in this paper are particular cases of the following general problem.

Let (X, d) be a complete metric space and $Y \subset X \times X$ be a nonempty subset. By definition, an operator $f : X \to X$ is called a Y-contraction if there exists $k \in]0, 1[$ such that

$$d(f(x), f(y)) \leq kd(x, y), \text{ for every } (x, y) \in Y.$$

The problem is to construct a fixed point theory of Y-contractions.

For other considerations and results, see [48, pp. 282–284] and the references therein.

References

1. V. Berinde, *Iterative Approximation of Fixed Points* (Springer, Berlin, 2007)
2. V. Berinde, Şt. Măruşter, I.A. Rus, An abstract point of view on iterative approximation of fixed points of nonself operators. J. Nonlinear Convex Anal. **15**, 851–865 (2014)
3. V. Berinde, Şt. Măruşter, I.A. Rus, Saturated contraction principle for non-self operators, generalizations and applications. Filomat **31**, 3391–3406 (2017)
4. V. Berinde, A. Petruşel, I.A. Rus, M.-A. Şerban, in *The Retraction-Displacement Condition in the Theory of Fixed Point Equation with a Convergent Iterative Algorithm*, ed. by T. Rassias, V. Gupta. Mathematical Analysis, Approximation Theory and Their Applications. Springer Optimization and Its Applications, vol. 111, (Springer, Cham, 2016), pp. 75–106
5. M. Bota-Boriceanu, A. Petruşel, Ulam-Hyers stability for operatorial equations. Analele Univ. Al.I. Cuza Iaşi **57**, 65–74 (2011)
6. A. Chiş-Novac, R. Precup, I.A. Rus, Data dependence of fixed points for nonself generalized contractions. Fixed Point Theory **10**, 73–87 (2009)
7. K. Deimling, *Nonlinear Functional Analysis* (Springer, Berlin, 1985)
8. J. Dieudonne, *Foundations of Modern Analysis* (Academic, New York, 1960)
9. A.-D. Filip, I.A. Rus, Fixed point theory for non-self generalized contractions in Kasahara spaces. Ann. West Univ. Timişoara Math-Inf. (to appear)

10. G.E. Hardy, T.D. Rogers, A generalization of a fixed point theorem of Reich. Canad. Math. Bull. **16**, 201–206 (1973)
11. T.L. Hicks, B.E. Rhoades, A Banach type fixed point theorem. Math. Japon. **24**, 327–330 (1979)
12. D.H. Hyers, G. Isac, T.M. Rassias, *Topics in Nonlinear Analysis and Applications* (World Scientific, Singapore, 1997)
13. L.V. Kantorovici, G.P. Akilov,*Analyse fonctionelle* (MIR, Moscou, 1981)
14. W.A. Kirk, Contraction mappings and extensions, in ed. by W.A. Kirk, B. Sims, *Handbook of Fixed Point Theory* (Kluwer Academic Publishers, Dordrecht, 2001)
15. Y.V. Konots, P.P. Zabreiko, The majorization fixed point principle and application to nonlinear integral equations. Fixed Point Theory **13**, 547–564 (2012)
16. M.A. Krasnoselskii, P.P. Zabrejko, *Geometrical Methods in Nonlinear Analysis* (Springer, Berlin, 1984)
17. T.A. Lazăr, A. Petruşel, N. Shahzad, Fixed points for non-self operators and domain invariance theorems. Nonlinear Anal. **70**, 117–125 (2009)
18. Şt. Măruşter, The stability of gradient-like methods. Appl. Math. Comput. **17**, 103–115 (2001)
19. D. O'Regan, R. Precup, *Theorems of Leray-Schauder Type and Applications* (Gordon and Breach Science Publishers, Amsterdam, 2001)
20. J.M. Ortega, *Numerical Analysis* (Academic, New York, 1972)
21. J.M. Ortega, W.C. Rheinboldt, *Iterative Solutions of Nonlinear Equation in Several Variables* (Academic, New York, 1970)
22. A. Petruşel, I.A. Rus, Fixed point theorems in ordered L-spaces. Proc. Amer. Math. Soc., **134**, 411–418 (2005)
23. A. Petruşel, I.A. Rus, in *A Class of Functional-Integral Equations with Applications to a Bilocal Problem*, ed. by T.M. Rassias, L. Toth. Topics in Mathematical Analysis and Applications (Springer, Berlin, 2014), pp. 609–631
24. A. Petruşel, I.A. Rus, M.-A. Şerban, Nonexpansive operators as graphic contractions. J. Nonlinear Convex Anal. **17**, 1409–1415 (2016)
25. A. Petruşel, G. Petruşel, B. Samet, J.-C. Yao, Coupled fixed point theorems for symmetric contractions in b-metric spaces with applications to operator equation systems. Fixed Point Theory **17**, 459–478 (2016)
26. R. Precup, *Methods in Nonlinear Integral Equations*, (Kluwer Academic Publishers, Dordrecht, 2002)
27. A.C.M. Ran, M.C.B. Reurings, A fixed point theorem in partially ordered sets and some applications to matrix equations. Proc. Amer. Math. Soc. **132**, 1435–1443 (2004)
28. T.M. Rassias (ed.), *Functional Equations, Inequalities and Applications* (Kluwer Academic Publishers, Dordrecht, 2003)
29. T.M. Rassias (ed.), *Handbook of Functional Equations: Stability Theory* (Springer, New York, 2014)
30. S. Reich, A.J. Zaslavski, Well-posedness of the fixed point problem. Far East J. Math. Sci. **3**, 393–401 (2001)
31. W.C. Rheinboldt, A unified convergence theory for a class of iterative processes. SIAM J. Numer. Anal. **5**, 42–63 (1968)
32. B.E. Rhoades, A comparison of various definitions of contractive mappings. Trans. Am. Math. Soc. **226**, 257–290 (1977)
33. I.A. Rus, On the method of successive approximations. Revue Roum. Math. Pures et Appl. **17**, 1433–1437 (1972)
34. I.A. Rus, Weakly Picard mappings. Comment. Math. Univ. Carol. **34**, 769–773 (1993)
35. I.A. Rus, *Generalized Contractions and Applications* (Transilvania Press, Cluj-Napoca, 2001)
36. I.A. Rus, Picard operators and applications. Sci. Math. Jpn. **58**, 191–219 (2003)
37. I.A. Rus, Metric space with fixed point property with respect to contractions. Studia Univ. Babeş-Bolyai Math. **51**, 115–121, (2006)
38. I.A. Rus, Iterates of Bernstein operators, via contraction principle. J. Math. Anal. Appl. **292**, 259–261 (2009)

39. I.A. Rus, Properties of the solutions of those equations for which the Krasnoselskii iteration converges. Carpathian J. Math. **28**, 329–336 (2010)
40. I.A. Rus, A müller's examples to Cauchy problem: an operatorial point of view. Creat. Math. Infor. **21**, 215–220 (2012)
41. I.A. Rus, An abstract point of view on iterative approximations of fixed points: impact on the theory of fixed points equations. Fixed Point Theory **13**, 179–192 (2012)
42. I.A. Rus, in *Results and Problems in Ulam Stability of Operatorial Equations and Inclusions*, ed. by T. Rassias. Handbook of Functional Equations. Springer Optimization and Its Applications, vol. 96 (Springer, New York, 2014), pp. 323–352
43. I.A. Rus, The generalized retraction method in fixed point theory for nonself operators. Fixed Point Theory **15**, 559–578 (2014)
44. I.A. Rus, Relevant classes of weakly Picard operators. Anal. Univ. de Vest Timişoara Math. Inform. **54**, 3–19 (2016)
45. I.A. Rus, Some variants of the contraction principle, generalizations and applications. Studia Univ. Babeş-Bolyai Math. **61**, 343–358 (2016)
46. I.A. Rus, M.A. Şerban, Operators on infinite dimensional Cartesian product. Anal. Univ. de Vest Timişoara, Math. Inform. **48**, 253–263 (2010)
47. I.A. Rus, M.A. Şerban, Basic problems of the metric fixed point theory and the relevance of a metric fixed point theorem. Carpathian J. Math. **29**, 239–258 (2013)
48. I.A. Rus, A. Petruşel, G. Petruşel, Fixed Point Theory (Cluj University Press, Cluj-Napoca 2008)
49. P.V. Subrahmanyam, Remarks on some fixed point theorems related to Banach's contraction principle. J. Malh. Phys. Sri. **8**, 445–457 (1974); Erratum, **9**, 195

A New Approach for the Inversion of the Attenuated Radon Transform

Nicholas E. Protonotarios, George A. Kastis, and Athanassios S. Fokas

Abstract One of the most well-known generalizations of the celebrated Radon transform is the so-called *attenuated Radon transform*. The inversion of the attenuated Radon transform provides the mathematical foundation of the important field of medical imaging, known as single photon emission computed tomography (SPECT). In this chapter, we present a novel mathematical formulation of the inversion of the attenuated Radon transform and the corresponding numerical implementation.

1 Introduction

The renowned Radon transform of a two-dimensional function is defined as the set of all its line integrals [1–3]. There exists a natural generalization of the Radon transform, the so-called *attenuated Radon transform*, defined as the set of all line integrals of a two-dimensional function attenuated with respect to an attenuation function. The attenuated Radon transform provides the mathematical foundation of one of the most important medical imaging techniques, referred to as single photon

N. E. Protonotarios (✉)
Research Center of Mathematics, Academy of Athens, Athens, Greece

Department of Mathematics, National Technical University of Athens, Athens, Greece

G. A. Kastis
Research Center of Mathematics, Academy of Athens, Athens, Greece
e-mail: gkastis@academyofathens.gr

A. S. Fokas
Research Center of Mathematics, Academy of Athens, Athens, Greece

Department of Applied Mathematics and Theoretical Physics, University of Cambridge, Cambridge, UK

Viterbi School of Engineering, University of Southern California, Los Angeles, CA, USA
e-mail: t.fokas@damtp.cam.ac.uk

T. M. Rassias, P. M. Pardalos (eds.), *Mathematical Analysis and Applications*, Springer Optimization and Its Applications 154,
https://doi.org/10.1007/978-3-030-31339-5_16

emission computerized tomography (SPECT) [4]. The attenuated Radon transform gives rise to an associated *inverse problem*, namely to "reconstruct" a function from its attenuated line integrals. The main task in SPECT imaging is the numerical implementation of the inversion of the attenuated Radon transform.

In [5], Novikov and Fokas rederived the well-known inversion of the Radon transform [1] by performing the so-called *spectral analysis* of the following eigenvalue equation:

$$\left[\frac{1}{2}\left(k+\frac{1}{k}\right)\partial_{x_1}+\frac{1}{2i}\left(k-\frac{1}{k}\right)\partial_{x_2}\right]\Theta=f,\quad k\in\mathbb{C}. \tag{1}$$

This specific analysis encompasses two certain problems in modern complex analysis known as the $\bar{d}$-problem and the scalar Riemann–Hilbert (RH) problem, respectively.

Although the inversion of the Radon transform can be obtained in a less complicated manner, namely by employing the two-dimensional Fourier transform, the advantage of the derivation of [5] was established 10 years later (2002) by Novikov [6]. Novikov demonstrated that the inverse attenuated Radon transform can be derived by applying an analysis similar to that performed in (1), to a slight generalization of equation (1), namely the equation

$$\left[\frac{1}{2}\left(k+\frac{1}{k}\right)\partial_{x_1}+\frac{1}{2i}\left(k-\frac{1}{k}\right)\partial_{x_2}-\mu\right]\Theta=f,\quad k\in\mathbb{C}. \tag{2}$$

Actually, by employing the results of the analysis of (1), one can derive the inverse attenuated Radon transform in a substantially simpler manner (see Section 3). In this chapter, we present a novel formula for the inversion of the attenuated Radon transform and the corresponding numerical implementation.

2 The Radon Transform

A line L on the plane can be specified by the signed distance from the origin (ρ, with $-\infty<\rho<\infty$), and the angle with the x_1-axis (θ, with $0\leq\theta<2\pi$), see Figure 1. We denote the corresponding unit vectors parallel and perpendicular to L by $\mathbf{e}^{\parallel}$ and $\mathbf{e}^{\perp}$, respectively. These vectors are given by

$$\mathbf{e}^{\parallel}=(\cos\theta,\sin\theta)\,,$$
$$\mathbf{e}^{\perp}=(-\sin\theta,\cos\theta)\,.$$

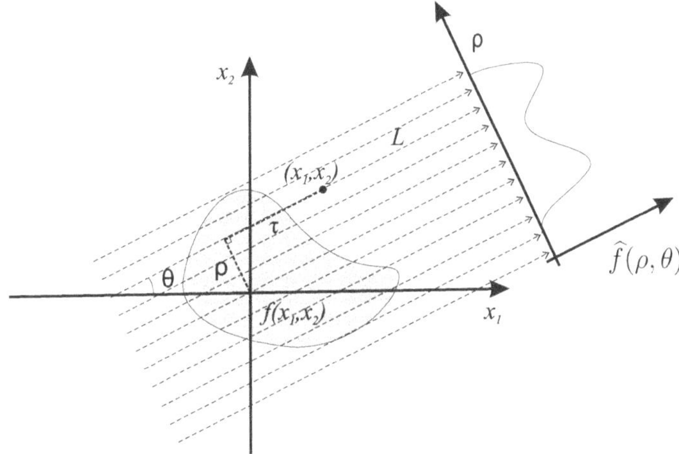

Fig. 1 A two-dimensional object $f(x_1, x_2)$ and its projections $\widehat{f}(\rho, \theta)$. Both Cartesian (x_1, x_2) and local (ρ, τ) coordinates are indicated

Every point $\mathbf{x} = (x_1, x_2)$ on L in Cartesian coordinates can be expressed in terms of the so-called *local coordinates* (ρ, τ) via

$$\mathbf{x} = \rho\, \mathbf{e}^{\perp} + \tau\, \mathbf{e}^{\parallel},$$

where τ denotes the arc length. Therefore,

$$x_1 = \tau \cos\theta - \rho \sin\theta, \tag{3a}$$

$$x_2 = \tau \sin\theta + \rho \cos\theta. \tag{3b}$$

Equation (3) can be inverted and expressed in the local coordinates (ρ, τ) in terms of the Cartesian coordinates (x_1, x_2) and the associated angle θ:

$$\rho = x_2 \cos\theta - x_1 \sin\theta, \tag{4a}$$

$$\tau = x_2 \sin\theta + x_1 \cos\theta. \tag{4b}$$

We define the line integral of a two-dimensional Schwartz function $f(x_1, x_2) \in S(\mathbb{R}^2)$ as the *Radon transform* of $f(x_1, x_2)$, denoted by $\widehat{f}(\rho, \theta)$. The Radon transform is usually stored in the form of the so-called *sinogram*, expressed as follows:

$$\widehat{f}(\rho, \theta) = \int_{-\infty}^{\infty} f(\tau \cos\theta - \rho \sin\theta, \tau \sin\theta + \rho \cos\theta) \mathrm{d}\tau,$$
$$0 \le \theta < 2\pi, \quad -\infty < \rho < \infty, \tag{5}$$

where $S(\mathbb{R}^2)$ denotes the space of Schwartz functions:

$$S(\mathbb{R}^2) = \left\{ f \in C^\infty(\mathbb{R}^2) : ||f||_{\alpha,\beta} < \infty \right\} \subset C^\infty(\mathbb{R}^2), \tag{6}$$

and

$$||f||_{\alpha,\beta} = \sup_{x\in\mathbb{R}^2} |x^\alpha D^\beta f(x)|, \quad \forall \text{ multi-index } \alpha, \beta,$$

$$|x^\alpha D^\beta f(x)| \to 0, \quad \text{as } |x| \to \infty. \tag{7}$$

The above transform (5) gives rise to the following *inverse problem*: Given the function $\widehat{f}(\rho, \theta)$, $0 \leq \theta < 2\pi$, $-\infty < \rho < \infty$, "reconstruct" the function $f(x_1, x_2)$.

3 The Inversion of the Radon Transform

3.1 Preliminaries

In order to solve the inverse problem defined in equation (5) and invert the Radon transform, it is essential to introduce appropriate mathematical machinery.

Lemma 1 (Generalized Cauchy Formula or Pompeiu's Formula) *Assume that the function $f(z, \bar{z})$ is continuous and has continuous partial derivatives in a finite region D and on the simple closed boundary ∂D.We denote by ∂D the closed boundary of D with counterclockwise direction. Then, $f(z, \bar{z})$ can be evaluated at any interior point z via the following formula*

$$f(z, \bar{z}) = \frac{1}{2\pi i} \left(\oint_{\partial D} f(\zeta, \bar{\zeta}) \frac{d\zeta}{\zeta - z} + \iint_D \frac{\partial f}{\partial \bar{\zeta}}(\zeta, \bar{\zeta}) \frac{d\zeta \wedge d\bar{\zeta}}{\zeta - z} \right), \tag{8}$$

where the so-called wedge product $d\zeta \wedge d\bar{\zeta}$, $\zeta = \xi + i\eta$, $\bar{\zeta} = \xi - i\eta$, *is defined as*

$$d\zeta \wedge d\bar{\zeta} = (d\xi + i d\eta) \wedge (d\xi - i d\eta) = -2i d\xi d\eta. \tag{9}$$

Proof Let the real functions $u(x, y)$ and $v(x, y)$, as well as their partial derivatives $u_x(x, y)$, $u_y(x, y)$, $v_x(x, y)$ and $v_y(x, y)$, be continuous inside a finite region D and on its simple closed boundary ∂D. Green's theorem yields

$$\oint_{\partial D} (u dx + v dy) = \iint_D (v_x - u_y) dx dy. \tag{10}$$

Replacing in equation (10) (x, y) by (ξ, η) and letting $g = u + iv$, we find

$$\oint_{\partial D} g \mathrm{d}\zeta = -\iint_{D} \frac{\partial g}{\partial \bar{\zeta}} \mathrm{d}\zeta \wedge \mathrm{d}\bar{\zeta}, \tag{11}$$

where $\mathrm{d}\zeta = \mathrm{d}\xi + i\mathrm{d}\eta$, $\partial g/\partial\bar{\zeta} = \frac{1}{2}(g_\xi + ig_\eta)$ and $\mathrm{d}\zeta \wedge \mathrm{d}\bar{\zeta}$ is defined in equation (9). We define the region D_ϵ as follows:

$$D_\epsilon = D \backslash \{|\zeta - z| \leq \epsilon\}, \quad \epsilon > 0, \tag{12}$$

i.e., D_ϵ is the region D without the circular area centered at $\zeta = z$ with radius ϵ. We denote the boundary of this circular area by ∂D_ϵ. The contour of D_ϵ comprises of two subcontours, namely ∂D and ∂D_ϵ. Noting that $(\zeta - z)^{-1}$ is analytic in D_ϵ, we may apply Green's theorem in the form (11) on the function $g(\zeta) = f(\zeta)(\zeta - z)^{-1}$

$$\oint_{\partial D} \frac{f(\zeta)}{\zeta - z} \mathrm{d}\zeta - \oint_{\partial D_\epsilon} \frac{f(\zeta)}{\zeta - z} \mathrm{d}\zeta = -\iint_{D} \frac{\partial f/\partial\bar{\zeta}}{\zeta - z} \mathrm{d}\zeta \wedge \mathrm{d}\bar{\zeta}. \tag{13}$$

However, we note that the points on the circle ∂D_ϵ can be expressed in the form $\zeta = z + \epsilon e^{i\theta}$, therefore

$$\oint_{\partial D_\epsilon} \frac{f(\zeta)}{\zeta - z} \mathrm{d}\zeta = \int_0^{2\pi} \frac{f(z + \epsilon e^{i\theta})}{\epsilon e^{i\theta}} i\epsilon e^{i\theta} \mathrm{d}\theta = \int_0^{2\pi} f(z + \epsilon e^{i\theta}) i \mathrm{d}\theta \xrightarrow{\epsilon \to 0} 2\pi i f(z). \tag{14}$$

The above is valid because (a) $f(z)$ is continuous, and (b) the integral of a continuous function inside a bounded region (∂D_ϵ) and the limit $(\epsilon \to 0)$ are interchangeable. Similarly, due to the fact that $(\zeta - z)^{-1}$ is integrable inside the region D_ϵ, together with the fact that $f_{\bar{\zeta}}$ is continuous, it follows that the double integral over D_ϵ converges to the double integral over the whole region D. Hence, their difference tends to zero as $\epsilon \to 0$:

$$\left| \iint_{D - D_\epsilon} \frac{\partial f/\partial\bar{\zeta}}{\zeta - z} \mathrm{d}\zeta \wedge \mathrm{d}\bar{\zeta} \right| \leq 2 \int_0^{\epsilon} \int_0^{2\pi} \frac{\partial f/\partial\bar{\zeta}}{r} r \mathrm{d}r \mathrm{d}\theta \leq 4\pi\Lambda\epsilon, \tag{15}$$

where we have used polar coordinates, i.e., $\zeta = z + \epsilon e^{i\theta}$, and the continuity of $\partial f/\partial\bar{\zeta}$ in a bounded region, namely

$$\left| \frac{\partial f}{\partial \bar{\zeta}} \right| \leq \Lambda. \tag{16}$$

As a consequence, we consider equation (13) in the limit $\epsilon \to 0$

$$\oint_{\partial D} \frac{f(\zeta)}{\zeta - z} \mathrm{d}\zeta - 2\pi i f(z) = -\iint_{D} \frac{\partial f/\partial\bar{\zeta}}{\zeta - z} \mathrm{d}\zeta \wedge \mathrm{d}\bar{\zeta}, \tag{17}$$

therefore, Pompeiu's formula (1) follows. □

Corollary 1 *If $f(z)$ is analytic in $\bar{D} = D \cup \partial D$, then Pompeiu's formula* (8) *reduces to Cauchy's integral formula*

$$f(z) = \frac{1}{2\pi i} \oint_{\partial D} \frac{f(\zeta)}{\zeta - z} \mathrm{d}\zeta. \tag{18}$$

Proof The proof is straightforward: if $f(z)$ is analytic in $\bar{D} = D \cup \partial D$, then $\partial f / \partial \bar{\zeta} = 0$. Hence, Cauchy's integral formula (18) follows from Pompeiu's formula (8). □

Lemma 2 (Plemelj Formulæ) *Let L be a smooth, finite, closed or open contour and let $\varphi(t)$ satisfy a Hölder condition on L. Then, the Cauchy-type integral*

$$\Phi(z) = \frac{1}{2\pi i} \int_L \frac{\varphi(\tau)}{\tau - z} \mathrm{d}\tau, \tag{19}$$

has the limiting values $\Phi^-(t)$ and $\Phi^+(t)$ as z approaches L from the right and the left, respectively, given that t is not an endpoint of L. The limits $\Phi^\pm(t)$ are given by

$$\Phi^\pm(t) = \pm\frac{1}{2}\varphi(t) + \frac{1}{2\pi i} ⨍_L \frac{\varphi(\tau)}{\tau - t} \mathrm{d}\tau, \tag{20}$$

where $⨍$ denotes the Cauchy principal value integral defined by

$$⨍_L g(\tau)\mathrm{d}\tau = \lim_{\epsilon \to 0} \int_{L - L_\epsilon} g(\tau)\mathrm{d}\tau, \tag{21}$$

and L_ε denotes the part of the contour L that is centered around t, with length 2ϵ.

Proof We shall derive the Plemelj formulæ only in the case that $\varphi(\tau)$ is analytic in the neighborhood of L. The derivation of these formulæ in the more general case of $\varphi(\tau)$ satisfying a Hölder condition on L is far too complicated for the purposes of the present study, see [7]. If $\varphi(\tau)$ is analytic at $\tau = t$, then we use the Cauchy theorem to deform L into two separate contours, namely $L - L_\epsilon$ and C_ϵ, where C_ϵ is the semicircle of radius $r = \epsilon$, centered at $\tau = t$. Taking the limit of equation (19) as $\epsilon \to 0$ yields

$$\Phi^+(t) = \lim_{\epsilon \to 0} \frac{1}{2\pi i} \int_{L - L_\epsilon} \frac{\varphi(\tau)}{\tau - t} \mathrm{d}\tau + \lim_{\epsilon \to 0} \int_{C_\epsilon} \frac{\varphi(\tau)}{\tau - t} \mathrm{d}\tau, \tag{22}$$

where we used the deformation $L = (L - L_\epsilon) \cup C_\epsilon$. However, if we use polar coordinates, namely $\tau = t + \epsilon e^{i\theta}$, then the second integral on the right-hand side of equation (22) becomes

$$\lim_{\epsilon \to 0} \frac{1}{2\pi i} \int_{C_\epsilon} \frac{\varphi(\tau)}{\tau - t} \mathrm{d}\tau = \frac{1}{2\pi i} \lim_{\epsilon \to 0} \int_{-\pi}^{0} \varphi(t + \epsilon e^{i\theta}) i \mathrm{d}\theta = \frac{1}{2}\varphi(\tau). \tag{23}$$

Hence, equation (23) reduces to equation (20) representing $\Phi^+(t)$. Similar considerations apply for the corresponding part of the equation (20) representing $\Phi^-(t)$. □

Lemma 3 *The solution of the following scalar Riemann–Hilbert problem*

$$\Phi^+(t) - \Phi^-(t) = g(t), \quad t \in L, \tag{24a}$$

$$\Phi(z) = O\left(\frac{1}{z}\right), \quad z \to \infty, \quad z \notin L, \tag{24b}$$

is given by

$$\Phi(z) = \frac{1}{2\pi i}\int_L \frac{g(\tau)}{\tau - z}d\tau. \tag{25}$$

Proof Equation (24a), together with Liouville's theorem, implies that the unique solution of the scalar Riemann–Hilbert problem (24) is given by (25). □

3.2 The Radon Transform Pair

In 1992, Roman Novikov and one of the authors rederived the well-known inversion of the Radon transform [5] by performing the so-called *spectral analysis* of the following eigenvalue equation:

$$\left[\frac{1}{2}\left(k + \frac{1}{k}\right)\partial_{x_1} + \frac{1}{2i}\left(k - \frac{1}{k}\right)\partial_{x_2}\right]\Theta(x_1, x_2; k, \bar{k}) = f(x_1, x_2), \quad k \in \mathbb{C}. \tag{26}$$

The Radon transform inversion can be obtained in a more straightforward manner, by employing the two-dimensional Fourier transform. However, the major advantage of the derivation of [5] was established 10 years later (2002) by Novikov [6]. Novikov demonstrated that the inverse attenuated Radon transform can be derived by applying a similar analysis to a slight generalization of equation (26), namely the equation

$$\left[\frac{1}{2}\left(k + \frac{1}{k}\right)\partial_{x_1} + \frac{1}{2i}\left(k - \frac{1}{k}\right)\partial_{x_2} - \mu(x_1, x_2)\right]\Theta(x_1, x_2; k, \bar{k}) = f(x_1, x_2), \quad k \in \mathbb{C}. \tag{27}$$

In the following pages, we present an algorithmic approach for the inversion of the Radon transform. The corresponding analysis, referred to as spectral analysis, consists of two steps:

(a) The direct problem. We solve the eigenvalue equation (26) in terms of the function f for all complex eigenvalues k. The solution of the eigenvalue equation must be bounded for all complex values of k in $\mathbb{C}$. The direct problem gives rise to a certain problem in complex analysis known as the $\bar{d}$-problem.

(b) The inverse problem. Given that the solution Θ of equation (26) is bounded for all complex eigenvalues k, we derive an equivalent representation of Θ which, instead of depending on f, depends on the Radon transform of f denoted by $\widehat{f}$, as in equation (5). The inverse problem gives rise to a certain problem in complex analysis known as the scalar Riemann–Hilbert (RH) problem.

Proposition 1 *The inversion of the Radon transform $\widehat{f}(\rho, \theta)$ of a function $f(x_1, x_2) \in S(\mathbb{R}^2)$, defined in equation* (5) *is given by*

$$f(x_1, x_2) = -\frac{1}{4\pi} \int_0^{2\pi} \left[\frac{\partial(\mathscr{H}\widehat{f})(\rho, \theta)}{\partial \rho} \right]_{\rho = x_2 \cos\theta - x_1 \sin\theta} \mathrm{d}\theta, \tag{28}$$

with $-\infty < x_1, x_2 < \infty$ and $\mathscr{H}$ denoting the Hilbert transform in the variable ρ, i.e.

$$(\mathscr{H}\widehat{f})(\rho, \theta) \equiv \frac{1}{\pi} ⨍_{-\infty}^{\infty} \frac{\widehat{f}(r, \theta)}{r - \rho} \mathrm{d}r, \quad -\infty < \rho < \infty, \quad 0 \leq \theta < 2\pi, \tag{29}$$

and $⨍$ denotes the Cauchy principal value integral defined in equation (21).

Proof We will invert the Radon transform by performing the spectral analysis of the eigenvalue equation (26).

3.2.1 The Direct Problem: A $\bar{d}$-Problem

In order to solve the direct problem, we solve equation (26) for all complex k values, assuming that the function f is known. We introduce the following change of variables from (x_1, x_2) to $(z, \bar{z})$:

$$z = \frac{1}{2i} \left(k - \frac{1}{k} \right) x_1 - \frac{1}{2} \left(k + \frac{1}{k} \right) x_2, \tag{30a}$$

$$\bar{z} = -\frac{1}{2i} \left(\bar{k} - \frac{1}{\bar{k}} \right) x_1 + \frac{1}{2} \left(\bar{k} + \frac{1}{\bar{k}} \right) x_2. \tag{30b}$$

Employing the chain rule we find

$$\partial_{x_1} = \frac{1}{2i} \left(k - \frac{1}{k} \right) \partial_z - \frac{1}{2i} \left(\bar{k} - \frac{1}{\bar{k}} \right) \partial_{\bar{z}}, \tag{31a}$$

$$\partial_{x_2} = -\frac{1}{2} \left(k + \frac{1}{k} \right) \partial_z - \frac{1}{2} \left(\bar{k} + \frac{1}{\bar{k}} \right) \partial_{\bar{z}}. \tag{31b}$$

Hence, we are able to rewrite equation (26) in the following form:

$$\nu(|k|)\frac{\partial \Theta(x_1, x_2, k)}{\partial \bar{z}} = f(x_1, x_2), \quad k \in \mathbb{C}, \quad |k| \neq 1, \tag{32}$$

where

$$\nu(|k|) = \frac{1}{2i}\left(\frac{1}{|k|^2} - |k|^2\right). \tag{33}$$

Equation (32) can be further simplified, namely

$$\Theta_{\bar{z}} = \frac{f}{\nu}, \quad |k| \neq 1. \tag{34}$$

It is important to note that if μ was analytic, then $\mu_{\bar{z}} = 0$, since the $\bar{z}$ derivative measures the "departure" from analyticity. Furthermore, we supplement equation (32) with the following boundary condition at infinity

$$\Theta = O\left(\frac{1}{z}\right), \quad z \to \infty, \quad \text{i.e.,} \quad \exists\, B > 0: \quad |\Theta| \leq \frac{B}{|z|}. \tag{35}$$

The solution of equation (32), equipped with the boundary condition (35), via Pompeiu's formula (see Lemma 1) is given by

$$\Theta = \frac{1}{2\pi i}\iint_{\mathbb{R}^2} \frac{f(x_1', x_2')}{\nu(|k|)}\frac{\mathrm{d}z' \wedge \mathrm{d}\bar{z}'}{z' - z}, \quad k \in \mathbb{C}, \quad |k| \neq 1. \tag{36}$$

However, by employing

$$\mathrm{d}z' \wedge \mathrm{d}\bar{z}' = \frac{1}{2i}\left|\frac{1}{|k|^2} - |k|^2\right| \mathrm{d}x_1'\mathrm{d}x_2',$$

we rewrite equation (36) as follows:

$$\Theta(x_1, x_2, k) = \frac{1}{2\pi i}\operatorname{sgn}\left(\frac{1}{|k|^2} - |k|^2\right)\iint_{\mathbb{R}^2} f(x_1', x_2')\frac{\mathrm{d}x_1'\mathrm{d}x_2'}{z' - z}. \tag{37}$$

It is evident from equation (37) that Θ depends on k only through $z - z'$, hence $\Theta(x_1, x_2, k)$ constitutes a *sectionally analytic* function with a "jump" across the unit circle $|k| = 1$ of the complex k-plane. Equation (37) represents the solution of the direct problem for all complex values of the eigenvalue k, in terms of the function f.

3.2.2 The Inverse Problem: A Riemann–Hilbert Problem

In order to solve the inverse problem, we solve equation (26) in terms of $\widehat{f}$, instead of f itself. We note that equation (37) implies

$$\Theta = O\left(\frac{1}{k}\right), \quad k \to \infty, \tag{38}$$

i.e., the solution Θ of equation (26) is bounded for all complex eigenvalues k.

We note that

$$z - z' = \frac{1}{2i}\left(k - \frac{1}{k}\right)(x_1 - x_1') - \frac{1}{2}\left(k + \frac{1}{k}\right)(x_2 - x_2'). \tag{39}$$

In order to investigate the behavior of Θ as k approaches the unit circle, let

$$k^{\pm} = (1 \mp \varepsilon)e^{i\theta}, \quad 0 \leq \theta < 2\pi, \quad \varepsilon > 0. \tag{40}$$

Hence,

$$k^{+} \mp \frac{1}{k^{+}} = (1 - \varepsilon)e^{i\theta} \mp (1 + \varepsilon)e^{-i\theta} + O(\varepsilon^2), \tag{41a}$$

$$k^{-} \mp \frac{1}{k^{-}} = (1 + \varepsilon)e^{i\theta} \mp (1 - \varepsilon)e^{-i\theta} + O(\varepsilon^2). \tag{41b}$$

Equation (30a) implies

$$\begin{aligned} z - z' &= (x_1 - x_1')\sin\theta - (x_2 - x_2')\cos\theta \pm i\varepsilon\left[(x_1 - x_1')\cos\theta - (x_2 - x_2')\sin\theta\right] + O(\varepsilon^2) \\ &= \rho - \rho' \pm i\varepsilon(\tau - \tau') + O(\varepsilon^2). \end{aligned} \tag{42}$$

Let $\Theta^{\pm}$ denote the limits of the function Θ as k approaches the unit circle $|k| = 1$ from inside and outside, respectively:

$$\Theta^{\pm} \equiv \lim_{\varepsilon \to 0} \Theta(x_1, x_2, (1 \mp \varepsilon)e^{i\theta}) = \lim_{\varepsilon \to 0} \Theta(x_1, x_2, k^{\mp}). \tag{43}$$

Replacing $z - z'$ in equation (37) by the corresponding representation (42) yields

$$\Theta^{\pm} = \mp\frac{1}{2\pi i} \lim_{\varepsilon \to 0} \iint_{\mathbb{R}^2} \frac{\varphi(\rho', \tau', \theta)\mathrm{d}\rho'\mathrm{d}\tau'}{\rho' - [\rho \pm i\varepsilon(\tau' - \tau)]}, \tag{44}$$

where

$$\varphi(\rho, \tau, \theta) = f\left(\tau\cos\theta - \rho\sin\theta, \tau\sin\theta + \rho\cos\theta\right), \tag{45}$$

i.e., φ is the function f expressed in the local coordinates. For the evaluation of the limit (44), we split the integral over $\mathrm{d}\tau'$, so as to control the sign of $\tau' - \tau$

$$\Theta^{\pm} = \mp\frac{1}{2\pi i}\lim_{\varepsilon\to 0}\int_{-\infty}^{\infty}\left\{\int_{-\infty}^{\tau}\frac{\varphi \mathrm{d}\tau'}{\rho' - [\rho \pm i\varepsilon(\tau'-\tau)]}\right.$$
$$\left. + \int_{\tau}^{\infty}\frac{\varphi \mathrm{d}\tau'}{\rho' - [\rho \pm i\varepsilon(\tau'-\tau)]}\right\}\mathrm{d}\rho'. \tag{46}$$

In the first integral of (46) $(\tau'-\tau)$ is negative, whereas in the second integral $(\tau'-\tau)$ is positive, thus

$$\Theta^{\pm} = \mp\frac{1}{2\pi i}\int_{-\infty}^{\tau}\left\{\mp\pi i\varphi(\rho,\tau',\theta) + (\mathscr{H}\varphi)(\rho,\tau',\theta)\right\}\mathrm{d}\tau'$$
$$\mp\frac{1}{2\pi i}\int_{\tau}^{\infty}\left\{\pm\pi i\varphi(\rho,\tau',\theta) + (\mathscr{H}\varphi)(\rho,\tau',\theta)\right\}\mathrm{d}\tau', \tag{47}$$

where we have employed the Plemelj formulæ, see Lemma 2, as well as the definition of the Hilbert transform, see equation (29). In the right-hand side of equation (47) we add and subtract $\mp\frac{1}{2\pi i}\int_{\tau}^{\infty}\pi i\varphi(\rho,\tau',\theta)\mathrm{d}\tau'$, to obtain

$$\Theta^{\pm} = \mp(P^{\mp}\hat{f})(\rho,\theta) - \int_{\tau}^{\infty}\varphi(\rho,\tau',\theta)\mathrm{d}\tau', \tag{48}$$

where

$$(P^{\mp}g)(\rho) = \pm\frac{g(\rho)}{2} + \frac{1}{2\pi i}\int_{-\infty}^{\infty}\frac{g(r)}{r-\rho}\mathrm{d}r = \pm\frac{g(\rho)}{2} + \frac{1}{2i}(\mathscr{H}g)(\rho), \tag{49}$$

denote a family of projectors in the variance ρ.

Equations (43) and (48) imply

$$\Theta^{+} - \Theta^{-} = i(\mathscr{H}\widehat{f})(\rho,\theta). \tag{50}$$

We supplement equation (50) with the boundary condition (38) and we construct a scalar Riemann–Hilbert problem. The solution to this problem is, according to Lemma 3,

$$\Theta = \frac{1}{2\pi i}\int_{|k'|=1}\frac{(\Theta^{+}-\Theta^{-})(\rho,\theta')\mathrm{d}k'}{k'-k}. \tag{51}$$

The equation $|k'| = 1$ can be rewritten as $k' = e^{i\theta'}$, implying that $\mathrm{d}k' = ie^{i\theta'}\mathrm{d}\theta'$, hence

$$\Theta = \frac{1}{2\pi i}\int_{0}^{2\pi}\frac{(\Theta^{+}-\Theta^{-})(\rho,\theta')ie^{i\theta'}\mathrm{d}\theta'}{e^{i\theta'}-k}. \tag{52}$$

Replacing in equation (52) $\Theta^+ - \Theta^-$ by the right-hand side of equation (50) yields:

$$\Theta = -\frac{1}{2\pi i}\int_0^{2\pi} \frac{e^{i\theta'}(\mathscr{H}\widehat{f})(\rho,\theta')\mathrm{d}\theta'}{e^{i\theta'}-k}, \; k \in \mathbb{C},\; |k| \neq 1,\; \rho \in \mathbb{R}. \tag{53}$$

Equation (53) represents the solution to the inverse problem. In order to express f as a function of $\widehat{f}$ we must utilize the equivalence of the solutions of the direct and inverse problems, provided by equations (37) and (53), respectively. We proceed with the asymptotic analysis of the behavior of Θ for large k. Equation (53) yields

$$\Theta = \left\{\frac{1}{2\pi i}\int_0^{2\pi} e^{i\theta'}(\mathscr{H}\widehat{f})(\rho,\theta')\mathrm{d}\theta'\right\}\frac{1}{k} + O\left(\frac{1}{k^2}\right), \quad k \to \infty. \tag{54}$$

Substituting in equation (26) the above expression, we find that the $O(1)$ term implies

$$f = \frac{1}{4\pi i}\left(\partial_{x_1} - i\partial_{x_2}\right)\int_0^{2\pi} e^{i\theta}(\mathscr{H}\widehat{f})(\rho,\theta)\Big|_{\rho = x_2\cos\theta - x_1\sin\theta} \mathrm{d}\theta. \tag{55}$$

Using the identity

$$(\partial_{x_1} - i\partial_{x_2}) = e^{-i\theta}(\partial_\tau - i\partial_\rho), \tag{56}$$

in equation (55), we obtain equation (28). □

An immediate consequence of the above is the following corollary.

Corollary 2 *Let $k^\pm$ denote the limits of k, defined in equation* (40)*, as k approaches the unit circle from inside and outside the unit circle* ($|k| = 1$)*, and let z and ν be defined in equations* (30a) *and* (33)*, respectively. Then,*

$$\lim_{k\to k^\pm}\left(\partial_{\bar z}^{-1}\left\{\frac{f(x_1,x_2)}{\nu(|k|)}\right\}\right) = \mp\left(P^\mp \widehat{f}\right)(\rho,\theta) - \int_\tau^\infty \varphi(\rho,s,\theta)\mathrm{d}s,$$
$$(\rho,\tau)\in\mathbb{R}^2, \quad \theta\in(0,2\pi), \tag{57}$$

where $\widehat{f}$ denotes the Radon transform of f (defined in equation (5)*), and $P^\pm$, (ρ,τ) and φ are defined in equations* (49), (4)*, and* (45)*, respectively.*

Proof Equation (34) implies that

$$\Theta(x_1,x_2,k) = \partial_{\bar z}^{-1}\left\{\frac{f(x_1,x_2)}{\nu(|k|)}\right\}. \tag{58}$$

Therefore, taking the limit of equation (58) as k approaches the unit circle $|k| = 1$ from inside and outside yields

$$\Theta^{\pm}(x_1, x_2) = \lim_{k \to k^{\pm}} \left(\partial_{\bar{z}}^{-1} \left\{ \frac{f(x_1, x_2)}{\nu(|k|)} \right\} \right), \tag{59}$$

where $\Theta^{\pm}$ is defined in equation (43). Inserting equation (48) in equation (59) yields equation (57). □

4 The Attenuated Radon Transform

We define the line integral of a two-dimensional function $f(x_1, x_2)$ attenuated with respect to the function $\mu(x_1, x_2)$ as the *attenuated Radon transform* of $f(x_1, x_2)$, denoted by $\widehat{f}_{\mu}(\rho, \theta)$, see Figure 2. The attenuated Radon transform is usually stored in the form of the so-called *attenuated sinogram*, expressed as follows:

$$\widehat{f}_{\mu}(\rho, \theta) = \int_{-\infty}^{\infty} e^{-\int_{\tau}^{\infty} \mu(s\cos\theta - \rho\sin\theta, s\sin\theta + \rho\cos\theta)ds} \times$$

$$f(\tau\cos\theta - \rho\sin\theta, \tau\sin\theta + \rho\cos\theta)d\tau, \; 0 \le \theta < 2\pi, \quad -\infty < \rho < \infty. \tag{60}$$

The above transform (60) gives rise to the following *inverse problem*: Given the functions $\widehat{f}_{\mu}(\rho, \theta)$, $0 \le \theta < 2\pi$, $-\infty < \rho < \infty$ and $\mu(x_1, x_2)$, $-\infty < x_1, x_2 < \infty$, "reconstruct" the function $f(x_1, x_2)$.

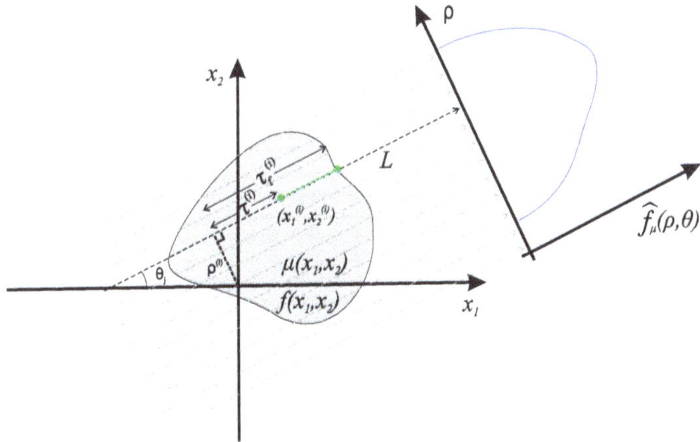

Fig. 2 A two-dimensional object, with attenuation coefficient $\mu(x_1, x_2)$, being imaged with parallel-beam projection geometry. Both Cartesian (x_1, x_2) and local (ρ, τ) coordinates are indicated

5 The Inversion of the Attenuated Radon Transform

5.1 The Attenuated Radon Transform Pair

In [6] Novikov demonstrated that the inverse attenuated Radon transform can be derived by performing the spectral analysis of the eigenvalue equation (27), which is a slight generalization of the eigenvalue equation (26). However, one can derive the inversion of the attenuated Radon transform in a simpler manner, using the results of the inversion of the non-attenuated Radon transform (see Section 3) as well as Corollary 2.

Proposition 2 *The inverse of the attenuated Radon transform $\widehat{f}_\mu(\rho,\theta)$ of a function $f(x_1,x_2)$, attenuated with respect to the function $\mu(x_1,x_2)$ (with $f,\mu \in S(\mathbb{R}^2)$), defined in equation* (60) *is given by*

$$f(x_1,x_2) = \frac{1}{4\pi}(\partial_{x_1} - i\partial_{x_2}) \int_0^{2\pi} e^{i\theta} J(x_1,x_2,\theta)\mathrm{d}\theta, \quad -\infty < x_1, x_2 < \infty, \tag{61a}$$

where the function J is defined by

$$J(x_1,x_2,\theta) = e^{M(\tau,\rho,\theta)} L_\mu(\rho,\theta)\widehat{f}_\mu(\rho,\theta)\Bigg|_{\substack{\tau=x_2\sin\theta+x_1\cos\theta \\ \rho=x_2\cos\theta-x_1\sin\theta}}, \tag{61b}$$

with M and L_μ defined by

$$M(\tau,\rho,\theta) = \int_\tau^\infty \mu\left(s\cos\theta - \rho\sin\theta, s\sin\theta + \rho\cos\theta\right)\mathrm{d}s, \tag{61c}$$

$$L_\mu(\rho,\theta) = e^{P^-\widehat{\mu}(\rho,\theta)} P^- e^{P^-\widehat{\mu}(\rho,\theta)} + e^{-P^+\widehat{\mu}(\rho,\theta)} P^+ e^{P^+\widehat{\mu}(\rho,\theta)}; \tag{61d}$$

in equation (61d)*, $\widehat{\mu}$ represents the Radon transform of the attenuation function μ, i.e.,*

$$\widehat{\mu}(\rho,\theta) = \int_{-\infty}^\infty \mu\left(\tau\cos\theta - \rho\sin\theta, \tau\sin\theta + \rho\cos\theta\right)\mathrm{d}\tau, \; 0 \le \theta < 2\pi, \; -\infty < \rho < \infty, \tag{61e}$$

whereas the projection operators $P^\pm$ are defined in equation (49).

Proof Rewriting equation (27) in the form

$$\Theta_{\bar{z}} + \frac{\mu}{\nu}\Theta = \frac{f}{\nu}, \tag{62}$$

and multiplying both sides of equation (62) by $e^{\partial_{\bar{z}}^{-1}\{\frac{\mu}{\nu}\}}$ yields

$$\Theta_{\bar{z}} e^{\partial_{\bar{z}}^{-1}\{\frac{\mu}{\nu}\}} + \frac{\mu}{\nu} \Theta e^{\partial_{\bar{z}}^{-1}\{\frac{\mu}{\nu}\}} = \frac{f}{\nu} e^{\partial_{\bar{z}}^{-1}\{\frac{\mu}{\nu}\}}. \tag{63}$$

Therefore,

$$\frac{\partial}{\partial \bar{z}} \left(\Theta e^{\partial_{\bar{z}}^{-1}\{\frac{\mu}{\nu}\}} \right) = \frac{f}{\nu} e^{\partial_{\bar{z}}^{-1}\{\frac{\mu}{\nu}\}}. \tag{64}$$

Hence,

$$e^{\partial_{\bar{z}}^{-1}\{\frac{\mu}{\nu}\}} \Theta = \partial_{\bar{z}}^{-1} \left\{ \frac{f}{\nu} e^{\partial_{\bar{z}}^{-1}\{\frac{\mu}{\nu}\}} \right\}, \quad (x_1, x_2) \in \mathbb{R}^2, \quad k \in \mathbb{C}. \tag{65}$$

Equation (65) represents the solution to the direct problem, which, following the same approach as with the inversion of the non-attenuated Radon transform, defines a sectionally analytic function Θ with a "jump" across the unit circle, $|k| = 1$, of the complex k-plane. Furthermore, Θ satisfies the estimate (38). Therefore, Θ is given by equation (52). However, this time the jump $\Theta^+ - \Theta^-$ will be different than the one given by equation (50). The determination of the jump involves the limits of $\partial_{\bar{z}}^{-1}(f/\nu)$ as k approaches $k^{\pm}$, hence can be computed via Corollary 2. In the limit $k \to k^{\pm}$, equation (65), where f is replaced by μ, implies

$$e^{(\mp P^{\mp}\widehat{\mu} - \int_{\tau}^{\infty} \Phi(\rho,s,\theta) \mathrm{d}s)} \Theta^{\pm} = \lim_{k \to k^{\pm}} \partial_{\bar{z}}^{-1} \left\{ \frac{f}{\nu} e^{(\mp P^{\mp}\widehat{\mu} - \int_{\tau}^{\infty} \Phi(\rho,s,\theta) \mathrm{d}s)} \right\}, \tag{66}$$

where $\widehat{\mu}$ denotes the Radon transform of μ, defined in equation (5), and Φ denotes the function μ expressed in the local coordinates:

$$\Phi(\rho, \tau, \theta) = \mu\left(\tau \cos\theta - \rho \sin\theta, \tau \sin\theta + \rho \cos\theta\right). \tag{67}$$

Applying Corollary 2 on the right-hand side of equation (66) and using $\widehat{f_{\mu}}$ defined in equation (60) yields

$$\mp P^{\mp} e^{\mp P^{\mp}\widehat{\mu}} \widehat{f_{\mu}} - \int_{\tau}^{\infty} \varphi(\rho, \tau', \theta) e^{\mp P^{\mp}\widehat{\mu} e^{-\int_{\tau}^{\infty} \Phi(\rho,s,\theta) \mathrm{d}s}} \mathrm{d}\tau', \tag{68}$$

where in equation (57), instead of f we used the function $(e^{\mp P^{\mp}\widehat{\mu} e^{-\int_{\tau}^{\infty} \Phi(\rho,s,\theta) \mathrm{d}s}}) f$. We note that inside the integral of the expression (68), the term $e^{\mp P^{\mp}\widehat{\mu}}$ is independent of τ', hence it can come outside the integral. This means that this specific term can be canceled throughout equation (66). Therefore, the jump is given by

$$\Theta^+ - \Theta^- = -J, \tag{69}$$

where J is defined in equation (61b). Hence, equation (52) implies

$$\Theta = -\frac{1}{2\pi}\int_0^{2\pi} \frac{J(\rho,\tau,\theta')e^{i\theta'}\mathrm{d}\theta'}{e^{i\theta'}-k}. \tag{70}$$

Therefore,

$$\Theta = \left[\frac{1}{2\pi}\int_0^{2\pi} e^{i\theta} J(\rho,\tau,\theta)\mathrm{d}\theta\right]\frac{1}{k} + O\left(\frac{1}{k^2}\right), \quad \text{as} \quad k \to \infty. \tag{71}$$

Substituting equation (71) in equation (27), we find that the $O(1)$ term in equation (27) implies equation (61a). □

5.2 A Novel Equivalent Formula

In what follows, it is useful to define F as half the Hilbert transform of $\widehat{\mu}$, i.e.,

$$F(\rho,\theta) \equiv \frac{1}{2}\mathcal{H}\{\widehat{\mu}(\rho,\theta)\} = \frac{1}{2\pi}\oint_{-\infty}^{\infty}\frac{\widehat{\mu}(r,\theta)}{r-\rho}\mathrm{d}r. \tag{72}$$

Proposition 3 *The inverse of the attenuated Radon transform $\widehat{f_\mu}(\rho,\theta)$ of a function $f(x_1,x_2)$, attenuated with respect to the function $\mu(x_1,x_2)$ (with $f,\mu \in S(\mathbb{R}^2)$), defined in equation* (61a) *is equivalent to the representation*

$$f(x_1,x_2) = -\frac{1}{2\pi}\int_0^{2\pi} e^{M(\tau,\rho,\theta)}\left[M_\rho(\tau,\rho,\theta)G(\rho,\theta) + G_\rho(\rho,\theta)\right]\Bigg|_{\substack{\rho=x_2\cos\theta-x_1\sin\theta\\ \tau=x_2\sin\theta+x_1\cos\theta}} d\theta, \tag{73}$$

where M is defined in equation (61c)*, the subscripts denote differentiation with respect to ρ, and G is defined as follows:*

$$G(\rho,\theta) = e^{-\frac{1}{2}\widehat{\mu}(\rho,\theta)}\left[\cos(F(\rho,\theta))G^C(\rho,\theta) + \sin(F(\rho,\theta))G^S(\rho,\theta)\right], \tag{74}$$

with the functions G^C and G^S defined by

$$G^C(\rho,\theta) = \frac{1}{2\pi}\oint_{-\infty}^{\infty} e^{\frac{1}{2}\widehat{\mu}(r,\theta)}\cos F(r,\theta)\frac{\widehat{f_\mu}(r,\theta)dr}{r-\rho}, \tag{75a}$$

$$G^S(\rho,\theta) = \frac{1}{2\pi}\oint_{-\infty}^{\infty} e^{\frac{1}{2}\widehat{\mu}(r,\theta)}\sin F(r,\theta)\frac{\widehat{f_\mu}(r,\theta)dr}{r-\rho}. \tag{75b}$$

Proof Applying the operator L_μ, defined in equation (61d), on the attenuated Radon transform $\widehat{f_\mu}$, defined in equation (60), yields

$$\left(L_\mu \widehat{f_\mu}\right)(\rho,\theta) = \left\{e^{P^-\widehat{\mu}(\rho,\theta)} P^- e^{P^-\widehat{\mu}(\rho,\theta)} + e^{-P^+\widehat{\mu}(\rho,\theta)} P^+ e^{P^+\widehat{\mu}(\rho,\theta)}\right\} \widehat{f_\mu}(\rho,\theta). \tag{76}$$

Furthermore, equations (49) and (72) imply

$$e^{P^\pm \widehat{\mu}} = e^{\pm\frac{\widehat{\mu}}{2} - iF}. \tag{77}$$

Therefore,

$$e^{P^-\widehat{\mu}} P^- \left\{e^{-P^-\widehat{\mu}} \widehat{f_\mu}\right\} = e^{-\frac{\widehat{\mu}}{2} - iF}\left[-\frac{1}{2} e^{\frac{\widehat{\mu}}{2}+iF} \widehat{f_\mu} + \frac{1}{2i}\mathscr{H}\left\{e^{\frac{\widehat{\mu}}{2}+iF}\widehat{f_\mu}\right\}\right], \tag{78a}$$

and

$$e^{-P^+\widehat{\mu}} P^+ \left\{e^{P^+\widehat{\mu}} \widehat{f_\mu}\right\} = e^{-\frac{\widehat{\mu}}{2} + iF}\left[\frac{1}{2} e^{\frac{\widehat{\mu}}{2}-iF} \widehat{f_\mu} + \frac{1}{2i}\mathscr{H}\left\{e^{\frac{\widehat{\mu}}{2}-iF}\widehat{f_\mu}\right\}\right]. \tag{78b}$$

Employing equations (78), and the fact that $e^{iF} = \cos F + i \sin F$, we are able to simplify equation (76) as follows: first, we combine equations (77) and (78) and rewrite L_μ in the form

$$(L_\mu \widehat{f_\mu})(\rho,\theta) = \frac{1}{2i} e^{-\frac{\widehat{\mu}}{2}}\left[e^{-iF}\mathscr{H}\left\{e^{\frac{\widehat{\mu}}{2}+iF}\widehat{f_\mu}\right\} + e^{iF}\mathscr{H}\left\{e^{\frac{\widehat{\mu}}{2}-iF}\widehat{f_\mu}\right\}\right]. \tag{79}$$

Using the fact that $e^{\pm iF} = \cos F \pm i \sin F$, equation (79) simplifies as follows:

$$\begin{aligned}
&(L_\mu \widehat{f_\mu})(\rho,\theta) \\
&= \frac{1}{2i} e^{-\frac{\widehat{\mu}}{2}}\left[(\cos F - i\sin F)\mathscr{H}\left\{e^{\frac{\widehat{\mu}}{2}+iF}\widehat{f_\mu}\right\} + (\cos F + i \sin F)\mathscr{H}\left\{e^{\frac{\widehat{\mu}}{2}-iF}\widehat{f_\mu}\right\}\right] \\
&= \frac{1}{2i} e^{-\frac{\widehat{\mu}}{2}}\left[\cos F\left(\mathscr{H}\left\{e^{\frac{\widehat{\mu}}{2}+iF}\widehat{f_\mu}\right\} + \mathscr{H}\left\{e^{\frac{\widehat{\mu}}{2}-iF}\widehat{f_\mu}\right\}\right)\right. \\
&\quad \left. - i\sin F\left(\mathscr{H}\left\{e^{\frac{\widehat{\mu}}{2}+iF}\widehat{f_\mu}\right\} - \mathscr{H}\left\{e^{\frac{\widehat{\mu}}{2}-iF}\widehat{f_\mu}\right\}\right)\right] \\
&= \frac{1}{2\pi i} e^{-\frac{\widehat{\mu}}{2}}\left[\cos F \oint_{-\infty}^{\infty} \frac{e^{\frac{\widehat{\mu}}{2}+iF} + e^{\frac{\widehat{\mu}}{2}-iF}}{r-\rho}\widehat{f_\mu}\mathrm{d}r - i \sin F \oint_{-\infty}^{\infty} \frac{e^{\frac{\widehat{\mu}}{2}+iF} - e^{\frac{\widehat{\mu}}{2}-iF}}{r-\rho}\widehat{f_\mu}\mathrm{d}r\right] \\
&= \frac{1}{2\pi i} e^{-\frac{\widehat{\mu}}{2}}\left[\cos F \oint_{-\infty}^{\infty} \frac{e^{\frac{\widehat{\mu}}{2}}(e^{iF} + e^{-iF})}{r-\rho}\widehat{f_\mu}\mathrm{d}r - i \sin F \oint_{-\infty}^{\infty} \frac{e^{\frac{\widehat{\mu}}{2}}(e^{iF} - e^{-iF})}{r-\rho}\widehat{f_\mu}\mathrm{d}r\right] \\
&= \frac{1}{2\pi i} e^{-\frac{\widehat{\mu}}{2}}\left[\cos F \oint_{-\infty}^{\infty} \frac{e^{\frac{\widehat{\mu}}{2}}(2\cos F)}{r-\rho}\widehat{f_\mu}\mathrm{d}r - i \sin F \oint_{-\infty}^{\infty} \frac{e^{\frac{\widehat{\mu}}{2}}(2i\sin F)}{r-\rho}\widehat{f_\mu}\mathrm{d}r\right]
\end{aligned}$$

$$= \frac{2}{i} e^{-\frac{\widehat{\mu}}{2}} \left[\cos F \left(\frac{1}{2\pi} \oint_{-\infty}^{\infty} \frac{e^{\frac{\widehat{\mu}}{2}} \cos F}{r - \rho} \widehat{f_\mu} \mathrm{d}r \right) + \sin F \left(\frac{1}{2\pi} \oint_{-\infty}^{\infty} \frac{e^{\frac{\widehat{\mu}}{2}} \sin F}{r - \rho} \widehat{f_\mu} \mathrm{d}r \right) \right]$$

$$= -2ie^{-\frac{\widehat{\mu}}{2}} \left[\cos(F) G^C + \sin(F) G^S \right],$$

with G, $G^C(\rho, \theta)$ and $G^S(\rho, \theta)$ defined in equations (74), (75a), and (75b), respectively. Therefore, equation (76) becomes

$$(L_\mu \widehat{f_\mu})(\rho, \theta) = -2iG(\rho, \theta), \tag{80}$$

where $G(\rho, \theta)$ is defined in equation (74). We emphasize the fact that equation (80) implies that the function $i(L_\mu \widehat{f_\mu})$ is real. Thus, equations (61b) and (80) yield

$$J(x_1, x_2, \theta) = -2i \left[e^{M(\tau, \rho, \theta)} G(\rho, \theta) \right]_{\substack{\tau = x_2 \sin\theta + x_1 \cos\theta \\ \rho = x_2 \cos\theta - x_1 \sin\theta}}. \tag{81}$$

Hence, using the identity (56), we can calculate the effect of the above operator on the function J:

$$\begin{aligned}(\partial_{x_1} - i\partial_{x_2}) J &= -2ie^{-i\theta} (\partial_\tau - i\partial_\rho) \left\{ e^M G \right\}_{\substack{\rho = x_2 \cos\theta - x_1 \sin\theta \\ \tau = x_2 \sin\theta + x_1 \cos\theta}} \\ &= -2ie^{-i\theta} \left[e^M (M_\tau - iM_\rho) G + e^M (G_\tau - iG_\rho) \right]_{\substack{\rho = x_2 \cos\theta - x_1 \sin\theta \\ \tau = x_2 \sin\theta + x_1 \cos\theta}} \\ &= -2e^{-i\theta} e^M \left[-i\mu G + M_\rho G + G_\rho \right]_{\substack{\rho = x_2 \cos\theta - x_1 \sin\theta \\ \tau = x_2 \sin\theta + x_1 \cos\theta}},\end{aligned} \tag{82}$$

where we have employed the identities

$$M_\tau(\tau, \rho, \theta) \Big|_{\substack{\rho = x_2 \cos\theta - x_1 \sin\theta \\ \tau = x_2 \sin\theta + x_1 \cos\theta}} = \mu(x_1, x_2) \quad \text{and} \quad G_\tau(\rho, \theta) = 0.$$

Inserting the operator $(\partial_{x_1} - i\partial_{x_2})$ inside the integral in the right-hand side of equation (61a), and combining equations (83) and (82), yields

$$f(x_1, x_2) = -\frac{1}{2\pi} \int_0^{2\pi} e^M \left[-i\mu G + M_\rho G + G_\rho \right] \Big|_{\substack{\rho = x_2 \cos\theta - x_1 \sin\theta \\ \tau = x_2 \sin\theta + x_1 \cos\theta}} \mathrm{d}\theta. \tag{83}$$

The first term of the integral on the right-hand side of equation (83) can be further simplified:

$$-i \int_0^{2\pi} \mu(x_1, x_2) \left[e^{M(\tau, \rho, \theta)} G(\rho, \theta) \right]_{\substack{\tau = x_2 \sin\theta + x_1 \cos\theta \\ \rho = x_2 \cos\theta - x_1 \sin\theta}} \mathrm{d}\theta = \frac{1}{2} \mu(x_1, x_2) \int_0^{2\pi} J(x_1, x_2, \theta) \mathrm{d}\theta. \tag{84}$$

Equation (2.9) of [4], with μ replaced by u, evaluated at $\lambda = 0$ implies

$$u(x_1, x_2, 0) = \frac{1}{2\pi} \int_0^{2\pi} J(x_1, x_2, \theta) \mathrm{d}\theta.$$

Moreover, the limit $\lambda \to 0$ of equation (2.2) of [4] yields

$$\frac{\partial u(x_1, x_2, 0)}{\partial \bar{z}} = 0,$$

which implies that u is analytic everywhere, including infinity, since u satisfies the boundary condition

$$u = O\left(\frac{1}{z}\right), \quad z \to \infty.$$

Thus, Liouville's theorem implies that the entire function u vanishes, hence

$$\int_0^{2\pi} J(x_1, x_2, \theta) \mathrm{d}\theta = 0. \tag{85}$$

Therefore, taking into account equation (85), equation (84) implies that

$$\int_0^{2\pi} \mu(x_1, x_2) \left[e^{M(\tau,\rho,\theta)} G(\rho, \theta) \right]_{\substack{\tau = x_2 \sin\theta + x_1 \cos\theta \\ \rho = x_2 \cos\theta - x_1 \sin\theta}} \mathrm{d}\theta = 0,$$

and hence, equation (61a) is equivalent to equation (73). □

6 The Attenuated Spline Reconstruction Technique

For the evaluation of all quantities involved in equation (73), we proceed as follows: for the computation of the function $M(\tau, \rho, \theta)$ we employ the Gauss-Legendre quadrature and for the computation of the functions $F(\rho, \theta)$ and $G(\rho, \theta)$ we employ splines [8]. For all the functions involved in the inversion of the attenuated Radon transform, we evaluate the solution to the inverse problem posed in (61) at a given reconstruction grid described by the points $(x_1^{(i)}, x_2^{(j)})$, for $i, j = 1, \ldots, n$.

6.1 The Evaluation of the Function M and Its Derivative M_ρ

Equation (61c) involves the computation of the integral of the given attenuation function $\mu(x_1, x_2)$ from $s = \tau^{(i)}$ to $s = \tau_f^{(i)}$, see Figure 2. However,

$$(\rho^{(i)})^2 + (\tau_f^{(i)})^2 \leq R^2,$$

where R denotes the radius of the support of functions f and μ. Therefore, M is given by

$$M(\tau^{(i)}, \rho^{(i)}, \theta_j) = \int_{\tau^{(i)}}^{\sqrt{R^2-(\rho^{(i)})^2}} \mu\left(s\cos\theta_j - \rho^{(i)}\sin\theta_j, s\sin\theta_j + \rho^{(i)}\cos\theta_j\right) ds. \tag{86}$$

The integral (86) can be computed using the Gauss-Legendre quadrature with two functional evaluations at every step, i.e.,

$$\int_\alpha^\beta f(s)ds \sim \frac{1}{2}(\beta - \alpha)[f(\tau_1) + f(\tau_2)],$$

where τ_1 and τ_2 are given by

$$\tau_1 = \alpha + \left(\frac{1}{2} - \frac{\sqrt{3}}{6}\right)(\beta - \alpha), \quad \tau_2 = \alpha + \left(\frac{1}{2} + \frac{\sqrt{3}}{6}\right)(\beta - \alpha).$$

For the evaluation of the derivative, $M_\rho(\tau, \rho, \theta)$, we employ a finite difference scheme, as in [9].

6.2 The Evaluation of the Function F

For the evaluation of the function F we adopt a similar approach as in [2], only this time, instead of evaluating the derivative of the Hilbert transform of $\widehat{\mu}(\rho, \theta)$, we evaluate the Hilbert transform of $\widehat{\mu}(\rho, \theta)$. Assuming that any function f involved in the inversion of the attenuated Radon transform, with $f(\rho, \theta)$, $-1 \leq \rho \leq 1$, $0 \leq \theta \leq 2\pi$, is given for every θ at the points $\{\rho_i\}_1^n$. We denote the value of f at ρ_i by f_i, i.e.,

$$f_i(\theta) = f(\rho_i, \theta), \quad \rho_i \in [-1, 1], \quad 0 \leq \theta < 2\pi, \quad i = 1, \ldots, n. \tag{87}$$

We also assume that the function f, along with its derivative with respect to ρ, f_ρ, vanish at the endpoints $\rho_1 = -1$ and $\rho_n = 1$:

$$f(-1,\theta)=f(1,\theta)=0,\quad 0\le\theta<2\pi, \tag{88a}$$

$$f_\rho(-1,\theta)=f_\rho(1,\theta)=0,\quad 0\le\theta<2\pi. \tag{88b}$$

For all $i=1,\ldots,n-1$ in the interval $\rho_i\le\rho\le\rho_{i+1}$ we approximate f by cubic splines, $S_i^{(3)}(\rho,\theta)$, in the variable ρ:

$$f(\rho,\theta)\sim S_i^{(3)}(\rho,\theta)\quad \rho_i\le\rho\le\rho_{i+1},\quad 0\le\theta<2\pi.$$

The cubic spline defined above interpolates the function $f(\rho,\theta)$ at the points (or knots) $\{\rho_i\}_{i=1}^n$

$$S_i^{(3)}(\rho_i,\theta)=f_i(\theta),\quad i=1,\ldots,n-1. \tag{89}$$

Furthermore,

$$S_i^{(3)}(\rho,\theta)=\sum_{j=0}^{3}c_i^{(j)}\rho^j\quad \rho_i\le\rho\le\rho_{i+1},\quad 0\le\theta<2\pi, \tag{90}$$

where the constants $\{c_i^{(j)}\}_{i=1}^n$ for $j=0,\ldots,3$, are given by, see [9]:

$$c_i^{(0)}=\frac{\rho_{i+1}f_i-\rho_i f_{i+1}}{\Delta_i}+\frac{f_i''}{6}\left(-\rho_{i+1}\Delta_i+\frac{\rho_{i+1}^3}{\Delta_i}\right)+\frac{f_{i+1}''}{6}\left(\rho_i\Delta_i-\frac{\rho_i^3}{\Delta_i}\right), \tag{91a}$$

$$c_i^{(1)}=\frac{f_{i+1}-f_i}{\Delta_i}-\frac{f_i''}{6}\left(-\Delta_i+\frac{3\rho_{i+1}^2}{\Delta_i}\right)+\frac{f_{i+1}''}{6}\left(-\Delta_i+\frac{3\rho_i^2}{\Delta_i}\right), \tag{91b}$$

$$c_i^{(2)}=\frac{1}{2\Delta_i}(\rho_{i+1}f_i''-\rho_i f_{i+1}''), \tag{91c}$$

$$c_i^{(3)}=\frac{f_{i+1}''-f_i''}{6\Delta_i}, \tag{91d}$$

$$\Delta_i=\rho_{i+1}-\rho_i, \tag{92}$$

with f_i'' denoting the second derivative of $f(\rho,\theta)$ with respect to ρ evaluated at ρ_i, i.e.,

$$f_i''=\left.\frac{\partial^2 f(\rho,\theta)}{\partial\rho^2}\right|_{\rho=\rho_i},\quad i=1,\ldots,n. \tag{93}$$

Equation (90) implies

$$\oint_{\rho_i}^{\rho_{i+1}} \frac{f(r,\theta)}{r-\rho}\mathrm{d}r \sim \sum_{j=0}^{3} c_i^{(j)} I_i^{(j)}(\rho), \tag{94}$$

where

$$I_i^{(j)}(\rho) = \oint_{\rho_i}^{\rho_{i+1}} \frac{r^j}{r-\rho}\mathrm{d}r, \quad j = 0, \ldots, 3. \tag{95}$$

For the integrals defined in equation (95), the following identities are valid:

$$I_i^{(0)}(\rho) = \ln\left|\frac{\rho_{i+1}-\rho}{\rho_i-\rho}\right|, \tag{96a}$$

$$I_i^{(1)}(\rho) = \Delta_i + \rho I_i^{(0)}(\rho), \tag{96b}$$

$$I_i^{(2)}(\rho) = \frac{1}{2}\left(\rho_{i+1}^2 - \rho_i^2\right) + \Delta_i \rho + \rho^2 I_i^{(0)}(\rho), \tag{96c}$$

$$I_i^{(3)}(\rho) = \frac{1}{3}\left(\rho_{i+1}^3 - \rho_i^3\right) + \frac{1}{2}\left(\rho_{i+1}^2 - \rho_i^2\right)\rho + \Delta_i \rho^2 + I_i^{(0)}(\rho)\rho^3. \tag{96d}$$

In equation (94), substituting equations (95) and (96) yields

$$\oint_{\rho_i}^{\rho_{i+1}} \frac{f(r,\theta)}{r-\rho}\mathrm{d}r \sim \alpha_i(\theta) + \beta_i(\theta)\rho + \gamma_i(\theta)\rho^2 + \left(\sum_{j=0}^{3} c_i^{(j)}\rho^j\right) I_i^{(0)}(\rho), \tag{97}$$

where

$$\alpha_i(\theta) = c_i^{(1)}(\theta)\Delta_i + \frac{1}{2}c_i^{(2)}\left(\rho_{i+1}^2 - \rho_i^2\right) + \frac{1}{3}c_i^{(3)}\left(\rho_{i+1}^3 - \rho_i^3\right), \tag{98}$$

$$\beta_i(\theta) = \left(c_i^{(2)}(\theta)\Delta_i + \frac{1}{2}c_i^{(3)}(\theta)\left(\rho_{i+1}^2 - \rho_i^2\right)\right), \tag{99}$$

$$\gamma_i(\theta) = c_i^{(3)}(\theta)\Delta_i. \tag{100}$$

The above expressions can be further simplified, taking into account equations (91), as follows:

$$\alpha_i(\theta) = (f_{i+1} - f_i) - \frac{1}{36}[17\rho_{i+1}^2 - 19\rho_{i+1}\rho_i + 8\rho_i^2]f_i'' - \frac{1}{36}[4\rho_{i+1}^2 - 5\rho_{i+1}\rho_i - 5\rho_i^2]f_{i+1}'', \tag{101}$$

$$\beta_i(\theta) = \frac{1}{12}\left[\left(5\rho_{i+1} - \rho_i\right) f_i'' - \left(5\rho_i - \rho_{i+1}\right) f_{i+1}''\right], \tag{102}$$

$$\gamma_i(\theta) = \frac{f_{i+1}'' - f_i''}{6}. \tag{103}$$

Employing the integral identity

$$\oint_{-1}^{1} \frac{f(r,\theta)}{r-\rho} \mathrm{d}r = \sum_{i=1}^{n-1} \oint_{\rho_i}^{\rho_{i+1}} \frac{f(r,\theta)}{r-\rho} \mathrm{d}r, \tag{104}$$

in equation (97) yields

$$\oint_{-1}^{1} \frac{f(r,\theta)}{r-\rho} \mathrm{d}r \sim A(\theta) + B(\theta)\rho + \frac{1}{6}(f_n'' - f_1'')\rho^2 + \sum_{i=1}^{n-2} \left[S_i^{(3)}(\rho,\theta) - S_{i+1}^{(3)}(\rho,\theta)\right] \ln|\rho_{i+1} - \rho|$$
$$+ S_{n-1}^{(3)}(\rho,\theta) \ln|\rho_n - \rho| - S_1^{(3)}(\rho,\theta) \ln|\rho_1 - \rho|, \tag{105}$$

where

$$A(\theta) = \sum_{i=1}^{n-1} \alpha_i(\theta) \quad \text{and} \quad B(\theta) = \sum_{i=1}^{n-1} \beta_i(\theta). \tag{106}$$

The right-hand side of equation (105) involves the known functions $\{f_i\}_1^n$ as well as the unknown functions $\{f_i''\}_1^n$. Denoting the first derivative of the cubic spline, $S_i^{(3)}$, by $S_i^{(2)}$, where the superscript denotes that $S_i^{(2)}$ is quadratic, implies

$$S_i^{(2)}(\rho,\theta) = \frac{\partial S_i^{(3)}(\rho,\theta)}{\partial \rho} = c_i^{(1)}(\theta) + 2c_i^{(2)}(\theta)\rho + 3c_i^{(3)}(\theta)\rho^2. \tag{107}$$

For the computation of $\{f_i''\}_1^n$ we follow the method of [2], based on the continuity of the first derivative of the cubic spline. More specifically, we solve the system of the following n equations:

$$S_i^{(2)}(\rho_{i+1},\theta) = S_{i+1}^{(2)}(\rho_{i+1},\theta), \ i = 1,\ldots,n-2, \ 0 \le \theta < 2\pi \tag{108a}$$

$$S_1^{(2)}(\rho_1,\theta) = S_{n-1}^{(2)}(\rho_n,\theta) = 0. \tag{108b}$$

The continuity of the spline (see equations (88a)) implies that the points $\{\rho_i\}_1^n$ are *removable* logarithmic singularities. The computation of $F(\rho,\theta)$ is concluded by using equation (105) with f replaced by $\widehat{\mu}$.

6.3 The Evaluation of the Function G and Its Derivative G_ρ

Define the functions f^C and f^S by

$$f^C(\rho,\theta) = e^{\frac{1}{2}\widehat{\mu}(\rho,\theta)}\cos(F(\rho,\theta))\widehat{f_\mu}(\rho,\theta), \quad 0 \leq \theta < 2\pi, \quad -1 \leq \rho \leq 1, \tag{109a}$$

$$f^S(\rho,\theta) = e^{\frac{1}{2}\widehat{\mu}(\rho,\theta)}\sin(F(\rho,\theta))\widehat{f_\mu}(\rho,\theta), \quad 0 \leq \theta \leq 2\pi, \quad -1 \leq \rho \leq 1, \tag{109b}$$

where the functions $\widehat{\mu}$, $\widehat{f_\mu}$, and F are defined in equations (61e), (60), and (72), respectively. We assume that the attenuated Radon transform of f, $\widehat{f_\mu}(\rho,\theta)$, is given at the points $\{\rho_i\}_1^n$. Computing the functions $\widehat{\mu}$ and F at these exact points enables us to compute the functions f^C and f^S at the same points. Therefore, using equation (105), and replacing f by f^C and f^S, we are able to compute G^C and G^S, respectively. Furthermore, if we require that both f^C and f^S vanish at the endpoints

$$f^C(-1,\theta) = f^C(1,\theta) = 0, \tag{110a}$$

$$f^S(-1,\theta) = f^S(1,\theta) = 0, \tag{110b}$$

then we may eliminate the logarithmic singularities of both G^C and G^S. The equations above are valid provided that the attenuated Radon transform vanishes at the endpoints in the variance ρ, i.e.,

$$\widehat{f_\mu}(-1,\theta) = \widehat{f_\mu}(1,\theta) = 0, \tag{111}$$

which is a valid conclusion, under the assumption that the attenuated Radon transform has finite support, which is the case in medical imaging. Combining G^C and G^S, we may compute $G(\rho,\theta)$ as in equation (74). For the numerical evaluation of the derivative of G, G_ρ, we employ an appropriate finite difference scheme.

Acknowledgements This work was partially supported by the research programme "Inverse Problems and Medical Imaging" (200/905) of the Research Committee of the Academy of Athens. G.A. Kastis acknowledges the financial support of the "Alexander S. Onassis Public Benefit Foundation" under grant N° R ZL 001-0. A.S. Fokas has been supported by EPSRC, UK in the form of a senior fellowship.

References

1. J. Radon, Über die bestimmung von funktionen durch ihre integralwerte längs gewisser mannigfaltigkeiten. Akad. Wiss. **69**, 262–277 (1917)
2. G.A. Kastis, D. Kyriakopoulou, A. Gaitanis, Y. Fernández, B.F. Hutton, A.S. Fokas, Evaluation of the spline reconstruction technique for PET. Med. Phys. **41**(4), 042501 (2014). https://doi.org/10.1118/1.4867862

3. P. Kuchment, *The Radon Transform and Medical Imaging*. CBMS-NSF Regional Conference Series in Applied Mathematics (Society for Industrial and Applied Mathematics, Philadelphia, 2014)
4. A. Fokas, A. Iserles, V. Marinakis, Reconstruction algorithm for single photon emission computed tomography and its numerical implementation. J. R. Soc. Interface **3**(6), 45–54 (2006). https://doi.org/10.1098/rsif.2005.0061
5. A. Fokas, R. Novikov, Discrete analogues of δ-equation and of Radon transform. Comptes Rendus Acad. Sci. Série 1, Mathématique **313**(2), 75–80 (1991)
6. R. Novikov, An inversion formula for the attenuated X-ray transformation. Ark. Mat. **40**(1), 145–167 (2002). https://doi.org/10.1007/BF02384507
7. N.I. Muskhelishvili, *Singular Integral Equations* (P. Nordhoff, Groningen, 1953)
8. M. Unser, J. Fageot, J.P. Ward, Splines are universal solutions of linear inverse problems with generalized TV regularization. SIAM Rev. **59**(4), 769–793 (2017). https://doi.org/10.1137/16M1061199
9. W.H. Press, S.A. Teukolsky, W.T. Vetterling, B.P. Flannery, *Numerical Recipes*, 3rd edn. The Art of Scientific Computing (Cambridge University Press, New York, 2007)

On Algorithms for Difference of Monotone Operators

Maede Ramazannejad, Mohsen Alimohammady, and Carlo Cattani

Abstract This review proposes a proximal algorithm for difference of two monotone operators in finite dimensional real Hilbert space. Our route begins with reviewing some properties of DC (difference of convex functions) programming and DCA (DC algorithms). Next, we recall some main results about a proximal point algorithm for DC programming.

1 Introduction

An important area of nonlinear analysis that emerged in the early 1960s [21–23, 40] is theory of monotone operators. During nearly six decades, this field has reached a high level of maturity. Application of maximal monotone operators in some branches such as optimization, variational analysis, algorithms, mathematical economics is the reason why it has grown dramatically.

One of the most challenging issues in theory of monotone operators is the problem of difference of monotone operators, because difference of two monotone operators is not always monotone. For this reason finding a zero of difference of two monotone operators has not been studied extensively. You can see the presence of this problem in DC programming, signal processing, machine learning, tomography, molecular biology, and optimization [8, 11, 25, 30, 32, 50, 57, 63].

M. Ramazannejad
Young Researchers and Elite Club, Islamic Azad University, Qaemshahr, Iran

M. Alimohammady
Department of Mathematics, Faculty of Mathematical Sciences, University of Mazandaran, Babolsar, Iran
e-mail: amohsen@umz.ac.ir

C. Cattani (✉)
Engineering School (DEIM), University of Tuscia, Viterbo, Italy
e-mail: cattani@unitus.it

T. M. Rassias, P. M. Pardalos (eds.), *Mathematical Analysis and Applications*, Springer Optimization and Its Applications 154,
https://doi.org/10.1007/978-3-030-31339-5_17

The main objective in this study is to offer a proximal algorithm for difference of two maximal monotone operators. It is the final section in our study but in the first step we try to review the features of the problem of minimizing difference of convex functions and the presented algorithms for this problem because finding the critical points of DC functions is a special case for finding the zeros of difference of two maximal monotone operators. The second step is about recalling a kind of DC programming and recommended proximal point algorithm for this.

2 DC Programming and DCA

This section is a small part of efforts of Pham Dinh Tao and Le Thi Hoai An in the way of development of DC programming and its applications in [50]. It is noteworthy that Pham Dinh Tao introduced DC programming and DCA in 1985 in their preliminary form.

The problem

$$\inf\{g(x) - h(x) : x \in \mathbb{R}^n\},$$

where g and h are convex functions, is called a *DC program.*

The importance of this problem in formulating of all most real-world optimization problems as DC programs cannot be exaggerated. The authors believe the main idea in convex analysis approach to DC programming is the use of the DC duality, which has been first studied by Toland in 1979 [62]. Though DC algorithms converge to a local solution generally, it is found from many literatures [3, 5–7, 9, 10, 49, 51] that DC algorithms converge quite often to a global solution. We see the different equivalent DC forms for primal and dual problems, because there is such an awareness that decompositions of the DC objective function may have an important influence on the qualities (robustness, stability, rate of convergence, and globability of sought solutions). In this regard, an interesting thing for guaranteeing of the convergence of DCA to a global solution is to check conditions on the choice of the DC decomposition and the initial point. We refer the reader to [3, 5–7, 9, 10, 49, 51] on applications of DCA in the study of many large-scale DC optimization problems and observing that DCA is more robust and efficient than related standard methods. Applying conjointly appropriate DC decompositions of convex functions and proximal regularization techniques [3, 47, 48] leads to proximal point algorithms [37, 55].

In this section, we want to express three outcomes of the main purposes in the paper of Pham and An [50]. To start, we will say the relationship between primal and dual solutions. We will continue argument by presenting of duality and local optimality conditions for DC optimization. Finally, the description of DCA will be reviewed.

We are now in a position to introduce some concepts and results about DC program, DCA, and their features.

Let $X = \mathbb{R}^n$, so the space X^* that is continuous dual of X can be identified with X. We assume X with norm $\|.\| = \langle ., . \rangle^{\frac{1}{2}}$ and that X and X^* are paired by $\langle ., . \rangle$. For an extended real value function $f : X \to \mathbb{R} \cup \{+\infty\}$, the domain of f is the set

$$\operatorname{dom} f = \{x \in \mathbb{R}^n : f(x) < +\infty\}.$$

The function f is said to be proper if its domain is nonempty.

If $B \subseteq X$, then the interior of B will be denoted by $\operatorname{int} B$ and if B is convex, then the relative interior of B will be denoted by $\operatorname{ri} B$.

Let $\Gamma_0(X)$ denote the set of all proper lower semicontinuous convex functions on X. The *indicator function* of subset C of X, written as $\chi_C(x)$, is defined at $x \in X$ by

$$\chi_C(x) = \begin{cases} 0 & x \in C, \\ +\infty & \text{otherwise.} \end{cases}$$

For a given function $f \in \Gamma_0(X)$,

$$f^*(x^*) = \sup\{\langle x, x^* \rangle - f(x) : x \in X\},$$

is *conjugate of the function* f which is a member of $\Gamma_0(X^*)$. Given a proper convex function f on X, $f : X \to \mathbb{R} \cup \{+\infty\}$, the *subdifferential* of f at x is given by

$$\partial f(x) = \{u \in X^* : f(y) \geq f(x) + \langle u, y - x \rangle \text{ for all } y \in X\}.$$

Also, $\operatorname{dom}\partial f = \{x \in X : \partial f(x) \neq \emptyset\}$ and $\operatorname{range}\partial f = \cup\{\partial f(x) : x \in \operatorname{dom}\partial f\}$. The *ε-subdifferential* of f was introduced by Brøndsted and Rockafellar in [20]. It is defined as:

$$\partial_\varepsilon f(x) := \{v \in X^* : f(y) \geq f(x) + \langle v, y - x \rangle - \varepsilon \text{ for all } y \in X\},$$

for any $\varepsilon \geq 0$, $x \in X$. Note that $\partial f(x) \subseteq \partial_\varepsilon f(x)$ for any $\varepsilon \geq 0$, $x \in X$. Here the convention $+\infty - (+\infty) = +\infty$ is followed.

Now we investigate the DC program

$$\text{(P)} \qquad \alpha = \inf\{f(x) := g(x) - h(x) : x \in X\},$$

where g and h belong to $\Gamma_0(X)$. Such a function f is called DC function on X and g, h are called its DC components. If g and h are in addition finite on all of X, then we say that $f = g - h$ is finite DC function on X.

In [61], we observe that

$$\inf\{g(x) - h(x) : x \in X\} = \inf\{h^*(x^*) - g^*(x^*) : x^* \in X^*\}.$$

With such an equality in hand we are ready to start main argument. The problem

$$\text{(D)} \qquad \alpha = \inf\{h^*(x^*) - g^*(x^*) : x^* \in X^*\},$$

is called *dual problem*. The attentive reader can see the perfect symmetry between primal and dual programs (P) and (D). Equivalence between problems (P) and (D) makes the reader relaxed to solve the easier problem.

Notice that since α is finite:

$$\mathrm{dom}g \subset \mathrm{dom}h \quad \text{and} \quad \mathrm{dom}h^* \subset \mathrm{dom}g^*. \tag{1}$$

We consider assumption (1) throughout this section.

A point $\bar{x}$ is called *local minimizer* of $g - h$ if $g(\bar{x}) - h(\bar{x})$ is finite, in other words $\bar{x} \in \mathrm{dom}g \cap \mathrm{dom}h$, and there exists a neighborhood U of $\bar{x}$ such that

$$g(\bar{x}) - h(\bar{x}) \leq g(x) - h(x), \quad \forall x \in U. \tag{2}$$

Under the convention $+\infty - (+\infty) = +\infty$, the property (2) is equivalent to

$$g(\bar{x}) - h(\bar{x}) \leq g(x) - h(x), \quad \forall x \in U \cap \mathrm{dom}g.$$

We say that $\bar{x}$ is a critical point of $g - h$ if $\partial g(\bar{x}) \cap \partial h(\bar{x}) \neq \emptyset$.

Here $\mathscr{P}$ and $\mathscr{D}$ denote the solution sets of problems (P) and (D), respectively. Let

$$\mathscr{P}_l = \{x \in X : \partial h(x) \subset \partial g(x)\},$$
$$\mathscr{D}_l = \{x^* \in X^* : \partial g^*(x^*) \subset \partial h^*(x^*)\}.$$

The first result to present is the following.

Theorem 1 ([3, 31, 46, 47])

(i) $x \in \mathscr{P}$ *if and only if* $\partial_\varepsilon h(x) \subset \partial_\varepsilon g(x) \quad \forall \varepsilon > 0$.
(ii) Dually, $x^* \in \mathscr{D}$ *if and only if* $\partial_\varepsilon g^*(x^*) \subset \partial_\varepsilon h^*(x^*) \quad \forall \varepsilon > 0$.
(iii) $\cup\{\partial h(x) : x \in \mathscr{P}\} \subset \mathscr{D} \subset \mathrm{dom}h^*$.
The first inclusion becomes equality if g^* *is subdifferentiable in* $\mathscr{D}$ *(in particular if* $\mathscr{D} \subset ri(\mathrm{dom}g^*)$ *or if* g^* *is subdifferentiable in dom* h^**).*
In this case $\mathscr{D} \subset (\mathrm{dom}\partial g^* \cap \mathrm{dom}\partial h^*)$.
(iv) $\cup\{\partial g^*(x^*) : x^* \in \mathscr{D}\} \subset \mathscr{P} \subset \mathrm{dom}g$.

The proof of the properties (i) and (ii) appears in Hiriart-Urruty's paper [31], based on the behavior of the ε-directional derivative of a convex function as a function of the parameters ε. Also the ideas behind the proofs of properties (iii) and (iv) are due to D.T. Pham and L. T. H. An. These proofs are based on the theory of subdifferential for convex functions [3, 46, 47].

The property (i) is a difficult way to reach the global solution to problem (P). You can analyze the meaningful relationships between primal and dual solutions in (iii) and (iv). One noteworthy conclusion to be drawn from (ii) and (iv) above is the fact that solving the problem (P) leads to solving the problem (D) and vice versa.

The fundamental results linking duality and local optimality conditions for DC program are the following; proofs will not be presented. These beautiful results are cited of Pham and An [50].

Theorem 2

(i) [3, 31, 46, 62] If $\bar{x}$ is a local minimizer of $g - h$, then $\bar{x} \in \mathscr{P}_l$.
(ii) Let $\bar{x}$ be a critical point of $g - h$ and $\bar{y}^* \in \partial g(\bar{x}) \cap \partial h(\bar{x})$. Let U be a neighborhood of $\bar{x}$ such that $U \cap domg \subset dom\partial h$. If for any $x \in U \cap domg$ there is $y^* \in \partial h(x)$ such that

$$h^*(y^*) - g^*(y^*) \geq h^*(\bar{y}^*) - g^*(\bar{y}^*),$$

then $\bar{x}$ is a local minimizer of $g - h$. More precisely,

$$g(x) - h(x) \geq g(\bar{x}) - h(\bar{x}), \quad \forall x \in U \cap domg.$$

Corollary 1 (Sufficient Local Optimality) *Let $\bar{x}$ be a point that admits a neighborhood U such that $\partial h(x) \cap \partial g(\bar{x}) \neq \emptyset$, $\forall x \in U \cap domg$. Then $\bar{x}$ is a local minimizer of $g - h$. More precisely,*

$$g(x) - h(x) \geq g(\bar{x}) - h(\bar{x}), \quad \forall x \in U \cap domg.$$

Corollary 2 (Sufficient Strict Local Optimality) *If $\bar{x} \in int(dom\, h)$ verifies $\partial h(\bar{x}) \subset int(\partial g(\bar{x}))$, then $\bar{x}$ is a strict local minimizer of $g - h$.*

Corollary 3 (DC Duality Transportation of a Local Minimizer) *Let $\bar{x} \in dom\partial h$ be a local minimizer of $g - h$ and let $\bar{y}^* \in \partial h(\bar{x})$, i.e. $\partial h(\bar{x})$ is nonempty, and $\bar{x}$ admits a neighborhood U such that*

$$g(x) - h(x) \geq g(\bar{x}) - h(\bar{x}), \quad \forall x \in U \cap domg.$$

If

$$\bar{y}^* \in int(domg^*) \quad and \quad \partial g^*(\bar{y}^*) \subset U, \tag{3}$$

((3) holds if g^* is differentiable at $\bar{y}^*$), then $\bar{y}^*$ is a local minimizer of $h^* - g^*$.

Now we are ready to describe DCA for general DC programs. How? Well consider the problem

$$(S(x)) \quad \inf\{h^*(y^*) - g^*(y^*) : \; y^* \in \partial h(x)\},$$

for any fixed $x \in X$. By the conjugate definition, this problem is equivalent to

$$\inf\{\langle x, y^* \rangle - g^*(y^*) : \ y^* \in \partial h(x)\}. \tag{4}$$

Also consider the problem

$$(T(y^*)) \qquad \inf\{g(x) - h(x) : \ x \in \partial g^*(y^*)\},$$

for any fixed $y^* \in X^*$. For the same reason as (4), the above problem is equivalent to

$$\inf\{\langle x, y^* \rangle - h(x) : \ x \in \partial g^*(y^*)\}. \tag{5}$$

Suppose $\mathscr{S}(x)$ and $\mathscr{T}(y^*)$ denote the solution sets of problems $(S(x))$ and $(T(y^*))$, respectively.

There is a face of DCA based on problems $(S(x))$ and $(T(y^*))$ which with an initial point $x_0 \in \mathrm{dom}g$ makes two sequences $\{x_k\}$ and $\{y_k^*\}$ such that

$$y_k^* \in \mathscr{S}(x_k), \quad x_{k+1} \in \mathscr{T}(y_k^*). \tag{6}$$

However the problems $S(x_k)$ and $(T(y_k^*))$ are simpler than (P) and (D), they are practically hard to solve. Therefore authors in [50] offered the simplified form of DCA as follows:

Start from any point $x_0 \in \mathrm{dom}g$ and consider the recursive process

$$y_k^* \in \partial h(x_k); \quad x_{k+1} \in \partial g^*(y_k^*).$$

This method aims at providing two sequences $\{x_k\}$ and $\{y_k^*\}$ which are easy to calculate and satisfy:

- the sequences $(g - h)(x_k)$ and $(h^* - g^*)(y_k^*)$ are decreasing;
- every limit point x (resp. y^*) of the sequence $\{x_k\}$ (resp. $\{y_k^*\}$) is a critical point of $g - h$ (resp. $h^* - g^*$).

Whenever DCA produces the sequences $\{x_k\}$ and $\{y_k^*\}$ as described above, so

$$\{x_k\} \subset \mathrm{range}\partial g^* = \mathrm{dom}\partial g \quad \text{and} \quad \{y_k^*\} \subset \mathrm{range}\partial h = \mathrm{dom}\partial h^*.$$

Hence obviously sequences $\{x_k\}$ and $\{y_k\}$ in DCA are well defined if and only if

$$\mathrm{dom}\partial g \subset \mathrm{dom}\partial h \quad \text{and} \quad \mathrm{dom}\partial h^* \subset \mathrm{dom}\partial g^*.$$

You can find the convergence of DCA for general DC programs in [50, Theorem 3].

3 Proximal Point Algorithm for DC Program

The heart and soul of this section are each devoted to recalling one kind of proximal point algorithm for minimizing the difference of a nonconvex function and a convex function introduced by An and Nam [4]. We quote several of their results and urge the reader to carefully study their original paper.

The structure of the problem of this section is flexible enough to include the problem of minimizing a smooth function on a closed set or minimizing a DC function. On doing so, we will study the convergence results of this algorithm.

The first work regarding the proximal minimization algorithm is due to Martinet [38, 39]. Proximal minimization was extended to the general proximal point algorithm for finding the zero of an arbitrary maximal monotone operator by Rockafellar [55]. There are many papers on applying different proximal algorithms to special problems, such as loss minimization in machine learning [15, 19, 29, 33], optimal control [44], energy management [34], and signal processing [28].

Neal Parikh and Stephen Boyd in their monograph [45] mentioned some reasons of many reasons to study proximal algorithms. Here, we recall them. First, these algorithms work under extremely general conditions, including cases where the functions are nonsmooth and extended real valued. Second, they can be fast, since there can be simple proximal operators for functions that are otherwise challenging to handle in an optimization problem. Third, they are amenable to distributed optimization, so they can be used to solve very large-scale problems. Finally, they are often conceptually and mathematically simple, so they are easy to understand, derive, and implement for a particular problem. Indeed, many proximal algorithms can be interpreted as generalizations of other well-known and widely used algorithms.

Consider concepts and notations as described in previous section but some notations, special to the present discussion, ought to be introduced.

Let f be a function from X into an extended real line $\mathbb{R} \cup \{+\infty\}$, finite at x. A set

$$\partial^F f(x) = \{x^* \in X^* : \liminf_{u \to x} \frac{f(u) - f(x) - \langle x^*, u - x\rangle}{\|u - x\|} \geq 0\}, \tag{7}$$

is called *Fréchet subdifferential* of f at x. Its elements are sometimes referred to as Fréchet subgradients. If $x \notin \operatorname{dom} f$, then we set $\partial^F f(x) = \emptyset$.

The set (7) is closed and convex, but Fréchet subdifferential mapping does not have a closed graph.

Employing a limiting "robust regularization" procedure over the subgradient mapping $\partial^F f(.)$ leads us to the subdifferential of f at x defined by

$$\partial^L f(x) := \limsup_{x_k \xrightarrow{f}} \partial^F f(x) = \{x^* \in X^* : \exists x_k \xrightarrow{f} x \text{ and } x_k^* \in \partial^F f(x_k) \text{ with } x_k^* \to x\} \tag{8}$$

where $x_k \xrightarrow{f} x$ means that $x_k \to x$ and $f(x_k) \to f(x)$. Set (8) which is closed is called *limiting/Mordukhovich subdifferential* of f at x.

It is easy to see that $\partial^F f(x) \subset \partial^L f(x)$ for every $x \in X$. If the function f is differentiable, then the Fréchet subdifferential reduces to the derivative. The limiting subdifferential of f at x reduces to $\{\nabla f(x)\}$ if f is continuously differentiable on a neighborhood of x. Also the Fréchet and the limiting subdifferential coincide with the subdifferential in the sense of convex analysis if f is convex.

It needs be remarked that for a nonempty subset K of X, the normal cone or normality operator N_K of K is defined as

$$N_K(x) = \begin{cases} \{x^* \in X^* : \langle x^*, u - x\rangle \leq 0 \;\; \forall u \in K\} & \text{if } x \in K, \\ \emptyset & \text{otherwise.} \end{cases}$$

We can find in [24, Proposition 3.6.2], for a nonempty, closed, and convex subset $K \subset X$, the following properties hold:

(a) $N_K = \partial \chi_K$.
(b) $N_K(x)$ is closed and convex for all $x \in X$.
(c) $N_K(x)$ is a cone for all $x \in K$.

For this K now let $P_K : X \to K$ the *orthogonal projection* onto K. It is easy to check that

$$P_K(x) = (I + N_K)^{-1}(x) = \{u \in K : \|x - u\| = \text{dist}(x; K)\},$$

where I is an identity map and $\text{dist}(x; K)$ is used to denote the distance from x to K, i.e. $\text{dist}(x; K) = \inf_{x \in K} \|x - u\|$.

Clarke, in [26], showed that for locally Lipschitz functions, the *Clarke subdifferential* admits the simple presentation

$$\partial^C f(x) = co\, \partial^L f(x),$$

where $co\ K$ is the convex hull of an arbitrary set $K \subset X$. Some important consequences of this are the following.

Proposition 1 ([56, Exercise 8.8]) *Let $f = g + h$, where g is lower semicontinuous and h is continuously differentiable on a neighborhood of x. Then*

$$\partial^F f(x) = \partial^F g(x) + \nabla h(x) \quad \text{and} \quad \partial^L f(x) = \partial^L g(x) + \nabla h(x).$$

Proposition 2 ([56, Theorem 10.1]) *If a lower semicontinuous function $f : X \to \mathbb{R} \cup \{+\infty\}$ has a local minimum at $\bar{x} \in dom f$, then $0 \in \partial^F f(\bar{x}) \subset \partial^L f(\bar{x})$. In the convex case, this condition is not only necessary for a local minimum but also sufficient for a global minimum.*

We state the definition of the *Kurdyka-Łojasiewicz property* from [12, 13].

The lower semicontinuous function $f : X \to \mathbb{R} \cup \{\infty\}$ has the Kurdyka-Łojasiewicz property at $\bar{x} \in \text{dom}\partial^L f$, if there exist $\eta \in (0, \infty]$, a neighborhood U of $\bar{x}$, and a continuous concave function $\varphi : [0, \eta) \to \mathbb{R}_+$ such that $\varphi(0) = 0$, $\varphi \in C^1(0, \eta)$, for all $s \in (0, \eta)$ it is $\varphi'(s) > 0$, and for all $x \in U \cap [f(\bar{x}) < f < f(\bar{x}) + \eta]$ the Kurdyka-Łojasiewicz inequality holds, i.e.,

$$\varphi'(f(x) - f(\bar{x}))\, \text{dist}(0, \partial^L f(x)) \geq 1.$$

We cannot present the results of [4] without recalling the concept of proximal mapping [56]. Let $g : X \to \mathbb{R} \cup \{+\infty\}$ be a proper lower semicontinuous function. The *Moreau proximal mapping* with regularization parameter $t > 0$, $\text{prox}_t^g(x)$, is defined by

$$\text{prox}_t^g = \text{argmin}_{u \in X}\{g(u) + \frac{t}{2}\|u - x\|^2\}.$$

When g is the indicator function χ_K, where K is a closed nonempty convex set, the proximal mapping of g reduces to projection mapping.

The following proposition shows us some conditions under which prox_t^g is well defined.

Proposition 3 ([18]) *If* $g : X \to \mathbb{R} \cup \{+\infty\}$ *be a proper lower semicontinuous function with* $\inf_{x \in X} g(x) > -\infty$, *then for every* $t \in (0, +\infty)$, *the set* $\text{prox}_t^g(x)$ *is nonempty and compact for every* $x \in X$.

Authors in [4] concentrated their attention on the convergence analysis of a proximal point algorithm for solving nonconvex optimization problems of the following type

$$\min\{f(x) = g_1(x) + g_2(x) - h(x) : x \in X\}, \tag{9}$$

where $g_1 : X \to \mathbb{R} \cup \{+\infty\}$ is proper and lower semicontinuous, $g_2 : X \to \mathbb{R}$ is differentiable with L-Lipschitz gradient, and $h : X \to \mathbb{R}$ is convex. The problem (9) is so flexible which it can be changed to the DC problem

$$\min\{f(x) = g(x) - h(x) : x \in X\}, \tag{10}$$

where $g \in \Gamma_0(X)$ and $h : X \to \mathbb{R}$ is convex. Also the problem (9) includes the following problem on a closed constraint set,

$$\min\{g(x) : x \in K\}. \tag{11}$$

There exists a meaningful relationship between the Moreau proximal operator $\text{prox}_{\frac{1}{\lambda}}^g$ and the subdifferential operator ∂g as follows:

$$\text{prox}_{\frac{1}{\lambda}}^g = (I + \lambda \partial g)^{-1}. \tag{12}$$

The right-hand side of (12) is called *resolvent operator* for ∂f with parameter $\lambda > 0$. You see that the proximal operator is the resolvent of the subdifferential operator. Regarding the problem (11), the *proximal point algorithm* which is also called *proximal minimization algorithm* or *proximal iteration* is

$$x_{k+1} := \text{prox}^{g}_{\frac{1}{\lambda}}(x_k), \tag{13}$$

where k is the iteration counter and x_k is the kth iterate of the algorithm.

We are ready for recalling a necessary optimality condition for minimizing the differences of functions in the nonconvex setting.

Proposition 4 ([41, Proposition 4.1]) *Consider the difference function $f = g - h$, where $g : X \to \mathbb{R} \cup \{+\infty\}$ and $h : X \to \mathbb{R}$ are lower semicontinuous functions. If $\bar{x} \in \text{dom} f$ is a local minimizer of f, then we have the inclusion*

$$\partial^F h(\bar{x}) \subset \partial^F g(\bar{x}).$$

If in addition, h is convex, then $\partial h(\bar{x}) \subset \partial^L g(\bar{x})$.

In sequel, you see the optimality condition associated with the problem (9).

Proposition 5 *If $\bar{x} \in \text{dom} f$ is a local minimizer of the function f considered in* (9), *then*

$$\partial h(\bar{x}) \subset \partial^L g_1(\bar{x}) + \nabla g_2(\bar{x}). \tag{14}$$

A stationary point is any point like $\bar{x} \in \text{dom} f$ that satisfies (14). Since the way to achieve condition (14) is tough, it can be relaxed to

$$[\partial^L g_1(\bar{x}) + \nabla g_2(\bar{x})] \cap \partial h(\bar{x}) \neq \emptyset. \tag{15}$$

The point $\bar{x}$ at the above condition is called *critical point*, so it is clear that every stationary point is a critical point. Therefore, if $0 \in \partial^L f(\bar{x})$, then $\bar{x}$ is a critical point of f, because in accordance with [41, Corollary 3.4] at any point $\bar{x}$ such that $g_1(\bar{x}) < +\infty$, we have

$$\partial^L (g_1 + g_2 - h)(\bar{x}) \subset \partial^L g_1(\bar{x}) + \nabla g_2(\bar{x}) - \partial h(\bar{x}).$$

It is time for recalling the main aim of this section. Authors in [4] introduced the generalized proximal point algorithm (GPPA) below to solve the problem (9):

1. Initialization: Choose $x_0 \in \text{dom} g_1$ and a tolerance $\varepsilon > 0$. Fix any $t > L$.
2. Find

$$y_k \in \partial h(x_k).$$

3. Find x_{k+1} as follows

$$x_{k+1} \in \text{prox}_t^{g_1}(x_k - \frac{\nabla g_2(x_k) - y_k}{t}). \tag{16}$$

4. If $\|x_k - x_{k+1}\| \leq \varepsilon$, then exit. Otherwise, increase k by 1 and go back to step 2.

The definition of proximal mapping tells us (16) is equivalent to

$$x_{k+1} \in \text{argmin}_{x \in X}\{g_1(x) - \langle y_k - \nabla g_2(x_k), x - x_k\rangle + \frac{t}{2}\|x - x_k\|^2\}. \tag{17}$$

In the following a few of the most attractive conditions assuring the convergence of GPPA are presented without proof.

Theorem 3 *Consider the GPPA for solving* (9) *in which* $g_1(x) : X \to \mathbb{R} \cup \{+\infty\}$ *is proper and lower semicontinuous with* $\inf_{x \in X} g_1(x) > -\infty$, $g_2(x) : X \to \mathbb{R}$ *is differentiable with L-Lipschitz gradient, and* $h : X \to \mathbb{R}$ *is convex. Then:*

(i) For any $k \geq 1$, *we have*

$$f(x_k) - f(x_{k+1}) \geq \frac{t - L}{2}\|x_k - x_{k+1}\|^2. \tag{18}$$

(ii) If $\alpha = \inf_{x \in X} f(x) > -\infty$, *then* $\lim_{k \to +\infty} f(x_k) = \bar{l} \geq \alpha$ *and* $\lim_{k \to +\infty} \|x_k - x_{k+1}\| = 0$.

(iii) If $\alpha = \inf_{x \in X} f(x) > -\infty$ *and* $\{x_k\}$ *is bounded, then every cluster point of* $\{x_k\}$ *is a critical point of* f.

Proposition 6 *Suppose that* $\inf_{x \in X} f(x) > -\infty$, f *is proper and lower semicontinuous. If the GPPA sequence* $\{x_k\}$ *has a cluster point* $\bar{x}$, *then* $\lim_{k \to +\infty} f(x_k) = f(\bar{x})$. *Thus,* f *has the same value at all cluster points of* $\{x_k\}$.

A few further remarks on the convergence of GPPA are in order.

- If $h(x) = 0$, then the GPPA coincides with the proximal forward–backward algorithm for minimizing $f = g_1 + g_2$ in [13]. When $h(x) = 0$ and g_1 is the indicator function χ_K where K is a nonempty closed set, then the GPPA turns into the projected gradient method (PGM) for minimizing the smooth function g_2 on a closed constraint set K:

$$x_{k+1} = P_K(x_k - \frac{1}{t}\nabla g_2(x_k)).$$

- Whenever $g_2 = 0$, the GPPA reduces to the PPA with constant stepsize suggested in [58, 59]. Authors by setting $t = \frac{L}{2}$ could recover the convergence result of the primal sequence $\{x_k\}$ generated by the DCA for minimizing the DC function $f = g_1 - h$ in [51, Theorem 3.7].

In the following, authors presented some sufficient conditions that warrant the convergence of the sequence $\{x_k\}$ generated by the GPPA. Denote by Ω the set of cluster points of the sequence $\{x_k\}$.

Theorem 4 *Suppose that* $\inf_{x\in X} f(x) > -\infty$, *and* f *is lower semicontinuous. Suppose further that* ∇h *is L(h)-Lipschitz continuous and* f *has the Kurdyka-Łojasiewicz property at any point* $x \in \mathrm{dom} f$. *If* $\Omega \neq \emptyset$, *then the GPPA sequence* $\{x_k\}$ *converges to a critical point of* f.

In the following theorem, you see some another sufficient conditions that warrant the convergence of the sequence $\{x_k\}$ generated by the GPPA.

Theorem 5 *Consider the difference of function* $f = g - h$ *with* $\inf_{x\in X} f(x) > -\infty$. *Suppose that* g *is differentiable and* ∇g *is L-Lipschitz continuous,* f *has the strong Kurdyka-Łojasiewicz property at any point* $x \in \mathrm{dom} f$, *and* h *is a finite convex function. If* $\Omega \neq \emptyset$, *then the GPPA sequence* $\{x_k\}$ *converges to a critical point of* f.

Authors in [35] investigated convergence of DC algorithm for DC programming with subanalytic data. From [16, Theorem 3.1] and [17, Corollary 16], it is obtained that a lower semicontinuous subanalytic function satisfies Kurdyka-Łojasiewicz property with a specific form of the function φ. From this you can find results in Theorems 4 and 5 as an extension of [35, Theorem 3.1].

Finally in the following proposition, authors exhibited some sufficient conditions for the set Ω to be nonempty.

Proposition 7 *Consider the function* $f = g - h$, *where* $g = g_1 + g_2$ *in* (9). *Let* $\{x_k\}$ *be the sequence generated by the GPPA for solving* (10). *The set of critical points* Ω *of* $\{x_k\}$ *is nonempty if one of the following conditions is satisfied:*

(i) *For any* α, *the lower level set* $L_{\leq\alpha} := \{x \in X : f(x) \leq \alpha\}$ *is bounded.*
(ii) $\liminf_{\|x\|\to+\infty} h(x) = +\infty$ *and* $\liminf_{\|x\|\to+\infty} \frac{g(x)}{h(x)} > 1$.

4 Inertial Proximal Algorithm for Difference of Two Maximal Monotone Operators

This section deals with presenting a proximal algorithm for difference of two maximal monotone operators. Before studying we investigate a bite of mathematical sociology.

Here our world will be extended to Hilbert space H. The notation $\langle ., . \rangle$ will be used for inner product in $H \times H$ and $\|.\|$ for the corresponding norm. A set valued operator $T : H \to 2^H$ is said to be *monotone* if

$$\langle x^* - y^*, x - y \rangle \geq 0, \quad \forall (x, x^*),\ (y, y^*) \in G(T),$$

wherein $G(T) := \{(x, y) \in H \times H;\ y \in Tx\}$ is graph of T. The domain of T is $D(T) := \{x \in H;\ T(x) \neq \emptyset\}$.

A monotone operator T is called *maximal monotone* if its graph is maximal in the sense of inclusion.

Associated with a given monotone operator T, the resolvent operator for T and parameter $\lambda > 0$ is $J_\lambda^T := (I + \lambda T)^{-1}$. The resolvent J_λ^T of a monotone operator T is a single valued nonexpansive map from $Im(I + \lambda T)$ to H [14, Proposition 3.5.3]. Moreover, the resolvent has full domain precisely when T is maximal monotone. For any $x \in H$, $\lim_{\lambda\to 0} J_\lambda^T(x) = Proj_{\overline{D(T)}}x$, wherein $Proj_{\overline{D(T)}}$ is the orthogonal projection on the closure of the domain of T. One of the best known approaches in the theory of optimization that is related to resolvent operators is *Yosida approximate* $T_\lambda := \frac{(I-J_\lambda^T)}{\lambda}$ of a maximal monotone operator T which satisfies in:

(i) For all $x \in H$, $T_\lambda(x) \in T(J_\lambda^T(x))$,
(ii) T_λ is Lipschitz with constant $\frac{1}{\lambda}$ and maximal monotone,
(iii) $T_\lambda(x)$ converges strongly to $T(x)$ as $\lambda \to 0$, for $x \in D(T)$,
(iv) $\|T_\lambda(x)\| \leq \|T^0(x)\|$ for every $x \in D(T)$, $\lambda > 0$, where T^0 is *minimal selection*

$$T^0(x) := \{y \in T(x);\ \|y\| = \min_{z\in T(x)} \|z\|\}, \qquad x \in D(T).$$

Consider the problem

$$\text{find } x \in H \text{ such that } 0 \in T(x) - S(x), \tag{19}$$

where $T, S : H \to 2^H$ are two maximal monotone operators on finite dimensional real Hilbert space H and it is equivalent to the problem

$$\text{find } x \in H \text{ such that } T(x) \cap S(x) \neq \emptyset. \tag{20}$$

Regarding the importance of this problem as mentioned, finding the critical points of the difference of two convex functions is the special case of finding the zeros of difference of two maximal monotone operators. Actually, an algorithm for difference of two maximal monotone operators plays a central role in the study of DC programming. The later studies are limited to Moudafi [42, 43]. By [43], a regularization of the problem (19) is

$$\text{find } x \in H \text{ such that } 0 \in T(x) - S_\lambda(x). \tag{21}$$

For finding a solution of (19) Moudafi [43] suggested a sequence $\{x_n\}$ by

$$x_{n+1} = J_{\mu_n}^T(x_n + \mu_n S_{\lambda_n} x_n) \quad \forall n \in \mathbb{N}, \tag{22}$$

where $\mu_n > 0$ and x_0 is an initial point.

Here, the problem (19) is studied via generalization of Moudafi's algorithm in [43] as the following:

$$x_{k+1} = J^T_{\beta_k}(x_k + \alpha_k(x_k - x_{k-1}) + \beta_k S_{\mu_k} x_k) \qquad \forall k \in \mathbb{N}, \tag{23}$$

with starting points $x_0, x_1 \in H$ and sequences $\{\mu_k\}$, $\{\alpha_k\}$ and $\{\beta_k\} \subset [0, +\infty)$ such that

(a) $\lim_{k\to+\infty} \mu_k = 0$;
(b) $\sum_{k=1}^{+\infty} \frac{\beta_k}{\mu_k} < +\infty$;
(c) $\lim_{k\to+\infty} \frac{\alpha_k}{\beta_k} = 0$;
also we suppose that
(d) $\sum_{k=1}^{+\infty} \alpha_k \|x_k - x_{k-1}\| < +\infty$;
(e) $\lim_{k\to+\infty} \frac{\|x_{k+1}-x_k\|}{\beta_k} = 0$.

We note that (23) is emanated from the evolution equation

$$x''(t) + \gamma x'(t) + \nabla f(x(t)) - \nabla g(x(t)) = 0, \tag{24}$$

where $\gamma > 0$ and algorithm (22) can be inspired from

$$x'(t) + \nabla f(x(t)) - \nabla g(x(t)) = 0, \tag{25}$$

in which both $f, g : H \to \mathbb{R}$ are differentiable convex functions and $\nabla f(x(t))$ and $\nabla g(x(t))$ are operators T and S in (19), respectively.

If $\nabla g(x(t)) = 0$, then (24) is *heavy ball with friction* system or (HBF) and (23) is equivalent to the standard gradient descent iteration (22) with an additional *inertia term* or *momentum term* $\alpha_k(x_k - x_{k-1})$. By the inertia term, convergence of the solution trajectories of the (HBF) system to a stationary point of f can be faster than those of the first order system (25) when $\nabla g(x(t)) = 0$ [52].

Another important advantage of algorithm (23) over algorithm (22) is using condition of local boundedness of S instead of boundedness in (22).

In this section, we present different conditions under which (23) converges to a solution of (19). Now, we recall some required results and definitions.

A set valued operator $T : H \to 2^H$ is *locally bounded* at $\bar{x}$ if there exists a neighborhood U of $\bar{x}$ such that the set $T(U)$ is bounded.

Lemma 1 ([53]) *A maximal monotone operator T is locally bounded at a point $\bar{x} \in D(T)$ if and only if $\bar{x}$ belongs to interior of $D(T)$.*

A set valued operator $T : H \to 2^H$ is upper semicontinuous at $\bar{x}$ if for any positive $\varepsilon > 0$ there exists $\delta > 0$ such that

$$\|x - \bar{x}\| \le \delta \Rightarrow T(x) \subseteq T(\bar{x}) + B(0, \varepsilon). \tag{26}$$

Lemma 2 ([1]) *Suppose that E is a Banach space. The maximal monotone operator $T : E \to 2^{E^*}$ where E^* is dual space of E is demiclosed, i.e., the following conditions hold.*

(1) If $\{x_k\} \subset E$ converges strongly to x_0 and $\{u_k \in T(x_k)\}$ converges weak to u_0 in E^*, then $u_0 \in T(x_0)$.*
(2) If $\{x_k\} \subset E$ converges weakly to x_0 and $\{u_k \in T(x_k)\}$ converges strongly to u_0 in E^*, then $u_0 \in T(x_0)$.

Lemma 3 ([36]) *Suppose that $\{a_n\}$, $\{b_n\}$, and $\{c_n\}$ are three sequences of nonnegative numbers such that*

$$a_{n+1} \leq (1 + b_n)a_n + c_n \quad for\ all\ n \geq 1.$$

If $\sum_{n=1}^{\infty} b_n < +\infty$ and $\sum_{n=1}^{\infty} c_n < +\infty$, then $\lim_{n\to\infty} a_n$ exists.

In the following, we improve the conditions of Theorem 2.1 in [43].

Theorem 6 *Assume that S is locally bounded on $\overline{D(S)}$ and the solution set Ω of problem (19) is nonempty. If the conditions $(a), \ldots, (e)$ satisfy and $D(T) \subset D(S)$, then the sequence $\{x_k\}$ generated by (23) converges to a solution of (19).*

Proof Take $x^* \in \Omega$. According to (20), there exists $y^* \in T(x^*) \cap S(x^*)$ and from (23), $x^* = J^T_{\beta_k}(x^* + \beta_k y^*)$. From the triangular inequality, (iv), nonexpansivity of $J^T_{\beta_k}$ and the fact that S_{μ_k} is also nonexpansive with constant $\frac{1}{\mu_k}$, one quickly deduces that

$$\begin{aligned}
\|x_{k+1} - x^*\| &= \|J^T_{\beta_k}(x_k + \alpha_k(x_k - x_{k-1}) + \beta_k S_{\mu_k} x_k) - J^T_{\beta_k}(x^* + \beta_k y^*)\| \\
&\leq \|x_k + \alpha_k(x_k - x_{k-1}) + \beta_k S_{\mu_k}(x_k) - x^* - \beta_k y^*\| \\
&\leq \|x_k - x^*\| + \alpha_k\|x_k - x_{k-1}\| + \beta_k\|S_{\mu_k}(x_k) - y^*\| \\
&\leq \|x_k - x^*\| + \alpha_k\|x_k - x_{k-1}\| \\
&+ \beta_k(\|S_{\mu_k}(x_k) - S_{\mu_k}(x^*)\| + \|S_{\mu_k}(x^*) - y^*\|) \\
&\leq (1 + \frac{\beta_k}{\mu_k})\|x_k - x^*\| + \alpha_k\|x_k - x_{k-1}\| + \beta_k(\|S^0 x^*\| + \|y^*\|).
\end{aligned}$$

Applying (a) and (b), $\sum_{k=0}^{\infty} \beta_k < \infty$. Also by (d) and Lemma 3, we have $\lim_{k\to+\infty} \|x_k - x^*\|$ exists. Hence, $\{x_k\}$ is bounded. Notice that there exist $\widetilde{x}$ and a subsequence $\{x_{k_\nu}\}$ such that $\lim_{\nu\to\infty} x_{k_\nu} = \widetilde{x}$, since H is a finite dimensional space. We see $J^S_{\mu_{k_\nu}} x_{k_\nu}$ tends to $\widetilde{x}$, because

$$\begin{aligned}
\|J^S_{\mu_{k_\nu}} x_{k_\nu} - \widetilde{x}\| &\leq \|J^S_{\mu_{k_\nu}} x_{k_\nu} - J^S_{\mu_{k_\nu}} \widetilde{x}\| + \|J^S_{\mu_{k_\nu}} \widetilde{x} - \widetilde{x}\| \\
&\leq \|x_{k_\nu} - \widetilde{x}\| + \|J^S_{\mu_{k_\nu}} \widetilde{x} - \widetilde{x}\|,
\end{aligned}$$

and $\lim_{\nu\to+\infty} J^S_{\mu_{k_\nu}}\widetilde{x} = Proj_{\overline{D(S)}}\widetilde{x} = \widetilde{x}$. This fact and the local boundedness of S imply that

$$\{S_{\mu_{k_\nu}} x_{k_\nu}\} \subseteq S(\{J^S_{\mu_{k_\nu}} x_{k_\nu}\}) \subseteq B, \tag{27}$$

where B is a bounded set. Therefore, $\{S_{\mu_{k_\nu}} x_{k_\nu}\}$ is bounded and there exist $\tilde{y}$ and a subsequence $\{S_{\mu_{k_{\nu'}}} x_{k_{\nu'}}\}$ such that $\lim_{\nu'\to\infty} S_{\mu_{k_{\nu'}}} x_{k_{\nu'}} = \tilde{y}$. Then $\tilde{y} \in S(\tilde{x})$ follows from

$$S_{\mu_{k_{\nu'}}} x_{k_{\nu'}} \in S(J^S_{\mu_{k_{\nu'}}} x_{k_{\nu'}}), \tag{28}$$

and Lemma 2. In sequel by (23), we have

$$S_{\mu_{k_{\nu'}}} x_{k_{\nu'}} - (\frac{x_{k_{\nu'}+1} - x_{k_{\nu'}}}{\beta_{k_{\nu'}}}) + \frac{\alpha_{k_{\nu'}}}{\beta_{k_{\nu'}}}(x_{k_{\nu'}} - x_{k_{\nu'}-1}) \in T x_{k_{\nu'}+1}, \tag{29}$$

tending ν' to $+\infty$ in (29) and using conditions (c), (e), boundedness of $\{x_k\}$ and Lemma 2, it is obtained that $\widetilde{y} \in T\widetilde{x}$. By similar procedure in the proof of Theorem 2.1 in [43], $\widetilde{x}$ is unique. Now proof is complete.

Example 1 The best example of Theorem 6 can be seen in digital halftoning which is a procedure for producing a sample of pixels when a limited number of colors are available with a binary system so that it is a continuous-tone image. In this context Teuber et al. [60] minimized difference of two functions that one is corresponding to attraction of the dots by the image gray values and the other corresponds to the repulsion between the dots. They signified black pixel with 0 and white pixel with 1 and investigated images $u : G \to [0, 1]$ on an integer grid $G := \{1, \ldots, n_x\} \times \{1, \ldots, n_y\}$. If m be the number of black pixels generated by the dithering procedure and $p := (p_k)_{k=1}^m = ((p_{k,x}, p_{k,y})^T)_{k=1}^m \in \mathbb{R}^{2m}$ be their position vector then $|p_k| := \sqrt{p_{k,x}^2 + p_{k,y}^2}$ is the Euclidean norm of the position of the k-th black pixel.

In [60], minimizer $\hat{p}$ is detected of the functional

$$E(p) = \underbrace{\sum_{k=1}^{m} \sum_{(i,j)\in G} w(i,j)|p_k - \binom{i}{j}|}_{F(p)} - \underbrace{\lambda \sum_{k=1}^{m} \sum_{l=k+1}^{m} |p_k - p_l|}_{G(p)}, \tag{30}$$

where $w := 1 - u$ is the corresponding weight distribution and $\lambda := \frac{1}{m}\sum_{(i,j)\in G} w(i,j)$. Given two functions $F(p)$ and $G(p)$, which are continuous and convex, since ∂F and ∂G are maximal monotone operators [54] and ∂G is locally bounded on $\mathbb{R}^{2m}$ [27], the problem of finding a minimizer of (30) is a special case of (19). If conditions (a),…,(e) satisfy and $D(\partial F) \subset D(\partial G)$, then by Theorem 6 the generated sequence $\{x_k\}$ of (23) converges to a minimizer of (30).

In next result, the condition of local boundedness of S in Theorem 6 is eliminated and domain of it will be entire H.

Corollary 4 *Assume that the solution set Ω of problem* (19) *is nonempty, conditions* $(a), \ldots, (e)$ *satisfy, and $D(S) = H$, then the sequence $\{x_k\}$ generated by* (23) *converges to a solution of* (21).

Proof *Since $D(S)$ is open, using Lemma 1 the operator S is locally bounded at any point of $D(S)$. The rest of proof is similar to Theorem 6.*

Remark 1 If $D(S) = H$ and $T - S$ is a monotone operator then by [2, Theorem 2.1], $T - S$ is maximal monotone. Hence, (19) reduces to find a zero point of maximal monotone operator $T - S$ and iteration algorithm (23) changes to $x_{k+1} = J_{\beta_k}^{T-S}(x_k + \alpha_k(x_k - x_{k-1}))$.

Corollary 5 *Assume that S is bounded value (i.e. for all $x \in H$, Sx is a bounded set) and upper semicontinuous at any point of $\overline{D(S)}$ and the solution set Ω of problem* (19) *is nonempty. If the conditions* $(a), \ldots, (e)$ *satisfy and $D(T) \subset D(S)$, then the sequence $\{x_k\}$ generated by* (23) *converges to a solution of* (19).

Proof Since S is bounded value and upper semicontinuous at any point of $\overline{D(S)}$, so it is locally bounded. The rest of proof is similar to Theorem 6.

Two types of interesting particular instances of (19) are:

$$\text{find} \quad x^* \in H \quad \text{such that} \quad y^* \in T(x^*), \tag{31}$$

and

$$\text{find} \quad x^* \in H \quad \text{such that} \quad x^* \in T(x^*). \tag{32}$$

It is assumed that $G(S) := H \times \{y\}$ for an arbitrary point $y \in H$ in (31) and $G(S) := \{(x, x); x \in H\}$ for any point $x \in H$ in (32).

In the following, we present the results of these types of problems.

Corollary 6 *Assume that the operator $S : H \to H$ is continuous and the solution set Ω of problem* (19) *is nonempty. If the conditions* $(a),\ldots,(e)$ *satisfy and $D(T) \subset D(S)$, then the sequence $\{x_k\}$ generated by* (23) *converges to a solution of* (19).

Proof It is easy to check that sequence $\{x_k\}$ is bounded and there exist $\widetilde{x}$ and a subsequence $\{x_{k_\nu}\}$ such that $\lim_{\nu\to\infty} x_{k_\nu} = \widetilde{x}$. In proof of Theorem 6 it has been shown that $\lim_{\nu\to\infty} J_{\mu_{k_\nu}} x_{k_\nu} = \widetilde{x}$. Consequently, from

$$S_{\mu_{k_\nu}} x_{k_\nu} - \left(\frac{x_{k_\nu+1} - x_{k_\nu}}{\beta_{k_\nu}}\right) + \frac{\alpha_{k_\nu}}{\beta_{k_\nu}}(x_{k_\nu} - x_{k_\nu-1}) \in T x_{k_\nu+1}, \tag{33}$$

$S_{\mu_{k_\nu}}(x_{k_\nu}) = S(J^S_{\mu_{k_\nu}}(x_{k_\nu}))$, continuity of S and by passing to a subsequence, we can arrange that left side of (33) converges to $S(\widetilde{x})$. By Lemma 2, we see that $S(\widetilde{x}) \in T(\widetilde{x})$, i.e. $0 \in T(\widetilde{x}) - S(\widetilde{x})$.

Corollary 7 *Assume that $S : H \to H$ is Lipschitz continuous, the solution set Ω of problem* (19) *is nonempty and $D(T) \subset D(S)$. If conditions (c),...,(e) satisfy and if one replaces condition $\sum_{k=1}^{\infty} \beta_k < \infty$ with (a) and (b), then the generated sequence $\{x_k\}$ of method*

$$x_{k+1} = J^T_{\beta_k}(x_k + \alpha_k(x_k - x_{k-1}) + \beta_k S(x_k))$$

converges to a solution of problem (19).

Remark 2 All results of this section have derived from Lemma 2. In an infinite dimensional real Hilbert space, boundedness of sequence $\{x_k\}$ in Theorem 6 implies that there exist subsequence $\{x_{k_\nu}\}$ and $\tilde{x} \in H$ such that $\{x_{k_\nu}\}$ converges weakly to $\tilde{x}$. The fundamental difficulties in proving $\tilde{y} \in S(\tilde{x})$ and $\tilde{y} \in T(\tilde{x})$ are showing strongly convergence of either $\{J^S_{\mu_{k_\nu}} x_{k_\nu}\}$ to $\tilde{x}$ or $\{S_{\mu_{k_\nu}} x_{k_\nu}\}$ to $\tilde{y}$ and the left side of (29) to $\tilde{y}$.

References

1. Y. Alber, I. Ryazantseva, *Nonlinear Ill-posed Problems of Monotone Type* (Springer, New York, 2006)
2. M. Alimohammady, M. Ramazannejad, M. Roohi, Notes on the difference of two monotone operators. Optim. Lett. **8**(1), 81–84 (2014)
3. L.T.H. An, Analyse numérique des algorithmes de l'optimisation d.c. Approches locales et globales. Code et simulations numériques en grande dimension. Applications, Thése de Doctorat de l'Université de Rouen, 1994
4. N.T. An, N.M. Nam, Convergence analysis of a proximal point algorithm for minimizing difference of functions. Optimization **66**(1), 129–147 (2017)
5. L.T.H. An, D.T. Pham, DCA with escaping procedure using a trust region algorithm for globally solving nonconvex quadratic programs, Technical Report, LMI-CNRS URA 1378, INSA-Rouen, 1996
6. L.T.H. An, D.T. Pham, Solving a class of linearly constrained indefinite quadratic problems by DC algorithms. J. Glob. Optim. **11**(3), 253–285 (1997)
7. L.T.H. An, D.T. Pham, A Branch-and-Bound method via D.C. optimization algorithm and ellipsoidal technique for box constrained nonconvex quadratic programming problems. J. Glob. Optim. **13**(2), 171–206 (1998)
8. L.T.H. An, D.T. Pham, D.C. programming approach for large-scale molecular optimization via the general distance geometry problem, in *Optimization in Computational Chemistry and Molecular Biology* (Springer, Boston, 2000), pp. 301–339
9. L.T.H. An, D.T. Pham, D.C. programming approach to the multidimensional scaling problem, in *From Local to Global Optimization*. Part of the Nonconvex Optimization and Its Applications Book Series (NOIA), vol. 53 (Springer, Dordrecht, 2001), pp. 231–276
10. L.T.H. An, D.T. Pham, A continuous approach for globally solving linearly constrained quadratic. Optimization **50**, 93–120 (2001)

11. L.T.H. An, D.T. Pham, The DC programming and DCA revisited with DC models of real world nonconvex optimization problems. Ann. Oper. Res. **133**, 25–46 (2005)
12. H. Attouch, J. Bolte, P. Redont et al., Proximal alternating minimization and projection methods for nonconvex problems: an approach based on the Kurdyka–Łojasiewicz inequality. Math. Oper. Res. **35**, 438–457 (2010)
13. H. Attouch, J. Bolte, B. Svaiter, Convergence of descent methods for semi-algebraic and tame problems: proximal algorithms, forward-backward splitting, and regularized Gauss–Seidel methods. Math. Program. A **137**, 91–124 (2011)
14. J.P. Aubin, H. Frankowska, *Set-valued Analysis* (Birkhäuser, Boston, 2009). Reprint of the 1990 Edition
15. F. Bach, R. Jenatton, J. Mairal, G. Obozinski, Optimization with sparsity-inducing penalties. Found. Trends Mach. Learn. **4**(1), 1–106 (2011)
16. J. Bolte, A. Daniilidis, A.S. Lewis, Lojasiewicz inequality for nonsmooth subanalytic functions with applications to subgradient dynamic systems. SIAM Optim. **17**, 1205–1223 (2007)
17. J. Bolte, A. Daniilidis, A.S. Lewis et al., Clarke subgradients of stratifiable functions. SIAM J. Optim. **18**, 556–572 (2007)
18. J. Bolte, S. Sabach, M. Teboulle, Proximal alternating linearized minimization for nonconvex and nonsmooth problems. Math. Program. A **146**, 459–494 (2014)
19. S. Boyd, N. Parikh, E. Chu, B. Peleato, J. Eckstein, Distributed optimization and statistical learning via the alternating direction method of multipliers. Found. Trends Mach. Learn. **3**(1), 1–122 (2011)
20. A. Brøndsted, R.T. Rockafellar, On the subdifferentiability of convex functions. Proc. Am. Math. Soc. **16**, 605–611 (1965)
21. F.E. Browder, Multi-valued monotone nonlinear mappings and duality mappings in Banach space. Trans. Am. Math. Soc. **118**, 338–551 (1965)
22. F.E. Browder, Nonlinear maximal monotone mappings in Banach spaces. Math. Ann. **175**, 89–113 (1968)
23. F.E. Browder, The fixed point theory of multi-valued mappings in topological vector spaces. Math. Ann. **177**, 283–301 (1968)
24. R.S. Burachik, A.N. Iusem, *Set-valued Mappings and Enlargements of Monotone Operators*. Optimization and its Applications (Springer, New York, 2008). ISBN: 978-0-387-69757-4
25. S. Chandra, Strong pseudo-convex programming. Indian J. Pure Appl. Math. **3**, 278–282 (1972)
26. F.H. Clarke, Generalized gradients and applications. Trans. Am. Math. Soc. **205**, 247–262 (1975)
27. F.H. Clarke, R.J. Stern, G. Sabidussi, *Nonlinear Analysis, Differential Equations and Control*. NATO Sciences Series, Series C: Mathematical and Physical Sciences, vol. 528 (Kluwer Academic Publishers, Dordrecht, 1999)
28. P. Combettes, J.-C. Pesquet, Proximal splitting methods in signal processing, in *Fixed-Point Algorithms for Inverse Problems in Science and Engineering* (Springer, New York, 2011), pp. 185–212
29. C. Do, Q. Le, C. Foo, Proximal regularization for online and batch learning, in *International Conference on Machine Learning* (2009), pp. 257–264
30. G. Gasso, A. Rakotomamonji, S. Canu, Recovering sparse signals with non-convex penalties and DC programming. IEEE Trans. Signal Process **57**(12), 4686–4698 (2009)
31. J.B. Hiriart-Urruty, From convex optimization to non convex optimization. Part I: Necessary and sufficient conditions for global optimality, in *Nonsmooth Optimization and Related Topics*. Ettore Majorana International Sciences, Series 43 (Plenum Press, New York, 1988)
32. S. Huda, R. Mukerjee, Minimax second-order designs over cuboidal regions for the difference between two estimated responses. Indian J. Pure Appl. Math. **41**(1), 303–312 (2010)
33. R. Jenatton, J. Mairal, G. Obozinski, F. Bach, Proximal methods for sparse hierarchical dictionary learning, in *International Conference on Machine Learning* (2010)

34. M. Kraning, E. Chu, J. Lavaei, S. Boyd, *Dynamic network energy management via proximal message passing*, Foundations and Trends in Optimization, **1**(2), 73–126 (2014)
35. H.A. Le Thi, V.N. Huynh, D.T. Pham, Convergence analysis of DC algorithm for DC programming with subanalytic data, Annals of Operations Research. Technical Report. LMI, INSA-Rouen, Rouen, 2013
36. D. Lei, L. Shenghong, Ishikawa iteration process with errors for nonexpansive mappings in uniformly convex Banach spaces. Int. J. Math. Math. Sci. **24**(1), 49–53 (2000)
37. P. Mahey, D.T. Pham, Proximal decomposition of the graph of maximal monotone operator. SIAM J. Optim. **5**, 454–468 (1995)
38. B. Martinet, Régularisation d'iné quations variationnelles par approximations successives, Revue Française de Informatique et Recherche Opérationelle (1970)
39. B. Martinet, Détermination approchée d'un point fixe d'une application pseudo-contractante. Rev C. R. Acad. Sci. Paris **274A**, 163–165 (1972)
40. G.J. Minty, Monotone networks. Proc. R. Soc. Lond. **257**, 194–212 (1960)
41. B.S. Mordukhovich, N.M. Nam, N.D. Yen, Fréchet subdifferential calculus and optimality conditions in nondifferentiable programming. Optimization **55**, 685–708 (2006)
42. A. Moudafi, On the difference of two maximal monotone operators: regularization and algorithmic approaches. Appl. Math. Comput. **202**, 446–452 (2008)
43. A. Moudafi, On critical points of the difference of two maximal monotone operators. Afr. Mat. **26**(3–4), 457–463 (2015)
44. B. O'Donoghue, G. Stathopoulos, S. Boyd, A splitting method for optimal control. IEEE Trans. Control Syst. Technol. **21**(6), 2432–2442 (2013)
45. N. Parikh, S. Boyd, Proximal algorithms. Found. Trends Optim. **1**(3), 123–231 (2013)
46. D.T. Pham, Duality in d.c. (difference of convex functions) optimization, in *Subgradient Methods, Trends in Mathematical Optimization*. International Series of Numerical Mathematics, vol. 84 (Birkhauser, Basel, 1988), pp. 277–293
47. D.T. Pham, L.T.H. An, Optimisation d.c. (différence de deux fonctions convexes). Dualité et Stabilité. Optimalités locale et globale. Algorithmes de l'optimisation d.c. (DCA), Technical Report, LMI-CNRS URA 1378, INSA Rouen, 1994
48. D.T. Pham, L.T.H. An, Polyhedral d.c. optimization. Theory, algorithms and applications, Technical Report, LMI-CNRS URA 1378, INSA-Rouen, 1994
49. D.T. Pham, L.T.H. An, Difference of convex functions optimization algorithms (DCA) for globally minimizing nonconvex quadratic forms on Euclidean balls and spheres. Oper. Res. Lett. **19**(5), 207–216 (1996)
50. D.T. Pham, L.T.H. An, Convex analysis approach to D.C. programming: theory, algorithms and applications. Acta Math. Vietnam. **22**(1), 289–355 (1997)
51. D.T. Pham, L.T.H. An, A D.C. optimization algorithms for solving the trust region subproblem. SIAM J. Optim. **8**(2), 476–505 (1998)
52. N. Qian, On the momentum term in gradient descent learning algorithms. Neural Netw. **12**(1), 145–151 (1999)
53. R.T. Rockafellar, Local boundedness of nonlinear monotone operators. Mich. Math. J. **16**, 397–407 (1969)
54. R.T. Rockafellar, On the maximal monotonicity of subdifferential mappings. Pac. J. Math. **33**, 209–216 (1970)
55. R.T. Rockafellar, Monotone operators and the proximal point algorithm. SIAM J. Control Optim. **14**(5), 877–898 (1976)
56. R.T. Rockafellar, R. Wets, *Variational analysis. Grundlehren der Mathematischen Wissenschaften [Variational analysis. Fundamental Principles of Mathematical Sciences]*, vol. 317 (Springer, Berlin, 1998)
57. T. Schüle, C. Schnörr, S. Weber, J. Hornegger, Discrete tomography by convex–concave regularization and D.C. programming. Discrete Appl. Math. **151**, 229–243 (2005)
58. J.C. Souza, P.R. Oliveira, A proximal point algorithm for DC functions on Hadamard manifolds. J. Glob. Optim. **63**, 797–810 (2015)

59. W. Sun, R.J.B. Sampaio, M.A.B. Candido, Proximal point algorithm for minimization of DC functions. J. Comput. Math. **21**, 451–462 (2003)
60. T. Teuber, G. Steidl, P. Gwosdek, Ch. Schmaltz, J. Weickert, Dithering by differences of convex functions. SIAM J. Imaging Sci. **4**(1), 79–108 (2011)
61. J.F. Toland, Duality in nonconvex optimization. J. Math. Anal. Appl. **66**, 399–415 (1978)
62. J.F. Toland, On subdifferential calculus and duality in nonconvex optimization. Bull. Soc. Math. France Mémoire **60**, 173–180 (1979)
63. Z. Zhang, J.T. Kwok, D.-Y. Yeung, Surrogate maximization/minimization algorithms and extensions. Mach. Learn. **69**, 1–33 (2007)

Finite Element Analysis in Fluid Mechanics

Anastasios Raptis, Konstantina Kyriakoudi, and Michail A. Xenos

Abstract In the last decades, the finite element method (FEM) in fluid mechanics applications has gained substantial momentum. FE analysis was initially introduced to solid mechanics. However, the progress in fluid mechanics problems was slower due to the non-linearities of the equations and inherent difficulties of the classical FEM to deal with instabilities in the solution of these problems. The main goal of this review is to analyze FEM and provide the theoretical basis of the approach mainly focusing on parabolic type of problems applied in fluid mechanics. Initially, we analyze the basics of FEM for the Stokes problem and we provide theorems for uniqueness and error estimates of the solution. We further discuss FE approaches for the solution of the advection–diffusion equation such as the stabilized FEM, the variational multiscale method, and the discontinuous Galerkin method. Finally, we extend the analysis on the non-linear Navier–Stokes equations and introduce recent FEM advancements.

1 Introduction

Finite element method (FEM) has gained substantial momentum in the last decades. FEM was initially introduced as an answer to solid mechanics problems that were difficult to solve until then. Most of them would be encountered in aeronautics or civil engineering due to the need of solving problems related to the construction of complicated structures. The method was extended to fluid mechanics applications where the convective terms play important role leading to a non-linear formulation of the problem. The progress in fluid mechanics was slower due to the non-linearities and instabilities of the solution of these problems.

The basic principles of the FEM were developed by the German mathematician Ritz in 1909. In 1915 Galerkin worked on the theoretical aspects of the method.

A. Raptis · K. Kyriakoudi · M. A. Xenos (✉)
Department of Mathematics, University of Ioannina, Ioannina, Greece
e-mail: anraptis@cc.uoi.gr; kkyriakoudi@cc.uoi.gr; mxenos@uoi.gr

T. M. Rassias, P. M. Pardalos (eds.), *Mathematical Analysis and Applications*, Springer Optimization and Its Applications 154,
https://doi.org/10.1007/978-3-030-31339-5_18

The absence of computers delayed further advancement of the method. Later on, when computers were introduced, the method was further developed. Hrenikoff, 1941, introduced the framework method, in which a plain elastic medium could be replaced by an equivalent system of sticks and rods. In 1943 Courant solved the torsion problem by using triangular elements based on the principle of minimum potential energy introducing the Rayleigh–Ritz method. Courant's theory could not be implemented due to the unavailability of computers at the time [82].

Argyris, 1955, in the book "Energy Theorems and Structural Analysis" introduced the principles of the finite element method [3, 85]. In 1956 Clough, Turner, Martin, and Top calculated the stiffness matrix of rod and other elements. Argyris and Kelsey, 1960, published their work which was based on the finite element principles. In the same year, the term finite element method was introduced by Clough in his paper and the term has been used extensively in the literature until today. Zienkiewicz and Chung wrote the first book on finite elements method, in 1967. Other notable researchers in the FEM field are Samuel Levy, Borje Langefors, Paul Denke, Baudoin Fraejis De Veubeke, L. Brandeis Wehle Jr., Theodore Pian, Warner Lansing, Bertran Klein, John Archer, Robert Melosh, John Przemieniecki, Ian Taig, Richard Gallagher, Bruce Irons, and others.

As mentioned before the progress of FEM in fluid mechanics applications had several drawbacks due to the non-linear convective terms and instabilities of the solution based on the element selection. For these reasons many researchers studied the advection–diffusion equation. The Galerkin method was introduced as a natural extension of the weak formulation of the PDEs under consideration. One of the reasons why finite elements have been less popular in the past than other numerical techniques such as finite differences is the lack of upwind techniques. However, accurate upwind methods have been constructed. The most popular of these upwind approaches is the Streamline Upwind Petrov–Galerkin method (SUPG) [89]. It can be shown that upwinding may increase the quality of the solution considerably. Another important aspect of upwinding is that it makes the systems of equations appropriate for the utilization of iterative methods. As a consequence both the number of iterations and the computation time substantially decrease.

The advection–diffusion equation represents diffusion of a scalar variable while convected by a velocity field. In this respect, the equation by itself applies in several physical phenomena and is a precursor to studying the non-linear Navier–Stokes equations that represents in a simplifying manner the transport of velocity itself. In any case, the development of accurate and stable numerical formulations for the advection–diffusion equation is quite challenging. For example, the classical Galerkin method is known to perform poorly for advection-dominated transport problems. Spurious oscillations emerge in the solution due to the truncation error inherently introduced in the discretized Galerkin approximation. The literature suggests numerous strategies to overcome this problem. The addition of artificial diffusion is a standard strategy, another is the employment of a non-centered discretization of the advection operator, the so-called upwind schemes [45]. Other strategies involve multiscale models using bubble functions or wavelets [72], while in many cases, these methods are equivalent [17]. In the relevant section of this chapter, more information is provided regarding some of the strategies in the context of finite element methods that have been developed to address the problems that standard discretizations face.

Studying the advection–diffusion equation helps in understanding more complicated problems such as the Navier–Stokes equations. For the discretization of the incompressible Navier–Stokes equations, since the pressure is an unknown in the momentum but not in the continuity equation, the discretization must satisfy some special requirements. In fact one is no longer free to choose any combination of pressure and velocity approximation but the finite elements must be constructed such that the Ladyzhenskaya–Brezzi–Babuska (LBB) condition is satisfied. This condition provides a relation between pressure and velocity approximation. In finite differences and finite volumes the equivalent of the LBB condition is satisfied if staggered grids are applied.

The solenoidal (divergence free) approach has been introduced where in this method, the elements are constructed in such a way that the approximate divergence freedom is satisfied explicitly. This method seems very attractive, however, the extension to three-dimensional problems is difficult. Stabilized and multiscale formulations are among the most fundamental method for fluid mechanics problems. The SUPG is one of the first finite element approaches for studying fluid mechanics applications. However, due to the advancement in research nowadays, new finite element approaches have emerged such as the variational multiscale method (VMS), the characteristic base split (CBS) method, the gradient smoothed method (GSM), discontinuous Galerkin (DG) and adaptive FEM.

In this review we initially present the basic analysis focused on the Stokes problem providing error estimates. We further analyze the advection–diffusion equation introducing several FEM advancements. We conclude this chapter with a brief analysis on FEM for the non-linear Navier–Stokes equations.

2 Preliminaries and Basic Theorems

We begin this chapter with the main steps of the finite element method. In advance, we formulate basic definitions and theorems about the existence and uniqueness of the solution in these problems. More details can be found in the textbooks by Brenner and Scott and Brezzi [15, 16].

Definition 1 Let $a(\cdot, \cdot)$ be a bilinear form on a normed linear space, H. The bilinear form is said to be *bounded* (or *continuous*) if exists $C < \infty$ such that

$$| a(u, v) | \leq C \|u\|_H \|v\|_H \quad \forall\, u, v \in H,$$

and *coercive* on subspace $V = \left\{v \in H^1(0, 1) : v(0) = 0\right\}$, $V \subset H$ if exists $\delta > 0$ such that

$$a(v, v) \geq \delta \, \|v\|_H^2 \,, \quad \forall \, v \in V,$$

where $a(u, v) = \int_0^1 u' \, v' \, dx$, $\|\cdot\|_H$ is the norm in space H.

Focusing our attention on the non-symmetric variational problem, that is more general, the following conditions are valid:

$$\begin{cases} (H, (\cdot, \cdot)) \text{ is a Hilbert space.} \\ V \text{ is a (closed) subspace of } H. \\ a(\cdot, \cdot) \text{ is a bilinear form on } V. \\ a(\cdot, \cdot) \text{ is continuous (bounded) on } V. \\ a(\cdot, \cdot) \text{ is coercive on } V. \end{cases}$$

Then the non-symmetric variational problem is the following, given $F \in V'$, find $u \in V$, such that $a(u, v) = F(v), \forall \, v \in V$, where V' is the dual space of V.

The discrete form or the Galerkin approximation of this problem is the following, given a finite dimensional subspace $V_h \subset V$ and $F \in V'$, find $u_h \in V_h$ such that

$$a(u_h, v) = F(v), \quad \forall \, v \in V_h. \tag{1}$$

Theorem 1 (Lax-Milgram) *Given a Hilbert space* $(V, (\cdot, \cdot))$, *a continuous, coercive bilinear form* $a(\cdot, \cdot)$ *and a continuous linear functional* $F \in V'$, *there exists a unique solution* $u \in V$, *such that*

$$a(u, v) = F(v), \quad \forall \, v \in V. \tag{2}$$

This theorem guarantees existence and uniqueness of the solution for both the variational and the approximation problems under the conditions mentioned previously and its proof can be found in [15, 16]. We define the *energy norm*, $\|\cdot\|_E$ as

$$\|v\|_E = \sqrt{a(v.v)}, \quad \forall \, v \in V. \tag{3}$$

Based on the above definition for the energy norm and with the use of the Schwartz' inequality the error estimate for the previous problem (2) is proven to be

$$\|u - u_h\|_E = \inf\{\|u - v\|_E : v \in S\}, \tag{4}$$

where u is the solution and u_h the approximate one and $v \in S$, S a finite dimensional subspace of V. This is the basic error estimate and is optimal in the energy norm. Moreover, in some cases it can be proved that we can replace "infimum" with "minimum," more details can be found elsewhere [15],

$$\|u - u_h\|_E = \min\{\|u - v\|_E : v \in S\}. \tag{5}$$

We further focus our attention on a specific linear parabolic problem the Stokes problem.

3 The Stokes Problem

Initially, we consider the stationary Stokes problem for incompressible flow. Ω is a bounded open set of $\mathbb{R}^n$ (where $n = 2, 3$) with regular boundary and $\mathbf{f}$ is a square integrable function on Ω. We seek a solution $(\mathbf{u}, p) \in H_0^1(\Omega)^2 \times (L^2(\Omega)/\mathbb{R})$ of the problem,

$$\begin{cases} -\Delta\mathbf{u} + \nabla p = \mathbf{f} \;\; in \;\; \Omega, \\ div\, \mathbf{u} = 0 \;\; in \;\; \Omega, \\ \mathbf{u} = 0 \;\; on \;\; \partial\Omega. \end{cases} \tag{6}$$

Based on this problem, we will introduce the error estimates (*a priori* and *a posteriori*) and we briefly discuss about the uniqueness of the solution for this problem [10]. Our goal is to extend these arguments for the non-stationary case.

According to the finite element analysis we end up with the following weak form:

$$\begin{cases} a\,(\mathbf{u}, \mathbf{v}) + b\,(p, \mathbf{v}) = (\mathbf{f}, \mathbf{v})\,, & \forall\, \mathbf{v} \in H_0^1(\Omega)^n,\;\; \mathbf{u} \in H_0^1(\Omega)^n, \\ b\,(\mathbf{u}, q) = 0 & \forall\, q \in H^1(\Omega),\;\; p \in H^1(\Omega), \end{cases} \tag{7}$$

where $a\,(\mathbf{u}, \mathbf{v}) = \displaystyle\int_\Omega \nabla\mathbf{u}\, \nabla\mathbf{v}\, d\Omega$ and $b\,(p, \mathbf{v}) = \displaystyle\int_\Omega p\, \nabla\mathbf{v}\, d\Omega$.

Given two finite dimensional subspaces $V_h \subset H^1(\Omega)^n$ and $Q_h \subset H^1(\Omega)$ the corresponding discrete form is

$$\begin{cases} a\,(\mathbf{u}_h, \mathbf{v}_h) + b\,(p_h, \mathbf{v}_h) = (\mathbf{f}, \mathbf{v}_h)\,, & \forall\, \mathbf{v}_h \in V_{0h},\;\; \mathbf{u}_h \in V_{0h}, \\ b\,(\mathbf{u}_h, q_h) = 0, & \forall\, q_h \in Q_h,\;\; p_h \in Q_h, \end{cases} \tag{8}$$

where $V_{0h} = \{\mathbf{v}_h \in V_h : \mathbf{v}_h\, |_{\partial\Omega} = 0\}$.

Two cases are analyzed for both triangular and quadrilateral elements depending on the number of nodes on each element [10]. We focus only on the Taylor–Hood method (six node triangular elements), second order polynomials for the velocity and first order polynomials for the pressure at each element ($P_2 - P_1$).

After finding a solution, for the problem under consideration, it is important to study whether the stability of the problem is affected by the input data. This can be done using the inf–sup condition, the Ladyzhenskaya–Babuska–Brezzi (LBB) condition. This is a condition for saddle point problems, i.e. problems arise in different types of discretization of equations. Convergence is ensured for most discretization schemes for positive definite problems but for saddle point problems there are still discretizations that are unstable, due to spurious oscillations [89]. In these cases a better approach is the adaptation of the computational grid [78]. We further discuss for the BB condition, introducing the following theorem.

Theorem 2 *If Ω is polygonal and $\Omega_h = \Omega$, $\Omega_h = \bigcup_i T_i$, where T_i are the triangles and h denotes the length of greatest triangle side, if all triangles have at least one vertex which is not on $\partial\Omega$, if V_h, Q_h are chosen as in the Taylor–Hood method, then there exists a constant C, independent of h, such that*

$$sup_{\mathbf{v}_h \in V_{0h}} \frac{(\mathbf{v}_h, \nabla q_h)}{(\mathbf{v}_h, \mathbf{v}_h)^{\frac{1}{2}}} \geq C \left(\nabla q_h, \nabla q_h\right)^{\frac{1}{2}}, \quad \forall q_h \in Q_h. \tag{9}$$

This theorem follows the idea of the BB condition and the proof depends on the choice of the elements and can be found in [10]. One of the most important questions in solving such a problem is that of existence and uniqueness of the solution. In this case we focus on the discrete form of the problem under consideration, (8) where we can ensure the previous with the following theorem.

Theorem 3 *Under the conditions of Theorem 2 the discrete form, Equation* (8), *has a unique solution $(\mathbf{u}_h, p_h)$ in $V_{0h} \times (Q_h/\mathbb{R})$.*

Additionally, we are interested in error estimates of the Stokes problem as discussed in the following sections.

3.1 A Priori Error Estimates

The *a priori* error estimates depend only on the exact solution, but not on the approximated one. On the other hand, the *a posteriori* error estimates require computation of the solution. *A posteriori* error estimates can also provide results on which element size gives a larger error contribution leading to conclusions about grid adaptation [78]. A theorem that provides *a priori* error estimates for the discrete form of the stationary Stokes problem using Taylor–Hood elements ($P_2 - P_1$) is as follows.

Theorem 4 *Let Ω be a polygon and $\Omega_h = \Omega$ for all h. We assume that each element of $\mathcal{T}_h$ (set of triangles) has at least one vertex not on the boundary. Then the following inequalities are valid:*

$$
\begin{aligned}
\|\nabla(\mathbf{u}-\mathbf{u}_h)\|_0 &\leqslant h^2K\left(\|\mathbf{u}\|_{H^3(\Omega)^N}+\|p\|_{H^2(\Omega)/R}\right), \\
\|\nabla(p-p_h)\|_0 &\leqslant hK\left(\|\mathbf{u}\|_{H^3(\Omega)^N}+\|p\|_{H^2(\Omega)/R}\right).
\end{aligned} \tag{10}
$$

Similar inequalities can be found in the case where we have quadrilaterals [10].

Expanding previous arguments for the non-stationary problem we find that there are not as many studies as in the previous case. According to Kemmochi [59] for the non-stationary Stokes problem,

$$
\begin{cases}
\mathbf{u}_t-\Delta\mathbf{u}+\nabla p=\mathbf{f} \;\; in \;\; \Omega\times[0,T], \\
div\,\mathbf{u}=0 \;\; in \;\; \Omega\times[0,T], \\
\mathbf{u}=0 \;\; on \;\; \partial\Omega\times[0,T], \\
\mathbf{u}(\cdot,0)=\mathbf{u}_0(\cdot) \;\; in \;\; \Omega,
\end{cases} \tag{11}
$$

the error estimates for the velocity $\mathbf{u}$ and pressure p are

$$
\begin{aligned}
\|\mathbf{u}-\mathbf{u}_h\|_H &\leqslant Ch^2t^{-1}\|\mathbf{u}_0\|_H, \\
\|p-p_h\|_Q &\leqslant Cht^{-1}\|\mathbf{u}_0\|_H.
\end{aligned} \tag{12}
$$

Remark 1 The difference between the *a priori* error estimates for the stationary and the non-stationary Stokes problem is the introduction of the time variable in the results. Additional results can be obtained for the time derivative for the non-stationary Stokes problem.

In many cases, of the classic finite element approach, the LBB condition is not satisfied, thus it is necessary to find a way to solve the problems and also satisfy this condition. An effective way to overcome this problem is to utilize the adaptive FEM. In the following, we analyze the method suggested by Arnold, Brezzi, and Fortin for the Stokes problem [4]. The discrete form is

$$
\begin{cases}
\sum_{i,j=1}^{2}\int_\Omega \epsilon_{ij}(u)\epsilon_{ij}(v)\,dx-\int_\Omega p\,\nabla\cdot u\,dx=\int_\Omega fv\,dx \;\; \forall v\in(H_0^1(\Omega))^2, \\
\int_\Omega q\,\nabla\cdot u\,dx=0 \;\; \forall q\in L^2(\Omega)/R,
\end{cases} \tag{13}
$$

where $\epsilon_{ij}(u)=\dfrac{(\partial_i u_j+\partial_j u_i)}{2}$. This method is based on using the MINI element as a way to satisfy the inf–sup condition introducing an operator $\Pi_h : (H_0^1(\Omega))^2 \to V_h$. Thus the second equation can be written as

$$
\int_\Omega q_h div\,(\Pi_h v-v)\,dx=0, \;\; \forall q_h\in Q_h, \forall v\in\left(H_0^1\right)^2 \tag{14}
$$

and

$$\|\Pi_h v\|_1 \leq c\,\|v\|_1 \quad \forall v \in \left(H_0^1\right)^2. \tag{15}$$

For the MINI element the space is

$$V_h = \left(\overset{\circ}{M}{}_0^1\right)^2 \oplus \left(B^3\right)^2, \quad Q_h = M_0^1, \tag{16}$$

where

$$M_0^k(\mathscr{T}_h) = \left\{v \mid v \in C^0(\Omega),\, v_{|T} \in P_k(T),\ \forall\, T \in T_h\right\},\ \overset{\circ}{M}{}_0^k(\mathscr{T}_h) = M_0^k(\mathscr{T}_h) \cap H_0^1(\Omega) \tag{17}$$

for $k \geq 1$ and

$$B^k(T_h) = \left\{v \mid v_{|T} \in P_k(T) \cap H_0^1(T),\ \ \forall\, T \in T_h\right\}, \tag{18}$$

for $k \geq 3$ and T the triangular elements of $\mathscr{T}_h$. For the problem based on the MINI elements, the following argument is valid:

$$\|u - u_h\|_1 + \|p + p_h\|_{0/R} \leq C \inf\left\{\|u - v\|_1 + \|p + q\|_{0/R}\right\} \leq Ch\,\|f\|_0, \tag{19}$$

where C is independent of h. These spaces can be further extended leading to other methods [4]. For example, there is a case where it can be seen as an enriched version of Taylor–Hood method where convergence is simpler than the classical Taylor–Hood method. In other methods discontinuous approximation of the pressure is used as mentioned in Crouzeix–Raviart [4, 30].

3.2 A Posteriori Error Estimates

In this section we focus our attention on *a posteriori* estimates for the approximation of time dependent Stokes equations. We introduce the notion of the Stokes reconstruction operator and present the error equation that satisfies the exact divergence-free condition described in detail in [57].

The energy technique for *a posteriori* error analysis of finite element discretizations of parabolic problems provides suboptimal rates in the $L^\infty(0, T; L^2(\Omega))$ norm. Makridakis and Nochetto in their study combine energy techniques with appropriate pointwise representation of the error based on an elliptic reconstruction operator which restores optimal order and regularity for piecewise polynomials of degree higher than one [68]. Additionally, Lakkis and Makridakis based on the previous work derive *a posteriori* error estimates for fully discrete approximations of the solutions of linear parabolic equations. The discretization uses finite element spaces that change in time [62]. Akrivis and collaborators presented a refined analysis for quasilinear parabolic problems applying implicit-explicit multistep finite element schemes [1]. Let us consider the non-stationary Stokes problem for incompressible flow. These equations are discretized in space by the finite elements or the finite volumes method. This problem is still open and directly related to Navier–Stokes equations. This is due to the fact that the *a posteriori* error theory is still in progress as reported by several researchers [11, 34, 57, 62, 68]. We assume the availability of *a posteriori* estimator for the Stokes problem, expressed by the following assumption.

Assumption Let $(\mathbf{w}, q) \in \mathbf{Z} \times \Pi$ and $(\mathbf{w}_h, q_h) \in \mathbf{Z}_h \times \Pi_h$ be the exact solution and its finite element approximation. For the space X (equal to $\mathbf{H} = (L^2(\Omega))^d$, $\mathbf{V} = (H_0^1(\Omega))^d$, $d = 2, 3$ or $\mathbf{V}'$ the dual space of $\mathbf{V}$), we assume that there exists *a posteriori* estimator function, $\mathscr{E}((\mathbf{w}_h, q_h)$ and $\mathscr{E}_{pres}((\mathbf{w}_h, q_h)$, which depend on $(\mathbf{w}_h, q_h)$, $\mathbf{g}$ and the corresponding norm, such that

$$\|\mathbf{w} - \mathbf{w}_h\|_X \leq \mathscr{E}((\mathbf{w}_h, q_h), \mathbf{g};\ X) \quad and \quad \|q - q_h\|_\Pi \leq \mathscr{E}_{pres}((\mathbf{w}_h, q_h), \mathbf{g};\ \Pi). \tag{20}$$

It can be shown that the discrete solution coincides with the continuous solution [57]. In order to define the Stokes reconstruction as introduced by Karakatsani and Makridakis, 2007, we provide the following definitions [46, 57],

Definition 2 (Stokes Operator) Let $\bar{\Delta} : \mathbf{H}^2 \cap \mathbf{Z} \subset \mathbf{J} \to \mathbf{J}$ be the Stokes operator, meaning, the L^2-projection of the Laplace operator onto $\mathbf{J}$. Then introducing the discrete version of the Stokes operator $\bar{\Delta}_h : \mathbf{Z}_h \to \mathbf{Z}_h$ by,

$$\left\langle \bar{\Delta}_h \mathbf{v}, \chi \right\rangle = -a(\mathbf{v}, \chi), \quad \forall \chi \in \mathbf{Z}_h. \tag{21}$$

Definition 3 (Stokes Reconstruction) For fixed $t \in [0, T]$, let $(\mathbf{U}, P) \in \mathbf{V} \times \Pi$ be the solution of the stationary Stokes problem,

$$\begin{cases} a(\mathbf{U}, \mathbf{v}) + b(\mathbf{v}, P) = \langle \mathbf{g}_h(t), \mathbf{v} \rangle, & \forall\ \mathbf{v} \in \mathbf{V}, \\ b(\mathbf{v}, P) = 0, & \forall\ q \in \Pi, \end{cases} \tag{22}$$

where

$$\mathbf{g}_h = -\Delta_h \mathbf{u}_h - \mathbf{f}_h + \mathbf{f}. \tag{23}$$

We call $(\mathbf{U}, P) = (\mathbf{U}(t), P(t))$ the Stokes reconstruction of the discrete velocity and pressure fields, $(\mathbf{u}_h(t), p_h(t))$.

Based on the above definitions Karakatsani and Makridakis, 2007, introduce the following theorem, where it provides the error equations based on the *a posteriori* estimator function introduced before [57].

Theorem 5 (Error Equation) *Let* $(\mathbf{U}, P)$ *be the Stokes reconstruction and* $(\mathbf{u}, p)$ *the solution of the Stokes problem which is assumed to be sufficiently regular. If* $\mathbf{e} = \mathbf{U} - \mathbf{u}$ *and* $\varepsilon = P - p$, *then* $(\mathbf{e}, \varepsilon)$ *is the weak solution of the problem,*

$$\begin{cases} \mathbf{e}_t - \Delta \mathbf{e} + \nabla \varepsilon = (\mathbf{U} - \mathbf{u}_h)_t, \\ div\ \mathbf{e} = 0. \end{cases} \tag{24}$$

Additionally, $\mathbf{U} - \mathbf{u}_h$ *and* $(\mathbf{U} - \mathbf{u}_h)_t$ *satisfy the following estimates:*

$$\left\| \partial_t^{(j)} (\mathbf{U} - \mathbf{u}_h) \right\|_X \leq \mathscr{E}((\partial_t^{(j)} \mathbf{u}_h, \partial_t^{(j)} p_h), \partial_t^{(j)} \mathbf{g}_h;\ X), \quad j = 0,\ 1, \tag{25}$$

where X *is one of the spaces,* $\mathbf{H}$, $\mathbf{V}$ *or* $\mathbf{V}'$, *discussed before and* $\mathscr{E}$ *is the a posteriori estimator function defined in previous assumption. The proof of this theorem can be found in [57].*

Theorem 6 ($L^\infty(\mathbf{H})$ **and** $L^2(\mathbf{V})$ **Norm Error Estimates)** *Let us assume that* $(\mathbf{u}, p)$ *is the solution of the time dependent Stokes problem, Equation* (11), *and* $(\mathbf{u}_h, p_h)$ *is the finite element approximation. Let* $(\mathbf{U}, P)$ *be the solution of the stationary Stokes problem and* $\mathscr{E}$ *is the a posteriori estimator function defined previously. Then the following a posteriori error bounds hold for,* $0 < t \leq T$,

$$\begin{aligned} &\|\mathbf{u}(t) - \mathbf{U}(t)\|_{\mathbf{H}}^2 + \int_0^t \|(\mathbf{u} - \mathbf{U})(s)\|_{\mathbf{V}}^2\, ds \\ &\qquad \leqslant \|\mathbf{u}(0) - \mathbf{U}(0)\|_{\mathbf{H}}^2 + \int_0^t \mathscr{E}((\mathbf{u}_{h,t}, p_{h,t}), \mathbf{g}_{h,t}; \mathbf{V}')^2 ds. \end{aligned} \tag{26}$$

Additional inequalities and the proof of this theorem can be found in [57].

They additionally provide a theorem for $L^\infty(\mathbf{V})$ norm error estimates and at the same study they discuss about estimates using the parabolic duality argument [34, 90]. In this study two related applications of the reconstruction of the Stokes problem are discussed [57].

3.3 Crouzeix–Raviart Finite Element Discretization and Finite Volume Scheme

An *a posteriori* bound for the time dependent Stokes problem under the Crouzeix–Raviart finite element approximation is derived. However, further detailed work is required related to the specific form of possible singularities of the exact solution for this problem [57]. The finite volume (FV) scheme approximations is the Crouzeix–Raviart couple $\mathbf{V}_h \times \Pi_h$. The FV methods rely on local conservation properties of the differential equations under consideration over the “control volume.” Integrating over a region $b \subset \Omega$ and utilizing the Green’s formula, we obtain the following system for the Stokes problem in the discrete form,

$$\begin{cases} \int_{b_e} \mathbf{u}_{h,t} - \int_{\partial b_e} \nabla \mathbf{u}_h \mathbf{n} + \int_{\partial b_e} p_h \mathbf{n} = \int_{b_e} \mathbf{f}, \quad \forall\, e \in E_h, \\ \int_K div\ \mathbf{u}_h = 0, \quad \forall\, K \in \mathscr{T}_h. \end{cases} \tag{27}$$

where z_K is an inner point of $K \in \mathscr{T}_h$, connecting the point with line segments to the vertices of the triangle K, we partition it into three segments K_e, where $e \in E_h(K)$, then each side e is associated with a quadrilateral, b_e, which is the union of the subregions K_e. Chatzipandelidis et al. have introduced *a priori* and *a posteriori* error estimates for the FV methods and for the stationary Stokes problem with the admission that FV scheme provides a variational formulation similar to the FE scheme [21]. These studies highlight the importance of *a posteriori* error estimates on a theoretical basis especially for parabolic problems such as the Stokes problem [12, 21].

We highlight the main finding from Karakatsani and Makridakis study for the FV scheme that is the following theorem,

Theorem 7 (Residual Based $L^2(H^1)$ and $L^\infty(H^1)$ Norm Error Estimates) *Let us assume that* $(\mathbf{u}, p)$ *is the solution of the time dependent Stokes problem and* $(\mathbf{u}_h, p_h)$ *is the finite volume approximation. The following a posteriori error bounds hold for,* $0 < t \leq T$,

$$\begin{aligned} \|\nabla(\mathbf{u} - \mathbf{u}_h)(t)\|_H \leq \left\|\mathbf{u}_0 - \mathbf{u}_h^0\right\|_V + C\left(\int_0^t \eta_1(\mathbf{u}_{h,t}(s))^2 ds\right)^{1/2} \\ + C\,\eta_1\,(\mathbf{u}_h(0)) + C\,\eta_1(\mathbf{u}_h(t)), \end{aligned} \tag{28}$$

Additional inequalities and the proof of this theorem can be found in [57].

Further, Larson and Malqvist derived a residual based *a posteriori* error estimates for parabolic problems on mixed form using Raviart-Thomas-Nedelec (RTN) finite elements in space and backward Euler in time [63]. In their study an *a posteriori* error estimate for the divergence of the flux in a weak norm is derived. The concept of elliptic reconstruction has been used to derive *a posteriori* error estimates for parabolic problems as briefly described before [57, 68]. In this framework, Larson and Malqvist use known *a posteriori* error estimates for the corresponding elliptic problem to derive error bounds for the parabolic problem [63]. However, the literature on FEM for parabolic problems on mixed form is less extensive and the development of the theory is still in progress [31, 90].

4 Advection–Diffusion Equation

The steady-state advection and diffusion of a scalar field is described by the partial differential equation (assuming homogeneous Dirichlet boundary condition),

$$\alpha \cdot \nabla u - \nabla \cdot (D\nabla u) = f \quad \text{in } \Omega, \tag{29}$$

$$u = 0 \quad \text{on } \partial\Omega, \tag{30}$$

where α is the velocity that the quantity, u, is moving with, which is considered to be divergent-free, $\nabla \cdot \alpha = 0$ [17]. For example, take as quantity the concentration of a chemical species that diffuses in a river while moving with its velocity α. The diffusion coefficient of the quantity is denoted with D and f represents sources or sinks.

The advection–diffusion problems are frequently treated as the point of departure for the study of the non-linear Navier–Stokes equations, at the level of developing discretization methods. The Peclet number, defined as the ratio of the advection and diffusion rates, $Pe = |a|h/D$, is a characteristic dimensionless number for such problems. A small Peclet number ($Pe \ll 1$) indicates diffusion-dominated flows while a large one ($Pe \gg 1$) indicates advection-dominated flows. In the diffusion-dominated regime, the standard Galerkin finite element method provides a good approximation of the solution [14].

The standard variational formulation arises by requesting the residual of Equation (29) to be orthogonal to a basis of the function space, H_0^1. The task is to find $u \in H_0^1(\Omega)$ such that

$$(\alpha \cdot \nabla u, v) - (\nabla \cdot (D\nabla u), v) = (f, v), \tag{31}$$

is satisfied for any test function $v \in H_0^1(\Omega)$. The Sobolev space, H_0^1, consists of functions that are one time weakly differentiable and also satisfy the zero Dirichlet boundary condition. In this respect, the second order term of the weak formulation can be integrated by parts, leading to,

$$(\alpha \cdot \nabla u, v) + (D\nabla u, \nabla v) = (f, v). \tag{32}$$

4.1 The Galerkin Formulation

To approximately solve Equation (32) using the Finite Element method, Ω is discretized in non-overlapping triangle element domains Ω_e with boundaries Γ_e, $e = 1, 2 \ldots K$, such that,

$$\Omega = \bigcup_{k=0}^{K} \overline{\Omega_k}.$$

The standard Galerkin formulation is retrieved by searching a solution in a finite-dimensional linear polynomial function space, $V_h \subset H_0^1(\Omega)$,

$$V_h = \{v_h \in H_0^1(\Omega) \mid v_h(\Omega_k) \in P_1(\Omega_k),\ \Omega_k \in \Omega\}$$

The problem now states, find $u_h \in V_h(\Omega)$ such that,

$$(\alpha \cdot \nabla u_h, v_h) + (D\nabla u_h, \nabla v_h) = (f, v_h),\ \ \forall\, v_h \in V_h(\Omega). \tag{33}$$

4.2 The Stabilized Finite Element Methods

It is well known that for advection-dominated flows, where the Peclet number is large, the solution involves non-physical oscillations [17]. To address the deficiency of the standard polynomial finite element method for advection-dominated flow problems, various approaches have been proposed, such as the streamline upwind Petrov–Galerkin (SUPG) method [18], Galerkin least squares (GLS) method [54], and the unusual stabilized FEM (USFEM) [38]. The common characteristic of the aforementioned methods is the introduction of artificial diffusion in the solution process while preserving the consistency of the discretization. Such methods are commonly referred to as stabilized finite element methods (SFEM).

The SFEM for the stationary advection–diffusion problem can be grouped as follows: find $u_h \in V_h(\Omega)$ such that,

$$B(u_h, v_h) = F(v_h),\ \ \forall\, v_h \in V_h(\Omega), \tag{34}$$

where,

$$B(u_h, v_h) = (\alpha \cdot \nabla u_h, v_h) + (D\nabla u_h, \nabla v_h) + Q(u_h, v_h), \tag{35}$$

$$F(v_h) = (f, v_h), \tag{36}$$

where $Q(u_h, v_h)$ indicates the additional terms added to the standard variational formulation. These are added to preserve consistency and enhance numerical stability. For instance, the stability term corresponding to the SUPG method is,

$$Q_{SUPG}(u_h, v_h) = \sum_K \tau_k(\alpha \cdot \nabla u_h - \nabla \cdot (k\nabla u_h) - f, \alpha \cdot \nabla u_h)_k, \tag{37}$$

where $(\cdot, \cdot)_k$ denotes element wise integration and τ_k is the stability coefficient for the SUPG method, as defined in [39],

$$
\begin{cases}
\tau_k = \dfrac{h_k}{2|\alpha|_p}\xi(Pe_k), \\
Pe_k = \dfrac{m_k|a|_p h_k}{2k}, \\
\xi(Pe_k) = \begin{cases} Pe_k, & 0 \le Pe_k < 1 \\ 1, & Pe_k \ge 1 \end{cases} \\
|\alpha|_p = \left(\sum_{i=1}^{N} |\alpha_i|_p\right)^{\frac{1}{p}}, \quad 1 \le p < \infty, \\
m_k = \min\left\{\dfrac{1}{3},\ 2C_k\right\}, \\
C_k \sum_K h_k^2 \left\|\Delta v_h\right\|_{0,K}^2 \leqslant \left\|\nabla v_h\right\|_0^2, \ v_h \in V_h.
\end{cases}
\tag{38}
$$

Accordingly, the stability terms added to the standard variational formulation for the GLS and the USFEM methods are,

$$
Q_{GLS}(u_h, v_h) = \sum_K \tau_k(\alpha \cdot \nabla u_h - \nabla \cdot (D\nabla u_h) - f, \alpha \cdot \nabla u_h - \nabla \cdot (D\nabla u_h))_k,
\tag{39}
$$

$$
Q_{USFEM}(u_h, v_h) = \sum_K \tau_k(\alpha \cdot \nabla u_h - \nabla \cdot (D\nabla u_h) - f, \alpha \cdot \nabla u_h + \nabla \cdot (D\nabla u_h))_k.
\tag{40}
$$

The stability of the SUPG method for transient convection–diffusion equations is studied in [13]. In the work by Onate [79], it was proven that the stabilization terms can be interpreted as a natural contribution to the governing differential equations of advection–diffusion problems. By considering the concept of flow equilibrium, the stabilization terms emerging in methods such as SUPG, subgrid scale (SS), GLS, Lax–Wendroff, characteristic Galerkin, Laplacian pressure operator, are not introduced as correction terms at the discretization level but rather derive naturally. For a comprehensive analysis of SFEM for the stationary or non-stationary advection-diffusion–reaction equation, the review by Codina [28] is recommended.

Writing the advection–diffusion equation in its first-order form via introduction of the flux of the scalar field as an additional unknown, is suited for many problems where higher accuracy of the flux is important, such as flow in porous media. Masud et al. studied the first-order form of the advection–diffusion equation in the framework of SFEM [74] .

Based on the partition of unity framework that is an instance of the generalized finite element method (GFEM), Turner et al. improved the performance of the Galerkin formulation designing enrichment functions using *a priori* knowledge about the qualitative behavior of solution to make better choices for the local approximation space [91]. The proposed method differs from the standard stabilization strategies as stability is not achieved by adding terms but by multiplying the polynomial with the enrichment functions.

4.3 The Variational Multiscale Method

Stabilized SUPG-type methods have several advantages, such as applicability to a wide range of problems and simplicity in computer implementations. However, spurious oscillations are often observed in regions around sharp layers even with the enhanced stability provided by the SUPG method [84]. To overcome the sharp features of the solution of advection-dominated problems, a higher resolution of the grid is usually employed which is, however, impractical in many cases. The multiscale approach comes at hand when fine scales cannot be captured by a given discretization in space [72].

The variational multiscale method was introduced in [49, 51] as a procedure for deriving numerical methods capable of dealing with multiscale phenomena that the straightforward application of the Galerkin's method with standard bases cannot address. It can be considered as a procedure to rebuild the error term in the weak form of the problem, yielding a stabilized form of the problem with higher accuracy on coarse grids.

The union of element interiors and element boundaries is denoted by Ω' and Γ', respectively,

$$\Omega' = \bigcup_{k=0}^{K} \Omega_k, \qquad \Gamma' = \bigcup_{k=0}^{K} \Gamma_k. \tag{41}$$

The appropriate function spaces for the coarse and the fine scale fields are introduced via a direct sum decomposition,

$$V = \overline{V} \oplus V', \tag{42}$$

where $\overline{V}$ is the space of trial and test functions for the coarse scale field,

$$\overline{V} = \{\overline{v} \in H_0^1(\Omega) \mid \overline{v}(\Omega_k) \in P_n(\Omega_k),\ \Omega_k \in \Omega\}, \tag{43}$$

where $P_n(\Omega_e)$ denotes polynomials of order n over the element interior.

In the discrete case, V' can contain various finite dimensional approximations such as bubble functions or p-refinements that further satisfy the assumption that the fine scales vanish identically over the element boundaries. Consequently,

$$V' = \left\{v' \mid v' = 0 \quad \text{on} \quad \Gamma'\right\}. \tag{44}$$

In this respect, the scalar field is decomposed in the coarse and fine scales denoted by $\overline{u}_h$ and u'_h, respectively,

$$u_h(x) = \overline{u_h}(x) + u'_h(x). \tag{45}$$

Likewise, the trial function is decomposed in its coarse and the fine scale components indicated as $\overline{v_h}$ and v'_h, respectively,

$$v_h(x) = \overline{v_h}(x) + v'_h(x). \tag{46}$$

Remark 2 Alternatively, the decomposition can be interpreted as the split of the solution in the part obtained on a given mesh and the part that is lost because its scale is smaller than the characteristic length of this mesh, representing the error in the solution.

The decomposed trial and test functions are substituted in the standard variational form (33), leading to,

$$(\alpha \cdot \nabla(\overline{u_h}+u'_h), (\overline{v_h}+v'_h))+(k\nabla(\overline{u_h}+u'_h), \nabla(\overline{v_h}+v'_h)) = (f, (\overline{v_h}+v'_h)). \tag{47}$$

Employing the linearity of the weighting function, the problem can be split into the coarse and the fine scale parts, indicated as $\overline{v_h}$ and v'_h. The coarse scale sub-problem can be written as

$$(\alpha \cdot \nabla(\overline{u_h} + u'_h), \overline{v_h}) + (D\nabla(\overline{u}_h + u'_h), \nabla\overline{v_h}) = (f, \overline{v_h}). \tag{48}$$

The fine scale sub-problem can be written as

$$(\alpha \cdot \nabla(\overline{u_h} + u'_h), v'_h) + (D\nabla(\overline{u_h} + u'_h), \nabla v'_h) = (f, v'_h). \tag{49}$$

When compared with the standard Galerkin method, the multiscale approach involves additional integrals that are evaluated element wise. These additional terms represent the effects of the subgrid scales in terms of the residuals of the coarse scales of the problem. The architecture of the method is simple: u'_h is determined analytically and is eliminated from the $\overline{u_h}$ problem that is computed numerically. $\overline{u_h}$ and u_h may overlap or be disjoint, and u_h may be globally or locally defined, while the effect of u_h on the $\overline{u_h}$ problem is nonlocal [49].

Hughes et al. generalized the problem working in the context of an abstract Dirichlet problem involving a second-order differential operator which enables the study of equations of practical interest, such as the advection–diffusion equation [55, 56]. After introducing the variational formulation of the Dirichlet problem, the authors took advantage of the multiscale approach.

An overview of finite element approximations to deal with oscillations near layers using the variational multiscale formulation is presented in [29], where the time-discretization of the sub-grid scales is also addressed. Recently, Sendur et al.

used the pseudo residual-free bubbles (PRFB) method to achieve discretization in space and the fractional-step θ-scheme for the discretization in time [84] . The discontinuous enrichment method augments the polynomial field by free-space solutions of the homogeneous differential equation differentiating from the standard bubble methods in enforcing continuity across element boundaries by Lagrange multipliers [91].

4.4 The Discontinuous Galerkin (DG) Method

Another class of important methods is the discontinuous Galerkin (DG) methods that are popular in convection-dominated advection–diffusion problems due to their good stability and local conservation properties [27, 76]. The DG methods have several advantages such as high order accuracy, local data structure, and high parallelization capacity, attracting the interest of several groups [6, 22, 25, 40]. Moreover, the DG methods can cover meshes with hanging nodes and/or locally varying polynomial degrees rendering them ideally suited for *hp*-adaptivity. In contrast with the continuous approach, in the discontinuous context, the local elemental bases can be chosen freely due to the lack of inter-element continuity requirements, yielding sparse mass matrices [43].

For advection and (advection-dominated) advection–diffusion equations, *hp*-finite element approximations have been investigated by Houston, Schwab, and Suli for interior penalty discontinuous finite elements [47], leading to the so-called *hp*-streamline diffusion method and the *hp*-discontinuous Galerkin method [19, 20].

To capture detailed features of the solution near singular points or sharp layers, a very fine mesh is required. However, the computation of the solution is very challenging due to the amount of computer memory and time needed. For quicker convergence and to reduce the computational cost, the mesh can be refined locally at suitable locations. For stationary convection–diffusion equations, the quest for robust *a posteriori* error estimators that are independent of the Peclet number has advanced in various contexts [83, 93, 97]. For instance, *a posteriori* estimates using the reconstruction of the flux term can be found in [36]. In non-stationary convection–diffusion equations, as time progresses, the nature of the solution may vary throughout the domain rendering the use of adaptive algorithms an attractive proposition for the accurate and efficient numerical approximation of such problems. As adaptive algorithms are usually based on suitable *a posteriori* error estimators, robust estimation of the temporal and spatial error depends on their formulation. For non-stationary linear convection–diffusion equations, *a posteriori* error estimators have been developed for various discretizations [2, 35, 42, 92].

Chung and Enquist [23] in 2006 conceived the staggered DG (SDG) method that is a sub-class of the DG method. The introduction of the staggered mesh approach automatically satisfies the preservation of the physical laws arising from the corresponding partial differential equations. The SDG method can be continuous along some of the faces and discontinuous along other faces. An SDG scheme for the convection-diffusion equation was proposed by Chung and Lee [24] in 2012. Recently, an adaptive SDG method to solve the steady state convection–diffusion equation was presented by Du et al. [33]. The study by Cockburn et al. [27] is devoted to some new DG methods for convection–diffusion–reaction problems, called local discontinuous Galerkin–hybridizable (LDG–H) methods. Three novel features render these methods attractive. Namely, the first is that they are hybridizable and hence efficiently implementable, the second is that they provide approximations for the flux which are optimally convergent when both the flux and the scalar variables are approximated by polynomials of the same degree on each element. Finally, the third feature is that the approximations to the scalar variable super converge.

5 The Navier–Stokes Problem

In finite element formulation and computation of incompressible flows there are two main sources of instabilities associated with the classical Galerkin formulation of the Navier–Stokes problem. One source of instabilities is due to the presence of advection terms leading to spurious oscillations mainly in the velocity field, as discussed in the previous section. The other source of instability is due to an inappropriate combination of interpolation functions for the velocity and pressure field. These instabilities usually appear as oscillations primarily in the pressure field [89]. Below, we present the most interesting FE methodologies for solving the Navier–Stokes problem.

5.1 Streamline–Upwind/Petrov–Galerkin (SUPG)

The most popular stabilized method, the Streamline–Upwind/Petrov–Galerkin (SUPG) formulation, was introduced in 1979 for the incompressible Navier–Stokes equations [9, 50]. By augmenting the Galerkin formulation with residual-based terms, the SUPG formulation addressed the instability of the Galerkin technique for convection dominated flows, leading to a stable method with optimal convergence properties. For compressible flows, the SUPG formulation was initially introduced in 1982 [87], but a more thorough presentation of the method with additional examples published in [52]. The compressible flow SUPG formulation was initially introduced for conservation variables, and later for primitive variables. For more details on these developments, the interested reader is referred to a recent paper on stabilized methods for compressible flows [9].

The Pressure-Stabilizing/Petrov–Galerkin (PSPG) formulation for the Navier–Stokes equations of incompressible flows in the framework of residual-based methods was introduced in [86, 89]. This method allowed the use of equal-order interpolation functions for the velocity and pressure variables and assured numerical stability and optimal accuracy. An earlier version of the PSPG formulation for the Stokes problem was introduced in [53]. The SUPG and PSPG stabilizations were combined under a single name, the SUPS stabilization method [8, 9].

5.2 Variational Multiscale Method (VMS) and Stabilized FEM

Stabilized and multiscale formulations are among the most fundamental and important methodologies for finite element computations of complex fluid mechanics problems. Tezduyar et al. have proposed certain stabilized formulations with bilinear and linear equal-order-interpolation elements for the computation of dynamic and steady incompressible flows [89]. In their study, the stabilization procedure involves a modified Galerkin/least-squares formulation of the steady-state equations. The results from the considered test problems show that the $Q_1 - Q_1$ element is slightly less dissipative than the $P_1 - P_1$ element. The solutions obtained with these elements compare well with the solutions obtained from other studies [88]. The incompressible Navier–Stokes equations are written as

$$\begin{cases} \dot{\mathbf{v}} + \mathbf{v} \cdot \nabla \mathbf{v} - 2\,\nu\,\nabla \cdot \varepsilon(\mathbf{v}) + \nabla p = \mathbf{f}, \quad in\ \Omega \times [0, \mathrm{T}], \\ div\,\mathbf{v} = 0, \quad in\ \Omega \times [0, \mathrm{T}], \\ \mathbf{v} = \mathbf{g}, \quad on\ \Gamma_{\mathbf{g}} \times [0, \mathrm{T}], \\ \boldsymbol{\sigma} \cdot \mathbf{n} = (2\,\nu\,\nabla^s \mathbf{v} - p\mathbf{I}) \cdot \mathbf{n} = \mathbf{h}, \quad on\ \Gamma_{\mathbf{h}} \times [0, \mathrm{T}], \\ \mathbf{v}(\mathbf{x}, 0) = \mathbf{v}_0(\mathbf{x}), \quad on\ \Omega_0, \end{cases} \tag{50}$$

where $\mathbf{v}$ is the velocity vector, p is the kinematic pressure, $\mathbf{f}$ is the body force vector, ν is the kinematic viscosity, $\nabla^s \mathbf{v}$ is the symmetric part of the velocity gradient, $\mathbf{I}$ is the identity tensor, and $\varepsilon(\mathbf{v})$ is the strain rate tensor which is defined as $\varepsilon(\mathbf{v}) = \frac{1}{2}(\nabla \mathbf{v} + \nabla \mathbf{v}^{\mathrm{T}})$. Equation (50) represents the momentum and continuity equations, with the Dirichlet and Neumann boundary conditions, and the initial condition, respectively.

Discretizing the bounded domain Ω into non-overlapping regions Ω^e with boundaries Γ^e, $e = 1, 2, \ldots, n_{el}$, such that $\Omega = \cup_{e=1}^{n_{el}} \Omega^e$. The union of element interiors and element boundaries are $\Omega' = \bigcup_{e=1}^{n_{el}} (\mathrm{int})\Omega^e$ and $\Gamma' = \bigcup_{e=1}^{n_{el}} \Gamma^e$, respectively. In variational multiscale method (VMS) the velocity field is decomposed into the sum of the coarse or resolved scales and the fine or subgrid scales [70, 73],

$$\mathbf{v}(\mathbf{x}, t) = \bar{\mathbf{v}}(\mathbf{x}, t) + \mathbf{v}'(\mathbf{x}, t), \tag{51}$$

and the weighting function is decomposed in its coarse and the fine scale components indicated as $\bar{\mathbf{w}}(\mathbf{x})$ and $\mathbf{w}'(\mathbf{x})$, respectively,

$$\mathbf{w}(\mathbf{x}) = \bar{\mathbf{w}}(\mathbf{x}) + \mathbf{w}'(\mathbf{x}). \tag{52}$$

Remark 3 The main goal of the VMS method is to solve the fine-scale problem, defined over the sum of element interiors to obtain the fine scale solution. This solution is then substituted in the coarse-scale problem, eliminating the explicit appearance of the fine scales while still modeling their effects. Both coarse and fine scale equations are nonlinear equations due to the convection term, and to solve them a linearization is taking place [73].

The resulting equation is expressed in terms of the coarse scales and for the sake of simplicity the superposed bars are dropped. So, the VMS residual-based stabilized form for the incompressible Navier–Stokes equations is

$$\begin{aligned} &(\mathbf{w}, \delta\mathbf{v}_t) + (\mathbf{w}, \delta\mathbf{v} \cdot \nabla\mathbf{v}^{(i)} + \mathbf{v}^{(i)} \cdot \nabla\delta\mathbf{v}) + \beta(\mathbf{w}, \mathbf{v}^{(i)}\nabla \cdot \delta\mathbf{v} + \delta\mathbf{v}\nabla \cdot \mathbf{v}^{(i)}) \\ &+(\nabla^S\mathbf{w}, 2\nu\, \nabla^S\delta\mathbf{v}) - (\nabla \cdot \mathbf{w}, \delta p) + (q, \nabla \cdot \delta\mathbf{v}) \\ &+(\mathbf{v}^{(i)} \cdot \nabla\mathbf{w} + 2\nu\, \Delta\mathbf{w} + \nabla q + (1-\beta)\mathbf{w}\nabla \cdot \mathbf{v}^{(i)}, \boldsymbol{\tau}\,\mathbf{r}_2) \\ &-(1-\beta)\,(\mathbf{w}, (\boldsymbol{\tau}\,\mathbf{r}_2) \cdot \nabla\mathbf{v}^{(i)}) + \beta((\boldsymbol{\tau}\,\mathbf{r}_2) \cdot \nabla\mathbf{w}, \mathbf{v}^{(i)}), \end{aligned} \tag{53}$$

where the last two lines of the equation correspond to the stabilization terms, $\beta \in [0, 1]$, $\mathbf{r}_2$ is the residual from the linearization of the non-linear fine-scale problem, $\boldsymbol{\tau}$ is the fine-scale variational operator, and Δ is the vector Laplacian operator. A significant contribution of the VMS method is the systematic and consistent derivation of the fine-scale variational operator, $\boldsymbol{\tau}$, termed as the stabilization tensor that possesses the right order in the advective and diffusive limits, and variationally projects the fine-scale solution on the coarse-scale space [73]. The stabilization operator can be defined as [70],

$$\begin{aligned} \boldsymbol{\tau} &= b^e \int b^e d\Omega \\ &\times \begin{bmatrix} \int (b^e)^2 \nabla^T \mathbf{v}^{(i)} d\Omega + \int b^e \mathbf{v}^{(i)} \cdot \nabla b^e d\Omega\, \mathbf{I} \\ +\beta \int b^e \mathbf{v}^{(i)} \otimes \nabla b^e d\Omega + \beta \int b^e (\nabla \cdot \mathbf{v}^{(i)})\, d\Omega\, \mathbf{I} \\ +\nu \int |\nabla b^e|^2\, d\Omega\, \mathbf{I} + \nu \int \nabla b^e \otimes \nabla b^e d\Omega \end{bmatrix}^{-1}, \end{aligned} \tag{54}$$

where $b^e(\xi)$ is a bubble function over Ω'. More details on the derivation and the obtained form of the VMS residual-based stabilized form and the fine-scale variational operator, τ, for the incompressible Navier–Stokes equations can be found in [70, 73]. Massud and collaborators have further extended the VMS methodology for shear-rate dependent non-Newtonian fluids and incompressible turbulent fluid flows [61, 71, 75].

5.3 Characteristic Based Split (CBS) Method and Two-Step Methods

The characteristic base split (CBS) method was first introduced by Zienkiewicz and Nithiarasu, 1995, in order to find a similar method to the Taylor–Galerkin, applicable in two or three dimensional problems. The algorithm is based on splitting the equations in two parts where the first would be a scalar convective-diffusion type of equations and the solution is derived from the characteristic Galerkin method [77, 98]. The second part constitutes of self-adjoined equations. There are four forms of the algorithm (fully explicit, semi-implicit, nearly implicit, fully implicit) depending on the problems we are called to solve. Here we focus only on the fully explicit and semi-implicit forms. Initially, we deal with the scalar convection–diffusion problem and the characteristic Galerkin explicit approximation. Assuming that the equation for this problem is

$$\frac{\partial V}{\partial t} = \frac{\partial F_i}{\partial x_i} + \frac{\partial G_i}{\partial x_i} + Q = 0, \tag{55}$$

where x_i is the i-th coordinate, F_i, G_i are the convected and the diffusion flux terms, respectively, and Q is the source term [98]. An alternative form of this equation is

$$\frac{\partial \phi}{\partial t} = -u_j \frac{\partial \phi}{\partial x_j} + \frac{\partial}{\partial x_i}\left(k \frac{\partial \phi}{\partial x_i}\right) - Q - \phi \frac{\partial u_j}{\partial x_j} = R(\phi). \tag{56}$$

The term, $-u_j \frac{\partial \phi}{\partial x_j}$, is not self-adjoined. Introducing a transformation we change the coordinate system, this term can be vanished and the equation will be a fully self-adjoined system. The stability condition for this problem is given as

$$\Delta t \leq \Delta t_{crit} = \frac{h}{\mid u \mid}\left(\sqrt{\frac{1}{Pe^2} + \frac{1}{3}} - \frac{1}{Pe}\right), \tag{57}$$

where Pe is the Peclet number defined as $Pe = \frac{|u|h}{2k}$. For multidimensional problems such as the two-dimensional Navier-Stokes the critical time step will be

$$\Delta t_{crit} = \frac{\Delta t_\sigma \Delta t_\nu}{\Delta t_\sigma + \Delta t_\nu}, \tag{58}$$

where Δt_σ is given by Equation (57) and $\Delta t_\nu = h^2/2k$. If $\Delta t = \Delta t_{crit}$ the steady state solutions are almost identical to that from the optimal streamline upwind methods [98]. The Navier–Stokes problem can be written in a form of the convection–diffusion problem as

$$\frac{\partial \mathbf{V}}{\partial t} = \frac{\partial \mathbf{F}_i}{\partial x_i} + \frac{\partial \mathbf{G}_i}{\partial x_i} + \mathbf{Q} = 0. \tag{59}$$

Then the basic steps for this problem are highlighted below:

- Solve momentum equation without pressure terms,
- Calculate pressure from Poisson equation,
- Correct the velocities,
- Calculate additional terms as temperature, concentration, energy, etc. from the corresponding equations.

The equations after the standard Galerkin discretization are

$$\Delta\overline{\mathbf{U}_\mathbf{i}^*} = -\mathbf{M}^{-1}\Delta t[(\mathbf{C}\overline{\mathbf{U}} + \mathbf{K}\overline{\mathbf{U}} - \mathbf{f}) - \Delta(\mathbf{K_u}\overline{\mathbf{U}} + \mathbf{f})]^n, \tag{60}$$

where $U_i = \mathbf{N}\overline{\mathbf{U}}_\mathbf{i}$, $\Delta U_i = \mathbf{N}\Delta\overline{\mathbf{U}}_\mathbf{i}$, $\Delta U_i^* = \mathbf{N}\Delta\overline{\mathbf{U}_\mathbf{i}^*}$ and $p = \mathbf{N}_p\overline{\mathbf{p}}$ in the first step. This gives the solution for $\overline{U*}_i$.

We can solve the following equation to find $\Delta\overline{\mathbf{p}}$,

$$(\tilde{\mathbf{M}} + \Delta t^2\theta_1\theta_2\mathbf{H})\Delta\overline{\mathbf{p}} = \Delta t[\mathbf{Q}(\overline{\mathbf{U}}^\mathbf{n} + \theta_1\Delta\overline{\mathbf{U}^*}) - \Delta t\theta_1\mathbf{H}\overline{\mathbf{p}} - \mathbf{f}_p]^n, \tag{61}$$

where $\mathbf{H}$, $\tilde{\mathbf{M}}$, $\mathbf{Q}$ are matrices, and this is step two.

Further $\overline{\mathbf{U}}^\mathbf{n+1}$, $\overline{\mathbf{p}}^\mathbf{n+1}$ can be computed from

$$\Delta\overline{\mathbf{U}} = \Delta\overline{\mathbf{U}^*} - \mathbf{M}^{-1}\Delta t\left[\mathbf{Q}^T(\overline{\mathbf{p}}^\mathbf{n} + \theta_2\Delta\overline{\mathbf{p}}) + \frac{\Delta t}{2}\mathbf{P}\overline{\mathbf{p}}^\mathbf{n}\right], \tag{62}$$

and this is the third step of the process .

Finally, the last step is to solve the equation for the energy,

$$\Delta\overline{\mathbf{E}} = -\Delta t\left[\mathbf{C}\overline{\mathbf{E}}^\mathbf{n} + \mathbf{K}\mathbf{T}^\mathbf{n} + \mathbf{f}_e^n - \Delta t(\mathbf{K}_u\overline{\mathbf{E}}^\mathbf{n} + \mathbf{f}_e^n)\right]. \tag{63}$$

More details about the method and the terms used can be found in [77, 98]. For incompressible problems the algorithm can be used in the semi-implicit form. This form is conditionally stable if $\theta_1, \theta_2 \in \left[\frac{1}{2}, 1\right]$, where θ_1, θ_2 are variables coming from the discretization at steps two and three. For the fully explicit form we can set $\theta_2 = 0$ and θ_1 will be the same as in the semi-implicit form.

Another method based on splitting is the two step algorithm [48, 94]. The idea comes from the two level method where two different type of meshes are used, a coarse mesh for solving a nonlinear system and a fine mesh for the linear system. The two-step method is based on solving Navier–Stokes equation in two different ways but using the same computational mesh. The first step is solving a Navier–Stokes problem using a lower order element pair ($P_1 - P_1$) and the projection of the pressure onto a piecewise constant space. In step two a general Stokes problem is solved with a higher order elements ($P_2 - P_2$) using the projection of the pressure gradient onto the same space. The purpose of the first step is to find a prediction of the solution and step two is a correction for the initial approximation. The convergence for both the velocity and pressure is of order $O(h^2)$. Huang et al. compare the method with the $P_1 - P_1$ and $P_2 - P_2$ stabilized method, analyzed before. They report that the two-step method timewise is between the other two but the error is similar with the $P_2 - P_2$ stabilized method [48].

5.4 Gradient Smoothed Method (GSM)

The gradient smoothed method (GSM) was developed by combining the meshfree methods with the FEM approach [64, 65]. The main idea in the GSM is to use a finite element mesh to construct numerical models of good performance. Liu and collaborators introduced the GSM for the solution of steady-state and transient incompressible fluid flow problems [95]. The proposed method is based on irregular cells and thus can be used for problems with arbitrarily complex geometrical boundaries.

In the GSM, derivatives at various locations, such as at nodes, cell centroids, and cell–edges midpoints, are approximated using gradient smoothing operation over relevant gradient smoothing domains. For a two dimensional problem the gradients of a field variable u, at a point of interest, $\mathbf{x}_i$, in the domain Ω_i can be approximated in the form,

$$\nabla u(\mathbf{x}_i) \approx \int_{\Omega_i} \nabla u(\mathbf{x})\, \bar{w}(\mathbf{x} - \mathbf{x}_i) dA, \tag{64}$$

where ∇ is the gradient operator and $\bar{w}$ is a smoothing function. For simplicity, the smoothing function can be set to be a piecewise constant over the smoothing domain. Integrating by parts or using Gauss divergence theorem and utilizing the properties of the smoothing function over the smoothing domain the following equations is obtained for the gradient,

$$\nabla u \approx \frac{1}{A_i} \oint_{\partial \Omega_i} u\, \bar{n}\, ds, \tag{65}$$

where $\bar{n}$ is the unit normal vector on $\partial \Omega_i$ and A_i is the area of the smoothing domain. Equation (65) provides a simple way to approximate gradients at a point by area-weighted integral along the boundary of a local smoothing domain, Ω_i. Similarly, by applying the gradient smoothing technique for the second-order derivatives the Laplace operator at a point of interest, $\mathbf{x}_i$, can be approximated as

$$\nabla \cdot (\Delta u_i) \approx \frac{1}{A_i} \oint_{\partial \Omega_i} \bar{n} \cdot \Delta u \, ds. \tag{66}$$

The spatial derivatives at any point of interest can be approximated over a smoothing domain that needs to be properly defined for a purpose, as presented above. The GSM can tackle the incompressible Navier–Stokes equations enhanced with artificial compressibility, in which the spatial derivatives are approximated by consistent and successive use of gradient smoothing operation over smoothing domains at various locations [66, 95, 96]. A favorable GSM scheme corresponding to a compact stencil with positive coefficients of influence has been derived in [95]. In this study, pseudo-time advancing approach is used for solving the governing equations with mixed hyperbolic–parabolic properties. The dual time stepping scheme and implicit five-stage Runge–Kutta method are implemented to enhance the efficiency and stability in the solution procedure. The obtained results show good agreement with literature [95].

5.5 *Discontinuous and Adaptive Galerkin Method*

In the last decades, discontinuous Galerkin (DG) methods form a class of numerical methods that combine features of the finite element and the finite volume framework, successfully applied to PDEs from a wide range of applications. An overview to DG method for elliptic problems and research directions can be found in [5, 26].

In order to use the equal order interpolation functions for velocity and pressure, the Navier–Stokes equations can be decoupled to distinct equations through the split method. The obtained equations are nonlinear hyperbolic, elliptic, and Helmholtz equations, respectively. The hybrid method combines DG and FE methods. Therefore, DG method is concerned to accomplish spatial discretization of the nonlinear hyperbolic equation to avoid using stabilization approaches in FEM. The split methods due to their decoupled schemes allows choosing equal order basis functions for velocity and pressure [32, 41, 44]. Marchandise and Remacle used an implicit pressure stabilized FEM to solve the Navier–Stokes equations, and DG method was employed to deal with the level-set equation [69]. They calculated the velocity and pressure in the coupled momentum equation together with adding stabilization terms for studying two-phase flows. Pandare and Luo proposed a coupled reconstructed discontinuous Galerkin (rDG) method and continuous Galerkin method for the solution of unsteady incompressible Navier–Stokes equations [80] .

In the paper by Gao et al., the main goal is to take full advantage of DG method and FEM on the basis of a split method [37, 58] to deal with the incompressible Navier–Stokes equations [41]. For the spatial discretization, they treat the nonlinear convection term through DG method, which can guarantee stability, accuracy and also avoid stabilization techniques used in FEM. Lomtev and Karniadakis in their study present a new DG method for simulating compressible viscous flows with shocks on standard unstructured grids [67]. This method is based on a discontinuous

Galerkin formulation both for the advective and the diffusive terms. High-order accuracy is achieved by using a recently developed hierarchical spectral basis. This basis is formed by combining Jacobi polynomials of high-order weights written in a new coordinate system. It retains a tensor-product property, and provides accurate numerical quadrature. Their formulation is conservative, and monotonicity is enforced by appropriately lowering the basis order and performing hp-refinement around discontinuities [67].

Bassi and Rebay introduce a high-order DG method for the numerical solution of the compressible flows [7]. The method combines two main ideas, the physics of wave propagation, accounted for by means of Riemann problems and accuracy being obtained by high-order polynomial approximations within elements. The method is suited to compute high-order accurate solution of the Navier–Stokes equations on unstructured grids. Klaij et al. in their study present a conservative arbitrary Lagrangian Eulerian (ALE) approach to deal with deforming meshes utilizing DG method for optimal flexibility on the local mesh refinement and adjustment of the polynomial order in each element (hp-adaptation) [60]. The numerical method allows for local grid adaptation as well as moving and deforming boundaries. Persson and colleagues introduced a method for computing time-dependent solutions to the compressible Navier–Stokes equations on variable geometries [81]. The transport equations are written as a conservation law for the independent variables in the reference configuration, the complexity introduced by variable geometry is reduced to solving a transformed conservation law in a fixed reference configuration. The spatial discretization is carried out using the DG method on unstructured meshes, while time integration is performed by a Runge–Kutta method. The problem under consideration was altered by adding an equation for the time evolution of the transformation Jacobian to the original conservation law and correcting for the accumulated metric errors. Results are discussed to present the capability of the approach to handle high-order approximations on complex geometries [81].

6 Conclusions

Finite element method (FEM) has gained substantial momentum in the last decades. FEM was initially introduced as an answer to solid mechanics problems while the progress in fluid mechanics was slower due to the non-linearities and instabilities in the solution. In this review we analyzed FEM providing the theoretical basis of the approach mainly focusing on parabolic type of problems, applied in fluid mechanics. Initially, we focused on basic FEM analysis for the Stokes problem. We further discussed FE approaches for the solution of the advection–diffusion equation such as the stabilized FEM, the variational multiscale method, and the discontinuous Galerkin method. Finally, we extended the analysis on the non-linear transport problems and discussed how FEM are utilized for the solution of the Navier–Stokes equations.

References

1. G. Akrivis, M. Crouzeix, C. Makridakis, Implicit-explicit multistep methods for quasilinear parabolic equations. Numer. Math. **82**(4), 521–541 (1999)
2. R. Araya, E. Behrens, R. Rodríguez, An adaptive stabilized finite element scheme for the advection–reaction–diffusion equation. Appl. Numer. Math. **54**(3–4), 491–503 (2005)
3. J.H. Argyris, S. Kelsey, *Energy Theorems and Structural Analysis*, vol. 960 (Springer, Berlin, 1960)
4. D.N. Arnold, F. Brezzi, M. Fortin, A stable finite element for the stokes equations. Calcolo **21**(4), 337–344 (1984)
5. D.N. Arnold, F. Brezzi, B. Cockburn, L.D. Marini, Unified analysis of discontinuous galerkin methods for elliptic problems. SIAM J. Numer. Anal. **39**(5), 1749–1779 (2002)
6. B. Ayuso, L.D. Marini, Discontinuous galerkin methods for advection–diffusion-reaction problems. SIAM J. Numer. Anal. **47**(2), 1391–1420 (2009)
7. F. Bassi, S. Rebay, A high-order accurate discontinuous finite element method for the numerical solution of the compressible Navier–Stokes equations. J. Comput. Phys. **131**(2), 267–279 (1997)
8. Y. Bazilevs, K. Takizawa, T.E. Tezduyar, *Computational Fluid-structure Interaction: Methods and Applications* (Wiley, Chichester, 2013)
9. Y. Bazilevs, K. Takizawa, T.E. Tezduyar, New directions and challenging computations in fluid dynamics modeling with stabilized and multiscale methods. Math. Models Methods Appl. Sci. **25**(12), 2217–2226 (2015). https://doi.org/10.1142/S0218202515020029. https://www.worldscientific.com/doi/abs/10.1142/S0218202515020029
10. M. Bercovier, O. Pironneau, Error estimates for finite element method solution of the stokes problem in the primitive variables. Numer. Math. **33**(2), 211–224 (1979). https://doi.org/10.1007/BF01399555
11. A. Bergam, C. Bernardi, Z. Mghazli, A posteriori analysis of the finite element discretization of some parabolic equations. Math. Comput. **74**(251), 1117–1138 (2005)
12. C. Bernardi, R. Verfürth, A posteriori error analysis of the fully discretized time-dependent stokes equations. ESAIM Math. Model. Numer. Anal. **38**(3), 437–455 (2004)
13. P.B. Bochev, M.D. Gunzburger, J.N. Shadid, Stability of the SUPG finite element method for transient advection–diffusion problems. Comput. Methods Appl. Mech. Eng. **193**(23–26), 2301–2323 (2004)
14. R. Borker, C. Farhat, R. Tezaur, A discontinuous galerkin method with Lagrange multipliers for spatially-dependent advectiondiffusion problems. Comput. Methods Appl. Mech. Eng. **327**, 93–117 (2017). https://doi.org/10.1016/j.cma.2017.08.024
15. S. Brenner, R. Scott, *The Mathematical Theory of Finite Element Methods*, vol. 15 (Springer Science & Business Media, New York, 2007)
16. F. Brezzi, M. Fortin, *Mixed and Hybrid Finite Element Methods*, vol. 15 (Springer Science & Business Media, New York, 2012)
17. F. Brezzi, M.O. Bristeau, L.P. Franca, M. Mallet, G. Rog, A relationship between stabilized finite element methods and the galerkin method with bubble functions. Comput. Methods Appl. Mech. Eng. **96**(1), 117–129 (1992). https://doi.org/10.1016/0045-7825(92)90102-p
18. A.N. Brooks, T.J. Hughes, Streamline upwind/Petrov–Galerkin formulations for convection dominated flows with particular emphasis on the incompressible Navier-Stokes equations. Comput. Methods Appl. Mech. Eng. **32**(1–3), 199–259 (1982)
19. E. Burman, A. Ern, Continuous interior penalty h-finite element methods for advection and advection-diffusion equations. Math. Comput. **76**(259), 1119–1140 (2007)
20. A. Cangiani, E.H. Georgoulis, S. Metcalfe, Adaptive discontinuous galerkin methods for nonstationary convection–diffusion problems. IMA J. Numer. Anal. **34**(4), 1578–1597 (2013)
21. P. Chatzipantelidis, R. Lazarov, V. Thomée, Error estimates for a finite volume element method for parabolic equations in convex polygonal domains. Numer. Methods Partial Differ. Equ. **20**(5), 650–674 (2004)

22. H. Chen, J. Li, W. Qiu, Robust a posteriori error estimates for HDG method for convection–diffusion equations. IMA J. Numer. Anal. **36**(1), 437–462 (2015)
23. E.T. Chung, B. Engquist, Optimal discontinuous galerkin methods for wave propagation. SIAM J. Numer. Anal. **44**(5), 2131–2158 (2006)
24. E. Chung, C.S. Lee, A staggered discontinuous galerkin method for the convection–diffusion equation. J. Numer. Math. **20**(1), 1–32 (2012)
25. E.T. Chung, W.T. Leung, A sub-grid structure enhanced discontinuous galerkin method for multiscale diffusion and convection-diffusion problems. Commun. Comput. Phys. **14**(2), 370–392 (2013)
26. B. Cockburn, G.E. Karniadakis, C.W. Shu (eds.), *Discontinuous Galerkin Methods* (Springer, Berlin, 2000). https://doi.org/10.1007/978-3-642-59721-3
27. B. Cockburn, B. Dong, J. Guzmán, M. Restelli, R. Sacco, A hybridizable discontinuous galerkin method for steady-state convection-diffusion-reaction problems. SIAM J. Sci. Comput. **31**(5), 3827–3846 (2009)
28. R. Codina, Comparison of some finite element methods for solving the diffusion-convection-reaction equation. Comput. Methods Appl. Mech. Eng. **156**(1–4), 185–210 (1998). https://doi.org/10.1016/s0045-7825(97)00206-5
29. R. Codina, Finite element approximation of the convection-diffusion equation: subgrid-scale spaces, local instabilities and anisotropic space-time discretizations, in *BAIL 2010-Boundary and Interior Layers, Computational and Asymptotic Methods* (Springer, Berlin, 2011), pp. 85–97
30. M. Crouzeix, P.A. Raviart, Conforming and nonconforming finite element methods for solving the stationary stokes equations I. Revue française d'automatique informatique recherche opérationnelle. Mathématique **7**(R3), 33–75 (1973)
31. C. Dawson, R. Kirby, Solution of parabolic equations by backward Euler-mixed finite element methods on a dynamically changing mesh. SIAM J. Numer. Anal. **37**(2), 423–442 (1999)
32. M.O. Deville, P.F. Fischer, E.H. Mund, *High-order Methods for Incompressible Fluid Flow*, vol. 9 (Cambridge University Press, Cambridge, 2002)
33. J. Du, E. Chung, An adaptive staggered discontinuous galerkin method for the steady state convection–diffusion equation. J. Sci. Comput., **77**(3), 1490–1518 (2018)
34. K. Eriksson, C. Johnson, Adaptive finite element methods for parabolic problems I: a linear model problem. SIAM J. Numer. Anal. **28**(1), 43–77 (1991)
35. A. Ern, J. Proft, A posteriori discontinuous galerkin error estimates for transient convection–diffusion equations. Appl. Math. Lett. **18**(7), 833–841 (2005)
36. A. Ern, A.F. Stephansen, M. Vohralík, Guaranteed and robust discontinuous galerkin a posteriori error estimates for convection–diffusion–reaction problems. J. Comput. Appl. Math. **234**(1), 114–130 (2010)
37. E. Ferrer, R. Willden, A high order discontinuous galerkin finite element solver for the incompressible Navier–Stokes equations. Comput. Fluids **46**(1), 224–230 (2011)
38. L.P. Franca, C. Farhat, Bubble functions prompt unusual stabilized finite element methods. Comput. Methods Appl. Mech. Eng. **123**(1–4), 299–308 (1995)
39. L.P. Franca, G. Hauke, A. Masud, Revisiting stabilized finite element methods for the advectivediffusive equation. Comput. Methods Appl. Mech. Eng. **195**(13–16), 1560–1572 (2006). https://doi.org/10.1016/j.cma.2005.05.028
40. G. Fu, W. Qiu, W. Zhang, An analysis of HDG methods for convection-dominated diffusion problems. ESAIM Math. Model. Numer. Anal. **49**(1), 225–256 (2015)
41. P. Gao, J. Ouyang, P. Dai, W. Zhou, A coupled continuous and discontinuous finite element method for the incompressible flows. Int. J. Numer. Methods Fluids **84**(8), 477–493 (2017)
42. E.H. Georgoulis, E. Hall, P. Houston, Discontinuous galerkin methods for advection-diffusion-reaction problems on anisotropically refined meshes. SIAM J. Sci. Comput. **30**(1), 246–271 (2007)
43. E.H. Georgoulis, O. Lakkis, J.M. Virtanen, A posteriori error control for discontinuous galerkin methods for parabolic problems. SIAM J. Numer. Anal. **49**(2), 427–458 (2011)

44. J.L. Guermond, P. Minev, J. Shen, An overview of projection methods for incompressible flows. Comput. Methods Appl. Mech. Eng. **195**(44–47), 6011–6045 (2006)
45. J. Heinrich, P. Huyakorn, O. Zienkiewicz, A. Mitchell, An upwindfinite element scheme for two-dimensional convective transport equation. Int. J. Numer. Methods Eng. **11**(1), 131–143 (1977)
46. J.G. Heywood, R. Rannacher, Finite element approximation of the nonstationary Navier–Stokes problem. I. Regularity of solutions and second-order error estimates for spatial discretization. SIAM J. Numer. Anal. **19**(2), 275–311 (1982)
47. P. Houston, C. Schwab, E. Süli, Discontinuous hp-finite element methods for advection-diffusion-reaction problems. SIAM J. Numer. Anal. **39**(6), 2133–2163 (2002)
48. P. Huang, X. Feng, Y. He, An efficient two-step algorithm for the incompressible flow problem. Adv. Comput. Math. **41**(6), 1059–1077 (2015)
49. T.J. Hughes, Multiscale phenomena: Green's functions, the Dirichlet-to-Neumann formulation, subgrid scale models, bubbles and the origins of stabilized methods. Comput. Methods Appl. Mech. Eng. **127**(1–4), 387–401 (1995)
50. T. Hughes, A. Brooks, A multidimensional upwind scheme with no crosswind diffusion, in *Finite Element Methods for Convection Dominated Flows*, ed. by T.J.R. Hughes (ASME, New York, 1979), pp. 19–35
51. T.J. Hughes, J.R. Stewart, A space-time formulation for multiscale phenomena. J. Comput. Appl. Math. **74**(1–2), 217–229 (1996)
52. T.J. Hughes, T. Tezduyar, Finite element methods for first-order hyperbolic systems with particular emphasis on the compressible Euler equations. Comput. Methods Appl. Mech. Eng. **45**(1–3), 217–284 (1984)
53. T.J. Hughes, L.P. Franca, M. Balestra, A new finite element formulation for computational fluid dynamics: V. Circumventing the babuška-brezzi condition: a stable Petrov–Galerkin formulation of the stokes problem accommodating equal-order interpolations. Comput. Methods Appl. Mech. Eng. **59**(1), 85–99 (1986)
54. T.J. Hughes, L.P. Franca, G.M. Hulbert, A new finite element formulation for computational fluid dynamics: VIII. The galerkin/least-squares method for advective-diffusive equations. Comput. Methods Appl. Mech. Eng. **73**(2), 173–189 (1989)
55. T.J. Hughes, G.R. Feijoo, L. Mazzei, J.B. Quincy, The variational multiscale method–a paradigm for computational mechanics. Comput. Methods Appl. Mech. Eng. **166**(1–2), 3–24 (1998). https://doi.org/10.1016/s0045-7825(98)00079-6
56. T.J.R. Hughes, G. Scovazzi, L.P. Franca, *Multiscale and Stabilized Methods*, chap. 2. (American Cancer Society, New York, 2004). https://doi.org/10.1002/0470091355.ecm051. https://onlinelibrary.wiley.com/doi/abs/10.1002/0470091355.ecm051
57. F. Karakatsani, C. Makridakis, A posteriori estimates for approximations of time-dependent stokes equations. IMA J. Numer. Anal. **27**(4), 741–764 (2006)
58. G.E. Karniadakis, M. Israeli, S.A. Orszag, High-order splitting methods for the incompressible Navier-Stokes equations. J. Comput. Phys. **97**(2), 414–443 (1991)
59. T. Kemmochi, On the finite element approximation for non-stationary saddle-point problems. Jpn. J. Ind. Appl. Math., **35**(2), 423–439 (2018)
60. C.M. Klaij, J.J. van der Vegt, H. van der Ven, Space–time discontinuous galerkin method for the compressible Navier–Stokes equations. J. Comput. Phys. **217**(2), 589–611 (2006)
61. J. Kwack, A. Masud, A stabilized mixed finite element method for shear-rate dependent non-Newtonian fluids: 3d benchmark problems and application to blood flow in bifurcating arteries. Comput. Mech. **53**(4), 751–776 (2014)
62. O. Lakkis, C. Makridakis, Elliptic reconstruction and a posteriori error estimates for fully discrete linear parabolic problems. Math. Comput. **75**(256), 1627–1658 (2006)
63. M.G. Larson, A. Målqvist, A posteriori error estimates for mixed finite element approximations of parabolic problems. Numer. Math. **118**(1), 33–48 (2011)
64. G.R. Liu, *Mesh Free Methods: Moving Beyond the Finite Element Method* (CRC Press, Boca Raton, 2002)
65. G.R. Liu, N.T. Trung, *Smoothed Finite Element Methods* (CRC Press, Boca Raton, 2016)

66. G. Liu, G.X. Xu, A gradient smoothing method (GSM) for fluid dynamics problems. Int. J. Numer. Methods Fluids **58**(10), 1101–1133 (2008)
67. I. Lomtev, G.E. Karniadakis, A discontinuous galerkin method for the Navier–Stokes equations. Int. J. Numer. Methods Fluids **29**(5), 587–603 (1999)
68. C. Makridakis, R.H. Nochetto, Elliptic reconstruction and a posteriori error estimates for parabolic problems. SIAM J. Numer. Anal. **41**(4), 1585–1594 (2003)
69. E. Marchandise, J.F. Remacle, A stabilized finite element method using a discontinuous level set approach for solving two phase incompressible flows. J. Comput. Phys. **219**(2), 780–800 (2006)
70. A. Masud, R. Calderer, A variational multiscale stabilized formulation for the incompressible Navier–Stokes equations. Comput. Mech. **44**(2), 145–160 (2009)
71. A. Masud, R. Calderer, A variational multiscale method for incompressible turbulent flows: bubble functions and fine scale fields. Comput. Methods Appl. Mech. Eng. **200**(33–36), 2577–2593 (2011)
72. A. Masud, R. Khurram, A multiscale/stabilized finite element method for the advectiondiffusion equation. Comput. Methods Appl. Mech. Eng. **193**(21–22), 1997–2018 (2004). https://doi.org/10.1016/j.cma.2003.12.047
73. A. Masud, R. Khurram, A multiscale finite element method for the incompressible Navier–Stokes equations. Comput. Methods Appl. Mech. Eng. **195**(13–16), 1750–1777 (2006)
74. A. Masud, J. Kwack, A stabilized mixed finite element method for the first-order form of advection–diffusion equation. Int. J. Numer. Methods Fluids **57**(9), 1321–1348 (2008)
75. A. Masud, J. Kwack, A stabilized mixed finite element method for the incompressible shear-rate dependent non-Newtonian fluids: variational multiscale framework and consistent linearization. Comput. Methods Appl. Mech. Eng. **200**(5–8), 577–596 (2011)
76. N.C. Nguyen, J. Peraire, B. Cockburn, An implicit high-order hybridizable discontinuous galerkin method for linear convection–diffusion equations. J. Comput. Phys. **228**(9), 3232–3254 (2009)
77. P. Nithiarasu, R. Codina, O. Zienkiewicz, The characteristic-based split (cbs) schemea unified approach to fluid dynamics. Int. J. Numer. Methods Eng. **66**(10), 1514–1546 (2006)
78. R.H. Nochetto, K.G. Siebert, A. Veeser, Theory of adaptive finite element methods: an introduction, in *Multiscale, Nonlinear and Adaptive Approximation* (Springer, Berlin, 2009), pp. 409–542
79. E. Oate, Derivation of stabilized equations for numerical solution of advective-diffusive transport and fluid flow problems. Comput. Methods Appl. Mech. Eng. **151**(1–2), 233–265 (1998). https://doi.org/10.1016/s0045-7825(97)00119-9
80. A.K. Pandare, H. Luo, A hybrid reconstructed discontinuous galerkin and continuous galerkin finite element method for incompressible flows on unstructured grids. J. Comput. Phys. **322**, 491–510 (2016)
81. P.O. Persson, J. Bonet, J. Peraire, Discontinuous galerkin solution of the Navier–Stokes equations on deformable domains. Comput. Methods Appl. Mech. Eng. **198**(17–20), 1585–1595 (2009)
82. S.S. Rao, *The Finite Element Method in Engineering* (Butterworth-Heinemann, Burlington, 2017)
83. G. Sangalli, Robust a-posteriori estimator for advection-diffusion-reaction problems. Math. Comput. **77**(261), 41–70 (2008)
84. A. Sendur, A. Nesliturk, Bubble-based stabilized finite element methods for time-dependent convection–diffusion–reaction problems. Int. J. Numer. Methods Fluids **82**(8), 512–538 (2016)
85. L.T. Tenek, J. Argyris, *Finite Element Analysis for Composite Structures*, vol. 59 (Springer Science & Business Media, New York, 2013)
86. T.E. Tezduyar, Stabilized finite element formulations for incompressible flow computations, in *Advances in Applied Mechanics*, vol. 28 (Elsevier, London, 1991), pp. 1–44
87. T. Tezduyar, T. Hughes, Development of time-accurate finite element techniques for first-order hyperbolic systems with particular emphasis on the compressible Euler equations. NASA Technical Report NASA-CR-204772, NASA, 1982

88. T. Tezduyar, S. Mittal, R. Shih, Time-accurate incompressible flow computations with quadrilateral velocity-pressure elements. Comput. Methods Appl. Mech. Eng. **87**(2–3), 363–384 (1991)
89. T.E. Tezduyar, S. Mittal, S. Ray, R. Shih, Incompressible flow computations with stabilized bilinear and linear equal-order-interpolation velocity-pressure elements. Comput. Methods Appl. Mech. Eng. **95**(2), 221–242 (1992)
90. V. Thomée, *Galerkin Finite Element Methods for Parabolic Problems* (Springer, New York, 2006)
91. D. Turner, K. Nakshatrala, K. Hjelmstad, A stabilized formulation for the advection–diffusion equation using the generalized finite element method. Int. J. Numer. Methods Fluids **66**(1), 64–81 (2011)
92. R. Verfürth, Robust a posteriori error estimates for nonstationary convection-diffusion equations. SIAM J. Numer. Anal. **43**(4), 1783–1802 (2005)
93. R. Verfürth, Robust a posteriori error estimates for stationary convection-diffusion equations. SIAM J. Numer. Anal. **43**(4), 1766–1782 (2005)
94. J. Wu, D. Liu, X. Feng, P. Huang, An efficient two-step algorithm for the stationary incompressible magnetohydrodynamic equations. Appl. Math. Comput. **302**, 21–33 (2017)
95. G.X. Xu, E. Li, V. Tan, G. Liu, Simulation of steady and unsteady incompressible flow using gradient smoothing method (GSM). Comput. Struct. **90**, 131–144 (2012)
96. J. Yao, G. Liu, A matrix-form GSM–CFD solver for incompressible fluids and its application to hemodynamics. Comput. Mech. **54**(4), 999–1012 (2014)
97. L. Zhu, D. Schötzau, A robust a posteriori error estimate for hp-adaptive DG methods for convection–diffusion equations. IMA J. Numer. Anal. **31**(3), 971–1005 (2010)
98. O. Zienkiewicz, P. Nithiarasu, R. Codina, M. Vazquez, P. Ortiz, The characteristic-based-split procedure: an efficient and accurate algorithm for fluid problems. Int. J. Numer. Methods Fluids **31**(1), 359–392 (1999)

On a Hilbert-Type Integral Inequality in the Whole Plane Related to the Extended Riemann Zeta Function

Michael Th. Rassias and Bicheng Yang

Abstract By the use of the methods of real analysis and the weight functions, a few equivalent conditions of a Hilbert-type integral inequality with the nonhomogeneous kernel in the whole plane are obtained. The best possible constant factor is related to the extended Riemann zeta function. As applications, a few equivalent conditions of a Hilbert-type integral inequality with the homogeneous kernel in the whole plane are deduced. We also consider the operator expressions and a few particular cases.

2000 Mathematics Subject Classification 26D15, 47A07, 65B10

1 Introduction

If $f(x), g(y) \geq 0$, $0 < \int_0^\infty f^2(x)dx < \infty$ and $0 < \int_0^\infty g^2(y)dy < \infty$, then we have the following well- known Hilbert's integral inequality (cf. [1]):

$$\int_0^\infty \int_0^\infty \frac{f(x)g(y)}{x+y} dxdy < \pi \left(\int_0^\infty f^2(x)dx \int_0^\infty g^2(y)dy \right)^{\frac{1}{2}}, \tag{1}$$

with the best possible constant factor π.

Recently, by means of the weight functions, a lot of extensions of (1) were given by Yang's books (cf. [2, 3]). Some Hilbert-type inequalities with the homogenous kernels of degree 0 and nonhomogenous kernels were provided by [4–9]. Some

M. Th. Rassias
Institute of Mathematics, University of Zurich, Zurich, Switzerland
Institute for Advanced Study Program in Interdisciplinary Studies, Princeton, NJ, USA
e-mail: michail.@math.uzh.ch

B. Yang (✉)
Department of Mathematics, Guangdong University of Education, Guangdong, People's Republic of China
e-mail: bcyang@gdei.edu.cn

T. M. Rassias, P. M. Pardalos (eds.), *Mathematical Analysis and Applications*, Springer Optimization and Its Applications 154,
https://doi.org/10.1007/978-3-030-31339-5_19

other kinds of Hilbert-type inequalities were obtained by [10–15]. Most of them are built in the quarter plane of the first quadrant.

In 2007, using the way of real analysis, Yang [16] gave a Hilbert-type integral inequality in the whole plane as follows:

$$\int_{-\infty}^{\infty}\int_{-\infty}^{\infty}\frac{f(x)g(y)}{(1+e^{x+y})^{\lambda}}dxdy$$

$$< B(\frac{\lambda}{2},\frac{\lambda}{2})\left(\int_{-\infty}^{\infty}e^{-\lambda x}f^2(x)dx\int_{-\infty}^{\infty}e^{-\lambda y}g^2(y)dy\right)^{\frac{1}{2}}, \tag{2}$$

with the best possible constant factor $B(\frac{\lambda}{2},\frac{\lambda}{2})(\lambda > 0,\ B(u,v)$ is the beta function) (cf. [17]). He et al. [18–32] proved some new Hilbert-type integral inequalities in the whole plane with the best possible constant factors.

In this paper, by means of the methods of real analysis and the weight functions, a few equivalent conditions of a Hilbert-type integral inequality with the nonhomogeneous kernel in the whole plane are obtained. The best possible constant factor is related to the extended Riemann zeta function. As applications, a few equivalent conditions of a Hilbert-type integral inequality with the homogeneous kernel in the whole plane are deduced. We also consider the operator expressions and a few particular cases.

2 An Example and Two Lemmas

Example 1 Setting $h(u) := \frac{|\ln|u||^{\beta}}{(\max\{|u|,1\})^{\lambda-1}|u-1|}$ $(u \in \mathbf{R} = (-\infty,\infty))$, then we have

$$h(xy) = \frac{|\ln|xy||^{\beta}}{(\max\{|xy|,1\})^{\lambda-1}|xy-1|}\quad (x,y\in\mathbf{R}), \tag{3}$$

and for $\beta,\sigma,\mu > 0, \sigma+\mu = \lambda$, it follows that

$$\begin{aligned}
K^{(1)}(\sigma) &:= \int_{-1}^{1}\frac{|\ln|u||^{\beta}|u|^{\sigma-1}}{(\max\{|u|,1\})^{\lambda-1}|u-1|}du\\
&= \int_{0}^{1}\frac{|\ln u|^{\beta}u^{\sigma-1}}{(\max\{u,1\})^{\lambda-1}}\left(\frac{1}{u+1}+\frac{1}{|u-1|}\right)du\\
&= \int_{0}^{1}(-\ln u)^{\beta}\left(\frac{1}{u+1}+\frac{1}{1-u}\right)u^{\sigma-1}du\\
&= 2\int_{0}^{1}(-\ln u)^{\beta}\frac{u^{\sigma-1}}{1-u^2}du\\
&= 2\int_{0}^{1}(-\ln u)^{\beta}\sum_{k=0}^{\infty}u^{2k+\sigma-1}du.
\end{aligned}$$

By Lebesgue term by term integration theorem (cf. [33]), we have

$$K^{(1)}(\sigma) = 2\sum_{k=0}^{\infty}\int_0^1 (-\ln u)^{\beta} u^{2k+\sigma-1} du$$
$$= 2\sum_{k=0}^{\infty}\frac{1}{(2k+\sigma)^{\beta+1}}\int_0^{\infty} v^{\beta} e^{-v} dv$$
$$= \frac{\Gamma(\beta+1)}{2^{\beta}}\zeta(\beta+1,\frac{\sigma}{2}) \in \mathbf{R}_+,$$

where

$$\zeta(s,a) = \sum_{k=0}^{\infty}\frac{1}{(k+a)^s} \quad (\mathrm{Re}\, s > 1; a > 0)$$

is the extended Riemann zeta function ($\zeta(s,1) = \sum_{k=1}^{\infty}\frac{1}{k^s}$ $(\mathrm{Re}\, s > 1)$ is the Riemann zeta function) (cf. [17]).

In the same way, setting $v = \frac{1}{u}$, we find

$$K^{(2)}(\sigma) := \int_{\{u;|u|\geq 1\}}\frac{|\ln|u||^{\beta}|u|^{\sigma-1}}{(\max\{|u|,1\})^{\lambda-1}|u-1|}du$$
$$= \int_1^{\infty}\frac{(\ln u)^{\beta}u^{\sigma-1}}{u^{\lambda-1}}\left(\frac{1}{u+1}+\frac{1}{u-1}\right)du$$
$$= \int_0^1 (-\ln v)^{\beta}\left(\frac{1}{v+1}+\frac{1}{1-v}\right)v^{\mu-1}du$$
$$= \frac{\Gamma(\beta+1)}{2^{\beta}}\zeta(\beta+1,\frac{\mu}{2}) \in \mathbf{R}_+,$$

and then

$$K(\sigma) := \int_{-\infty}^{\infty} h(u)|u|^{\sigma-1}du = K^{(1)}(\sigma) + K^{(2)}(\sigma)$$
$$= \frac{\Gamma(\beta+1)}{2^{\beta}}\left(\zeta(\beta+1,\frac{\sigma}{2})+\zeta(\beta+1,\frac{\mu}{2})\right). \tag{4}$$

In the following, we agree that $p > 1, \frac{1}{p}+\frac{1}{q} = 1, \beta, \sigma, \mu > 0, \sigma+\mu = \lambda$, and $\sigma_1 \in \mathbf{R}$.

For $n \in \mathbf{N} = \{1, 2, \cdots\}$, we define the following two expressions:

$$I_1 := \int_{\{y;|y|\geq 1\}} \left(\int_{\{x;|x|\leq 1\}} h(xy)|x|^{\sigma+\frac{1}{pn}-1} dx \right) |y|^{\sigma_1-\frac{1}{qn}-1} dy, \tag{5}$$

$$I_2 := \int_{\{y;|y|\leq 1\}} \left(\int_{\{x;|x|\geq 1\}} h(xy)|x|^{\sigma-\frac{1}{pn}-1} dx \right) |y|^{\sigma_1+\frac{1}{qn}-1} dy. \tag{6}$$

Setting $u = xy$ in (5), by Fubini theorem (cf. [33]), it follows that

$$\begin{aligned} I_1 &= \int_{\{y;|y|\geq 1\}} \left[\int_{-1}^{0} h(xy)(-x)^{\sigma+\frac{1}{pn}-1} dx \right. \\ &\quad \left. + \int_0^1 h(xy)x^{\sigma+\frac{1}{pn}-1} dx \right] |y|^{\sigma_1-\frac{1}{qn}-1} dy \\ &= 2\int_1^\infty \left[\int_0^y (h(-u)+h(u))(\frac{u}{y})^{\sigma+\frac{1}{pn}-1} \frac{1}{y} du \right] y^{\sigma_1-\frac{1}{qn}-1} dy \\ &= 2\int_1^\infty \left[\int_0^y (h(-u)+h(u))u^{\sigma+\frac{1}{pn}-1} du \right] y^{\sigma_1-\sigma-\frac{1}{n}-1} dy \\ &= 2\int_1^\infty \left[\int_0^1 (h(-u)+h(u))u^{\sigma+\frac{1}{pn}-1} du \right] y^{\sigma_1-\sigma-\frac{1}{n}-1} dy \\ &\quad +2\int_1^\infty \left[\int_1^y (h(-u)+h(u))u^{\sigma+\frac{1}{pn}-1} du \right] y^{\sigma_1-\sigma-\frac{1}{n}-1} dy \\ &= 2\int_1^\infty y^{\sigma_1-\sigma-\frac{1}{n}-1} dy \int_0^1 (h(-u)+h(u))u^{\sigma+\frac{1}{pn}-1} du \\ &\quad +2\int_1^\infty \left(\int_u^\infty y^{\sigma_1-\sigma-\frac{1}{n}-1} dy \right) (h(-u)+h(u))u^{\sigma+\frac{1}{pn}-1} du. \end{aligned} \tag{7}$$

In the same way, we find

$$\begin{aligned} I_2 &= \int_{\{x;|x|\geq 1\}} \left(\int_{\{y;|y|\leq 1\}} h(xy)|y|^{\sigma_1+\frac{1}{qn}-1} dy \right) |x|^{\sigma-\frac{1}{pn}-1} dx \\ &= 2\int_1^\infty \left[\int_0^x (h(-u)+h(u))u^{\sigma_1+\frac{1}{qn}-1} du \right] x^{\sigma-\sigma_1-\frac{1}{n}-1} dx \\ &= 2\int_1^\infty x^{\sigma-\sigma_1-\frac{1}{n}-1} dx \int_0^1 (h(-u)+h(u))u^{\sigma_1+\frac{1}{qn}-1} du \\ &\quad +2\int_1^\infty \left(\int_u^\infty x^{\sigma-\sigma_1-\frac{1}{n}-1} dx \right) (h(-u)+h(u))u^{\sigma_1+\frac{1}{qn}-1} du. \end{aligned} \tag{8}$$

Lemma 1 *There exists a constant M, such that for any nonnegative measurable functions $f(x)$ and $g(y)$ in* **R**, *the following inequality*

$$I := \int_{-\infty}^{\infty}\int_{-\infty}^{\infty}\frac{|\ln|xy||^{\beta}f(x)g(y)}{(\max\{|xy|,1\})^{\lambda-1}|xy-1|}dxdy$$
$$\leq M\left[\int_{-\infty}^{\infty}x^{p(1-\sigma)-1}f^{p}(x)dx\right]^{\frac{1}{p}}\left[\int_{\infty}^{\infty}y^{q(1-\sigma_1)-1}g^{q}(y)dy\right]^{\frac{1}{q}} \tag{9}$$

holds true, then we have $\sigma_1=\sigma$.

Proof If $\sigma_1<\sigma$, then for $n>\frac{1}{\sigma-\sigma_1}$ $(n\in\mathbf{N})$, we set the following functions:

$$f_n(x):=\begin{cases}0, & |x|<1\\ |x|^{\sigma-\frac{1}{pn}-1}, & |x|\geq 1\end{cases},$$
$$g_n(y):=\begin{cases}|y|^{\sigma_1+\frac{1}{qn}-1}, & |y|\leq 1\\ 0, & |y|>1\end{cases},$$

and find

$$J_2 := \left[\int_{-\infty}^{\infty}|x|^{p(1-\sigma)-1}f_n^{p}(x)dx\right]^{\frac{1}{p}}\left[\int_{-\infty}^{\infty}|y|^{q(1-\sigma_1)-1}g_n^{q}(y)dy\right]^{\frac{1}{q}}$$
$$=\left(\int_{\{x;|x|\geq1\}}|x|^{-\frac{1}{n}-1}dx\right)^{\frac{1}{p}}\left(\int_{\{y;|y|\leq1\}}|y|^{\frac{1}{n}-1}dy\right)^{\frac{1}{q}}=2n.$$

By (8), we have

$$2\int_{1}^{\infty}\left(\int_{u}^{\infty}x^{\sigma-\sigma_1-\frac{1}{n}-1}dx\right)(h(-u)+h(u))u^{\sigma_1+\frac{1}{qn}-1}du$$
$$\leq I_2=\int_{-\infty}^{\infty}\int_{-\infty}^{\infty}\frac{|\ln|xy||^{\beta}f_n(x)g_n(y)}{(\max\{|xy|,1\})^{\lambda-1}|xy-1|}dxdy$$
$$\leq MJ_2=2Mn. \tag{10}$$

Since $\sigma-\sigma_1-\frac{1}{n}>0$, it follows that for any $u\in[1,\infty)$, $\int_u^{\infty}x^{\sigma-\sigma_1-\frac{1}{n}-1}dx=\infty$. By (10), in view of

$$(h(-u)+h(u))u^{\sigma_1+\frac{1}{qn}-1}>0\ (u\in[1,\infty)),$$

we find that $\infty\leq 2Mn<\infty$, which is a contradiction.

If $\sigma_1 > \sigma$, then for $n > \frac{1}{\sigma_1 - \sigma}$ $(n \in \mathbf{N})$, we set the following functions:

$$\widetilde{f}_n(x) := \begin{cases} |x|^{\sigma + \frac{1}{pn} - 1}, |x| \leq 1 \\ 0, |x| > 1 \end{cases},$$

$$\widetilde{g}_n(y) := \begin{cases} 0, |y| < 1 \\ |y|^{\sigma_1 - \frac{1}{qn} - 1}, |y| \geq 1 \end{cases},$$

and find

$$\begin{aligned} \widetilde{J}_2 &:= \left[\int_{-\infty}^{\infty} |x|^{p(1-\sigma)-1} \widetilde{f}_n^p(x) dx\right]^{\frac{1}{p}} \left[\int_{-\infty}^{\infty} |y|^{q(1-\sigma_1)-1} \widetilde{g}_n^q(y) dy\right]^{\frac{1}{q}} \\ &= \left(\int_{-1}^{1} |x|^{\frac{1}{n}-1} dx\right)^{\frac{1}{p}} \left(\int_{\{y;|y|\geq 1\}} |y|^{-\frac{1}{n}-1} dy\right)^{\frac{1}{q}} = 2n. \end{aligned}$$

By (7), we have

$$\begin{aligned} & 2\int_{1}^{\infty} y^{\sigma_1 - \sigma - \frac{1}{n} - 1} dy \int_{0}^{1} (h(-u) + h(u)) u^{\sigma + \frac{1}{pn} - 1} du \\ & \leq I_1 = \int_{0}^{\infty} \int_{0}^{\infty} \frac{|\ln|xy||^{\beta} \widetilde{f}_n(x) \widetilde{g}_n(y) dx dy}{(\max\{|xy|, 1\})^{\lambda - 1} |xy - 1|} \\ & \leq M \widetilde{J}_2 = 2Mn. \end{aligned} \tag{11}$$

Since $(\sigma_1 - \sigma) - \frac{1}{n} > 0$, it follows that $\int_1^{\infty} y^{(\sigma_1 - \sigma) - \frac{1}{n} - 1} dy = \infty$. By (11), in view of

$$\int_{0}^{1} (h(-u) + h(u)) u^{\sigma + \frac{1}{pn} - 1} du > 0,$$

we have $\infty \leq 2Mn < \infty$, which is a contradiction.

Hence, we conclude that $\sigma_1 = \sigma$.

The lemma is proved.

For $\sigma_1 = \sigma$, we still have

Lemma 2 *If there exists a constant M, such that for any nonnegative measurable functions $f(x)$ and $g(y)$ in $\mathbf{R}$, the following inequality*

$$\begin{aligned} & \int_{-\infty}^{\infty} \int_{-\infty}^{\infty} \frac{|\ln|xy||^{\beta} f(x) g(y)}{(\max\{|xy|, 1\})^{\lambda - 1} |xy - 1|} dx dy \\ & \leq M \left[\int_{-\infty}^{\infty} |x|^{p(1-\sigma)-1} f^p(x) dx\right]^{\frac{1}{p}} \left[\int_{-\infty}^{\infty} |y|^{q(1-\sigma)-1} g^q(y) dy\right]^{\frac{1}{q}} \end{aligned} \tag{12}$$

holds true, then we have $M \geq K(\sigma) > 0$.

Proof For $\sigma_1 = \sigma$, we reduce (7) and then use inequality $I_1 \leq M\widetilde{J}_2$ (when $\sigma_1 = \sigma$) as

$$\frac{1}{2n}I_1 = \int_0^1 (h(-u) + h(u))u^{\sigma+\frac{1}{pn}-1}du + \int_1^\infty (h(-u) + h(u))u^{\sigma-\frac{1}{qn}-1}du \leq M. \tag{13}$$

By Fatou lemma (cf. [33]) and (13), we have

$$\begin{aligned} 0 < K(\sigma) &= \int_0^1 \lim_{n\to\infty} (h(-u) + h(u))u^{\sigma+\frac{1}{pn}-1}du \\ &+ \int_1^\infty \lim_{n\to\infty} (h(-u) + h(u))u^{\sigma-\frac{1}{qn}-1}du \\ &\leq \underline{\lim}_{n\to\infty}\frac{1}{2n}I_1 \leq M. \end{aligned}$$

The lemma is proved.

3 Main Results and a Few Corollaries

Theorem 1 *The following conditions are equivalent:*

(i) There exists a constant M*, such that for any* $f(x) \geq 0$*, satisfying*

$$0 < \int_{-\infty}^\infty |x|^{p(1-\sigma)-1} f^p(x)dx < \infty,$$

we have the following inequality:

$$J := \left[\int_{-\infty}^\infty |y|^{p\sigma_1-1}\left(\int_{-\infty}^\infty \frac{|\ln|xy||^\beta f(x)dx}{(\max\{|xy|, 1\})^{\lambda-1}|xy-1|}\right)^p dy\right]^{\frac{1}{p}} < M\left[\int_{-\infty}^\infty |x|^{p(1-\sigma)-1} f^p(x)dx\right]^{\frac{1}{p}}. \tag{14}$$

(ii) There exists a constant M*, such that for any* $f(x), g(y) \geq 0$*, satisfying*

$$0 < \int_{-\infty}^\infty |x|^{p(1-\sigma)-1} f^p(x)dx < \infty,$$

and $0 < \int_{-\infty}^{\infty} |y|^{q(1-\sigma_1)-1} g^q(y)dy < \infty$, *we have the following inequality:*

$$I = \int_{-\infty}^{\infty} \int_{-\infty}^{\infty} \frac{|\ln|xy||^{\beta} f(x)g(y)}{(\max\{|xy|, 1\})^{\lambda-1}|xy-1|} dxdy$$
$$< M\left[\int_{-\infty}^{\infty} |x|^{p(1-\sigma)-1} f^p(x)dx\right]^{\frac{1}{p}} \left[\int_{-\infty}^{\infty} |y|^{q(1-\sigma_1)-1} g^q(y)dy\right]^{\frac{1}{q}}. \quad (15)$$

(iii) $\sigma_1 = \sigma$, *and* $M \geq K(\sigma) > 0$.

Proof *(i)* $\Rightarrow$ *(ii)*. By Hölder's inequality (cf. [34]), we have

$$I = \int_{-\infty}^{\infty} \left(|y|^{\sigma_1 - \frac{1}{p}} \int_{-\infty}^{\infty} \frac{|\ln|xy||^{\beta} f(x)dx}{(\max\{|xy|, 1\})^{\lambda-1}|xy-1|}\right)\left(|y|^{\frac{1}{p}-\sigma_1} g(y)\right) dy$$
$$\leq J\left[\int_{-\infty}^{\infty} |y|^{q(1-\sigma_1)-1} g^q(y)dy\right]^{\frac{1}{q}}. \quad (16)$$

Then by (14), we have (15).

(ii) $\Rightarrow$ *(iii)*. By Lemma 1, we have $\sigma_1 = \sigma$. Then by Lemma 2, we have $M \geq K(\sigma) > 0$.

(iii) $\Rightarrow$ *(i)*. Setting $u = xy$, we obtain the following weight function: For $y \in (-\infty, 0) \cup (0, \infty)$,

$$\omega(\sigma, y) := |y|^{\sigma} \int_{-\infty}^{\infty} \frac{|\ln|xy||^{\beta} |x|^{\sigma-1}}{(\max\{|xy|, 1\})^{\lambda-1}|xy-1|} dx$$
$$= \int_0^{\infty} (h(-u) + h(u))u^{\sigma-1} du = K(\sigma). \quad (17)$$

By Hölder's inequality with weight and (17), we have

$$\left(\int_{-\infty}^{\infty} \frac{|\ln|xy||^{\beta} f(x)}{(\max\{|xy|, 1\})^{\lambda-1}|xy-1|} dx\right)^p$$
$$= \left\{\int_{-\infty}^{\infty} h(xy)\left[\frac{|y|^{(\sigma-1)/p}}{|x|^{(\sigma-1)/q}} f(x)\right]\left[\frac{|x|^{(\sigma-1)/q}}{|y|^{(\sigma-1)/p}}\right] dx\right\}^p$$
$$\leq \int_{-\infty}^{\infty} h(xy)\frac{|y|^{\sigma-1} f^p(x)}{|x|^{(\sigma-1)p/q}} dx \left[\int_{-\infty}^{\infty} h(xy)\frac{|x|^{\sigma-1}dx}{|y|^{(\sigma-1)q/p}}\right]^{p/q}$$
$$= \left[\omega(\sigma, y)|y|^{q(1-\sigma)-1}\right]^{p-1} \int_{=\infty}^{\infty} h(xy)\frac{|y|^{\sigma-1}}{|x|^{(\sigma-1)p/q}} f^p(x)dx$$
$$= (K(\sigma))^{p-1}|y|^{-p\sigma+1} \int_0^{\infty} h(xy)\frac{|y|^{\sigma-1}}{|x|^{(\sigma-1)p/q}} f^p(x)dx. \quad (18)$$

If (18) takes the form of equality for a $y \in (-\infty, 0) \cup (0, \infty)$, then (cf. [34]), there exist constants A and B, such that they are not all zero, and

$$A\frac{|y|^{\sigma-1}}{|x|^{(\sigma-1)p/q}} f^p(x) = B\frac{|x|^{\sigma-1}}{|y|^{(\sigma-1)q/p}} \ a.e. \ in \ \mathbf{R}.$$

We suppose that $A \neq 0$ (otherwise $B = A = 0$). Then it follows that

$$|x|^{p(1-\sigma)-1} f^p(x) = |y|^{q(1-\sigma)} \frac{B}{A|x|} \ a.e. \ in \ \mathbf{R},$$

which contradicts the fact that

$$0 < \int_{-\infty}^{\infty} |x|^{p(1-\sigma)-1} f^p(x)dx < \infty.$$

Hence, (18) takes the form of strict inequality.

For $\sigma_1 = \sigma$, by Fubini theorem (cf. [33]) and (18), we have

$$\begin{aligned} J &< (K(\sigma))^{\frac{1}{q}} \left[\int_{-\infty}^{\infty} \int_{-\infty}^{\infty} h(xy) \frac{|y|^{\sigma-1}}{|x|^{(\sigma-1)p/q}} f^p(x)dxdy \right]^{\frac{1}{p}} \\ &= (K(\sigma))^{\frac{1}{q}} \left\{ \int_{-\infty}^{\infty} \left[\int_{-\infty}^{\infty} h(xy) \frac{|y|^{\sigma-1}}{|x|^{(\sigma-1)(p-1)}} dy \right] f^p(x)dx \right\}^{\frac{1}{p}} \\ &= (K(\sigma))^{\frac{1}{q}} \left[\int_{-\infty}^{\infty} \omega(\sigma, x)|x|^{p(1-\sigma)-1} f^p(x)dx \right]^{\frac{1}{p}} \\ &= K(\sigma) \left[\int_{-\infty}^{\infty} |x|^{p(1-\sigma)-1} f^p(x)dx \right]^{\frac{1}{p}}. \end{aligned}$$

For $K(\sigma) \in \mathbf{R}_+$, and $M \geq K(\sigma)$, (14) follows.

Therefore, the conditions (i)–(iii) are equivalent.

The theorem is proved.

For $\sigma_1 = \sigma$, we still have

Theorem 2 *The following conditions are equivalent:*

(i) There exists a constant M, such that for any $f(x) \geq 0$, satisfying

$$0 < \int_{-\infty}^{\infty} |x|^{p(1-\sigma)-1} f^p(x)dx < \infty,$$

we have the following inequality:

$$\left[\int_{-\infty}^{\infty}|y|^{p\sigma-1}\left(\int_{-\infty}^{\infty}\frac{|\ln|xy||^{\beta}f(x)dx}{(\max\{|xy|,1\})^{\lambda-1}|xy-1|}\right)^{p}dy\right]^{\frac{1}{p}}$$

$$< M\left[\int_{-\infty}^{\infty}|x|^{p(1-\sigma)-1}f^{p}(x)dx\right]^{\frac{1}{p}}. \tag{19}$$

(ii) There exists a constant M, such that for any $f(x), g(y) \geq 0$, satisfying

$$0 < \int_{-\infty}^{\infty}|x|^{p(1-\sigma)-1}f^{p}(x)dx < \infty,$$

and $0 < \int_{-\infty}^{\infty}|y|^{q(1-\sigma)-1}g^{q}(y)dy < \infty$, we have the following inequality:

$$\int_{-\infty}^{\infty}\int_{-\infty}^{\infty}\frac{|\ln|xy||^{\beta}f(x)g(y)}{(\max\{|xy|,1\})^{\lambda-1}|xy-1|}dxdy$$

$$< M\left[\int_{-\infty}^{\infty}|x|^{p(1-\sigma)-1}f^{p}(x)dx\right]^{\frac{1}{p}}\left[\int_{-\infty}^{\infty}|y|^{q(1-\sigma)-1}g^{q}(y)dy\right]^{\frac{1}{q}}. \tag{20}$$

Moreover, the constant factor

$$M = K(\sigma) = \frac{\Gamma(\beta+1)}{2^{\beta}}\left(\zeta(\beta+1,\frac{\sigma}{2})+\zeta(\beta+1,\frac{\mu}{2})\right)$$

in (19) and (20) is the best possible.

Proof For $\sigma_1 = \sigma$ in Theorem 1, we still can conclude that the conditions (i) and (ii) in Theorem 2 are equivalent.

If there exists a constant $M \leq K(\sigma)$, such that (20) is valid, then by Lemma 2, we have $M \geq K(\sigma)$. Hence, the constant factor $M = K(\sigma) \in \mathbf{R}_{+}$ in (20) is the best possible.

The constant factor $M = K(\sigma)$ in (19) is still the best possible. Otherwise, by (16) (for $\sigma_1 = \sigma$), we would reach a conclusion that the constant factor $M = K(\sigma)$ in (20) is not the best possible.

The theorem is proved.

In particular, for $\sigma_1 = \sigma = \frac{1}{p}, \mu = \lambda - \frac{1}{p} > 0$ in Theorem 2, we have

Corollary 1 *If $\lambda > \frac{1}{p}$, then the following conditions are equivalent:*

(i) There exists a constant M, such that for any $f(x) \geq 0$, satisfying

$$0 < \int_{-\infty}^{\infty}|x|^{p-2}f^{p}(x)dx < \infty,$$

we have the following inequality:

$$\left[\int_{-\infty}^{\infty}\left(\int_{-\infty}^{\infty}\frac{|\ln|xy||^{\beta}f(x)dx}{(\max\{|xy|,1\})^{\lambda-1}|xy-1|}\right)^{p}dy\right]^{\frac{1}{p}}$$
$$< M\left(\int_{-\infty}^{\infty}|x|^{p-2}f^{p}(x)dx\right)^{\frac{1}{p}}. \tag{21}$$

(ii) *There exists a constant* M, *such that for any* $f(x), g(y) \geq 0$, *satisfying*

$$0 < \int_{-\infty}^{\infty}|x|^{p-2}f^{p}(x)dx < \infty,$$

and $0 < \int_{-\infty}^{\infty} g^{q}(y)dy < \infty$, *we have the following inequality:*

$$\int_{-\infty}^{\infty}\int_{-\infty}^{\infty}\frac{|\ln|xy||^{\beta}f(x)g(y)}{(\max\{|xy|,1\})^{\lambda-1}|xy-1|}dxdy$$
$$< M\left(\int_{-\infty}^{\infty}|x|^{p-2}f^{p}(x)dx\right)^{\frac{1}{p}}\left(\int_{-\infty}^{\infty}g^{q}(y)dy\right)^{\frac{1}{q}}. \tag{22}$$

Moreover, the constant factor

$$M = K(\frac{1}{p}) = \frac{\Gamma(\beta+1)}{2^{\beta}}\left(\zeta(\beta+1,\frac{1}{2p}) + \zeta(\beta+1,\frac{p\lambda-1}{2p})\right)$$

in (21) and (22) is the best possible.

Setting $y = \frac{1}{Y}$, $G(Y) = g(\frac{1}{Y})\frac{1}{Y^2}$ in Theorem 1–2, then replacing Y by y, we have

Corollary 2 *The following conditions are equivalent:*

(i) *There exists a constant* M, *such that for any* $f(x) \geq 0$, *satisfying*

$$0 < \int_{0}^{\infty}|x|^{p(1-\sigma)-1}f^{p}(x)dx < \infty,$$

we have the following inequality:

$$\left[\int_{-\infty}^{\infty}|y|^{-p\sigma_1-1}\left(\int_{-\infty}^{\infty}\frac{|\ln|x/y||^{\beta}f(x)dx}{(\max\{|x/y|,1\})^{\lambda-1}|(x/y)-1|}\right)^{p}dy\right]^{\frac{1}{p}}$$
$$< M\left[\int_{-\infty}^{\infty}|x|^{p(1-\sigma)-1}f^{p}(x)dx\right]. \tag{23}$$

(ii) *There exists a constant M, such that for any $f(x), G(y) \geq 0$, satisfying*

$$0 < \int_{-\infty}^{\infty} |x|^{p(1-\sigma)-1} f^p(x)dx < \infty,$$

and $0 < \int_{-\infty}^{\infty} |y|^{q(1+\sigma_1)-1} G^q(y)dy < \infty$, *we have the following inequality:*

$$\int_{-\infty}^{\infty} \int_{-\infty}^{\infty} \frac{|\ln |x/y||^{\beta} f(x)G(y)}{(\max\{|x/y|, 1\})^{\lambda-1}|(x/y) - 1|} dxdy$$

$$< M \left[\int_{-\infty}^{\infty} |x|^{p(1-\sigma)-1} f^p(x)dx \right]^{\frac{1}{p}} \left[\int_{-\infty}^{\infty} y^{q(1+\sigma_1)-1} G^q(y)dy \right]^{\frac{1}{q}}. \quad (24)$$

(iii) $\sigma_1 = \sigma$, *and* $M \geq K(\sigma) > 0$.

If Condition (iii) follows, then the constant factor $M = K(\sigma)$ in (23) and (24) (for $\sigma_1 = \sigma$) is the best possible.

For $g(y) = |y|^{\lambda} G(y)$ and $\mu_1 = \lambda - \sigma_1$ in Corollary 2, we have

Theorem 3 *The following conditions are equivalent:*

(i) *There exists a constant M, such that for any $f(x) \geq 0$, satisfying*

$$0 < \int_{-\infty}^{\infty} |x|^{p(1-\sigma)-1} f^p(x)dx < \infty,$$

we have the following inequality:

$$\left[\int_{-\infty}^{\infty} |y|^{p\mu_1 - 1} \left(\int_{-\infty}^{\infty} \frac{|\ln |x/y||^{\beta} f(x)dx}{(\max\{|x|, |y|\})^{\lambda-1}|x - y|} \right)^p dy \right]^{\frac{1}{p}}$$

$$< M \left[\int_{-\infty}^{\infty} |x|^{p(1-\sigma)-1} f^p(x)dx \right]^{\frac{1}{p}}. \quad (25)$$

(ii) *There exists a constant M, such that for any $f(x), g(y) \geq 0$, satisfying*

$$0 < \int_{-\infty}^{\infty} |x|^{p(1-\sigma)-1} f^p(x)dx < \infty,$$

and $0 < \int_{-\infty}^{\infty} |y|^{q(1-\mu_1)-1} g^q(y)dy < \infty$, *we have the following inequality:*

$$\int_{-\infty}^{\infty} \int_{-\infty}^{\infty} \frac{|\ln |x/y||^{\beta} f(x)g(y)}{(\max\{|x|, |y|\})^{\lambda-1}|x - y|} dxdy$$

$$< M \left[\int_{-\infty}^{\infty} |x|^{p(1-\sigma)-1} f^p(x)dx \right]^{\frac{1}{p}} \left[\int_{-\infty}^{\infty} |y|^{q(1-\mu_1)-1} g^q(y)dy \right]^{\frac{1}{q}}. \quad (26)$$

(iii) $\mu_1 = \mu(= \lambda - \sigma)$, *and* $M \geq K(\sigma) > 0$.

If Condition (iii) follows, then the constant factor

$$M = K(\sigma) = \frac{\Gamma(\beta+1)}{2^\beta}\left(\zeta(\beta+1, \frac{\sigma}{2}) + \zeta(\beta+1, \frac{\mu}{2})\right)$$

in (25) and (26) is the best possible.

In particular, for $\lambda = 1, \sigma = \frac{1}{q}, \mu = \frac{1}{p}$ in Theorem 3 (also refer to Theorem 2), we have

Corollary 3 *The following conditions are equivalent:*

(i) There exists a constant M, *such that for any* $f(x) \geq 0$, *satisfying*

$$0 < \int_{-\infty}^{\infty} f^p(x)dx < \infty,$$

we have the following inequality:

$$\left[\int_{-\infty}^{\infty}\left(\int_{-\infty}^{\infty}\frac{|\ln|x/y||^\beta}{|x-y|}f(x)dx\right)^p dy\right]^{\frac{1}{p}} < M\left(\int_{-\infty}^{\infty} f^p(x)dx\right). \tag{27}$$

(ii) There exists a constant M, *such that for any* $f(x), g(y) \geq 0$, *satisfying*

$$0 < \int_{-\infty}^{\infty} f^p(x)dx < \infty,$$

and $0 < \int_{-\infty}^{\infty} g^q(y)dy < \infty$, *we have the following inequality:*

$$\int_{-\infty}^{\infty}\int_{-\infty}^{\infty}\frac{|\ln|x/y||^\beta}{|x-y|}f(x)g(y)dxdy$$
$$< M\left(\int_{-\infty}^{\infty} f^p(x)dx\right)^{\frac{1}{p}}\left(\int_{-\infty}^{\infty} g^q(y)dy\right)^{\frac{1}{q}}. \tag{28}$$

The constant factor

$$M = \frac{\Gamma(\beta+1)}{2^\beta}\left(\zeta(\beta+1, \frac{1}{2q}) + \zeta(\beta+1, \frac{1}{2p})\right)$$

in (27) and (28) is the best possible.

For $\lambda = 1, \sigma = \frac{1}{p}, \mu = \frac{1}{q}$ in Theorem 3 (also refer to Theorem 2), we have the dual form of Corollary 3 as follows:

Corollary 4 *The following conditions are equivalent:*

(i) There exists a constant M, such that for any $f(x) \geq 0$, satisfying

$$0 < \int_{-\infty}^{\infty} |x|^{p-2} f^p(x)dx < \infty,$$

we have the following inequality:

$$\left[\int_{-\infty}^{\infty} |y|^{p-2} \left(\int_{-\infty}^{\infty} \frac{|\ln|x/y||^{\beta}}{|x-y|} f(x)dx\right)^p dy\right]^{\frac{1}{p}}$$
$$< M \left(\int_{-\infty}^{\infty} |x|^{p-2} f^p(x)dx\right)^{\frac{1}{p}}. \quad (29)$$

(ii) There exists a constant M, such that for any $f(x), g(y) \geq 0$, satisfying

$$0 < \int_{-\infty}^{\infty} |x|^{p-2} f^p(x)dx < \infty,$$

and $0 < \int_{-\infty}^{\infty} |y|^{q-2} g^q(y)dy < \infty$, we have the following inequality:

$$\int_{-\infty}^{\infty} \int_{-\infty}^{\infty} \frac{|\ln|x/y||^{\beta}}{|x-y|} f(x)g(y)dxdy$$
$$< M \left(\int_{-\infty}^{\infty} |x|^{p-2} f^p(x)dx\right)^{\frac{1}{p}} \left(\int_{-\infty}^{\infty} |y|^{q-2} g^q(y)dy\right)^{\frac{1}{q}}. \quad (30)$$

The constant factor

$$M = \frac{\Gamma(\beta+1)}{2^{\beta}} \left(\zeta(\beta+1, \frac{1}{2q}) + \zeta(\beta+1, \frac{1}{2p})\right)$$

in (29) and (30) is the best possible.

4 Operator Expressions

We set the following functions: $\varphi(x) := |x|^{p(1-\sigma)-1}$, $\psi(y) := |y|^{q(1-\sigma)-1}$, $\phi(y) := |y|^{q(1-\mu)-1}$, wherefrom, $\psi^{1-p}(y) = |y|^{p\sigma-1}$,

$$\phi^{1-p}(y) = |y|^{p\mu-1} (x, y \in \mathbf{R}).$$

Define the following real normed linear spaces:

$$L_{p,\varphi}(\mathbf{R}) := \left\{ f : ||f||_{p,\varphi} := \left(\int_{-\infty}^{\infty} \varphi(x)|f(x)|^p dx \right)^{\frac{1}{p}} < \infty \right\},$$

wherefrom,

$$L_{q,\psi}(\mathbf{R}) = \left\{ g : ||g||_{q,\psi} = \left(\int_{-\infty}^{\infty} \psi(y)|g(y)|^q dy \right)^{\frac{1}{q}} < \infty \right\},$$

$$L_{q,\phi}(\mathbf{R}) = \left\{ g : ||g||_{q,\phi} = \left(\int_{-\infty}^{\infty} \phi(y)|g(y)|^q dy \right)^{\frac{1}{q}} < \infty \right\},$$

$$L_{p,\psi^{1-p}}(\mathbf{R}) = \left\{ h : ||h||_{p,\psi^{1-p}} = \left(\int_{-\infty}^{\infty} \psi^{1-p}(y)|h(y)|^p dy \right)^{\frac{1}{p}} < \infty \right\},$$

$$L_{q,\phi^{1-p}}(\mathbf{R}) = \left\{ h : ||h||_{p,\phi^{1-p}} = \left(\int_{-\infty}^{\infty} \phi^{1-p}(y)|h(y)|^p dy \right)^{\frac{1}{p}} < \infty \right\}.$$

(a) In view of Theorem 2, for $f \in L_{p,\varphi}(\mathbf{R})$, setting

$$h_1(y) := \int_{-\infty}^{\infty} \frac{|\ln|x/y||^{\beta} f(x)}{(\max\{|x|,|y|\})^{\lambda-1}|x-y|} dx \ (y \in \mathbf{R}),$$

by (19), we have

$$||h_1||_{p,\psi^{1-p}} = \left[\int_{-\infty}^{\infty} \psi^{1-p}(y) h_1^p(y) dy \right]^{\frac{1}{p}} < M||f||_{p,\varphi} < \infty. \quad (31)$$

Definition 1 Define a Hilbert-type integral operator with the nonhomogeneous kernel $T^{(1)} : L_{p,\varphi}(\mathbf{R}) \to L_{p,\psi^{1-p}}(\mathbf{R})$ as follows: For any $f \in L_{p,\varphi}(\mathbf{R})$, there exists a unique representation $T^{(1)} f = h_1 \in L_{p,\psi^{1-p}}(\mathbf{R})$, satisfying for any $y \in \mathbf{R}$, $T^{(1)} f(y) = h_1(y)$.

In view of (31), it follows that

$$||T^{(1)} f||_{p,\psi^{1-p}} = ||h_1||_{p,\psi^{1-p}} \le M||f||_{p,\varphi},$$

and then the operator $T^{(1)}$ is bounded satisfying

$$||T^{(1)}|| := \sup_{f(\neq\theta)\in L_{p,\varphi}(\mathbf{R})} \frac{||T^{(1)} f||_{p,\psi^{1-p}}}{||f||_{p,\varphi}} \le M.$$

If we define the formal inner product of $T^{(1)}f$ and g as follows:

$$(T^{(1)}f, g) := \int_{-\infty}^{\infty} \left(\int_{-\infty}^{\infty} \frac{|\ln|x/y||^{\beta} f(x)dx}{(\max\{|x|, |y|\})^{\lambda-1}|x-y|} \right) g(y)dy,$$

then we can rewrite Theorem 2 as follows:

Theorem 4 *The following conditions are equivalent:*

(i) There exists a constant M, such that for any $f(x) \geq 0, f \in L_{p,\varphi}(\mathbf{R})$, $||f||_{p,\varphi} > 0$, we have the following inequality:

$$||T^{(1)}f||_{p,\psi^{1-p}} < M||f||_{p,\varphi}. \tag{32}$$

(ii) There exists a constant M, such that for any $f(x), g(y) \geq 0, f \in L_{p,\varphi}(\mathbf{R}), g \in L_{q,\psi}(\mathbf{R}), ||f||_{p,\varphi}, ||g||_{q,\psi} > 0$, we have the following inequality:

$$(T^{(1)}f, g) < M||f||_{p,\varphi}||g||_{q,\psi}. \tag{33}$$

Moreover, the constant factor $M = K(\sigma)(\in \mathbf{R}_+)$ in (32) and (33) is the best possible, namely, $0 < ||T^{(1)}|| = K(\sigma) \leq M$.

(b) In view of Theorem 3 (for $\mu_1 = \mu$), for $f \in L_{p,\varphi}(\mathbf{R})$, setting

$$h_2(y) := \int_{-\infty}^{\infty} \frac{|\ln|x/y||^{\beta} f(x)dx}{(\max\{|x|, |y|\})^{\lambda-1}|x-y|} \ (y \in \mathbf{R}),$$

by (25), we have

$$||h_2||_{p,\phi^{1-p}} = \left[\int_{-\infty}^{\infty} \phi^{1-p}(y)h_2^p(y)dy \right]^{\frac{1}{p}} < M||f||_{p,\varphi} < \infty. \tag{34}$$

Definition 2 Define a Hilbert-type integral operator with the homogeneous kernel $T^{(2)} : L_{p,\varphi}(\mathbf{R}) \to L_{p,\phi^{1-p}}(\mathbf{R})$ as follows: For any $f \in L_{p,\varphi}(\mathbf{R})$, there exists a unique representation $T^{(2)}f = h_2 \in L_{p,\phi^{1-p}}(\mathbf{R})$, satisfying for any $y \in \mathbf{R}$, $T^{(2)}f(y) = h_2(y)$.

In view of (34), it follows that

$$||T^{(2)}f||_{p,\phi^{1-p}} = ||h_2||_{p,\phi^{1-p}} \leq M||f||_{p,\varphi},$$

and then the operator $T^{(2)}$ is bounded satisfying

$$||T^{(2)}|| := \sup_{f(\neq\theta)\in L_{p,\varphi}(\mathbf{R})} \frac{||T^{(2)}f||_{p,\phi^{1-p}}}{||f||_{p,\varphi}} \leq M.$$

If we define the formal inner product of $T^{(2)}f$ and g as follows:

$$(T^{(2)}f, g) := \int_{-\infty}^{\infty}\left(\int_{-\infty}^{\infty}\frac{|\ln|x/y||^{\beta}f(x)dx}{(\max\{|x|, |y|\})^{\lambda-1}|x-y|}\right)g(y)dy,$$

then we can rewrite Theorem 3 (for $\mu_1 = \mu$) as follows:

Theorem 5 *The following conditions are equivalent:*

(i) There exists a constant M, such that for any $f(x) \geq 0, f \in L_{p,\varphi}(\mathbf{R})$, $||f||_{p,\varphi} > 0$, we have the following inequality:

$$||T^{(2)}f||_{p,\phi^{1-p}} < M||f||_{p,\varphi}. \tag{35}$$

(ii) There exists a constant M, such that for any $f(x), g(y) \geq 0$, $f \in L_{p,\varphi}(\mathbf{R})$, $g \in L_{q,\phi}(\mathbf{R})$, $||f||_{p,\varphi}, ||g||_{q,\phi} > 0$, we have the following inequality:

$$(T^{(2)}f, g) < M||f||_{p,\varphi}||g||_{q,\phi}. \tag{36}$$

Moreover, the constant factor $M = K_{\lambda}(\sigma)(\in \mathbf{R}_{+})$ in (35) and (36) is the best possible, namely, $0 < ||T^{(2)}|| = K_{\lambda}(\sigma) \leq M$.

Acknowledgements This work is supported by the National Natural Science Foundation (Nos. 61370186, 61640222), and Appropriative Researching Fund for Professors and Doctors, Guangdong University of Education (No. 2015ARF25). we are grateful for this help.

References

1. G.H. Hardy, J.E. Littlewood, G. Pólya, *Inequalities* (Cambridge University Press, Cambridge, 1934)
2. B.C. Yang, *The Norm of Operator and Hilbert-Type Inequalities* (Science Press, Beijing, 2009)
3. B.C. Yang, *Hilbert-Type Integral Inequalities* (Bentham Science Publishers, Sharjah, 2009)
4. B.C. Yang, On the norm of an integral operator and applications. J. Math. Anal. Appl. **321**, 182–192 (2006)
5. J.S. Xu, Hardy-Hilbert's inequalities with two parameters. Adv. Math. **36**(2), 63–76 (2007)
6. B.C. Yang, On the norm of a Hilbert's type linear operator and applications. J. Math. Anal. Appl. **325**, 529–541 (2007)
7. D.M. Xin, A Hilbert-type integral inequality with the homogeneous kernel of zero degree. Math. Theory Appl. **30**(2), 70–74 (2010)
8. B.C. Yang, A Hilbert-type integral inequality with the homogenous kernel of degree 0. J. Shandong Univ. (Natural Science) **45**(2), 103–106 (2010)
9. L. Debnath, B.C. Yang, Recent developments of Hilbert-type discrete and integral inequalities with applications. Int. J. Math. Math. Sci. **2012**, 871845 (2012)
10. M.T. Rassias, B.C. Yang, On half-discrete Hilbert's inequality. Appl. Math. Comput. **220**, 75–93 (2013)

11. B.C. Yang, M. Krnic, A half-discrete Hilbert-type inequality with a general homogeneous kernel of degree 0. J. Math. Inequal. **6**(3), 401–417 (2012)
12. T.M. Rassias, B.C. Yang, A multidimensional half-discrete Hilbert-type inequality and the Riemann zeta function. Appl. Math. Comput. **225**, 263–277 (2013)
13. M.T. Rassias, B.C. Yang, On a multidimensional half-discrete Hilbert-type inequality related to the hyperbolic cotangent function. Appl. Math. Comput. **242**, 800–813 (2013)
14. M.T. Rassias, B.C. Yang, A multidimensional Hilbert-type integral inequality related to the Riemann zeta function, in *Applications of Mathematics and Informatics in Science and Engineering*, ed. by N.J. Daras (Springer, New York, 2014), pp. 417–433
15. Q. Chen, B.C. Yang, A survey on the study of Hilbert-type inequalities. J. Inequal. Appl. **2015**, 302 (2015)
16. B.C. Yang, A new Hilbert-type integral inequality. Soochow J. Math. **33**(4), 849–859 (2007)
17. Z.Q. Wang, D.R. Guo, *Introduction to Special Functions* (Science Press, Beijing, 1979)
18. B. He, B.C. Yang, On a Hilbert-type integral inequality with the homogeneous kernel of 0-degree and the hypergeometrc function. Math. Practice Theory **40**(18), 105–211 (2010)
19. B.C. Yang, A new Hilbert-type integral inequality with some parameters. J. Jilin Univ. **46**(6), 1085–1090 (2008)
20. B.C. Yang, A Hilbert-type integral inequality with a non-homogeneous kernel. J. Xiamen Univ. **48**(2), 165–169 (2008)
21. Z. Zeng, Z.T. Xie, On a new Hilbert-type integral inequality with the homogeneous kernel of degree 0 and the integral in whole plane. J. Inequal. Appl. **2010**, 256796 (2010)
22. B.C. Yang, A reverse Hilbert-type integral inequality with some parameters. J. Xinxiang Univ. **27**(6), 1–4 (2010)
23. A.Z. Wang, B.C. Yang, A new Hilbert-type integral inequality in whole plane with the non-homogeneous kernel. J. Inequal. Appl. **2011**, 123 (2011)
24. D.M. Xin, B.C. Yang, A Hilbert-type integral inequality in whole plane with the homogeneous kernel of degree-2. J. Inequal. Appl. **2011**, 401428 (2011)
25. B. He, B.C. Yang, On an inequality concerning a non-homogeneous kernel and the hypergeometric function. Tamsul Oxford J. Inform. Math. Sci. **27**(1), 75–88 (2011)
26. B. Yang, A reverse Hilbert-type integral inequality with a non-homogeneous kernel. J. Jilin Univ. **49**(3), 437–441 (2011)
27. Z.T. Xie, Z. Zeng, Y.F. Sun, A new Hilbert-type inequality with the homogeneous kernel of degree-2. Adv. Appl. Math. Sci. **12**(7), 391–401 (2013)
28. Q.L. Huang, S.H. Wu, B.C. Yang, Parameterized Hilbert-type integral inequalities in the whole plane. Sci. World J. **2014**, 169061 (2014)
29. Z. Zhen, K. Raja Rama Gandhi, Z.T. Xie, A new Hilbert-type inequality with the homogeneous kernel of degree-2 and with the integral. Bull. Math. Sci. Appl. **3**(1), 11–20 (2014)
30. M.T. Rassias, B.C. Yang, A Hilbert-type integral inequality in the whole plane related to the hyper geometric function and the beta function. J. Math. Anal. Appl. **428**(2), 1286–1308 (2015)
31. X.Y. Huang, J.F. Cao, B. He, B.C. Yang, Hilbert-type and Hardy-type integral inequalities with operator expressions and the best constants in the whole plane. J. Inequal. Appl. **2015**, 129 (2015)
32. Z.H. Gu, B.C. Yang, A Hilbert-type integral inequality in the whole plane with a non-homogeneous kernel and a few parameters. J. Inequal. Appl. **2015**, 314 (2015)
33. J.C. Kuang, *Real and Functional Analysis (Continuation)* vol. 2 (Higher Education Press, Beijing, 2015)
34. J.C. Kuang, *Applied Inequalities* (Shangdong Science and Technology Press, Jinan, 2004)

On Metric Structures of Normed Gyrogroups

Teerapong Suksumran

Abstract In this article, we indicate that the open unit ball in n-dimensional Euclidean space $\mathbb{R}^n$ admits norm-like functions compatible with the Poincaré and Beltrami–Klein metrics. This leads to the notion of a normed gyrogroup, similar to that of a normed group in the literature. We then examine topological and geometric structures of normed gyrogroups. In particular, we prove that the normed gyrogroups are homogeneous and form left invariant metric spaces and derive a version of the Mazur–Ulam theorem. We also give certain sufficient conditions, involving the right-gyrotranslation inequality and Klee's condition, for a normed gyrogroup to be a topological gyrogroup.

1 Introduction

Roughly speaking, a normed group is a group that comes with a compatible norm (also called a length function), similar to the case of normed linear spaces. A prominent example of a normed group is a finitely generated group with the word metric, which is one of the main ingredients in geometric group theory. The normed groups abound as an integral part in the theory of topological groups [5]. They are blended objects that have significance in group theory, geometry, analysis, and topology, to name a few.

In [27], Ungar studies a parametrization of the Lorentz transformation group. This leads to the formation of gyrogroup theory, a rich subject in mathematics [3, 9, 23–26, 30]. Loosely speaking, a gyrogroup is a group-like structure in which the associative law fails to satisfy. However, it obeys the *gyroassociative* law, a weak form of associativity, as well as the *loop property*, an algebraic rule equivalent to the *Bol identity* in loop theory. One of the virtues of studying gyrogroups is

T. Suksumran (✉)
Department of Mathematics, Faculty of Science, Chiang Mai University, Chiang Mai, Thailand
e-mail: teerapong.suksumran@cmu.ac.th

T. M. Rassias, P. M. Pardalos (eds.), *Mathematical Analysis and Applications*, Springer Optimization and Its Applications 154,
https://doi.org/10.1007/978-3-030-31339-5_20

an application in non-Euclidean geometry [28], where Ungar examines analytic hyperbolic geometry using the gyrolanguage.

Atiponrat [4] and Cai et al. [6] pave the way for studying topological gyrogroups. In particular, Atiponrat proves that in the class of topological gyrogroups, being a T_0-space is equivalent to being a T_3-space [4, Theorem 3]. Further, she attempts to extend the famous Birkhoff–Kakutani theorem by proving that every first-countable Hausdorff topological gyrogroup is premetrizable [4, Theorem 4]. The latter result is strengthened when Cai et al. prove that every first-countable Hausdorff topological gyrogroup is metrizable [6, Theorem 2.3]. The achieved results inspire us to investigate topological properties of gyrogroups, which eventually bring us to the notion of a normed gyrogroup. This provides a large class of gyrogroups with a left invariant metric, and some of them are indeed topological gyrogroups.

2 Preliminaries

Standard terminology and notation in algebra, topology, and geometry used throughout the article are defined as usual. In this section, we collect relevant definitions and elementary properties of gyrogroups for reference [22, 28]. The reader familiar with gyrogroup theory may skip this section.

Let G be a nonempty set equipped with a binary operation $\oplus$ on G. Denote by $\operatorname{Aut} G$ the group of automorphisms of $(G, \oplus)$.

Definition 1 (Gyrogroups) A nonempty set G, together with a binary operation $\oplus$ on G, is called a *gyrogroup* if it satisfies the following axioms:

(G1) There exists an element $e \in G$ such that $e \oplus a = a$ for all $a \in G$.

(G2) For each $a \in G$, there exists an element $b \in G$ such that $b \oplus a = e$.

(G3) For all $a, b \in G$, there is an automorphism $\operatorname{gyr}[a, b] \in \operatorname{Aut} G$ such that

$$a \oplus (b \oplus c) = (a \oplus b) \oplus \operatorname{gyr}[a, b]c \qquad \text{(left gyroassociative law)}$$

for all $c \in G$.

(G4) For all $a, b \in G$, $\operatorname{gyr}[a \oplus b, b] = \operatorname{gyr}[a, b]$. (left loop property)

Note that the axioms in Definition 1 imply the right counterparts. In particular, any gyrogroup has a unique two-sided identity e, and an element a of the gyrogroup has a unique two-sided inverse $\ominus a$. The automorphism $\operatorname{gyr}[a, b]$ is called the *gyroautomorphism* generated by a and b. It is clear that every group satisfies the gyrogroup axioms (the gyroautomorphisms are the identity map) and hence is a gyrogroup. Conversely, any gyrogroup with *trivial* gyroautomorphisms forms a group. From this point of view, gyrogroups naturally generalize groups.

The table below summarizes some algebraic properties of gyrogroups [22, 28], which will prove useful in studying topological and geometric aspects of gyrogroups in Sections 3 and 4. We remark that gyroautomorphisms play an essential role in gyrogroup theory; for example, they appear as part of generic algebraic rules extended from group-theoretic identities.

3 Normed Gyrogroups and Concrete Examples

In this section, we establish that any gyrogroup with an appropriate *length function*, called a *gyronorm*, has the metric structure and hence is a Hausdorff space. We also exhibit a few examples of well-known gyrogroups that have gyronorms.

3.1 The Definition and Basic Properties

Definition 2 (Gyronorms) Let G be a gyrogroup. A function $\|\cdot\|\colon G \to \mathbb{R}$ is called a *gyronorm* on G if the following properties hold:

1. $\|x\| \geq 0$ for all $x \in G$ and $\|x\| = 0$ if and only if $x = e$; (positivity)
2. $\|\ominus x\| = \|x\|$ for all $x \in G$; (invariant under taking inverses)
3. $\|x \oplus y\| \leq \|x\| + \|y\|$ for all $x, y \in G$; (subadditivity)
4. $\|\mathrm{gyr}[a, b]x\| = \|x\|$ for all $a, b, x \in G$. (invariant under gyrations)

Any gyrogroup with a gyronorm is called a *normed* gyrogroup. We remark that the term "gyronorm" is quite different from what Ungar used in Chapter 4 of [28]. In Section 3.2, we give several concrete examples of gyrogroups with a gyronorm. Clearly, Definition 2 is a generalization of the notion of a *group-norm* [5, p. 8], which in turn is motivated by norms on linear spaces. Furthermore, any gyrogroup may be viewed as a normed gyrogroup with a gyronorm defined by

$$\|x\| = \begin{cases} 0 & \text{if } x = e; \\ 1 & \text{if } x \neq e. \end{cases}$$

Theorem 1 *Let G be a normed gyrogroup. Define*

$$d(x, y) = \|\ominus x \oplus y\| \tag{1}$$

for all $x, y \in G$. Then d is a metric on G and so (G, d) forms a metric space.

Proof By definition, $d(x, y) = \|\ominus x \oplus y\| \geq 0$ for all $x, y \in G$. Clearly, $d(x, x) = \|e\| = 0$ for all $x \in G$. Suppose that $d(x, y) = 0$. Then $\|\ominus x \oplus y\| = 0$. By definition, $\ominus x \oplus y = e$. Hence, $x = y$ by the left cancellation law.

Let $x, y, z \in G$. Using appropriate properties of gyrogroups in Table 1, together with the defining properties of a gyronorm, we obtain

$$\begin{aligned} d(y, x) &= \|\ominus y \oplus x\| = \|\ominus(\ominus y \oplus x)\| = \|\mathrm{gyr}[\ominus y, x](\ominus x \oplus y)\| \\ &= \|\ominus x \oplus y\| = d(x, y). \end{aligned}$$

Table 1 Algebraic properties of gyrogroups (cf. [22, 28])

Gyrogroup identity	Name/reference
$\ominus(\ominus a) = a$	Involution of inversion
$\ominus a \oplus (a \oplus x) = x$	Left cancellation law
$\mathrm{gyr}[a, b]c = \ominus(a \oplus b) \oplus (a \oplus (b \oplus c))$	Gyrator identity
$\ominus(a \oplus b) = \mathrm{gyr}[a, b](\ominus b \ominus a)$	cf. $(ab)^{-1} = b^{-1}a^{-1}$ in a group
$(\ominus a \oplus b) \oplus \mathrm{gyr}[\ominus a, b](\ominus b \oplus c) = \ominus a \oplus c$	cf. $(a^{-1}b)(b^{-1}c) = a^{-1}c$ in a group
$\mathrm{gyr}[\ominus a, \ominus b] = \mathrm{gyr}[a, b]$	Even property
$\mathrm{gyr}[b, a] = \mathrm{gyr}^{-1}[a, b]$, the inverse of $\mathrm{gyr}[a, b]$	Inversive symmetry
$\varphi(\mathrm{gyr}[a, b]c) = \mathrm{gyr}[\varphi(a), \varphi(b)]\varphi(c)$	Gyration preserving under a gyrogroup homomorphism φ
$L_a \circ L_b = L_{a\oplus b} \circ \mathrm{gyr}[a, b]$	Composition law for left gyrotranslations

Furthermore, we obtain

$$\begin{aligned} d(x, z) &= \| \ominus x \oplus z\| \\ &= \|(\ominus x \oplus y) \oplus \mathrm{gyr}[\ominus x, y](\ominus y \oplus z)\| \\ &\leq \| \ominus x \oplus y\| + \|\mathrm{gyr}[\ominus x, y](\ominus y \oplus z)\| \\ &= \| \ominus x \oplus y\| + \| \ominus y \oplus z\| \\ &= d(x, y) + d(y, z). \end{aligned}$$

This proves that d satisfies the defining properties of a metric. □

The metric d induced by a gyronorm on G in Theorem 1 is called a *gyronorm* metric. Whenever we say that G is a normed gyrogroup, we assume that G is endowed with the corresponding gyronorm metric and that G carries the topology induced by this metric, unless mentioned otherwise. It is clear that every isometry of a normed gyrogroup to itself is a homeomorphism for the inverse of an isometry is again an isometry. Next, we show that known gyrogroups in the literature possess gyronorms.

3.2 *Concrete Examples*

3.2.1 An n-Dimensional Euclidean Version of the Einstein Gyrogroup

Let $\mathbb{B}$ denote the open unit ball in n-dimensional Euclidean space $\mathbb{R}^n$, that is,

$$\mathbb{B} = \{\mathbf{v} \in \mathbb{R}^n : \|\mathbf{v}\| < 1\}, \tag{2}$$

where $\|\cdot\|$ denotes the *Euclidean* norm on $\mathbb{R}^n$. The *Einstein* gyrogroup consists of $\mathbb{B}$, together with Einstein addition $\oplus_E$ given by

$$\mathbf{u} \oplus_E \mathbf{v} = \frac{1}{1 + \langle \mathbf{u}, \mathbf{v} \rangle} \left(\mathbf{u} + \frac{1}{\gamma_{\mathbf{u}}} \mathbf{v} + \frac{\gamma_{\mathbf{u}}}{1 + \gamma_{\mathbf{u}}} \langle \mathbf{u}, \mathbf{v} \rangle \mathbf{u} \right) \tag{3}$$

for all $\mathbf{u}, \mathbf{v} \in \mathbb{B}$, where $\gamma_{\mathbf{u}}$ is the *Lorentz factor* given by $\gamma_{\mathbf{u}} = \dfrac{1}{\sqrt{1 - \|\mathbf{u}\|^2}}$. The zero vector $\mathbf{0}$ acts as the identity of $\mathbb{B}$ under $\oplus_E$. For each $\mathbf{v} \in \mathbb{B}$, the negative vector $-\mathbf{v}$ acts as the inverse of $\mathbf{v}$ with respect to Einstein addition. By Proposition 2.4 of [14], the gyroautomorphisms of $(\mathbb{B}, \oplus_E)$ are orthogonal in the sense that

$$\|\mathrm{gyr}[\mathbf{u}, \mathbf{v}]\mathbf{w}\| = \|\mathbf{w}\|$$

for all $\mathbf{u}, \mathbf{v}, \mathbf{w} \in \mathbb{B}$.

Define a function $\|\cdot\|_E$ on $\mathbb{B}$ by the equation

$$\|\mathbf{v}\|_E = \tanh^{-1} \|\mathbf{v}\|, \qquad \mathbf{v} \in \mathbb{B}, \tag{4}$$

where $\tanh^{-1}$ denotes the inverse of the hyperbolic tangent function on $\mathbb{R}$.

Theorem 2 *The function* $\|\cdot\|_E$ *defined by* (4) *is a gyronorm on the Einstein gyrogroup.*

Proof Since $\tanh r \geq 0$ if and only $r \geq 0$, it follows that $\|\mathbf{v}\|_E \geq 0$ for all $\mathbf{v} \in \mathbb{B}$. Note that $\|\mathbf{0}\|_E = \tanh^{-1} 0 = 0$. Suppose that $\|\mathbf{v}\|_E = 0$. Then $0 = \tanh^{-1} \|\mathbf{v}\|$, which implies $\|\mathbf{v}\| = 0$. Hence, $\mathbf{v} = \mathbf{0}$. Let $\mathbf{v} \in \mathbb{B}$. Then

$$\|\ominus \mathbf{v}\|_E = \|-\mathbf{v}\|_E = \tanh^{-1} \|-\mathbf{v}\| = \tanh^{-1} \|\mathbf{v}\| = \|\mathbf{v}\|_E.$$

Let $\mathbf{u}, \mathbf{v} \in \mathbb{B}$. Applying Proposition 3.3 of [14] and Lemma 3.2 (iv) of [14] gives

$$\|\mathbf{u} \oplus_E \mathbf{v}\|_E \leq \|\mathbf{u}\|_E + \|\mathbf{v}\|_E.$$

Let $\mathbf{u}, \mathbf{v}, \mathbf{w} \in \mathbb{B}$. Direct computation shows that

$$\|\mathrm{gyr}[\mathbf{u}, \mathbf{v}]\mathbf{w}\|_E = \tanh^{-1} \|\mathrm{gyr}[\mathbf{u}, \mathbf{v}]\mathbf{w}\| = \tanh^{-1} \|\mathbf{w}\| = \|\mathbf{w}\|_E.$$

□

It follows from Theorems 1 and 2 that

$$d_E(\mathbf{u}, \mathbf{v}) = \tanh^{-1} \|-\mathbf{u} \oplus_E \mathbf{v}\| \tag{5}$$

defines a metric on $\mathbb{B}$. This metric is called the *rapidity* metric on the Einstein gyrogroup [14, 28]. It is known that the rapidity metric on the Einstein gyrogroup agrees with the *Cayley–Klein* metric on the Beltrami–Klein model of n-dimensional hyperbolic geometry [14, p. 1233].

Note that the Euclidean norm is indeed a gyronorm on the Einstein gyrogroup. This follows from the fact that

$$\|\mathbf{u} \oplus_E \mathbf{v}\| \leq \|\mathbf{u}\| \oplus \|\mathbf{v}\| = \frac{\|\mathbf{u}\| + \|\mathbf{v}\|}{1 + \|\mathbf{u}\|\|\mathbf{v}\|} \leq \|\mathbf{u}\| + \|\mathbf{v}\|,$$

where the first inequality is worked out in Proposition 3.3 of [14] and $\oplus$ is the restricted Einstein addition on the open interval $(-1, 1)$ given by $r \oplus s = \dfrac{r+s}{1+rs}$ for all $r, s \in (-1, 1)$. Denote by d_e the gyronorm metric induced by the Euclidean norm. That is,

$$d_e(\mathbf{u}, \mathbf{v}) = \|-\mathbf{u} \oplus_E \mathbf{v}\| \tag{6}$$

for all $\mathbf{u}, \mathbf{v} \in \mathbb{B}$. In fact, d_e is known as the *Einstein gyrometric* [28, p. 222]. Our results provide an elegant proof that the Einstein gyrometric is indeed a metric on the open unit ball of $\mathbb{R}^n$.

Note that $d_e(\mathbf{u}, \mathbf{v}) \leq d_E(\mathbf{u}, \mathbf{v})$ for all $\mathbf{u}, \mathbf{v} \in \mathbb{B}$ because $x \mapsto x - \tanh^{-1} x$ defines a strictly decreasing function on the open interval $(0, 1)$. This implies that the topology generated by d_E is finer than the topology generated by d_e. Next, we prove that the topology generated by d_e is finer than the topology generated by d_E. Let $\mathbf{u} \in \mathbb{B}$ and let $\epsilon > 0$. Choose $\delta = \tanh \epsilon$. Let $\mathbf{v} \in B_{d_e}(\mathbf{u}, \delta)$. Then $d_e(\mathbf{u}, \mathbf{v}) < \delta$, that is, $\|-\mathbf{u} \oplus_E \mathbf{v}\| < \tanh \epsilon$. It follows that

$$d_E(\mathbf{u}, \mathbf{v}) = \tanh^{-1} \|-\mathbf{u} \oplus_E \mathbf{v}\| < \epsilon$$

for $\tanh^{-1}$ is a strictly increasing function on its domain. Thus, $\mathbf{v} \in B_{d_E}(\mathbf{u}, \epsilon)$. This proves that $B_{d_e}(\mathbf{u}, \delta) \subseteq B_{d_E}(\mathbf{u}, \epsilon)$. Therefore, d_e and d_E generate the same topology on $\mathbb{B}$.

3.2.2 An n-Dimensional Euclidean Version of the Möbius Gyrogroup

The *Möbius* gyrogroup consists of the same underlying set as the Einstein gyrogroup, but its binary operation, called *Möbius addition*, is defined by

$$\mathbf{u} \oplus_M \mathbf{v} = \frac{(1 + 2\langle \mathbf{u}, \mathbf{v}\rangle + \|\mathbf{v}\|^2)\mathbf{u} + (1 - \|\mathbf{u}\|^2)\mathbf{v}}{1 + 2\langle \mathbf{u}, \mathbf{v}\rangle + \|\mathbf{u}\|^2\|\mathbf{v}\|^2} \tag{7}$$

for all $\mathbf{u}, \mathbf{v} \in \mathbb{B}$. The zero vector acts as the identity of $\mathbb{B}$ under $\oplus_M$. For each $\mathbf{v} \in \mathbb{B}$, the negative vector $-\mathbf{v}$ acts as the inverse of $\mathbf{v}$ with respect to Möbius addition. By

Proposition 2.4 of [14], the gyroautomorphisms of $(\mathbb{B}, \oplus_M)$ are orthogonal in the sense that

$$\|\mathrm{gyr}[\mathbf{u}, \mathbf{v}]\mathbf{w}\| = \|\mathbf{w}\|$$

for all $\mathbf{u}, \mathbf{v}, \mathbf{w} \in \mathbb{B}$.

By Proposition 2.3 of [14], the map Φ defined by

$$\Phi(\mathbf{v}) = \frac{2}{1 + \|\mathbf{v}\|^2}\mathbf{v}, \qquad \mathbf{v} \in \mathbb{B} \tag{8}$$

is a gyrogroup isomorphism from $(\mathbb{B}, \oplus_M)$ to $(\mathbb{B}, \oplus_E)$. In particular, Φ is a bijection from $\mathbb{B}$ to itself. Let $\|\cdot\|_E$ be the gyronorm on the Einstein gyrogroup defined by (4). Define a function $\|\cdot\|_M$ by the equation

$$\|\mathbf{v}\|_M = \frac{1}{2}\|\Phi(\mathbf{v})\|_E, \qquad \mathbf{v} \in \mathbb{B}. \tag{9}$$

Theorem 3 *The function* $\|\cdot\|_M$ *defined by* (9) *is a gyronorm on the Möbius gyrogroup.*

Proof The theorem follows directly from the fact that $\|\cdot\|_E$ defines a gyronorm on $(\mathbb{B}, \oplus_E)$ and that Φ is a gyrogroup isomorphism. □

By Theorems 1 and 3,

$$d_M(\mathbf{u}, \mathbf{v}) = \frac{1}{2}\tanh^{-1}\|\Phi(-\mathbf{u} \oplus_M \mathbf{v})\| \tag{10}$$

defines a metric on $\mathbb{B}$. This metric is called the *rapidity* metric on the Möbius gyrogroup [14, 28]. It is known that the rapidity metric on the Möbius gyrogroup is half of the *Poincaré* metric with curvature -1 on the Poincaré model of n-dimensional hyperbolic geometry [14, Theorem 3.7].

4 Topological and Geometric Structures

Normed gyrogroups have nice topological and geometric structures. Further, they share certain remarkable analogies with normed groups. In fact, several of the results proved in this section are inspired by the expository article of Bingham and Ostaszewski [5] that treats normed and topological groups. Roughly speaking, normed gyrogroups are left invariant, homogeneous, and isotropic.

4.1 Topological and Geometric Properties

Theorem 4 *Let G be a normed gyrogroup. Then the gyronorm metric is invariant under left gyrotranslation,*

$$d(a \oplus x, a \oplus y) = d(x, y) \tag{11}$$

for all $a, x, y \in G$. Hence, every left gyrotranslation of G is an isometry of G with respect to the gyronorm metric.

Proof Let $a \in G$. Recall that the *left gyrotranslation* by a, denoted by L_a, is defined by $L_a(x) = a \oplus x$ for all $x \in G$. By Theorem 18 (1) of [22], L_a is a bijection from G to itself.

Next, we prove that the gyronorm metric d is invariant under L_a. Let $x, y \in G$. Using appropriate properties of gyrogroups in Table 1, together with the defining properties of a gyronorm, we obtain

$$\begin{aligned} d(L_a(x), L_a(y)) &= \| \ominus (a \oplus x) \oplus (a \oplus y)\| \\ &= \|\mathrm{gyr}[a, x](\ominus x \ominus a) \oplus (a \oplus y)\| \\ &= \|(\ominus x \ominus a) \oplus \mathrm{gyr}[x, a](a \oplus y)\| \\ &= \|(\ominus x \ominus a) \oplus \mathrm{gyr}[\ominus x, \ominus a](a \oplus y)\| \\ &= \| \ominus x \oplus y\| \\ &= d(x, y). \end{aligned}$$

□

Corollary 1 *If G is a normed gyrogroup, then every left gyrotranslation of G is a homeomorphism.*

Theorem 5 *Let G be a normed gyrogroup. If $\tau \in \mathrm{Aut}\, G$ and $\|\tau(x)\| = \|x\|$ for all $x \in G$, then τ is an isometry of G with respect to the gyronorm metric.*

Proof By assumption,

$$d(\tau(x), \tau(y)) = \| \ominus \tau(x) \oplus \tau(y)\| = \|\tau(\ominus x \oplus y)\| = \| \ominus x \oplus y\| = d(x, y)$$

and so τ defines an isometry of G. □

Corollary 2 *If G is a normed gyrogroup, then the gyroautomorphisms of G are isometries (and also homeomorphisms) of G.*

Theorem 6 (Homogeneity) *If G is a normed gyrogroup, then G is homogeneous in the sense that if x and y are arbitrary points of G, then there is an isometry $T: G \to G$ (and also a homeomorphism of G) such that $T(x) = y$.*

Proof Let $x, y \in G$. Define $T = L_y \circ L_{\ominus x}$. By Theorem 4, T is an isometry of G. Further,

$$T(x) = (L_y \circ L_{\ominus x})(x) = L_y(L_{\ominus x}(x)) = L_y(\ominus x \oplus x) = L_y(e) = y \oplus e = y.$$

Thus, G is homogeneous. □

Theorem 7 (Isotropy) *If G is a nondegenerate normed gyrogroup; that is, G has a nonidentity gyroautomorphism, then G is isotropic in the sense that for each point $p \in G$, there exists a nonidentity isometry T of G such that $T(p) = p$.*

Proof Let p be an arbitrary point of G. Let τ be a *nonidentity* gyroautomorphism of G. Then $\tau(e) = e$. Define $T = L_p \circ \tau \circ L_{\ominus p}$. Note that T is an isometry of G, being the composite of isometries of G. Further, $T(p) = p$. Note that T is not the identity transformation of G; otherwise, we would have $I = L_p \circ \tau \circ L_{\ominus p} = L_p \circ \tau \circ L_p^{-1}$ and would have $\tau = I$, a contradiction. □

Recall that the famous Mazur–Ulam theorem states that any isometry between normed linear spaces over $\mathbb{R}$ that fixes the zero vector must be linear; see, for instance, [10, Theorem 1.3.5]. Extensions of the Mazur–Ulam theorem are studied by Rassias [18, 19] and by Rassias et al. [20, 21]. Further, the Mazur–Ulam theorem is examined in the setting of *gyrovector spaces* by Abe [1] and by Abe and Hatori [2]. Here, we prove a normed-gyrogroup version of the Mazur–Ulam theorem.

Theorem 8 *Let G be a normed gyrogroup. If f is an isometry of G with respect to the gyronorm metric, then*

$$f = L_{f(e)} \circ \rho, \tag{12}$$

where ρ is an isometry of G that leaves the gyrogroup identity fixed.

Proof Suppose that f is an isometry of G. By definition, f is a permutation of G. By Proposition 19 of [22], $f = L_{f(e)} \circ \rho$, where ρ is a permutation of G that fixes e. As in the proof of Theorem 18 of [22], $L_{f(e)}^{-1} = L_{\ominus f(e)}$ and so $\rho = L_{\ominus f(e)} \circ f$. Hence, ρ is an isometry of G, being the composite of isometries of G. □

4.2 A Characterization of Normed Gyrogroups

Note that the gyronorm of an arbitrary normed gyrogroup can be recovered by its corresponding metric,

$$\|x\| = d(e, x), \qquad x \in G. \tag{13}$$

It turns out that Theorem 4 provides a characterizing property of normed gyrogroups, as shown in the following theorem.

Theorem 9 *Let G be a gyrogroup with a metric d. If d is invariant under left gyrotranslation, that is,*

$$d(a \oplus x, a \oplus y) = d(x, y)$$

for all $a, x, y \in G$, then $\|x\| = d(e, x)$ defines a gyronorm on G that generates the same metric.

Proof It is clear that $\|x\| \geq 0$ and $\|x\| = 0$ if and only if $x = e$. Let $x \in G$. By the left gyrotranslation invariant, $\|\ominus x\| = d(e, \ominus x) = d(x \oplus e, x \ominus x) = d(x, e) = \|x\|$.

Let $x, y \in G$. Direct computation shows that

$$\|x \oplus y\| = d(x \oplus y, e) = d(y, \ominus x) \leq d(y, e) + d(e, \ominus x) = \|x\| + \|y\|.$$

Let $a, b, x \in G$. By the gyrator identity,

$$\begin{aligned}\|\text{gyr}[a, b]x\| &= d(\text{gyr}[a, b]x, e) \\ &= d(\ominus(a \oplus b) \oplus (a \oplus (b \oplus x)), e) \\ &= d(a \oplus (b \oplus x), a \oplus b)) \\ &= d(b \oplus x, b) \\ &= d(x, e) \\ &= \|x\|.\end{aligned}$$

□

In view of Theorems 1, 4, and 9, there is a one-to-one correspondence between the class of normed gyrogroups and the class of gyrogroups with a left-gyrotranslation-invariant metric:

$$\{\text{Normed gyrogroups}\} \quad \longleftrightarrow \quad \{(G, d), d \text{ left-gyrotranslation-invariant metric}\}.$$

One of the advantages of Theorem 9 is illustrated in the example below.

Example 1 (The Complex Möbius Gyrogroup) The Poincaré disk model consists of the open unit disk in the complex plane,

$$\mathbb{D} = \{z \in \mathbb{C} : |z| < 1\}, \tag{14}$$

and the (complex version of) Poincaré metric defined by

$$d_P(w, z) = 2 \tanh^{-1} \left| \frac{w - z}{1 - \overline{w} z} \right| \tag{15}$$

for all $w, z \in \mathbb{D}$. Here, the factor 2 is added to (15) so that the metric corresponds to a curvature of -1.

A *complex* version of Möbius addition is defined by

$$a \oplus_M b = \frac{a+b}{1+\overline{a}b}, \qquad a, b \in \mathbb{D}, \tag{16}$$

which gives $\mathbb{D}$ the gyrogroup structure [29]. It is not difficult to check that 0 is the identity of $\mathbb{D}$, that the inverse of a is $-a$, and that the gyroautomorphism generated by a and b is a disk rotation corresponding to the unimodular complex $\dfrac{1+a\overline{b}}{1+\overline{a}b}$. Using (15) and (16), we have by inspection that

$$d_P(a \oplus_M w, a \oplus_M z) = d_P(w, z)$$

for all $a, w, z \in \mathbb{D}$. Hence, by Theorem 9, $\mathbb{D}$ forms a normed gyrogroup whose gyronorm is given by $\|z\| = 2\tanh^{-1}|z|$ for all $z \in \mathbb{D}$. This leads to the well-known result that any Möbius transformation (also called a *conformal* self-map) of $\mathbb{D}$ of the form

$$z \mapsto \frac{a+z}{1+\overline{a}z}, \qquad z \in \mathbb{D},$$

where a is a fixed element in $\mathbb{D}$, is an isometry of $\mathbb{D}$ with respect to the Poincaré metric. The study of Möbius transformations is an important topic in mathematics. This is evidenced by characterizations of Möbius transformations found in the literature; see, for instance, [7, 8, 11–13, 16, 17].

4.3 Sufficient Conditions to be a Topological Gyrogroup

In [15], Klee shows that in the class of groups with a metric d the condition that $d(xy, ab) \le d(x, a) + d(y, b)$ is equivalent to the bi-invariant of d. This motivates the following theorem for normed gyrogroups.

Theorem 10 *Let G be a normed gyrogroup with the corresponding metric d. Then the following conditions are equivalent.*

(I) Right-gyrotranslation inequality: $d(x \oplus a, y \oplus a) \le d(x, y)$ for all $a, x, y \in G$;
(II) Klee's condition: $d(x \oplus y, a \oplus b) \le d(x, a) + d(y, b)$ for all $a, b, x, y \in G$.

Proof Assume that the right-gyrotranslation inequality holds. Then we have

$$\begin{aligned} d(x \oplus y, a \oplus b) &\le d(x \oplus y, x \oplus b) + d(x \oplus b, a \oplus b) \\ &= d(y, b) + d(x \oplus b, a \oplus b) \\ &\le d(y, b) + d(x, a) \\ &= d(x, a) + d(y, b) \end{aligned}$$

for all $a, b, x, y \in G$. Conversely, if Klee's condition holds, then

$$d(x \oplus a, y \oplus a) \leq d(x, y) + d(a, a) = d(x, y)$$

for all $a, x, y \in G$. □

Recall that a gyrogroup G endowed with a topology is called a *topological* gyrogroup if (i) the gyroaddition map $(x, y) \mapsto x \oplus y$ is jointly continuous and (ii) the inversion map $x \mapsto \ominus x$ is continuous [4, Definition 1]. In general, a normed gyrogroup need not be a topological gyrogroup. The conditions mentioned in Theorem 10 are sufficient conditions for a normed gyrogroup to be a topological gyrogroup. It is still an open question whether these conditions are necessary.

Theorem 11 *Let G be a normed gyrogroup. If one of the conditions in Theorem 10 holds, then G is a topological gyrogroup with respect to the topology induced by the gyronorm metric.*

Proof Denote by A the gyroaddition map, $A(x, y) = x \oplus y$. Let (x, y) be an arbitrary point of $G \times G$ and let V be a neighborhood of $A(x, y) = x \oplus y$. By definition, there is an $\epsilon > 0$ such that $B(x \oplus y, \epsilon) \subseteq V$. Define $S = B(x, \epsilon/2)$ and $T = B(y, \epsilon/2)$. Set $U = S \times T$. Since S and T are open in G, it follows that U is a neighborhood of (x, y) in $G \times G$. Let $(a, b) \in U$. Then $a \in S$ and $b \in T$. By assumption,

$$d(x \oplus y, a \oplus b) \leq d(x, a) + d(y, b) < \epsilon.$$

Hence, $A(a, b) = a \oplus b \in B(x \oplus y, \epsilon) \subseteq V$ and so $A(U) \subseteq V$. This proves that A is continuous.

Denote by ι the inversion map of G. By the right-gyrotranslation inequality,

$$\begin{aligned} d(\iota(x), \iota(y)) &= d(\ominus x, \ominus y) \\ &\leq d(e, \ominus y \oplus x) \\ &= d(y \oplus e, y \oplus (\ominus y \oplus x)) \\ &= d(x, y) \end{aligned}$$

for all $x, y \in G$. This also implies that $d(x, y) \leq d(\iota(x), \iota(y))$ for all $x, y \in G$ because $\iota = \iota^{-1}$. Thus, $d(\iota(x), \iota(y)) = d(x, y)$ for all $x, y \in G$ and so ι is an isometry of G. Hence, ι is continuous. This proves that G is a topological gyrogroup. □

According to Theorem 2.18 of [5], the group-norm of a group Γ is *abelian*, that is,

$$\|gh\| = \|hg\|$$

for all $g, h \in \Gamma$ if and only if the metric induced by this group-norm is bi-invariant. This motivates the following theorem for normed gyrogroups.

Theorem 12 *Let G be a normed gyrogroup with the corresponding metric d. Then the following conditions are equivalent.*

(I) Commutative-like condition: $\|(a \oplus x) \oplus \mathrm{gyr}[a, x](y \ominus a)\| = \|x \oplus y\|$ *for all* $a, x, y \in G$.
(II) Bi-gyrotranslation invariant: $d(x \oplus a, y \oplus a) = d(x, y) = d(a \oplus x, a \oplus y)$ *for all* $a, x, y \in G$.

Proof Let $a, x, y \in G$. Direct computation shows that

$$\begin{aligned} d(x \oplus a, y \oplus a) &= \| \ominus (x \oplus a) \oplus (y \oplus a)\| \\ &= \|\mathrm{gyr}[x, a](\ominus a \ominus x) \oplus (y \oplus a)\| \\ &= \|(\ominus a \ominus x) \oplus \mathrm{gyr}[a, x](y \oplus a)\| \\ &= \|(\ominus a \ominus x) \oplus \mathrm{gyr}[\ominus a, \ominus x](y \ominus (\ominus a))\| \\ &= \| \ominus x \oplus y\| \\ &= d(x, y). \end{aligned} \tag{17}$$

Hence, d is invariant under right gyrotranslation. By Theorem 4, d is invariant under left gyrotranslation as well. Conversely, computation as in (17) with $\ominus a$ in place of a and $\ominus x$ in place of x gives

$$\|x \oplus y\| = d(\ominus x, y) = d(\ominus x \ominus a, y \ominus a) = \|(a \oplus x) \oplus \mathrm{gyr}[a, x](y \ominus a)\|.$$

□

Acknowledgement The author would like to thank Themistocles M. Rassias and Watchareepan Atiponrat for their collaboration.

References

1. T. Abe, Gyrometric preserving maps on Einstein gyrogroups, Möbius gyrogroups and proper velocity gyrogroups. Nonlinear Funct. Anal. Appl. **19**, 1–17 (2014)
2. T. Abe, O. Hatori, Generalized gyrovector spaces and a Mazur–Ulam theorem. Publ. Math. Debrecen **87**(3–4), 393–413 (2015)
3. T. Abe, K. Watanabe, Finitely generated gyrovector subspaces and orthogonal gyrodecomposition in the Möbius gyrovector space. J. Math. Anal. Appl. **449**, 77–90 (2017)
4. W. Atiponrat, Topological gyrogroups: generalization of topological groups. Topol. Appl. **224**, 73–82 (2017)
5. N. Bingham, A. Ostaszewski, Normed versus topological groups: dichotomy and duality. Diss. Math. **472**, 1–138 (2010)
6. Z. Cai, S. Lin, W. He, A note on paratopological loops. Bull. Malays. Math. Sci. Soc. **42**(5), 2535–2547 (2019). https://doi.org/10.1007/s40840-018-0616-y
7. O. Demirel, A characterization of Möbius transformations by use of hyperbolic triangles. J. Math. Anal. Appl. **398**(2), 457–461 (2013)

8. O. Demirel, E.S. Seyrantepe, A characterization of Möbius transformations by use of hyperbolic regular polygons. J. Math. Anal. Appl. **374**(2), 566–572 (2011)
9. M. Ferreira, Harmonic analysis on the Möbius gyrogroup. J. Fourier Anal. Appl. **21**(2), 281–317 (2015)
10. R. Fleming, J. Jamison, Isometries on Banach spaces: function spaces, in *Monographs and Surveys in Pure and Applied Mathematics*, vol. 129 (Chapman & Hall/CRC, Boca Raton, 2003)
11. H. Haruki, T.M. Rassias, A new characteristic of Möbius transformations by use of Apollonius points of triangles. J. Math. Anal. Appl. **197**, 14–22 (1996)
12. H. Haruki, T.M. Rassias, A new characteristic of Möbius transformations by use of Apollonius quadrilaterals. Proc. Am. Math. Soc. **126**(10), 2857–2861 (1998)
13. H. Haruki, T.M. Rassias, A new characterization of Möbius transformations by use of Apollonius hexagons. Proc. Am. Math. Soc. **128**(7), 2105–2109 (2000)
14. S. Kim, J. Lawson, Unit balls, Lorentz boosts, and hyperbolic geometry. Results Math. **63**, 1225–1242 (2013)
15. V. Klee, Invariant metrics in groups (solution of a problem of Banach). Proc. Am. Math. Soc. **3**, 484–487 (1952)
16. P. Niamsup, A note on the characteristics of Möbius transformations. J. Math. Anal. Appl. **248**, 203–215 (2000)
17. P. Niamsup, A note on the characteristics of Möbius transformations. II. J. Math. Anal. Appl. **261**, 151–158 (2001)
18. T.M. Rassias, On the A. D. Aleksandrov problem of conservative distances and the Mazur–Ulam theorem, in *Proceedings of the Third World Congress of Nonlinear Analysts, Part 4 (Catania, 2000)*, vol. 47 (Elsevier, Amsterdam, 2001), pp. 2597–2608
19. T.M. Rassias, On the Aleksandrov problem for isometric mappings. Appl. Anal. Discrete Math. **1**, 18–28 (2007)
20. T.M. Rassias, P. Šemrl, On the Mazur–Ulam theorem and the Aleksandrov problem for unit distance preserving mappings. Proc. Am. Math. Soc. **118**(3), 919–925 (1993)
21. T.M. Rassias, S. Xiang, On mappings with conservative distances and the Mazur–Ulam theorem. Univ. Beograd. Publ. Elektrotehn. Fak. Ser. Mat. **11**, 1–8 (2000)
22. T. Suksumran, Essays in mathematics and its applications: in *Honor of Vladimir Arnold*, ed. by T.M. Rassias, P.M. Pardalos. The Algebra of Gyrogroups: Cayley's Theorem, Lagrange's Theorem, and Isomorphism Theorems (Springer, Cham, 2016), pp. 369–437
23. T. Suksumran, Gyrogroup actions: a generalization of group actions. J. Algebra **454**, 70–91 (2016)
24. T. Suksumran, Involutive groups, unique 2-divisibility, and related gyrogroup structures. J. Algebra Appl. **16**(6), 1750 (2017)
25. T. Suksumran, Modern discrete mathematics and analysis, in *Springer Optimization and Its Applications*, ed. by N.J. Daras, T.M. Rassias, vol. 131. Cauchy's Functional Equation, Schur's Lemma, One-Dimensional Special Relativity, and Möbius's Functional Equation (Springer, Cham, 2018), pp. 389–396
26. T. Suksumran, K. Wiboonton, Möbius's functional equation and Schur's lemma with applications to the complex unit disk. Aequat. Math. **91**(3), 491–503 (2017)
27. A. Ungar, Thomas rotation and the parametrization of the Lorentz transformation group. Found. Phys. Lett. **1**, 57–89 (1988)
28. A. Ungar, *Analytic Hyperbolic Geometry and Albert Einstein's Special Theory of Relativity* (World Scientific, Hackensack, 2008)
29. A. Ungar, From Möbius to gyrogroups. Am. Math. Monthly **115**(2), 138–144 (2008)
30. A. Ungar, Parametric realization of the Lorentz transformation group in pseudo-Euclidean spaces. J. Geom. Symmetry Phys. **38**, 39–108 (2015)

Birelator Spaces Are Natural Generalizations of Not Only Bitopological Spaces, But Also Ideal Topological Spaces

Árpád Száz

Abstract In 1962, W. J. Pervin proved that every topology $\mathscr{T}$ on a set X can be derived from the quasi-uniformity $\mathscr{U}$ on X generated by the preorder relations $R_A = A^2 \cup A^c \times X$ with $A \in \mathscr{T}$.

Thus, a quasi-uniform space $X(\mathscr{U})$ is a generalization of a topological space $X(\mathscr{T})$, and a bi-quasi-uniform space $X(\mathscr{U}, \mathscr{V})$ is a generalization of a bitopological space $X(\mathscr{P}, \mathscr{Q})$, studied first by J. C. Kelly in 1963.

Now, we shall show that a bi-quasi-uniform space $X(\mathscr{U}, \mathscr{V})$ is also a certain generalization of an ideal topological space $X(\mathscr{T}, \mathscr{I})$ studied first by K. Kuratowski in 1933.

Actually, instead of a bi-quasi-uniform space $X(\mathscr{U}, V)$, we shall use a birelator space $(X, Y)(\mathscr{R}, \mathscr{S})$, where X and Y are sets and $\mathscr{R}$ and $\mathscr{S}$ are relators (families of relations) on X to Y.

Much more general results could be achieved by using corelations $\big($functions of $\mathscr{P}(X)$ to $\mathscr{P}(Y)\big)$ instead relations on X to Y. However, a detailed theory of corelators has not been worked out yet.

1 Some Basic Facts on Topological Spaces

If $\mathscr{T}$ is a family of subsets of a set X such that $\mathscr{T}$ is closed under finite intersections and arbitrary unions, then the family $\mathscr{T}$ is called a *topology* on X, and the ordered pair $X(\mathscr{T}) = (X, \mathscr{T})$ is called a *topological space*.

Note that $\emptyset \subseteq \mathscr{T}$ such that $\emptyset = \bigcup \emptyset$ and $X = \bigcap \emptyset$. Therefore, $\emptyset, X \in \mathscr{T}$. Thus, a family $\mathscr{T} \subseteq \mathscr{P}(X)$ may also be called a topology on X if it contains both $\emptyset$ and X, and it is closed under nonvoid unions and pairwise intersections.

Á. Száz (✉)
Department of Mathematics, University of Debrecen, Debrecen, Hungary
e-mail: szaz@science.unideb.hu

T. M. Rassias, P. M. Pardalos (eds.), *Mathematical Analysis and Applications*, Springer Optimization and Its Applications 154,
https://doi.org/10.1007/978-3-030-31339-5_21

The members of a topology $\mathscr{T}$ on X are called the *open subsets* of X. By Thron [117], the use of open sets as a starting point was first suggested by Tietze [119] and Alexandroff [2], and later standardized by Bourbaki [7] and Kelley [42].

The notion of a topology has been generalized by a surprisingly great number of mathematicians. For instance, if $\mathscr{T} \subseteq \mathscr{P}(X)$ such that $X \in \mathscr{T}$ and $\mathscr{T}$ is closed under arbitrary unions, then $\mathscr{T}$ is called a *generalized topology* on X.

While, if $\tau \subseteq \mathscr{P}(X)$ such that $\emptyset \in \tau$ and $X \in \tau$, then τ is called a *minimal structure* on X. Generalized topologies and minimal structures were introduced by Lugojan [54] and Maki [55], respectively.

Recently, even the conditions $X \in \mathscr{T}$ and $X \in \tau$ have also been omitted by Császár [10, 12, 14]. Several authors, publishing in the same Acta, considered this to be "the one of the most important developments of general topology."

On the other hand, if $\mathscr{P}$ and $\mathscr{Q}$ are topologies on X, then the ordered pair $X(\mathscr{P}, \mathscr{Q}) = \big(X, (\mathscr{P}, \mathscr{Q})\big)$ was called a *bitopological space* by Kelly [43]. (See also [3, 24, 49, 50, 73, 123].) Thus, $X(\mathscr{P})$ may be considered as $X(\mathscr{P}, \mathscr{P})$.

A less straightforward generalization of topological spaces is the *ideal topological spaces* studied by Kuratowski [47], Vaidyanathaswamy [121] and Janković and Hamlett [39] and many others [26, 30, 31, 33, 38, 58, 59, 62–64, 79, 125].

If $\mathscr{T}$ is a topology and $\mathscr{I}$ is an *ideal* on X (in the sense that it is a nonvoid, descending family of subsets of X which is closed under pairwise unions), then the ordered pair $X(\mathscr{T}, \mathscr{I}) = \big(X, (\mathscr{T}, \mathscr{I})\big)$ is called an *ideal topological space*.

In such a space $X(\mathscr{T}, \mathscr{I})$, for any $A \subseteq X$, one defines

$$A^* = \big\{ x \in X : \quad \forall\, U \in \mathscr{T}(x) : \quad U \cap A \notin \mathscr{I} \big\},$$

where $\mathscr{T}(x) = \{ U \in \mathscr{T} : \ x \in U \}$.

Moreover, by using this closure-like operation $*$, one also defines

$$\mathrm{cl}^*(A) = A \cup A^*.$$

Thus, cl^* is already a *Kuratowski closure* on X such that $\mathrm{cl}^* = \mathrm{cl}$ if $\mathscr{I} = \{\emptyset\}$.

Bitopological spaces and ideal topological spaces have also been generalized by Császár [11, 13]. He used generalized topologies and *hereditary classes* (descending systems) instead of topologies and ideals, respectively.

For us, instead of ideals, it will be more convenient to use *filters* and *stacks* (ascending systems) [118]. Namely, if $\mathscr{I}$ is an ideal on X, then its elementwise complement $\mathscr{F} = \mathscr{I}^{\,c} = \{ A^c : \ A \in \mathscr{I} \}$ is a filter on X such that $\mathscr{I} = \mathscr{F}^{\,c}$.

Therefore, for any $V \subseteq X$, we have $V \notin \mathscr{I}$ if and only if $V \notin \mathscr{F}^{\,c}$, i.e., $V^c \notin \mathscr{F}$. That is, V is in the generated *grill* $\mathscr{G} = \{ A \subseteq X : \ A^c \notin \mathscr{F} \}$, which can also be written in the form $\mathscr{G} = \{ A \subseteq X : \ \ \forall\, F \in \mathscr{F} : \ \ A \cap F \neq \emptyset \}$.

2 Some Basic Facts on Relator Spaces

Instead of open sets, Hausdorff [32], Kuratowski [47], Weil [122], Tukey [120], Efremovič and Švarc [18, 19], Kowalsky [45], Császár [9], Doičinov [17], Herrlich [34] and others [8, 37, 66, 81] offered some more powerful tools.

For instance, from the works of Davis [16], Pervin [75] and Hunsaker and Lindgren [36], it should have been completely clear that topologies, closures and proximities should not be studied without generalized uniformities.

Considering several papers and some books on generalized uniformities and their induced structures, the present author in [85] offered *relators* (families of relations) as the most suitable basic term on which abstract analysis should be based on.

Later, reading the work of Höhle and Kubiak [35], he has observed that *corelations* on X to Y $\big($functions of $\mathcal{P}(X)$ to $\mathcal{P}(Y)\big)$ are somewhat less flexible, but much more powerful tools than relations on X to Y.

Therefore, the extensive theory of relators has to be generalized to *corelators* (families of corelations). However, in the present paper, to make our recent ideas more easily accessible to the reader, we shall restrict ourselves to relators.

Thus, if $\mathcal{R}$ is a family of relations on X to Y (i.e., $\mathcal{R} \subseteq \mathcal{P}(X \times Y)$), then $\mathcal{R}$ is called a *relator* on X to Y, and the ordered pair $(X, Y)(\mathcal{R}) = \big((X, Y), \mathcal{R}\big)$ is called a *relator space*.

If in particular $\mathcal{R}$ is a relator on X to itself, then we may simply say that $\mathcal{R}$ is a relator on X. Moreover, by identifying singletons with their elements, we may naturally write $X(\mathcal{R})$ instead of $(X, X)(\mathcal{R})$.

A relator $\mathcal{R}$ on X to Y, or a relator space $(X, Y)(\mathcal{R})$, is called *simple* if $\mathcal{R} = \{R\}$ for some relation R. Simple relator spaces $(X, Y)(R)$ and $X(R)$ were called *formal context* and *generalized ordered sets* in [27] and [107], respectively.

Moreover, a relator $\mathcal{R}$ on X, or a relator space $X(\mathcal{R})$, may, for instance, be naturally called *reflexive* if each member of $\mathcal{R}$ is reflexive on X. Thus, we may also naturally speak of *preorder, tolerance*, and *equivalence relators*.

For instance, for a family $\mathcal{A}$ of subsets of X, the family $\mathcal{R}_{\mathcal{A}} = \{ R_A : \ A \in \mathcal{A}\}$, where $R_A = A^2 \cup A^c \times X$, is an important preorder relator on X. Such relators were first used by Davis [16], Pervin [75] and Levine [53].

While, for a family $\mathcal{D}$ of *pseudo-metrics* on X, the family $\mathcal{R}_{\mathcal{D}} = \{B_r^d : \ r > 0, d \in \mathcal{D}\}$, where $B_r^d = \{(x, y) : \ d(x, y) < r\}$, is an important tolerance relator on X. Such relators were first considered by Weil [122].

Moreover, if $\mathfrak{S}$ is a family of *partitions* of X, then the family $\mathcal{R}_{\mathfrak{S}} = \{ S_{\mathcal{A}} : \mathcal{A} \in \mathfrak{S}\}$, where $S_{\mathcal{A}} = \bigcup_{A \in \mathcal{A}} A^2$, is an equivalence relator on X. Such practically important relators were first investigated by Levine [52].

3 Some Important Structures Derived from Relators

A function $\mathfrak{F}$ of the class of all relator spaces to some other class is called a *structure for relators* if, for any relator $\mathscr{R}$ on X to Y, the value $\mathfrak{F}_{\mathscr{R}} = \mathfrak{F}_{\mathscr{R}}^{XY} = \mathfrak{F}\big((X, Y)(\mathscr{R})\big)$ is in a power set depending only on X and Y.

A structure $\mathfrak{F}$ for relators is called *increasing* if $\mathscr{R} \subseteq \mathscr{S}$ implies $\mathfrak{F}_{\mathscr{R}} \subseteq \mathfrak{F}_{\mathscr{S}}$ for any two relators $\mathscr{R}$ and S. And, $\mathfrak{F}$ is called *quasi-increasing* if $R \in \mathscr{R}$ implies $\mathfrak{F}_R \subseteq \mathfrak{F}_{\mathscr{R}}$ for any relator $\mathscr{R}$. Note that here $\mathfrak{F}_R = \mathfrak{F}_{\{R\}}$.

Moreover, the structure $\mathfrak{F}$ is called *union-preserving* if $\mathfrak{F}_{\bigcup_{i\in I} \mathscr{R}_i} = \bigcup_{i\in I} \mathfrak{F}_{\mathscr{R}_i}$ for any family $(\mathscr{R}_i)_{i\in I}$ of relators. It can be shown that $\mathfrak{F}$ is union-preserving if and only if $\mathfrak{F}_{\mathscr{R}} = \bigcup_{R\in\mathscr{R}} \mathfrak{F}_R$ for any relator $\mathscr{R}$ [104].

For instance, if $\mathscr{R}$ is a relator on X to Y, then for any $a \in X$, $A \subseteq X$ and $B \subseteq Y$ we write:

(1) $A \in \operatorname{Int}_{\mathscr{R}}(B)$ if $R[A] \subseteq B$ for some $R \in \mathscr{R}$;
(2) $A \in \operatorname{Cl}_{\mathscr{R}}(B)$ if $R[A] \cap B \neq \emptyset$ for all $R \in \mathscr{R}$;
(3) $a \in \operatorname{int}_{\mathscr{R}}(B)$ if $\{a\} \in \operatorname{Int}_{\mathscr{R}}(B)$;
(4) $a \in \operatorname{cl}_{\mathscr{R}}(B)$ if $\{a\} \in \operatorname{Cl}_{\mathscr{R}}(B)$;
(5) $B \in \mathscr{E}_{\mathscr{R}}$ if $\operatorname{int}_{\mathscr{R}}(B) \neq \emptyset$;
(6) $B \in \mathscr{D}_{\mathscr{R}}$ if $\operatorname{cl}_{\mathscr{R}}(B) = X$.

Moreover, if in particular $\mathscr{R}$ is a relator on X, then for any $A \subseteq X$ we also write:

(7) $A \in \tau_{\mathscr{R}}$ if $A \in \operatorname{Int}_{\mathscr{R}}(A)$;
(8) $A \in \bar{\tau}_{\mathscr{R}}$ if $A^c \notin \operatorname{Cl}_{\mathscr{R}}(A)$;
(9) $A \in \mathscr{T}_{\mathscr{R}}$ if $A \subseteq \operatorname{int}_{\mathscr{R}}(A)$;
(10) $A \in \mathscr{F}_{\mathscr{R}}$ if $\operatorname{cl}_{\mathscr{R}}(A) \subseteq A$.

The relations $\operatorname{Int}_{\mathscr{R}}$ and $\operatorname{int}_{\mathscr{R}}$ are called *the proximal and topological interiors* generated by $\mathscr{R}$, respectively. While, the members of the families, $\tau_{\mathscr{R}}$, $\mathscr{T}_{\mathscr{R}}$ and $\mathscr{E}_{\mathscr{R}}$ are called the *proximally open, topologically open, and fat sets* generated by $\mathscr{R}$, respectively.

The origins of the relations $\operatorname{Cl}_{\mathscr{R}}$ and $\operatorname{Int}_{\mathscr{R}}$ go back to Efremović's proximity δ [18] and Smirnov's strong inclusion $\Subset$ [82], respectively. The notations $\operatorname{Cl}_{\mathscr{R}}$ and $\operatorname{Int}_{\mathscr{R}}$, and the families $\tau_{\mathscr{R}}$ and $\mathscr{E}_{\mathscr{R}}$ were first explicitly used by the present author [85–88]. While, the notation $\bar{\tau}_{\mathscr{R}}$ has been suggested by J. Kurdics.

The proximal closure and proximally open sets are usually more convenient tools than the topological closure (proximal interior) and topologically open sets, respectively. Namely, for instance, we have $\operatorname{Cl}_{\mathscr{R}^{-1}} = \operatorname{Cl}_{\mathscr{R}}^{-1}$ and $\bar{\tau}_{\mathscr{R}} = \tau_{\mathscr{R}^{-1}}$.

While, the topological interior and fat sets are usually more convenient tools than the topologically open sets. Namely, in general, we only have $\bigcup \mathscr{T}_{\mathscr{R}} \cap \mathscr{P}(A) \subseteq \operatorname{int}_{\mathscr{R}}(A)$, and the mapping $\mathscr{R} \mapsto \mathscr{T}_{\mathscr{R}}$ is not union-preserving.

In this respect, we can also note that if $\leq$ is a certain order relation on X, then $\mathscr{E}_{\leq}$ and $\mathscr{D}_{\leq}$ are just the families of all *residual and cofinal subsets* of the ordered set $X(\leq)$, respectively. While, $\mathscr{T}_{\leq}$ is the family of all *ascending subsets* of $X(\leq)$.

Moreover, we can also note that if in particular $R(x) = \{x-1\} \cup [\,x\,, +\infty\,[$ for all $x \in \mathbb{R}$, then $\mathscr{T}_R = \{\emptyset\,, \mathbb{R}\}$, but $\mathscr{E}_R$ is quite large family. Namely, the supersets of each $R(x)$, with $x \in \mathbb{R}$, are also in $\mathscr{E}_R$.

Furthermore, we can also note that the best tools in a relator spaces $(X\,, Y)(\mathscr{R})$, the *convergence* $\operatorname{Lim}_{\mathscr{R}}$ and *adherence* $\operatorname{Adh}_{\mathscr{R}}$ of a net of points or sets to another can also be most naturally defined with the help of the fat and dense sets.

However, it is now more important to note that *minimal structures, generalized topologies, and proper stacks* (ascending systems) can always be derived from the corresponding Pervin relators [98].

In this respect, it is also worth mentioning that all reasonable generalizations of *closures, proximities, and convergences* can also be derived from relators [84, 88]. Therefore, they need not also be studied separately.

Hence, it is clear that *relator and birelator spaces* are generalizations of not only *topological and bitopological spaces*, but also their natural generalizations. Moreover, we can note that if $\mathscr{R} = \emptyset$, then $\tau_{\mathscr{R}} = \mathscr{E}_{\mathscr{R}} = \emptyset$, but $\mathscr{T}_{\mathscr{R}} = \{\emptyset\}$.

In the present paper, we shall show that a *birelator space* $X(\mathscr{R}, \mathscr{S}) = \big(X, (\mathscr{R}, \mathscr{S})\big)$ is also a certain generalization of an *ideal topological space* $X(\mathscr{T}, \mathscr{I})$, where $\mathscr{T}$ is a topology and $\mathscr{I}$ is an ideal on X.

For this, for any $A \subseteq X$, we shall define

$$\operatorname{cl}_{(\mathscr{R},\mathscr{S})}(A) = \big\{ x \in X: \quad \forall\ R \in \mathscr{R}: \quad R(x) \cap A \in \mathscr{D}_{\mathscr{S}} \big\}$$

instead of $A^* = \big\{ x \in X: \ \forall\ U \in \mathscr{T}_{\mathscr{R}}(x): \ U \cap A \in \mathscr{D}_{\mathscr{S}} \}$.

Finally, we note that if $\mathscr{R}$ is a relator on X to Y, then according to [94] for any $A \subseteq X$ and $B \subseteq Y$, we may also naturally write:

(a) $B \in \operatorname{Ub}_{\mathscr{R}}(A)$ and $A \in \operatorname{Lb}_{\mathscr{R}}(B)$ if $A \times B \subseteq R$ for some $R \in \mathscr{R}$.

Moreover, in particular $\mathscr{R}$ is a relator on X, then for any $A \subseteq X$ we may also naturally define:

(b) $\operatorname{Max}_{\mathscr{R}}(A) = \mathscr{P}(A) \cap \operatorname{Ub}_{\mathscr{R}}(A)$;
(c) $\operatorname{Min}_{\mathscr{R}}(A) = \mathscr{P}(A) \cap \operatorname{Lb}_{\mathscr{R}}(A)$,
(d) $\operatorname{Sup}_{\mathscr{R}}(A) = \operatorname{Min}_{\mathscr{R}}\big[\operatorname{Lb}_{\mathscr{R}}(A)\big]$;
(e) $\operatorname{Inf}_{\mathscr{R}}(A) = \operatorname{Max}_{\mathscr{R}}\big[\operatorname{Ub}_{\mathscr{R}}(A)\big]$.

However, these algebraic tools, and their specializations, are not independent from the former topological ones. Namely, if R is a relation on X to Y, then for any $A \subseteq X$ and $B \subseteq Y$ we have

$$\begin{aligned} A \times B \subseteq R \iff \forall\, a \in A: \ B \subseteq R(a) \iff \forall\, a \in A: \ R(a)^c \subseteq B^c \\ \iff \forall\, a \in A: \ R^c(a) \subseteq B^c \iff R^c[\,A\,] \subseteq B^c. \end{aligned}$$

Therefore, if $\mathscr{R}$ is a relator on X to Y, then by the corresponding definitions, for any $A \subseteq X$ and $B \subseteq Y$, we have

$$A \in \mathrm{Lb}_{\mathscr{R}}(B) \iff A \in \mathrm{Int}_{\mathscr{R}^c}(B^c) \iff A \in \big(\mathrm{Int}_{\mathscr{R}^c} \circ \mathscr{C}_Y\big)(B).$$

Thus, $\mathrm{Lb}_{\mathscr{R}} = \mathrm{Int}_{\mathscr{R}^c} \circ \mathscr{C}_Y$ and $\mathrm{Int}_{\mathscr{R}} = \mathrm{Lb}_{\mathscr{R}^c} \circ \mathscr{C}_Y$.

Therefore, in contrast to a common belief, some algebraic and topological structures are just as closely related to each other by the above equalities, and their particular cases, as the exponential and the trigonometric functions are so by the celebrated Euler formula $\exp(i\,z) = \cos(z) + i\,\sin(z)$.

4 Some Important Unary Operations for Relators

A function $\square$ of the class of all relator spaces to the class of all relators is called a *direct (indirect) unary operation for relators* if, for any relator $\mathscr{R}$ on X to Y, the value $\mathscr{R}^{\square} = \mathscr{R}^{\square_{XY}} = \square\big((X, Y)(\mathscr{R})\big)$ is a relator on X to Y (on Y to X).

A unary operation $\square$ for relators is called *extensive, intensive, involutive*, and *idempotent* if, for any relator $\mathscr{R}$ on X to Y, we have $\mathscr{R} \subseteq \mathscr{R}^{\square}$, $\mathscr{R}^{\square} \subseteq \mathscr{R}$, $\mathscr{R}^{\square\square} = \mathscr{R}$ and $\mathscr{R}^{\square\square} = \mathscr{R}^{\square}$, respectively.

In particular, an increasing involutive (idempotent) operation for relators will be called an *involution (projection or modification) operation*. While, an extensive projection operation for relators will be called a *closure or refinement operation*.

For instance, the functions c and -1, defined by

$$\mathscr{R}^c = \{R^c : \ R \in \mathscr{R}\} \qquad \text{and} \qquad \mathscr{R}^{-1} = \{R^{-1} : \ R \in \mathscr{R}\}$$

for all relator $\mathscr{R}$ on X to Y, are involutive operations such that, for any relator $\mathscr{R}$ on X to Y, we have $\big(\mathscr{R}^c\big)^{-1} = \big(\mathscr{R}^{-1}\big)^c$. Therefore, the operation c is *inversion compatible*.

While, the functions ∞ and ∂, defined by

$$\mathscr{R}^{\infty} = \{R^{\infty} : \ R \in \mathscr{R}\} \qquad \text{and} \qquad \mathscr{R}^{\partial} = \{S \subseteq X^2 : \ S^{\infty} \in \mathscr{R}\}$$

for all relator $\mathscr{R}$ on X, are projection operations such that, for any two relators $\mathscr{R}$ and $\mathscr{S}$ on X, we have

$$\mathscr{R}^{\infty} \subseteq \mathscr{S} \iff \mathscr{R} \subseteq \mathscr{S}^{\partial}.$$

Therefore, the operations ∞ and ∂ form a *Galois connection* [15, p. 155]. Thus, in particular $\infty\,\partial$ is a closure operation for relators such that $\infty = \infty\,\partial\,\infty$ [96].

To investigate inclusions between generalized topologies derived from relations and relators, the operations ∞ and ∂ were first introduced by my former PhD students Mala [56] and Pataki [71], respectively.

Moreover, by using several more powerful structures, the present author [89] and Pataki [71] defined several important closure operations for relators. Some of them were already used by Kenyon [44] and Nakano and Nakano [66].

For instance, for any relator $\mathscr{R}$ on X to Y, the relators

$$\mathscr{R}^{*} = \big\{ S \subseteq X \times Y : \quad \exists\ R \in \mathscr{R} : \quad R \subseteq S \big\} ;$$

$$\mathscr{R}^{\#} = \big\{ S \subseteq X \times Y : \quad \forall\ A \subseteq X : \quad \exists\ R \in \mathscr{R} : \quad R[A] \subseteq S[A] \big\} ;$$

$$\mathscr{R}^{\wedge} = \big\{ S \subseteq X \times Y : \quad \forall\ x \in X : \quad \exists\ R \in \mathscr{R} : \quad R(x) \subseteq S(x) \big\}$$

and

$$\mathscr{R}^{\triangle} = \big\{ S \subseteq X \times Y : \quad \forall\ x \in X : \quad \exists\ u \in X : \quad \exists\ R \in \mathscr{R} : \quad R(u) \subseteq S(x) \big\}$$

are called the *uniform, proximal, topological, and paratopological closures (refinements)* of the relator $\mathscr{R}$, respectively.

Thus, we evidently have $\mathscr{R} \subseteq \mathscr{R}^{*} \subseteq \mathscr{R}^{\#} \subseteq \mathscr{R}^{\wedge} \subseteq \mathscr{R}^{\triangle}$. Moreover, if in particular $\mathscr{R}$ is a relator on X, then we can easily see that $\mathscr{R}^{\infty} \subseteq \mathscr{R}^{*\infty} \subseteq \mathscr{R}^{\infty *} \subseteq \mathscr{R}^{*}$. However, it is now more important to note that, because of the corresponding definitions of Section 3, for any relator $\mathscr{R}$ on X to Y, we also have

$$\mathscr{R}^{\triangle} = \big\{ S \subseteq X \times Y : \quad \forall\ x \in X : \quad S(x) \in \mathscr{E}_{\mathscr{R}} \big\} ;$$

$$\mathscr{R}^{\wedge} = \big\{ S \subseteq X \times Y : \quad \forall\ x \in X : \quad x \in \operatorname{int}_{\mathscr{R}} \big(S(x) \big) \big\} ;$$

$$\mathscr{R}^{\#} = \big\{ S \subseteq X \times Y : \quad \forall\ A \subseteq X : \quad A \in \operatorname{Int}_{\mathscr{R}} \big(S[A] \big) \big\} .$$

Therefore, the operations $\triangle$, $\wedge$ and $\#$ are closely related to the structures $\mathscr{E}$, int and Int. More concretely, if $\mathfrak{F}$ is a quasi-increasing structure for relators, then following the ideas of Pataki [71] we may naturally define an operation $\square_{\mathfrak{F}}$ for relators such that

$$\mathscr{R}^{\square_{\mathfrak{F}}} = \big\{ S \subseteq X \times Y : \quad \mathfrak{F}_{S} \subseteq \mathfrak{F}_{\mathscr{R}} \big\}$$

for any relator $\mathscr{R}$ on X to Y.

Thus, as an applicable particular case of a more general result, we can state that if $\mathfrak{F}$ is a *union-preserving structure* for relators, then $\mathfrak{F}$ is *increasingly* $\square_{\mathfrak{F}}$*-regular* in the sense that, for any two relators $\mathscr{R}$ and $\mathscr{S}$ on X to Y we have

$$\mathfrak{F}_{\mathscr{R}} \subseteq \mathfrak{F}_{\mathscr{S}} \quad \Longleftrightarrow \quad \mathscr{R} \subseteq \mathscr{S}^{\square_{\mathfrak{F}}} .$$

That is, the structure $\mathfrak{F}$ and the operation $\square_{\mathfrak{F}}$ form an *increasing Pataki connection* which is a certain particular case of a corresponding *increasing Galois connection*. (See [97, 99, 106].)

From the above fact, it is not very hard to derive that if $\mathfrak{F}$ is a *union-preserving structure* for relators, then $\square_{\mathfrak{F}}$ is a closure (refinement) operation for relators such that, for any relator $\mathcal{R}$ on X to Y, just $\mathcal{S} = \mathcal{R}^{\square_{\mathfrak{F}}}$ is the largest relator on X to Y such that $\mathfrak{F}_{\mathcal{S}} = \mathfrak{F}_{\mathcal{R}}$ $\big(\mathfrak{F}_{\mathcal{S}} \subseteq \mathfrak{F}_{\mathcal{R}}\big)$.

Unfortunately, the increasing structure $\mathscr{T}$ is not union-preserving. Namely, if for instance $x_1 \in X$ and $x_2 \in X \setminus \{x_1\}$, and

$$R_i = \{x_i\}^2 \cup \big(X \setminus \{x_i\}\big)^2$$

for all $i = 1, 2$, then $\mathcal{R} = \{R_1, R_2\}$ is an equivalence relator on X such that $\{x_1, x_2\} \in \mathscr{T}_{\mathcal{R}} \setminus \big(\mathscr{T}_{R_1} \cup \mathscr{T}_{R_2}\big)$, and thus $\mathscr{T}_{\mathcal{R}} \not\subseteq \mathscr{T}_{R_1} \cup \mathscr{T}_{R_2}$.

Mala [56, Example 5.3] proved that if $\operatorname{card}(X) > 2$, then even for the equivalence relator $\mathcal{R} = \{X^2\}$ there does not exist a largest relator $\mathcal{S}$ on X such that $\mathscr{T}_{\mathcal{S}} = \mathscr{T}_{\mathcal{R}}$. Moreover Pataki [71, Example 7.2] proved that the operation $\square_{\mathscr{T}} = \wedge\,\partial$ is not idempotent. Therefore, in connection with the structure $\mathscr{T}$, the projection (modification) operation $\wedge\infty$, introduced by Mala [56, 57], has to be used.

In the light of the several disadvantages of the structure $\mathscr{T}$, it is rather curious that most of the works in topology and analysis have been based on open sets. This seems to be the greatest mistake in the history of mathematics. Namely, a topology and its natural generalizations can be easily derived from preorder relators. Moreover, in relator spaces, we can still use most of the techniques of metric and uniform spaces.

For instance, in relator spaces, we can naturally define well-chainedness [48], connectedness [72], continuity [92, 102, 109, 110, 114, 116], completeness [90], compactness [91] and Baire properties [93, 95]. For instance, if $\mathcal{R}$ is a relator on X to Y and $\square$ is a unary operation for relators, then a net in Y may be naturally called *$\square$-Cauchy* with respect to $\mathcal{R}$ if it properly converges with respect to each $R \in \mathcal{R}^{\square}$. Thus, "$\wedge$-Cauchy" coincides with "convergent."

Moreover, $\mathcal{R}$ may be naturally called *$\square$-compact* with respect to $\mathcal{R}$ if forevery $R \in \mathcal{R}^{\square}$ there exists a finite subset A of X such that $Y = R[A]$. Thus,"$*$-compact" and "$\wedge$-compact" coincide with the usual "precompactness" and "compactness." However, #-compactness and Δ-compactness seem to be some new properties.

5 Some Basic Theorems on Fat and Dense Sets

Notation 1 *In this and the next section, we shall assume that $\mathcal{R}$ is a relator on X to Y.*

The subsequent simple, but important theorems have been mostly proved in our former papers. Therefore, their proofs will usually be omitted.

Theorem 1 *For any* $B \subseteq Y$, *we have*

(1) $B \in \mathscr{E}_{\mathscr{R}}$ *if and only if* $R(x) \subseteq B$ *for some* $x \in X$ *and* $R \in \mathscr{R}$;
(2) $B \in \mathscr{D}_{\mathscr{R}}$ *if and only if* $R(x) \cap B \neq \emptyset$ *for all* $x \in X$ *and* $R \in \mathscr{R}$.

Proof This is immediate from the definitions of the families $\mathscr{E}_{\mathscr{R}}$ and $\mathscr{D}_{\mathscr{R}}$ and the relations $\operatorname{int}_{\mathscr{R}}$ and $\operatorname{cl}_{\mathscr{R}}$.

Remark 1 Because of this theorem, the families $\mathscr{D}_{\mathscr{R}}$ and $\mathscr{E}_{\mathscr{R}}$ can also be easily introduced without using the relations $\operatorname{int}_{\mathscr{R}}$ and $\operatorname{cl}_{\mathscr{R}}$.

Theorem 2 *For any* $B \subseteq Y$, *we have*

(1) $B \in \mathscr{D}_{\mathscr{R}}$ *if and only if* $X = R^{-1}[B]$ *for all* $R \in \mathscr{R}$;
(2) $B \in \mathscr{E}_{\mathscr{R}}$ *if and only if* $X \neq R^{-1}[B^c]$ *for some* $R \in \mathscr{R}$.

Proof To prove this, one can use the equality

$$\operatorname{cl}_{\mathscr{R}}(B) = \bigcap_{R \in \mathscr{R}} R^{-1}[B]$$

and the following theorem instead of Theorem 1.

Theorem 3 *For any* $B \subseteq Y$, *we have*

(1) $B \in \mathscr{D}_{\mathscr{R}} \iff B^c \notin \mathscr{E}_{\mathscr{R}}$;
(2) $B \in \mathscr{E}_{\mathscr{R}} \iff B^c \notin \mathscr{D}_{\mathscr{R}}$.

Proof To prove this, instead of Theorem 1, one can use the equalities

$$\operatorname{cl}_{\mathscr{R}}(B) = \operatorname{int}_{\mathscr{R}}(B^c)^c \qquad \text{and} \qquad \operatorname{int}_{\mathscr{R}}(B) = \operatorname{cl}_{\mathscr{R}}(B^c)^c.$$

Theorem 4 *For any* $B \subseteq Y$, *we have*

(1) $B \in \mathscr{D}_{\mathscr{R}}$ *if and only if* $B \cap E \neq \emptyset$ *for all* $E \in \mathscr{E}_{\mathscr{R}}$;
(2) $B \in \mathscr{E}_{\mathscr{R}}$ *if and only if* $B \cap D \neq \emptyset$ *for all* $D \in \mathscr{D}_{\mathscr{R}}$.

Proof In principle this theorem can be derived from Theorem 3. However, it can be more easily proved with the help of Theorem 1.

Theorem 5 *The families* $\mathscr{D}_{\mathscr{R}}$ *and* $\mathscr{E}_{\mathscr{R}}$ *are stacks (ascending systems) in* Y.

Proof This is quite obvious from Theorem 1.

Remark 2 In addition, we can also note that $\mathscr{B} = \{ R(x) : \ x \in X, \ R \in \mathscr{R} \}$ is a base for $\mathscr{E}_{\mathscr{R}}$ whose cardinality can be estimated.

The importance of the investigation of the cardinalities of the bases of $\mathscr{E}_{\mathscr{R}}$ was first recognized by Jenő Deák and Gergely Pataki. (See [70].)

By using Theorems 1 and 3, in addition to Theorem 5, we can also easily prove the following two theorems.

Theorem 6 *The following assertions are equivalent:*

(1) $\emptyset \notin \mathscr{D}_{\mathscr{R}}$;
(2) $\mathscr{E}_{\mathscr{R}} \neq \emptyset$;
(3) $Y \in \mathscr{E}_{\mathscr{R}}$;
(4) $\mathscr{D}_{\mathscr{R}} \neq \mathscr{P}(Y)$;
(5) $X \neq \emptyset$ *and* $\mathscr{R} \neq \emptyset$.

Theorem 7 *The following assertions are equivalent:*

(1) $\emptyset \notin \mathscr{E}_{\mathscr{R}}$;
(2) $\mathscr{D}_{\mathscr{R}} \neq \emptyset$;
(3) $Y \in \mathscr{D}_{\mathscr{R}}$;
(4) $\mathscr{E}_{\mathscr{R}} \neq \mathscr{P}(Y)$;
(5) $X = R^{-1}[Y]$ *if* $R \in \mathscr{R}$.

Proof Note that assertion (5), in a detailed form, means only that for any $x \in X$ and $R \in \mathscr{R}$ we have $x \in R^{-1}[Y]$. That is, there exists $y \in Y$ such that $x \in R^{-1}(y)$, i.e., $y \in R(x)$. Consequently, $R(x) \neq \emptyset$ for all $x \in X$ and $R \in \mathscr{R}$. That is, X is the domain of each member of $\mathscr{R}$.

Remark 3 If the assertions (5) of Theorems 6 and 7 hold, then the relator $\mathscr{R}$ on X to Y, or the relator space $(X, Y)(\mathscr{R})$, may be called *non-degenerated* and *non-partial*, respectively.

Now, we can also easily prove the following simple extension of our former [98, Theorem 3.15].

Theorem 8 *For a family $\mathscr{B}$ of subsets of Y, the following assertion are equivalent:*

(1) $\mathscr{B} = \mathscr{E}_{\mathscr{R}_{\mathscr{B}}}$;
(2) $\mathscr{B}$ *is a proper stack on* Y;
(3) $\mathscr{B} = \mathscr{E}_{\mathscr{R}}$ *for some preorder relator* $\mathscr{R}$ *on* Y;
(4) $\mathscr{B} = \mathscr{E}_{\mathscr{R}}$ *for some non-partial relator* $\mathscr{R}$ *on* X *to* Y.

Proof To prove the implication (2) $\Longrightarrow$ (1), note that if $B \in \mathscr{B}$, then by the definition of the Pervin relator $\mathscr{R}_{\mathscr{B}}$ we can state that $R_B = B^2 \cup B^c \times Y$ is in $\mathscr{R}_{\mathscr{B}}$. Moreover, we can note that $R_B(y) = B$ for all $y \in B$. Thus, by Theorem 1, we have $B \in \mathscr{E}_{\mathscr{R}_{\mathscr{B}}}$ if $B \neq \emptyset$. Therefore, $\mathscr{B} \subseteq \mathscr{E}_{\mathscr{R}_{\mathscr{B}}}$ if $\emptyset \notin \mathscr{B}$. That is, $\mathscr{B} \neq \mathscr{P}(Y)$ whenever $\mathscr{B}$ is ascending.

On the other hand, if $E \in \mathscr{E}_{\mathscr{R}_{\mathscr{B}}}$, then by Theorem 1 there exist $y \in Y$ and $B \in \mathscr{B}$ such that $R_B(y) \subseteq E$. Hence, if $E \neq Y$, we can infer that $y \in B$, and thus $B \subseteq E$. Therefore, $E \in \mathscr{B}$ if $E \neq Y$ and $\mathscr{B}$ is ascending. Moreover, if $\mathscr{B}$ is nonvoid and ascending, then we necessarily have $Y \in \mathscr{B}$. Therefore, $\mathscr{E}_{\mathscr{R}_{\mathscr{B}}} \subseteq \mathscr{B}$ if $\mathscr{B}$ is nonvoid and ascending.

While, if $\mathscr{B} = \emptyset$, then by the definition of $\mathscr{R}_{\mathscr{B}}$ we also have $\mathscr{R}_{\mathscr{B}} = \emptyset$. Thus, by Theorem 6, the equality $\mathscr{E}_{\mathscr{R}_{\mathscr{B}}} = \emptyset$, and thus $\mathscr{E}_{\mathscr{R}_{\mathscr{B}}} = \mathscr{B}$ also holds.

Remark 4 This theorem shows that stacks, and thus also their particular cases filters and grills, should not be studied without generalized uniformities.

Concerning the families $\mathscr{E}_{\mathscr{R}}$ and $\mathscr{D}_{\mathscr{R}}$, we can also easily establish the following

Theorem 9 *We have*

(1) $\mathscr{E}_{\mathscr{R}} = \bigcup_{R \in \mathscr{R}} \mathscr{E}_R$;
(2) $\mathscr{D}_{\mathscr{R}} = \bigcap_{R \in \mathscr{R}} \mathscr{D}_R$.

Corollary 1 *The mapping*

(1) $\mathscr{R} \mapsto \mathscr{E}_{\mathscr{R}}$ *is union-preserving;*
(2) $\mathscr{R} \mapsto \mathscr{D}_{\mathscr{R}}$ *is intersection-preserving.*

Remark 5 In Section 4, we have seen that the increasing mappings $\mathscr{R} \mapsto \mathscr{T}_{\mathscr{R}}$ and $\mathscr{R} \mapsto \mathscr{F}_{\mathscr{R}}$ are not union-preserving.

However, in the next section we shall see that, in some particular cases, the families $\mathscr{T}_{\mathscr{R}}$ and $\mathscr{F}_{\mathscr{R}}$ are still better tools than $\mathscr{E}_{\mathscr{R}}$ and $\mathscr{D}_{\mathscr{R}}$.

For this, it is convenient to prove first the following two theorems.

Theorem 10 *For any relator* $\mathscr{R}$ *on* X, *we have*

(1) $\mathscr{T}_{\mathscr{R}} \setminus \{\emptyset\} \subseteq \mathscr{E}_{\mathscr{R}}$;
(2) $\mathscr{D}_{\mathscr{R}} \cap \mathscr{F}_{\mathscr{R}} \subseteq \{X\}$.

Hence, by using global complementations, we can easily infer

Corollary 2 *For any relator* $\mathscr{R}$ *on* X, *we have*

(1) $\mathscr{F}_{\mathscr{R}} \subseteq \left(\mathscr{D}_{\mathscr{R}}\right)^c \cup \{X\}$;
(2) $\mathscr{D}_{\mathscr{R}} \subseteq \left(\mathscr{F}_{\mathscr{R}}\right)^c \cup \{X\}$.

Proof To prove (1), note that if (1) does not hold, then there exists $A \in \mathscr{F}_{\mathscr{R}}$ such that $A \notin \left(\mathscr{D}_{\mathscr{R}}\right)^c \cup \{X\}$. Hence, since now $\left(\mathscr{D}_{\mathscr{R}}\right)^c = \mathscr{P}(Y) \setminus \mathscr{D}_{\mathscr{R}}$, we can infer that $A \in \mathscr{D}_{\mathscr{R}}$ and $A \neq X$. However, by Theorem 10, we now also have $A \in \mathscr{D}_{\mathscr{R}} \cap \mathscr{F}_{\mathscr{R}} \subseteq \{X\}$. Therefore, $A = X$ which is a contradiction.

Theorem 11 *If* $\mathscr{R}$ *is a relator on* X, *then for any* $A \subseteq X$ *we have*

(1) $A \in \mathscr{E}_{\mathscr{R}}$ *if* $V \subseteq A$ *for some* $V \in \mathscr{T}_R \setminus \{\emptyset\}$;
(2) $A \in \mathscr{D}_{\mathscr{R}}$ *only if* $A \setminus W \neq \emptyset$ *for all* $W \in \mathscr{F}_{\mathscr{R}} \setminus \{X\}$.

Proof To prove (2), note that if $W \in \mathscr{F}_{\mathscr{R}} \setminus \{X\}$, then $W \in \mathscr{F}_{\mathscr{R}}$ and $W \neq X$. Therefore, $W^c \in \mathscr{T}_{\mathscr{R}}$ and $W^c \neq \emptyset$. Hence, by using Theorem 10, we can infer that $W^c \in \mathscr{E}_{\mathscr{R}}$. Therefore, if $A \in \mathscr{D}_{\mathscr{R}}$, then by Theorem 4 we necessarily have $A \setminus W = A \cap W^c \neq \emptyset$.

Remark 6 In the next section, we shall see that if the relator $\mathscr{R}$ is *topological*, in the sense that for any $x \in X$ and $R \in \mathscr{R}$, there exists $V \in \mathscr{T}_{\mathscr{R}}$ such that $x \in V \subseteq R(x)$, then the converses of the above assertions (1) and (2) are also true.

6 Some Further Theorems on Fat and Dense Sets

Having in mind *gosets* (generalized ordered sets), we may also naturally introduce the following

Definition 1 We say that the relator

(1) $\mathcal{R}$ is *semi-directed* if $R_1(x_1) \cap R_2(x_2) \neq \emptyset$ for all $x_1, x_2 \in X$ and $R_1, R_2 \in \mathcal{R}$;
(2) $\mathcal{R}$ is *quasi-directed* if $R_1(x_1) \cap R_2(x_2) \in \mathcal{E}_{\mathcal{R}}$ for all $x_1, x_2 \in X$ and $R_1, R_2 \in \mathcal{R}$;

Remark 7 Note that if $\mathcal{R}$ is semi-directed, then by Remark 3 we can state that $\mathcal{R}$ is non-partial.

While, if $\mathcal{R}$ is *directed* in the sense that its both non-partial and quasi-directed, then by Theorem 7 we can state that $\mathcal{R}$ is semi-directed.

By using the corresponding definitions, we can also easily establish the following

Theorem 12 *The following assertions are equivalent:*

(1) $\mathcal{R}^{-1} \circ \mathcal{R} \subseteq \{X^2\}$*;*
(2) $\mathcal{R}$ *is semi-directed;*
(3) $R_2^{-1} \circ R_1 = X^2$ *for all* $R_1, R_2 \in \mathcal{R}$.

Remark 8 If $R_1, R_2 \in \mathcal{R}$, then by using two important properties of the *boxproduct* [103], defined by

$$(R_1 \boxtimes R_2)(x_1, x_2) = R_1(x_1) \times R_2(x_2),$$

for all $x_1, x_2 \in X$, it can be seen that

$$R_2^{-1} \circ R_1 = R_2^{-1} \circ \Delta_Y \circ R_1 = \big(R_1^{-1} \boxtimes R_2^{-1}\big)(\Delta_Y) = (R_1 \boxtimes R_2)^{-1}(\Delta_Y),$$

and thus

$$\mathrm{cl}_{\mathcal{R} \boxtimes \mathcal{R}}(\Delta_Y) = \bigcap_{R_1, R_2 \in \mathcal{R}} (R_1 \boxtimes R_2)^{-1}(\Delta_Y) = \bigcap_{R_1, R_2 \in \mathcal{R}} R_2^{-1} \circ R_1.$$

Therefore, by Theorem 12, we can also state that $\mathcal{R}$ is semi-directed if and only if $\mathrm{cl}_{\mathcal{R} \boxtimes \mathcal{R}}(\Delta_Y) = X^2$, or equivalently $\Delta_Y \in \mathcal{E}_{\mathcal{R} \boxtimes \mathcal{R}}$.

However, it is now more important to note that, by using Definition 1 and the results of Section 5, we can also easily prove the following three theorems.

Theorem 13 *The following assertions are equivalent:*

(1) $\mathcal{E}_{\mathcal{R}} \subseteq \mathcal{D}_{\mathcal{R}}$*;*
(2) $\mathcal{R}$ *is semi-directed;*
(3) $R(x) \in \mathcal{D}_{\mathcal{R}}$ *for all* $x \in X$ *and* $R \in \mathcal{R}$.

Remark 9 This theorem shows that semi-directedness is a generalization of the *hyperconnectedness* of Steen and Seebach [83, p. 29], studied first in [51] by Levine, and later in [1, 4, 5, 20, 22, 28, 46, 60, 67–69, 77, 78, 80] by others.

Moreover, from Theorem 13, we can also see that, in the $X = Y$, $\operatorname{card}(X) > 1$ and $\mathscr{R} \neq \emptyset$ particular case, semi-directedness coincides with the *paratopological connectedness* introduced and investigated by Pataki and Száz [72].

Theorem 14 *The following assertions are equivalent:*

(1) $\mathscr{R}$ *is semi-directed;*
(2) $B \in \mathscr{E}_{\mathscr{R}}$ *implies* $B^c \notin \mathscr{E}_{\mathscr{R}}$*;*
(3) $B \subseteq Y$ *implies either* $B \in \mathscr{D}_{\mathscr{R}}$ *or* $B^c \in \mathscr{D}_{\mathscr{R}}$*;*
(4) if $B \subseteq Y$ *such that* $B \cap D \neq \emptyset$ *for all* $D \in \mathscr{D}_{\mathscr{R}}$*, then* $B \in \mathscr{D}_{\mathscr{R}}$.

Theorem 15 *The following assertions are equivalent:*

(1) $\mathscr{R}$ *is semi-directed;*
(2) $B_1,\ B_2 \in \mathscr{E}_{\mathscr{R}}$ *implies* $B_1 \cap B_2 \neq \emptyset$*;*
(3) $X = B_1 \cup B_2$ *implies either* $B_1 \in \mathscr{D}_{\mathscr{R}}$ *or* $B_2 \in \mathscr{D}_{\mathscr{R}}$*;*

Proof To prove the implication (2) $\Longrightarrow$ (3), note that if (3) does not hold, then there exist $B_1,\ B_2 \subseteq Y$ such that $X = B_1 \cup B_2$, but $B_1 \notin \mathscr{D}_{\mathscr{R}}$ and $B_2 \notin \mathscr{D}_{\mathscr{R}}$. Hence, by using Theorem 3, we can infer that $B_1^c \in \mathscr{E}_{\mathscr{R}}$ and $B_2^c \in \mathscr{E}_{\mathscr{R}}$. Moreover, we can also note that

$$B_1^c \cap B_2^c = \big(B_1 \cup B_2 \big)^c = X^c = \emptyset .$$

Therefore, (2) does not also hold.

Concerning quasi-directedness, we can only prove the following

Theorem 16 *The following assertions are equivalent:*

(1) $\mathscr{R}$ *is quasi-directed;*
(2) $B_1,\ B_2 \in \mathscr{E}_{\mathscr{R}}$ *implies* $B_1 \cap B_2 \in \mathscr{E}_{\mathscr{R}}$*;*
(3) $B_1 \in \mathscr{E}_{\mathscr{R}}$ *and* $B_2 \in \mathscr{D}_{\mathscr{R}}$ *imply* $B_1 \cap B_2 \in \mathscr{D}_{\mathscr{R}}$*;*
(4) $B_1 \cup B_2 \in \mathscr{D}_{\mathscr{R}}$ *implies either* $B_1 \in \mathscr{D}_{\mathscr{R}}$ *or* $B_2 \in \mathscr{D}_{\mathscr{R}}$.

Proof To prove the implication (2) $\Longrightarrow$ (3), note that if $B_1 \in \mathscr{E}_{\mathscr{R}}$ and $B_2 \in \mathscr{D}_{\mathscr{R}}$ and (2) holds, then by Theorem 4 we have

$$\big(B_1 \cap B_2 \big) \cap E = \big(B_1 \cap E \big) \cap B_2 \neq \emptyset$$

for all $E \in \mathscr{E}_{\mathscr{R}}$. Therefore, by Theorem 4, we have $B_1 \cap B_2 \in \mathscr{D}_{\mathscr{R}}$.

Finally, we note that, in addition to Theorem 11, we can also prove

Theorem 17 *If in particular $\mathscr{R}$ is a topological relator on X, then for any $A \subseteq X$ we have*

(1) $A \in \mathscr{E}_{\mathscr{R}}$ *if and only if* $V \subseteq A$ *for some* $V \in \mathscr{T}_R \setminus \{\emptyset\}$*;*
(2) $A \in \mathscr{D}_{\mathscr{R}}$ *if and only if* $A \setminus W \neq \emptyset$ *for all* $W \in \mathscr{F}_{\mathscr{R}} \setminus \{X\}$.

Remark 10 A relator $\mathscr{R}$ on X has been called *topological* if for any $x \in X$ and $R \in \mathscr{R}$ there exists $V \in \mathscr{T}_{\mathscr{R}}$ such that $x \in V \subseteq R(x)$.

Topological relators have several useful characterizations. For instance, it can be shown that $\mathscr{R}$ is topological if and only if $\operatorname{int}_{\mathscr{R}}(A) = \bigcup \mathscr{T}_{\mathscr{R}} \cap \mathscr{P}(A)$ for all $A \subseteq X$. Therefore, if $\mathscr{R}$ is topological, then $\mathscr{T}_{\mathscr{R}}$ and $\operatorname{int}_{\mathscr{R}}$ are equivalent tools.

Moreover, it can be shown that $\mathscr{R}$ is topological if and only if $\mathscr{R}$ is reflexive and *quasi-topological* in the sense that $x \in \operatorname{int}_{\mathscr{R}}\big(\operatorname{int}_{\mathscr{R}}\big(R(x)\big)\big)$ for all $x \in X$ and $R \in \mathscr{R}$. (Note that $x \in \operatorname{int}_{\mathscr{R}}\big(R(x)\big)$ is always true.)

Furthermore, we can also note that $\mathscr{R}$ is topological if and only if its topological refinement $\mathscr{R}^{\wedge}$ is *proximal* in the sense that for any $A \subseteq X$ and $R \in \mathscr{R}^{\wedge}$ there exists $V \in \tau_{\mathscr{R}}$ such that $A \subseteq V \subseteq R[A]$.

In this respect, it is also worth mentioning that, by using global complementation, in addition to Theorem 10 we can also prove the following

Theorem 18 *For any relator $\mathscr{R}$ on X, we have*

(1) $\mathscr{T}_{\mathscr{R}^{\triangle}} = \mathscr{E}_{\mathscr{R}} \cup \{\emptyset\}$,
(2) $\mathscr{F}_{\mathscr{R}^{\triangle}} = \big(\mathscr{D}_{\mathscr{R}}\big)^c \cup \{X\}$.

Hence, by using Theorem 7, we can immediately derive

Corollary 3 *For any non-partial relator $\mathscr{R}$ on X, we have*

(1) $\mathscr{E}_{\mathscr{R}} = \mathscr{T}_{\mathscr{R}^{\triangle}} \setminus \{\emptyset\}$,
(2) $\mathscr{D}_{\mathscr{R}} = \big(\mathscr{F}_{\mathscr{R}^{\triangle}}\big)^c \cup \{X\}$.

Remark 11 From Theorem 17, we can see that if in particular $\mathscr{R}$ is a topological relator on X, then not only the relation $\operatorname{int}_{\mathscr{R}}$, but also the family $\mathscr{T}_{\mathscr{R}}$ is also a stronger tool than $\mathscr{E}_{\mathscr{R}}$.

While, from Theorem 17 and Corollary 3, we can see that if $\mathscr{R}$ is not only a topological, but also a *paratopologically fine* relator on X in the sense that $\mathscr{R}^{\triangle} = \mathscr{R}$, then the families $\mathscr{T}_{\mathscr{R}}$ and $\mathscr{E}_{\mathscr{R}}$ are already equivalent tools.

7 Proximal Closures and Interiors Derived from Ordered Pairs of Relators

Notation 2 *In the sequel, we shall assume that $\mathscr{R}$ and $\mathscr{S}$ are relators on X to Y.*

Remark 12 Here, instead of relators, we should also rather use corelators [104, 112, 115]. However, to clarify our present ideas, relators are more convenient than corelators.

Definition 2 For any $A \subseteq X$ and $B \subseteq Y$, we write

(1) $A \in \mathrm{Cl}_{(\mathscr{R},\mathscr{S})}(B)$ if $R[A] \cap B \in \mathscr{D}_{\mathscr{S}}$ for all $R \in \mathscr{R}$;
(2) $A \in \mathrm{Int}_{(\mathscr{R},\mathscr{S})}(B)$ if $R[A]^c \cup B \in \mathscr{E}_{\mathscr{S}}$ for some $R \in \mathscr{R}$.

The relations $\mathrm{Cl}_{(\mathscr{R},\mathscr{S})}$ and $\mathrm{Int}_{(\mathscr{R},\mathscr{S})}$ will be called the *proximal closure and interior* generated by the pair $(\mathscr{R}, \mathscr{S})$ of relators, respectively.

Remark 13 This definition has been mainly motivated by our former definitions of the relations $\mathrm{Cl}_{\mathscr{R}}$ and $\mathrm{Int}_{\mathscr{R}}$, and the definition of the operation $*$ used in the theory of ideal topological spaces [39].

Note that if X and $\mathscr{S}$ are nonvoid and $\mathscr{S}$ is quasi-directed, then by Theorems 5, 6, and 16 the family $\mathscr{E}_{\mathscr{S}}$ is already a filter on Y. Thus, $\mathscr{I} = \{B^c : B \in \mathscr{E}_{\mathscr{S}}\}$ is an ideal on Y such that $\mathscr{P}(Y) \setminus \mathscr{I} = \mathscr{D}_{\mathscr{S}}$.

Modifying Definition 2, by taking some other important families derived from the relator $\mathscr{S}$, one may also get some reasonable notions. However, in this paper we shall only be interested in the present setting.

The appropriateness of our present definitions is also apparent from the following analogue of the corresponding basic theorem for the relations $\mathrm{Cl}_{\mathscr{R}}$ and $\mathrm{Int}_{\mathscr{R}}$.

Theorem 19 *For any $B \subseteq Y$, we have*

(1) $\mathrm{Int}_{(\mathscr{R},\mathscr{S})}(B) = \mathscr{P}(X) \setminus \mathrm{Cl}_{(\mathscr{R},\mathscr{S})}(Y \setminus B)$;
(2) $\mathrm{Cl}_{(\mathscr{R},\mathscr{S})}(B) = \mathscr{P}(X) \setminus \mathrm{Int}_{(\mathscr{R},\mathscr{S})}(Y \setminus B)$.

Proof If $A \in \mathrm{Int}_{(\mathscr{R},\mathscr{S})}(B)$, then by Definition 2 there exists $R \in \mathscr{R}$ such that

$$R[A]^c \cup B \in \mathscr{E}_{\mathscr{S}}.$$

Hence, by using Theorem 3, we can infer that

$$\left(R[A]^c \cup B\right)^c \notin \mathscr{D}_{\mathscr{S}}, \qquad \text{and thus} \qquad R[A] \cap B^c \notin \mathscr{D}_{\mathscr{S}}.$$

Therefore, by Definition 2, we have

$$A \notin \mathrm{Cl}_{(\mathscr{R},\mathscr{S})}\left(B^c\right), \qquad \text{and thus} \qquad A \in \mathscr{P}(X) \setminus \mathrm{Cl}_{(\mathscr{R},\mathscr{S})}\left(B^c\right).$$

Thus, we have proved that $\mathrm{Int}_{(\mathscr{R},\mathscr{S})}(B) \subseteq \mathscr{P}(X) \setminus \mathrm{Cl}_{(\mathscr{R},\mathscr{S})}(Y \setminus B)$.

Now, the converse inclusion can be proved quite similarly by reversing the above argument. Moreover, assertion (2) can be derived from (1) by writing $Y \setminus B$ in place of B and taking complement with respect to $\mathscr{P}(X)$.

By using appropriate complementations, the above theorem can be reformulated in a more concise form.

Corollary 4 *We have*

(1) $\operatorname{Int}_{(\mathscr{R},\mathscr{S})} = \big(\operatorname{Cl}_{(\mathscr{R},\mathscr{S})} \circ \mathscr{C}_Y\big)^c = \big(\operatorname{Cl}_{(\mathscr{R},\mathscr{S})}\big)^c \circ \mathscr{C}_Y;$
(2) $\operatorname{Cl}_{(\mathscr{R},\mathscr{S})} = \big(\operatorname{Int}_{(\mathscr{R},\mathscr{S})} \circ \mathscr{C}_Y\big)^c = \big(\operatorname{Int}_{(\mathscr{R},\mathscr{S})}\big)^c \circ \mathscr{C}_Y.$

Remark 14 The second equalities in assertions (1) and (2) follow from the general fact that if f is a function on X to Y and G is a relation on Y to Z, then $(G \circ f)^c = G^c \circ f$.

Theorem 20 *We have*

(1) $\operatorname{Cl}_{(\mathscr{R},\mathscr{S})}(\emptyset) = \emptyset$ *if* $X, \mathscr{R}, \mathscr{S} \neq \emptyset;$
(2) $\operatorname{Cl}_{(\mathscr{R},\mathscr{S})}(B_1) \subseteq \operatorname{Cl}_{(\mathscr{R},\mathscr{S})}(B_2)$ *if* $B_1 \subseteq B_2 \subseteq Y.$

Proof To prove (1), note that, by Definition 2 and Theorem 6, for some $A \subseteq X$ we have

$$A \in \operatorname{Cl}_{(\mathscr{R},\mathscr{S})}(\emptyset) \iff \forall\, R \in \mathscr{R}: \quad R[A] \cap \emptyset \in \mathscr{D}_{\mathscr{S}}$$
$$\iff \forall\, R \in \mathscr{R}: \quad \emptyset \in \mathscr{D}_{\mathscr{S}} \iff \forall\, R \in \mathscr{R}: \quad X = \emptyset \text{ or } \mathscr{S} = \emptyset.$$

Therefore, assertion (1) is true.

Remark 15 Note that if any one of the families X, $\mathscr{R}$ and $\mathscr{S}$ is empty, then in contrast to (1) we have $\operatorname{Cl}_{(\mathscr{R},\mathscr{S})}(B) = \mathscr{P}(X)$ for all $B \subseteq Y$.

Theorem 21 *We have*

(1) $\operatorname{Cl}^{-1}_{(\mathscr{R},\mathscr{S})}(\emptyset) = \emptyset$ *if* $X, \mathscr{R}, \mathscr{S} \neq \emptyset;$
(2) $\operatorname{Cl}^{-1}_{(\mathscr{R},\mathscr{S})}(A_1) \subseteq \operatorname{Cl}^{-1}_{(\mathscr{R},\mathscr{S})}(A_2)$ *if* $A_1 \subseteq A_2 \subseteq X.$

Proof To prove (2), note that if $B \in \operatorname{Cl}^{-1}_{(\mathscr{R},\mathscr{S})}(A_1)$, then by the definition of the inverse relation we have $A_1 \in \operatorname{Cl}_{(\mathscr{R},\mathscr{S})}(B)$. Thus, by Definition 2, for any $R \in \mathscr{R}$ we have

$$R[A_1] \cap B \in \mathscr{D}_{\mathscr{S}}.$$

Moreover, we can also note that $R[A_1] \subseteq R[A_2]$, and thus

$$R[A_1] \cap B \subseteq R[A_2] \cap B.$$

Hence, by Theorem 5, we can already see that

$$R[A_2] \cap B \in \mathscr{D}_{\mathscr{S}}$$

also holds. Therefore, by the corresponding definitions, we can state that $A_2 \in \operatorname{Cl}_{(\mathscr{R},\mathscr{S})}(B)$, and thus $B \in \operatorname{Cl}^{-1}_{(\mathscr{R},\mathscr{S})}(A_2)$. Therefore, assertion (2) is true.

Remark 16 Now, an analogue of the important fact that $\mathrm{Cl}_{\mathscr{R}^{-1}} = \mathrm{Cl}_{\mathscr{R}}^{-1}$ cannot be proved. Therefore, Theorem 21 cannot be derived from Theorem 20.

Theorem 22 *We have*

(1) $\mathrm{Int}_{(\mathscr{R},\mathscr{S})}(X) = \mathscr{P}(X)$ *if* $X, \mathscr{R}, \mathscr{S} \neq \emptyset$;
(2) $\mathrm{Int}_{(\mathscr{R},\mathscr{S})}(B_1) \subseteq \mathrm{Int}_{(\mathscr{R},\mathscr{S})}(B_2)$ *if* $B_1 \subseteq B_2 \subseteq Y$.

Remark 17 Note that if any one of the families X, $\mathscr{R}$ and $\mathscr{S}$ is empty, then by Remark 15 and Theorem 19 we have $\mathrm{Int}_{(\mathscr{R},\mathscr{S})}(B) = \emptyset$ for all $B \subseteq Y$.

Theorem 23 *We have*

(1) $\mathrm{Int}_{(\mathscr{R},\mathscr{S})}^{-1}(\emptyset) = \mathscr{P}(X)$ *if* $X, \mathscr{R}, \mathscr{S} \neq \emptyset$;
(2) $\mathrm{Int}_{(\mathscr{R},\mathscr{S})}^{-1}(A_2) \subseteq \mathrm{Int}_{(\mathscr{R},\mathscr{S})}^{-1}(A_1)$ *if* $A_1 \subseteq A_2 \subseteq X$.

Remark 18 Note that the proof of (2) also need only that, for any relation R on X to Y, the union-preserving corelation $R^{\triangleright}$, defined by $R^{\triangleright}(A) = R[A]$ for all $A \subseteq X$, is increasing.

Theorem 24 *We have*

(1) $\mathrm{Cl}_{(\mathscr{R},\mathscr{S})} = \bigcap_{R \in \mathscr{R}} \mathrm{Cl}_{(R,\mathscr{S})}$;
(2) $\mathrm{Int}_{(\mathscr{R},\mathscr{S})} = \bigcup_{R \in \mathscr{R}} \mathrm{Int}_{(R,\mathscr{S})}$.

Corollary 5 *The mapping*

(1) $\mathscr{R} \mapsto \mathrm{Int}_{(\mathscr{R},\mathscr{S})}$ *is union-preserving;*
(2) $\mathscr{R} \mapsto \mathrm{Cl}_{(\mathscr{R},\mathscr{S})}$ *is intersection-preserving.*

Theorem 25 *We have*

(1) $\mathrm{Cl}_{(\mathscr{R},\mathscr{S})} = \bigcap_{S \in \mathscr{S}} \mathrm{Cl}_{(\mathscr{R},S)}$;
(2) $\mathrm{Int}_{(\mathscr{R},\mathscr{S})} = \bigcup_{S \in \mathscr{S}} \mathrm{Int}_{(\mathscr{R},S)}$.

Proof To prove (1), note that, by Definition 2 and Theorem 9, for any $A \subseteq X$ and $B \subseteq Y$, we have

$$A \in \mathrm{Cl}_{(\mathscr{R},\mathscr{S})}(B) \iff \forall\, R \in \mathscr{R}: \quad R[A] \cap B \in \mathscr{D}_{\mathscr{S}} \iff$$

$$\forall\, R \in \mathscr{R}: \quad R[A] \cap B \in \bigcap_{S \in \mathscr{S}} \mathscr{D}_S \iff \forall\, R \in \mathscr{R}: \ \forall\, S \in \mathscr{S}: \quad R[A] \cap B \in \mathscr{D}_S$$

$$\iff \forall\, S \in \mathscr{S}: \ \forall\, R \in \mathscr{R}: \quad R[A] \cap B \in \mathscr{D}_S \iff \forall\, S \in \mathscr{S}: \quad A \in \mathrm{Cl}_{(\mathscr{R},S)}(B)$$

$$\iff A \in \bigcap_{S \in \mathscr{S}} \mathrm{Cl}_{(\mathscr{R},S)}(B) \iff A \in \Big(\bigcap_{S \in \mathscr{S}} \mathrm{Cl}_{(\mathscr{R},S)} \Big)(B).$$

Therefore, assertion (1) is true.

Corollary 6 *The mapping*

(1) $\mathscr{S} \mapsto \mathrm{Int}_{(\mathscr{R},\mathscr{S})}$ *is union-preserving;*
(2) $\mathscr{S} \mapsto \mathrm{Cl}_{(\mathscr{R},\mathscr{S})}$ *is intersection-preserving.*

8 Some Further Theorems on the Induced Proximal Closures and Interiors

By using the corresponding definitions, we can easily prove the following

Theorem 26 *For any* $A \subseteq X$ *and* $B \subseteq Y$, *the following assertions are equivalent:*

(1) $A \in \mathrm{Cl}_{(\mathscr{R},\mathscr{S})}(B)$;
(2) *for all* $R \in \mathscr{R}$, *we have* $\mathrm{cl}_{\mathscr{S}}\big(R[A] \cap B\big) = X$;
(3) *for all* $x \in X$, $R \in \mathscr{R}$ *and* $S \in \mathscr{S}$ *we have* $S(x) \cap R[A] \cap B \neq \emptyset$;
(4) *for all* $x \in X$, $R \in \mathscr{R}$ *and* $S \in \mathscr{S}$ *there exist* $a \in A$ *and* $b \in B$ *such that* $b \in S(x)$ *and* $b \in R(a)$.

Proof If (1) holds, then by Definition 2, for any $R \in \mathscr{R}$, we have $R[A] \cap B \in \mathscr{D}_{\mathscr{S}}$. Hence, by the definition of the family $\mathscr{D}_{\mathscr{S}}$, we can infer that $\mathrm{cl}_{\mathscr{S}}\big(R[A] \cap B\big) = X$. Therefore, for any $x \in X$, we have $x \in \mathrm{cl}_{\mathscr{S}}\big(R[A] \cap B\big)$. Thus, by the definition of the relation $\mathrm{cl}_{\mathscr{S}}$, for any $S \in \mathscr{S}$, we have $S(x) \cap R[A] \cap B \neq \emptyset$. Consequently, there exists $b \in B$ such that $b \in S(x)$ and $b \in R[A]$. Moreover, there exists $a \in A$ such that $b \in R(a)$.

Hence, it is clear that (1) $\Longrightarrow$ (2) $\Longrightarrow$ (3) $\Longrightarrow$ (4). The converse implications can be proved by reversing the above argument. Note that equivalence of (1) and (3) is also apparent from Theorem 1 by Definition 2.

From this theorem, by using the definition of $\mathrm{Cl}_{\mathscr{R}}$, we can immediately derive

Corollary 7 *For any* $B \subseteq Y$, *we have*

$$\mathrm{Cl}_{(\mathscr{R},\mathscr{S})}(B) = \bigcap_{x \in X} \bigcap_{S \in \mathscr{S}} \mathrm{Cl}_{\mathscr{R}}\big(S(x) \cap B\big).$$

Analogously to Theorem 26, we can also easily prove the following theorem which can also be proved with the help of Theorems 19 and 26.

Theorem 27 *For any* $A \subseteq X$ *and* $B \subseteq Y$, *the following assertions are equivalent:*

(1) $A \in \mathrm{Int}_{(\mathscr{R},\mathscr{S})}(B)$;
(2) *there exists* $R \in \mathscr{R}$ *such that* $\mathrm{int}_{\mathscr{S}}\big(R[A]^c \cup B\big) \neq \emptyset$;
(3) *there exist* $x \in X$, $R \in \mathscr{R}$ *and* $S \in \mathscr{S}$ *such that* $S(x) \subseteq R[A]^c \cup B$;
(4) *there exist* $x \in X$, $R \in \mathscr{R}$ *and* $S \in \mathscr{S}$ *such that for all* $v \in S(x)$ *we have either* $v \in B$ *or* $v \notin R(u)$ *for all* $u \in A$.

Remark 19 Because of Theorems 26 and 27, the relations $\mathrm{Cl}_{(\mathcal{R},\mathcal{S})}$ and $\mathrm{Int}_{(\mathcal{R},\mathcal{S})}$ can be easily introduced without using the families $\mathcal{D}_{\mathcal{S}}$ and $\mathcal{E}_{\mathcal{S}}$.

Now, by using Theorems 26 and 19, we can also easily establish the following theorems which also show the appropriateness of Definition 2.

Theorem 28 *If in particular* $\mathcal{S}$ *is non-degenerated (i.e.,* $X \neq \emptyset$ *and* $\mathcal{S} \neq \emptyset$*), then*

(1) $\mathrm{Cl}_{(\mathcal{R},\mathcal{S})} \subseteq \mathrm{Cl}_{\mathcal{R}}$*;*
(2) $\mathrm{Int}_{\mathcal{R}} \subseteq \mathrm{Int}_{(\mathcal{R},\mathcal{S})}$.

Proof To prove (1), note that if $A \subseteq X$ and $B \subseteq Y$ such that $A \in \mathrm{Cl}_{(\mathcal{R},\mathcal{S})}(B)$, then by Theorem 26, for any $x \in X$, $R \in \mathcal{R}$ and $S \in \mathcal{S}$, we have

$$S(x) \cap R[A] \cap B \neq \emptyset.$$

Hence, by choosing $x \in X$ and $S \in \mathcal{S}$, we can infer that

$$R[A] \cap B \neq \emptyset$$

for all $R \in \mathcal{R}$. Therefore, by the definition of the relation $\mathrm{Cl}_{\mathcal{R}}$, we have $A \in \mathrm{Cl}_{\mathcal{R}}(B)$. Thus, (1) is true.

Theorem 29 *If in particular* $\mathcal{S} \subseteq \{X \times Y\}$ *(i.e.,* $Y = S(x)$ *for all* $x \in X$ *and* $S \in \mathcal{S}$*), then*

(1) $\mathrm{Cl}_{\mathcal{R}} \subseteq \mathrm{Cl}_{(\mathcal{R},\mathcal{S})}$*;*
(2) $\mathrm{Int}_{(\mathcal{R},\mathcal{S})} \subseteq \mathrm{Int}_{\mathcal{R}}$.

Proof To prove (1), note that if $A \subseteq X$ and $B \subseteq Y$ such that $A \in \mathrm{Cl}_{\mathcal{R}}(B)$, then by the definition of the relation $\mathrm{Cl}_{\mathcal{R}}$, for any $R \in \mathcal{R}$, we have

$$R[A] \cap B \neq \emptyset.$$

Hence, by the assumption of the theorem, it is clear that, for any $x \in X$ and $S \in \mathcal{S}$, we also have

$$S(x) \cap R[A] \cap B = Y \cap R[A] \cap B = R[A] \cap B \neq \emptyset.$$

Therefore, by Theorem 26, we have $A \in \mathrm{Cl}_{(\mathcal{R},\mathcal{S})}(B)$. Thus, assertion (1) is true.

Now, as an immediate consequence of Theorems 28 and 29, we can also state

Corollary 8 *If in particular* $X \neq \emptyset$ *and* $\mathcal{S} = \{X \times Y\}$*, then*

(1) $\mathrm{Cl}_{\mathcal{R}} = \mathrm{Cl}_{(\mathcal{R},\mathcal{S})}$*;*
(2) $\mathrm{Int}_{\mathcal{R}} = \mathrm{Int}_{(\mathcal{R},\mathcal{S})}$.

Remark 20 To see the necessity of the assumption $X \neq \emptyset$, note that if $X = \emptyset$, then by Theorem 26 we necessarily have $\mathrm{Cl}_{(\mathscr{R},\mathscr{S})}(B) = \mathscr{P}(X) = \mathscr{P}(\emptyset) = \{\emptyset\}$ for all $B \subseteq Y$.

On the other hand, if $X = \emptyset$, then $\mathscr{R} \subseteq \mathscr{P}(X \times Y) = \mathscr{P}(\emptyset \times Y) = \mathscr{P}(\emptyset) = \{\emptyset\}$. Therefore, either $\mathscr{R} = \emptyset$ or $\mathscr{R} = \{\emptyset\}$. Moreover, by the definition of $\mathrm{Cl}_{\mathscr{R}}$, we have $\mathrm{Cl}_{\emptyset}(B) = \mathscr{P}(X) = \mathscr{P}(\emptyset) = \{\emptyset\}$ and $\mathrm{Cl}_{\{\emptyset\}}(B) = \emptyset$ for all $B \subseteq Y$.

In addition to Theorems 28 and 29, we can also easily prove the following two theorems.

Theorem 30 *If in particular, for every $y \in Y$, there exist $x \in X$ and $S \in \mathscr{S}$ such that $S(x) \subseteq \{y\}$, then for any $A \subseteq X$ and $B \subseteq X$ the following assertions hold:*

(1) if $A \in \mathrm{Cl}_{(\mathscr{R},\mathscr{S})}(B)$, then $A \in \mathscr{D}_{\mathscr{R}^{-1}}$ and $B = Y$;
(2) if either $A^c \in \mathscr{E}_{\mathscr{R}^{-1}}$ or $B \neq \emptyset$, then $A \in \mathrm{Int}_{(\mathscr{R},\mathscr{S})}(B)$.

Proof To prove (1), note that if $A \in \mathrm{Cl}_{(\mathscr{R},\mathscr{S})}(B)$, then by Theorem 26, for any $x \in X$, $R \in \mathscr{R}$ and $S \in \mathscr{S}$, we have

$$S(x) \cap R[A] \cap B \neq \emptyset.$$

This implies, in particular, that $S(x) \neq \emptyset$ for all $x \in X$ and $S \in \mathscr{S}$. Thus, by the assumption of the theorem, for every $y \in Y$, there exist $x_y \in X$ and $S_y \in \mathscr{S}$ such that $S_y(x_y) = \{y\}$.

Hence, it is clear that, for any $y \in Y$ and $R \in \mathscr{R}$, we have

$$\{y\} \cap R[A] \cap B = S_y(x_y) \cap R[A] \cap B \neq \emptyset.$$

Therefore, $y \in R[A] \cap B$, and thus $y \in B$ and $y \in R[A]$. Therefore, we necessarily have

$$B = Y \qquad \text{and} \qquad R[A] = Y.$$

Hence, by Theorem 2, we can already see that $A \in \mathscr{D}_{\mathscr{R}^{-1}}$. Therefore, assertion (1) is true.

Theorem 31 *If in particular $\mathscr{S}$ is non-partial (i.e., $S(x) \neq \emptyset$ for all $x \in X$ and $S \in \mathscr{S}$), then for any $A \subseteq X$ and $B \subseteq Y$ the following assertions hold:*

(1) if $A \in \mathscr{D}_{\mathscr{R}^{-1}}$ and $B = Y$, then $A \in \mathrm{Cl}_{(\mathscr{R},\mathscr{S})}(B)$;
(2) if $A \in \mathrm{Int}_{(\mathscr{R},\mathscr{S})}(B)$, then either $A^c \in \mathscr{E}_{\mathscr{R}^{-1}}$ or $B \neq \emptyset$.

Proof To prove (1), note that if $A \in \mathscr{D}_{\mathscr{R}^{-1}}$, then by Theorem 2, for any $R \in \mathscr{R}$, we have $R[A] = Y$. Hence, if $B = Y$ also holds, it is clear that, for any $x \in X$ and $S \in \mathscr{S}$, we have

$$S(x) \cap R[A] \cap B = S(x) \cap Y \cap Y = S(x) \neq \emptyset.$$

Therefore, by Theorem 26, we have $A \in \mathrm{Cl}_{(\mathscr{R},\mathscr{S})}(B)$. Thus, assertion (1) is true.

Now, as an immediate consequence of Theorems 30 and 31, we can also state

Corollary 9 *If in particular $\mathscr{S}$ is non-partial and for every $y \in Y$, there exist $x \in X$ and $S \in \mathscr{S}$ such that $S(x) \subseteq \{y\}$, then for any $A \subseteq X$ and $B \subseteq X$ we have*

(1) $A \in \mathrm{Cl}_{(\mathscr{R},\mathscr{S})}(B)$ if and only if $A \in \mathscr{D}_{\mathscr{R}^{-1}}$ and $B = Y$;
(2) $A \in \mathrm{Int}_{(\mathscr{R},\mathscr{S})}(B)$ if and only if either $A^c \in \mathscr{E}_{\mathscr{R}^{-1}}$ or $B \neq \emptyset$.

Remark 21 By using some further natural specializations of the relator $\mathscr{S}$, and also those of $\mathscr{R}$, one can also easily establish some other interesting consequences of Theorems 26 and 19.

However, it is now more important to note that, by using Theorems 26, 2, 3 and 19, we can also easily prove the following two theorems.

Theorem 32 *If in particular $\mathscr{R} \subseteq \mathscr{S}$, then for any $B \subseteq Y$*

(1) $\mathrm{Cl}_{(\mathscr{R},\mathscr{S})}(B) \neq \emptyset$ implies $B \in \mathscr{D}_{\mathscr{R}}$;
(2) $B \in \mathscr{E}_{\mathscr{R}}$ implies $\mathrm{Int}_{(\mathscr{R},\mathscr{S})}(B) = \mathscr{P}(X)$.

Proof To prove (1), note that if $\mathrm{Cl}_{(\mathscr{R},\mathscr{S})}(B) \neq \emptyset$, then there exists $A \subseteq X$ such that $A \in \mathrm{Cl}_{(\mathscr{R},\mathscr{S})}(B)$. Thus, by Theorem 26, for any $x \in X$, $R \in \mathscr{R}$ and $S \in \mathscr{S}$, we have

$$S(x) \cap R[A] \cap B \neq \emptyset.$$

Hence, since $\mathscr{R} \subseteq \mathscr{S}$, it is clear that, in particular, for any $x \in X$ and $R \in \mathscr{R}$ we also have

$$R(x) \cap R[A] \cap B \neq \emptyset, \qquad \text{and thus} \qquad R(x) \cap B \neq \emptyset.$$

Hence, by the definition of the relation $\mathrm{cl}_{\mathscr{R}}$, we can see that $x \in \mathrm{cl}_{\mathscr{R}}(B)$. Therefore, $\mathrm{cl}_{\mathscr{R}}(B) = X$, and thus by Theorem 2 $B \in \mathscr{D}_{\mathscr{R}}$ also holds.

Theorem 33 *If in particular $\mathscr{R} \subseteq \mathscr{S}$, then for any $A \subseteq X$*

(1) $\mathrm{Cl}^{-1}_{(\mathscr{R},\mathscr{S})}(A) \neq \emptyset$ implies $A \in \mathscr{D}_{\mathscr{R}^{-1} \circ \mathscr{R}}$;
(2) $A^c \in \mathscr{E}_{\mathscr{R}^{-1} \circ \mathscr{R}}$ implies $\mathrm{Int}^{-1}_{(\mathscr{R},\mathscr{S})}(A) = \mathscr{P}(Y)$.

Proof To prove (1), note that if $\mathrm{Cl}^{-1}_{(\mathscr{R},\mathscr{S})}(A) \neq \emptyset$, then there exists $B \subseteq Y$ such that $B \in \mathrm{Cl}^{-1}_{(\mathscr{R},\mathscr{S})}(A)$, and thus $A \in \mathrm{Cl}_{(\mathscr{R},\mathscr{S})}(B)$. Therefore, by Theorem 26, for any $x \in X$, $R \in \mathscr{R}$ and $S \in \mathscr{S}$, we have

$$S(x) \cap R[A] \cap B \neq \emptyset.$$

Hence, since $\mathscr{R} \subseteq \mathscr{S}$, it is clear that, in particular, for any $x \in X$ and $R, U \in \mathscr{R}$ we have

$$U(x) \cap R[A] \cap B \neq \emptyset, \qquad \text{and thus} \qquad U(x) \cap R[A] \neq \emptyset.$$

This implies that

$$x \in U^{-1}\big[R[A]\big] = \big(U^{-1} \circ R\big)[A].$$

Therefore, for any $R, U \in \mathscr{R}$, we actually have

$$\big(U^{-1} \circ R\big)[A] = X, \qquad \text{and thus} \qquad \big(R^{-1} \circ U\big)^{-1}[A] = X.$$

Hence, by using the notation

$$\mathscr{R}^{-1} \circ \mathscr{R} = \big\{ R^{-1} \circ U : \ R, U \in \mathscr{R} \big\}$$

and Theorem 2, we can already infer that $A \in \mathscr{D}_{\mathscr{R}^{-1} \circ \mathscr{R}}$.

Moreover, quite similarly, we can also prove the following two theorems.

Theorem 34 *If in particular* $\mathscr{S} \subseteq \mathscr{R}$, *then for any* $B \subseteq Y$ *then*

(1) $\operatorname{Cl}_{(\mathscr{R}, \mathscr{S})}(B) \neq \emptyset$ *implies* $B \in \mathscr{D}_{\mathscr{S}}$;
(2) $B \in \mathscr{E}_{\mathscr{S}}$ *implies* $\operatorname{Int}_{(\mathscr{R}, \mathscr{S})}(B) = \mathscr{P}(X)$.

Theorem 35 *If in particular* $\mathscr{S} \subseteq \mathscr{R}$, *then for any* $A \subseteq X$

(1) $\operatorname{Cl}^{-1}_{(\mathscr{R}, \mathscr{S})}(A) \neq \emptyset$ *implies* $A \in \mathscr{D}_{\mathscr{S}^{-1} \circ \mathscr{S}}$;
(2) $A^c \in \mathscr{E}_{\mathscr{S}^{-1} \circ \mathscr{S}}$ *implies* $\operatorname{Int}^{-1}_{(\mathscr{R}, \mathscr{S})}(A) = \mathscr{P}(Y)$.

9 Topological Closures and Interiors Derived from Ordered Pairs of Relators

Now, by using Definition 2, we may also naturally introduce the following

Definition 3 For any $a \in X$ and $B \subseteq Y$, we write

(1) $a \in \operatorname{cl}_{(\mathscr{R}, \mathscr{S})}(B)$ if $\{a\} \in \operatorname{Cl}_{(\mathscr{R}, \mathscr{S})}(B)$;
(2) $a \in \operatorname{int}_{(\mathscr{R}, \mathscr{S})}(B)$ if $\{a\} \in \operatorname{Int}_{(\mathscr{R}, \mathscr{S})}(B)$.

The relations $\operatorname{cl}_{(\mathscr{R}, \mathscr{S})}$ and $\operatorname{int}_{(\mathscr{R}, \mathscr{S})}$ will be called the *topological closure and interior* generated by the pair $(\mathscr{R}, \mathscr{S})$ of relators, respectively.

Thus, by specializing the results of Sections 7 and 8, we can easily establish the following theorems.

Theorem 36 *For any* $a \in X$ *and* $B \subseteq Y$, *the following assertions are equivalent:*

(1) $a \in \mathrm{cl}_{(\mathscr{R},\mathscr{S})}(B)$;
(2) *for all* $R \in \mathscr{R}$ *we have* $R(a) \cap B \in \mathscr{D}_{\mathscr{S}}$;
(3) *for all* $R \in \mathscr{R}$ *we have* $\mathrm{cl}_{\mathscr{S}}\big(R(a) \cap B\big) = X$;
(4) *for all* $x \in X$, $R \in \mathscr{R}$ *and* $S \in \mathscr{S}$ *we have* $S(x) \cap R(a) \cap B \neq \emptyset$;
(5) *for all* $x \in X$, $R \in \mathscr{R}$ *and* $S \in \mathscr{S}$ *there exists* $b \in B$ *such that* $b \in S(x)$ *and* $b \in R(a)$.

Corollary 10 *For any* $B \subseteq Y$, *we have*

$$\mathrm{cl}_{(\mathscr{R},\mathscr{S})}(B) = \bigcap_{x \in X} \bigcap_{S \in \mathscr{S}} \mathrm{cl}_{\mathscr{R}}\big(S(x) \cap B\big).$$

Theorem 37 *For any* $a \in X$ *and* $B \subseteq Y$, *the following assertions are equivalent:*

(1) $a \in \mathrm{int}_{(\mathscr{R},\mathscr{S})}(B)$;
(2) *there exists* $R \in \mathscr{R}$ *such that* $R(a)^c \cup B \in \mathscr{E}_{\mathscr{S}}$;
(3) *there exists* $R \in \mathscr{R}$ *such that* $\mathrm{int}_{\mathscr{S}}\big(R(a)^c \cup B\big) \neq \emptyset$;
(4) *there exist* $x \in X$, $R \in \mathscr{R}$ *and* $S \in \mathscr{S}$ *such that* $S(x) \subseteq R(a)^c \cup B$;
(5) *there exist* $x \in X$, $R \in \mathscr{R}$ *and* $S \in \mathscr{S}$ *such that for all* $v \in S(x)$ *we have either* $v \in B$ *or* $v \notin R(a)$.

Theorem 38 *For any* $A \subseteq X$ *and* $B \subseteq Y$,

(1) $A \in \mathrm{Int}_{(\mathscr{R},\mathscr{S})}(B)$ *implies* $A \subseteq \mathrm{int}_{(\mathscr{R},\mathscr{S})}(B)$;
(2) $A \cap \mathrm{cl}_{(\mathscr{R},\mathscr{S})}(B) \neq \emptyset$ *implies* $A \in \mathrm{Cl}_{(\mathscr{R},\mathscr{S})}(B)$.

Proof To prove (1), note that if $A \in \mathrm{Int}_{(\mathscr{R},\mathscr{S})}(B)$, then by Theorem 27 there exist $x \in X$, $R \in \mathscr{R}$ and $S \in \mathscr{S}$ such that

$$S(x) \subseteq R[A]^c \cup B.$$

Moreover, if $a \in A$, then we can note that $R(a) \subseteq R[A]$, and thus

$$R[A]^c \subseteq R(a)^c.$$

Therefore, we also have

$$S(x) \subseteq R(a)^c \cup B.$$

Hence, by Theorem 37, we can already see that $a \in \mathrm{int}_{(\mathscr{R},\mathscr{S})}(B)$ also holds. Therefore, $A \subseteq \mathrm{int}_{(\mathscr{R},\mathscr{S})}(B)$, and thus (1) is true.

Theorem 39 *For any* $B \subseteq Y$, *we have*

(1) $\mathrm{int}_{(\mathscr{R},\mathscr{S})}(B) = X \setminus \mathrm{cl}_{(\mathscr{R},\mathscr{S})}(Y \setminus B)$;
(2) $\mathrm{cl}_{(\mathscr{R},\mathscr{S})}(B) = X \setminus \mathrm{int}_{(\mathscr{R},\mathscr{S})}(Y \setminus B)$.

Corollary 11 *We have*

(1) $\operatorname{int}_{(\mathscr{R},\mathscr{S})} = \big(\operatorname{cl}_{(\mathscr{R},\mathscr{S})} \circ \mathscr{C}_Y\big)^c = \big(\operatorname{cl}_{(\mathscr{R},\mathscr{S})}\big)^c \circ \mathscr{C}_Y$;
(2) $\operatorname{cl}_{(\mathscr{R},\mathscr{S})} = \big(\operatorname{int}_{(\mathscr{R},\mathscr{S})} \circ \mathscr{C}_Y\big)^c = \big(\operatorname{int}_{(\mathscr{R},\mathscr{S})}\big)^c \circ \mathscr{C}_Y$.

Theorem 40 *We have*

(1) $\operatorname{cl}_{(\mathscr{R},\mathscr{S})}(\emptyset) = \emptyset$ *if* $X,\ \mathscr{R},\ \mathscr{S} \neq \emptyset$;
(2) $\operatorname{cl}_{(\mathscr{R},\mathscr{S})}(B_1) \subseteq \operatorname{cl}_{(\mathscr{R},\mathscr{S})}(B_2)$ *if* $B_1 \subseteq B_2 \subseteq Y$.

Theorem 41 *We have*

(1) $\operatorname{int}_{(\mathscr{R},\mathscr{S})}(X) = X$ *if* $X,\ \mathscr{R},\ \mathscr{S} \neq \emptyset$;
(2) $\operatorname{int}_{(\mathscr{R},\mathscr{S})}(B_1) \subseteq \operatorname{int}_{(\mathscr{R},\mathscr{S})}(B_2)$ *if* $B_1 \subseteq B_2 \subseteq Y$.

Remark 22 Note that if any one of the families X, $\mathscr{R}$ and $\mathscr{S}$ is empty, then by Theorems 37 and 36 we have $\operatorname{int}_{(\mathscr{R},\mathscr{S})}(B) = \emptyset$ and $\operatorname{cl}_{(\mathscr{R},\mathscr{S})}(B) = X$ for all $B \subseteq Y$.

Theorem 42 *We have*

(1) $\operatorname{cl}_{(\mathscr{R},\mathscr{S})} = \bigcap_{R \in \mathscr{R}} \operatorname{cl}_{(R,\mathscr{S})}$;
(2) $\operatorname{int}_{(\mathscr{R},\mathscr{S})} = \bigcup_{R \in \mathscr{R}} \operatorname{int}_{(R,\mathscr{S})}$.

Corollary 12 *The mapping*

(1) $\mathscr{R} \mapsto \operatorname{int}_{(\mathscr{R},\mathscr{S})}$ *is union-preserving;*
(2) $\mathscr{R} \mapsto \operatorname{cl}_{(\mathscr{R},\mathscr{S})}$ *is intersection-preserving.*

Theorem 43 *We have*

(1) $\operatorname{cl}_{(\mathscr{R},\mathscr{S})} = \bigcap_{S \in \mathscr{S}} \operatorname{cl}_{(\mathscr{R},S)}$;
(2) $\operatorname{int}_{(\mathscr{R},\mathscr{S})} = \bigcup_{S \in \mathscr{S}} \operatorname{int}_{(\mathscr{R},S)}$.

Corollary 13 *The mapping*

(1) $\mathscr{S} \mapsto \operatorname{int}_{(\mathscr{R},\mathscr{S})}$ *is union-preserving;*
(2) $\mathscr{S} \mapsto \operatorname{cl}_{(\mathscr{R},\mathscr{S})}$ *is intersection-preserving.*

Theorem 44 *If in particular* $\mathscr{S} \neq \emptyset$, *then*

(1) $\operatorname{cl}_{(\mathscr{R},\mathscr{S})} \subseteq \operatorname{cl}_{\mathscr{R}}$;
(2) $\operatorname{int}_{\mathscr{R}} \subseteq \operatorname{int}_{(\mathscr{R},\mathscr{S})}$.

Remark 23 To see that now the assumption $X \neq \emptyset$ could be omitted, note that if $X = \emptyset$, then by Remark 20 and Definition 3 we have $\operatorname{cl}_{(\mathscr{R},\mathscr{S})}(B) = X = \emptyset$ for all $B \subseteq Y$.

On the other hand, if $X = \emptyset$, then by Remark 20 we have either $\mathscr{R} = \emptyset$ or $\mathscr{R} = \{\emptyset\}$. Moreover, by the definition of the relation $\operatorname{cl}_{\mathscr{R}}$, we have $\operatorname{cl}_{\emptyset}(B) = X = \emptyset$ and $\operatorname{cl}_{\{\emptyset\}}(B) = \emptyset$.

Theorem 45 *If in particular* $\mathscr{S} \subseteq \{X \times Y\}$, *then*

(1) $\operatorname{cl}_{\mathscr{R}} \subseteq \operatorname{cl}_{(\mathscr{R},\mathscr{S})}$;
(2) $\operatorname{int}_{(\mathscr{R},\mathscr{S})} \subseteq \operatorname{int}_{\mathscr{R}}$.

Corollary 14 *If in particular* $\mathscr{S} = \{X \times Y\}$, *then*

(1) $\operatorname{cl}_{\mathscr{R}} = \operatorname{cl}_{(\mathscr{R},\mathscr{S})}$;
(2) $\operatorname{int}_{\mathscr{R}} = \operatorname{int}_{(\mathscr{R},\mathscr{S})}$.

Theorem 46 *If in particular for every* $y \in Y$, *there exist* $x \in X$ *and* $S \in \mathscr{S}$ *such that* $S(x) \subseteq \{y\}$, *then for any* $a \in X$ *and* $B \subseteq Y$ *the following assertions hold:*

(1) if $a \in \operatorname{cl}_{(\mathscr{R},\mathscr{S})}(B)$, *then* $B = Y$ *and* $R(a) = Y$ *for all* $R \in \mathscr{R}$;
(2) if either $B \neq \emptyset$ *or* $R(a) \neq Y$ *for some* $R \in \mathscr{R}$, *then* $a \in \operatorname{Int}_{(\mathscr{R},\mathscr{S})}(B)$.

Proof To prove (1), note that by Definition 3 and Theorem 30 we have

$$a \in \operatorname{cl}_{(\mathscr{R},\mathscr{S})}(B) \iff \{a\} \in \operatorname{Cl}_{(\mathscr{R},\mathscr{S})}(B) \iff B = Y \quad \text{and} \quad \{a\} \in \mathscr{E}_{\mathscr{R}^{-1}}.$$

Moreover, by Theorem 2, we have $\{a\} \in \mathscr{E}_{\mathscr{R}^{-1}}$ if and only if $R\big[\{a\}\big] = Y$ for all $R \in \mathscr{R}$. Therefore, assertion (1) is true.

Theorem 47 *If in particular* $\mathscr{S}$ *is non-partial (i.e.,* $S(x) \neq \emptyset$ *for all* $x \in X$ *and* $S \in \mathscr{S}$*), then for any* $a \in X$ *and* $B \subseteq Y$ *the following assertions hold:*

(1) if $B = Y$ *and* $R(a) = Y$ *for all* $R \in \mathscr{R}$, *then* $a \in \operatorname{cl}_{(\mathscr{R},\mathscr{S})}(B)$;
(2) if $a \in \operatorname{int}_{(\mathscr{R},\mathscr{S})}(B)$, *then either* $B \neq \emptyset$ *or* $R(a) \neq Y$ *for some* $R \in \mathscr{R}$.

Corollary 15 *If in particular for every* $x \in X$ *there exist* $x \in X$ *and* $S \in \mathscr{S}$ *such that* $S(x) = \{y\}$, *then for any* $a \in X$ *and* $B \subseteq Y$ *we have*

(1) $a \in \operatorname{cl}_{(\mathscr{R},\mathscr{S})}(B)$ *if and only if* $B = Y$ *and* $R(a) = Y$ *for all* $R \in \mathscr{R}$;
(2) $a \in \operatorname{int}_{(\mathscr{R},\mathscr{S})}(B)$ *if and only if either* $B \neq \emptyset$ *or* $R(a) \neq Y$ *for some* $R \in \mathscr{R}$.

Theorem 48 *If in particular* $\mathscr{R} \subseteq \mathscr{S}$, *then for any* $B \subseteq Y$

(1) $\operatorname{cl}_{(\mathscr{R},\mathscr{S})}(B) \neq \emptyset$ *implies* $B \in \mathscr{D}_{\mathscr{R}}$;
(2) $B \in \mathscr{E}_{\mathscr{R}}$ *implies* $\operatorname{int}_{(\mathscr{R},\mathscr{S})}(B) = X$.

Theorem 49 *If in particular* $\mathscr{R} \subseteq \mathscr{S}$, *then for any* $a \in X$ *we can state that*

(1) if $a \in \operatorname{cl}_{(\mathscr{R},\mathscr{S})}(B)$ *for some* $B \subseteq Y$, *then* $\big(R^{-1} \circ U\big)(a) = X$ *for all* $R, U \in \mathscr{R}$;
(2) if $\big(R^{-1} \circ U\big)(a) \neq X$ *for some* $R, U \in \mathscr{R}$, *then* $a \in \operatorname{int}_{(\mathscr{R},\mathscr{S})}(B)$ *for all* $B \subseteq Y$.

Proof To prove (1), note that if $B \subseteq Y$ such that $a \in \operatorname{cl}_{(\mathscr{R},\mathscr{S})}(B)$, then by Definition 3 we have $\{a\} \in \operatorname{Cl}_{(\mathscr{R},\mathscr{S})}(B)$, and thus $B \in \operatorname{Cl}_{(\mathscr{R},\mathscr{S})}^{-1}(\{a\})$. Therefore, $\operatorname{Cl}_{(\mathscr{R},\mathscr{S})}^{-1}\big(\{a\}\big) \neq \emptyset$. Thus, by Theorem 33, we have $\{a\} \in \mathscr{D}_{\mathscr{R}^{-1}\circ\mathscr{R}}$. Hence, by using Theorem 2, we can already infer that $\left(U^{-1}\circ R\right)^{-1}[\{a\}] = X$, and thus $\left(R^{-1}\circ U\right)(a) = X$ for all $R, U \in \mathcal{R}$.

Theorem 50 *If in particular $\mathcal{S} \subseteq \mathcal{R}$, then for any $B \subseteq Y$*

(1) $\mathrm{cl}_{(\mathcal{R},\mathcal{S})}(B) \neq \emptyset$ *implies* $B \in \mathcal{D}_{\mathcal{S}}$*;*
(2) $B \in \mathcal{E}_{\mathcal{S}}$ *implies* $\mathrm{int}_{(\mathcal{R},\mathcal{S})}(B) = X$.

Theorem 51 *If in particular $\mathcal{S} \subseteq \mathcal{R}$, then for any $a \in X$ we can state that*

(1) if $a \in \mathrm{cl}_{(\mathcal{R},\mathcal{S})}(B)$ *for some* $B \subseteq Y$*, then* $\left(S^{-1}\circ V\right)(a) = X$ *for all* $S, V \in \mathcal{S}$*;*
(2) if $\left(S^{-1}\circ V\right)(a) \neq X$ *for some* $S, V \in \mathcal{S}$*, then* $a \in \mathrm{int}_{(\mathcal{R},\mathcal{S})}(B)$ *for all* $B \subseteq X$.

10 Fat and Dense Sets Derived from Ordered Pairs of Relators

Now, by using Definition 3, we can also naturally introduce the following

Definition 4 For any $B \subseteq Y$, we write

(1) $B \in \mathcal{D}_{(\mathcal{R},\mathcal{S})}$ if $\mathrm{cl}_{(\mathcal{R},\mathcal{S})}(B) = X$;
(2) $B \in \mathcal{E}_{(\mathcal{R},\mathcal{S})}$ if $\mathrm{int}_{(\mathcal{R},\mathcal{S})}(B) \neq \emptyset$.

The members of the families $\mathcal{E}_{(\mathcal{R},\mathcal{S})}$ and $\mathcal{D}_{(\mathcal{R},\mathcal{S})}$ will be called the *fat and dense sets* generated by the pair $(\mathcal{R}, \mathcal{S})$ of relators, respectively.

Thus, by using the results of Section 9, we can easily prove the following analogues of the results of Sections 5 and 6.

Theorem 52 *For any $B \subseteq Y$, the following assertions are equivalent:*

(1) $B \in \mathcal{D}_{(\mathcal{R},\mathcal{S})}$*;*
(2) for all $x \in X$ *we have* $x \in \mathrm{cl}_{(\mathcal{R},\mathcal{S})}(B)$*;*
(3) for all $x \in X$ *and* $R \in \mathcal{R}$ *we have* $R(x) \cap B \in \mathcal{D}_{\mathcal{S}}$*;*
(4) for all $x \in X$ *and* $R \in \mathcal{R}$ *we have* $\mathrm{cl}_{\mathcal{S}}\left(R(x) \cap B\right) = X$*;*
(5) for all $x, u \in X$*,* $R \in \mathcal{R}$ *and* $S \in \mathcal{S}$ *we have*

$$S(u) \cap R(x) \cap B \neq \emptyset ;$$

(6) for all $x, u \in X$*,* $R \in \mathcal{R}$ *and* $S \in \mathcal{S}$ *there exists* $y \in B$ *such that* $y \in R(x)$ *and* $y \in S(u)$.

Corollary 16 *For any $B \subseteq Y$, the following assertions are equivalent:*

(1) $B \in \mathcal{D}_{(\mathcal{R},\mathcal{S})}$*;*
(2) $X = S^{-1}[R(x) \cap B]$ *for all* $x \in X$*,* $R \in \mathcal{R}$ *and* $S \in \mathcal{S}$*;*

(3) $X = R^{-1}[\, S(x) \cap B\,]$ *for all* $x \in X$, $R \in \mathscr{R}$ *and* $S \in \mathscr{S}$.

Theorem 53 *For any* $B \subseteq Y$, *the following assertions are equivalent:*

(1) $B \in \mathscr{E}_{(\mathscr{R},\mathscr{S})}$;
(2) there exists $x \in X$ *such that* $x \in \operatorname{int}_{(\mathscr{R},\mathscr{S})}(B)$;
(3) there exist $x \in X$ *and* $R \in \mathscr{R}$ *such that* $R(x)^c \cup B \in \mathscr{E}_{\mathscr{S}}$;
(4) there exist $x \in X$ *and* $R \in \mathscr{R}$ *such that* $\operatorname{int}_{\mathscr{S}}\big(R(x)^c \cup B\big) \neq \emptyset$;
(5) there exist $x, u \in X$, $R \in \mathscr{R}$ *and* $S \in \mathscr{S}$ *such that* $S(u) \subseteq R(x)^c \cap B$;
(6) there exist $x, u \in X$, $R \in \mathscr{R}$ *and* $S \in \mathscr{S}$ *such that for all* $y \in S(u)$ *we have* $y \notin R(x)$ *and* $y \in B$.

Theorem 54 *For any* $B \subseteq Y$, *we have*

(1) $B \in \mathscr{D}_{(\mathscr{R},\mathscr{S})} \iff B^c \notin \mathscr{E}_{(\mathscr{R},\mathscr{S})}$;
(2) $B \in \mathscr{E}_{(\mathscr{R},\mathscr{S})} \iff B^c \notin \mathscr{D}_{(\mathscr{R},\mathscr{S})}$.

Theorem 55 *For any* $B \subseteq Y$, *we have*

(1) $B \in \mathscr{D}_{(\mathscr{R},\mathscr{S})}$ *if and only if* $B \cap E \neq \emptyset$ *for all* $E \in \mathscr{E}_{(\mathscr{R},\mathscr{S})}$;
(2) $B \in \mathscr{E}_{(\mathscr{R},\mathscr{S})}$ *if and only if* $B \cap D \neq \emptyset$ *for all* $D \in \mathscr{D}_{(\mathscr{R},\mathscr{S})}$.

Theorem 56 *We have*

(1) $\mathscr{D}_{(\mathscr{S},\mathscr{R})} = \mathscr{D}_{(\mathscr{R},\mathscr{S})}$;
(2) $\mathscr{E}_{(\mathscr{S},\mathscr{R})} = \mathscr{E}_{(\mathscr{R},\mathscr{S})}$.

Proof From Theorem 52, it is clear that assertion (1) is true. Now, by using Theorem 54, we can also see that

$$B \in \mathscr{E}_{(\mathscr{S},\mathscr{R})} \iff B^c \notin \mathscr{D}_{(\mathscr{S},\mathscr{R})} \iff B^c \notin \mathscr{D}_{(\mathscr{R},\mathscr{S})} \iff B \in \mathscr{E}_{(\mathscr{R},\mathscr{S})}.$$

Therefore, assertion (2) is also true.

Remark 24 To give a direct proof of (2), one can note that if $B \in \mathscr{E}_{(\mathscr{S},\mathscr{R})}$, then by Theorem 53 there exist $x, u \in X$, $R \in \mathscr{R}$ and $S \in \mathscr{S}$ such that $R(u) \subseteq S(x)^c \cap B$. This implies that

$$S(x) \cup B^c = \big(S(x)^c \cap B\big)^c \subseteq R(u)^c.$$

Hence, we can infer that

$$S(x) = S(x) \cap Y = S(x) \cap (B^c \cup \mathscr{B}) = \big(S(x) \cap B^c\big) \cup \big(S(x) \cap B\big) \subseteq R(u)^c \cap B.$$

Therefore, by Theorem 53, we also have $B \in \mathscr{E}_{(\mathscr{R},\mathscr{S})}$.

Theorem 57 *The families* $\mathscr{D}_{(\mathscr{R},\mathscr{S})}$ *and* $\mathscr{E}_{(\mathscr{R},\mathscr{S})}$ *are stacks (ascending systems) in* Y.

Theorem 58 *The following assertions are equivalent:*

(1) $\emptyset \notin \mathscr{D}_{(\mathscr{R},\mathscr{S})}$;
(2) $\mathscr{E}_{(\mathscr{R},\mathscr{S})} \neq \emptyset$;
(3) $Y \in \mathscr{E}_{(\mathscr{R},\mathscr{S})}$;
(4) $\mathscr{D}_{(\mathscr{R},\mathscr{S})} \neq \mathscr{P}(Y)$;
(5) $X, \mathscr{R}, \mathscr{S} \neq \emptyset$.

Proof To prove the equivalence of (1) and (5), note that, by Theorem 52, we have $\emptyset \notin \mathscr{D}_{(\mathscr{R},\mathscr{S})}$ if and only if there exist $x, u \in X$, $R \in \mathscr{R}$ and $S \in \mathscr{S}$ such that $S(u) \cap R(x) \cap \emptyset = \emptyset$, or equivalently $\emptyset = \emptyset$. That is, each of the families X, $\mathscr{R}$ and $\mathscr{S}$ is nonempty.

Theorem 59 *The following assertions are equivalent:*

(1) $\emptyset \notin \mathscr{E}_{(\mathscr{R},\mathscr{S})}$;
(2) $\mathscr{D}_{(\mathscr{R},\mathscr{S})} \neq \emptyset$;
(3) $Y \in \mathscr{D}_{(\mathscr{R},\mathscr{S})}$;
(4) $\mathscr{E}_{(\mathscr{R},\mathscr{S})} \neq \mathscr{P}(Y)$;
(5) $\mathscr{S}^{-1} \circ \mathscr{R} \subseteq \{X^2\}$.

Proof To prove the equivalence of (1), (3), and (5), note that, by Theorem 54, we have

$$\emptyset \notin \mathscr{E}_{(\mathscr{R},\mathscr{S})} \iff Y^c \notin \mathscr{E}_{(\mathscr{R},\mathscr{S})} \iff Y \in \mathscr{D}_{(\mathscr{R},\mathscr{S})}.$$

Moreover, by Theorem 52, we have $Y \in \mathscr{D}_{(\mathscr{R},\mathscr{S})}$ if and only if for any $x, u \in X$, $R \in \mathscr{R}$ and $S \in \mathscr{S}$ we have

$$S(u) \cap R(x) \cap Y \neq \emptyset, \qquad \text{or equivalently} \qquad S(u) \cap R(x) \neq \emptyset.$$

That is, $u \in S^{-1}[R(x)]$, which can be written in the form that $u \in \big(S^{-1} \circ R\big)(x)$, or equivalently $(x, u) \in S^{-1} \circ R$. Hence, we can see that assertion (1) is equivalent to the requirement that $S^{-1} \circ R = X^2$ for all $R \in \mathscr{R}$ and $S \in \mathscr{S}$. That is, (5) holds.

Remark 25 If the assertions (5) of Theorems 58 and 59 hold, then analogously to Remark 3 the pair $(\mathscr{R}, \mathscr{S})$, or the birelator space $(\mathscr{R}, Y)(\mathscr{R}, \mathscr{S})$, may be called *non-degenerated* and *non-partial*, respectively.

Note that if $\mathscr{S}^{-1} \circ \mathscr{R} \subseteq \{X^2\}$, then by Theorem 12 the pair $(\mathscr{R}, \mathscr{S})$ may also be naturally called *semi-directed*. Thus, as an extension of Theorem 13, we can prove that $(\mathscr{R}, \mathscr{S})$ is semi-directed if and only if $\mathscr{E}_{\mathscr{R}} \subseteq \mathscr{D}_{\mathscr{S}}$.

Theorem 60 *We have*

(1) $\mathscr{E}_{(\mathscr{R},\mathscr{S})} = \bigcup_{R \in \mathscr{R}} \mathscr{E}_{(R,\mathscr{S})}$;
(2) $\mathscr{D}_{(\mathscr{R},\mathscr{S})} = \bigcap_{R \in \mathscr{R}} \mathscr{D}_{(R,\mathscr{S})}$.

Corollary 17 *The mapping*

(1) $\mathscr{R} \mapsto \mathscr{E}_{(\mathscr{R},\mathscr{S})}$ *is union-preserving;*
(2) $\mathscr{R} \mapsto \mathscr{D}_{(\mathscr{R},\mathscr{S})}$ *is intersection-preserving.*

Theorem 61 *We have*

(1) $\mathscr{E}_{(\mathscr{R},\mathscr{S})} = \bigcup_{S \in \mathscr{S}} \mathscr{E}_{(\mathscr{R},S)}$;
(2) $\mathscr{D}_{(\mathscr{R},\mathscr{S})} = \bigcap_{S \in \mathscr{S}} \mathscr{D}_{(\mathscr{R},S)}$.

Corollary 18 *The mapping*

(1) $\mathscr{S} \mapsto \mathscr{E}_{(\mathscr{R},\mathscr{S})}$ *is union-preserving;*
(2) $\mathscr{S} \mapsto \mathscr{D}_{(\mathscr{R},\mathscr{S})}$ *is intersection-preserving.*

Remark 26 Note that, Theorem 61 and Corollary 18 can also be derived from Theorem 60 and Corollary 17 by using Theorem 56.

Theorem 62 *For any* $B \subseteq Y$,

(1) $\mathrm{Int}_{(\mathscr{R},\mathscr{S})}(B) \setminus \{\emptyset\} \neq \emptyset$ *implies* $B \in \mathscr{E}_{(\mathscr{R},\mathscr{S})}$;
(2) $B \in \mathscr{D}_{(\mathscr{R},\mathscr{S})}$ *implies* $\mathscr{P}(X) \setminus \{\emptyset\} \subseteq \mathrm{Cl}_{(\mathscr{R},\mathscr{S})}(B)$.

Proof To prove (1), note that if $\mathrm{Int}_{(\mathscr{R},\mathscr{S})}(B) \setminus \{\emptyset\} \neq \emptyset$, then there exists $A \subseteq X$ such that $A \neq \emptyset$ and $A \in \mathrm{Int}_{(\mathscr{R},\mathscr{S})}(B)$. Hence, by using Theorem 38, we can infer that $A \subseteq \mathrm{int}_{(\mathscr{R},\mathscr{S})}(B)$. Thus, in particular $\mathrm{int}_{(\mathscr{R},\mathscr{S})}(B) \neq \emptyset$, and thus $B \in \mathscr{E}_{(\mathscr{R},\mathscr{S})}$ also holds.

Now, for instance, by using Corollaries 14 and 15, we can also prove the following two theorems.

Theorem 63 *If in particular* $\mathscr{S} = \{X \times Y\}$, *then*

(1) $\mathscr{D}_{\mathscr{R}} = \mathscr{D}_{(\mathscr{R},\mathscr{S})}$;
(2) $\mathscr{E}_{\mathscr{R}} = \mathscr{E}_{(\mathscr{R},\mathscr{S})}$.

Theorem 64 *If in particular for every* $v \in Y$ *there exist* $u \in X$ *and* $S \in \mathscr{S}$ *such that* $S(u) = \{v\}$, *then for any* $B \subseteq Y$ *we have*

(1) $B \in \mathscr{D}_{(\mathscr{R},\mathscr{S})}$ *if and only if* $B = X$ *and* $\mathscr{R} \subseteq \{X \times Y\}$;
(2) $A \in \mathscr{E}_{(\mathscr{R},\mathscr{S})}(A)$ *if and only if either* $B \neq \emptyset$ *or* $\mathscr{R} \setminus \{X \times Y\} \neq \emptyset$.

Proof To prove the "only if part" of assertion (1), note that if $B \in \mathscr{D}_{(\mathscr{R},\mathscr{S})}$, then for any $x \in X$, we have $x \in \mathrm{cl}_{(\mathscr{R},\mathscr{S})}(B)$. Hence, by using Corollary 15, we can infer that $B = Y$ and $R(x) = Y$ for all $R \in \mathscr{R}$. Therefore, we actually have $R = X \times Y$ for all $R \in \mathscr{R}$, and thus $\mathscr{R} \subseteq \{X \times Y\}$ also holds.

Moreover, for instance, by using Theorem 51, we can also prove the following

Theorem 65 *If in particular $\mathscr{R} \subseteq \mathscr{S}$, then*

(1) $\mathscr{D}_{(\mathscr{R},\mathscr{S})} \neq \emptyset$ *implies* $\mathscr{R}^{-1} \circ \mathscr{R} \subseteq \{X^2\}$;
(2) $\mathscr{E}_{(\mathscr{R},\mathscr{S})} = \mathscr{P}(Y)$ *implies* $\mathscr{R}^{-1} \circ \mathscr{R} \setminus \{X^2\} \neq \emptyset$.

Proof To prove (1), note that if $\mathscr{D}_{(\mathscr{R},\mathscr{S})} \neq \emptyset$, then there exists $B \subseteq Y$ such that $B \in \mathscr{D}_{(\mathscr{R},\mathscr{S})}$, and thus $\mathrm{cl}_{(\mathscr{R},\mathscr{S})}(B) = X$. Therefore, $x \in \mathrm{cl}_{(\mathscr{R},\mathscr{S})}(B)$ for all $x \in X$. Hence, by Theorem 51, we can see that $\big(R^{-1} \circ U\big)(x) = X$ for all $x \in X$ and $R,\, U \in \mathscr{R}$. Therefore, $R^{-1} \circ U = X^2$ for all $R,\, U \in \mathscr{R}$, and thus $\mathscr{R}^{-1} \circ \mathscr{R} \subseteq \{X^2\}$ also holds.

Remark 27 Note that from the latter three theorems, by using Theorem 56, we can also derive some further reasonable theorems.

11 Proximally Open and Closed Sets Derived from Ordered Pairs of Relators

Notation 3 *In this and the next section, we shall assume that $\mathscr{R}$ and $\mathscr{S}$ are relators on X.*

Now, by using Definition 2, we may also naturally introduce the following

Definition 5 For any $A \subseteq X$, we write

(1) $A \in \tau_{(\mathscr{R},\mathscr{S})}$ if $A \in \mathrm{Int}_{(\mathscr{R},\mathscr{S})}(A)$;
(2) $A \in \tau\!\!\!-_{(\mathscr{R},\mathscr{S})}$ if $A^c \notin \mathrm{Cl}_{(\mathscr{R},\mathscr{S})}(A)$.

The members of the families $\tau_{(\mathscr{R},\mathscr{S})}$ and $\tau\!\!\!-_{(\mathscr{R},\mathscr{S})}$ will be called the *proximally open and closed sets* generated by the pair $(\mathscr{R},\, \mathscr{S})$ of relators, respectively.

Thus, by using the results of Sections 7 and 8, we can easily prove the following theorems.

Theorem 66 *For any $A \subseteq X$, the following assertions are equivalent:*

(1) $A \in \tau_{(\mathscr{R},\mathscr{S})}$;
(2) *there exists* $R \in \mathscr{R}$ *such that* $R[A]^c \cup A \in \mathscr{E}_{\mathscr{S}}$;
(3) *there exists* $R \in \mathscr{R}$ *such that* $\mathrm{int}_{\mathscr{S}}\big(R[A]^c \cup A\big) \neq \emptyset$;
(4) *there exist* $x \in X$, $R \in \mathscr{R}$ *and* $S \in \mathscr{S}$ *such that* $S(x) \subseteq R[A]^c \cup A$;
(5) *there exist* $x \in X$, $R \in \mathscr{R}$ *and* $S \in \mathscr{S}$ *such that for all* $v \in S(x)$ *we have either* $v \in A$ *or* $v \notin R(u)$ *for all* $u \in A$.

Theorem 67 *For any $A \subseteq X$, the following assertions are equivalent:*

(1) $A \in \tau\!\!\!-_{(\mathscr{R},\mathscr{S})}$;
(2) *there exists* $R \in \mathscr{R}$ *such that* $R[A^c] \cap A \notin \mathscr{D}_{\mathscr{S}}$;

(3) there exists $R \in \mathscr{R}$ *such that* $\operatorname{cl}_{\mathscr{S}}\big(R[A^c] \cap A\big) \neq X$;
(4) there exist $x \in X$, $R \in \mathscr{R}$ *and* $S \in \mathscr{S}$ *such that* $S(x) \cap R[A^c] \cap A = \emptyset$;
(5) there exist $x \in X$, $R \in \mathscr{R}$ *and* $S \in \mathscr{S}$ *such that for all* $v \in S(x)$ *we have either* $v \notin A$ *or* $v \notin R(u)$ *for all* $u \in A^c$.

Theorem 68 *For any* $A \subseteq X$, *we have*

(1) $A \in \tau\!\!\!\!-_{(\mathscr{R},\mathscr{S})} \iff A^c \in \tau_{(\mathscr{R},\mathscr{S})}$;
(2) $A \in \tau_{(\mathscr{R},\mathscr{S})} \iff A^c \in \tau\!\!\!\!-_{(\mathscr{R},\mathscr{S})}$.

Theorem 69 *If in particular* X, $\mathscr{R}$ *and* $\mathscr{S}$ *are not empty, then*

(1) $\{\emptyset,\, X\} \subseteq \tau_{(\mathscr{R},\mathscr{S})}$;
(2) $\{\emptyset,\, X\} \subseteq \tau\!\!\!\!-_{(\mathscr{R},\mathscr{S})}$.

Remark 28 Note that if any one of the families X, $\mathscr{R}$ and $\mathscr{S}$ is empty, then by Theorems 66 and 67 we have $\tau_{(\mathscr{R},\mathscr{S})} = \emptyset$ and $\tau\!\!\!\!-_{(\mathscr{R},\mathscr{S})} = \emptyset$.

Therefore, if τ is a weak structure on X by Császár [14] (i.e., $\emptyset \in \tau \subseteq \mathscr{P}(X)$) such that $X \notin \tau$, then $\tau \neq \tau_{(\mathscr{R},\mathscr{S})}$ for any two relators $\mathscr{R}$ and $\mathscr{S}$ on X.

Theorem 70 *We have*

(1) $\tau_{(\mathscr{R},\mathscr{S})} = \bigcup_{R \in \mathscr{R}} \tau_{(R,\mathscr{S})}$;
(2) $\tau\!\!\!\!-_{(\mathscr{R},\mathscr{S})} = \bigcup_{R \in \mathscr{R}} \tau\!\!\!\!-_{(R,\mathscr{S})}$.

Corollary 19 *The mappings* $\mathscr{R} \mapsto \tau_{(\mathscr{R},\mathscr{S})}$ *and* $\mathscr{R} \mapsto \tau\!\!\!\!-_{(\mathscr{R},\mathscr{S})}$ *are union-preserving.*

Theorem 71 *We have*

(1) $\tau_{(\mathscr{R},\mathscr{S})} = \bigcup_{S \in \mathscr{S}} \tau_{(\mathscr{R},S)}$;
(2) $\tau\!\!\!\!-_{(\mathscr{R},\mathscr{S})} = \bigcup_{S \in \mathscr{S}} \tau\!\!\!\!-_{(\mathscr{R},S)}$.

Corollary 20 *The mappings* $\mathscr{S} \mapsto \tau_{(\mathscr{R},\mathscr{S})}$ *and* $\mathscr{S} \mapsto \tau\!\!\!\!-_{(\mathscr{R},\mathscr{S})}$ *are union-preserving.*

Now, for instance, by Corollaries 8 and 9, we can also state the following two theorems.

Theorem 72 *If in particular* $X \neq \emptyset$ *and* $\mathscr{S} = \{X^2\}$, *then*

(1) $\tau_{\mathscr{R}} = \tau_{(\mathscr{R},\mathscr{S})}$;
(2) $\tau\!\!\!\!-_{\mathscr{R}} = \tau\!\!\!\!-_{(\mathscr{R},\mathscr{S})}$.

Remark 29 To see the necessity of the condition $X \neq \emptyset$, note that if $X = \emptyset$, then by Remark 28 we have $\tau\!\!\!\!-_{(\mathscr{R},\mathscr{S})} = \emptyset$.

On the other hand, if $X = \emptyset$, then by Remark 8.8, we have either $\mathscr{R} = \emptyset$ or $\mathscr{R} = \{\emptyset\}$. Moreover, by the definition of the family $\tau\!\!\!\!-_{\mathscr{R}}$, we have $\tau\!\!\!\!-_{\emptyset} = \emptyset$ and $\tau\!\!\!\!-_{\{\emptyset\}} = \mathscr{P}(X) = \mathscr{P}(\emptyset) = \{\emptyset\}$.

Theorem 73 *If in particular* $\mathscr{S}$ *is non-partial and for every* $y \in X$*, there exist* $x \in X$ *and* $S \in \mathscr{S}$ *such that* $S(x) \subseteq \{y\}$*, then for any* $A \subseteq X$ *we have*

(1) $A \in \tau_{(\mathscr{R},\mathscr{S})}$ *if and only if either* $A \neq \emptyset$ *or* $A \notin \mathscr{D}_{\mathscr{R}^{-1}}$*;*
(2) $A \in \tau\!\!\!-_{(\mathscr{R},\mathscr{S})}$ *if and only if either* $A \neq X$ *or* $A \in \mathscr{E}_{\mathscr{R}^{-1}}$*.*

Moreover, for instance, by using Theorems 33 and 68, we can prove the following

Theorem 74 *If in particular* $\mathscr{R} \subseteq \mathscr{S}$*, then*

(1) $\mathscr{E}_{\mathscr{R}^{-1}\circ\mathscr{R}} \subseteq \tau\!\!\!-_{(\mathscr{R},\mathscr{S})}$*;*
(2) $\mathscr{P}(X) \setminus \mathscr{D}_{\mathscr{R}^{-1}\circ\mathscr{R}} \subseteq \tau_{(\mathscr{R},\mathscr{S})}$*.*

Proof To prove (1), note that if $A \in \mathscr{E}_{\mathscr{R}^{-1}\circ\mathscr{R}}$, then by Theorem 33 we have $\operatorname{Int}^{-1}_{(\mathscr{R},\mathscr{S})}(A^c) = \mathscr{P}(X)$. Hence, in particular it follows that $A^c \in \operatorname{Int}^{-1}_{(\mathscr{R},\mathscr{S})}(A^c)$, and thus $A^c \in \operatorname{Int}_{(\mathscr{R},\mathscr{S})}(A^c)$. Therefore, by Definition 5 and Theorem 68, we have $A^c \in \tau_{(\mathscr{R},\mathscr{S})}$, and thus $A \in \tau\!\!\!-_{(\mathscr{R},\mathscr{S})}$.

12 Topologically Open and Closed Sets Derived from Ordered Pairs of Relators

By using Definition 3, we may also naturally introduce the following

Definition 6 For any $A \subseteq X$, we write

(1) $A \in \mathscr{F}_{(\mathscr{R},\mathscr{S})}$ if $\operatorname{cl}_{(\mathscr{R},\mathscr{S})}(A) \subseteq A$;
(2) $A \in \mathscr{T}_{(\mathscr{R},\mathscr{S})}$ if $A \subseteq \operatorname{int}_{(\mathscr{R},\mathscr{S})}(A)$.

The members of the families $\mathscr{T}_{(\mathscr{R},\mathscr{S})}$ and $\mathscr{F}_{(\mathscr{R},\mathscr{S})}$ will be called the *topologically open and closed sets* generated by the pair $(\mathscr{R},\,\mathscr{S})$ of relators, respectively.

Thus, by using the results of Section 9, we can easily prove the following theorems.

Theorem 75 *For any* $A \subseteq X$*, the following assertions are equivalent:*

(1) $A \in \mathscr{T}_{(\mathscr{R},\mathscr{S})}$*;*
(2) for all $x \in A$ *we have* $x \in \operatorname{int}_{(\mathscr{R},\mathscr{S})}(A)$*;*
(3) for all $x \in A$ *there exists* $R \in \mathscr{R}$ *such that* $R(x)^c \cup A \in \mathscr{E}_{\mathscr{S}}$*;*
(4) for all $x \in A$ *there exists* $R \in \mathscr{R}$ *such that* $\operatorname{int}_{\mathscr{S}}\big(R(x)^c \cup A\big) \neq \emptyset$*;*
(5) for all $x \in A$ *there exist* $u \in X$*,* $R \in \mathscr{R}$ *and* $S \in \mathscr{S}$ *such that* $S(u) \subseteq R(x)^c \cup A$*;*
(6) for all $x \in A$ *there exist* $u \in X$*,* $R \in \mathscr{R}$ *and* $S \in \mathscr{S}$ *such that for all* $v \in S(u)$ *we have either* $v \in A$ *or* $v \notin R(x)$*.*

Theorem 76 *For any* $A \subseteq Y$, *the following assertions are equivalent:*

(1) $A \in \mathscr{F}_{(\mathscr{R},\mathscr{S})}$;
(2) for all $x \in \operatorname{cl}_{(\mathscr{R},\mathscr{S})}(A)$ *we have* $x \in A$;
(3) if $x \in X$ *such that for all* $R \in \mathscr{R}$ *we have* $R(x) \cap A \in \mathscr{D}_{\mathscr{S}}$, *then* $x \in A$;
(4) if $x \in X$ *such that for all* $R \in \mathscr{R}$ *we have* $\operatorname{cl}_{\mathscr{S}}\big(R(x) \cap A\big) = X$, *then* $x \in A$;
(5) if $x \in X$ *such that for all* $u \in X$, $R \in \mathscr{R}$ *and* $S \in \mathscr{S}$ *we have* $S(u) \cap R(x) \cap A \neq \emptyset$, *then* $x \in A$;
(6) if $x \in X$ *such that for all* $u \in X$, $R \in \mathscr{R}$ *and* $S \in \mathscr{S}$ *there exists* $v \in S(u)$ *such that* $v \in A$ *and* $v \in R(x)$, *then* $x \in A$.

Theorem 77 *For any* $A \subseteq X$, *we have*

(1) $A \in \mathscr{F}_{(\mathscr{R},\mathscr{S})} \iff A^c \in \mathscr{T}_{(\mathscr{R},\mathscr{S})}$;
(2) $A \in \mathscr{T}_{(\mathscr{R},\mathscr{S})} \iff A^c \in \mathscr{F}_{(\mathscr{R},\mathscr{S})}$.

Theorem 78 *We have*

(1) $X \in \mathscr{T}_{(\mathscr{R},\mathscr{S})}$ *if* $\mathscr{R}, \mathscr{S} \neq \emptyset$;
(2) $\mathscr{A} \subseteq \mathscr{T}_{(\mathscr{R},\mathscr{S})}$ *implies* $\bigcup \mathscr{A} \in \mathscr{T}_{(\mathscr{R},\mathscr{S})}$.

Theorem 79 *We have*

(1) $\emptyset \in \mathscr{F}_{(\mathscr{R},\mathscr{S})}$ *if* $\mathscr{R}, \mathscr{S} \neq \emptyset$;
(2) $\mathscr{A} \subseteq \mathscr{F}_{(\mathscr{R},\mathscr{S})}$ *implies* $\bigcap \mathscr{A} \in \mathscr{F}_{(\mathscr{R},\mathscr{S})}$.

From the latter two theorems, by taking $\mathscr{A} = \emptyset$, we can infer

Corollary 21 *We have*

(1) $\emptyset \in \mathscr{T}_{(\mathscr{R},\mathscr{S})}$;
(2) $X \in \mathscr{F}_{(\mathscr{R},\mathscr{S})}$.

Remark 30 In this respect, it is also worth noticing that if either $\mathscr{S}$ or $\mathscr{R}$ is empty, then $\mathscr{T}_{(\mathscr{R},\mathscr{S})} = \{\emptyset\}$ and $\mathscr{F}_{(\mathscr{R},\mathscr{S})} = \{X\}$.

Theorem 80 *We have*

(1) $\tau_{(\mathscr{R},\mathscr{S})} \subseteq \mathscr{T}_{(\mathscr{R},\mathscr{S})}$;
(2) $\bar{\tau}_{(\mathscr{R},\mathscr{S})} \subseteq \mathscr{F}_{(\mathscr{R},\mathscr{S})}$.

Proof To prove (1), note that, by Theorem 38 and Definitions 5 and 6,

$$A \in \tau_{(\mathscr{R},\mathscr{S})} \implies A \in \operatorname{Int}_{(\mathscr{R},\mathscr{S})}(A) \implies A \subseteq \operatorname{int}_{(\mathscr{R},\mathscr{S})}(A) \implies A \in \mathscr{T}_{(\mathscr{R},\mathscr{S})}.$$

Therefore, assertion (1) is true.

Remark 31 To feel the difference between proximally and topologically open sets, note that by Theorems 75 and 67, for any $A \subseteq X$, we have

(1) $A \in \mathcal{T}_{(\mathcal{R}, \mathcal{S})}$ if and only if for all $x \in A$ there exist $u_x \in X$, $R_x \in \mathcal{R}$ and $S_x \in \mathcal{S}$ such that $S_x(u_x) \subseteq R_x(x)^c \cup A$;

(2) $A \in \tau_{(\mathcal{R}, \mathcal{S})}$ if and only if there exist $u \in X$, $R \in \mathcal{R}$ and $S \in \mathcal{S}$ such that $S(u) \subseteq R[A]^c \cup A$, or equivalently $S(u) \subseteq R(x)^c \cup A$ for all $x \in A$.

Namely, by the corresponding definitions and De Morgan's law, we have

$$R[A]^c \cup A = \Big(\bigcup_{x \in A} R(x) \Big)^c \cup A = \Big(\bigcap_{x \in A} R(x)^c \Big) \cup A = \bigcap_{x \in A} R(x)^c \cup A.$$

Now, analogously to the Theorems 10 and 11, we can also easily establish the following two theorems.

Theorem 81 *We have*

(1) $\mathcal{T}_{(\mathcal{R}, \mathcal{S})} \setminus \{\emptyset\} \subseteq \mathcal{E}_{(\mathcal{R}, \mathcal{S})}$;
(2) $\mathcal{D}_{(\mathcal{R}, \mathcal{S})} \cap \mathcal{F}_{(\mathcal{R}, \mathcal{S})} \subseteq \{X\}$.

Corollary 22 *We have*

(1) $\mathcal{F}_{(\mathcal{R}, \mathcal{S})} \subseteq \big(\mathcal{D}_{(\mathcal{R}, \mathcal{S})} \big)^c \cup \{X\}$;
(2) $\mathcal{D}_{(\mathcal{R}, \mathcal{S})} \subseteq \big(\mathcal{F}_{(\mathcal{R}, \mathcal{S})} \big)^c \cup \{X\}$.

Theorem 82 *For any* $A \subseteq X$ *we have*

(1) $A \in \mathcal{E}_{(\mathcal{R}, \mathcal{S})}$ *if* $V \subseteq A$ *for some* $V \in \mathcal{T}_{(\mathcal{R}, \mathcal{S})} \setminus \{\emptyset\}$;
(2) $A \in \mathcal{D}_{(\mathcal{R}, \mathcal{S})}$ *only if* $A \setminus W \neq \emptyset$ *for all* $W \in \mathcal{F}_{(\mathcal{R}, \mathcal{S})} \setminus \{X\}$.

However, instead of the analogues of Corollaries 19 and 20, we can only prove the following two theorems.

Theorem 83 *The mappings* $\mathcal{R} \mapsto \mathcal{T}_{(\mathcal{R}, \mathcal{S})}$ *and* $\mathcal{R} \mapsto \mathcal{F}_{(\mathcal{R}, \mathcal{S})}$ *are increasing.*

Corollary 23 *We have*

(1) $\bigcup_{R \in \mathcal{R}} \mathcal{T}_{(R, \mathcal{S})} \subseteq \mathcal{T}_{(\mathcal{R}, \mathcal{S})}$;
(2) $\bigcup_{R \in \mathcal{R}} \mathcal{F}_{(R, \mathcal{S})} \subseteq \mathcal{F}_{(\mathcal{R}, \mathcal{S})}$.

Theorem 84 *The mappings* $\mathcal{S} \mapsto \mathcal{T}_{(\mathcal{R}, \mathcal{S})}$ *and* $\mathcal{S} \mapsto \mathcal{F}_{(\mathcal{R}, \mathcal{S})}$ *are increasing.*

Corollary 24 *We have*

(1) $\bigcup_{S \in \mathcal{S}} \mathcal{T}_{(\mathcal{R}, S)} \subseteq \mathcal{T}_{(\mathcal{R}, \mathcal{S})}$;
(2) $\bigcup_{S \in \mathcal{S}} \mathcal{F}_{(\mathcal{R}, S)} \subseteq \mathcal{F}_{(\mathcal{R}, \mathcal{S})}$.

Now, for instance, by using Corollaries 14 and 15 we can also prove the following two theorems.

Theorem 85 *If in particular* $\mathscr{S} = \{X^2\}$, *then*

(1) $\mathscr{T}_{\mathscr{R}} = \mathscr{T}_{(\mathscr{R}, \mathscr{S})}$;
(2) $\mathscr{F}_{\mathscr{R}} = \mathscr{F}_{(\mathscr{R}, \mathscr{S})}$.

Theorem 86 *If in particular for every* $v \in X$ *there exist* $u \in X$ *and* $S \in \mathscr{S}$ *such that* $S(u) = \{v\}$, *then for any* $A \subseteq X$ *we have*

(1) $A \in \mathscr{T}_{(\mathscr{R}, \mathscr{S})}$ *if and only if either* $A \neq \emptyset$ *or for all* $x \in X$ *there exists* $R \in \mathscr{R}$ *such that* $R(x) \neq X$;
(2) $A \in \mathscr{F}_{(\mathscr{R}, \mathscr{S})}$ *if and only if either* $A \neq X$ *or for all* $x \in A^c$ *there exists* $R \in \mathscr{R}$ *such that* $R(x) \neq X$.

Moreover, for instance, by using Theorem 51, we can also prove the following

Theorem 87 *If in particular* $\mathscr{R} \subseteq \mathscr{S}$, *then for any* $A \subseteq X$ *the following assertions hold:*

(1) *if for all* $x \in A$ *there exist* $R, U \in \mathscr{R}$ *such that* $\left(R^{-1} \circ U\right)(x) \neq X$, *then* $A \in \mathscr{T}_{(\mathscr{R}, \mathscr{S})}$;
(2) *if* $A \notin \mathscr{F}_{(\mathscr{R}, \mathscr{S})}$, *then there exists* $x \in A^c$ *such that for all* $R, U \in \mathscr{R}$ *we have* $\left(R^{-1} \circ U\right)(x) = X$.

13 The Real Line as a Prime Example for Birelator Spaces

Notation 4 *Let* $X = \mathbb{R}$, *and for all* $r > 0$ *define*

$$B_r = \left\{(x, y) \in X^2 : \ d(x, y) < r\right\} \quad \text{and} \quad S = \left\{(x, y) \in X^2 : \ x \leq y\right\}.$$

Moreover, define

$$\mathscr{R} = \left\{B_r\right\}_{r>0} \qquad \text{and} \qquad \mathscr{S} = \{S\}.$$

Remark 32 Then, for all $x \in X$ and $r > 0$, we have

$$B_r(x) = \left]x - r, \ x + r\right[\qquad \text{and} \qquad S(x) = \left[x, \ +\infty\right[.$$

Moreover, we can note that $X(\mathscr{R}, \mathscr{S})$ is the most plausible birelator space for the illustration of our present ideas.

However, to establish some basic properties of the pair $(\mathscr{R}, \mathscr{S})$ of relators, it is convenient to list first some of those of the relators $\mathscr{R}$ and $\mathscr{S}$.

First of all, by using the notation

$$d(A, B) = \inf\left\{d(a, b) : \ a \in A, \ b \in B\right\}$$

for $A, B \subseteq X$, we shall prove the following

Theorem 88 *For any* $A, B \subseteq X$, *the following assertions are equivalent:*

(1) $d(A, B) = 0$;
(2) $A \in \mathrm{Cl}_{\mathscr{R}}(B)$.

Proof To prove the implication (1) $\Longrightarrow$ (2), note that if (1) holds, then for any $r > 0$ we have $d(A, B) < r$. Therefore, by the corresponding property of the infimum, there exist $a \in A$ and $b \in B$ such that $d(a, b) < r$, and thus $b \in B_r(a)$. Therefore, $B_r[A] \cap B \neq \emptyset$, and thus assertion (2) also holds.

Now, because of the corresponding definitions, we can also state

Corollary 25 *For any* $a \in X$ *and* $B \subseteq X$, *the following assertions are equivalent:*

(1) $d(a, B) = 0$;
(2) $a \in \mathrm{cl}_{\mathscr{R}}(B)$.

To prove some analogous results for the relator $\mathscr{S}$, we have to use the first type induced relation of Mitrinović and Berković [61].

Definition 7 For any $A, B \subseteq X$, we write $A \preceq B$ if $a \leq b$ for some $a \in A$ and $b \in B$.

Remark 33 Thus, $\preceq$ is closely related to a more natural inequality $\leq$ defined on $\mathscr{P}(X)$ such that, for any $A, B \subseteq X$, we write $A \leq B$ if $a \leq b$ for all $a \in A$ and $b \in B$.

More concretely, we can see that if $A \leq B$ and $A, B \neq \emptyset$, then $A \preceq B$. Moreover, if $A \not\leq B$, then there exist $a \in A$ and $b \in B$ such that $a \not\leq b$, and thus $b \leq a$. Therefore, $B \preceq A$.

However, it is now more important to note that thus we can also prove

Theorem 89 *For any* $A, B \subseteq X$, *the following assertions are equivalent:*

(1) $A \preceq B$;
(2) $A \in \mathrm{Cl}_{\mathscr{S}}(B)$.

Proof By the corresponding definitions, it is clear that

$$A \preceq B \iff \exists\, a \in A,\ b \in B:\ a \leq b$$

$$\iff \exists\, a \in A,\ b \in B:\ b \in S(a) \iff S[A] \cap B \neq \emptyset \iff A \in \mathrm{Cl}_{\mathscr{S}}(B).$$

Remark 34 In this respect, it is also worth noticing that, by the corresponding definitions, for any $A, B \subseteq X$ we have

$$A \leq B \iff \forall\, a \in A,\ b \in B:\ a \leq b$$

$$\Longleftrightarrow \ \forall\ a \in A\,,\ \ b \in B: \quad (a,\,b) \in S \Longleftrightarrow A \times B \subseteq S \Longleftrightarrow A \in \mathrm{Lb}_{\mathscr{S}}(B).$$

Now, because of the corresponding definitions, we can also state

Corollary 26 *For any $a \in X$ and $B \subseteq X$, the following assertions are equivalent:*

(1) $a \preceq B$*;*
(2) $a \in \mathrm{cl}_{\mathscr{S}}(B)$.

In the sequel, we shall also need the following

Notation 5 *For any $A \subseteq X$, define*

$$\mathrm{ub}\,(A) = \big\{ x \in X : \quad \forall\ a \in A : \quad a \leq x \big\}.$$

Moreover, define

$$\mathscr{U} = \big\{ A \subseteq X : \quad \mathrm{ub}\,(A) \neq \emptyset \big\}.$$

Remark 35 Thus, $\mathrm{ub}\,(A)$ is the family of all upper bounds of A in X. And, $\mathscr{U}$ is the family of all upper-bounded subsets of X.

Concerning the family $\mathscr{U}$, we can easily establish the following

Theorem 90 *We have*

(1) $\emptyset \in \mathscr{U}$ *and* $X \notin \mathscr{U}$*;*
(2) $A \subseteq B \in \mathscr{U}$ *implies* $A \in \mathscr{U}$*;*
(3) $A,\ B \in \mathscr{U}$ *implies* $A \cup B \in \mathscr{U}$.

Remark 36 Therefore, $\mathscr{U}$ is a proper ideal on X, and thus

$$\mathscr{U}^{\,c} = \{ A^c : \ \ A \in \mathscr{U} \},$$

which should not be confused now with $\mathscr{P}(X) \setminus \mathscr{U}$, is a proper filter on X.

Moreover, having in mind the extended real line $\overline{\mathbb{R}} = \mathbb{R} \cup \{-\infty,\ +\infty\}$, we can also note that

$$\mathscr{U} = \big\{ A \subseteq X : \quad \sup\,(A) < +\infty \}.$$

Now, for instance, by using Theorem 26, we can easily prove the following

Theorem 91 *For any $A,\ B \subseteq X$, the following assertions are equivalent:*

(1) $A \in \mathrm{Cl}_{(\mathscr{R},\,\mathscr{S})}(B)$*;*
(2) for all $x \in X$ and $r > 0$, there exist $a \in A$ and $b \in B$ such that

$$d\,(a,\,b) < r \qquad\qquad \text{and} \qquad\qquad x \leq b.$$

Proof To prove the implication (1) $\Longrightarrow$ (2), note that if (1) holds, then by Theorem 26, for any $x \in X$ and $r > 0$, there exist $a \in A$ and $b \in B$ such that

$$b \in B_r(a) \qquad \text{and} \qquad b \in S(x).$$

Therefore, $d(a, b) < r$ and $x \leq b$. Thus, assertion (2) also holds.

Now, by using Theorems 91 and 19, we can also prove

Corollary 27 *For any* $A \in \mathscr{D}_{\mathscr{R}}$ *and* $B \in \mathscr{P}(X) \setminus \mathscr{U}$, *we have*

(1) $A \in \mathrm{Cl}_{(\mathscr{R}, \mathscr{S})}(B)$;
(2) $A \notin \mathrm{Int}_{(\mathscr{R}, \mathscr{S})}(B^c)$.

Proof To prove (1), note that if $x \in X$, then because of $B \notin \mathscr{U}$ there exists $b \in B$ such that $x \leq b$. Moreover, if $r > 0$, then because of $A \in \mathscr{D}_{\mathscr{R}}$, there exists $a \in A$ such that $d(a, b) < r$. Therefore, by Theorem 91, $A \in \mathrm{Cl}_{(\mathscr{R}, \mathscr{S})}(B)$ also holds.

Remark 37 Thus, in particular, we can state that

$$\mathbb{Q} \in \mathrm{Cl}_{(\mathscr{R}, \mathscr{S})}(\mathbb{N}) \qquad \text{and} \qquad \mathbb{Q}^c \in \mathrm{Cl}_{(\mathscr{R}, \mathscr{S})}(\mathbb{Q}).$$

Moreover, by using Theorem 91, we can also easily prove the following

Theorem 92 *If* $A, B \subseteq X$, *such that* $A \in \mathrm{Cl}_{(\mathscr{R}, \mathscr{S})}(B)$, *then*

(1) $A \in \mathrm{Cl}_{\mathscr{R}}(B)$;
(2) $d(A, B) = 0$;
(3) $B \notin \mathscr{U}$.

Proof From Theorem 88, we know that assertions (1) and (2) are equivalent. Moreover, if $A \in \mathrm{Cl}_{(\mathscr{R}, \mathscr{S})}(B)$ holds, then by Theorem 91, for any $x \in X$ and $r > 0$, there exist $a \in A$ and $b \in B$ such that

$$d(a, b) < r \qquad \text{and} \qquad x \leq b.$$

Hence, by the definitions of $d(A, B)$ and $\sup(B)$, it is clear that

$$d(A, B) < r \qquad \text{and} \qquad x \leq \sup(B).$$

Therefore, we actually have

$$d(A, B) = 0 \qquad \text{and} \qquad \sup(B) = +\infty.$$

Thus, assertions (2) and (3) also hold.

Remark 38 Unfortunately, assertions (2) and (3) do not, in general, imply that $A \in \mathrm{Cl}_{(\mathscr{R}, \mathscr{S})}(B)$.

Namely, if they hold, then we can only state that, for any $x \in X$ and $r > 0$, there exist $a \in A$ and $b_1, b_2 \in B$ such that $d(a, b_1) < r$ and $x \leq b_2$. Therefore, to infer the required conclusion, by Theorem 91, we had to assume that B is a singleton, and thus $b_1 = b_2$.

To be more concrete, for instance, we can note that $d(\{1\}, \mathbb{N}) = 0$ and $\mathbb{N} \notin \mathscr{U}$, but $\{1\} \notin \mathrm{Cl}_{(\mathscr{R}, \mathscr{S})}(\mathbb{N})$. Namely, if this not the case, then by Theorem 91 there exists $n \in \mathbb{N}$ such that $d(1, n) < 1/2$ and $2 \leq n$. Therefore, $n = 1$, and thus $2 \leq 1$, which is a contradiction.

Now, as an immediate consequence of Theorems 92 and 19, we can state

Corollary 28 *For any $B \in \mathscr{U}$, we have*

(1) $\mathrm{Cl}_{(\mathscr{R}, \mathscr{S})}(B) = \emptyset$;
(2) $\mathrm{Int}_{(\mathscr{R}, \mathscr{S})}\left(B^c\right) = \mathscr{P}(X)$.

Remark 39 Thus, in particular, we can also state that

$$\mathrm{Cl}_{(\mathscr{R}, \mathscr{S})}\left(B_r(x)\right) = \emptyset \qquad \text{and} \qquad \mathrm{Int}_{(\mathscr{R}, \mathscr{S})}\left(S(x)\right) = \mathscr{P}(X)$$

for all $x \in X$ and $r > 0$. Namely, the sets $B_r(x)$ and $S(x)^c$ are in $\mathscr{U}$.

By using Theorems 91 and 19, we can also easily prove the following

Theorem 93 *For any $B \subseteq X$, we have*

(1) $\mathrm{cl}_{(\mathscr{R}, \mathscr{S})}(B) = \emptyset$;
(2) $\mathrm{int}_{(\mathscr{R}, \mathscr{S})}(B) = X$.

Proof To prove (1), note that if $a \in \mathrm{cl}_{(\mathscr{R}, \mathscr{S})}(B)$, then by Definition 3 we have $\{a\} \in \mathrm{Cl}_{(\mathscr{R}, \mathscr{S})}(B)$. Thus, by Theorem 91, there exists $b \in B$ such that

$$d(a, b) < 1 \qquad \text{and} \qquad a + 1 \leq b.$$

This implies that $b < b$, which is contradiction.

From this theorem, by Definitions 4 and 6, we can immediately derive

Corollary 29 *We have*

(1) $\mathscr{D}_{(\mathscr{R}, \mathscr{S})} = \emptyset$;
(2) $\mathscr{E}_{(\mathscr{R}, \mathscr{S})} = \mathscr{P}(X)$;
(3) $\mathscr{F}_{(\mathscr{R}, \mathscr{S})} = \mathscr{P}(X)$;
(4) $\mathscr{T}_{(\mathscr{R}, \mathscr{S})} = \mathscr{P}(X)$.

From Theorem 91, by using Definition 5, we can also immediately derive

Theorem 94 *For any* $A \subseteq X$, *the following assertions are equivalent:*

(1) $A \in \tau\!\!\!-_{(\mathscr{R},\,\mathscr{S})}$;
(2) there exist $x \in X$ *and* $r > 0$ *such that for any* $a \in A$ *and* $b \in A^c$ *we have either*

$$d(a,\,b) \geq r \qquad \text{or} \qquad a < x.$$

Remark 40 From Remark 37, by Definition 5 and Theorem 68, we can once see that

$$\mathbb{Q} \notin \tau\!\!\!-_{(\mathscr{R},\,\mathscr{S})} \qquad \text{and} \qquad \mathbb{Q}^c \notin \tau_{(\mathscr{R},\,\mathscr{S})}.$$

However, to derive the latter statements from Theorem 94, it seems necessary to use that $\mathbb{Q}$ is countable.

Moreover, from Corollary 28, by using Definition 5 and Theorems 69 and 68, we can immediately derive the following

Theorem 95 *We have*

(1) $\mathscr{U} \cup \{X\} \subseteq \tau\!\!\!-_{(\mathscr{R},\,\mathscr{S})}$;
(2) $\{\emptyset\} \cup \mathscr{U}^{\,c} \subseteq \tau_{(\mathscr{R},\,\mathscr{S})}$.

Proof To prove (1), note that by Theorem 69 we have $X \in \tau\!\!\!-_{(\mathscr{R},\,\mathscr{S})}$. Moreover, if $B \in \mathscr{U}$, then by Corollary 28 we have $\mathrm{Cl}_{(\mathscr{R},\,\mathscr{S})}(B) = \emptyset$. Thus, in particular $B^c \notin \mathrm{Cl}_{(\mathscr{R},\,\mathscr{S})}(B)$. Therefore, by Definition 5, $B \in \tau\!\!\!-_{(\mathscr{R},\,\mathscr{S})}$ also holds.

Remark 41 From the proper well-chainedness of $\mathscr{R}$ and [72, Theorem 12.8], we know that $\tau_{\mathscr{R}} = \{\emptyset,\, X\}$.

Moreover, by using the corresponding definitions, we can see that for any $A \subseteq X$ we have $A \in \tau_{\mathscr{S}}$ if and only if $S(a) \subseteq A$, i.e., $[\,a, +\infty\,[\, \subseteq A$ for all $a \in A$. Therefore, A is either the empty set or an interval with right endpoint $+\infty$.

Acknowledgements The work of the author has been supported by the Hungarian Scientific Research Fund (OTKA) Grant K-111651.

The author is indebted to Amr Zakaria for calling his attention to the theory of ideal topological spaces and their modifications. (His joint papers [40, 41] contain closely related ideas.)

References

1. N. Ajmal, J.K. Kohli, Properties of hyperconnected spaces, their mappings into Hausdorff spaces and embeddings into hyperconnected spaces. Acta Math. Hungar. **60**, 41–49 (1992)
2. P. Alexandroff, Zur Begründung der n-dimensionalen mengentheoretischen Topologie. Math. Ann. **94**, 296–308 (1925)

3. Z. Balogh, Relative compactness and recent common generalizations of metric and locally compact spaces. Fund. Math. **100**, 165–177 (1978). Proceedings of the Fourth Prague Topological Symposium, 1976, 37–44
4. K. Bhavani, D. Sivaraj, On $\mathfrak{I}$-hyperconnected spaces. Bull. Allahabad Math. Soc. **29**, 15–25 (2014)
5. C.R. Borges, Hyperconnectivity of hyperspaces. Math. Jap. **30**, 757–761 (1985)
6. M.K. Bose, R. Tiwari, (ω)Topological connectedness and hyperconnectedness. Note Mat. **31**, 93–101 (2011)
7. N. Bourbaki, *General Topology*, Chapters 1–4 (Springer, Berlin, 1989)
8. E. Čech, *Topological Spaces* (Academia, Prague, 1966)
9. Á. Császár, *Foundations of General Topology* (Pergamon Press, London, 1963)
10. Á. Császár, Generalized topology, generalized continuity. Acta Math. Hungar. **96**, 351–357 (2002)
11. Á. Császár, Modification of generalized topologies via hereditary classes. Acta Math. Hungar. **115**, 29–36 (2007)
12. Á. Császár, Remarks on quasi-topologies. Acta Math. Hungar. **119**, 197–200 (2008)
13. Á. Császár, Mixed constructions for generalized topologies. Acta Math. Hungar. **122**, 153–159 (2009)
14. Á. Császár, Weak structures. Acta Math. Hungar. **131**, 193–195 (2011)
15. B.A. Davey, H.A. Priestley, *Introduction to Lattices and Order* (Cambridge University Press, Cambridge, 2002)
16. A.S. Davis, Indexed systems of neighborhoods for general topological spaces. Am. Math. Mon. **68**, 886–893 (1961)
17. D. Doičinov, A unified theory of topological spaces, proximity spaces and uniform spaces. Dokl. Acad. Nauk SSSR **156**, 21–24 (1964) (Russian)
18. V.A. Efremovič, The geometry of proximity. Mat. Sb. **31**, 189–200 (1952) (Russian)
19. V.A. Efremović, A.S. Švarc, A new definition of uniform spaces. Metrization of proximity spaces. Dokl. Acad. Nauk. SSSR **89**, 393–396 (1953) (Russian)
20. E. Ekici, Generalized hyperconnectedness. Acta Math. Hungar. **133**, 140–147 (2011)
21. E. Ekici, Further new generalized topologies via mixed constructions due to Császár. Math. Bohem. **140**, 1–9 (2015)
22. E. Ekici, T. Noiri, $*$-Hyperconnected ideal topological spaces. An. Stiint. Univ. Al. I. Cuza Iasi **58**, 121–129 (2012)
23. P. Fletcher, W.F. Lindgren, *Quasi-Uniform Spaces* (Marcel Dekker, New York, 1982)
24. P. Flether, H.B. Hoyle, C.W. Patty, The comparison of topologies. Duke Math. J. **36**, 325–331 (1969)
25. G.B. Folland, A tale of topology. Am. Math. Mon. **117**, 663–672 (2010)
26. G. Freud, Ein Beitrag zu dem Satze von Cantor und Bendixson. Acta Math. Hungar. **9**, 333–336 (1958)
27. B. Ganter, R. Wille, *Formal Concept Analysis* (Springer, Berlin, 1999)
28. B. Garai, C. Bandyopadhyay, On pairwise hyperconnected spaces. Soochow J. Math. **27**, 391–399 (2001)
29. T. Glavosits, Generated preorders and equivalences. Acta Acad. Paed. Agrienses, Sect. Math. **29**, 95–103 (2002)
30. T.R. Hamlett, D. Janković, Ideals in topological spaces and the set operator ψ. Bollettino U.M.I. **(7) 4-B**, 863–874 (1990)
31. H. Hashimoto, On the $*$topology and its applications. Fund. Math. **91**, 5–10 (1976)
32. F. Hausdorff, *Grundzüge der Mengenlehre* (De Gruyter, Berlin, 1914)
33. E. Hayashi, Topologies defined by local properties. Math. Ann. **156**, 205–215 (1964)
34. H. Herrlich, Topological structures. Math. Centre Tracts **52**, 59–122 (1974)
35. U. Höhle, T. Kubiak, On regularity of sup-preserving maps: generalizing Zareckiĭ's theorem. Semigroup Forum **83**, 313–319 (2011)
36. W. Hunsaker, W. Lindgren, Construction of quasi-uniformities. Math. Ann. **188**, 39–42 (1970)
37. J.R. Isbell, *Uniform Spaces* (American Mathematical Society, Providence, 1964)

38. S. Jafari, N. Rajesh, Generalized closed sets with respect to an ideal. Eur. J. Pure Appl. Math. **4**, 147–151 (2011)
39. D. Janković, T.R. Hamlett, New topologies from old via ideals. Am. Math. Mon. **97**, 295–310 (1990)
40. A. Kandil, M.M. Yakout, A. Zakaria, Generalized rough sets via ideals. Ann. Fuzzy Math. Inf. **5**, 525–532 (2013)
41. A. Kandil, O.A. El-Tantawy, S.A. El-Sheikh, A. Zakaria, On $\mathscr{I}$-proximity spaces. Appl. Math. Inf. Sci. Lett. **4**, 79–84 (2016)
42. J.L. Kelley, *General Topology* (Van Nostrand Reinhold Company, New York, 1955)
43. J.C. Kelly, Bitopological spaces. Proc. Lond. Math. Soc. **13**, 71–89 (1963)
44. H. Kenyon, Two theorems about relations. Trans. Am. Math. Soc. **107**, 1–9 (1963)
45. H.-J. Kowalsky, *Topologische Räumen* (Birkhäuser, Basel, 1960)
46. M.K.R.S.V. Kumar, Hyperconnected type spaces. Acta Cienc. Indica Math. **31**, 273–275 (2005)
47. K. Kuratowski, *Topologie I* (Warszawa, 1933) (Revised and augmented edition: Topology I. Academic, New York, 1966)
48. J. Kurdics, Á. Száz, Well-chained relator spaces. Kyungpook Math. J. **32**, 263–271 (1992)
49. E.P. Lane, Bitopological spaces and quasi-uniform spaces. Proc. Lond. Math. Soc. **17**, 241–256 (1967)
50. E.P. Lane, Quasi-proximities and bitopological spaces. Portugal. Math. **28**, 151–159 (1969)
51. N. Levine, Dense topologies. Am. Math. Mon. **75**, 847–852 (1968)
52. N. Levine, On uniformities generated by equivalence relations. Rend. Circ. Mat. Palermo **18**, 62–70 (1969)
53. N. Levine, On Pervin's quasi uniformity. Math. J. Okayama Univ. **14**, 97–102 (1970)
54. S. Lugojan, Topologii generalizate. Stud. Cerc. Mat. **34**, 348–360 (1982)
55. H. Maki, On generalizing semi-open and preopen sets. Report for Meeting on Topological Spaces Theory and its Applications, Yatsushiro College of Technology (1996), pp. 13–18
56. J. Mala, Relators generating the same generalized topology. Acta Math. Hungar. **60**, 291–297 (1992)
57. J. Mala, Á. Száz, Modifications of relators. Acta Math. Hungar. **77**, 69–81 (1997)
58. R. Manoharan, P. Thangevalu, Some new sets and topologies in ideal topological spaces. Chin. J. Math. **2013**, 6 pp. (2013)
59. N.F.G. Martin, Generalized condensation point. Duke Math. J. **28**, 507–514 (1961)
60. P.M. Mathew, On hyperconnected spaces. Indian J. Pure Appl. Math. **19**, 1180–1184 (1988)
61. Ž.M. Mitrinović, I.F. Berković, The induced relations on a power set. Filomat (Niš) **9**, 857–865 (1995)
62. S. Modak, Some new topologies on ideal topological spaces. Proc. Natl. Acad. Sci. India **82**, 233–243 (2012)
63. S. Modak, Topology of grill filter space and continuity. Bol. Soc. Paran. Mat. **31**, 219–230 (2013) .
64. S. Modak, C. Bandyopadhyay, A note on ψ-operator. Bull. Malays. Math. Sci. Soc. **30**, 43–48 (2007)
65. S.A. Naimpally, B.D. Warrack, *Proximity Spaces* (Cambridge University Press, Cambridge, 1970)
66. H. Nakano, K. Nakano, Connector theory. Pac. J. Math. **56**, 195–213 (1975)
67. T. Noiri, A note on hyperconnected sets. Mat. Vesn. **3**, 53–60 (1979)
68. T. Noiri, Hyperconnectedness and preopen sets. Rev. Roum. Math. Pures Appl. **29**, 329–334 (1984)
69. T. Noiri, Properties of hyperconnected spaces. Acta Math. Hungar. **66**, 147–154 (1995)
70. G. Pataki, Supplementary notes to the theory of simple relators. Radovi Mat. **9**, 101–118 (1999)
71. G. Pataki, On the extensions, refinements and modifications of relators. Math. Balk. **15**, 155–186 (2001)

72. G. Pataki, Á. Száz, A unified treatment of well-chainedness and connectedness properties. Acta Math. Acad. Paedagog. Nyházi. (N.S.) **19**, 101–165 (2003)
73. C.W. Patty, Bitopological spaces. Duke Math. J. **34**, 387–391 (1967)
74. W.J. Pervin, Uniformizations of neighborhood axioms. Math. Ann. **147**, 313–315 (1962)
75. W.J. Pervin, Quasi-uniformization of topological spaces. Math. Ann. **147**, 316–317 (1962)
76. W.J. Pervin, Quasi-proximities for topological spaces. Math. Ann. **150**, 325–326 (1963)
77. V. Renukadevi, On generalizations of hyperconnected spaces. J. Adv. Res. Pure Math. **4**, 46–58 (2012)
78. V. Renukadevi, Remarks on generalized hypperconnectedness. Acta Math. Hung. **136**, 157–164 (2012)
79. P. Samuels, A topology formed from a given topology and ideal. J. Lond. Math. Soc. **10**, 409–416 (1975)
80. A.K. Sharma, On some properties of hyperconnected spaces. Mat. Vesnik **14**, 25–27 (1977)
81. W. Sierpinski, *General Topology*. Mathematical Expositions, vol. 7 (University of Toronto Press, Toronto, 1956)
82. Yu.M. Smirnov, On proximity spaces. Math. Sb. **31**, 543–574 (1952) (Russian)
83. L.A. Steen, J.A. Seebach, *Counterexamples in Topology* (Springer, New York, 1970)
84. Á. Száz, Coherences instead of convergences, in *Proceedings of the Conference on Convergence and Generalized Functions (Katowice, 1983)* (Institute of Mathematics, Polish Academy of Sciences, Warsaw, 1984), pp. 141–148
85. Á. Száz, Basic tools and mild continuities in relator spaces. Acta Math. Hungar. **50**, 177–201 (1987)
86. Á. Száz, The fat and dense sets are more important than the open and closed ones, in *Abstracts of the Seventh Prague Topological Symposium*, Institute of Mathematics, Czechoslovak Academy of Science, 1991, p. 106
87. Á. Száz, Relators, Nets and Integrals. Unfinished doctoral thesis, Debrecen, 1991, 126 pp.
88. Á. Száz, Structures derivable from relators. Singularité **3**, 14–30 (1992)
89. Á. Száz, Refinements of relators. Tech. Rep., Inst. Math., Univ. Debrecen, 76 (1993), 19 pp.
90. Á. Száz, Cauchy nets and completeness in relator spaces. Colloq. Math. Soc. János Bolyai **55**, 479–489 (1993)
91. Á. Száz, Uniformly, proximally and topologically compact relators. Math. Pannon. **8**, 103–116 (1997)
92. Á. Száz, Somewhat continuity in a unified framework for continuities of relations. Tatra Mt. Math. Publ. **24**, 41–56 (2002)
93. Á. Száz, An extension of Baire's category theorem to relator spaces. Math. Morav. **7**, 73–89 (2003)
94. Á. Száz, Upper and lower bounds in relator spaces. Serdica Math. J. **29**, 239–270 (2003)
95. Á. Száz, Rare and meager sets in relator spaces. Tatra Mt. Math. Publ. **28**, 75–95 (2004)
96. Á. Száz, Galois-type connections on power sets and their applications to relators. Tech. Rep., Inst. Math., Univ. Debrecen, 2005/2 (2005), 38 pp.
97. Á. Száz, Supremum properties of Galois-type connections. Comment. Math. Univ. Carolin. **47**, 569–583 (2006)
98. Á. Száz, Minimal structures, generalized topologies, and ascending systems should not be studied without generalized uniformities. Filomat **21**, 87–97 (2007)
99. Á. Száz, Galois type connections and closure operations on preordered sets. Acta Math. Univ. Comen. **78**, 1–21 (2009)
100. Á. Száz, Applications of relations and relators in the extensions of stability theorems for homogeneous and additive functions. Aust. J. Math. Anal. Appl. **6**, 1–66 (2009)
101. Á. Száz, Foundations of the theory of vector relators. Adv. Stud. Contemp. Math. **20**, 139–195 (2010)
102. Á. Száz, Lower semicontinuity properties of relations in relator spaces. Adv. Stud. Contemp. Math. (Kyungshang) **23**, 107–158 (2013)
103. Á. Száz, Inclusions for compositions and box products of relations. J. Int. Math. Virtual Inst. **3**, 97–125 (2013)

104. Á. Száz, A particular Galois connection between relations and set functions. Acta Univ. Sapientiae, Math. **6**, 73–91 (2014)
105. Á. Száz, Generalizations of Galois and Pataki connections to relator spaces. J. Int. Math. Virtual Inst. **4**, 43–75 (2014)
106. Á. Száz, Galois and Pataki connections on generalized ordered sets. Tech. Rep., Inst. Math., Univ. Debrecen, 2014/3 (2014), 27 pp.
107. Á. Száz, Basic tools, increasing functions, and closure operations in generalized ordered sets, in *Contributions in Mathematics and Engineering: In Honor of Constantin Caratheodory*, ed. by P.M. Pardalos, Th.M. Rassias (Springer, Berlin, 2016), pp. 551–616
108. Á. Száz, A natural Galois connection between generalized norms and metrics. Acta Univ. Sapientiae Math. **9**, 360–373 (2017)
109. Á. Száz, Four general continuity properties, for pairs of functions, relations and relators, whose particular cases could be investigated by hundreds of mathematicians. Tech. Rep., Inst. Math., Univ. Debrecen, 2017/1 (2017), 17 pp.
110. Á. Száz, Contra continuity properties of relations in relator spaces. Tech. Rep., Inst. Math., Univ. Debrecen, 2017/5 (2017), 48 pp.
111. Á. Száz, Ideal topological spaces and their immediate generalizations should not also be studied without generalized uniformities, in *Summary of a Lecture at the Seminar on Analysis*, Debrecen, 2017, 4 pp.
112. Á. Száz, Relationships between inclusions for relations and inequalities for corelations. Math. Pannon. **26**, 15–31 (2017–2018)
113. Á. Száz, The closure-interior Galois connection and its applications to relational inclusions and equations. J. Int. Math. Virt. Inst. **8**, 181–224 (2018)
114. Á. Száz, A unifying framework for studying continuity, increasingness, and Galois connections. MathLab J. **1**, 154–173 (2018)
115. Á. Száz, Corelations are more powerful tools than relations, in *Applications of Nonlinear Analysis*, ed. by Th.M. Rassias. Springer Optimizations and Its Applications, vol. 134 (Springer, Berlin, 2018), pp. 711–779
116. Á. Száz, A. Zakaria, Mild continuity properties of relations and relators in relator spaces, in *Essays in Mathematics and Its Applications: In Honor of Vladimir Arnold*, ed. by P.M. Pardalos, Th.M. Rassias (Springer, Berlin, 2016), pp. 439–511
117. W.J. Thron, *Topological Structures* (Holt, Rinehart and Winston, New York, 1966)
118. W.J. Thron, Proximity structures and grills. Math. Ann. **206**, 35–62 (1973)
119. H. Tietze, Beiträge zur allgemeinen Topologie I. Axiome für verschiedene Fassungen des Umgebungsbegriffs. Math. Ann. **88**, 290–312 (1923)
120. J.W. Tukey, *Convergence and Uniformity in Topology*. Annals of Mathematics Studies, vol. 2 (Princeton University Press, Princeton, 1940)
121. R. Vaidyanathaswamy, The localization theory in set-topology. Proc. Indian Acad. Sci. **20**, 51–61 (1945)
122. A. Weil, Sur les espaces á structure uniforme et sur la topologie générale. Actual. Sci. Ind. vol. 551 (Herman and Cie, Paris, 1937)
123. J.D. Weston, On the comparison of topologies. J. Lond. Math. Soc. **32**, 342–354 (1957)
124. I. Zuzčák, Generalized topological spaces. Math. Slovaca **33**, 249–256 (1983)
125. I. Zvina, On i-topological spaces: generalization of the concept of a topological space via ideals. Appl. Gen. Top. **7**, 51–66 (2006)

PPF Dependent Fixed Points in Razumikhin Metrical Chains

Mihai Turinici

Abstract A functional extension is given for the (metrical) PPF dependent fixed point statement in Drici et al. (Nonlin. Anal. 67:641–647, 2007). Some technical aspects of the result in question are also being discussed.

AMS Subject Classification 47H10 (Primary), 54H25 (Secondary)

1 Introduction

Let X be a nonempty set. Call the subset Y of X, *almost singleton* (in short: *asingleton*), provided $[y_1, y_2 \in Y$ implies $y_1 = y_2]$; and *singleton* if, in addition, Y is nonempty; note that in this case, $Y = \{y\}$ for some $y \in X$. Take a metric $d : X \times X \to R_+ := [0, \infty[$ over X; as well as a selfmap $T \in \mathscr{F}(X)$. [Here, for each couple A, B of nonempty sets, $\mathscr{F}(A, B)$ denotes the class of all functions from A to B; when $A = B$, we write $\mathscr{F}(A)$ in place of $\mathscr{F}(A, A)$]. Denote $\text{Fix}(T) = \{x \in X; x = Tx\}$; each point of this set is referred to as *fixed* under T. The determination of such elements is to be performed under the directions below, comparable with the ones in Rus [38, Ch 2, Sect. 2.2]:

(pic-0) We say that T is *fix-asingleton* if $\text{Fix}(T)$ is an asingleton; and *fix-singleton* if $\text{Fix}(T)$ is a singleton

(pic-1) We say that T is a *Picard operator* (modulo d) if, for each $x \in X$, the iterative sequence $(T^n x; n \geq 0)$ is d-Cauchy

(pic-2) We say that T is a *strong Picard operator* (modulo d) if, for each $x \in X$, the iterative sequence $(T^n x; n \geq 0)$ is d-convergent with $\lim_n(T^n x) \in \text{Fix}(T)$.

M. Turinici (✉)
A. Myller Mathematical Seminar, A. I. Cuza University, Iaşi, Romania
e-mail: mturi@uaic.ro

T. M. Rassias, P. M. Pardalos (eds.), *Mathematical Analysis and Applications*, Springer Optimization and Its Applications 154,
https://doi.org/10.1007/978-3-030-31339-5_22

A basic result in this area is the 1922 one due to Banach [3]. Call T, $(d;k)$-*contractive* (where $k \geq 0$), if

$d(Tx, Ty) \leq kd(x, y)$, for all $x, y \in X$.

Theorem 1 *Assume that T is $(d;k)$-contractive, for some $k \in [0, 1[$. In addition, let X be d-complete. Then,*

(11-a) *T is fix-singleton:* $\mathrm{Fix}(T) = \{z\}$, *for some $z \in X$*
(11-b) *T is strong Picard (modulo d):* $\lim_n(T^n x) = z$, *for each $x \in X$.*

This result (referred to as: Banach's fixed point principle; in short: (B-fpp)) found some basic applications to the operator equations theory. As a consequence, many extensions of it were proposed. From the perspective of this exposition, the following ones are of interest:

(I) Structural contractive extensions: the initial (Banach) contractive property is taken in a generalized (implicit) way, as

(i-contr) $F(d(Tx, Ty), d(x, y), d(x, Tx), d(y, Ty), d(x, Ty), d(Tx, y)) \leq 0$, for all $x, y \in X, x\nabla y$;

where $F : R^6_+ \to R$ is a function and $\nabla \subseteq X \times X$ is a *relation* over X. Note that in the case of

$\nabla = X \times X$ (the *trivial relation* over X),

a consistent list of such extensions may be found in the survey paper by Rhoades [36]; further aspects were developed in Leader [26] and Turinici [47]. On the other hand, the case of

∇ is *reflexive* and *transitive* (hence, a *quasi-order*)

was considered in the 1986 paper by Turinici [49]. Two decades later, these results have been re-discovered— in a (partially) ordered context— by Ran and Reurings [35]; see also Nieto and Rodriguez-Lopez [32]. Finally, the case

∇ is *amorphous* (i.e., it has no regularity properties at all)

was discussed via graph techniques by Jachymski [18]; and (in a general context) by Samet and Turinici [40]. Some other aspects involving additional convergence structures may be found in Kasahara [22].

(II) Nonself extensions: T is no longer a selfmap. In 1977, Bernfeld et al. [5] introduced the concept of *PPF (past-present-future) dependent fixed point* for nonself-mappings (whose domain is distinct from their range). Furthermore, the quoted authors established some PPF dependent fixed point theorems for certain contraction mappings of this type. As precise there, the obtained results are useful tools in the study of existence and uniqueness questions for solutions of nonlinear functional differential/integral equations which may depend upon the past history, present data, and future evolution. As a consequence, this theory attracted a lot of contributors in the area; see, for instance, Chuasuk and Kaewcharoen [7], Drici et al. [12], Dhage [10, 11], Hussain et al. [16], Kaewcharoen [21], Kutbi and Sintunavarat [25], to name only a few. However, according to a recent paper by Turinici [53], in all results of this type— based on the algebraical closeness of associated Razumikhin class— we may arrange for these PPF dependent fixed points belonging to the constant class of ambient functional space. It is our aim of the present exposition to show that a similar conclusion is valid at the level of subsequent metrical setting attached to these problems. Further aspects will be discussed elsewhere.

2 Preliminaries

Throughout this exposition, the axiomatic system in use is Zermelo-Fraenkel's (abbreviated: ZF), as described by Cohen [9, Ch 2]. The notations and basic facts to be considered in this system are more or less standard. Some important ones are discussed below.

(A) Let X be a nonempty set. By a *relation* over X, we mean any (nonempty) part $\mathscr{R} \subseteq X \times X$. For simplicity, we sometimes write $(x, y) \in \mathscr{R}$ as $x\mathscr{R}y$. Note that $\mathscr{R}$ may be regarded as a mapping between X and exp$[X]$ (=the class of all subsets in X). In fact, denote for $x \in X$:

$X(x, \mathscr{R}) = \{y \in X; x\mathscr{R}y\}$ (the *section* of $\mathscr{R}$ through x);

then, the announced mapping representation is $[\mathscr{R}(x) = X(x, \mathscr{R}), x \in X]$. A basic example of such object is

$\mathscr{I} = \{(x, x); x \in X\}$ [the *identical relation* over X].

Given the relations $\mathscr{R}$, $\mathscr{S}$ over X, define their *product* $\mathscr{R} \circ \mathscr{S}$ as

$(x, z) \in \mathscr{R} \circ \mathscr{S}$, if there exists $y \in X$ with $(x, y) \in \mathscr{R}$, $(y, z) \in \mathscr{S}$.

Also, for each relation $\mathscr{R}$ in X, denote

$\mathscr{R}^{-1} = \{(x, y) \in X \times X; (y, x) \in \mathscr{R}\}$ (the *inverse* of $\mathscr{R}$).

Finally, given the relations $\mathscr{R}$ and $\mathscr{S}$ on X, let us say that $\mathscr{R}$ is *coarser* than $\mathscr{S}$ (or, equivalently: $\mathscr{S}$ is *finer* than $\mathscr{R}$), provided

$\mathscr{R} \subseteq \mathscr{S}$; i.e., $x\mathscr{R}y$ implies $x\mathscr{S}y$.

Given a relation $\mathscr{R}$ on X, the following properties are to be discussed here:

(P1) $\mathscr{R}$ is *reflexive*: $\mathscr{I} \subseteq \mathscr{R}$
(P2) $\mathscr{R}$ is *irreflexive*: $\mathscr{R} \cap \mathscr{I} = \emptyset$
(P3) $\mathscr{R}$ is *transitive*: $\mathscr{R} \circ \mathscr{R} \subseteq \mathscr{R}$
(P4) $\mathscr{R}$ is *symmetric*: $\mathscr{R}^{-1} = \mathscr{R}$
(P5) $\mathscr{R}$ is *antisymmetric*: $\mathscr{R}^{-1} \cap \mathscr{R} \subseteq \mathscr{I}$.

This yields the classes of relations to be used; the following ones are important for our developments:

(C0) $\mathscr{R}$ is *amorphous* (i.e., it has no specific properties)
(C1) $\mathscr{R}$ is a *quasi-order* (reflexive and transitive)
(C2) $\mathscr{R}$ is a *strict order* (irreflexive and transitive)
(C3) $\mathscr{R}$ is an *equivalence* (reflexive, transitive, symmetric)
(C4) $\mathscr{R}$ is a *(partial) order* (reflexive, transitive, antisymmetric)
(C5) $\mathscr{R}$ is *trivial* (i.e., $\mathscr{R} = X \times X$).

A basic relational structure is $(N, \leq)$; here, $N = \{0, 1, \ldots\}$ is the set of *natural* numbers and $(\leq)$ is defined as

$m \leq n$ iff $m + p = n$, for some $p \in N$.

The basic property of this structure may be written as:

$(N, \leq)$ is *well ordered* (each nonempty part of N has a first element); hence, in particular: $(N, \leq)$ is (partially) ordered.

For each $n \in N(1, \leq)$, let $N(n, >) := \{0, \ldots, n-1\}$ stand for the *initial interval* (in N) induced by n. Any set P with $P \sim N$ (in the sense: there exists a bijection from P to N) will be referred to as *effectively denumerable*. In addition, given some natural number $n \in N(1, \leq)$, any set Q with $Q \sim N(n, >)$ will be said to be *n-finite*; if n is generic here, we say that Q is *finite*. Then, the (nonempty) set Y is called (at most) *denumerable* iff it is either effectively denumerable or finite. Finally, by a *sequence* in X, we mean any mapping $x : N \to X$. For simplicity reasons, it will be useful to denote it as $(x(n); n \geq 0)$, or $(x_n; n \geq 0)$; moreover, when no confusion can arise, we further simplify this notation as $(x(n))$ or (x_n), respectively. Also, any sequence $(y_n := x_{i(n)}; n \geq 0)$ with

$(i(n); n \geq 0)$ is *strictly ascending* (hence: $i(n) \to \infty$ as $n \to \infty$)

will be referred to as a *subsequence* of $(x_n; n \geq 0)$. Note that, under such a convention, the relation "subsequence of" is transitive; i.e.,

(z_n)=subsequence of (y_n) and (y_n)=subsequence of (x_n)
imply (z_n)=subsequence of (x_n).

(B) Remember that an outstanding part of (ZF) is the *Axiom of Choice* (abbreviated: AC), which, in a convenient manner, may be written as

(AC) For each couple (J, X) of nonempty sets and each function $F : J \to \exp(X)$, there exists a (selective) function $f : J \to X$, with $f(\nu) \in F(\nu)$, for each $\nu \in J$.

(Here, $\exp(X)$ stands for the class of all nonempty elements in $\exp[X]$). Sometimes, when the ambient set X is endowed with denumerable type structures, the case of $J = N$ will suffice for all involved choice reasonings; and existence of a selective function may be determined by using a weaker form of (AC), called: *Dependent Choice* principle (in short: DC). Call the relation $\mathscr{R}$ over X, *proper* when

$(X(x, \mathscr{R}) =)\mathscr{R}(x)$ is nonempty, for each $x \in X$.

Note that, in this case, $\mathscr{R}$ is to be viewed as a mapping between X and $\exp(X)$; the couple $(X, \mathscr{R})$ will be then referred to as a *proper relational structure*. Given $a \in X$, let us say that the sequence $(x_n; n \geq 0)$ in X is $(a; \mathscr{R})$*-iterative*, provided

$x_0 = a$ and $[x_n \mathscr{R} x_{n+1}$ (i.e., $x_{n+1} \in \mathscr{R}(x_n)), \forall n]$.

Proposition 1 *Let the relational structure $(X, \mathscr{R})$ be proper. Then, for each $a \in X$ there is at least an $(a, \mathscr{R})$-iterative sequence in X.*

This principle— proposed, independently, by Bernays [4] and Tarski [46]— is deductible from (AC), but not conversely; cf. Wolk [57]. Moreover, by the developments in Moskhovakis [31, Ch 8], and Schechter [42, Ch 6], the *reduced system* (ZF-AC+DC) is comprehensive enough so as to cover the "usual" mathematics; see also Moore [30, Appendix 2].

A basic consequence of the underlying statement is the so-called *Denumerable Axiom of Choice* [in short: (AC(N))]. Precisely, we have, in (DC-AC+DC):

Proposition 2 *Let $F : N \to \exp(X)$ be a function. Then, for each $a \in F(0)$ there exists a function $f : N \to X$ with $f(0) = a$ and $f(n) \in F(n)$, $\forall n \in N$.*

Proof Denote $P = N \times X$; and introduce a relation $\mathscr{R}$ over P, according to:

$\mathscr{R}(n, x) = \{n + 1\} \times F(n + 1),\ n \in N, x \in X.$

By the Dependent Choice principle (DC) it follows that, for the starting element $b := (0, a) \in P$, there must be a sequence $(b_n := (i(n), a_n); n \geq 0)$ in P, such that

$b_0 = b$; [that is: $i(0) = 0, a_0 = a$],
$(\forall n)$: $b_{n+1} \in \mathscr{R}(b_n)$ [that is: $i(n + 1) = i(n) + 1, a_{n+1} \in F(i(n) + 1)$].

By the first half of this last relation, we get $(i(n) = n, \forall n)$; and this, combined with the second half of the same, yields $(a_n \in F(n), \forall n)$. It will suffice taking $(f(n) = a_n; n \in N)$ to derive our desired conclusion.

As a consequence of the above fact, (DC) $\Longrightarrow$ (AC(N)) in (ZF-AC). The reciprocal inclusion is not true; see Moskhovakis [31, Ch 8, Sect 8.25] for details.

(C) Let (X, d) be a metric space. We introduce a d-convergence and d-Cauchy structure on X as follows. Given the sequence (x_n) in X and the point $x \in X$, we say that (x_n), d-converges to x (written as: $x_n \xrightarrow{d} x$), provided $d(x_n, x) \to 0$ as $n \to \infty$; that is,

$\forall \varepsilon > 0, \exists i = i(\varepsilon)\text{: } i \leq n \Longrightarrow d(x_n, x) < \varepsilon;$

the set of all such points x will be denoted $\lim_n(x_n)$. By this definition, we have

(d-conv-1) $((\xrightarrow{d})$ is hereditary)
$x_n \xrightarrow{d} x$ implies $y_n \xrightarrow{d} x$, for each subsequence (y_n) of (x_n)
(d-conv-2) $((\xrightarrow{d})$ is reflexive)
$(\forall u \in X)$: the constant sequence $(x_n = u; n \geq 0)$ fulfills $x_n \xrightarrow{d} u$.

As a consequence, $(\xrightarrow{d})$ is a sequential convergence on X, under the general conventions in Kasahara [22]; with in addition (as d is triangular and symmetric)

(d-conv-3) $(\xrightarrow{d})$ is separated (referred to as: d is *separated*):
$\lim_n(x_n)$ is an asingleton, for each sequence $(x_n; n \geq 0)$ in X.

If $\lim_n(x_n)$ is nonempty, then (x_n) is called *d-convergent*; in this case, $\lim_n(x_n) = \{z\}$ is written as $\lim_n(x_n) = z$.

Further, call the sequence (x_n), *d-Cauchy* when $d(x_m, x_n) \to 0$ as $m, n \to \infty$, $m < n$; that is,

$$\forall \varepsilon > 0, \exists j = j(\varepsilon):\ j \leq m < n \Longrightarrow d(x_m, x_n) < \varepsilon;$$

the class of all these will be denoted as $Cauchy(d)$. As before, we have

(d-Cauchy-1) ($Cauchy(d)$ is hereditary)
(x_n) is d-Cauchy implies (y_n) is d-Cauchy,
for each subsequence (y_n) of (x_n)
(d-Cauchy-2) ($Cauchy(d)$ is reflexive)
$(\forall u \in X)$: the constant sequence $(x_n = u; n \geq 0)$ is d-Cauchy;

hence, $Cauchy(d)$ is a Cauchy structure on X under the conventions in Turinici [51]. Moreover, we get (again via d=triangular, symmetric)

(d-Cauchy-3) $((\xrightarrow{d}), Cauchy(d))$ is *regular* (referred to as: d is *regular*):
each d-convergent sequence is d-Cauchy.

Finally, let us say that $(x_n; n \geq 0)$ is *d-semi-Cauchy*, provided

$$\lim_n d(x_n, x_{n+1}) = 0; \text{ or, equivalently: } \lim_n d(x_n, x_{n+i}) = 0, \forall i \geq 1.$$

Clearly, each d-Cauchy sequence is d-semi-Cauchy too; the reciprocal of this is not in general true.

The introduced concepts allow us to give a useful property. (Since its verification is immediate, we do not give further details).

Proposition 3 *The mapping $(x, y) \mapsto d(x, y)$ is d-Lipschitz, in the sense*

$$|d(x, y) - d(u, v)| \leq d(x, u) + d(y, v), \text{ for all } (x, y), (u, v) \in X \times X.$$

As a consequence, this map is d-continuous; i.e.,

$$x_n \xrightarrow{d} x,\ y_n \xrightarrow{d} y \text{ imply } d(x_n, y_n) \to d(x, y).$$

Note, finally, that an extended setting of these developments is possible, under the lines described by the 2001 PhD Thesis in Hitzler [15, Ch 1, Sect 1.2]; we do not give details.

(D) In the following, some auxiliary facts involving convergent real sequences are given. For each sequence (r_n) in R, and each element $r \in R$, denote

$$r_n \to r+ \text{ (resp., } r_n \to r-\text{), when } r_n \to r \text{ and } [r_n > r \text{ (resp., } r_n < r\text{)}, \forall n].$$

Proposition 4 *Let the sequence $(r_n; n \geq 0)$ in R and the number $\varepsilon \in R$ be such that $r_n \to \varepsilon+$. There exists then a subsequence $(r_n^* := r_{i(n)}; n \geq 0)$ of $(r_n; n \geq 0)$, endowed with the properties*

$(r_n^*; n \geq 0)$ is strictly descending and $r_n^* \to \varepsilon+$.

Proof Put $i(0) = 0$. As $\varepsilon < r_{i(0)}$ and $r_n \to \varepsilon+$, we have that

$A(i(0)) := \{n > i(0); r_n < r_{i(0)}\}$ is not empty;
 hence, $i(1) := \min(A(i(0)))$ is an element of it, and $r_{i(1)} < r_{i(0)}$.

Likewise, as $\varepsilon < r_{i(1)}$ and $r_n \to \varepsilon+$, we have that

$A(i(1)) := \{n > i(1); r_n < r_{i(1)}\}$ is not empty;
 hence, $i(2) := \min(A(i(1)))$ is an element of it, and $r_{i(2)} < r_{i(1)}$.

This procedure may continue indefinitely; and yields (without any choice technique) a strictly ascending rank sequence $(i(n); n \geq 0)$ (hence, $i(n) \to \infty$ as $n \to \infty$) for which the attached subsequence $(r_n^* := r_{i(n)}; n \geq 0)$ of $(r_n; n \geq 0)$ fulfills

$r_{n+1}^* < r_n^*$, for all n; hence, (r_n^*) is (strictly) descending.

On the other hand, by this very subsequence property,

$(r_n^* > \varepsilon, \forall n)$, and $\lim_n r_n^* = \lim_n r_n = \varepsilon$.

Putting these together, we get the desired fact.

A bi-dimensional counterpart of these facts may be given along the lines below. Let $\pi(t, s)$ (where $t, s \in R$) be a logical property involving couples of real numbers. Given the couple of real sequences $(t_n; n \geq 0)$ and $(s_n; n \geq 0)$, call the subsequences $(t_n^*; n \geq 0)$ of (t_n) and $(s_n^*; n \geq 0)$ of (s_n), *compatible* when

$(t_n^* = t_{i(n)}; n \geq 0)$, and $(s_n^* = s_{i(n)}; n \geq 0)$,
 for the same strictly ascending rank sequence $(i(n); n \geq 0)$.

Proposition 5 *Let the couple of real sequences $(t_n; n \geq 0)$, $(s_n; n \geq 0)$ and the couple of numbers $a, b \in R$ be such that*

$t_n \to a+$, $s_n \to b+$ as $n \to \infty$ and $(\pi(t_n, s_n)$ is true, for all $n)$.

There exists then a couple of subsequences $(t_n^*; n \geq 0)$ of $(t_n; n \geq 0)$ and $(s_n^*; n \geq 0)$ of $(s_n; n \geq 0)$, respectively, with

(25-1) *$(t_n^*; n \geq 0)$ and $(s_n^*; n \geq 0)$ are strictly descending, compatible*
(25-2) *$(t_n^* \to a+$, $s_n^* \to b+$, as $n \to \infty)$, and $(\pi(t_n^*, s_n^*)$ is true, for all $n)$.*

Proof By the preceding statement, the sequence $(t_n; n \geq 0)$ admits a subsequence $(T_n := t_{i(n)}; n \geq 0)$, with

$(T_n; n \geq 0)$ is strictly descending, and $(T_n \to a+$, as $n \to \infty)$.

Denote $(S_n := s_{i(n)}; n \geq 0)$; clearly,

$(S_n; n \geq 0)$ is a subsequence of $(s_n; n \geq 0)$ with $S_n \to b+$ as $n \to \infty$.

Moreover, by this very construction

$\pi(T_n, S_n)$ holds, for all n.

Again by that statement, there exists a subsequence $(s_n^* := S_{j(n)} = s_{i(j(n))}; n \geq 0)$ of $(S_n; n \geq 0)$ (hence, of $(s_n; n \geq 0)$ as well), with

$(s_n^*; n \geq 0)$ is strictly descending, and $(s_n^* \to b+$, as $n \to \infty)$.

Denote further $(t_n^* := T_{j(n)} = t_{i(j(n))}; n \geq 0)$; this is a subsequence of $(T_n; n \geq 0)$ (hence, of $(t_n; n \geq 0)$ as well), with

$(t_n^*; n \geq 0)$ is strictly descending, and $(t_n^* \to a+$, as $n \to \infty)$;

Finally, by this very construction (and a previous relation)

$\pi(t_n^*, s_n^*)$ holds, for all n.

Summing up, the couple of subsequences $(t_n^*; n \geq 0)$ and $(s_n^*; n \geq 0)$ has all needed properties; and the conclusion follows.

3 Meir-Keeler Relations

Let $R_+^0 :=]0, \infty[$ stand for the class of all strictly positive real numbers; and $\Omega \subseteq R_+^0 \times R_+^0$ be a relation over R_+^0; usually, we write $(t, s) \in \Omega$ as $t\Omega s$. The following basic (global) properties upon this object are considered

(u-diag) Ω is *upper diagonal*:
$t\Omega s$ implies $t < s$

(1-decr) Ω is *first variable decreasing*:
$t_1, t_2, s \in R_+^0$, $t_1 \geq t_2$ and $t_1\Omega s$ imply $t_2\Omega s$

(2-incr) Ω is *second variable increasing*:
$t, s_1, s_2 \in R_+^0$, $s_1 \leq s_2$ and $t\Omega s_1$ imply $t\Omega s_2$.

The class of all upper diagonal relations will be denoted as udiag(R_+^0). Note that all subsequent constructions are being connected to this setting. In particular, the following basic property for upper diagonal relations Ω is considered:

(M-adm) Ω in *Matkowski admissible*:
$(t_n; n \geq 0)$ in R_+^0 and $(t_{n+1}\Omega t_n, \forall n)$ imply $\lim_n t_n = 0$.

In parallel to this, the following local conditions involving the same class of relations will be useful for us:

(g-MK) Ω has the *geometric Meir-Keeler property*:
$\forall \varepsilon > 0, \exists \delta > 0: t\Omega s, \varepsilon < s < \varepsilon + \delta \Longrightarrow t \leq \varepsilon$

(g-left-s) Ω is *geometric left separable*:
$\forall \beta > 0, \exists \gamma \in]0, \beta[: \gamma < t < \beta \Longrightarrow (t, \beta) \notin \Omega$.

Note that, by the upper diagonal property, the former of these local conditions may be also written as

(g-MK-var) $\forall \varepsilon > 0, \exists \delta > 0: t\Omega s, s < \varepsilon + \delta \Longrightarrow t \leq \varepsilon$.

It is the main aim of this section to study the above defined concepts. Most of these investigations were already carried out; see, in this direction, Turinici [55] and the references therein. However, to make this exposition self-contained, we provide an argument for the underlying facts.

To start with, note that— technically speaking— the geometric Meir-Keeler property is strongly related to the Matkowski admissible property we just introduced. Precisely, the following auxiliary fact is available.

Proposition 6 *The following are valid, in (ZF-AC+DC):*

(31-1) *For each $\Omega \in \mathrm{udiag}(R_+^0)$, we have*

Ω has the geometric Meir-Keeler property implies
Ω is Matkowski admissible.

(31-2) *For each first variable decreasing $\Omega \in \mathrm{udiag}(R_+^0)$, we have*

Ω is Matkowski admissible implies
Ω has the geometric Meir-Keeler property.

Hence, summing up

(31-3) *For each first variable decreasing $\Omega \in \mathrm{udiag}(R_+^0)$, we have*

Ω has the geometric Meir-Keeler property iff Ω is Matkowski admissible.

Proof

(i) Suppose that $\Omega \in \mathrm{udiag}(R_+^0)$ has the geometric Meir-Keeler property; we have to establish that Ω is Matkowski admissible. Let $(t_n; n \geq 0)$ be a sequence in R_+^0, with

$t_{n+1}\Omega t_n$, for all n.

By the upper diagonal property, we get

$t_{n+1} < t_n$, for all n;

i.e., (t_n) is strictly descending. As a consequence, $\tau = \lim_n t_n$ exists in R_+; with in addition ($t_n > \tau$, for all n). Let $\sigma > 0$ be the number assured by the geometric Meir-Keeler property. By definition, there exists an index $n(\sigma)$, with

$(t_{n+1}\Omega t_n$ and$)$ $\tau < t_n < \tau + \sigma$, for all $n \geq n(\sigma)$.

This, by the quoted property, gives (for the same ranks)

$\tau < t_{n+1} \leq \tau$; contradiction.

Hence, necessarily, $\tau = 0$; and the conclusion follows.

(ii) Suppose that the first variable decreasing $\Omega \in \mathrm{udiag}(R_+^0)$ has the Matkowski property; we have to establish that Ω has the geometric Meir-Keeler property. Suppose by contradiction that this is not true; i.e. (for some $\varepsilon > 0$)

$H(\delta) := \{(t, s) \in \Omega; \varepsilon < s < \varepsilon + \delta, t > \varepsilon\} \neq \emptyset$, for each $\delta > 0$.

Taking a zero converging sequence $(\delta_n; n \geq 0)$ in R_+^0, we get by the Denumerable Axiom of Choice (AC(N)) [deductible, as precise, in (ZF-AC+DC)], a sequence $((t_n, s_n); n \geq 0)$ in $R_+^0 \times R_+^0$, so as

$(\forall n)$: (t_n, s_n) is an element of $H(\delta_n)$;

or, equivalently (by definition and upper diagonal property)

$((t_n, s_n) \in \Omega$ and$)$ $\varepsilon < t_n < s_n < \varepsilon + \delta_n$, for all n.

Note that, as a direct consequence,

$((t_n, s_n) \in \Omega$, for all $n)$, and $t_n \to \varepsilon+$, $s_n \to \varepsilon+$, as $n \to \infty$.

Put $i(0) = 0$. As

$\varepsilon < t_{i(0)}$ and $s_n \to \varepsilon+$ as $n \to \infty$,

we have that

$A(i(0)) := \{n > i(0); s_n < t_{i(0)}\}$ is not empty;
hence, $i(1) := \min(A(i(0)))$ is an element of it, and $s_{i(1)} < t_{i(0)}$;
so that $s_{i(1)} \Omega s_{i(0)}$ (as $t_{i(0)} \Omega s_{i(0)}$ and Ω is first variable decreasing).

Likewise, as

$\varepsilon < t_{i(1)}$ and $s_n \to \varepsilon+$ as $n \to \infty$,

we have that

$A(i(1)) := \{n > i(1); s_n < t_{i(1)}\}$ is not empty;
hence, $i(2) := \min(A(i(1)))$ is an element of it, and $s_{i(2)} < t_{i(1)}$;
so that $s_{i(2)} \Omega s_{i(1)}$ (as $t_{i(1)} \Omega s_{i(1)}$ and Ω is first variable decreasing).

This procedure may continue indefinitely; and yields (without any choice technique) a strictly ascending rank sequence $(i(n); n \geq 0)$ in N (hence, $i(n) \to \infty$ as $n \to \infty$) for which the attached subsequence $(r_n := s_{i(n)}; n \geq 0)$ of (s_n) fulfills

$r_{n+1} \Omega r_n$, for all n; whence $r_n \to 0$ (as Ω is Matkowski admissible).

On the other hand, by our subsequence property,

$(r_n > \varepsilon, \forall n)$ and $\lim_n r_n = \lim_n s_n = \varepsilon$; i.e., $r_n \to \varepsilon+$.

The obtained relation is in contradiction with the previous one. Hence, the working condition cannot be true; and we are done.

(iii) Evident, by the above.

In the following, sufficient (sequential) conditions are given for the geometric Meir-Keeler property and geometric left separable property above. Given the upper diagonal relation $\Omega \in \mathrm{udiag}(R^0_+)$, let us introduce the (asymptotic type) conventions

(a-MK) Ω has the *asymptotic Meir-Keeler property*:
there are no strictly descending sequences (t_n) and (s_n) in R^0_+
and no elements ε in R^0_+, with
$((t_n, s_n) \in \Omega, \forall n)$ and $(t_n \to \varepsilon+, s_n \to \varepsilon+)$

(a-left-s) Ω is *asymptotic left separable*:
there are no strictly ascending sequences (t_n) in R^0_+
and no elements β in R^0_+, with $((t_n, \beta) \in \Omega, \forall n)$ and $(t_n \to \beta-)$.

Here, as precise, for each sequence (r_n) in R, and each element $r \in R$,

$$r_n \to r+ \text{ (resp., } r_n \to r-\text{), when } r_n \to r \text{ and } [r_n > r \text{ (resp., } r_n < r\text{)}, \forall n].$$

Proposition 7 *The following generic relationships are valid (for an arbitrary upper diagonal relation $\Omega \subseteq R^0_+ \times R^0_+$), in the reduced system (ZF-AC+DC):*

(32-1) asymptotic Meir-Keeler implies geometric Meir-Keeler

(32-2) asymptotic left separable is equivalent with geometric left separable.

Proof

(i) Let $\Omega \in \mathrm{udiag}(R^0_+)$ be asymptotic Meir-Keeler; but— contrary to the conclusion— assume that Ω does not have the geometric Meir-Keeler property; i.e., (for some $\varepsilon > 0$)

$$H(\delta) := \{(t, s) \in \Omega; s < \varepsilon + \delta, t > \varepsilon\} \neq \emptyset, \text{ for each } \delta > 0.$$

Taking a zero converging sequence $(\delta_n; n \geq 0)$ in R^0_+, we get by the Denumerable Axiom of Choice (AC(N)) [deductible, as precise, in (ZF-AC+DC)], a sequence $((t_n, s_n); n \geq 0)$ in $R^0_+ \times R^0_+$, so as

$$(\forall n)\text{: } (t_n, s_n) \text{ is an element of } H(\delta_n);$$

or, equivalently (by definition and upper diagonal property)

$$(t_n \Omega s_n \text{ and}) \ \varepsilon < t_n < s_n < \varepsilon + \delta_n, \text{ for all } n.$$

Note that, as a direct consequence,

$$(t_n \Omega s_n, \text{ for all } n), \text{ and } t_n \to \varepsilon+, s_n \to \varepsilon+, \text{ as } n \to \infty.$$

By a previous result, there exists a compatible couple of subsequences $(t^*_n := t_{i(n)}; n \geq 0)$ of $(t_n; n \geq 0)$ and $(s^*_n := s_{i(n)}; n \geq 0)$ of $(s_n; n \geq 0)$, with

$$(t^*_n \Omega s^*_n, \text{ for all } n), (t^*_n; n \geq 0), (s^*_n; n \geq 0) \text{ are strictly descending, and } (t^*_n \to \varepsilon+, s^*_n \to \varepsilon+, \text{ as } n \to \infty).$$

This, however, is in contradiction with respect to the posed hypothesis; wherefrom, our assertion follows.

(ii-1) Let $\Omega \in \text{udiag}(R^0_+)$ be an asymptotic left separable relation; we have to establish that Ω is geometric left separable. Suppose— contrary to this conclusion— that Ω is not endowed with such a property; that is, (for some $\beta > 0$)

$$K(\gamma) := \{t \in]\gamma, \beta[; (t, \beta) \in \Omega\} \neq \emptyset, \text{ for each } \gamma \in]0, \beta[.$$

Taking a strictly ascending sequence $(\gamma_n; n \geq 0)$ in $]0, \beta[$ with $\gamma_n \to \beta-$, we get by the Denumerable Axiom of Choice (AC(N)) [deductible, as precise, in (ZF-AC+DC)], a sequence $(t_n; n \geq 0)$ in R^0_+, so as

$$(\forall n): t_n \text{ is an element of } K(\gamma_n);$$

or, equivalently (by the very definition above)

$$(\forall n): \gamma_n < t_n < \beta, \text{ and } (t_n, \beta) \in \Omega.$$

By the former part, we must have $(t_n \to \beta-)$; and this, along with an auxiliary fact, tells us that there exists a subsequence $(t^*_n := t_{i(n)}; n \geq 0)$ such that

$$(t^*_n; n \geq 0) \text{ is strictly ascending, and } t^*_n \to \beta-.$$

On the other hand, by the latter part of our previous relation,

$$(t^*_n, \beta) \in \Omega, \text{ for all } n.$$

This, however, contradicts the asymptotic left separable property. Hence, Ω is geometric left separable; and the assertion follows.

(ii-2) Let $\Omega \in \text{udiag}(R^0_+)$ be geometric left separable; we have to establish that Ω is asymptotic left separable. Suppose— contrary to this conclusion— that Ω is not asymptotic left separable:

there exist a strictly ascending sequence (t_n) in R^0_+ and an element β in R^0_+, with $((t_n, \beta) \in \Omega, \forall n)$ and $(t_n \to \beta-)$.

As Ω is geometric left separable, it follows that given this $\beta > 0$ there exists $\gamma \in]0, \beta[$, such that

$$\gamma < t < \beta \Longrightarrow (t, \beta) \notin \Omega.$$

On the other hand, from the above convergence property, we have that for this $\gamma \in]0, \beta[$, there exists some index $n(\gamma)$, such that

$$(\forall n \geq n(\gamma)): \gamma < t_n < \beta \text{ and } (t_n, \beta) \in \Omega.$$

This contradicts the preceding relation (involving the number γ). Hence, our working assumption cannot be accepted; and the assertion follows.

Under these general facts, it is our objective to give some basic examples of (upper diagonal) Matkowski admissible relations. The general scheme of constructing these may be described as follows.

Let $R(\pm\infty) := R \cup \{-\infty, \infty\}$ stand for the set of all *extended real numbers*. For each relation Ω over R^0_+, let us associate a function $\xi : R^0_+ \times R^0_+ \to R(\pm\infty)$, via

$\xi(t, s) = 0$, if $(t, s) \in \Omega$; $\xi(t, s) = -\infty$, if $(t, s) \notin \Omega$.

It will be referred to as the *function* generated by Ω; clearly,

$(t, s) \in \Omega$ iff $\xi(t, s) \geq 0$.

Conversely, given a function $\xi : R^0_+ \times R^0_+ \to R(\pm\infty)$, we may associate it a relation Ω over R^0_+, according to

$\Omega = \{(t, s) \in R^0_+ \times R^0_+; \xi(t, s) \geq 0\}$; in short: $\Omega = [\xi \geq 0]$;
referred to as: the *the positive section* of ξ.

Note that the correspondence between the function ξ and its associated relation $\Omega = [\xi \geq 0]$ is not injective; because, for the function $\eta := \lambda\xi$ (where $\lambda > 0$), its associated relation $[\eta \geq 0]$ is identical with the relation $[\xi \geq 0]$ attached to ξ.

Now, call the function $\xi : R^0_+ \times R^0_+ \to R(\pm\infty)$, *upper diagonal* provided:

(u-diag-fct) $\xi(t, s) \geq 0$ implies $t < s$.

Note that all subsequent constructions are being considered within this setting. In particular, the following basic property (condition) for upper diagonal functions ξ is considered:

(M-adm-fct) ξ in *Matkowski admissible*:
$(t_n; n \geq 0)$ in R^0_+ and $(\xi(t_{n+1}, t_n) \geq 0, \forall n)$ imply $\lim_n t_n = 0$.

The following local conditions involving our functions are to be considered:

(g-MK-fct) ξ has the *geometric Meir-Keeler property*:
$\forall \varepsilon > 0, \exists \delta > 0$: $\xi(t, s) \geq 0, \varepsilon < s < \varepsilon + \delta \Longrightarrow t \leq \varepsilon$

(g-left-s-fct) ξ is *geometric left separable*: for each $\beta > 0$,
there exists $\gamma \in]0, \beta[$, such that: $\gamma < t < \beta \Longrightarrow \xi(t, \beta) < 0$.

The relationship between this geometric Meir-Keeler property and the Matkowski admissible one attached to upper diagonal functions is nothing else than a simple translation of the one involving upper diagonal relations; we do not give details.

Summing up, any concept (like the ones above) about (upper diagonal) relations over R^0_+ may be written as a concept about (upper diagonal) functions in $\mathcal{F}(R^0_+ \times R^0_+, R(\pm\infty))$. For the rest of our exposition, it will be convenient working with relations over R^0_+, and not with functions in $\mathcal{F}(R^0_+ \times R^0_+, R(\pm\infty))$; this, however, is but a methodology question.

Having these precise, we may now pass to the description of some basic objects in this area.

Part-Case I The former of these corresponds to the choice

$(\xi(t, s) = \chi(s) - t; t, s \in R^0_+)$, where $\chi \in \mathcal{F}(R^0_+, R)$.

Precisely, denote by $\mathcal{F}(re)(R^0_+, R)$ the family of all functions $\varphi \in \mathcal{F}(R^0_+, R)$ with

φ is *regressive*: $\varphi(t) < t$, for all $t \in R^0_+$.

Then, let χ be a function in $\mathscr{F}(re)(R^0_+, R)$. Define the *Matkowski* relation $\Omega := \Omega[\chi]$ over R^0_+, as

$(t, s \in R^0_+)$: $(t, s) \in \Omega$ iff $t \leq \chi(s)$;

note that, as χ=regressive, Ω is upper diagonal and first variable decreasing. The introduced terminology is suggested by the developments in Matkowski [28].

In the following, sufficient conditions upon χ are given so as its attached relation $\Omega[\chi]$ be Matkowski admissible (or, equivalently: geometric Meir-Keeler).

For each $\varphi \in \mathscr{F}(re)(R^0_+, R)$, let us introduce the sequential property

(M-a) φ is *Matkowski admissible*:
for each $(t_n; n \geq 0)$ in R^0_+ with $(t_{n+1} \leq \varphi(t_n), \forall n)$ we have $\lim_n t_n = 0$.

To get concrete circumstances under which this property holds, we need some conventions. Given $\varphi \in \mathscr{F}(re)(R^0_+, R)$, let us introduce the global property

(MK-a) φ is *Meir-Keeler admissible*:
for each $\varepsilon > 0$, there exists $\delta > 0$, such that
$\varepsilon < s < \varepsilon + \delta \Longrightarrow \varphi(s) \leq \varepsilon$.

Proposition 8 *For each $\varphi \in \mathscr{F}(re)(R^0_+, R)$, we have in (ZF-AC+DC)*

(33-1) φ is Meir-Keeler admissible implies φ is Matkowski admissible
(33-2) φ is Matkowski admissible implies φ is Meir-Keeler admissible
(33-3) φ is Meir-Keeler admissible iff φ is Matkowski admissible.

Proof

(i) Suppose that φ is Meir-Keeler admissible; we have to establish that φ is Matkowski admissible. Let $(t_n; n \geq 0)$ be a sequence in R^0_+ with the property $(t_{n+1} \leq \varphi(t_n); n \geq 0)$. Clearly, (t_n) is strictly descending in R^0_+; hence, $\tau := \lim_n t_n$ exists in R_+. Suppose by contradiction that $\tau > 0$; and let $\sigma > 0$ be given by the Meir-Keeler property of φ. By the above convergence property, there exists some rank $n(\sigma)$, such that

$n \geq n(\sigma)$ implies $\tau < t_n < \tau + \sigma$.

But then, from the very choice of (t_n), we get (for the same ranks)

$\tau < t_{n+1} \leq \varphi(t_n) \leq \tau$; contradiction.

Hence, necessarily, $\tau = 0$; and conclusion follows.

(ii) Suppose that φ is Matkowski admissible; we claim that φ is Meir-Keeler admissible. For, if φ is not endowed with such a property, one has (for some $\gamma > 0$)

$H(\beta) := \{t \in R^0_+; \gamma < t < \gamma + \beta, \varphi(t) > \gamma\}$ is not empty, for each $\beta > 0$.

Taking a strictly descending sequence $(\beta_n; n \geq 0)$ in R^0_+ with $\beta_n \to 0$, we get by the Denumerable Axiom of Choice (AC(N)) [deductible, as precise, in (ZF-AC+DC)], a sequence $(t_n; n \geq 0)$ in R^0_+, so as

$(\forall n)$: t_n is an element of $H(\beta_n)$;

or, equivalently (by the very definition above and φ=regressive)

$(\forall n)$: $\gamma < \varphi(t_n) < t_n < \gamma + \beta_n$;
hence, in particular: $\varphi(t_n) \to \gamma+$ and $t_n \to \gamma+$.

Put $i(0) = 0$. As

$\gamma < \varphi(t_{i(0)})$ and $t_n \to \gamma+$,

we have that

$A(i(0)) := \{n > i(0); t_n < \varphi(t_{i(0)})\}$ is not empty;
hence, $i(1) := \min(A(i(0)))$ is an element of it, and $t_{i(1)} < \varphi(t_{i(0)})$.

Likewise, as

$\gamma < \varphi(t_{i(1)})$ and $t_n \to \gamma+$,

we have that

$A(i(1)) := \{n > i(1); t_n < \varphi(t_{i(1)})\}$ is not empty;
hence, $i(2) := \min(A(i(1)))$ is an element of it, and $t_{i(2)} < \varphi(t_{i(1)})$.

This procedure may continue indefinitely; and yields (without any choice technique) a strictly ascending rank sequence $(i(n); n \geq 0)$ in N (hence, $i(n) \to \infty$ as $n \to \infty$), for which the attached subsequence $(s_n := t_{i(n)}; n \geq 0)$ of (t_n) fulfills

$s_{n+1} < \varphi(s_n)(< s_n)$, for all n;
wherefrom, $s_n \to 0$ as $n \to \infty$ (as φ is Matkowski admissible).

On the other hand, $t_n \to \gamma+$ implies (by this subsequence property)

$(s_n > \gamma, \forall n)$ and $\lim_n s_n = \lim_n t_n = \gamma$.

The obtained relations contradict the preceding properties of $(s_n; n \geq 0)$; hence, the working condition cannot be true; and we are done.

(iii) Evident.

Summing up, the concept of Meir-Keeler admissible function is a basic one for the precise context; so that a list of basic examples of such functions is welcomed from a practical perspective.

For any $\varphi \in \mathscr{F}(re)(R^0_+, R)$ and any $s \in R^0_+$, put

$$\Lambda^+\varphi(s) = \inf_{\varepsilon>0} \Phi(s+)(\varepsilon), \text{ where } \Phi(s+)(\varepsilon) = \sup \varphi(]s, s+\varepsilon[), \varepsilon > 0;$$

referred to as: the *right superior limit* of φ at s. From the regressiveness of φ,

$$-\infty \leq \Lambda^+\varphi(s) \leq s, \forall s \in R^0_+;$$

but the case of these extremal values being attained cannot be avoided.

Call $\varphi \in \mathscr{F}(re)(R^0_+, R)$, *Boyd-Wong admissible* [6], if

(BW-adm) $\quad \Lambda^+\varphi(s) < s$, for all $s > 0$.

In particular, $\varphi \in \mathscr{F}(re)(R_+^0, R)$ is Boyd-Wong admissible provided it is *upper semicontinuous at the right* on R_+^0:

$\Lambda^+\varphi(s) \leq \varphi(s)$, for each $s \in R_+^0$.

This, e.g., is fulfilled when φ is *continuous at the right* on R_+^0; for, in such a case,

$\Lambda^+\varphi(s) = \varphi(s)$, for each $s \in R_+^0$.

Further, let $\mathscr{F}(re, in)(R_+^0, R)$ stand for the class of all $\varphi \in \mathscr{F}(re)(R_+^0, R)$, with

φ is increasing on R_+^0 ($0 < t_1 \leq t_2$ implies $\varphi(t_1) \leq \varphi(t_2)$).

The following characterization of our previous Matkowski property imposed to φ is available. For each $t > 0$, let the iterates sequence of φ at t be introduced as

$\varphi^0(t) = t, \varphi^1(t) = \varphi(t), \ldots, \varphi^{n+1}(t) = \varphi(\varphi^n(t)), n \geq 0$;

Note that, by the choice of φ, this iterates sequence may be not effective; for, e.g.,

$\varphi^2(t) = \varphi(\varphi(t))$ is not defined if $\varphi(t) \leq 0$.

Proposition 9 *For each $\varphi \in \mathscr{F}(re, in)(R_+)$, we have*

φ is Matkowski admissible, iff
($\forall t > 0$): $\lim_n \varphi^n(t) = 0$, whenever $(\varphi^n(t); n \geq 0)$ exists.

The proof is immediate, by the increasing property of φ; we do not give details.
As before, sufficient conditions (involving the class $\mathscr{F}(re, in)(R_+^0, R)$) under which this property holds are needed. For each $\varphi \in \mathscr{F}(re, in)(R_+^0, R)$, denote

$\varphi(s + 0) := \lim_{t \to s+} \varphi(t), s \in R_+^0$ (the *right limit* of φ at s).

It is not hard to see that the following evaluation holds:

$\varphi(s) \leq \varphi(s + 0) \leq s$, for all $s > 0$.

Proposition 10 *Suppose that the function $\varphi \in \mathscr{F}(re, in)(R_+^0, R)$ fulfills φ is strongly regressive: $\varphi(s + 0) < s$, for each $s > 0$. Then, φ is Matkowski admissible.*

Proof Given $s_0 > 0$, suppose that the iterative sequence $(s_n = \varphi^n(s_0); n \geq 0)$ exists in R_+^0. By the regressive property of φ, (s_n) is strictly descending; hence, $s := \lim_n s_n$ exists, with (in addition) $s_n > s$, for all n. Suppose by contradiction that $s > 0$. Combining with

$\varphi(s + 0) = \lim_n \varphi(s_n) = \lim_n s_{n+1}$,

yields $\varphi(s + 0) = s$; contradiction. Hence, $s = 0$; and we are done.

Remark 1 Concerning the reverse inclusion, let $(r_n; n \geq 0)$ be a strictly descending sequence in R_+^0 with $r_n \to 0$; and take the function $\varphi \in \mathscr{F}(re, in)(R_+^0)$, according to

$$\varphi(t) = r_0, \text{ if } r_0 < t$$

$$\varphi(t) = r_{n+1}, \text{ if } r_{n+1} < t \leq r_n, \text{ for all } n.$$

Clearly, φ is Matkowski admissible, via $(\varphi(r_n) = r_{n+1}, \forall n)$. On the other hand,

$\varphi(r_n + 0) = r_n$, for all n;

so that φ is not strongly regressive; and this proves our claim. For a slightly different example of this type, see Turinici [48] and the references therein.

Now, it is natural to establish the connection between these introduced classes and the Meir-Keeler one. An appropriate answer to this is contained in

Proposition 11 *Under these conventions, the following inclusions are valid:*

(36-1) *If $\varphi \in \mathscr{F}(re)(R^0_+, R)$ is Boyd-Wong admissible, then φ is Meir-Keeler admissible (or, equivalently: Matkowski admissible)*

(36-2) *The function $\varphi \in \mathscr{F}(re, in)(R^0_+, R)$ is Matkowski admissible, if and only if it is Meir-Keeler admissible.*

Proof

(i) (cf. Meir and Keeler [29]). Suppose that $\varphi \in \mathscr{F}(re)(R^0_+, R)$ is Boyd-Wong admissible; and fix $\gamma > 0$. As $\Lambda^+\varphi(\gamma) < \gamma$, there exists $\beta = \beta(\gamma) > 0$ such that $\Phi(\gamma+)(\beta) < \gamma$; wherefrom,

$\gamma < t < \gamma + \beta$ implies $\varphi(t) < \gamma$;

hence the claim.

(ii) The verification of this fact is entirely deductible from the one developed in a related statement; however, for completeness reasons, we shall give a (different) reasoning for it.

(ii-1) (cf. Jachymski [17]). Assume that $\varphi \in \mathscr{F}(re, in)(R^0_+, R)$ is Matkowski admissible; we have to establish that it is Meir-Keeler admissible. If the underlying property fails, then (for some $\gamma > 0$):

$\forall \beta > 0, \exists t \in]\gamma, \gamma + \beta[$, such that $\varphi(t) > \gamma$.

As φ is increasing, this yields (by the arbitrariness of β)

$(\varphi(t) > \gamma, \forall t > \gamma)$; whence, by induction: $(\varphi^n(t) > \gamma, \forall n, \forall t > \gamma)$.

Taking some $t > \gamma$ and passing to limit as $n \to \infty$, one gets $0 \geq \gamma$; contradiction; and the claim follows.

(ii-2) Assume that $\varphi \in \mathscr{F}_0(re, in)(R^0_+, R)$ is Meir-Keeler admissible; we have to establish that it is Matkowski admissible. Given $s_0 > 0$, suppose that $(s_n = \varphi^n(s_0); n \geq 0)$ exists in R^0_+. By the regressive property of φ, (s_n) is strictly descending; hence, $s := \lim_n s_n$ exists, with $s_n > s$, for all n. Suppose by contradiction that $s > 0$; and let $r > 0$ be the number assured by the Meir-Keeler admissible property of φ. By definition, there exists a rank $n(r) \geq 0$, such that

$n \geq n(r)$ implies $s < s_n < s + r$.

This, by the underlying property, gives (for the same ranks)

$s < s_{n+1} = \varphi(s_n) \leq s$; contradiction.

Hence, $s = 0$; wherefrom φ is Matkowski admissible.

Remark 2 Concerning the reverse of our first inclusion, the answer is negative. In fact (cf. a previous example) let $(r_n; n \geq 0)$ be a strictly descending sequence in R^0_+ with $r_n \to 0$; and take the function $\varphi \in \mathscr{F}(re, in)(R^0_+, R)$, according to

$$\varphi(t) = r_0, \text{ if } r_0 < t$$

$$\varphi(t) = r_{n+1}, \text{ if } r_{n+1} < t \leq r_n, \text{ for all } n.$$

Clearly,

$(\forall n)$: $\varphi(r_n + 0) = r_n$; whence $\Lambda^+\varphi(r_n) = r_n$;

which tells us that φ is not Boyd-Wong admissible. On the other hand,

φ is Matkowski admissible; since $\lim_n \varphi^n(t) = \lim_n r_n = 0$, $\forall t \in R^0_+$;

hence (cf. the previous result), φ Meir-Keeler admissible.

With these preliminaries, we may now return to the posed question. Precisely, let χ be a function in $\mathscr{F}(re)(R^0_+, R)$. Define the *Matkowski* relation $\Omega := \Omega[\chi]$ over R^0_+, as

$(t, s \in R^0_+)$: $(t, s) \in \Omega$ iff $t \leq \chi(s)$;

note that, as χ=regressive, Ω is upper diagonal.

Proposition 12 *The following properties (involving our upper diagonal relation $\Omega := \Omega[\chi]$) are valid:*

(37-1) *Ω is first variable decreasing*
(37-2) *Ω is second variable increasing, when χ is increasing*
(37-3) *Ω is asymptotic Meir-Keeler (hence, geometric Meir-Keeler; hence, Matkowski admissible) when χ is Meir-Keeler admissible (or, equivalently: Matkowski admissible)*
(37-4) *Ω is geometric left separable (or, equivalently: asymptotic left separable).*

Proof

(i), (ii) Evident, by definition.
(iii) Assume that χ is Meir-Keeler admissible; we claim that Ω is asymptotic Meir-Keeler. Suppose that this is not true

there exist strictly descending sequences (t_n) and (s_n) in R_+^0
and elements ε in R_+^0, with
$((t_n, s_n) \in \Omega, \forall n)$ and $(t_n \to \varepsilon+, s_n \to \varepsilon+)$.

By the very definition of our relation, this means

there exist strictly descending sequences (t_n) and (s_n) in R_+^0
and elements ε in R_+^0, with
$(t_n \leq \chi(s_n), \forall n)$ and $(t_n \to \varepsilon+, s_n \to \varepsilon+)$.

Let $\delta > 0$ be the number given by the Meir-Keeler property of χ. By the above relations, there exists some index $n(\delta)$, such that

$$\varepsilon < t_n \leq \chi(s_n) < s_n < \varepsilon + \delta, \text{ for all } n \geq n(\delta).$$

But then (according to the underlying property) we have (for the same ranks)

$$\chi(s_n) \leq \varepsilon; \text{ whence, } t_n \leq \varepsilon;$$

in contradiction with the above property. Hence, our working assumption is not acceptable; and the conclusion follows.

(iv) Let $\beta > 0$ be arbitrary fixed. As χ is regressive,

$$\chi(\beta) < \beta; \text{ whence, } \theta := \max\{0, \chi(\beta)\} < \beta.$$

Denote $\gamma := (1/2)(\theta + \beta)$; clearly, $0 \leq \theta < \gamma < \beta$. By definition, we have

$$t \in]\gamma, \beta[\text{ implies } t > \chi(\beta) \text{ (i.e., } (t, \beta) \notin \Omega);$$

and, from this, Ω appears as geometric left separable.

Part-Case II The second particular case corresponds to the choice

$$\xi(t, s) = \psi(s) - \psi(t) - \varphi(s), \text{ where } \psi, \varphi \in \mathscr{F}(R_+^0, R).$$

So, let (ψ, φ) be a pair of functions in $\mathscr{F}(R_+^0, R)$, fulfilling

(Rhoades) (ψ, φ) is a *Rhoades couple*:
ψ is increasing, and φ is *strictly positive* $(\varphi(t) > 0, \forall t > 0)$.

Define the *Rhoades* associated relation $\Omega := \Omega[\psi, \varphi]$ in $R_+^0 \times R_+^0$, as

$$(t, s \in R_+^0): (t, s) \in \Omega \text{ iff } \psi(t) \leq \psi(s) - \varphi(s).$$

Note that, by the Rhoades couple condition above, Ω is upper diagonal. In fact, let $t, s \in R_+^0$ be such that

$$(t, s) \in \Omega; \text{ i.e., } \psi(t) \leq \psi(s) - \varphi(s).$$

By the strict positivity of φ, one gets $\psi(t) < \psi(s)$; and this, along with the increasing property of ψ, shows that $t < s$; whence, the assertion. The introduced convention is related to the developments in Rhoades [37]. (Some related aspects may be found in Wardowski [56]; see also Dutta and Choudhury [13]).

Now, further properties of this associated relation are available under some extra sequential conditions upon the Rhoades couple (ψ, φ) to be considered; precisely,

(a-pos) φ is *asymptotic positive*:
for each strictly descending sequence $(t_n; n \geq 0)$ in R^0_+
and each $\varepsilon > 0$ with $t_n \to \varepsilon+$, we must have $\limsup_n(\varphi(t_n)) > 0$

(bd-left-osc) ψ is *φ-bounded left oscillating*:
for each $\beta > 0$, we have $\varphi(\beta) > \psi(\beta) - \psi(\beta - 0)$.

Proposition 13 *Under the precise framework, the following conclusions hold:*

(38-1) *Ω is first variable decreasing*
(38-2) *Ω is second variable increasing if $\psi - \varphi$ is increasing*
(38-3) *Ω is asymptotic Meir-Keeler (hence, geometric Meir-Keeler; hence, Matkowski admissible), if φ is asymptotic positive*
(38-4) *Ω is geometric left separable (or, equivalently: asymptotic left separable) when ψ is φ-bounded left oscillating.*

Proof

(i), (ii) Evident.

(iii) Suppose that Ω does not have the asymptotic Meir-Keeler property: there exist strictly descending sequences (t_n), (s_n) in R^0_+ and elements ε in R^0_+, with

$$((t_n, s_n) \in \Omega, \forall n) \text{ and } (t_n \to \varepsilon+, s_n \to \varepsilon+).$$

By the former of these, we get

$$(0 <)\varphi(s_n) \leq \psi(s_n) - \psi(t_n), \forall n.$$

Passing to the superior limit as $n \to \infty$, and noting that

$$\lim_n \psi(s_n) = \lim_n \psi(t_n) = \psi(\varepsilon + 0),$$

one gets $\limsup_n \varphi(t_n) = 0$; in contradiction with the asymptotic positive property of φ. So, necessarily, Ω has the asymptotic Meir-Keeler property.

(iv) Suppose, by contradiction, that Ω is not asymptotic left separable; i.e., there exist strictly ascending sequences (t_n) in R^0_+ and elements β in R^0_+ with

$$((t_n, \beta) \in \Omega, \forall n) \text{ and } t_n \to \beta-.$$

By the former of these, we get (via φ=strictly positive)

$$(0 <)\varphi(\beta) \leq \psi(\beta) - \psi(t_n), \forall n.$$

Passing to limit as $n \to \infty$ yields (as ψ=increasing)

$$\varphi(\beta) \leq \psi(\beta) - \psi(\beta - 0);$$

or, in other words: ψ is not φ-bounded left oscillating at β; contradiction. So, Ω is asymptotic left separable; as asserted.

In the following, some practical examples of (Rhoades) couples (ψ, φ) fulfilling the imposed extra conditions in the statement above are to be discussed.

Part-Case II-a The construction in the preceding step (involving a certain $\chi \in \mathscr{F}(re)(R^0_+, R)$) is nothing else than a particular case of this one, corresponding to the choice

$$\psi(t) = t, \varphi(t) = t - \chi(t), t \in R^0_+.$$

Since the verification is immediate, we do not give details.

Part-Case II-b Let $\psi \in \mathscr{F}(R^0_+, R)$ be an increasing function and $\tau > 0$ be a constant. Taking $\varphi \in \mathscr{F}(R^0_+, R)$ as

$$\varphi(t) = \tau, \text{ for all } t \in R^0_+,$$

the couple (ψ, τ) of functions over $\mathscr{F}(R^0_+, R)$ appears as a Rhoades one, with φ=asymptotic positive. As a consequence, the relation $\Omega := \Omega[\psi, \tau]$ fulfills

Ω is first variable decreasing, second variable increasing, and asymptotic Meir-Keeler (hence, Matkowski admissible).

Note that, a direct proof of the last affirmation above is available; we do not give details. Further aspects may be found in Suzuki and Vetro [45].

Part-Case II-c Let $\lambda : R^0_+ \to]1, \infty[$ and $\mu : R^0_+ \to]0, 1[$ be a couple of functions, with λ=increasing. Define a relation $\Omega := \Omega[[\lambda, \mu]]$ over R^0_+ as

$$t\Omega s \text{ iff } \lambda(t) \leq [\lambda(s)]^{\mu(s)}.$$

This will be referred to as the *Jleli-Samet relation* attached to $\lambda(.)$ and $\mu(.)$. [The proposed convention comes from the developments in Jleli and Samet [20], corresponding to $\mu(.)$=constant). By a direct calculation, it is evident that

$$t\Omega s \text{ iff } t\Omega[\psi, \varphi]s; \text{ where } \psi(t) = \log[\log(\lambda(t))], \varphi(t) = -\log(\mu(t)), t > 0.$$

Hence, this construction is entirely reducible to the standard one in this series.

In particular, when ψ and φ [appearing at the beginning of this construction] are continuous, our statements reduce to the geometric one in Jachymski [19]; see also Suzuki [44]. Nevertheless, it is to be stressed that the proposed techniques cannot help us— in general— to handle contractive conditions like in Khan et al. [24]; so, we may ask of under which conditions is this removable. Further aspects will be delineated elsewhere.

4 Geometric Contractions in OMS

Let X be a nonempty set, $d : X \times X \to R_+$ be a metric on X, and $(\leq)$ be a *quasi-order* on X; the triple $(X, d, \leq)$ will be referred to as a *quasi-ordered metric space*. Call the subset Y of X, $(\leq)$-*asingleton*, if $[y_1, y_2 \in Y, y_1 \leq y_2$ imply $y_1 = y_2]$; and $(\leq)$-*singleton*, if in addition Y is nonempty.

(A) Further, take some $T \in \mathscr{F}(X)$. Assume in the following that

(s-prog) T is semi-progressive: $X(T, \leq) := \{x \in X; x \leq Tx\} \neq \emptyset$
(incr) T is increasing: $x \leq y$ implies $Tx \leq Ty$.

The basic directions under which the fixed points of T are to be determined are shown in the list below, comparable with the one in Turinici [52]:

(pic-0) We say that T is *fix-($\leq$)-asingleton*, when $\text{Fix}(T)$ is an ($\leq$)-asingleton; and *fix-($\leq$)-singleton*, when $\text{Fix}(T)$ is a ($\leq$)-singleton
(pic-1) We say that $x \in X(T, \leq)$ is a *Picard point* (modulo $(d, \leq; T)$) if the iterative sequence $(T^n x; n \geq 0)$ is d-Cauchy; when this property holds for all $x \in X(T, \leq)$, then T is called a *Picard operator* (modulo $(d, \leq)$)
(pic-2) We say that $x \in X(T, \leq)$ is a *strong Picard point* (modulo $(d, \leq; T)$) if $(T^n x; n \geq 0)$ is d-convergent and $\lim_n(T^n x) \in \text{Fix}(T)$; when this property holds for all $x \in X(T, \leq)$, then T is called a *strong Picard operator* (modulo $(d, \leq)$)
(pic-3) We say that $x \in X(T, \leq)$ is a *Bellman Picard point* (modulo $(d, \leq; T)$) if $(T^n x; n \geq 0)$ is d-convergent and $[T^n x \leq \lim_n(T^n x) \in \text{Fix}(T), \forall n]$; when this property holds for all $x \in X(T, \leq)$, then T is called a *Bellman Picard operator* (modulo $(d, \leq)$).

The sufficient (regularity) conditions for such properties are being founded on *ascending orbital full* concepts (in short: (a-o-f)-concepts). Namely, let us say that the sequence $(z_n; n \geq 0)$ in X is

ascending, if $(\forall i, \forall j)$: $i \leq j$ implies $z_i \leq z_j$
orbital, when $(z_n = T^n x; n \geq 0)$, for some $x \in X(T, \leq)$
full, provided $n \mapsto z_n$ is injective ($i \neq j$ implies $z_i \neq z_j$);

the intersection of these is just the precise concept.

(reg-1) Call X, *(a-o-f,d)-complete* provided (for each (a-o-f)-sequence) d-Cauchy $\Longrightarrow$ d-convergent
(reg-2) We say that T is *(a-o-f,d)-continuous*, if [(z_n)=(a-o-f)-sequence and $z_n \xrightarrow{d} z$] imply $Tz_n \xrightarrow{d} Tz$
(reg-3) Call ($\leq$), *(a-o-f,d)-selfclosed* when [(z_n)=(a-o-f)-sequence and $z_n \xrightarrow{d} z$] imply $[z_n \leq z, \forall n]$.

Finally, as a basic completion of these facts, we have to formulate the metrical conditions to be used. Let ($<$) stand for the relation

$x < y$ iff $x \leq y$ and $x \neq y$.

Clearly, ($<$) is *irreflexive*; but not transitive, as long as ($\leq$) is not antisymmetric. Further, denote for each $x, y \in X$,

$$A_1(x, y) = d(x, y),\ M(x, y) = \text{diam}\{x, Tx, y, Ty\}$$
$$A_2(x, y) = (1/2)[d(x, Tx) + d(y, Ty)],\ A_3(x, y) = \max\{d(x, Tx), d(y, Ty)\}$$
$$A_4(x, y) = (1/2)[d(x, Ty) + d(Tx, y)],\ \mathscr{A} = \{A_1, A_2, A_3, A_4\}.$$

By taking all possible maxima between these, one gets $2^4 - 1 = 15$ functions of this type (including the ones we just listed) that may be collected in the class

$$\mathscr{B} := \{\max(\mathscr{K}); \mathscr{K} \in \exp(\mathscr{A})\}.$$

Note that, as $A_2 \leq A_3$, some of these maxima are identical. Taking the practical importance of these into account, we arrive at the following basic maps

$$A_1, (B_i = \max\{A_1, A_i\}, i \in \{2, 3, 4\}), (C_j = \max\{A_1, A_j, A_4\}, j \in \{2, 3\}).$$

Denote, for simplicity,

$$\mathscr{G}_1 = \{A_1, B_2, B_4, C_2\}, \mathscr{G}_2 = \{B_3, C_3\}, \mathscr{G} = \mathscr{G}_1 \cup \mathscr{G}_2.$$

Note that, for each $G \in \mathscr{G}$ we have

$A_1(x, y) \leq G(x, y), \forall x, y \in X$ (G is *separated*).

Moreover, for each $G \in \mathscr{G}$ we have

$G(x, y) \leq M(x, y), \forall x, y \in X$ (G is *diametral*).

In fact, by the definitions above, we have

$$A_k(x, y) \leq M(x, y), \forall x, y \in X, \forall k \in \{1, 2, 3, 4\};$$

and, from this, all is clear. Finally, for each $G \in \mathscr{G}$, one has

$G(x, Tx) \leq A_3(x, Tx), \forall x \in X$ (G is *orbitally bounded*).

Indeed, the case $G = A_1$ is clear; and— for the remaining ones— we have

$$A_4(x, Tx) = (1/2)d(x, T^2x) \leq A_2(x, Tx) \leq A_3(x, Tx), \forall x \in X;$$

hence the claim.

Given some $G \in \mathscr{G}$, let us say that T is *Meir-Keeler* $(d, \leq; G)$*-contractive*, if

(mk-1) $x < y$ imply $d(Tx, Ty) < G(x, y)$
(T is *strictly nonexpansive* (modulo $(d, \leq; G)$))

(mk-2) $\forall \varepsilon > 0, \exists \delta > 0$: $[x < y, \varepsilon < G(x, y) < \varepsilon + \delta] \Longrightarrow d(Tx, Ty) \leq \varepsilon$
(T has the *Meir-Keeler property* (modulo $(d, \leq; G)$)). Note that, by the former of these, the Meir-Keeler property may be written as

(mk-3) $\forall \varepsilon > 0, \exists \delta > 0$: $[x < y, G(x, y) < \varepsilon + \delta] \Longrightarrow d(Tx, Ty) \leq \varepsilon$.

In particular, when $G = A_1$, this convention is comparable with the one due to Meir and Keeler [29].

A geometric version of the above concept may be given along the lines below. Remember that the relation $\Omega \subseteq R_+^0 \times R_+^0$ is called *upper diagonal*, provided

(u-diag) $(t, s) \in \Omega$ implies $t < s$;

the class of all these will be denoted as $\text{udiag}(R_+^0)$. Further, let us introduce the conditions (over the class $\text{udiag}(R_+^0)$)

(g-mkr-1) (Ω has the *geometric Meir-Keeler property*)
$\forall \varepsilon > 0, \exists \delta > 0$, such that: $(t, s) \in \Omega, \varepsilon < s < \varepsilon + \delta \Longrightarrow t \leq \varepsilon$

(g-mkr-2) (Ω is *geometric left separable*)
$\forall \beta > 0, \exists \gamma \in]0, \beta[$, such that: $\gamma < t < \beta \Longrightarrow (t, \beta) \notin \Omega$.

Now, given $G \in \mathscr{G}$ and the relation $\Omega \subseteq R^0_+ \times R^0_+$, let us say that the selfmap T is $(d, \leq; G; \Omega)$-*contractive*, provided

(G-Om-con) $(d(Tx, Ty), G(x, y)) \in \Omega, \forall x, y \in X, x < y, Tx < Ty$.

Proposition 14 *Suppose that the selfmap T is $(d, \leq; G; \Omega)$-contractive where $\Omega \subseteq R^0_+ \times R^0_+$ is upper diagonal and geometric Meir-Keeler. Then, necessarily, T is Meir-Keeler $(d, \leq; G)$-contractive.*

Proof

(i) Let $x, y \in X$ be such that $x < y$; hence, $G(x, y) > 0$ and (as T is increasing), $Tx \leq Ty$. If $Tx = Ty$, then $d(Tx, Ty) = 0 < G(x, y)$. Suppose now that $Tx \neq Ty$; hence, $Tx < Ty$. As a consequence of this,

$$(t, s) \in \Omega; \text{ where } t := d(Tx, Ty), s := G(x, y).$$

Combining with the upper diagonal property of Ω, one gets $t < s$; i.e., $d(Tx, Ty) < G(x, y)$. Summing up, T is strictly nonexpansive (modulo $(d, \leq; G)$).

(ii) Let $\varepsilon > 0$ be arbitrary fixed; and $\delta > 0$ be the number assured by the geometric Meir-Keeler property for Ω. Further, let $x, y \in X$ be such that $x < y$ (hence, $Tx \leq Ty$, $s := G(x, y) > 0$) and $\varepsilon < G(x, y) < \varepsilon + \delta$; i.e., $\varepsilon < s < \varepsilon + \delta$. If $Tx = Ty$, then $d(Tx, Ty) = 0 < \varepsilon$. Suppose now that $Tx \neq Ty$; whence, $Tx < Ty$, $t := d(Tx, Ty) > 0$. By definition, we must have $(t, s) \in \Omega$; and this along with $\varepsilon < s < \varepsilon + \delta$ gives (by the geometric Meir-Keeler property for Ω), $t \leq \varepsilon$; i.e., $d(Tx, Ty) \leq \varepsilon$. Putting these alternatives together, it follows that T has the Meir-Keeler property (modulo $(d, \leq; G)$).

In the following, a kind of reciprocal for this result is formulated. Given the selfmap T (endowed with the general properties we already specified) and the map $G \in \mathscr{G}$, let $\Omega = \Omega[d, \leq; G; T]$ stand for the associated relation over R^0_+:

$\Omega = \{(d(Tx, Ty), G(x, y)); x < y, Tx < Ty\}$; or, in other words:
$(t, s) \in \Omega$ iff $t = d(Tx, Ty), s = G(x, y)$, where $x < y, Tx < Ty$.

Proposition 15 *Let the couple T and G be taken as before. Then,*

(42-1) *If T is Meir-Keeler $(d, \leq; G)$-contractive, then the attached relation $\Omega = \Omega[d, \leq; G; T]$ over R^0_+ is upper diagonal and geometric Meir-Keeler*

(42-2) *T is Meir-Keeler $(d, \leq; G)$-contractive if and only if the attached relation $\Omega = \Omega[d, \leq; G; T]$ over R^0_+ is upper diagonal and geometric Meir-Keeler.*

Proof

(i) Suppose that T is Meir-Keeler $(d, \leq; G)$-contractive. There are two steps to be passed.

(i-1) Let $(t, s) \in R_+^0 \times R_+^0$ be such that $(t, s) \in \Omega$; hence (by definition)

$t = d(Tx, Ty), s = G(x, y)$, where $x < y, Tx < Ty$.

From the strict nonexpansive property of T, we must have $d(Tx, Ty) < G(x, y)$; or, equivalently, $t < s$, which shows that Ω is upper diagonal.

(i-2) Let $\varepsilon > 0$ be arbitrary fixed; and $\delta > 0$ be the number associated by the Meir-Keeler property for T. Further, let $(t, s) \in R_+^0 \times R_+^0$ be such that $(t, s) \in \Omega$ and $\varepsilon < s < \varepsilon + \delta$; hence (see above)

$t = d(Tx, Ty), s = G(x, y)$, where $x < y, Tx < Ty$;
so that (by definition): $x < y$, and $\varepsilon < G(x, y) < \varepsilon + \delta$.

By the Meir-Keeler-property for T, we get

$d(Tx, Ty) \leq \varepsilon$; i.e. (under our notation): $t \leq \varepsilon$;

so that Ω has the geometric Meir-Keeler property. Hence, by simply combining these, it results that Ω is upper diagonal and geometric Meir-Keeler.

(ii) Suppose that the associated relation $\Omega = \Omega[d, \leq; G; T]$ over R_+^0 is upper diagonal and geometric Meir-Keeler. By the very definition of this object, T is $(d, \leq; G; \Omega)$-contractive. Combining with the preceding result, one derives that T appears as Meir-Keeler $(d, \leq; G)$-contractive; and the conclusion follows.

As a consequence of this, it follows that the Meir-Keeler $(d, \leq; G)$-contractive properties of T are finally reducible to the upper diagonal and geometric Meir-Keeler properties for the associated relation $\Omega[d, \leq; G; T]$. Concerning this aspect, remember that various examples of such objects were treated in a previous place; see also Lim [27].

Let $(X, d, \leq)$ be a quasi-ordered metric space. Further, take some selfmap $T \in \mathscr{F}(X)$; supposed to satisfy

T is $(\leq)$-semi-progressive and $(\leq)$-increasing.

As precise, we have to determine conditions assuring that $\mathrm{Fix}(T)$ is nonempty. The specific directions under which this problem is to be solved were already listed. Sufficient requirements for getting such properties involve the *ascending orbital full concepts* (in short: (a-o-f)-concepts) and *contractive properties* we just introduced.

The following fixed point result (referred to as: Meir-Keeler geometric fixed point principle; in short: (MK-g-fpp)) will be useful in our next developments.

Theorem 2 *Assume that the selfmap T is $(d, \leq; G; \Omega)$-contractive, for some $G \in \mathscr{G}$ and some upper diagonal geometric Meir-Keeler relation $\Omega \subseteq R_+^0 \times R_+^0$. In addition, let X be (a-o-f,d)-complete. Then, the following conclusions hold:*

(41-a) *T is fix-$(\leq)$-asingleton*

(41-b) *T is a strong Picard operator (modulo $(d, \leq)$) when, in addition, T is (a-o-f,d)-continuous*

(41-c) *T is a Bellman Picard operator (modulo $(d, \leq)$) when, in addition, $(\leq)$ is (a-o-f,d)-selfclosed and*

either ($G \in \mathscr{G}_1$) or ($G \in \mathscr{G}_2$ and Ω is geometric left separable).

Proof There are several steps to be passed.

Part 1 Let us firstly check the fix-$(\leq)$-asingleton property. Take a couple of points $z_1, z_2 \in \text{Fix}(T)$ with $z_1 \leq z_2$; and assume by contradiction that $z_1 \neq z_2$; whence $z_1 < z_2$ (as well as $Tz_1 < Tz_2$). The contractive condition is thus applicable to (z_1, z_2); and yields (via Ω=upper diagonal)

$(d(z_1, z_2) =)d(Tz_1, Tz_2)\Omega G(z_1, z_2)$; whence, $d(z_1, z_2) < G(z_1, z_2)$.

On the other hand, as G is diametral, we derive

$G(z_1, z_2) \leq M(z_1, z_2) = d(z_1, z_2)$.

The obtained facts are thus contradictory; and this proves the assertion.

Part 2 We claim that, under the conditions in our statement

(2-iter) $d(Tx, T^2x)\Omega G(x, Tx)(= d(x, Tx))$ and
$d(Tx, T^2x) < G(x, Tx)(= d(x, Tx))$, whenever $x < Tx, Tx < T^2x$.

In fact, let $x \in X$ be as in the premise above. By the contractive property (and Ω=upper diagonal),

$d(Tx, T^2x)\Omega G(x, Tx)$ and $d(Tx, T^2x) < G(x, Tx)$.

On the other hand, as $G \in \mathscr{G}$ is orbitally bounded,

$(d(x, Tx) \leq)\ G(x, Tx) \leq A_3(x, Tx) = \max\{d(x, Tx), d(Tx, T^2x)\}$.

Combining with the preceding relation gives

$d(Tx, T^2x) < \max\{d(x, Tx), d(Tx, T^2x)\}$;

and this, in turn, yields $d(Tx, T^2x) < d(x, Tx)$. Note that, as a consequence,

$A_3(x, Tx) = d(x, Tx)$; whence: $d(x, Tx) \leq G(x, Tx) \leq d(x, Tx)$;

and our claim follows.

It remains now to establish the strong/Bellman Picard property (modulo $(d, \leq)$). Take some $x_0 \in X(T, \leq)$, and put $(x_n = T^n x_0; n \geq 0)$; this is an ascending orbital sequence. If $x_n = x_{n+1}$ for some $n \geq 0$, we are done (via $x_n \in \text{Fix}(T)$); so, without loss, one may assume that, for all $n \geq 0$,

(tele-dist) $x_n \neq x_{n+1}$; hence, $x_n < x_{n+1}, \rho_n := d(x_n, x_{n+1}) > 0$.

Part 3 By the preceding developments, we have

$\rho_{n+1}\Omega\rho_n, \rho_{n+1} < \rho_n, \forall n$;

so, the sequence $(\rho_n; n \geq 0)$ is strictly descending in R_+; whence, $\rho := \lim_n \rho_n$ exists as an element of R_+. Assume by contradiction that $\rho > 0$; and let $\sigma > 0$

be the number given by the geometric Meir-Keeler property upon Ω. By definition, there exists a rank $n(\sigma) \geq 0$, such that

$n \geq n(\sigma)$ implies $\rho < \rho_n < \rho + \sigma$.

Taking a previous relation into account, one derives (for all $n \geq n(\sigma)$)

$(x_n < x_{n+1}$ and) $\rho < G(x_n, x_{n+1}) = \rho_n < \rho + \sigma$;

and this, combined with the geometric Meir-Keeler property, gives

$(\forall n \geq n(\sigma))$: $\rho < \rho_{n+1} \leq \rho$;

a contradiction. Hence, $\rho = 0$; so that

(x_n) is d-semi-Cauchy [$\rho_n := d(x_n, x_{n+1}) = d(x_n, Tx_n) \to 0$, as $n \to \infty$].

Part 4 Suppose that

(non-inj) there exist $i, j \in N$ such that $i < j$, $x_i = x_j$.

Denoting $p = j - i$, we thus have $p > 0$ and $x_i = x_{i+p}$; so that

$x_{i+1} = x_{i+p+1}$; whence $\rho_i = \rho_{i+p}$;

in contradiction with the strict descending property of $(\rho_n; n \geq 0)$. Hence, our working hypothesis cannot hold; wherefrom

$i < j \Longrightarrow x_i < x_j$; so that $(x_n; n \geq 0)$ is a full sequence.

Part 5 We now establish that $(x_n; n \geq 0)$ is a d-Cauchy sequence. Let $\gamma > 0$ be arbitrary fixed; and $\beta > 0$ be the number associated by the Meir-Keeler property of Ω; without loss, one may assume that $\beta < \gamma$. As $(x_n; n \geq 0)$ is d-semi-Cauchy, there exists $n(\beta) \geq 0$, with

(d-s-C) $d(x_n, x_{n+1}) < \beta/4$, for all $n \geq n(\beta)$;
whence: $d(x_n, x_{n+i}) < \beta/2$, $\forall i \in \{1, 2\}$, $\forall n \geq n(\beta)$.

We now claim that

(d-C) $(\forall p \geq 1)$: $[d(x_n, x_{n+p}) < \gamma + \beta/2, \forall n \geq n(\beta)]$;
wherefrom, the desired property of (x_n) follows. To do this, an induction argument upon p is applied. The case $p \in \{1, 2\}$ is clear, by the above. Assume that our evaluation holds, for $p \in \{1, \ldots, q\}$, where $q \geq 2$; we show that it holds as well for $p = q + 1$. So, let $n \geq n(\beta)$ be arbitrary fixed. By a previous stage, we have

$x_n < x_{n+q}, x_{n+1} < x_{n+q+1}$;

which tells us that the contractive condition applies to (x_n, x_{n+q}); and yields

$(d(x_{n+1}, x_{n+q+1}) =) d(Tx_n, Tx_{n+q}) \Omega G(x_n, x_{n+q})$.

On the other hand, by the d-semi-Cauchy property and inductive hypothesis,

$$d(x_{n+i}, x_{n+q}) < \gamma + \beta/2, \ \forall i \in \{0, 1\}$$
$$d(x_{n+1}, x_{n+q+1}) < \gamma + \beta/2 < \gamma + \beta,$$
$$d(x_n, x_{n+1}), d(x_{n+q}, x_{n+q+1}) < \beta/2.$$

Moreover, from the triangular inequality,

$$d(x_n, x_{n+q+1}) \leq d(x_n, x_{n+q}) + d(x_{n+q}, x_{n+q+1}) < \gamma + \beta;$$

and this, by the diametral property of G, yields

$$G(x_n, x_{n+q}) \leq M(x_n, x_{n+q}) < \gamma + \beta.$$

Taking the contractive condition into account yields [by the (variant of) geometric Meir-Keeler property upon Ω]

$$d(x_{n+1}, x_{n+q+1}) = d(Tx_n, Tx_{n+q}) \leq \gamma;$$

so that, by the triangular inequality,

$$d(x_n, x_{n+q+1}) \leq d(x_n, x_{n+1}) + d(x_{n+1}, x_{n+q+1}) < \gamma + \beta/2;$$

and the assertion follows.

Part 6 As X is (a-o-f,d)-complete, $x_n \xrightarrow{d} z$, for some $z \in X$. There are two cases to discuss.

Case 6a Suppose that T is (a-o-f,d)-continuous. Then

$$y_n := Tx_n \xrightarrow{d} Tz, \text{ as } n \to \infty.$$

On the other hand, $(y_n = x_{n+1}; n \geq 0)$ is a subsequence of $(x_n; n \geq 0)$; whence $y_n \xrightarrow{d} z$; and this yields (as d is separated), $z = Tz$.

Case 6b Now, let us assume that $(\leq)$ is (a-o-f,d)-selfclosed. By the convergence property above, it results that (via T=increasing)

$(\forall n)$: $x_n \leq z$, and $Tx_n \leq Tz$.

As $(x_n; n \geq 0)$ and $(Tx_n = x_{n+1}; n \geq 0)$ are full sequences,

$H_1 := \{n \geq 0; x_n = z\}$ and $H_2 := \{n \geq 0; Tx_n = Tz\}$ are asingletons;

so, necessarily, there exists some index $h \geq 0$, such that

(h-dist) $\quad n \geq h$ implies $x_n < z$, $Tx_n < Tz$.

We now intend to show that the alternative $z \neq Tz$ (hence $\beta := d(z, Tz) > 0$) yields a contradiction.

Define the subsequence (u_n) of (x_n) according to

$u_n := x_{h+n} = T^n x_h$ (hence, $Tu_n = x_{h+n+1} = u_{n+1}$), $n \geq 0$;

clearly, $(u_n; n \geq 0)$ is an (a-o-f)-sequence. Combining with our previous parts and the metrical continuous property of $d(.,.)$ (see above), one gets

$$d(u_n, Tu_n) \to 0, d(u_n, z), d(Tu_n, z) \to 0, d(u_n, Tz), d(Tu_n, Tz) \to \beta;$$

and this, in turn, yields $M(u_n, z) \to \beta$.

This, by definition, gives

$A_1(u_n, z) \to 0$, $A_2(u_n, z) \to \beta/2$, $A_3(u_n, z) \to \beta$, $A_4(u_n, z) \to \beta/2$;

so that, necessarily,

(lim-G1) $(\exists) \lim_n G(u_n, z) \in \{0, \beta/2\}$, for each map $G \in \mathscr{G}_1$.

Further, the convergence properties of $(u_n; n \geq 0)$ and $(Tu_n; n \geq 0)$ relative to z tell us that, for a certain rank $p(z) \geq 0$, we must have

$d(u_n, Tu_n) < \beta/2$, and $d(u_n, z), d(Tu_n, z) < \beta/2, \forall n \geq p(z)$.

Moreover, the same convergence properties of $(u_n; n \geq 0)$ and $(Tu_n; n \geq 0)$ relative to z give us (along with $\beta > 0$), some other rank $q(z) \geq 0$, with

$|d(u_n, Tz) - \beta|, |d(Tu_n, Tz) - \beta| < \beta/2, \forall n \geq q(z)$;

wherefrom (for the same ranks)

$\beta/2 < d(u_n, Tz), d(Tu_n, Tz) < 3\beta/2$.

Combining these yields, for all $n \geq r(z) := \max\{p(z), q(z)\}$

$$A_1(u_n, z) < \beta/2, \ A_2(u_n, z) < 3\beta/4, \ A_3(u_n, z) = \beta, \ A_4(u_n, z) < \beta;$$

whence, for the subsequence $(v_n := u_{r(z)+n}; n \geq 0)$ of $(u_n; n \geq 0)$, one has

$G(v_n, z) = \beta$, for all $n \geq 0$.

Two alternatives come into discussion.

Alter 1 Suppose that $G \in \mathscr{G}_1$. By the very construction of our subsequence,

$u_n < z, Tu_n < Tz$, for all $n \geq 0$.

Hence, the contractive condition is applicable to (u_n, z), for each $n \geq 0$; and yields (for these ranks)

(u-contr) $d(Tu_n, Tz) \Omega G(u_n, z)$, and $d(Tu_n, Tz) < G(u_n, z) \leq M(u_n, z)$.

Passing to limit yields $\lim_n G(u_n, z) = \beta$; in contradiction with a previous limit relation deduced from the choice of G. So, necessarily, $\beta = 0$; i.e.; $z = Tz$.

Alter 2 Suppose that $G \in \mathscr{G}_2$. From the above subsequence constructions,

$v_n < z, Tv_n < Tz$, for all $n \geq 0$.

Hence, the contractive condition is applicable to (v_n, z), for all n; and yields

$d(Tv_n, Tz) \Omega G(v_n, z) (= \beta)$, and $d(Tv_n, Tz) < G(v_n, z) (= \beta), \forall n$;

that is, (under the notation $(t_n := d(Tv_n, Tz); n \geq 0)$)

$(t_n, \beta) \in \Omega, t_n < \beta$, for all n.

This along with

$\lim_n t_n = \lim_n d(Tv_n, Tz) = \beta$ (see above)

gives us a contradiction with respect to the geometric left separable property of Ω. Hence, the alternative $\beta > 0$ is (again) impossible; wherefrom $\beta = 0$ [i.e., $z = Tz$]; and conclusion follows.

In the following, two basic particular cases of this result will be stated along the choice $G = A_1$. By a $(\leq, d)$-type regularity condition we shall mean any (a-o-f,d)-regularity condition, where the orbital and full properties are ignored.

Given some function $\chi \in \mathcal{F}(R^0_+, R)$, let us say that the selfmap T is $(d, \leq; \chi)$-*contractive*, provided

$d(Tx, Ty) \leq \chi(d(x, y))$, for all $x, y \in X, x < y, Tx < Ty$.

Likewise, given a couple of functions $\psi, \varphi \in \mathcal{F}(R^0_+, R)$, let us say that the selfmap T is $(d, \leq; (\psi, \varphi))$-*contractive*, provided

$\psi(d(Tx, Ty)) \leq \psi(d(x, y)) - \varphi(d(x, y))$, for all $x, y \in X, x < y, Tx < Ty$.

Now, as a consequence of Meir-Keeler geometric fixed point principle (MK-g-fpp), the following synthetic fixed point result (referred to as: Matkowski-Rhoades fixed point principle; in short: (MR-fpp)) is now available.

Theorem 3 *Suppose that T is semi-progressive, increasing, and fulfills one of the contractive conditions below*

(41-i) *T is $(d, \leq; \chi)$-contractive, for some Meir-Keeler admissible function $\chi \in \mathcal{F}(re)(R^0_+, R)$*

(41-ii) *T is $(d, \leq; (\psi, \varphi))$-contractive, for some Rhoades couple of functions $\psi, \varphi \in \mathcal{F}(R^0_+, R)$, with φ=asymptotic positive.*

In addition, let X be (a-o-f,d)-complete. Then,

(42-a) *T is fix-$(\leq)$-asingleton*

(42-b) *T is a strong Picard operator when T is $(\leq, d)$-continuous*

(42-c) *T is a Bellman Picard operator when $(\leq)$ is $(\leq, d)$-selfclosed.*

Proof There are two parts to be passed.

Part 1 Let the relation $\Omega := \Omega[\chi]$ over R^0_+ be introduced as

$(t, s \in R^0_+)$: $t\Omega s$ iff $t \leq \chi(s)$.

By the imposed upon χ conditions, Ω appears as upper diagonal and geometric Meir-Keeler. This, along with $G \in \mathcal{G}_1$ yields, via Meir-Keeler geometric fixed point principle (MK-g-fpp), all needed conclusions.

Part 2 Let the relation $\Omega := \Omega[\psi, \varphi]$ over R^0_+ be introduced as

$(t, s \in R^0_+)$: $t\Omega s$ iff $\psi(t) \leq \psi(s) - \varphi(s)$.

By the imposed upon (ψ, φ) conditions, Ω is upper diagonal and geometric Meir-Keeler. Combining with $G \in \mathscr{G}_1$ and Meir-Keeler geometric fixed point principle (MK-g-fpp), we are done.

Note that a different argument for proving the first half of Matkowski-Rhoades fixed point principle (MR-fpp) was provided in Turinici [50]; see also Agarwal et al. [1]. On the other hand, the second half of Matkowski-Rhoades fixed point principle (MR-fpp) yields, under a continuity condition upon ψ and φ, the main result in Dutta and Choudhury [13]. Further aspects involving these facts may be found in Turinici [54].

5 Razumikhin Chains

Let (E, d) be a metric space. Further, let $I = [a, b]$ (where $a < b$) be a bounded closed real interval; and $E_0 := C(I, E)$ stand for the class of all continuous functions $\varphi : I \to E$, endowed with the supremum distance

$D(\varphi, \xi) = \sup\{d(\varphi(t), \xi(t)); t \in I\}, \varphi, \xi \in E_0;$

clearly, (E_0, D) is a metric space.

Let in the following $\mathscr{G} : E_0 \to E$ be an operator. Define a relation $(\perp)$ on E_0 as

(Raz-r) $\varphi \perp \xi$ iff $D(\varphi, \xi) \leq d(\mathscr{G}\varphi, \mathscr{G}\xi)$;

referred to as: the *Razumikhin relation* attached to $\mathscr{G}$. Clearly,

(Raz-r-1) $(\perp)$ is *reflexive*: $\varphi \perp \varphi, \forall \varphi \in E_0$
(Raz-r-2) $(\perp)$ is *symmetric*: $\varphi \perp \xi$ iff $\xi \perp \varphi$;

however, it is not in general transitive, as simple examples show. Note that, by the very definition above, the following *sufficiency* property is valid:

(Raz-suf) $\varphi \perp \xi$ and $\mathscr{G}\varphi = \mathscr{G}\xi$ imply $\varphi = \xi$.

Having these precise, denote for each $\varphi \in E_0$

$E_0(\varphi, \perp) = \{\xi \in E_0; \varphi \perp \xi\}$ [the *section* of $(\perp)$ at φ].

A natural problem is that of determining those elements $\varphi \in E_0$ with "large" sections. The following answer to this is useful in the sequel.

Proposition 16 *Let $\varphi \in E_0$ and the (nonempty) subset M_0 of E_0 be such that*

$\mathscr{G}\varphi \in \mathscr{G}(M_0)$, $E_0(\varphi, \perp) \supseteq M_0$.

Then, necessarily, $\varphi \in M_0$.

Proof As $\mathscr{G}\varphi \in \mathscr{G}(M_0)$, there exists an element $\xi \in M_0$, such that $\mathscr{G}\varphi = \mathscr{G}\xi$. On the other hand, by the imposed hypothesis,

$\xi \in E(\varphi, \perp)$; whence, $\varphi \perp \xi$.

Combining these gives (by the sufficiency relation above) $\varphi = \xi$; so that $\varphi \in M_0$. This completes the argument.

In particular, call the subset $L_0 \subseteq E_0$, *full* provided $\mathscr{G}(L_0) = E$. As a consequence of the above result, we have that

L_0=full, $\varphi \in E_0 \setminus L_0$ implies [$E_0(\varphi, \perp) \supseteq L_0$ is false];

hence, for any $\varphi \in E_0 \setminus L_0$, the associated section $E(\varphi, \perp)$ is not very large.

The following "global" consequence of this fact is to be noted. Call the subset L_0 of E_0, a *Razumikhin* $\mathscr{G}$*-chain*, provided

(Raz-chn) $\varphi, \xi \in L_0 \Longrightarrow \varphi \perp \xi$;

i.e., any two elements of L_0 are ($\perp$)-comparable. The following fact about such objects— referred to as *Reduction statement*— is available.

Proposition 17 *Under the precise conventions,*

(52-1) *If the Razumikhin* $\mathscr{G}$*-chain* L_0 *of* E_0 *and the (nonempty) subset* M_0 *of* E_0 *are such that* [$\mathscr{G}(L_0) = \mathscr{G}(M_0)$, $L_0 \supseteq M_0$], *then,* $L_0 = M_0$

(52-2) *Let* L_0 *and* M_0 *be a couple of full Razumikhin* $\mathscr{G}$*-chains of* E_0*. Then, either* [$L_0 = M_0$], *or* [L_0 *and* M_0 *are not inclusion comparable*].

Proof

(i) Suppose that L_0 and M_0 are like in the premise above; and let $\varphi \in L_0$ be arbitrary fixed; hence, $\mathscr{G}\varphi \in \mathscr{G}(L_0) = \mathscr{G}(M_0)$. As L_0 is a Razumikhin $\mathscr{G}$-chain, we have (by the imposed hypothesis)

$$E_0(\varphi, \perp) \supseteq L_0 \supseteq M_0;$$

so that, from the previous result, $\varphi \in M_0$. As φ was arbitrarily chosen in L_0, we necessarily have $L_0 \subseteq M_0$; whence, $L_0 = M_0$.

(ii) Evident, by $\mathscr{G}(L_0) = \mathscr{G}(M_0) = E$ and the preceding step.

Some concrete examples of such objects will be given a bit further.

6 Conditional Result

Let E be a nonempty set. Take a metric $d : E \times E \to R_+$ on E, and a quasi-order ($\leq$) over E; then, $(E, d, \leq)$ will be referred to as a *quasi-ordered metric space*. Further, let $I = [a, b]$ (where $a < b$) be a bounded closed real interval; and $E_0 := C(I, E)$ denote the class of all continuous functions $\varphi : I \to E$. The mapping

$D(\varphi, \xi) = \sup\{d(\varphi(t), \xi(t)); t \in I\}$, $\varphi, \xi \in E_0$

is a metric on E_0 as it can be directly seen. Moreover, the relation introduced as

$(\varphi, \xi \in E_0)$: $\varphi \preceq \xi$ iff $\varphi(t) \leq \xi(t)$, for all $t \in I$

is a quasi-order on E_0; so that $(E_0, D, \preceq)$ is a quasi-ordered metric space.

Finally, let $\mathscr{G} : E_0 \to E$ be a (nonself) map. Remember that the *Razumikhin* relation $(\perp)$ attached to $\mathscr{G}$ writes

$$\varphi \perp \xi \text{ iff } D(\varphi, \xi) \leq d(\mathscr{G}\varphi, \mathscr{G}\xi);$$

it is reflexive and symmetric, but not in general transitive.

Remark 3 In particular, assume that

(non-exp) $\mathscr{G}$ is (D, d)-nonexpansive: $d(\mathscr{G}\varphi, \mathscr{G}\xi) \leq D(\varphi, \xi)$, $\forall \varphi, \xi \in E_0$.

Note that, in this case, we necessarily have

$\mathscr{G}$ is (D, d)-continuous [$\varphi_n \xrightarrow{D} \varphi$ implies $\mathscr{G}\varphi_n \xrightarrow{d} \mathscr{G}\varphi$].

Moreover, under such a condition, the associated Razumikhin relation becomes

$$\varphi \perp \xi \text{ iff } D(\varphi, \xi) = d(\mathscr{G}\varphi, \mathscr{G}\xi).$$

A basic example of such objects is given as below. Let $c \in I$ be arbitrary fixed. Then, the operator $\mathscr{G}_c : E_0 \to E$ introduced as

$$\mathscr{G}_c(\varphi) = \varphi(c), \varphi \in E_0,$$

is (D, d)-nonexpansive, as it can be directly seen; hence, all the more, (D, d)-continuous. Moreover, the associated Razumikhin relation $(\perp_c)$ writes (by the property above)

$$\varphi \perp_c \xi \text{ iff } D(\varphi, \xi) = d(\mathscr{G}_c(\varphi), \mathscr{G}_c(\xi)).$$

Returning to the general case, let $\mathscr{T} : E_0 \to E$ be a (nonself) mapping. We say that $\varphi \in E_0$ is a *coincidence point* of $\mathscr{G}$ and $\mathscr{T}$, when $\mathscr{G}\varphi = \mathscr{T}\varphi$; the class of all these will be denoted as $\mathrm{Cop}(\mathscr{G}, \mathscr{T})$. In particular, when $\mathscr{G} = \mathscr{G}_c$ (for some $c \in I$), the coincidence point relation for the couple $(\mathscr{G}_c, \mathscr{T})$ becomes: $\varphi(c) = \mathscr{T}\varphi$; and φ is then called a *PPF dependent fixed point* of T.

To state our main result concerning such points, the following concepts and constructions are needed [under the reduced axiomatic system (ZF-AC+DC)].

(I) Suppose that

(cond-1) $\mathscr{G} : E_0 \to E$ is surjective: $\mathscr{G}(E_0) = E$.

As a consequence, the map $\mathscr{G}^{-1} \circ \mathscr{T} : E_0 \to \exp[E_0]$ fulfills

$$\mathscr{G}^{-1} \circ \mathscr{T}(\varphi) \neq \emptyset, \text{ for each } \varphi \in E_0;$$

so, it is a proper relation over E_0. By the Dependent Choice principle [available in (ZF-AC+DC)], it follows that, for each $\varphi_0 \in E_0$, there exists a sequence $(\varphi_n; n \geq 0)$ in E_0 with

$$\varphi_{n+1} \in \mathscr{G}^{-1} \circ \mathscr{T}(\varphi_n) \text{ [i.e.; } \mathscr{T}\varphi_n = \mathscr{G}\varphi_{n+1}], n \geq 0;$$

it will be referred to as the $(\mathscr{G}, \mathscr{T})$-*iterative sequence* generated by φ_0.

(II) Let $(\leq_{\mathscr{G}})$ stand for the quasi-order on E_0

$$(\varphi, \xi \in E_0): \varphi \leq_{\mathscr{G}} \xi \text{ iff } \mathscr{G}\varphi \leq \mathscr{G}\xi.$$

Further, let us assume that

(cond-2) $(\mathscr{G}, \mathscr{T})$ is E_0-semi-progressive:
$E_0(\mathscr{G}, \mathscr{T}; \leq) := \{\varphi \in E_0; \mathscr{G}\varphi \leq \mathscr{T}\varphi\}$ is nonempty

(cond-3) $\mathscr{T}$ is $(\leq_{\mathscr{G}}, \leq)$-increasing: $\varphi \leq_{\mathscr{G}} \xi$ implies $\mathscr{T}\varphi \leq \mathscr{T}\xi$.

Then, for each $\varphi_0 \in E_0(\mathscr{G}, \mathscr{T}; \leq)$, the $(\mathscr{G}, \mathscr{T})$-iterative sequence $(\varphi_n; n \geq 0)$ generated by φ_0 fulfills

$$\varphi_n \leq_{\mathscr{G}} \varphi_{n+1}, \text{ (hence, } \mathscr{T}\varphi_n \leq \mathscr{T}\varphi_{n+1}\text{), for all } n \geq 0;$$

since the verification is immediate, we do not give details.

(III) We say that $\varphi_0 \in E_0(\mathscr{G}, \mathscr{T}; \leq)$ is $(\perp)$-*starting* when there exists a $(\mathscr{G}, \mathscr{T})$-iterative sequence $(\varphi_n; n \geq 0)$ generated by φ_0, with

$$(\varphi_n) \text{ is } (\perp)\text{-ascending } [\varphi_n \perp \varphi_{n+1}, \forall n].$$

This means that the iterative sequence in question $(\varphi_n; n \geq 0)$ fulfills

$$\varphi_n \leq_{\mathscr{G}} \varphi_{n+1}, \mathscr{T}\varphi_n = \mathscr{G}\varphi_{n+1}, \text{ and } \varphi_n \perp \varphi_{n+1}, \text{ for all } n;$$

in this case, $(\varphi_n; n \geq 0)$ will be referred to as a $(\mathscr{G}, \mathscr{T}; \perp)$-*iterative* sequence (in E_0) generated by φ_0; or, simply: a $(\varphi_0, \perp)$-*iterative* sequence.

(IV) The regularity conditions to be used are the following.

(reg-1) Let us say that E_0 is $(\leq_{\mathscr{G}}, D)$-*complete*, provided

(for each $(\leq_{\mathscr{G}})$-ascending sequence): D-Cauchy implies D-convergent.

(reg-2) Let us say that $(\leq_{\mathscr{G}})$ is $(\leq_{\mathscr{G}}, D)$-*selfclosed*, provided:

$(\varphi_n; n \geq 0)$ is $(\leq_{\mathscr{G}})$-ascending, $\varphi_n \xrightarrow{D} \varphi$ imply $\varphi_n \leq_{\mathscr{G}} \varphi$, for all n.

(V) Finally, as a basic completion of all these, we are passing to the contractive conditions to be considered.

Given $k \geq 0$, call $\mathscr{T}$, $(D, d; \leq_{\mathscr{G}}; k)$-*contractive* provided

(contr) $d(\mathscr{T}\varphi, \mathscr{T}\xi) \leq kD(\varphi, \xi)$, for all $\varphi, \xi \in E_0, \varphi \leq_{\mathscr{G}} \xi$.

Note that, as a consequence of this,

$\mathscr{T}$ is $(\leq_{\mathscr{G}}; D, d)$-continuous:
$\varphi_n \xrightarrow{D} \varphi$ and $(\varphi_n \leq_{\mathscr{G}} \varphi, \forall n)$ imply $\mathscr{T}\varphi_n \xrightarrow{d} \mathscr{T}\varphi$.

The following coincidence point statement (referred to as: Conditional BLR theorem) is now available.

Theorem 4 *Suppose that $\mathscr{T}$ is $(D, d; \leq_{\mathscr{G}}; k)$-contractive, for some $k \in [0, 1[$. Further, let $\mathscr{G}$ be a surjective (D, d)-continuous operator, E_0 be $(\leq_{\mathscr{G}}, D)$-complete, and $(\leq_{\mathscr{G}})$ be $(\leq_{\mathscr{G}}, D)$-selfclosed. Then,*

(61-a) *Given the $(\perp)$-starting point $\varphi_0 \in E_0(\mathscr{G}, \mathscr{T}; \leq)$, any $(\varphi_0, \perp)$-iterative sequence $(\varphi_n; n \geq 0)$ in E_0 is D-convergent, and $\varphi := \lim_n \varphi_n$ is a coincidence point of $\mathscr{G}$ and $\mathscr{T}$ (i.e., $\varphi \in \mathrm{Cop}(\mathscr{G}, \mathscr{T})$).*

(61-b) *Let $\varphi_0, \xi_0 \in E_0(\mathscr{G}, \mathscr{T}; \leq)$ be a couple of $(\perp)$-starting points, and $(\varphi_n; n \geq 0)$, $(\xi_n; n \geq 0)$ be a $(\varphi_0, \perp)$-iterative sequence and $(\xi_0, \perp)$-iterative sequence in E_0, respectively. Then,*

(61-b-1) $D(\varphi_n, \xi_n) \leq (1/(1-k))[D(\varphi_0, \varphi_1) + D(\xi_0, \xi_1)] + D(\varphi_0, \xi_0)$, $\forall n$;

(61-b-2) *if, in addition, $(\varphi_n \leq_{\mathscr{G}} \xi_n, \varphi_n \perp \xi_n, \forall n)$, then $\lim_n \varphi_n = \lim_n \xi_n$.*

(61-c) *Let $\varphi_0 = \xi_0 \in E_0(\mathscr{G}, \mathscr{T}; \leq)$ be a $(\perp)$-starting point, and $(\varphi_n; n \geq 0)$, $(\xi_n; n \geq 0)$ be a couple of $(\varphi_0, \perp)$-iterative sequence and $(\xi_0 = \varphi_0, \perp)$-iterative sequence in E_0, respectively. Then,*

(61-c-1) $D(\varphi_n, \xi_n) \leq (2/(1-k))d(\mathscr{G}\varphi_0, \mathscr{T}\varphi_0)$, *for all* $n \geq 0$.

In particular, if $\mathscr{G}$ is (D, d)-nonexpansive, the above relation writes

(61-c-2) $D(\varphi_n, \xi_n) \leq (2/(1-k))D(\varphi_0, \varphi_1)$, $\forall n$.

(61-d) *If φ^*, ξ^* are two elements in $\mathrm{Cop}(\mathscr{G}, \mathscr{T})$ with $(\varphi^* \leq_{\mathscr{G}} \xi^*, \varphi^* \perp \xi^*)$ then, necessarily, $\varphi^* = \xi^*$.*

Proof There are several steps to be considered.

Part 1 Take a $(\perp)$-starting point $\varphi_0 \in E_0(\mathscr{G}, \mathscr{T}; \leq)$; and let $(\varphi_n; n \geq 0)$ be an associated $(\varphi, \perp)$-iterative sequence in E_0. In particular, this means

$$\varphi_0 \leq_{\mathscr{G}} \varphi_1, \ \mathscr{T}\varphi_0 = \mathscr{G}\varphi_1, \ D(\varphi_0, \varphi_1) \leq d(\mathscr{G}\varphi_0, \mathscr{G}\varphi_1);$$

as well as

$$\varphi_1 \leq_{\mathscr{G}} \varphi_2, \ \mathscr{T}\varphi_1 = \mathscr{G}\varphi_2, \ D(\varphi_1, \varphi_2) \leq d(\mathscr{G}\varphi_1, \mathscr{G}\varphi_2);$$

note that, as a combination of these, we have (by the contractive property of T)

$$D(\varphi_1, \varphi_2) \leq d(\mathscr{G}\varphi_1, \mathscr{G}\varphi_2) = d(\mathscr{T}\varphi_0, \mathscr{T}\varphi_1) \leq kD(\varphi_0, \varphi_1).$$

This procedure may continue indefinitely; and gives the iterative relations

$$\varphi_n \leq_{\mathscr{G}} \varphi_{n+1}, \ D(\varphi_{n+1}, \varphi_{n+2}) \leq kD(\varphi_n, \varphi_{n+1}), \ \forall n.$$

By a finite induction procedure, one gets

$$\varphi_n \leq_{\mathscr{G}} \varphi_{n+1},\ D(\varphi_n, \varphi_{n+1}) \leq k^n D(\varphi_0, \varphi_1),\ \forall n;$$

and since the series $\sum_n k^n$ converges, the sequence $(\varphi_n; n \geq 0)$ is $(\leq_{\mathscr{G}})$-ascending, D-Cauchy. As E_0 is $(\leq_{\mathscr{G}}, D)$-complete and $(\leq_{\mathscr{G}})$ is $(\leq_{\mathscr{G}}, D)$-selfclosed, there must be some $\varphi^* \in E_0$, with

$(\varphi_n \xrightarrow{D} \varphi^*$ as $n \to \infty)$, and $(\varphi_n \leq_{\mathscr{G}} \varphi^*$, for all $n)$.

Now, by the very definition of our $(\varphi_0, \perp)$-iterative sequence (φ_n), one has

$(\forall n)$: $\varphi_n \leq_{\mathscr{G}} \varphi_{n+1}, \mathscr{T}\varphi_n = \mathscr{G}\varphi_{n+1}, \varphi_n \perp \varphi_{n+1}$.

Passing to limit as $n \to \infty$ in the operator part of this relation, we get (as $\mathscr{G}$ is (D, d)-continuous and [by the contractive condition] $\mathscr{T}$ is $(\leq_{\mathscr{G}}; D, d)$-continuous)

$\mathscr{T}\varphi^* = \mathscr{G}\varphi^*$; i.e., φ^* is a coincidence point of $\mathscr{G}$ and $\mathscr{T}$.

Part 2 Take a couple of $(\perp)$-starting points $\varphi_0, \xi_0 \in E_0(\mathscr{G}, \mathscr{T}; \leq)$, and let $(\varphi_n; n \geq 0)$, $(\xi_n; n \geq 0)$, be a $(\varphi_0, \perp)$-iterative sequence and $(\xi_0, \perp)$-iterative sequence in E_0, respectively. By the preceding part, $(\varphi_n; n \geq 0)$ fulfills the inductive relations we just listed. Likewise, $(\xi_n; n \geq 0)$ fulfills the iterative relations

$$\xi_n \leq_{\mathscr{G}} \xi_{n+1},\ D(\xi_{n+1}, \xi_{n+2}) \leq kD(\xi_n, \xi_{n+1}),\ \forall n \geq 0;$$

wherefrom (by a finite induction procedure)

$$\xi_n \leq_{\mathscr{G}} \xi_{n+1},\ D(\xi_n, \xi_{n+1}) \leq k^n D(\xi_0, \xi_1),\ \forall n \geq 0.$$

Combining with the triangle inequality, one gets for all $n \geq 1$,

$$D(\varphi_n, \xi_n) \leq D(\varphi_{n-1}, \varphi_n) + D(\xi_{n-1}, \xi_n) + D(\varphi_{n-1}, \xi_{n-1}) \leq$$
$$k^{n-1}[D(\varphi_0, \varphi_1) + D(\xi_0, \xi_1)] + D(\varphi_{n-1}, \xi_{n-1});$$

wherefrom— by repeating $(n - 1)$ times this procedure— the first part of our conclusion follows. Further, suppose that (in addition) $(\varphi_n \leq_{\mathscr{G}} \xi_n, \varphi_n \perp \xi_n, \forall n)$. By the very definition of these points (and the contractive condition), we have

$$D(\varphi_n, \xi_n) \leq d(\mathscr{G}\varphi_n, \mathscr{G}\xi_n) = d(\mathscr{T}\varphi_{n-1}, \mathscr{T}\xi_{n-1}) \leq kD(\varphi_{n-1}, \xi_{n-1}),\ \forall n \geq 1.$$

This, by a finite induction procedure, gives

$$D(\varphi_n, \xi_n) \leq k^n D(\varphi_0, \xi_0),\ \forall n.$$

Passing to limit as $n \to \infty$, and noting that both $(\varphi_n; n \geq 0)$ and $(\xi_n; n \geq 0)$ are D-convergent (see above), the second part of our conclusion is established.

Part 3 Let $\varphi_0 = \xi_0 \in E_0(\mathcal{G}, \mathcal{T}; \leq)$ be a $(\perp)$-starting point; and $(\varphi_n; n \geq 0)$, $(\xi_n; n \geq 0)$ be a couple of $(\varphi_0, \perp)$-iterative sequence and $(\xi_0 = \varphi_0, \perp)$-iterative sequence in E_0, respectively. Then,

$$D(\varphi_0, \varphi_1) \leq d(\mathcal{G}\varphi_0, \mathcal{G}\varphi_1) = d(\mathcal{G}\varphi_0, \mathcal{T}\varphi_0),$$
$$D(\xi_0, \xi_1) \leq d(\mathcal{G}\xi_0, \mathcal{G}\xi_1) = d(\mathcal{G}\xi_0, \mathcal{T}\xi_0) = d(\mathcal{G}\varphi_0, \mathcal{T}\varphi_0);$$

and so, the first part of our assertion follows. Moreover, assume that $\mathcal{G}$ is (D, d)-nonexpansive. Then, from $\varphi_0 = \xi_0$, we get $\mathcal{T}\varphi_0 = \mathcal{T}\xi_0$; whence $\mathcal{G}\varphi_1 = \mathcal{G}\xi_1$. This (according with the representation of $(\perp)$) gives

$$D(\varphi_0, \varphi_1) = d(\mathcal{G}\varphi_0, \mathcal{G}\varphi_1) = d(\mathcal{G}\varphi_0, \mathcal{T}\varphi_0),$$
$$D(\xi_0, \xi_1) = d(\mathcal{G}\xi_0, \mathcal{G}\xi_1) = d(\mathcal{G}\varphi_0, \mathcal{G}\varphi_1) = D(\mathcal{G}\varphi_0, \mathcal{T}\varphi_0);$$

wherefrom, the second part of our assertion follows.

Part 4 Let φ^*, ξ^* be two elements in $\text{Cop}(\mathcal{G}, \mathcal{T})$ $\varphi \leq_{\mathcal{G}} \xi$, $\varphi^* \perp \xi^*$. By the working condition and contractive property of T,

$$D(\varphi^*, \xi^*) \leq d(\mathcal{G}\varphi^*, \mathcal{G}\xi^*) = d(\mathcal{T}\varphi^*, \mathcal{T}\xi^*) \leq kD(\varphi^*, \xi^*);$$

and this yields (as D=metric), $\varphi^* = \xi^*$. The proof is thereby complete.

Remark 4 Given the map $\mathcal{G} : E_0 \to E$ as before, let $d_{\mathcal{G}} : E_0 \times E_0 \to R_+$ stand for the associated semimetric

$d_{\mathcal{G}}(\varphi, \xi) = d(\mathcal{G}\varphi, \mathcal{G}\xi), \varphi, \xi \in E_0.$

Then, call $\mathcal{T}$, $(D, d_{\mathcal{G}}; \leq_{\mathcal{G}}, \perp; k)$-*contractive* (where $k \geq 0$), provided

(contr) $\quad d(\mathcal{T}\varphi, \mathcal{T}\xi) \leq kd_{\mathcal{G}}(\varphi, \xi)$, for all $\varphi, \xi \in E_0, \varphi \leq_{\mathcal{G}} \xi, \varphi \perp \xi$.

Clearly,

$\mathcal{T}$ is $(D, d; \leq_{\mathcal{G}}; k)$-*contractive* implies $\mathcal{T}$ is $(D, d_{\mathcal{G}}; \leq_{\mathcal{G}}, \perp; k)$-*contractive*;

but the converse is not in general true. Note that, replacing the initial contractive property with the above one gives us, practically, the same amount of information as the one provided by the above result. This may be of some avail in certain practical circumstances; we do not give details.

In particular, choose $(\leq)$ as

$(\leq) = E \times E$ (the trivial quasi-ordering on E)

and let $c \in I$ be arbitrary fixed. Then, the operator

$(\mathcal{G}_c : E_0 \to E)$: $\mathcal{G}_c(\varphi) = \varphi(c), \varphi \in E_0$,

is nonexpansive (see above). The associated to $\mathcal{G}_c$ relation $(\perp_c)$ writes

$(\varphi, \xi \in E_0)$: $\varphi \perp_c \xi$ iff $D(\varphi, \xi) = d(\varphi(c), \xi(c))$;

as precise, it is reflexive and symmetric but not in general transitive. Note that the corresponding version of Conditional BLR theorem above is just the result in Bernfeld et al. [5] proved via "normed" type methods. Further extensions of this statement may be found in Pathak [34] and Som [43].

7 Full Razumikhin Chains

The above result is, ultimately, a relative (conditional) one. Precisely, the existence of coincidence points introduced there is essentially depending on the possibility of effectively constructing iterative sequences according to the described conventions. So, it is natural to ask whether such iterative sequences exist. It is our aim in the following to show that, if we restrict our developments to certain classes of full Razumikhin chains, this always happens; a number of appropriate examples will be also given so as to support the obtained (general) results.

Let $(E, d, \leq)$ be a quasi-ordered metric space. As usual, denote by $(<)$ the associated irreflexive order

$x < y$ iff $x \leq y$ and $x \neq y$.

Further, let $I = [a, b]$ (where $a < b$) be a bounded closed real interval; and $E_0 := C(I, E)$ stand for the class of all continuous functions in $\mathcal{F}(I, E)$, endowed with the supremum distance and related quasi-order

$[D(\varphi, \xi) = \sup\{d(\varphi(t), \xi(t)); t \in I\}]$, $[\varphi \preceq \xi$ iff $\varphi(t) \leq \xi(t), t \in I]$;

clearly, $(E_0, D, \preceq)$ is a quasi-ordered metric space. As before, denote by $(\prec)$ the associated irreflexive order

$\varphi \prec \xi$ iff $\varphi \preceq \xi$ and $\varphi \neq \xi$.

Remember that, given $\mathcal{G} : E_0 \to E$, the attached Razumikhin relation

$\varphi \perp \xi$ iff $D(\varphi, \xi) \leq d(\mathcal{G}\varphi, \mathcal{G}\xi)$

is reflexive and symmetric, but not in general transitive. Further, by a *Razumikhin $\mathcal{G}$-chain* we mean any part $\mathcal{L}$ of E_0, with:

$\varphi \perp \xi, \forall \varphi, \xi \in \mathcal{L}$ [i.e., any two elements of $\mathcal{L}$ are $(\perp)$-comparable];

if in addition $\mathcal{G}(\mathcal{L}) = E$, then $\mathcal{L}$ is called *full*.

A basic example of couples $(\mathcal{G}, \mathcal{L})$ like before may be constructed by starting from the constant functions subclass of E_0. Precisely, for each $u \in E$, let $K[u]$ denote the constant function of E_0, introduced as

$K[u](t) = u, t \in I$ (hence, $K[u](I) = \{u\}$).

Denote, for simplicity,

$\mathcal{K} = \{K[u]; u \in E\}$ (the *constant class* of E_0).

The following properties of the constant class $\mathscr{K}$ will be useful for us.

Proposition 18 *Under the above conventions,*

(71-1) $D(K[u], K[v]) = d(u, v), \forall u, v \in E$
(71-2) *the mapping* $u \mapsto K[u]$ *is an isometry (hence, a topological isomorphism) between* (E, d) *and* $(\mathscr{K}, D)$
(71-3) $u \leq v$ *iff* $K[u] \preceq K[v]$.
(71-4) $\mathscr{K}$ *is* D*-closed in* E_0.

Proof

(i), (ii), (iii) Evident.
(iv) Let $(\varphi_n := K[u_n]; n \geq 0)$ be a sequence in $\mathscr{K}$, with

$$\varphi_n \xrightarrow{D} \varphi, \text{ for some } \varphi \in E_0.$$

By the very definition of our metric D, we must have

$$(u_n =)\varphi_n(t) \xrightarrow{d} \varphi(t), \text{ for each } t \in I;$$

and this yields directly

$$u = \lim_n(u_n) \text{ exists in } E \text{ and } \varphi = K[u].$$

The proof is complete.

Having these precise, let $\mathscr{G} : E_0 \to E$ be an operator with

(met-K-1) $\mathscr{G}$ is $(\preceq, \leq)$*-increasing*: $\varphi \preceq \xi$ implies $\mathscr{G}(\varphi) \leq \mathscr{G}(\xi)$
(met-K-2) $\mathscr{G}$ is K-invertible: $\mathscr{G}(K[u]) = u, \forall u \in E$;

we then say that $\mathscr{G}$ is a *metrical* K*-subordinated Razumikhin map*. If, in addition,

(met-K-3) $\mathscr{G}$ is (D, d)*-nonexpansive*: $d(\mathscr{G}\varphi, \mathscr{G}\xi) \leq D(\varphi, \xi), \forall \varphi, \xi \in E_0$,

then $\mathscr{G}$ is called a *nonexpansive-metrical* K*-subordinated Razumikhin map*. For example, all these properties (including the last one) hold under the choice

$\mathscr{G} = \mathscr{G}_c$ (see above), for some $c \in I$;

but this is not the only possible example of (nonexpansive-) metrical K-subordinated Razumikhin map; some other examples will be discussed a bit further.

By the imposed properties, we have

(i-Raz) $\mathscr{G}(\mathscr{K}) = E$ and $[D(\varphi, \xi) = d(\mathscr{G}\varphi, \mathscr{G}\xi), \forall \varphi, \xi \in \mathscr{K}]$;

referred to as: the constant class $\mathscr{K}$ is an *isometric full Razumikhin* $\mathscr{G}$*-chain* of E_0; hence, in particular: $\mathscr{K}$ is a full Razumikhin $\mathscr{G}$-chain of E_0. Note that, by these properties, $\mathscr{G}$ is a bijection between $\mathscr{K}$ and $E = \mathscr{G}(\mathscr{K})$, with $\mathscr{G}^{-1} = K$. As a consequence, K is bijective (from E to $\mathscr{K}$), with

$\mathscr{G} = K^{-1}$; whence, $K[\mathscr{G}\varphi] = \varphi, \forall \varphi \in \mathscr{K}$.

Further, let $\mathscr{T} : E_0 \to E$ be a (nonself) mapping. Remember that $\varphi \in E_0$ is a *coincidence point* of $\mathscr{G}$ and $\mathscr{T}$, when $\mathscr{G}\varphi = \mathscr{T}\varphi$. To determine such points, the following construction is useful. Let T be the selfmap of E introduced as

$Tu = \mathscr{T}(K[u]), u \in E.$

Proposition 19 *Under these conventions, the following are valid:*

(72-1) *If $x \in E$ is a fixed point of T, then $\xi := K[x] \in \mathscr{K}$ is a coincidence point of $\mathscr{G}$ and $\mathscr{T}$*

(72-2) *Conversely, if $\zeta := K[z] \in \mathscr{K}$ is a coincidence point of $\mathscr{G}$ and $\mathscr{T}$, then $z \in E$ is a fixed point of T.*

Proof

(i) If $x \in E$ is a fixed point of T, we have

$$\mathscr{G}(K[x]) = x = Tx = \mathscr{T}(K[x]); \text{so, } \mathscr{G}(\xi) = \mathscr{T}(\xi), \text{where } \xi := K[x] \in \mathscr{K};$$

which tells us that $\xi \in \mathscr{K}$ is a coincidence point of $\mathscr{G}$ and $\mathscr{T}$.

(ii) Suppose that $\zeta = K[z] \in \mathscr{K}$ is a coincidence point of $\mathscr{G}$ and $\mathscr{T}$; i.e., $\mathscr{G}(\zeta) = \mathscr{T}(\zeta)$. This yields (by the properties of $\mathscr{G}$)

$$z = \mathscr{G}(K[z]) = \mathscr{T}(K[z]) = Tz;$$

so that $z \in E$ is a fixed point of T.

Having these precise, we may now proceed to the formulation of our main result. Some basic concepts and auxiliary facts are needed. These, roughly speaking, consist in determining the appropriate conditions upon $(E_0, D, \preceq)$ and $(\mathscr{G}, \mathscr{T})$, under which the needed regularity properties involving $(E, d, \leq)$ and T are fulfilled. The following technical aspect will be of some avail for us. Remember that, by $(\leq_{\mathscr{G}})$ we denoted the quasi-order on E_0

$(\varphi, \xi \in E_0)$: $\varphi \leq_{\mathscr{G}} \xi$ iff $\mathscr{G}\varphi \leq \mathscr{G}\xi$.

Clearly, by the $(\preceq, \leq)$-increasing property of $\mathscr{G}$, we have

$(\varphi, \xi \in E_0)$: $\varphi \preceq \xi$ implies $\varphi \leq_{\mathscr{G}} \xi$.

On the other hand, by definition,

$(\varphi, \xi \in \mathscr{K})$: $\varphi \preceq \xi$ iff $\varphi \leq_{\mathscr{G}} \xi$.

Now, it is worth noting that, by a simple substitution of $(\preceq)$ by $(\leq_{\mathscr{G}})$ in the underlying conditions to be stated we get extended versions of these. This along with the $(\preceq, \leq)$-increasing property of $\mathscr{G}$ being no more needed leads us to the question of to what extent is the initial approach (modulo $(\preceq)$) motivated. The answer to be provided is practical in nature; we do not give details.

(I) Let us start these conditions with a remark about continuity properties.

Proposition 20 *Suppose that*

$\mathscr{T}$ *is* $(\preceq; D, d)$*-continuous:*
$(\varphi_n; n \geq 0)$ *is* $(\preceq)$*-ascending and* $\varphi_n \xrightarrow{D} \varphi$ *imply* $\mathscr{T}\varphi_n \xrightarrow{d} \mathscr{T}\varphi$.

Then, necessarily,

$T = \mathscr{T} \circ K$ *is* $(\leq, d)$*-continuous:*
$(x_n; n \geq 0)$ *is* $(\leq)$*-ascending and* $x_n \xrightarrow{d} x$ *imply* $Tx_n \xrightarrow{d} Tx$.

Proof Let (x_n) and x be as in this premise. The associated sequence $(\varphi_n := K[x_n]; n \geq 0)$ is $(\preceq)$-ascending and (with the notation $\varphi = K[x]$)

$$D(\varphi_n, \varphi) = D(K[x_n], K[x]) = d(x_n, x), \forall n,$$

if one takes the isometric property of K into account. As a consequence, $\varphi_n \xrightarrow{D} \varphi$ as $n \to \infty$; wherefrom (combining with posed hypothesis)

$$\mathscr{T}(\varphi_n) \xrightarrow{d} \mathscr{T}(\varphi); \text{i.e., } Tx_n \xrightarrow{d} Tx;$$

and the assertion follows.

(II) Further, the following completeness and selfcloseness result is available.

Proposition 21 *Under the precise conventions,*

(74-1) *If* E_0 *is* $(\preceq, D)$*-complete, then* E *is necessarily* $(\leq, d)$*-complete*
(74-2) *If* $(\preceq)$ *is* $(\preceq, D)$*-selfclosed (over the space* E_0*), then,* $(\leq)$ *is* $(\leq, d)$*-selfclosed (over the space* E*).*

Proof

(i) Let the sequence $(u_n; n \geq 0)$ in E be $(\leq)$-ascending and d-Cauchy. By the above observation and K=isometry,

$(\varphi_n := K[u_n]; n \geq 0)$ is $(\preceq)$-ascending, D-Cauchy in $\mathscr{K}$.

From the completeness of E_0 hypothesis, there exists an element $\varphi \in E_0$, with

$\varphi_n \xrightarrow{D} \varphi$ as $n \to \infty$; whence, $\varphi_n(t) \xrightarrow{d} \varphi(t)$ as $n \to \infty$, for each $t \in I$.

This, along with $\mathscr{K}$ being D-closed (see above), yields

$u = \lim_n u_n$ exists in E, and $\varphi = K[u]$;

wherefrom, the assertion follows.

(ii) Let the sequence $(u_n; n \geq 0)$ in E and the point $u \in E$ be such that

(u_n) is $(\leq)$-ascending and $u_n \xrightarrow{d} u$.

By the above observation and K=isometry,

$(\varphi_n := K[u_n]; n \geq 0)$ is $(\preceq)$-ascending in $\mathscr{K}$, with $\varphi_n \xrightarrow{D} \varphi := K[u]$.

Combining with the imposed hypothesis gives

$(\varphi_n \preceq \varphi, \forall n)$; wherefrom (see above) $u_n \le u$, for all n;

and conclusion follows.

(III) Remember that $(\mathscr{G}, \mathscr{T})$ is E_0-semi-progressive, provided

$E_0(\mathscr{G}, \mathscr{T}; \le) := \{\varphi \in E_0; \mathscr{G}\varphi \le \mathscr{T}\varphi\}$ is nonempty.

Also, let us say that $(\mathscr{G}, \mathscr{T})$ is $\mathscr{K}$-semi-progressive, provided

$\mathscr{K}(\mathscr{G}, \mathscr{T}; \le) = \{\varphi \in \mathscr{K}; \mathscr{G}(\varphi) \le \mathscr{T}(\varphi)\}$ is nonempty.

Finally, let us introduce the condition

$\mathscr{T}$ is $(\le_{\mathscr{G}}, \le)$-increasing: $\varphi \le_{\mathscr{G}} \xi$ implies $\mathscr{T}\varphi \le \mathscr{T}\xi$.

This convention may be viewed as a stronger version of

$\mathscr{T}$ is $(\preceq, \le)$-increasing: $\varphi \preceq \xi$ implies $\mathscr{T}\varphi \le \mathscr{T}\xi$.

Precisely, by the $(\preceq, \le)$-increasing property of $\mathscr{G}$, we have

$(\varphi, \xi \in E_0)$: $\varphi \preceq \xi$ implies $\varphi \le_{\mathscr{G}} \xi$;

so that (by these very definitions)

$\mathscr{T}$ is $(\le_{\mathscr{G}}, \le)$-increasing implies $\mathscr{T}$ is $(\preceq, \le)$-increasing.

The reciprocal is not in general true, as simple examples show. But, if we restrict this operator to the constant class $\mathscr{K}$, this happens; we do not give details. In particular, with $\varphi = K[u]$, $\xi = K[v]$, where $u, v \in E$, either of these conditions yields

$u, v \in E, u \le v \Longrightarrow Tu \le Tv$;

i.e., the selfmap T is $(\le)$-increasing on E.

The usefulness of this (rather strong) condition is to be judged from the following relative type statement (with a basic role in the sequel).

Proposition 22 *Assume that $(\mathscr{G}, \mathscr{T})$ is E_0-semi-progressive and the operator $\mathscr{T}$ is $(\le_{\mathscr{G}}, \le)$-increasing. Then,*

(75-1) *$(\mathscr{G}, \mathscr{T})$ is $\mathscr{K}$-semi-progressive*
(75-2) *the associated mapping T is semi-progressive and increasing.*

Proof

(i) By the imposed condition, there exists some $\varphi_0 \in E_0$ with $\mathscr{G}(\varphi_0) \le \mathscr{T}(\varphi_0)$. As $u_0 := \mathscr{T}(\varphi_0) \in E$, we may consider the point $\xi_0 = K[u_0]$ in the class $\mathscr{K} := K[E]$; note that, by this definition

$$\mathscr{G}(\xi_0) = \mathscr{G}(K[u_0]) = u_0 = \mathscr{T}(\varphi_0).$$

The E_0-semi-progressive property of $(\mathscr{G}, \mathscr{T})$ may then be written as $\varphi_0 \leq_{\mathscr{G}} \xi_0$. Since $\mathscr{T}$ is $(\leq_{\mathscr{G}}, \leq)$-increasing, this yields $\mathscr{T}(\varphi_0) \leq \mathscr{T}(\xi_0)$. Combining with a preceding relation, we thus have

$\mathscr{G}(\xi_0) \leq \mathscr{T}(\xi_0)$; i.e., $(\mathscr{G}, \mathscr{T})$ is $\mathscr{K}$-semi-progressive.

(ii) As $\xi_0 = K[u_0]$, where $u_0 := \mathscr{T}(\varphi_0) \in E$, the last relation gives $u_0 \leq Tu_0$; whence $u_0 \in E(T, \leq)$. The proof is thereby complete.

(IV) Given $f \in \mathscr{F}(R^0_+, R)$, let us say that $\mathscr{T}$ is $(D, d; \preceq; f)$*-contractive*, when

(E0-contr) $d(\mathscr{T}\varphi, \mathscr{T}\xi) \leq f(D(\varphi, \xi)), \forall \varphi, \xi \in E_0, \varphi \prec \xi, \mathscr{T}_\varphi < \mathscr{T}_\xi$

In particular, taking $x, y \in E$ with $x < y$, $T_x < T_y$, the functions $\varphi = K[x]$ and $\xi = K[y]$ satisfy $\varphi \prec \xi$, $\mathscr{T}_\varphi < \mathscr{T}_\xi$. This yields (under the properties of K)

(E-contr) $d(Tx, Ty) \leq f(D(K[x], K[y])) = f(d(x, y)), \forall x, y \in E, x < y, T_x < T_y$;
whence: $d(Tx, Ty) \leq f(d(x, y)), \forall x, y \in E, x < y, Tx < Ty$;

which tells us that the associated selfmap $T : E \to E$ is $(d, \leq; f)$-contractive (according to a previous convention).

(V) Let the mapping $\mathscr{T}$ be $(\leq_{\mathscr{G}}, \leq)$-increasing. Given $\varphi_0 = K[x_0] \in \mathscr{K}(\mathscr{G}, \mathscr{T}; \leq)$, we say that the sequence $(\varphi_n = K[x_n]; n \geq 0)$ in $\mathscr{K}$ is $(\mathscr{G}, \mathscr{T}; \leq)$*-iterative*, provided

$\mathscr{T}\varphi_n = \mathscr{G}\varphi_{n+1}$, for all $n \geq 0$;

note that, by a previous observation, this $(\mathscr{G}, \mathscr{T}; \leq)$-iterative sequence fulfills

$\varphi_n \preceq \varphi_{n+1}$ (or, equivalently: $\varphi_n \leq_{\mathscr{G}} \varphi_{n+1}$), and $\mathscr{T}\varphi_n \leq \mathscr{T}\varphi_{n+1}$, for all $n \geq 0$.

The family of such sequences is nonempty. In fact, for the starting $x_0 \in E(T, \leq)$, the T-iterative sequence $(x_n; n \geq 0)$ in E is well defined (and ascending) according to the formula $(Tx_n = x_{n+1}; n \geq 0)$. But then,

$$\mathscr{G}\varphi_n = x_n \leq x_{n+1} = \mathscr{G}\varphi_{n+1}, \ \mathscr{T}\varphi_n = Tx_n = x_{n+1} = \mathscr{G}\varphi_{n+1}, \ \forall n \geq 0.$$

Conversely, if $\varphi_0 \in \mathscr{K}(\mathscr{G}, \mathscr{T}; \leq)$, and $(\varphi_n := K[x_n]; n \geq 0)$ is $(\mathscr{G}, \mathscr{T}; \leq)$-iterative, then (by the same formula), the sequence $(x_n; n \geq 0)$ of E is T-iterative, ascending.

Putting these together, it follows, from the Matkowski-Rhoades fixed point principle (MR-fpp), the following coincidence point result involving such data (referred to as: Constant BLR theorem):

Theorem 5 *Suppose that $\mathscr{T}$ is $(D, d; \preceq; f)$-contractive, for some Meir-Keeler admissible function $f \in \mathscr{F}(R^0_+, R)$. In addition, suppose that $(\mathscr{G}, \mathscr{T})$ is $\mathscr{K}$-semi-progressive, $\mathscr{T}$ is $(\leq_{\mathscr{G}}, \leq)$-increasing, and E_0 is $(\preceq, D)$-complete. Then,*

(71-a) *if φ^* and ξ^* are coincidence points in $\mathscr{K} := K[E]$ for the couple $(\mathscr{G}, \mathscr{T})$ with $\varphi^* \preceq \xi^*$, then $\varphi^* = \xi^*$*

(71-b) *for each starting point* $\varphi_0 \in \mathscr{K}(\mathscr{G}, \mathscr{T}; \leq)$*, the* $(\mathscr{G}, \mathscr{T}; \leq)$*-iterative sequence* $(\varphi_n; n \geq 0)$ *in* $\mathscr{K}$ *introduced as* $(\mathscr{T}\varphi_n = \mathscr{G}\varphi_{n+1}; n \geq 0)$*, is*

D-convergent towards some coincidence point $\varphi^* \in \mathscr{K}$ *of* $(\mathscr{G}, \mathscr{T})$*, when* $\mathscr{T}$ *is* $(\preceq; D, d)$*-continuous*
D-convergent towards some coincidence point $\varphi^* \in \mathscr{K}$ *of* $(\mathscr{G}, \mathscr{T})$*, with* $(\varphi_n \preceq \varphi^*, \forall n)$*, when* $(\preceq)$ *is* $(\preceq, D)$*-selfclosed.*

Note that, by a previous fact, any full Razumikhin $\mathscr{G}$-chain $\mathscr{A}$ that includes $\mathscr{K}$ fulfills $\mathscr{A} = \mathscr{K}$; so that all versions of Constant BLR theorem constructed with such Razumikhin $\mathscr{G}$-chains are identical with Constant BLR theorem.

In the following, we shall discuss a certain particular case of Constant BLR theorem due to Drici et al. [12].

Let $(E, d, \leq)$ be a partially ordered metric space. Further, let $I = [a, b]$ (where $a < b$) be a bounded closed real interval; and $E_0 = C(I, E)$ stand for the space of all continuous functions in $\mathscr{F}(I, E)$, endowed with the supremum metric and the partial order

$$(D(\varphi, \xi) = \sup\{d(\varphi(t), \xi(t)); t \in I\}), (\varphi \preceq \xi \text{ iff } \varphi(t) \leq \xi(t), t \in I).$$

Remember that $(\preceq)$ is called *selfclosed*, provided

$$(\varphi_n) \text{ is } (\preceq)\text{-increasing and } \varphi_n \xrightarrow{D} \varphi \text{ imply } \varphi_n \preceq \varphi, \text{ for all } n.$$

Finally, take some point $c \in I$; and let $\mathscr{T} : E_0 \to E$ be a map. The following contractive condition will be used

(contr) $\mathscr{T}$ is $(\preceq, k)$-contractive (for some $k \geq 0$):
$d(\mathscr{T}\varphi, \mathscr{T}\xi) \leq kD(\varphi, \xi)$, when $\varphi, \xi \in E_0, \varphi \preceq \xi$.

Theorem 6 *Suppose that* $\mathscr{T}$ *is* $(\preceq, k)$*-contractive, for some* $k \in [0, 1[$*; and*

(72-i) there exists a (lower solution) $\varphi_0 \in E_0$ *with* $\varphi_0(c) \leq \mathscr{T}\varphi_0$
(72-ii) $\mathscr{T}$ *is increasing:* $\varphi \preceq \xi$ *implies* $\mathscr{T}\varphi \leq \mathscr{T}\xi$.

Further, let E be d-complete and the alternative type condition holds

(72-iii) either $\mathscr{T}$ *is continuous or* $(\preceq)$ *is selfclosed.*

Then, there exists some $\varphi^* \in E_0$ *with* $\varphi^*(c) = \mathscr{T}\varphi^*$.

For technical reasons, we shall provide the basic lines of authors' proof, with some modifications (imposed by our notations).

Proof Essentially, two steps must be considered.

Step 1 Let $\mathscr{T}\varphi_0 = x_1 \in E$. Choose $\varphi_1 \in E_0$ such that

(R1) $\varphi_1(c) = x_1, \varphi_0 \preceq \varphi_1, D(\varphi_0, \varphi_1) = d(\varphi_0(c), \varphi_1(c))$;
as well as (via $\mathscr{T}$=increasing) $\mathscr{T}\varphi_0 = \varphi_1(c) \leq \mathscr{T}\varphi_1$.

Further, let $\mathscr{T}\varphi_1 = x_2 \in E$. Choose $\varphi_2 \in E_0$ such that

(R2) $\varphi_2(c) = x_2, \varphi_1 \preceq \varphi_2, D(\varphi_1, \varphi_2) = d(\varphi_1(c), \varphi_2(c))$;

as well as (via $\mathscr{T}$=increasing) $\mathscr{T}\varphi_1 = \varphi_2(c) \leq \mathscr{T}\varphi_2$.

By induction we get, for all $n \geq 1$,

(R3) $\varphi_{n-1} \leq \varphi_n$, $D(\varphi_{n-1}, \varphi_n) = d(\varphi_{n-1}(c), \varphi_n(c))$,
and $\mathscr{T}\varphi_{n-1} = \varphi_n(c) \leq \mathscr{T}\varphi_n$.

Step 2 Again by induction, we show that

$d(\mathscr{T}\varphi_{n-1}, \mathscr{T}\varphi_n) \leq k^n D(\varphi_0, \varphi_1)$, for all $n \geq 1$.

As E_0 is D-complete, there exists $\varphi^* \in E_0$ such that

(R4) $\varphi_n \xrightarrow{D} \varphi^*$, as $n \to \infty$. To establish that $\varphi^*(c) = \mathscr{T}(\varphi)$, we pass to limit in (R3) (the second part). Two alternatives are possible.

Alternative 1 Suppose that $\mathscr{T}$ is continuous. We then have

(R5) $\lim_n \mathscr{T}\varphi_n = \mathscr{T}\varphi^*$, $\lim_n \varphi_n(c) = \varphi^*(c)$;

so that (by the underlying iterative relation) $\mathscr{T}\varphi^* = \varphi^*(c)$.

Alternative 2 Suppose that $(\preceq)$ is selfclosed. Combining with (R3) (the first part) we have ($\varphi_n \preceq \varphi^*$, for all n). This, along with the contractive hypothesis, gives again (R5); wherefrom (see above) $\mathscr{T}\varphi^* = \varphi^*(c)$.

Comments

(I) The basic property used by the authors in **Step 1** is

(P) for each $\varphi \in E_0$ with $\varphi(c) \leq \mathscr{T}\varphi$ there exists $\xi \in E_0$ with
$\varphi \leq \xi$, $D(\varphi, \xi) = d(\varphi(c), \xi(c))$, $\mathscr{T}\varphi = \xi(c) \leq \mathscr{T}(\xi)$.

As already precise in Constant BLR theorem, the hypotheses about $(E, d, \leq)$, $(E_0, D, \preceq)$ and $(c, \mathscr{T})$ are not in general sufficient for the property (P) being available. Hence, for the moment, **Step 1** above is not acceptable.

(II) A possibility of avoiding this trouble is that of replacing (72-i) with

(72-i-1) there exists a (lower solution) $\varphi_0 \in \mathscr{K}$ with $\varphi_0(c) \leq \mathscr{T}\varphi_0$.

Then, the Matkowski-Rhoades fixed point principle (MR-fpp) applies to $(E, d, \leq)$ and operator $T := \mathscr{T} \circ K$ so as to get the desired fact. For example, this relation holds if the monotone property of $\mathscr{T}$ is to be taken in the stronger way:

(s-incr) $\mathscr{T}$ is *strong increasing*: $\varphi(c) \leq \xi(c)$ implies $\mathscr{T}\varphi \leq \mathscr{T}\xi$;

the detailed argument for this is just the one appearing in Constant BLR theorem.

In conclusion, the methods proposed in the paper by Drici et al. [12] are not in general suitable for the PPF dependent fixed point problem to be solved. This is also true for the application to functional differential equations developed by the quoted authors. Further aspects will be discussed elsewhere.

8 Linear Aspects

Let $(E, ||.||)$ be a normed space; note that, with respect to the associated map

$d(x, y) = ||x - y||, x, y \in E,$

the structure (E, d) is a metric space. Fix a *cone* $E^+ \subseteq E$, in the sense

$\alpha E^+ + \beta E^+ \subseteq E^+$, for all $\alpha, \beta \geq 0$;

so, if $(\leq)$ stands for the attached quasi-order

$x \leq y$ iff $y - x \in E^+$ (whence, $E^+ = \{x \in E; x \geq 0\}$),

the triple $(E, d, \leq)$ is a quasi-ordered metric space.

Further, let $I = [a, b]$ be a bounded closed real interval, and $E_0 := C(I, E)$ denote the class of all continuous functions in $\mathscr{F}(I, E)$, endowed with the supremum norm

$||\varphi||_0 = \sup\{||\varphi(t)||; t \in I\}, \varphi \in E_0;$

note that with respect to the associated map

$D(\varphi, \xi) = ||\varphi - \xi||_0, \varphi, \xi \in E_0,$

the structure (E_0, D) becomes a metric space. On the other hand,

$E_0^+ = \{\varphi \in E_0; \varphi(t) \geq 0, t \in I\}$ is a cone in E_0;

so, denoting by $(\preceq)$ its attached quasi-order

$\varphi \preceq \xi$ iff $\xi - \varphi \in E_0^+$ (whence, $E_0^+ = \{\varphi \in E_0; \varphi \succeq 0\}$),

the triple $(E_0, D, \preceq)$ is a quasi-ordered metric space. In fact, some additional properties of these ordering structures are also available; we do not give details.

Having these precise, the following linear counterpart of a previous metric type construction is our starting point.

Define a mapping $K : E \to E_0$ according to: for each $u \in E$, $K[u]$ denote the constant function of E_0, defined as

$K[u](t) = u, t \in I$ (whence $K[u](I) = \{u\}$).

Denote, for simplicity, $\mathscr{K} = K[E]$; i.e.,

$\mathscr{K} = \{K[u]; u \in E\}$; referred to as: the *constant class* of E_0.

The following properties of the mapping K and its associated constant class $\mathscr{K}$ are almost immediate; so, we do not give details.

Proposition 23 *The mapping K is linear isometric, in the sense*

(81-1) *K is linear: $K[u + v] = K[u] + K[v]$, $K[\lambda u] = \lambda K[u]$, for each $u, v \in E$ and each $\lambda \in R$*

(81-2) *K is isometric: $||K[u]||_0 = ||u||$, for each $u \in E$*

(81-3) *$K[u] \succeq 0$ iff $u \geq 0$.*

As a consequence of these properties, the following conclusions hold:

(81-4) $\mathscr{K}$ *is a D-closed linear subspace of* E_0

(81-5) $u \mapsto K[u]$ *is an increasing linear isomorphism and isometry (hence, topological isomorphism as well) between* $(E, d, \leq)$ *and* $(\mathscr{K}, D, \preceq)$.

Given the linear isometric application K, let $\mathscr{G} : E_0 \to E$ be a mapping with

(lin-K-0) $\mathscr{G}$ is a *linear operator*:
$\mathscr{G}(\varphi + \xi) = \mathscr{G}(\varphi) + \mathscr{G}(\xi), \mathscr{G}(r\varphi) = r\mathscr{G}(\varphi), \varphi, \xi \in E_0, r \in R$

(lin-K-1) $\mathscr{G}$ is *positive*: $\varphi \succeq 0$ implies $\mathscr{G}(\varphi) \geq 0$

(lin-K-2) $\mathscr{G}$ is K-invertible: $\mathscr{G}(K[u]) = u$, for each $u \in E$;

we then say that $\mathscr{G}$ is a *linear K-subordinated Razumikhin map*. If, in addition,

(lin-K-3) $\mathscr{G}$ is *unitary*: $||\mathscr{G}(\varphi)|| \leq ||\varphi||_0$, for each $\varphi \in E_0$,

then $\mathscr{G}$ is called a *unitary-linear K-subordinated Razumikhin map*.

Note that, by the linear property of $\mathscr{G}$ (and our previous conventions), we have

(met-K-1) $\mathscr{G}$ is $(\preceq, \leq)$*-increasing*: $\varphi \preceq \xi$ implies $\mathscr{G}(\varphi) \leq \mathscr{G}(\xi)$

(met-K-2) $\mathscr{G}$ is $\mathscr{K}$-invertible: $\mathscr{G}(K[u]) = u, \forall u \in E$;

or, in other words: $\mathscr{G}$ is a metrical $\mathscr{K}$-subordinated Razumikhin map. On the other hand, under the additional unitary property,

(met-K-3) $\mathscr{G}$ is (d, D)*-nonexpansive*: $d(\mathscr{G}\varphi, \mathscr{G}\xi) \leq D(\varphi, \xi), \varphi, \xi \in E_0$;

wherefrom: $\mathscr{G}$ is a nonexpansive-metrical $\mathscr{K}$-subordinated Razumikhin map.

Returning to the general case, note that, by the definitions above

$||\varphi||_0 = ||\mathscr{G}(\varphi)||$, for each $\varphi \in \mathscr{K}$.

Concerning this aspect, let the *$\mathscr{G}$-attached Razumikhin class* be defined as

$\mathscr{R}(\mathscr{G}) = \{\varphi \in E_0; ||\varphi||_0 = ||\mathscr{G}(\varphi)||\}$.

Note that, by these observations,

$\mathscr{K} \subseteq \mathscr{R}(\mathscr{G})$; hence, in particular, $0 \in \mathscr{R}(\mathscr{G})$.

The converse inclusion is true under certain regularity assumptions about $\mathscr{R}(\mathscr{G})$. To verify this, a lot of preliminary facts involving our class are needed.

Proposition 24 *The $\mathscr{G}$-attached Razumikhin class $\mathscr{R}(\mathscr{G})$ is homogeneous; i.e.,*

$\varphi \in \mathscr{R}(\mathscr{G}) \Longrightarrow \mu\varphi \in \mathscr{R}(\mathscr{G})$, *for all* $\mu \in R$.

Proof Denote for simplicity $\xi = \mu\varphi$. By definition (and $\mathscr{G}$=linear operator)

$||\xi||_0 = |\mu| \cdot ||\varphi||_0, ||\mathscr{G}(\xi)|| = |\mu| \cdot ||\mathscr{G}(\varphi)||$.

This, along with the choice of φ, gives $\xi \in \mathscr{R}(\mathscr{G})$; and we are done.

Now, call the $\mathscr{G}$-attached Razumikhin class $\mathscr{R}(\mathscr{G})$, *difference closed* provided

(dif-clo) $\varphi, \xi \in \mathscr{R}(\mathscr{G}) \Longrightarrow \varphi - \xi \in \mathscr{R}(\mathscr{G})$.

The following characterization of this concept is almost immediate by the above auxiliary facts; so, we do not give details.

Proposition 25 *The following conditions are equivalent;*

(83-1) $\mathscr{R}(\mathscr{G})$ *is difference closed*
(83-2) $\mathscr{R}(\mathscr{G})$ *is additive:* $\varphi, \psi \in \mathscr{R}(\mathscr{G}) \Longrightarrow \varphi + \psi \in \mathscr{R}(\mathscr{G})$
(83-3) $\mathscr{R}(\mathscr{G})$ *is a linear subspace of* E_0.

Having these precise, the following answer to the posed question is available.

Proposition 26 *Suppose that the* $\mathscr{G}$*-attached Razumikhin class* $\mathscr{R}(\mathscr{G})$ *is difference closed; or, equivalently, additive. Then, necessarily,* $\mathscr{R}(\mathscr{G}) = \mathscr{K}$.

Proof Suppose that $\mathscr{R}(\mathscr{G})$ includes strictly $\mathscr{K}$; and let φ be some function with

$\varphi \in \mathscr{R}(\mathscr{G}) \setminus \mathscr{K}$; i.e., $\varphi \neq K[u]$, for all $u \in E$.

The function $\xi := K[u]$ where $u := \mathscr{G}(\varphi)$ is an element of the class $\mathscr{K}$; hence, of the Razumikhin class $\mathscr{R}(\mathscr{G})$ as well. As $\mathscr{R}(\mathscr{G})$ is difference closed, $\delta := \varphi - \xi$ is in $\mathscr{R}(\mathscr{G})$; with, in addition (as $\mathscr{G}$=linear operator)

$$\mathscr{G}(\delta) = \mathscr{G}(\varphi) - \mathscr{G}(K[u]) = \mathscr{G}(\varphi) - u = \mathscr{G}(\varphi) - \mathscr{G}(\varphi) = 0.$$

This, by definition, gives

$||\delta||_0 = ||\mathscr{G}(\delta)|| = 0$; or, equivalently: $\varphi = K[u] \in \mathscr{K}$, where $u = \mathscr{G}(\varphi) \in E$;

in contradiction with the choice of φ. Hence, $\mathscr{R}(\mathscr{G}) = \mathscr{K}$, as claimed.

Let $\mathscr{G}$ be a linear $\mathscr{K}$-subordinated Razumikhin map; and $\mathscr{T} : E_0 \to E$ be a (nonself) mapping. We say that $\varphi \in E_0$ is a *coincidence point* of $(\mathscr{G}, \mathscr{T})$, when $\mathscr{G}\varphi = \mathscr{T}\varphi$. From the same technical motivations as the ones discussed in our previous metrical context, such fixed points are to be sought in the constant class $\mathscr{K} := K[E]$. Concerning this aspect, the following general reduction principle is to be noted. Define an associated to $\mathscr{T}$ selfmap $T : E \to E$, according to

$$Tu = \mathscr{T}(K[u]),\ u \in E.$$

Proposition 27 *Under these conventions, the following are valid:*

(85-1) *If* $z \in E$ *is a fixed point of* T*, then* $\zeta := K[z] \in \mathscr{K}$ *is a coincidence point of* $(\mathscr{G}, \mathscr{T})$
(85-2) *Conversely, if* $\psi = K[u] \in \mathscr{K}$ *is a coincidence point of* $(\mathscr{G}, \mathscr{T})$*, then* $u \in E$ *is a fixed point of* T.

The proof is identical with the related one we already presented in our metrical setting; so, further details are not given.

Finally, let us give some basic examples of linear K-subordinated Razumikhin maps $\mathscr{G}$ (introduced as before).

Example 1 Let $c \in I$ be an arbitrary fixed point. The linear operator $\mathcal{G}_c : E_0 \to E$ introduced as

$$\mathcal{G}_c(\varphi) = \varphi(c), \varphi \in E_0,$$

is unitary, positive, and K-invertible; so, it is a unitary-linear K-subordinated Razumikhin map. Moreover, the Razumikhin class of functions in E_0 attached to $\mathcal{G}_c$ is given by the (known) representation formula

$$\text{Raz}(\mathcal{G}_c) = \{\varphi \in E_0; ||\varphi||_0 = ||\varphi(c)||\}.$$

Example 2 Let $C = (c_n; n \geq 0)$ be a sequence in I; and $\Lambda = (\lambda_n; n \geq 0)$ be a sequence in R_+ with $\sum_n \lambda_n = 1$. The linear operator $\mathcal{G}_{(C,\Lambda)} : E_0 \to E$ given as

$$\mathcal{G}_{(C,\Lambda)}(\varphi) = \sum_n \lambda_n \varphi(c_n), \varphi \in E_0$$

is well defined, because φ is bounded on I. In addition, $\mathcal{G}_{(C,\Lambda)}$ is unitary, positive, and K-invertible; so, it is a unitary-linear K-subordinated Razumikhin map. Moreover, the Razumikhin class of functions in E_0 attached to $\mathcal{G}_{(C,\Lambda)}$ is expressed as

$$\text{Raz}(\mathcal{G}_{(C,\Lambda)}) = \{\varphi \in E_0; ||\varphi||_0 = ||\mathcal{G}_{(C,\Lambda)}(\varphi)||\}.$$

In particular, when the sequence $(c_n; n \geq 0)$ is constant and $(\lambda_0 = 1; \lambda_n = 0, \forall n \geq 1)$ this operator reduces to the preceding one.

Example 3 Define the linear operator $\mathcal{G}_\int$ over E_0 as

$$\mathcal{G}_\int(\varphi) = (1/(b-a)) \int_a^b \varphi(t)dt, \varphi \in E_0;$$

where $\int$ stands for the Bochner integral. By the elementary properties of this object, $\mathcal{G}_\int$ is unitary, positive, and K-invertible; so, it is a unitary-linear K-subordinated Razumikhin map. Moreover, the Razumikhin class of functions in E_0 attached to $\mathcal{G}_\int$ is given by the (representation) formula

$$\text{Raz}(\mathcal{G}_\int) = \{\varphi \in E_0; ||\varphi||_0 = ||\mathcal{G}_\int(\varphi)||\}.$$

Starting from these observations, we may reduce the PPF dependent fixed point theory over such structures to the (standard) fixed point theory over quasi-ordered metric spaces. Some elements of this reduction principle have been sketched in Turinici [53]. Precisely, it was established there that a recent PPF dependent fixed point result in Agarwal et al. [2] is reducible to a fixed point problem involving a class of contractions over standard metric structures introduced under the lines proposed by Samet et al. [41]. Note that the same reduction technique is applicable to some other statements in this area due to Dhage [10, 11], Chuasuk and Kaewcharoen [7], Ćirić et al. [8], Farajzadeh and Kaewcharoen [14]. Kutbi and Sintunavarat [25], Khan and Jhade [23], Paknazar et al. [33], or Saluja et al. [39]; we do not give details. Further aspects will be discussed in a separate paper.

References

1. R.P. Agarwal, M.A. El-Gebeily, D. O'Regan, Generalized contractions in partially ordered metric spaces. Appl. Anal. **87**, 109–116 (2008)
2. R.P. Agarwal, P. Kumam, W. Sintunavarat, PPF dependent fixed point theorems for an α_c-admissible non-self mapping in the Razumikhin class. Fixed Point Theory Appl. **2013**, 280 (2013)
3. S. Banach, Sur les opérations dans les ensembles abstraits et leur application aux équations intégrales. Fundam. Math. **3**, 133–181 (1922)
4. P. Bernays, A system of axiomatic set theory: Part III. Infinity and enumerability analysis. J. Symb. Log. **7**, 65–89 (1942)
5. S.R. Bernfeld, V. Lakshmikantham, Y.M. Reddy, Fixed point theorems of operators with PPF dependence in Banach spaces. Appl. Anal. **6**, 271–280 (1977)
6. D.W. Boyd, J.S.W. Wong, On nonlinear contractions. Proc. Am. Math. Soc. **20**, 458–464 (1969)
7. P. Chuasuk, A. Kaewcharoen, Fixed point results for generalized rational type contractions with PPF dependence. Int. J. Pure Appl. Math. **96**, 197–212 (2014)
8. L.B. Cirić, S.M. Alsulami, P. Salimi, P. Vetro, PPF dependent fixed point results for triangular α_c -admissible mappings. Sci. World J. **2014**, Article ID 673647 (2014)
9. P.J. Cohen, *Set Theory and the Continuum Hypothesis* (Benjamin, New York, 1966)
10. B.C. Dhage, Fixed point theorems with PPF dependence and functional differential equations. Fixed Point Theory **13**, 439–452 (2012)
11. B. Dhage, On some common fixed point theorems with PPF dependence in Banach spaces. J. Nonlin. Sci. Appl. **5**, 220–232 (2012)
12. Z. Drici, F.A. McRae, J. Vasundara Devi, Fixed point theorem in partially ordered metric spaces for operators with PPF dependence. Nonlin. Anal. **67**, 641–647 (2007)
13. P.N. Dutta, B.S. Choudhury, A generalization of contraction principle in metric spaces. Fixed Point Theory Appl. **2008**, Article ID 406368 (2008)
14. A. Farajzadeh, A. Kaewcharoen, On fixed point theorems for mappings with PPF dependence. J. Ineq. Appl. **2014**, 372 (2014)
15. P. Hitzler, Generalized metrics and topology in logic programming semantics, PhD Thesis, Natl. Univ. Ireland, Univ. College Cork, 2001
16. N. Hussain, S. Khaleghizadeh, P. Salimi, F. Akbar, New fixed point results with PPF dependence in Banach spaces endowed with a graph. Abstr. Appl. Anal. **2013**, Article ID 827205 (2013)
17. J. Jachymski, Common fixed point theorems for some families of mappings. Indian J. Pure Appl. Math. **25**, 925–937 (1994)
18. J. Jachymski, The contraction principle for mappings on a metric space with a graph. Proc. Am. Math. Soc. **136**, 1359–1373 (2008)
19. J. Jachymski, Equivalent conditions for generalized contractions on (ordered) metric spaces. Nonlin. Anal. **74**, 768–774 (2011)
20. M. Jleli, B. Samet, A new generalization of the Banach contraction principle. J. Inequal. Appl. **2014**, 38 (2014)
21. A. Kaewcharoen, PPF dependent common fixed point theorems for mappings in Banach spaces. J. Inequal. Appl. **2013**, 287 (2013)
22. S. Kasahara, On some generalizations of the Banach contraction theorem. Publ. Res. Inst. Math. Sci. Kyoto Univ. **12**, 427–437 (1976)
23. M.S. Khan, P.K. Jhade, On a fixed point theorem with PPF dependence in the Razumikhin class. Gazi Univ. J. Sci. **28**, 211–219 (2015)
24. M.S. Khan, M. Swaleh, S. Sessa, Fixed point theorems by altering distances between the points. Bull. Aust. Math. Soc. **30**, 1–9 (1984)
25. M.A. Kutbi, W. Sintunavarat, On sufficient conditions for the existence of past-present-future dependent fixed point in the Razumikhin class and application. Abstr. Appl. Anal. **2014**, Article ID 342687 (2014)

26. S. Leader, Fixed points for general contractions in metric spaces. Jpn. J. Math. **24**, 17–24 (1979)
27. T.-C. Lim, On characterizations of Meir-Keeler contractive maps. Nonlinear Anal. **46**, 113–120 (2001)
28. J. Matkowski, Integrable solutions of functional equations. *Dissertationes Math.* vol. 127 (Polish Scientific Publishers, Warsaw, 1975)
29. A. Meir, E. Keeler, A theorem on contraction mappings. J. Math. Anal. Appl. **28**, 326–329 (1969)
30. G.H. Moore, *Zermelo's Axiom of Choice: Its Origin, Development and Influence* (Springer, New York, 1982)
31. Y. Moskhovakis, *Notes on Set Theory* (Springer, New York, 2006)
32. J.J. Nieto, R. Rodriguez-Lopez, Contractive mapping theorems in partially ordered sets and applications to ordinary differential equations. Order **22**, 223–239 (2005)
33. M. Paknazar, M.A. Kutbi, M. Demma, P. Salimi, PPF dependent fixed point in modified Razumikhin class with applications. J. Nonlin. Sci. Appl. **9**, 270–286 (2016)
34. H.K. Pathak, A fixed point theorem on Banach space and its applications. Commun. Fac. Sci. Univ. Ank. (Ser. A) **42**, 61–67 (1993)
35. A.C.M. Ran, M.C. Reurings, A fixed point theorem in partially ordered sets and some applications to matrix equations. Proc. Am. Math. Soc. **132**, 1435–1443 (2004)
36. B.E. Rhoades, A comparison of various definitions of contractive mappings. Trans. Am. Math. Soc. **226**, 257–290 (1977)
37. B.E. Rhoades, Some theorems on weakly contractive maps. Nonlinear Anal. **47**, 2683–2693 (2001)
38. I.A. Rus, *Generalized Contractions and Applications* (Cluj University Press, Cluj-Napoca, 2001)
39. A.S. Saluja, R.A. Rashwan, D. Magarde, P.K. Jhade, Some result in ordered metric spaces for rational type expressions. Facta Univ. Niš (Ser. Math. Inform.) **31**, 125–138 (2016)
40. B. Samet, M. Turinici, Fixed point theorems on a metric space endowed with an arbitrary binary relation and applications. Commun. Math. Anal. **13**, 82–97 (2012)
41. B. Samet, C. Vetro, P. Vetro, Fixed point theorems for $\alpha - \psi$-contractive type mappings. Nonlinear Anal. **75**, 2154–2165 (2012)
42. E. Schechter, *Handbook of Analysis and Its Foundation* (Academic Press, New York, 1997)
43. T. Som, Some fixed point theorems of metric and Banach spaces. Indian J. Pure Appl. Math. **16**, 575–585 (1985)
44. T. Suzuki, Discussion of several contractions by Jachymski's approach. Fixed Point Theory Appl. **2016**, 91 (2016)
45. T. Suzuki, C. Vetro, Three existence theorems for weak contractions of Matkowski type. Int. J. Math. Stat. **6**, 110–120 (2010)
46. A. Tarski, Axiomatic and algebraic aspects of two theorems on sums of cardinals. Fundam. Math. **35**, 79–104 (1948)
47. M. Turinici, Fixed points of implicit contraction mappings. An. Şt. Univ. "A. I. Cuza" Iaşi (S I-a, Mat) **22**, 177–180 (1976)
48. M. Turinici, Nonlinear contractions and applications to Volterra functional equations. An. Şt. Univ. "Al. I. Cuza" Iaşi (S I-a, Mat) **23**, 43–50 (1977)
49. M. Turinici, Fixed points for monotone iteratively local contractions. Demonstratio Math. **19**, 171–180 (1986)
50. M. Turinici, Abstract comparison principles and multivariable Gronwall-Bellman inequalities. J. Math. Anal. Appl. **117**, 100–127 (1986)
51. M. Turinici, Function pseudometric VP and applications. Bul. Inst. Polit. Iaşi (S. Mat., Mec. Teor., Fiz.) **53**(57), 393–411 (2007)
52. M. Turinici, Ran-Reurings theorems in ordered metric spaces. J. Indian Math. Soc. (N.S.) **78**, 207–214 (2011)
53. M. Turinici, PPF dependent fixed points in A-closed Razumikhin classes. ROMAI J. **10**(No. 1), 205–220 (2014)

54. M. Turinici, Contraction maps in ordered metrical structures. *Mathematics Without Boundaries*, ed. by P.M. Pardalos, T.M. Rassias (Springer, New York, 2014), pp. 533–575
55. M. Turinici, Diagonal fixed points of geometric contractions. *Modern Discrete Mathematics and Analysis*, ed. by N.J. Daras, Th.M. Rassias (Springer, New York, 2017)
56. D. Wardowski, Fixed points of a new type of contractive mappings in complete metric spaces. J. Fixed Point Theory Appl. **2012**, 94 (2012)
57. E.S. Wolk, On the principle of dependent choices and some forms of Zorn's lemma. Can. Math. Bull. **26**, 365–367 (1983)

Equivalent Properties of Parameterized Hilbert-Type Integral Inequalities

Bicheng Yang

Abstract By the use of the techniques of real analysis and the weight functions, a few equivalent statements of a general Hilbert-type integral inequality with the nonhomogeneous kernel related to another inequality, the parameters and the integral of kernel are obtained. The best possible constant factor is given. As a corollary, a few equivalent statements of a general Hilbert-type integral inequality with the homogeneous kernel and a best possible constant factor are deduced. Moreover, we also study the case of the reverses. The operator expressions, a few particular cases and some examples are considered.

Mathematics Subject Classification 26D15, 47A07

1 Introduction

If $0 < \int_0^\infty f^2(x)dx < \infty$ and $0 < \int_0^\infty g^2(y)dy < \infty$, then we have the following Hilbert's integral inequality (cf. [1]):

$$\int_0^\infty \int_0^\infty \frac{f(x)g(y)}{x+y} dxdy < \pi \left(\int_0^\infty f^2(x)dx \int_0^\infty g^2(y)dy \right)^{\frac{1}{2}}, \tag{1}$$

where the constant factor π is the best possible. In 1925, by introducing one pair of conjugate exponents (p, q), Hardy [2] gave an extension of (1) as follows: For $p > 1$, $\frac{1}{p} + \frac{1}{q} = 1$, $f(x), g(y) \geq 0$, $0 < \int_0^\infty f^p(x)dx < \infty$ and $0 < \int_0^\infty g^q(y) dy < \infty$, we have the following Hardy-Hilbert's inequality:

B. Yang (✉)
Department of Mathematics, Guangdong University of Education, Guangzhou, Guangdong, People's Republic of China
e-mail: bcyang@gdei.edu.cn

T. M. Rassias, P. M. Pardalos (eds.), *Mathematical Analysis and Applications*, Springer Optimization and Its Applications 154,
https://doi.org/10.1007/978-3-030-31339-5_23

$$\int_0^\infty \int_0^\infty \frac{f(x)g(y)}{x+y} dxdy$$
$$< \frac{\pi}{\sin(\pi/p)} \left(\int_0^\infty f^p(x)dx \right)^{\frac{1}{p}} \left(\int_0^\infty g^q(y)dy \right)^{\frac{1}{q}}, \quad (2)$$

where the constant factor $\frac{\pi}{\sin(\pi/p)}$ is the best possible. Inequalities (1) and (2) are important in analysis and its applications (cf. [3, 4]).

In 1934, Hardy et al. gave an extension of (2) as follows: If $k_1(x, y)$ is a nonnegative homogeneous function of degree -1, $k_p = \int_0^\infty k_1(u, 1)u^{\frac{-1}{p}} du \in \mathbf{R}_+ = (0, \infty)$, then we have the following Hardy-Hilbert-type integral inequality:

$$\int_0^\infty \int_0^\infty k_1(x, y) f(x)g(y)dxdy$$
$$< k_p \left(\int_0^\infty f^p(x)dx \right)^{\frac{1}{p}} \left(\int_0^\infty g^q(y)dy \right)^{\frac{1}{q}}, \quad (3)$$

where the constant factor k_p is the best possible (cf. [3, Theorem 319]). Also a general Hilbert-type integral inequality with the nonhomogeneous kernel is proved as follows: If $h(u) > 0$, $\phi(\sigma) = \int_0^\infty h(u)u^{\sigma-1}du \in \mathbf{R}_+$, then

$$\int_0^\infty \int_0^\infty h(xy) f(x)g(y)dxdy$$
$$< \phi(\frac{1}{p}) \left(\int_0^\infty x^{p-2} f^p(x)dx \right)^{\frac{1}{p}} \left(\int_0^\infty g^q(y)dy \right)^{\frac{1}{q}}, \quad (4)$$

where the constant factor $\phi(\frac{1}{p})$ is the best possible (cf. [3, Theorem 350]).

In 1998, by introducing an independent parameter $\lambda > 0$, Yang gave an extension of (1) as follows (cf. [5, 6]):

$$\int_0^\infty \int_0^\infty \frac{f(x)g(y)}{(x+y)^\lambda} dxdy$$
$$< B(\frac{\lambda}{2}, \frac{\lambda}{2}) \left(\int_0^\infty x^{1-\lambda} f^2(x)dx \int_0^\infty y^{1-\lambda} g^2(y)dy \right)^{\frac{1}{2}}, \quad (5)$$

where the constant factor $B(\frac{\lambda}{2}, \frac{\lambda}{2})$ is the best possible ($B(u, v)$ is the beta function).

In 2004, by introducing another pair of conjugate exponents (r, s), Yang [7] gave an extension of (2) as follows: If $\lambda > 0, r > 1, \frac{1}{r} + \frac{1}{s} = 1$, $f(x), g(y) \geq 0, 0 < \int_0^\infty x^{p(1-\frac{\lambda}{r})-1} f^p(x)dx < \infty$ and $0 < \int_0^\infty y^{q(1-\frac{\lambda}{s})-1} g^q(y)dy < \infty$, then

$$\int_0^\infty \int_0^\infty \frac{f(x)g(y)}{x^\lambda + y^\lambda} dxdy$$
$$< \frac{\pi}{\lambda \sin(\pi/r)} \left[\int_0^\infty x^{p(1-\frac{\lambda}{r})-1} f^p(x)dx\right]^{\frac{1}{p}} \left[\int_0^\infty y^{q(1-\frac{\lambda}{s})-1} g^q(y)dy\right]^{\frac{1}{q}}, \quad (6)$$

where the constant factor $\frac{\pi}{\lambda \sin(\pi/r)}$ is the best possible. For $\lambda = 1, r = q, s = p$, (6) reduces to (2); for $\lambda = 1, r = p, s = q$, (6) reduces to the dual form of (2) as follows:

$$\int_0^\infty \int_0^\infty \frac{f(x)g(y)}{x + y} dxdy$$
$$< \frac{\pi}{\sin(\pi/p)} \left(\int_0^\infty x^{p-2} f^p(x)dx\right)^{\frac{1}{p}} \left(\int_0^\infty y^{q-2} g^q(y)dy\right)^{\frac{1}{q}}, \quad (7)$$

where the constant factor $\frac{\pi}{\sin(\pi/p)}$ is still the best possible. For $p = q = 2$, both (2) and (7) reduce to (1).

In 2005, Yang et al. [8] also gave an extension of (2) and (5) as follows:

$$\int_0^\infty \int_0^\infty \frac{f(x)g(y)}{(x + y)^\lambda} dxdy$$
$$< B(\frac{\lambda}{r}, \frac{\lambda}{s}) \left[\int_0^\infty x^{p(1-\frac{\lambda}{r})-1} f^p(x)dx\right]^{\frac{1}{p}} \left[\int_0^\infty y^{q(1-\frac{\lambda}{s})-1} g^q(y)dy\right]^{\frac{1}{q}}, \quad (8)$$

where the constant factor $B(\frac{\lambda}{r}, \frac{\lambda}{s})(\lambda > 0)$ is the best possible. Krnić et al. [9–19] provided some extensions and particular cases of (2)–(4) with parameters.

In 2009, Yang gave an extension of (3), (6), and (8) as follows (cf. [20, 21]): If $\lambda_1 + \lambda_2 = \lambda \in \mathbf{R} = (-\infty, \infty)$, $k_\lambda(x, y)$ is a nonnegative homogeneous function of degree $-\lambda$, satisfying

$$k_\lambda(ux, uy) = u^{-\lambda} k_\lambda(x, y)(u, x, y > 0),$$

with $k(\lambda_1) = \int_0^\infty k_\lambda(u, 1)u^{\lambda_1 - 1}du \in \mathbf{R}_+$, then we have

$$\int_0^\infty \int_0^\infty k_\lambda(x, y) f(x)g(y)dxdy$$
$$< k(\lambda_1) \left(\int_0^\infty x^{p(1-\lambda_1)-1} f^p(x)dx\right)^{\frac{1}{p}} \left(\int_0^\infty y^{q(1-\lambda_2)-1} g^q(y)dy\right)^{\frac{1}{q}}, \quad (9)$$

where the constant factor $k(\lambda_1)$ is the best possible. For $\lambda = 1, \lambda_1 = \frac{1}{q}, \lambda_2 = \frac{1}{p}$, (9) reduces to (3); for $\lambda > 0, k_\lambda(x, y) = \frac{1}{(x+y)^\lambda}, \lambda_1 = \frac{\lambda}{r}, \lambda_2 = \frac{\lambda}{s}$, (9) reduces to (8).

Also an extension of (4) with a parameter was given as follows:

$$\int_0^\infty \int_0^\infty h(xy)f(x)g(y)dxdy$$
$$< \phi(\sigma)\left(\int_0^\infty x^{p(1-\sigma)-1}f^p(x)dx\right)^{\frac{1}{p}}\left(\int_0^\infty y^{q(1-\sigma)-1}g^q(y)dy\right)^{\frac{1}{q}}, \quad (10)$$

where the constant factor $\phi(\sigma)$ is the best possible (cf. [22]). For $\sigma = \frac{1}{p}$, (10) reduces to (4).

Some equivalent inequalities of (9) and (10) are considered by Yang [21]. In 2013, Yang [22] also studied the equivalency of (9) and (10) under an adding condition. In 2017, Hong [23] studied an equivalent condition between (9) and the parameters.

In this chapter, by the use of the techniques of real analysis and the weight functions, a few equivalent statements of (10) related to another inequality, the parameters and the integral of kernel are obtained. The best possible constant factor is given. As a corollary, a few equivalent statements of (9) and a best possible constant factor are deduced. Moreover, we also study the case of the reverses. The operator expressions, a few particular cases, and some examples are considered.

2 Some Lemmas

In the following, we agree that $p > 0$ $(p \neq 1)$, $\frac{1}{p} + \frac{1}{q} = 1, \sigma_1, \sigma \in \mathbf{R}$, $h(u)$ is a nonnegative measurable function in $(0, \infty)$, such that

$$k(\sigma) := \int_0^\infty h(u)u^{\sigma-1}du \ (\geq 0). \quad (11)$$

For $n \in \mathbf{N} = \{1, 2, \cdots\}$, we define the following two expressions:

$$I_1 := \int_1^\infty \left(\int_0^1 h(xy)x^{\sigma+\frac{1}{pn}-1}dx\right) y^{\sigma_1-\frac{1}{qn}-1}dy, \quad (12)$$

$$I_2 := \int_0^1 \left(\int_1^\infty h(xy)x^{\sigma-\frac{1}{pn}-1}dx\right) y^{\sigma_1+\frac{1}{qn}-1}dy. \quad (13)$$

Setting $u = xy$ in (12) and (13), by Fubini theorem (cf. [24]), it follows that

$$I_1 = \int_1^\infty \left[\int_0^y h(u)\left(\frac{u}{y}\right)^{\sigma+\frac{1}{pn}-1}\frac{1}{y}du\right] y^{\sigma_1-\frac{1}{qn}-1}dy$$
$$= \int_1^\infty y^{(\sigma_1-\sigma)-\frac{1}{n}-1}\left(\int_0^y h(u)u^{\sigma+\frac{1}{pn}-1}du\right)dy$$

$$= \int_1^\infty y^{(\sigma_1-\sigma)-\frac{1}{n}-1}dy \int_0^1 h(u)u^{\sigma+\frac{1}{pn}-1}du$$
$$+ \int_1^\infty y^{(\sigma_1-\sigma)-\frac{1}{n}-1} \int_1^y h(u)u^{\sigma+\frac{1}{pn}-1}dudy$$
$$= \int_1^\infty y^{(\sigma_1-\sigma)-\frac{1}{n}-1}dy \int_0^1 h(u)u^{\sigma+\frac{1}{pn}-1}du$$
$$+ \int_1^\infty \left[\int_u^\infty y^{(\sigma_1-\sigma)-\frac{1}{n}-1}dy\right] h(u)u^{\sigma+\frac{1}{pn}-1}du, \tag{14}$$

$$I_2 = \int_0^1 \left[\int_y^\infty h(u)\left(\frac{u}{y}\right)^{\sigma-\frac{1}{pn}-1}\frac{1}{y}du\right] y^{\sigma_1+\frac{1}{qn}-1}dy$$
$$= \int_0^1 y^{(\sigma_1-\sigma)+\frac{1}{n}-1}\left(\int_y^\infty h(u)u^{\sigma-\frac{1}{pn}-1}du\right)dy$$
$$= \int_0^1 y^{(\sigma_1-\sigma)+\frac{1}{n}-1}dy \int_y^1 h(u)u^{\sigma-\frac{1}{pn}-1}du$$
$$+ \int_0^1 y^{(\sigma_1-\sigma)+\frac{1}{n}-1} \int_1^\infty h(u)u^{\sigma-\frac{1}{pn}-1}dudy$$
$$= \int_0^1 \left[\int_0^u y^{(\sigma_1-\sigma)+\frac{1}{n}-1}dy\right] h(u)u^{\sigma-\frac{1}{pn}-1}du$$
$$+ \int_0^1 y^{(\sigma_1-\sigma)+\frac{1}{n}-1}dy \int_1^\infty h(u)u^{\sigma-\frac{1}{pn}-1}du. \tag{15}$$

Lemma 1 *If $p > 1, k(\sigma) > 0$, there exists a constant M, such that for any nonnegative measurable functions $f(x)$ and $g(y)$ in $(0, \infty)$, the following inequality*

$$I := \int_0^\infty \int_0^\infty h(xy)f(x)g(y)dxdy$$
$$\leq M\left[\int_0^\infty x^{p(1-\sigma)-1}f^p(x)dx\right]^{\frac{1}{p}}\left[\int_0^\infty y^{q(1-\sigma_1)-1}g^q(y)dy\right]^{\frac{1}{q}} \tag{16}$$

holds true, then we have $\sigma_1 = \sigma$.

Proof Setting $k_1(\sigma) := \int_0^1 h(u)u^{\sigma-1}du$ and $k_2(\sigma) := \int_1^\infty h(u)u^{\sigma-1}du$, it follows that $k(\sigma) = k_1(\sigma) + k_2(\sigma) > 0$. Since $k_i(\sigma) \geq 0 (i = 1, 2)$, without loss of generality, we assume that $k_1(\sigma) > 0$, namely, $h(u) > 0$ a.e. in an interval $(a, b) \subset (0, 1)$ $(a < b)$.

If $\sigma_1 < \sigma$, then for $n > \frac{1}{\sigma - \sigma_1}$ ($n \in \mathbf{N}$), we set the following two functions:

$$f_n(x) := \begin{cases} 0, 0 < x < 1 \\ x^{\sigma - \frac{1}{pn} - 1}, x \geq 1 \end{cases},$$

$$g_n(y) := \begin{cases} y^{\sigma_1 + \frac{1}{qn} - 1}, 0 < y \leq 1 \\ 0, y > 1 \end{cases}.$$

Hence, we find

$$\begin{aligned} J_2 &:= \left[\int_0^\infty x^{p(1-\sigma)-1} f_n^p(x)dx\right]^{\frac{1}{p}} \left[\int_0^\infty y^{q(1-\sigma_1)-1} g_n^q(y)dy\right]^{\frac{1}{q}} \\ &= \left[\int_1^\infty x^{p(1-\sigma)-1} x^{p(\sigma - \frac{1}{pn} - 1)} dx\right]^{\frac{1}{p}} \left[\int_0^1 y^{q(1-\sigma_1)-1} y^{q(\sigma_1 + \frac{1}{qn} - 1)} dy\right]^{\frac{1}{q}} \\ &= \left(\int_1^\infty x^{-\frac{1}{n}-1} dx\right)^{\frac{1}{p}} \left(\int_0^1 y^{\frac{1}{n}-1} dy\right)^{\frac{1}{q}} = n. \end{aligned}$$

By (15), we have

$$\begin{aligned} &\int_0^1 \left[\int_0^u y^{(\sigma_1 - \sigma) + \frac{1}{n} - 1} dy\right] h(u) u^{\sigma - \frac{1}{pn} - 1} du \\ &\leq I_2 = \int_0^\infty \int_0^\infty h(xy) f_n(x) g_n(y) dx dy \\ &\leq M J_2 = Mn < \infty. \end{aligned} \tag{17}$$

Since $(\sigma_1 - \sigma) + \frac{1}{n} < 0$, it follows that for any $u \in (0, 1)$, $\int_0^u y^{(\sigma_1 - \sigma) + \frac{1}{n} - 1} dy = \infty$. By (17), in view of $h(u) u^{\sigma - \frac{1}{pn} - 1} > 0$ a.e. in $(a, b) \subset (0, 1)$, we find that $\infty \leq Mn < \infty$, which is a contradiction.

If $\sigma_1 > \sigma$, then for $n > \frac{1}{\sigma_1 - \sigma}$ ($n \in \mathbf{N}$), we set the following two functions:

$$\widetilde{f}_n(x) := \begin{cases} x^{\sigma + \frac{1}{pn} - 1}, 0 < x \leq 1 \\ 0, x > 1 \end{cases},$$

$$\widetilde{g}_n(y) := \begin{cases} 0, 0 < y < 1 \\ y^{\sigma_1 - \frac{1}{qn} - 1}, y \geq 1 \end{cases}.$$

Hence, we find

$$\begin{aligned}\widetilde{J}_2 &:= \left[\int_0^\infty x^{p(1-\sigma)-1}\widetilde{f}_n^p(x)dx\right]^{\frac{1}{p}}\left[\int_0^\infty y^{q(1-\sigma_1)-1}\widetilde{g}_n^q(y)dy\right]^{\frac{1}{q}} \\ &= \left[\int_0^1 x^{p(1-\sigma)-1}x^{p(\sigma+\frac{1}{pn}-1)}dx\right]^{\frac{1}{p}}\left[\int_1^\infty y^{q(1-\sigma_1)-1}y^{q(\sigma_1-\frac{1}{qn}-1)}dy\right]^{\frac{1}{q}} \\ &= \left(\int_0^1 x^{\frac{1}{n}-1}dx\right)^{\frac{1}{p}}\left(\int_1^\infty y^{-\frac{1}{n}-1}dy\right)^{\frac{1}{q}} = n.\end{aligned}$$

By (14), we have

$$\begin{aligned}&\int_1^\infty y^{(\sigma_1-\sigma)-\frac{1}{n}-1}dy\int_0^1 h(u)u^{\sigma+\frac{1}{pn}-1}du \\ &\leq I_1 = \int_0^\infty\int_0^\infty h(xy)\widetilde{f}_n(x)\widetilde{g}_n(y)dxdy \\ &\leq M\widetilde{J}_2 = Mn < \infty. \qquad (18)\end{aligned}$$

Since $(\sigma_1 - \sigma) - \frac{1}{n} > 0$, it follows that $\int_1^\infty y^{(\sigma_1-\sigma)-\frac{1}{n}-1}dy = \infty$. By (18), in view of $\int_0^1 h(u)u^{\sigma+\frac{1}{pn}-1}du > 0$, we have $\infty \leq Mn < \infty$, which is a contradiction.

Hence, we conclude that $\sigma_1 = \sigma$.

The lemma is proved.

Lemma 2 *If $p > 1, k(\sigma) > 0$, there exists a constant M, such that for any nonnegative measurable functions $f(x)$ and $g(y)$ in $(0, \infty)$, the following inequality*

$$\begin{aligned}&\int_0^\infty\int_0^\infty h(xy)f(x)g(y)dxdy \\ &\leq M\left[\int_0^\infty x^{p(1-\sigma)-1}f^p(x)dx\right]^{\frac{1}{p}}\left[\int_0^\infty y^{q(1-\sigma)-1}g^q(y)dy\right]^{\frac{1}{q}} \qquad (19)\end{aligned}$$

holds true, then we have $k(\sigma) \leq M < \infty$, namely, $k(\sigma) \in \mathbf{R}_+$.

Proof For $\sigma_1 = \sigma$, we reduce (14) and then use inequality $I_1 \leq M\widetilde{J}_2$ (when $\sigma_1 = \sigma$) as follows:

$$\begin{aligned}\frac{1}{n}I_1 = \frac{1}{n}\Bigg[&\int_1^\infty y^{-\frac{1}{n}-1}dy\int_0^1 h(u)u^{\sigma+\frac{1}{pn}-1}du \\ &+\int_1^\infty\left(\int_u^\infty y^{-\frac{1}{n}-1}dy\right)h(u)u^{\sigma+\frac{1}{pn}-1}du\Bigg]\end{aligned}$$

$$
\begin{aligned}
&= \int_0^1 h(u) u^{\sigma+\frac{1}{pn}-1} du + \int_1^\infty h(u) u^{\sigma-\frac{1}{qn}-1} du \\
&\le \frac{1}{n} M \widetilde{J}_2 = M.
\end{aligned} \tag{20}
$$

By Fatou lemma (cf. [24]) and (20), we have

$$
\begin{aligned}
k(\sigma) &= \int_0^1 \lim_{n\to\infty} h(u) u^{\sigma+\frac{1}{pn}-1} du + \int_1^\infty \lim_{n\to\infty} h(u) u^{\sigma-\frac{1}{qn}-1} du \\
&\le \underline{\lim}_{n\to\infty} \left[\int_0^1 h(u) u^{\sigma+\frac{1}{pn}-1} du + \int_1^\infty h(u) u^{\sigma-\frac{1}{qn}-1} du \right] \\
&\le M < \infty.
\end{aligned}
$$

The lemma is proved.

Lemma 3 *If $0 < p < 1, k(\sigma) < \infty$, there exist constants $\delta_0, M > 0$, such that $k(\sigma \pm \delta_0) < \infty$, and for any nonnegative measurable functions $f(x)$ and $g(y)$ in $(0, \infty)$, the following inequality*

$$
\begin{aligned}
I &= \int_0^\infty \int_0^\infty h(xy) f(x) g(y) dx dy \\
&\ge M \left[\int_0^\infty x^{p(1-\sigma)-1} f^p(x) dx \right]^{\frac{1}{p}} \left[\int_0^\infty y^{q(1-\sigma_1)-1} g^q(y) dy \right]^{\frac{1}{q}}
\end{aligned} \tag{21}
$$

holds true, then we have $\sigma_1 = \sigma$ and $k(\sigma) \ge M > 0$.

Proof If $\sigma_1 > \sigma$, then for $n > \frac{1}{\delta_0 p} (n \in \mathbf{N}, 0 < p < 1)$, we set the following two functions as in Lemma 1:

$$
f_n(x) = \begin{cases} 0, 0 < x < 1 \\ x^{\sigma-\frac{1}{pn}-1}, x \ge 1 \end{cases},
$$

$$
g_n(y) = \begin{cases} y^{\sigma_1+\frac{1}{qn}-1}, 0 < y \le 1 \\ 0, y > 1 \end{cases}.
$$

We find

$$
J_2 = \left[\int_0^\infty x^{p(1-\sigma)-1} f_n^p(x) dx \right]^{\frac{1}{p}} \left[\int_0^\infty y^{q(1-\sigma_1)-1} g_n^q(y) dy \right]^{\frac{1}{q}} = n.
$$

By (15), we have

$$
\begin{aligned}
I_2 &\leq \int_0^1 y^{(\sigma_1-\sigma)+\frac{1}{n}-1} dy \int_0^\infty h(u) u^{\sigma-\frac{1}{pn}-1} du \\
&= \frac{1}{\sigma_1-\sigma+\frac{1}{n}} \left(\int_0^1 h(u) u^{\sigma-\frac{1}{pn}-1} du + \int_1^\infty h(u) u^{\sigma-\frac{1}{pn}-1} du \right) \\
&\leq \frac{1}{\sigma_1-\sigma} \left(\int_0^1 h(u) u^{(\sigma-\delta_0)-1} du + \int_1^\infty h(u) u^{\sigma-1} du \right) \\
&\leq \frac{1}{\sigma_1-\sigma} \left(k(\sigma-\delta_0) + k(\sigma) \right),
\end{aligned}
$$

and then by (21), we find

$$
\begin{aligned}
&\frac{1}{\sigma_1-\sigma} \left(k(\sigma-\delta_0) + k(\sigma) \right) \\
&\geq I_2 = \int_0^\infty \int_0^\infty h(xy) f_n(x) g_n(y) dx dy \\
&\geq M J_2 = Mn.
\end{aligned} \tag{22}
$$

By (22), in view of $\sigma_1 - \sigma > 0$, $0 \leq k(\sigma-\delta_0) + k(\sigma) < \infty$, for $n \to \infty$, we find

$$
\infty > \frac{1}{\sigma_1-\sigma} \left(k(\sigma-\delta_0) + k(\sigma) \right) \geq \infty,
$$

which is a contradiction.

If $\sigma_1 < \sigma$, then for $n \in \mathbf{N}, n > \frac{1}{\delta_0 p}$, we set the following two functions as in Lemma 1:

$$
\widetilde{f}_n(x) = \begin{cases} x^{\sigma+\frac{1}{pn}-1}, 0 < x \leq 1 \\ 0, x > 1 \end{cases},
$$

$$
\widetilde{g}_n(y) = \begin{cases} 0, 0 < y < 1 \\ y^{\sigma_1-\frac{1}{qn}-1}, y \geq 1 \end{cases}.
$$

We find

$$
\widetilde{J}_2 := \left[\int_0^\infty x^{p(1-\sigma)-1} \widetilde{f}_n^p(x) dx \right]^{\frac{1}{p}} \left[\int_0^\infty y^{q(1-\sigma_1)-1} \widetilde{g}_n^q(y) dy \right]^{\frac{1}{q}} = n.
$$

By (14), we have

$$I_1 \leq \int_1^\infty y^{(\sigma_1-\sigma)-\frac{1}{n}-1}dy \int_0^\infty h(u)u^{\sigma+\frac{1}{pn}-1}du$$
$$= \frac{1}{\sigma-\sigma_1+\frac{1}{n}}\left(\int_0^1 h(u)u^{\sigma+\frac{1}{pn}-1}du + \int_1^\infty h(u)u^{\sigma+\frac{1}{pn}-1}du\right)$$
$$\leq \frac{1}{\sigma-\sigma_1}\left(\int_0^1 h(u)u^{\sigma-1}du + \int_1^\infty h(u)u^{\sigma+\delta_0-1}du\right)$$
$$\leq \frac{1}{\sigma-\sigma_1}\left(k(\sigma)+k(\sigma+\delta_0)\right),$$

and then by (21), we find

$$\frac{1}{\sigma-\sigma_1}\left(k(\sigma)+k(\sigma+\delta_0)\right)$$
$$\geq I_1 = \int_0^\infty \int_0^\infty h(xy)\widetilde{f}_n(x)\widetilde{g}_n(y)dxdy$$
$$\geq M\widetilde{J}_2 = Mn. \tag{23}$$

By (23), for $n \to \infty$, we still find that

$$\infty > \frac{1}{\sigma-\sigma_1}\left(k(\sigma)+k(\sigma+\delta_0)\right) \geq \infty,$$

which is a contradiction.

Hence, we conclude that $\sigma_1 = \sigma$.

For $\sigma_1 = \sigma$, we still have $nM = MJ_2 \leq I_2$. Then we find

$$M = \frac{1}{n}MJ_2 \leq \frac{1}{n}I_2 = \frac{1}{n}\int_0^1 y^{\frac{1}{n}-1}\left(\int_y^\infty h(u)u^{\sigma-\frac{1}{pn}-1}du\right)dy$$
$$= \frac{1}{n}\int_0^1 y^{\frac{1}{n}-1}\left(\int_y^1 h(u)u^{\sigma-\frac{1}{pn}-1}du\right)dy + \int_1^\infty h(u)u^{\sigma-\frac{1}{pn}-1}du$$
$$= \frac{1}{n}\int_0^1 \left(\int_0^u y^{\frac{1}{n}-1}dy\right)h(u)u^{\sigma-\frac{1}{pn}-1}du + \int_1^\infty h(u)u^{\sigma-\frac{1}{pn}-1}du$$
$$\leq \int_0^1 h(u)u^{\sigma+\frac{1}{qn}-1}du + \int_1^\infty h(u)u^{\sigma-1}du. \tag{24}$$

Since for $n > \frac{1}{\delta_0|q|}(n \in \mathbf{N})$, we have

$$h(u)u^{\sigma+\frac{1}{qn}-1} \leq h(u)u^{\sigma-\delta_0-1}(0 < u \leq 1)$$

and

$$\int_0^1 h(u)u^{\sigma-\delta_0-1}du \leq k(\sigma-\delta_0) < \infty,$$

then by (24) and Lebesgue control convergence theorem (cf. [24]), we have

$$\begin{aligned} k(\sigma) &= \int_0^1 \lim_{n\to\infty} h(u)u^{\sigma+\frac{1}{qn}-1}du + \int_1^\infty h(u)u^{\sigma-1}du \\ &= \lim_{n\to\infty}\left[\int_0^1 h(u)u^{\sigma+\frac{1}{qn}-1}du + \int_1^\infty h(u)u^{\sigma-1}du\right] \\ &\geq M > 0. \end{aligned}$$

The lemma is proved.

For $\sigma_1 = \sigma$, we still have

Lemma 4 *If* $0 < p < 1, k(\sigma) < \infty$, *there exist constants* $\delta_0, M > 0$, *such that* $k(\sigma-\delta_0) < \infty$ *(or* $k(\sigma+\delta_0) < \infty$*), and for any nonnegative measurable functions* $f(x)$ *and* $g(y)$ *in* $(0,\infty)$, *the following inequality*

$$\begin{aligned} I &:= \int_0^\infty \int_0^\infty h(xy)f(x)g(y)dxdy \\ &\geq M\left[\int_0^\infty x^{p(1-\sigma)-1}f^p(x)dx\right]^{\frac{1}{p}}\left[\int_0^\infty y^{q(1-\sigma)-1}g^q(y)dy\right]^{\frac{1}{q}} \end{aligned} \tag{25}$$

holds true, then we have $k(\sigma) \geq M > 0$, *namely,* $k(\sigma) \in \mathbf{R}_+$.

Proof If $k(\sigma-\delta_0) < \infty$, then for $\sigma_1 = \sigma$, by (24) and the proof of Lemma 3, we have $k(\sigma) \geq M$; if $k(\sigma+\delta_0) < \infty$, then we have $nM = M\widetilde{J}_2 \leq I_1$ (for $\sigma_1 = \sigma$), and by (14), we find

$$\begin{aligned} M &= \frac{1}{n}M\widetilde{J}_2 \leq \frac{1}{n}I_1 = \frac{1}{n}\int_1^\infty y^{-\frac{1}{n}-1}\left(\int_0^y h(u)u^{\sigma+\frac{1}{pn}-1}du\right)dy \\ &= \int_0^1 h(u)u^{\sigma+\frac{1}{pn}-1}du + \frac{1}{n}\int_1^\infty y^{-\frac{1}{n}-1}\left(\int_1^y h(u)u^{\sigma+\frac{1}{pn}-1}du\right)dy \\ &= \int_0^1 h(u)u^{\sigma+\frac{1}{pn}-1}du + \frac{1}{n}\int_1^\infty \left(\int_u^\infty y^{-\frac{1}{n}-1}dy\right)h(u)u^{\sigma+\frac{1}{pn}-1}du \\ &\leq \int_0^1 h(u)u^{\sigma-1}du + \int_1^\infty h(u)u^{\sigma-\frac{1}{qn}-1}du. \end{aligned} \tag{26}$$

Since for $n > \frac{1}{\delta_0|q|}(n \in \mathbf{N})$, we have

$$h(u)u^{\sigma-\frac{1}{qn}-1} \leq h(u)u^{\sigma+\delta_0-1}(u \geq 1)$$

and

$$\int_1^\infty h(u)u^{\sigma+\delta_0-1}du \leq k(\sigma+\delta_0) < \infty,$$

then by (26) and Lebesgue control convergence theorem (cf. [24]), we have

$$\begin{aligned} k(\sigma) &= \int_0^1 h(u)u^{\sigma-1}du + \int_1^\infty \lim_{n\to\infty} h(u)u^{\sigma-\frac{1}{qn}-1}du \\ &= \lim_{n\to\infty}\left[\int_0^1 h(u)u^{\sigma-1}du + \int_1^\infty h(u)u^{\sigma-\frac{1}{qn}-1}du\right] \\ &\geq M > 0. \end{aligned}$$

The lemma is proved.

3 The Case of the Conjugate Exponents $p > 1(q > 1)$

Theorem 1 *For $p > 1, k(\sigma) > 0, M \in \mathbf{R}$, the following statements are equivalent:*

(i) If $f(x) \geq 0$, satisfying

$$0 < \int_0^\infty x^{p(1-\sigma)-1}f^p(x)dx < \infty,$$

then we have the following inequality:

$$\begin{aligned} J &:= \left[\int_0^\infty y^{p\sigma_1-1}\left(\int_0^\infty h(xy)f(x)dx\right)^p dy\right]^{\frac{1}{p}} \\ &< M\left[\int_0^\infty x^{p(1-\sigma)-1}f^p(x)dx\right]^{\frac{1}{p}}. \end{aligned} \tag{27}$$

(ii) If $f(x), g(y) \geq 0$, satisfying

$$0 < \int_0^\infty x^{p(1-\sigma)-1}f^p(x)dx < \infty,$$

and $0 < \int_0^\infty y^{q(1-\sigma_1)-1} g^q(y)dy < \infty$, then we have the following inequality:

$$I = \int_0^\infty \int_0^\infty h(xy)f(x)g(y)dxdy$$
$$< M\left[\int_0^\infty x^{p(1-\sigma)-1} f^p(x)dx\right]^{\frac{1}{p}} \left[\int_0^\infty y^{q(1-\varsigma)-1} g^q(y)dy\right]^{\frac{1}{q}}. \quad (28)$$

(iii) $\sigma_1 = \sigma$, and $k(\sigma) \leq M < \infty$.

Proof $(i) => (ii)$. By Hölder's inequality (cf. [25]), we have

$$I = \int_0^\infty \left(y^{\sigma_1 - \frac{1}{p}} \int_0^\infty h(xy)f(x)dx\right)\left(y^{\frac{1}{p}-\sigma_1} g(y)\right) dy$$
$$\leq J\left[\int_0^\infty y^{q(1-\sigma_1)-1} g^q(y)dy\right]^{\frac{1}{q}}. \quad (29)$$

Then by (27), we have (28).

$(ii) => (iii)$. Since $k(\sigma) > 0$, by Lemma 1, we have $\sigma_1 = \sigma$. Then by Lemma 2, we have $k(\sigma) \leq M < \infty$.

$(iii) => (i)$. Setting $u = xy$, we obtain the following weight function: For $y > 0$,

$$\omega(\sigma, y) := y^\sigma \int_0^\infty h(xy)x^{\sigma-1}dx = \int_0^\infty h(u)u^{\sigma-1}du = k(\sigma). \quad (30)$$

By Hölder's inequality with weight and (30), we have

$$\left(\int_0^\infty h(xy)f(x)dx\right)^p$$
$$= \left\{\int_0^\infty h(xy)\left[\frac{y^{(\sigma-1)/p}}{x^{(\sigma-1)/q}} f(x)\right]\left[\frac{x^{(\sigma-1)/q}}{y^{(\sigma-1)/p}}\right] dx\right\}^p$$
$$\leq \int_0^\infty h(xy)\frac{y^{\sigma-1}}{x^{(\sigma-1)p/q}} f^p(x)dx \left[\int_0^\infty h(xy)\frac{x^{\sigma-1}}{y^{(\sigma-1)q/p}} dx\right]^{p/q}$$
$$= \left[\omega(\sigma, y)y^{q(1-\sigma)-1}\right]^{p-1} \int_0^\infty h(xy)\frac{y^{\sigma-1}}{x^{(\sigma-1)p/q}} f^p(x)dx$$
$$= (k(\sigma))^{p-1} y^{-p\sigma+1} \int_0^\infty h(xy)\frac{y^{\sigma-1}}{x^{(\sigma-1)p/q}} f^p(x)dx. \quad (31)$$

If (31) takes the form of equality for a $y \in (0, \infty)$, then (cf. [25]), there exists constants A and B, such that they are not all zero, and

$$A \frac{y^{\sigma-1}}{x^{(\sigma-1)p/q}} f^p(x) = B \frac{x^{\sigma-1}}{y^{(\sigma-1)q/p}} \; a.e. \; in \; \mathbf{R}_+.$$

We suppose that $A \neq 0$ (otherwise $B = A = 0$). Then it follows that

$$x^{p(1-\sigma)-1} f^p(x) = y^{q(1-\sigma)} \frac{B}{Ax} \; a.e. \; in \; \mathbf{R}_+,$$

which contradicts the fact that $0 < \int_0^\infty x^{p(1-\sigma)-1} f^p(x)dx < \infty$. Hence, (31) takes the form of strict inequality.

For $\sigma_1 = \sigma$, by Fubini theorem (cf. [24]) and (31), we have

$$\begin{aligned} J &< (k(\sigma))^{\frac{1}{q}} \left[\int_0^\infty \int_0^\infty h(xy) \frac{y^{\sigma-1}}{x^{(\sigma-1)p/q}} f^p(x)dxdy \right]^{\frac{1}{p}} \\ &= (k(\sigma))^{\frac{1}{q}} \left\{ \int_0^\infty \left[\int_0^\infty h(xy) \frac{y^{\sigma-1}}{x^{(\sigma-1)(p-1)}} dy \right] f^p(x)dx \right\}^{\frac{1}{p}} \\ &= (k(\sigma))^{\frac{1}{q}} \left[\int_0^\infty \omega(\sigma, x) x^{p(1-\sigma)-1} f^p(x)dx \right]^{\frac{1}{p}} \\ &= k(\sigma) \left[\int_0^\infty x^{p(1-\sigma)-1} f^p(x)dx \right]^{\frac{1}{p}}. \end{aligned}$$

Since $k(\sigma) \in \mathbf{R}_+$, setting $M \geq k(\sigma)$ in the above inequality, (27) follows.

Therefore, the statements (i)–(iii) are equivalent.

The theorem is proved.

Remark 1 If $k(\sigma) = 0$, then $h(u) = 0$ a.e. in $(0, \infty)$, we still can show that (i) is equivalent to (ii), but (ii) does not deduce to $\sigma_1 = \sigma$ in (iii).

For $\sigma_1 = \sigma$, we have

Theorem 2 *For $p > 1, k(\sigma) > 0, M \in \mathbf{R}$, the following statements are equivalent:*

(i) If $f(x) \geq 0$, satisfying

$$0 < \int_0^\infty x^{p(1-\sigma)-1} f^p(x)dx < \infty,$$

then we have the following inequality:

$$\left[\int_0^\infty y^{p\sigma-1}\left(\int_0^\infty h(xy)f(x)dx\right)^p dy\right]^{\frac{1}{p}}$$
$$< M\left[\int_0^\infty x^{p(1-\sigma)-1}f^p(x)dx\right]^{\frac{1}{p}}. \tag{32}$$

(ii) If $f(x), g(y) \geq 0$, satisfying

$$0 < \int_0^\infty x^{p(1-\sigma)-1}f^p(x)dx < \infty,$$

and $0 < \int_0^\infty y^{q(1-\sigma)-1}g^q(y)dy < \infty$, then we have the following inequality:

$$\int_0^\infty \int_0^\infty h(xy)f(x)g(y)dxdy$$
$$< M\left[\int_0^\infty x^{p(1-\sigma)-1}f^p(x)dx\right]^{\frac{1}{p}}\left[\int_0^\infty y^{q(1-\sigma)-1}g^q(y)dy\right]^{\frac{1}{q}}. \tag{33}$$

(iii) $k(\sigma) \leq M < \infty$.

Moreover, if (iii) follows, then the constant factor $M = k(\sigma) \in \mathbf{R}_+$ in (32) and (33) is the best possible.

Proof For $\sigma_1 = \sigma$ in Theorem 1, we still can conclude that the conditions (i)–(iii) in Theorem 2 are equivalent.

When Condition (iii) follows, if there exists a constant $M \leq k(\sigma)$, such that (33) is valid, then in view of $k(\sigma) \leq M$, the constant factor $M = k(\sigma)$ in (33) is the best possible.

The constant factor $k(\sigma)$ in (32) is still the best possible. Otherwise, by (29) (for $\sigma_1 = \sigma$), we can conclude that the constant factor $k(\sigma)$ in (33) is not the best possible.

The theorem is proved.

In particular, for $\sigma = \sigma_1 = \frac{1}{p}$ in Theorem 2, we have

Corollary 1 *For* $p > 1, k(\frac{1}{p}) > 0, M \in \mathbf{R}$, *the following conditions are equivalent:*

(i) If $f(x) \geq 0$, satisfying

$$0 < \int_0^\infty x^{p-2}f^p(x)dx < \infty,$$

then we have the following inequality:

$$\left[\int_0^\infty \left(\int_0^\infty h(xy)f(x)dx\right)^p dy\right]^{\frac{1}{p}} < M\left(\int_0^\infty x^{p-2}f^p(x)dx\right)^{\frac{1}{p}}. \tag{34}$$

(ii) If $f(x), g(y) \geq 0$, satisfying

$$0 < \int_0^\infty x^{p-2}f^p(x)dx < \infty,$$

and $0 < \int_0^\infty g^q(y)dy < \infty$, then we have the following inequality:

$$\int_0^\infty \int_0^\infty h(xy)f(x)g(y)dxdy$$
$$< M\left(\int_0^\infty x^{p-2}f^p(x)dx\right)^{\frac{1}{p}}\left(\int_0^\infty g^q(y)dy\right)^{\frac{1}{q}}. \tag{35}$$

(iii) $k(\frac{1}{p}) \leq M < \infty$.

If (iii) follows, then the constant $M = k(\frac{1}{p})$ $(\in \mathbf{R}_+)$ in (34) and (35) is the best possible.

Setting $y = \frac{1}{Y}$, $G(Y) = g(\frac{1}{Y})\frac{1}{Y^2}$ in Theorems 1–2, then replacing Y by y, we have

Corollary 2 *For* $p > 1, k(\sigma) > 0, M \in \mathbf{R}$, *the following statements are equivalent:*

(i) If $f(x) \geq 0$, satisfying

$$0 < \int_0^\infty x^{p(1-\sigma)-1}f^p(x)dx < \infty,$$

then we have the following inequality:

$$\left[\int_0^\infty y^{-p\sigma_1-1}\left(\int_0^\infty h(\frac{x}{y})f(x)dx\right)^p dy\right]^{\frac{1}{p}}$$
$$< M\left[\int_0^\infty x^{p(1-\sigma)-1}f^p(x)dx\right]^{\frac{1}{p}}. \tag{36}$$

(ii) If $f(x), G(y) \geq 0$, satisfying

$$0 < \int_0^\infty x^{p(1-\sigma)-1}f^p(x)dx < \infty,$$

and $0 < \int_0^\infty y^{q(1+\sigma_1)-1} G^q(y)dy < \infty$, then we have the following inequality:

$$\int_0^\infty \int_0^\infty h(\frac{x}{y}) f(x)G(y)dxdy$$

$$< M\left[\int_0^\infty x^{p(1-\sigma)-1} f^p(x)dx\right]^{\frac{1}{p}} \left[\int_0^\infty y^{q(1+\sigma_1)-1} G^q(y)dy\right]^{\frac{1}{q}}. \quad (37)$$

(iii) $\sigma_1 = \sigma$, and $k(\sigma) \leq M < \infty$.

If (iii) follows, then the constant $M = k(\sigma)\,(\in \mathbf{R}_+)$ in (36) and (37) (for $\sigma_1 = \sigma$) is the best possible.

Note $h(\frac{x}{y})$ is a homogeneous function of degree 0, namely, $h(\frac{x}{y}) = k_0(x, y)$.

Setting $h(u) = k_\lambda(u, 1)$, where $k_\lambda(x, y)$ $(x, y > 0)$ is the homogeneous function of degree $-\lambda \in \mathbf{R}$, then for $g(y) = y^\lambda G(y)$ and $\mu = \lambda - \sigma_1$ in Corollary 2, we have

Corollary 3 *For* $p > 1$, $k_\lambda(\sigma) := \int_0^\infty k_\lambda(u, 1)u^{\sigma-1}du > 0$, $M \in \mathbf{R}$, *the following statements are equivalent:*

(i) If $f(x) \geq 0$, satisfying

$$0 < \int_0^\infty x^{p(1-\sigma)-1} f^p(x)dx < \infty,$$

then we have the following inequality:

$$\left[\int_0^\infty y^{p\mu-1}\left(\int_0^\infty k_\lambda(x, y) f(x)dx\right)^p dy\right]^{\frac{1}{p}}$$

$$< M\left[\int_0^\infty x^{p(1-\sigma)-1} f^p(x)dx\right]^{\frac{1}{p}}. \quad (38)$$

(ii) If $f(x), g(y) \geq 0$, satisfying

$$0 < \int_0^\infty x^{p(1-\sigma)-1} f^p(x)dx < \infty,$$

and $0 < \int_0^\infty y^{q(1-\mu)-1} g^q(y)dy < \infty$, then we have the following inequality:

$$\int_0^\infty \int_0^\infty k_\lambda(x, y) f(x)g(y)dxdy$$

$$< M\left[\int_0^\infty x^{p(1-\sigma)-1} f^p(x)dx\right]^{\frac{1}{p}} \left[\int_0^\infty y^{q(1-\mu)-1} g^q(y)dy\right]^{\frac{1}{q}}. \quad (39)$$

(iii) $\mu + \sigma = \lambda$, and $k_\lambda(\sigma) \leq M < \infty$.

If (iii) follows, then the constant $M = k_\lambda(\sigma)\ (\in \mathbf{R}_+)$ in (38) and (39) is the best possible.

Remark 2 If $\lambda = 0, \mu = -\sigma_1, k_0(x, y) = h(\frac{x}{y})$, then Corollary 3 reduces to Corollary 2.

In particular, for $\lambda = 1, \sigma = \frac{1}{q}, \mu = \frac{1}{p}$ in Corollary 3, we have

Corollary 4 *For* $p > 1, k_1(\frac{1}{q}) > 0, M \in \mathbf{R}$, *the following statements are equivalent:*

(i) If $f(x) \geq 0$, satisfying $0 < \int_0^\infty f^p(x)dx < \infty$, then we have the following inequality:

$$\left[\int_0^\infty \left(\int_0^\infty k_1(x, y)f(x)dx\right)^p dy\right]^{\frac{1}{p}} < M\left(\int_0^\infty f^p(x)dx\right)^{\frac{1}{p}}. \tag{40}$$

(ii) If $f(x), g(y) \geq 0$, satisfying $0 < \int_0^\infty f^p(x)dx < \infty$, and $0 < \int_0^\infty g^q(y)dy < \infty$, then we have the following inequality:

$$\int_0^\infty \int_0^\infty k_1(x, y)f(x)g(y)dxdy$$
$$< M\left(\int_0^\infty f^p(x)dx\right)^{\frac{1}{p}}\left(\int_0^\infty g^q(y)dy\right)^{\frac{1}{q}}. \tag{41}$$

(iii) $k_1(\frac{1}{q}) \leq M < \infty$.

If (iii) follows, then the constant $M = k_1(\frac{1}{q})\ (\in \mathbf{R}_+)$ in (40) and (41) is the best possible.

For $\lambda = 1, \sigma = \frac{1}{p}, \mu = \frac{1}{q}$ in Corollary 3, we have

Corollary 5 *For* $p > 1, k_\lambda(\frac{1}{p}) > 0, M \in \mathbf{R}$, *the following statements are equivalent:*

(i) If $f(x) \geq 0$, satisfying

$$0 < \int_0^\infty x^{p-2} f^p(x)dx < \infty,$$

then we have the following inequality:

$$\left[\int_0^\infty y^{p-2}\left(\int_0^\infty k_1(x, y)f(x)dx\right)^p dy\right]^{\frac{1}{p}}$$
$$< M\left(\int_0^\infty x^{p-2} f^p(x)dx\right)^{\frac{1}{p}}. \tag{42}$$

(ii) If $f(x), g(y) \geq 0$, satisfying

$$0 < \int_0^\infty x^{p-2} f^p(x)dx < \infty,$$

and $0 < \int_0^\infty y^{q-2} g^q(y)dy < \infty$, then we have the following inequality:

$$\int_0^\infty \int_0^\infty k_1(x, y) f(x) g(y) dx dy$$
$$< M \left(\int_0^\infty x^{p-2} f^p(x)dx \right)^{\frac{1}{p}} \left(\int_0^\infty y^{q-2} g^q(y)dy \right)^{\frac{1}{q}}. \tag{43}$$

(iii) $k_1(\frac{1}{q}) \leq M < \infty$.

If (iii) follows, then the constant $M = k(\frac{1}{q}) (\in \mathbf{R}_+)$ in (42) and (43) is the best possible.

4 The Case of the Conjugate Exponents $0 < p < 1 (q < 0)$

Theorem 3 *For* $0 < p < 1, k(\sigma) < \infty, M > 0$, *the following conditions are equivalent:*

(i) If $f(x) \geq 0$, satisfying

$$0 < \int_0^\infty x^{p(1-\sigma)-1} f^p(x)dx < \infty,$$

then we have the following inequality:

$$J := \left[\int_0^\infty y^{p\sigma_1 - 1} \left(\int_0^\infty h(xy) f(x)dx \right)^p dy \right]^{\frac{1}{p}}$$
$$> M \left[\int_0^\infty x^{p(1-\sigma)-1} f^p(x)dx \right]^{\frac{1}{p}}. \tag{44}$$

(ii) If $g(y) \geq 0$, satisfying

$$0 < \int_0^\infty y^{q(1-\sigma_1)-1} g^q(y)dy < \infty,$$

then we have the following inequality:

$$K := \left[\int_0^\infty x^{q\sigma-1}\left(\int_0^\infty h(xy)g(y)dy\right)^q dx\right]^{\frac{1}{q}}$$
$$> M\left[\int_0^\infty y^{q(1-\sigma_1)-1}g^q(y)dy\right]^{\frac{1}{q}}. \quad (45)$$

(iii) If $f(x), g(y) \geq 0$, satisfying

$$0 < \int_0^\infty x^{p(1-\sigma)-1}f^p(x)dx < \infty,$$

and $0 < \int_0^\infty y^{q(1-\sigma_1)-1}g^q(y)dy < \infty$, then we have the following inequality:

$$I = \int_0^\infty \int_0^\infty h(xy)f(x)g(y)dxdy$$
$$> M\left[\int_0^\infty x^{p(1-\sigma)-1}f^p(x)dx\right]^{\frac{1}{p}}\left[\int_0^\infty y^{q(1-\sigma_1)-1}g^q(y)dy\right]^{\frac{1}{q}}. \quad (46)$$

(iv) If there exists a constant $\delta_0 > 0$, such that $k(\sigma \pm \delta_0) < \infty$, then $\sigma_1 = \sigma$, and $k(\sigma) \geq M > 0$.

Proof $(i) => (iii)$. By the reverse Hölder's inequality (cf. [25]), we have

$$I = \int_0^\infty \left(y^{\sigma_1-\frac{1}{p}}\int_0^\infty h(xy)f(x)dx\right)\left(y^{\frac{1}{p}-\sigma_1}g(y)\right)dy$$
$$\geq J\left[\int_0^\infty y^{q(1-\sigma_1)-1}g^q(y)dy\right]^{\frac{1}{q}}. \quad (47)$$

Then by (44), we have (46).
$(ii) => (iii)$. Still by the reverse Hölder's inequality, we have

$$I = \int_0^\infty \left(y^{\frac{1}{q}-\sigma}f(x)\right)\left(x^{\sigma-\frac{1}{q}}\int_0^\infty h(xy)g(y)dy\right)dx$$
$$\geq \left[\int_0^\infty x^{p(1-\sigma)-1}f^p(x)dx\right]^{\frac{1}{p}}K. \quad (48)$$

Then by (45), we have (46).
"$(iii) => (iv)$". By Lemma 3, we have $\sigma_1 = \sigma$, and $k(\sigma) \geq M > 0$.
"$(iv) => (i)$". Setting $u = xy$, we obtain the following weight function: For $y > 0$,

$$\omega(\sigma, y) = y^{\sigma} \int_0^{\infty} h(xy)x^{\sigma-1}dx = \int_0^{\infty} h(u)u^{\sigma-1}du = k(\sigma). \tag{49}$$

By the reverse Hölder's inequality with weight and (49), we have

$$\begin{aligned}
&\left(\int_0^{\infty} h(xy)f(x)dx\right)^p \\
&= \left\{\int_0^{\infty} h(xy)\left[\frac{y^{(\sigma-1)/p}}{x^{(\sigma-1)/q}}f(x)\right]\left[\frac{x^{(\sigma-1)/q}}{y^{(\sigma-1)/p}}\right]dx\right\}^p \\
&\geq \int_0^{\infty} h(xy)\frac{y^{\sigma-1}}{x^{(\sigma-1)p/q}}f^p(x)dx\left[\int_0^{\infty} h(xy)\frac{x^{\sigma-1}}{y^{(\sigma-1)q/p}}dx\right]^{p/q} \\
&= \left[\omega(\sigma, y)y^{q(1-\sigma)-1}\right]^{p-1}\int_0^{\infty} h(xy)\frac{y^{\sigma-1}}{x^{(\sigma-1)p/q}}f^p(x)dx \\
&= (k(\sigma))^{p-1}y^{-p\sigma+1}\int_0^{\infty} h(xy)\frac{y^{\sigma-1}}{x^{(\sigma-1)p/q}}f^p(x)dx
\end{aligned} \tag{50}$$

In the same way, we can prove that (50) takes the form of strict inequality.
For $\sigma_1 = \sigma$, by (50) and Fubini theorem, we have

$$\begin{aligned}
J &> (k(\sigma))^{\frac{1}{q}}\left[\int_0^{\infty}\int_0^{\infty} h(xy)\frac{y^{\sigma-1}}{x^{(\sigma-1)p/q}}f^p(x)dxdy\right]^{\frac{1}{p}} \\
&= (k(\sigma))^{\frac{1}{q}}\left\{\int_0^{\infty}\left[\int_0^{\infty} h(xy)\frac{y^{\sigma-1}}{x^{(\sigma-1)(p-1)}}dy\right]f^p(x)dx\right\}^{\frac{1}{p}} \\
&= (k(\sigma))^{\frac{1}{q}}\left[\int_0^{\infty} \omega(\sigma, x)x^{p(1-\sigma)-1}f^p(x)dx\right]^{\frac{1}{p}} \\
&= k(\sigma)\left[\int_0^{\infty} x^{p(1-\sigma)-1}f^p(x)dx\right]^{\frac{1}{p}}.
\end{aligned} \tag{51}$$

Setting $0 < M \leq k(\sigma)$, then (44) follows.

$(iv) => (ii)$. In the same way, we can obtain (45).

Therefore, the statements (i)–(iv) are equivalent.

The theorem is proved.

For $\sigma_1 = \sigma$, we still have

Theorem 4 *If* $0 < p < 1, 0 < k(\sigma) < \infty$, *then the following statements are equivalent:*

(i) If $f(x) \geq 0$, satisfying

$$0 < \int_0^{\infty} x^{p(1-\sigma)-1}f^p(x)dx < \infty,$$

then we have the following inequality:

$$\left[\int_0^\infty y^{p\sigma-1}\left(\int_0^\infty h(xy)f(x)dx\right)^p dy\right]^{\frac{1}{p}}$$
$$> k(\sigma)\left[\int_0^\infty x^{p(1-\sigma)-1}f^p(x)dx\right]^{\frac{1}{p}}. \tag{52}$$

(ii) If $g(y) \geq 0$, satisfying

$$0 < \int_0^\infty y^{q(1-\sigma)-1}g^q(y)dy < \infty,$$

then we have the following inequality:

$$\left[\int_0^\infty x^{q\sigma-1}\left(\int_0^\infty h(xy)g(y)dy\right)^q dx\right]^{\frac{1}{q}}$$
$$> k(\sigma)\left[\int_0^\infty y^{q(1-\sigma)-1}g^q(y)dy\right]^{\frac{1}{q}}. \tag{53}$$

(iii) If $f(x), g(y) \geq 0$, satisfying

$$0 < \int_0^\infty x^{p(1-\sigma)-1}f^p(x)dx < \infty,$$

and $0 < \int_0^\infty y^{q(1-\sigma)-1}g^q(y)dy < \infty$, then we have the following inequality:

$$\int_0^\infty \int_0^\infty h(xy)f(x)g(y)dxdy$$
$$> k(\sigma)\left[\int_0^\infty x^{p(1-\sigma)-1}f^p(x)dx\right]^{\frac{1}{p}}\left[\int_0^\infty y^{q(1-\sigma)-1}g^q(y)dy\right]^{\frac{1}{q}}. \tag{54}$$

Moreover, if there exists a constant $\delta_0 > 0$, such that $k(\sigma - \delta_0) < \infty$ (or $k(\sigma + \delta_0) < \infty$), then the constant factor $k(\sigma)$ in (52)–(54) is the best possible.

Proof For $\sigma_1 = \sigma$ in Theorem 3, since $0 < k(\sigma) < \infty$, setting $M = k(\sigma)$ in (44)–(46), in the same way, we still can prove that the statements (i)–(iii) are equivalent in Theorem 4. If there exists a constant $M \geq k(\sigma)$, such that (54) is valid, then by Lemma 4, we have $M \leq k(\sigma)$. Hence, the constant factor $M = k(\sigma)$ in (54) is the best possible.

The constant factor $k(\sigma)$ in (52) (53) is still the best possible. Otherwise, by (47) (48) for $\sigma_1 = \sigma$, we can conclude that the constant factor $M = k(\sigma)$ in (54) is not the best possible.

The theorem is proved.

In particular, for $\sigma = \frac{1}{p}$ in Theorem 4, we have

Corollary 6 *For* $0 < p < 1, 0 < k(\frac{1}{p}) < \infty$, *the following statements are equivalent:*

(i) If $f(x) \geq 0$, satisfying

$$0 < \int_0^\infty x^{p-2} f^p(x)dx < \infty,$$

then we have the following inequality:

$$\left[\int_0^\infty \left(\int_0^\infty h(xy)f(x)dx\right)^p dy\right]^{\frac{1}{p}} > k(\frac{1}{p})\left(\int_0^\infty x^{p-2} f^p(x)dx\right)^{\frac{1}{p}}. \quad (55)$$

(ii) If $g(y) \geq 0$, satisfying

$$0 < \int_0^\infty g^q(y)dy < \infty,$$

then we have the following inequality:

$$\left[\int_0^\infty x^{q-2}\left(\int_0^\infty h(xy)g(y)dy\right)^q dx\right]^{\frac{1}{q}} > k(\frac{1}{p})\left(\int_0^\infty g^q(y)dy\right)^{\frac{1}{q}}. \quad (56)$$

(iii) If $f(x), g(y) \geq 0$,

$$0 < \int_0^\infty x^{p-2} f^p(x)dx < \infty,$$

and $0 < \int_0^\infty g^q(y)dy < \infty$, then we have the following inequality:

$$\int_0^\infty \int_0^\infty h(xy)f(x)g(y)dxdy > k(\frac{1}{p})\left(\int_0^\infty x^{p-2} f^p(x)dx\right)^{\frac{1}{p}}\left(\int_0^\infty g^q(y)dy\right)^{\frac{1}{q}}. \quad (57)$$

Moreover, if there exists a constant $\delta_0 > 0$, such that $k(\frac{1}{p} - \delta_0) < \infty$ (or $k(\frac{1}{p} + \delta_0) < \infty$), then the constant factor $k(\frac{1}{p})$ in (55)–(57) is the best possible.

In particular, for $\sigma = \frac{1}{q}$ in Theorem 4, we have

Corollary 7 *If* $0 < p < 1, 0 < k(\frac{1}{q}) < \infty$, *then the following statements are equivalent:*

(i) If $f(x) \geq 0$, satisfying

$$0 < \int_0^\infty f^p(x)dx < \infty,$$

then we have the following inequality:

$$\left[\int_0^\infty y^{p-2}\left(\int_0^\infty h(xy)f(x)dx\right)^p dy\right]^{\frac{1}{p}} > k(\frac{1}{q})\left(\int_0^\infty f^p(x)dx\right)^{\frac{1}{p}}. \tag{58}$$

(ii) If $g(y) \geq 0$, satisfying

$$0 < \int_0^\infty y^{q-2}g^q(y)dy < \infty,$$

then we have the following inequality:

$$\left[\int_0^\infty \left(\int_0^\infty h(xy)g(y)dy\right)^q dx\right]^{\frac{1}{q}} > k(\frac{1}{q})\left(\int_0^\infty y^{q-2}g^q(y)dy\right)^{\frac{1}{q}}. \tag{59}$$

(iii) If $f(x), g(y) \geq 0$, satisfying

$$0 < \int_0^\infty f^p(x)dx < \infty,$$

and $0 < \int_0^\infty y^{q-2}g^q(y)dy < \infty$, then we have the following inequality:

$$\int_0^\infty \int_0^\infty h(xy)f(x)g(y)dxdy > k(\frac{1}{q})\left(\int_0^\infty f^p(x)dx\right)^{\frac{1}{p}}\left(\int_0^\infty y^{q-2}g^q(y)dy\right)^{\frac{1}{q}}. \tag{60}$$

Moreover, if there exists a constant $\delta_0 > 0$, such that $k(\frac{1}{q} - \delta_0) < \infty$ (or $k(\frac{1}{q} + \delta_0) < \infty$), then the constant factor $k(\frac{1}{q})$ in (58)–(60) is the best possible.

Setting $y = \frac{1}{Y}$, $G(Y) = g(\frac{1}{Y})\frac{1}{Y^2}$ in Theorem 3, then replacing Y by y, we have

Corollary 8 *For* $0 < p < 1, k(\sigma) < \infty, M > 0$, *the following statements are equivalent:*

(i) If $f(x) \geq 0$, satisfying

$$0 < \int_0^\infty x^{p(1-\sigma)-1} f^p(x)dx < \infty,$$

then we have the following inequality:

$$\left[\int_0^\infty y^{-p\sigma_1-1}\left(\int_0^\infty h(\frac{x}{y})f(x)dx\right)^p dy\right]^{\frac{1}{p}}$$
$$> M\left[\int_0^\infty x^{p(1-\sigma)-1} f^p(x)dx\right]^{\frac{1}{p}}. \tag{61}$$

(ii) If $G(y) \geq 0$, satisfying

$$0 < \int_0^\infty y^{q(1+\sigma_1)-1} G^q(y)dy < \infty,$$

then we have the following inequality:

$$\left[\int_0^\infty x^{q\sigma-1}\left(\int_0^\infty h(\frac{x}{y})G(y)dy\right)^q dx\right]^{\frac{1}{q}}$$
$$> M\left[\int_0^\infty y^{q(1+\sigma_1)-1} G^q(y)dy\right]^{\frac{1}{q}}. \tag{62}$$

(iii) If $f(x), G(y) \geq 0$, satisfying

$$0 < \int_0^\infty x^{p(1-\sigma)-1} f^p(x)dx < \infty,$$

and $0 < \int_0^\infty y^{q(1+\sigma_1)-1} G^q(y)dy < \infty$, then we have the following inequality:

$$\int_0^\infty \int_0^\infty h(\frac{x}{y}) f(x)G(y)dxdy$$
$$> M\left[\int_0^\infty x^{p(1-\sigma)-1} f^p(x)dx\right]^{\frac{1}{p}} \left[\int_0^\infty y^{q(1+\sigma_1)-1} G^q(y)dy\right]^{\frac{1}{q}}. \tag{63}$$

(iv) If there exists a constant $\delta_0 > 0$, such that $k(\sigma \pm \delta_0) < \infty$, then $\sigma_1 = \sigma$, and $k(\sigma) \geq M > 0$.

Setting $h(u) = k_\lambda(u, 1)$, where $k_\lambda(x, y)$ is the homogeneous function of degree $-\lambda \in \mathbf{R}$, then for $g(y) = y^\lambda G(y)$ and $\mu_1 = \lambda - \sigma_1$ in Corollary 8, we have

Corollary 9 *For* $0 < p < 1, k_\lambda(\sigma) < \infty, M > 0$, *the following statements are equivalent:*

(i) If $f(x) \geq 0$, satisfying

$$0 < \int_0^\infty x^{p(1-\sigma)-1} f^p(x)dx < \infty,$$

then we have the following inequality:

$$\left[\int_0^\infty y^{p\mu_1-1}\left(\int_0^\infty k_\lambda(x, y)f(x)dx\right)^p dy\right]^{\frac{1}{p}} > M\left[\int_0^\infty x^{p(1-\sigma)-1} f^p(x)dx\right]^{\frac{1}{p}}. \quad (64)$$

(ii) If $g(y) \geq 0$, satisfying

$$0 < \int_0^\infty y^{q(1-\mu_1)-1} g^q(y)dy < \infty,$$

then we have the following inequality:

$$\left[\int_0^\infty x^{q\sigma-1}\left(\int_0^\infty k_\lambda(x, y)g(y)dy\right)^q dx\right]^{\frac{1}{q}} > M\left[\int_0^\infty y^{q(1-\mu_1)-1} g^q(y)dy\right]^{\frac{1}{q}}; \quad (65)$$

(iii) If $f(x), g(y) \geq 0$, satisfying

$$0 < \int_0^\infty x^{p(1-\sigma)-1} f^p(x)dx < \infty,$$

and $0 < \int_0^\infty y^{q(1-\mu_1)-1} g^q(y)dy < \infty$, then we have the following inequality:

$$\int_0^\infty \int_0^\infty k_\lambda(x, y)f(x)g(y)dxdy > M\left[\int_0^\infty x^{p(1-\sigma)-1} f^p(x)dx\right]^{\frac{1}{p}} \left[\int_0^\infty y^{q(1-\mu_1)-1} g^q(y)dy\right]^{\frac{1}{q}}. \quad (66)$$

(iv) If there exists a constant $\delta_0 > 0$, such that $k_\lambda(\sigma \pm \delta_0) < \infty$, then $\sigma + \mu_1 = \lambda$, and $k_\lambda(\sigma) \geq M > 0$.

For $\mu_1 = \mu = \lambda - \sigma$, in Corollary 9, we still have

Corollary 10 *For* $0 < p < 1, 0 < k_\lambda(\sigma) < \infty$, *the following statements are equivalent:*

(i) If $f(x) \geq 0$, satisfying

$$0 < \int_0^\infty x^{p(1-\sigma)-1} f^p(x)dx < \infty,$$

then we have the following inequality:

$$\left[\int_0^\infty y^{p\mu-1} \left(\int_0^\infty k_\lambda(x,y) f(x)dx\right)^p dy\right]^{\frac{1}{p}} > k_\lambda(\sigma)\left[\int_0^\infty x^{p(1-\sigma)-1} f^p(x)dx\right]^{\frac{1}{p}}. \tag{67}$$

(ii) If $g(y) \geq 0$, satisfying

$$0 < \int_0^\infty y^{q(1-\mu)-1} g^q(y)dy < \infty,$$

then we have the following inequality:

$$\left[\int_0^\infty x^{q\sigma-1} \left(\int_0^\infty k_\lambda(x,y) g(y)dy\right)^q dx\right]^{\frac{1}{q}} > k_\lambda(\sigma)\left[\int_0^\infty y^{q(1-\mu)-1} g^q(y)dy\right]^{\frac{1}{q}}. \tag{68}$$

(iii) If $f(x), g(y) \geq 0$, satisfying

$$0 < \int_0^\infty x^{p(1-\sigma)-1} f^p(x)dx < \infty,$$

and $0 < \int_0^\infty y^{q(1-\mu)-1} g^q(y)dy < \infty$, then we have the following inequality:

$$\int_0^\infty \int_0^\infty k_\lambda(x,y) f(x)g(y)dxdy > k_\lambda(\sigma)\left[\int_0^\infty x^{p(1-\sigma)-1} f^p(x)dx\right]^{\frac{1}{p}} \left[\int_0^\infty y^{q(1-\mu)-1} g^q(y)dy\right]^{\frac{1}{q}}. \tag{69}$$

Moreover, if there exists a constant $\delta_0 > 0$, such that $k_\lambda(\sigma - \delta_0) < \infty$ (or $k_\lambda(\sigma + \delta_0) < \infty$), then the constant factor $k_\lambda(\sigma)$ in (67)–(69) is the best possible.

In particular, for $\lambda = 1, \sigma = \frac{1}{q}, \mu = \frac{1}{p}$ in Corollary 10, we have

Corollary 11 *For $0 < p < 1$, $0 < k_1(\frac{1}{q}) < \infty$, the following statements are equivalent:*

(i) If $f(x) \geq 0$, satisfying

$$0 < \int_0^\infty f^p(x)dx < \infty,$$

then we have the following inequality:

$$\left[\int_0^\infty \left(\int_0^\infty k_1(x, y) f(x)dx\right)^p dy\right]^{\frac{1}{p}} > k_1(\frac{1}{q}) \left(\int_0^\infty f^p(x)dx\right)^{\frac{1}{p}}. \tag{70}$$

(ii) If $g(y) \geq 0$, satisfying

$$0 < \int_0^\infty g^q(y)dy < \infty,$$

then we have the following inequality:

$$\left[\int_0^\infty \left(\int_0^\infty k_1(x, y) g(y)dy\right)^q dx\right]^{\frac{1}{q}} > k_1(\frac{1}{q}) \left(\int_0^\infty g^q(y)dy\right)^{\frac{1}{q}}. \tag{71}$$

(iii) If $f(x), g(y) \geq 0$, satisfying

$$0 < \int_0^\infty f^p(x)dx < \infty,$$

and $0 < \int_0^\infty g^q(y)dy < \infty$, then we have the following inequality:

$$\int_0^\infty \int_0^\infty k_1(x, y) f(x) g(y)dxdy > k_1(\frac{1}{q}) \left(\int_0^\infty f^p(x)dx\right)^{\frac{1}{p}} \left(\int_0^\infty g^q(y)dy\right)^{\frac{1}{q}}. \tag{72}$$

Moreover, if there exists a constant $\delta_0 > 0$, such that $k_1(\frac{1}{q} - \delta_0) < \infty$ (or $k_1(\frac{1}{q} + \delta_0) < \infty$,), then the constant factor $k_1(\frac{1}{q})$ in (70)–(72) is the best possible.

For $\lambda = 1, \sigma = \frac{1}{p}, \mu = \frac{1}{q}$ in Corollary 11, we have

Corollary 12 *For* $0 < p < 1$, $0 < k_1(\frac{1}{p}) < \infty$, *the following statements are equivalent:*

(i) If $f(x) \geq 0$, satisfying

$$0 < \int_0^\infty x^{p-2} f^p(x)dx < \infty,$$

then we have the following inequality:

$$\left[\int_0^\infty y^{p-2} \left(\int_0^\infty k_1(x, y) f(x)dx\right)^p dy\right]^{\frac{1}{p}} > k_1(\frac{1}{p}) \left(\int_0^\infty x^{p-2} f^p(x)dx\right)^{\frac{1}{p}}. \tag{73}$$

(ii) If $g(y) \geq 0$, satisfying

$$0 < \int_0^\infty y^{q-2} g^q(y)dy < \infty,$$

then we have the following inequality:

$$\left[\int_0^\infty x^{q-2} \left(\int_0^\infty k_1(x, y)g(y)dy\right)^q dx\right]^{\frac{1}{q}} > k_1(\frac{1}{p}) \left(\int_0^\infty y^{q-2} g^q(y)dy\right)^{\frac{1}{q}}. \tag{74}$$

(iii) If $f(x), g(y) \geq 0$, satisfying

$$0 < \int_0^\infty x^{p-2} f^p(x)dx < \infty,$$

and $0 < \int_0^\infty y^{q-2} g^q(y)dy < \infty$, then we have the following inequality:

$$\int_0^\infty \int_0^\infty k_1(x, y) f(x)g(y)dxdy > k_1(\frac{1}{p}) \left(\int_0^\infty x^{p-2} f^p(x)dx\right)^{\frac{1}{p}} \left(\int_0^\infty y^{q-2} g^q(y)dy\right)^{\frac{1}{q}}. \tag{75}$$

Moreover, if there exists a constant $\delta_0 > 0$, such that $k_1(\frac{1}{p} - \delta_0) < \infty$ (or $k_1(\frac{1}{p} + \delta_0) < \infty$), then the constant factor $k_1(\frac{1}{p})$ in (73)–(75) is the best possible.

5 Operator Expressions and a Few Examples

For $p > 1$, $\mu + \sigma = \lambda$, we set the following functions: $\varphi(x) := x^{p(1-\sigma)-1}$, $\psi(y) := y^{q(1-\sigma)-1}$, $\phi(y) := y^{q(1-\mu)-1}$, wherefrom, $\psi^{1-p}(y) = y^{p\sigma-1}$,

$$\phi^{1-p}(y) = y^{p\mu-1} (x, y \in \mathbf{R}_+).$$

Define the following real normed linear spaces:

$$L_{p,\varphi}(\mathbf{R}_+) := \left\{ f : ||f||_{p,\varphi} := \left(\int_0^\infty \varphi(x) |f(x)|^p dx \right)^{\frac{1}{p}} < \infty \right\},$$

wherefrom,

$$L_{q,\psi}(\mathbf{R}_+) = \left\{ g : ||g||_{q,\psi} = \left(\int_0^\infty \psi(y) |g(y)|^q dy \right)^{\frac{1}{q}} < \infty \right\},$$

$$L_{q,\phi}(\mathbf{R}_+) = \left\{ g : ||g||_{q,\phi} = \left(\int_0^\infty \phi(y) |g(y)|^q dy \right)^{\frac{1}{q}} < \infty \right\},$$

$$L_{p,\psi^{1-p}}(\mathbf{R}_+) = \left\{ h : ||h||_{p,\psi^{1-p}} = \left(\int_0^\infty \psi^{1-p}(y) |h(y)|^p dy \right)^{\frac{1}{p}} < \infty \right\},$$

$$L_{q,\phi^{1-p}}(\mathbf{R}_+) = \left\{ h : ||h||_{p,\phi^{1-p}} = \left(\int_0^\infty \phi^{1-p}(y) |h(y)|^p dy \right)^{\frac{1}{p}} < \infty \right\}.$$

(a) In view of Theorem 2, for $f \in L_{p,\varphi}(\mathbf{R}_+)$, setting

$$h_1(y) := \int_0^\infty h(xy) f(x) dx \ (y \in \mathbf{R}_+),$$

by (32), we have

$$||h_1||_{p,\psi^{1-p}} = \left[\int_0^\infty \psi^{1-p}(y) h_1^p(y) dy \right]^{\frac{1}{p}} < M ||f||_{p,\varphi} < \infty. \tag{76}$$

Definition 1 Define a Hilbert-type integral operator with the nonhomogeneous kernel $T^{(1)} : L_{p,\varphi}(\mathbf{R}_+) \to L_{p,\psi^{1-p}}(\mathbf{R}_+)$ as follows: For any $f \in L_{p,\varphi}(\mathbf{R}_+)$, there exists a unique representation $T^{(1)}f = h_1 \in L_{p,\psi^{1-p}}(\mathbf{R}_+)$, satisfying for any $y \in \mathbf{R}_+$, $T^{(1)}f(y) = h_1(y)$.

In view of (76), it follows that

$$||T^{(1)}f||_{p,\psi^{1-p}} = ||h_1||_{p,\psi^{1-p}} \leq M||f||_{p,\varphi},$$

and then the operator $T^{(1)}$ is bounded satisfying

$$||T^{(1)}|| := \sup_{f(\neq\theta)\in L_{p,\varphi}(\mathbf{R}_+)} \frac{||T^{(1)}f||_{p,\psi^{1-p}}}{||f||_{p,\varphi}} \leq M.$$

If we define the formal inner product of $T^{(1)}f$ and g as follows:

$$(T^{(1)}f, g) := \int_0^\infty \left(\int_0^\infty h(xy)f(x)dx \right) g(y)dy,$$

then we can rewrite Theorem 2 as follows:

Theorem 5 *For $p > 1, k(\sigma) > 0, M \in \mathbf{R}$, the following statements are equivalent:*

(i) If $f(x) \geq 0, f \in L_{p,\varphi}(\mathbf{R}_+)$, satisfying $||f||_{p,\varphi} > 0$, then we have the following inequality:

$$||T^{(1)}f||_{p,\psi^{1-p}} < M||f||_{p,\varphi}. \tag{77}$$

(ii) If $f(x), g(y) \geq 0, f \in L_{p,\varphi}(\mathbf{R}_+), g \in L_{q,\psi}(\mathbf{R}_+)$, satisfying $||f||_{p,\varphi}$, $||g||_{q,\psi} > 0$, then we have the following inequality:

$$(T^{(1)}f, g) < M||f||_{p,\varphi}||g||_{q,\psi}. \tag{78}$$

(iii) $k(\sigma) \leq M < \infty$.

Moreover, if (iii) follows, then the constant factor $M = k(\sigma)(\in \mathbf{R}_+)$ in (77) and (78) is the best possible, namely, $0 < ||T^{(1)}|| = k(\sigma) \leq M$.

(b) In view of Corollary 3 ($\sigma + \mu = \lambda$), for $f \in L_{p,\varphi}(\mathbf{R}_+)$, setting

$$h_2(y) := \int_0^\infty k_\lambda(x, y)f(x)dx \ (y \in \mathbf{R}_+),$$

by (38), we have

$$||h_2||_{p,\phi^{1-p}} = \left[\int_0^\infty \phi^{1-p}(y)h_2^p(y)dy \right]^{\frac{1}{p}} < M||f||_{p,\varphi} < \infty. \tag{79}$$

Definition 2 Define a Hilbert-type integral operator with the homogeneous kernel $T^{(2)} : L_{p,\varphi}(\mathbf{R}_+) \to L_{p,\phi^{1-p}}(\mathbf{R}_+)$ as follows: For any $f \in L_{p,\varphi}(\mathbf{R})$, there exists a unique representation $T^{(2)}f = h_2 \in L_{p,\phi^{1-p}}(\mathbf{R}_+)$, satisfying for any $y \in \mathbf{R}_+$, $T^{(2)}f(y) = h_2(y)$.

In view of (79), it follows that

$$||T^{(2)}f||_{p,\phi^{1-p}} = ||h_2||_{p,\phi^{1-p}} \leq M||f||_{p,\varphi},$$

and then the operator $T^{(2)}$ is bounded satisfying

$$||T^{(2)}|| := \sup_{f(\neq\theta)\in L_{p,\varphi}(\mathbf{R}_+)} \frac{||T^{(2)}f||_{p,\phi^{1-p}}}{||f||_{p,\varphi}} \leq M.$$

If we define the formal inner product of $T^{(2)}f$ and g as follows:

$$(T^{(2)}f, g) := \int_0^\infty \left(\int_0^\infty k_\lambda(x, y) f(x) dx \right) g(y) dy,$$

then we can rewrite Corollary 3 (for $\mu + \sigma = \lambda$) as follows:

Corollary 13 *For $p > 1, k_\lambda(\sigma) > 0, M \in \mathbf{R}$, the following statements are equivalent:*

(i) If $f(x) \geq 0, f \in L_{p,\varphi}(\mathbf{R}_+)$, satisfying $||f||_{p,\varphi} > 0$, then we have the following inequality:

$$||T^{(2)}f||_{p,\phi^{1-p}} < M||f||_{p,\varphi}. \tag{80}$$

(ii) If $f(x), g(y) \geq 0, f \in L_{p,\varphi}(\mathbf{R}_+), g \in L_{q,\phi}(\mathbf{R}_+)$, satisfying $||f||_{p,\varphi}$, $||g||_{q,\phi} > 0$, then we have the following inequality:

$$(T^{(2)}f, g) < M||f||_{p,\varphi}||g||_{q,\phi}. \tag{81}$$

(iii) $k_\lambda(\sigma) \leq M < \infty$.

If (iii) follows, then the constant factor $M = k_\lambda(\sigma)(\in \mathbf{R}_+)$ in (80) and (81) is the best possible, namely, $0 < ||T^{(2)}|| = k_\lambda(\sigma) \leq M$.

Example 1 Setting

$$h(u) = k_0(u, 1) = \operatorname{csc} h(u) = \frac{2}{e^u - e^{-u}} \; (u > 0),$$

where $\csc h(u)$ is the hyperbolic cosecant function (cf. [26]), then we find $h(xy) = \csc h(xy) = \frac{2}{e^{xy}-e^{-xy}}$,

$$k_0(x, y) = \csc h(\frac{x}{y}) = \frac{2}{e^{x/y} - e^{-x/y}}(x, y > 0),$$

and for $\sigma > 1$, it follows that

$$\begin{aligned}
k(\sigma) = k_0(\sigma) &= \int_0^\infty \csc h(u)u^{\sigma-1}du \\
&= \int_0^\infty \frac{2u^{\sigma-1}}{e^u - e^{-u}}du = \int_0^\infty \frac{2u^{\sigma-1}e^{-u}}{1 - e^{-2u}}du \\
&= 2\int_0^\infty u^{\sigma-1}\sum_{k=0}^\infty e^{-(2k+1)u}du = 2\sum_{k=0}^\infty \int_0^\infty u^{\sigma-1}e^{-(2k+1)u}du.
\end{aligned}$$

Setting $v = (2k + 1)u$ in the above integral, we have

$$\begin{aligned}
k(\sigma) = k_0(\sigma) &= 2\int_0^\infty v^{\sigma-1}e^{-v}dv\sum_{k=0}^\infty \frac{1}{(2k+1)^\sigma} \\
&= 2\Gamma(\sigma)\left[\sum_{k=1}^\infty \frac{1}{k^\sigma} - \sum_{k=1}^\infty \frac{1}{(2k)^\sigma}\right] \\
&= 2\Gamma(\sigma)\left(1 - \frac{1}{2^\sigma}\right)\zeta(\sigma) \in \mathbf{R}_+,
\end{aligned}$$

where $\Gamma(u) := \int_0^\infty v^{u-1}e^{-u}dv\ (u > 0)$ is the Gamma function, and

$$\zeta(s) := \sum_{k=1}^\infty \frac{1}{k^s}\ (\mathrm{Re}\, s > 1)$$

is the Riemann zeta function (cf. [27]). Then by Theorem 5 and Corollary 13, we have

$$||T^{(1)}|| = ||T^{(2)}|| = 2\Gamma(\sigma)\left(1 - \frac{1}{2^\sigma}\right)\zeta(\sigma). \tag{82}$$

Setting $\delta_0 = \frac{\sigma-1}{2}(> 0)$, then $\sigma \pm \delta_0 > 1$ and $k(\sigma \pm \delta_0) = k_0(\sigma \pm \delta_0) \in \mathbf{R}_+$.We can use Example 1 to Theorems 4, 5 and Corollaries 9, 10 as particular kernels.

Example 2 Setting

$$h(u) = k_\lambda(u, 1) = \frac{(\min\{u, 1\})^\alpha}{(\max\{u, 1\})^{\lambda+\alpha}} |\ln u|^\beta \ (u > 0),$$

then we find $h(xy) = \frac{(\min\{xy, 1\})^\alpha}{(\max\{xy, 1\})^{\lambda+\alpha}} |\ln xy|^\beta$,

$$k_\lambda(x, y) = \frac{(\min\{x, y\})^\alpha}{(\max\{x, y\})^{\lambda+\alpha}} |\ln \frac{x}{y}|^\beta \ (x, y > 0),$$

and for $\beta > -1, \sigma, \mu > -\alpha, \sigma + \mu = \lambda \in \mathbf{R}$, it follows that

$$\begin{aligned} k(\sigma) = k_\lambda(\sigma) &= \int_0^\infty \frac{(\min\{u, 1\})^\alpha |\ln u|^\beta}{(\max\{u, 1\})^{\lambda+\alpha}} u^{\sigma-1} du \\ &= \int_0^1 u^{\alpha+\sigma-1}(-\ln u)^\beta du + \int_1^\infty u^{-\mu-\alpha-1}(\ln^\beta u) du \\ &= \frac{1}{(\sigma+\alpha)^{\beta+1}} \int_0^\infty e^{-v} v^\beta dv + \frac{1}{(\mu+\alpha)^{\beta+1}} \int_0^\infty e^{-v} v^\beta dv \\ &= \left[\frac{1}{(\sigma+\alpha)^{\beta+1}} + \frac{1}{(\mu+\alpha)^{\beta+1}}\right] \Gamma(\beta+1) \in \mathbf{R}_+. \end{aligned}$$

Then by Theorem 5 and Corollary 13, we have

$$||T^{(1)}|| = ||T^{(2)}|| = \left[\frac{1}{(\sigma+\alpha)^{\beta+1}} + \frac{1}{(\mu+\alpha)^{\beta+1}}\right] \Gamma(\beta+1). \tag{83}$$

Setting $\delta_0 = \min\{\frac{\sigma+\alpha}{2}, \frac{\mu+\alpha}{2}\}(> 0)$, then $\sigma \pm \delta_0 > -\alpha, \mu \pm \delta_0 > -\alpha$, and $k(\sigma \pm \delta_0) = k_0(\sigma \pm \delta_0) \in \mathbf{R}_+$. We can use Example 2 to Theorems 4, 5 and Corollaries 9, 10 as particular kernels.

Example 3 Setting

$$h(u) = k_\lambda(u, 1) = \frac{|\ln u|^\beta}{|u^\lambda - 1|} \ (u > 0),$$

then we find $h(xy) = \frac{|\ln xy|^\beta}{|(xy)^\lambda - 1|}$,

$$k_\lambda(x, y) = \frac{|\ln x/y|^\beta}{|x^\lambda - y^\lambda|} \ (x, y > 0),$$

and for $\beta, \sigma, \mu > 0, \sigma + \mu = \lambda$, it follows that

$$k(\sigma) = k_\lambda(\sigma) = \int_0^\infty \frac{|\ln u|^\beta}{|u^\lambda - 1|} u^{\sigma-1} du$$
$$= \int_0^1 \frac{(-\ln u)^\beta}{1 - u^\lambda} u^{\sigma-1} du + \int_1^\infty \frac{\ln^\beta u}{u^\lambda - 1} u^{\sigma-1} du$$
$$= \int_0^1 \frac{(-\ln u)^\beta}{1 - u^\lambda} (u^{\sigma-1} + u^{\mu-1}) du$$
$$= \int_0^1 (-\ln u)^\beta \sum_{k=0}^\infty u^{\lambda k} (u^{\sigma-1} + u^{\mu-1}) du.$$

By Lebesgue term by term integration theorem (cf. [24]), we have

$$k(\sigma) = k_\lambda(\sigma) = \sum_{k=0}^\infty \int_0^1 (-\ln u)^\beta (u^{\lambda k+\sigma-1} + u^{\lambda k+\mu-1}) du$$
$$= \sum_{k=0}^\infty \left[\frac{1}{(\lambda k + \sigma)^{\beta+1}} + \frac{1}{(\lambda k + \mu)^{\beta+1}} \right] \int_0^\infty v^\beta e^{-v} dv$$
$$= \frac{\Gamma(\beta+1)}{\lambda^{\beta+1}} \left(\zeta(\beta+1, \frac{\sigma}{\lambda}) + \zeta(\beta+1, \frac{\mu}{\lambda}) \right) \in \mathbf{R}_+,$$

where

$$\zeta(s, a) = \sum_{k=0}^\infty \frac{1}{(k+a)^s} (\mathrm{Re}\, s > 1; 0 < a \le 1)$$

is the Hurwitz zeta function (cf. [27]). Then by Theorem 5 and Corollary 13, we have

$$||T^{(1)}|| = ||T^{(2)}|| = \frac{\Gamma(\beta+1)}{\lambda^{\beta+1}} \left(\zeta(\beta+1, \frac{\sigma}{\lambda}) + \zeta(\beta+1, \frac{\mu}{\lambda}) \right). \tag{84}$$

Setting $\delta_0 = \min\{\frac{\sigma}{2}, \frac{\mu}{2}\}(> 0)$, then $\sigma \pm \delta_0 > 0$, $\mu \pm \delta_0 > 0$, and $k(\sigma \pm \delta_0) = k_\lambda(\sigma \pm \delta_0) \in \mathbf{R}_+$.We can use Example 3 to Theorems 4, 5 and Corollaries 9, 10 as particular kernels.

Example 4 Setting

$$h(u) = k_\lambda(u, 1) = \frac{1}{(u^\alpha + 1)^{\lambda/\alpha}} \ (u > 0),$$

then we find $h(xy) = \frac{1}{[(xy)^\alpha+1]^{\lambda/\alpha}}$,

$$k_\lambda(x, y) = \frac{1}{(x^\alpha + y^\alpha)^{\lambda/\alpha}} \ (x, y > 0),$$

and for $\alpha, \sigma, \mu > 0, \sigma + \mu = \lambda$, it follows that

$$\begin{aligned} k(\sigma) = k_\lambda(\sigma) &= \int_0^\infty \frac{1}{(u^\alpha + 1)^{\lambda/\alpha}} u^{\sigma-1} du \\ &= \frac{1}{\alpha} \int_0^\infty \frac{1}{(v + 1)^{\lambda/\alpha}} v^{\frac{\sigma}{\alpha}-1} dv \\ &= \frac{1}{\alpha} B(\frac{\sigma}{\alpha}, \frac{\mu}{\alpha}) \in \mathbf{R}_+. \end{aligned}$$

Then by Theorem 5 and Corollary 13, we have

$$||T^{(1)}|| = ||T^{(2)}|| = \frac{1}{\alpha} B(\frac{\sigma}{\alpha}, \frac{\mu}{\alpha}). \tag{85}$$

Setting $\delta_0 = \min\{\frac{\sigma}{2}, \frac{\mu}{2}\}(> 0)$, then $\sigma \pm \delta_0 > 0$, $\mu \pm \delta_0 > 0$, and $k(\sigma \pm \delta_0) = k_\lambda(\sigma \pm \delta_0) \in \mathbf{R}_+$.We can use Example 3 to Theorems 4, 5 and Corollaries 9, 10 as particular kernels.

Example 5 Setting

$$h(u) = k_\lambda(u, 1) = \frac{1}{|u - 1|^\lambda} \ (u > 0),$$

then we find $h(xy) = \frac{1}{|xy-1|^\lambda|}$,

$$k_\lambda(x, y) = \frac{1}{|x - y|^\lambda} \ (x, y > 0),$$

and for $\beta, \sigma, \mu > 0, \sigma + \mu = \lambda < 1$, it follows that

$$\begin{aligned} k(\sigma) = k_\lambda(\sigma) &= \int_0^\infty \frac{1}{|u - 1|^\lambda} u^{\sigma-1} du \\ &= \int_0^1 \frac{u^{\sigma-1}}{(1 - u)^\lambda} du + \int_1^\infty \frac{u^{\sigma-1}}{(u - 1)^\lambda} du \\ &= \int_0^1 \frac{1}{(1 - u)^\lambda} (u^{\sigma-1} + u^{\mu-1}) du \\ &= B(1 - \lambda, \sigma) + B(1 - \lambda, \mu) \in \mathbf{R}_+. \end{aligned}$$

Then by Theorem 5 and Corollary 13, we have

$$||T^{(1)}|| = ||T^{(2)}|| = B(1-\lambda, \sigma) + B(1-\lambda, \mu). \tag{86}$$

Setting $\delta_0 = \min\{\frac{\sigma}{2}, \frac{\mu}{2}\}(> 0)$, then $\sigma \pm \delta_0 > 0$, $\mu \pm \delta_0 > 0$, and $k(\sigma \pm \delta_0) = k_\lambda(\sigma \pm \delta_0) \in \mathbf{R}_+$.We can use Example 3 to Theorems 4, 5 and Corollaries 9, 10 as particular kernels.

Acknowledgements This work is supported by the National Natural Science Foundation (Nos. 61370186, 61640222), and Appropriative Researching Fund for Professors and Doctors, Guangdong University of Education (No. 2015ARF25). I'm grateful for this help.

References

1. I. Schur, Bernerkungen sur Theorie der beschrankten Billnearformen mit unendlich vielen Veranderlichen. J. Math. **140**, 1–28 (1911)
2. G.H. Hardy, Note on a theorem of Hilbert concerning series of positive terms. Proc. Lond. Math. Soc. **23**(2), Records of Proc. xlv–xlvi (1925)
3. G.H. Hardy, J.E. Littlewood, G. Pólya, *Inequalities* (Cambridge University Press, Cambridge, 1934)
4. D.S. Mitrinović, J.E. Pečarić, A.M. Fink, *Inequalities Involving Functions and Their Integrals and Derivatives* (Kluwer Academic, Boston, 1991)
5. B.C. Yang, On Hilbert's integral inequality. J. Math. Anal. Appl. **220**, 778–785 (1998)
6. B.C. Yang, A note on Hilbert's integral inequality. Chin. Q. J. Math. **13**(4), 83–86 (1998)
7. B.C. Yang, On an extension of Hilbert's integral inequality with some parameters. Aust. J. Math. Anal. Appl. **1**(1), Art.11, 1–8 (2004)
8. B.C. Yang, I. Brnetić, M. Krnić, J.E. Pečarić, Generalization of Hilbert and Hardy-Hilbert integral inequalities. Math. Inequal. Appl. **8**(2), 259–272 (2005)
9. M. Krnić, J.E. Pečarić, Hilbert's inequalities and their reverses. Publ. Math. Debrecen **67**(3–4), 315–331 (2005)
10. Y. Hong, On Hardy-Hilbert integral inequalities with some parameters. J. Inequal. Pure Appl. Math. **6**(4), Art. 92: 1–10 (2005)
11. B. Arpad, O. Choonghong, Best constant for certain multi linear integral operator. J. Inequal. Appl. **2006**, no. 28582 (2006)
12. Y.J. Li, B. He, On inequalities of Hilbert's type. Bull. Aust. Math. Soc. **76**(1), 1–13 (2007)
13. W.Y. Zhong, B.C. Yang, On multiple Hardy-Hilbert's integral inequality with kernel. J. Inequal. Appl. **2007**, Art.ID 27962, 17 p. (2007)
14. J.S. Xu, Hardy-Hilbert's Inequalities with two parameters. Adv. Math. **36**(2), 63–76 (2007)
15. M. Krnić, M.Z. Gao, J.E. Pečarić, et al., On the best constant in Hilbert's inequality. Math. Inequal. Appl, **8**(2), 317–329 (2005)
16. Y. Hong, On Hardy-type integral inequalities with some functions. Acta Mathmatica Sinica **49**(1), 39–44 (2006)
17. M.T. Rassias, B.C. Yang, On a multidimensional half C discrete Hilbert C type inequality related to the hyperbolic cotangent function. Appl. Math. Comput. **242**, 800–813 (2014)
18. M.T. Rassias, B.C. Yang, A Hilbert C type integral inequality in the whole plane related to the hyper geometric function and the beta function. J. Math. Anal. Appl. **428**(2), 1286–1308 (2015)
19. M.T. Rassias, B.C. Yang, On a Hardy-Hilbert-type inequality with a general homogeneous kernel. Int. J. Nonlinear Anal. Appl. 2.16, **7**(1), 249–269 (2015)

20. B.C. Yang, *The Norm of Operator and Hilbert-Type Inequalities* (Science Press, Beijing, 2009)
21. B.C. Yang, *Hilbert-Type Integral Inequalities* (Bentham Science Publishers Ltd., The United Emirates, 2009)
22. B.C. Yang, On Hilbert-type integral inequalities and their operator expressions. J. Guangaong Univ. Educ. **33**(5), 1–17 (2013)
23. Y. Hong, On the structure character of Hilbert's type integral inequality with homogeneous kernel and applications. J. Jilin Univ. (Sci. Ed.) **55**(2), 189–194 (2017)
24. J.C. Kuang, *Real and Functional Analysis (Continuation)*, 2nd vol. (Higher Education Press, Beijing, 2015)
25. J.C. Kuang, *Applied Inequalities* (Shangdong Science and Technology Press, Jinan, 2004)
26. Y.Q. Zhong, *Introduction to Complex Functions*, 3rd vol. (Higher Education Press, Beijing, 2003)
27. Z.X. Wang, D.R. Guo, *Introduction to Special Functions* (Science Press, Beijing, 1979)

Trotter Product Formula for Non-self-Adjoint Gibbs Semigroups

Valentin A. Zagrebnov

Abstract In this note we present the *analytic extension method* for holomorphic families of the Trotter product formula *approximants*. It allows to prove the trace-norm convergence of this formula for Gibbs semigroups for the case when the involved non-self-adjoint generators are not subordinated.

1 Introduction

Let $\mathscr{H}$ be a separable Hilbert space and $\mathscr{L}(\mathscr{H})$ be algebra of bounded operators on $\mathscr{H}$. Let $\mathscr{C}_\infty(\mathscr{H})$ denote the ideal of compact operators in $\mathscr{L}(\mathscr{H})$. Then the eigenvalues $\{\lambda_k(|C|)\}_{k=0}^\infty$ of the compact operator $|C| = \sqrt{C^*C}$ are known as the singular values $\{s_k(C)\}_{k=0}^\infty$ of C. We denote by $\mathscr{C}_p(\mathscr{H})$ $(1 \leq p < \infty)$ the ideals in $\mathscr{L}(\mathscr{H})$ consisting of all compact operators C for which

$$\sum_{n=1}^{\infty} s_n(C)^p < \infty ,$$

where $s_n(C)$ are the singular values of the compact operator C (see, e.g., [4, 12]). Then $\mathscr{C}_p(\mathscr{H})$ is the Banach space endowed with the norm:

$$\|C\|_p = \left(\sum_{n=0}^{\infty} s_n(C)^p\right)^{1/p} .$$

The space $\mathscr{C}_1(\mathscr{H})$ is the set of the trace-class operators, and one has $\|C\|_1 = \mathrm{Tr}\,|C|$.

V. A. Zagrebnov (✉)
Institut de Mathématiques de Marseille (UMR 7373) - AMU, Centre de Mathématiques et Informatique, Technopôle Château-Gombert, Marseille Cedex 13, France
e-mail: Valentin.Zagrebnov@univ-amu.fr

T. M. Rassias, P. M. Pardalos (eds.), *Mathematical Analysis and Applications*, Springer Optimization and Its Applications 154,
https://doi.org/10.1007/978-3-030-31339-5_24

Definition 1 ([14]) A strongly continuous one-parameter semigroup $\{G_t\}_{t\geq 0}$ is called the (immediate) Gibbs semigroup if $G_t \in \mathscr{C}_1(\mathscr{H})$ for any $t > 0$.

For self-adjoint Gibbs semigroups the Trotter-Kato product formula in the trace-norm topology of $\mathscr{C}_1(\mathscr{H})$ is known since the papers [9, 10] for generators subordinated by a relative smallness. In the present note we describe a generalisation of the trace-norm convergence for the (exponential) Trotter product formula to the non-self-adjoint semigroups. This is a subtle problem since the *self-adjointness* was used in the papers mentioned above for some important estimates. Moreover, in contrast to [9, 10] our method allows also to extend the trace-norm convergent Trotter product formula to the case of non-subordinated generators.

We consider the Trotter product formula when the involved strongly continuous (C_0-)semigroups have *m-sectorial* generators A and B. The strategy of the proof of the Trotter product formula convergence is based on the *lifting up* the corresponding results for the self-adjoint generators. To this aim we propose the *analytic extension method* for holomorphic families of sectorial forms (for generators) and of the corresponding semigroups. We develop details in Section 2.

In Section 3 we use this method to prove the trace-norm compactness of holomorphic families of the Trotter *approximants.*

This allows to prove in Section 4 a statement (Theorem 4.4) about the Trotter product formula convergence in the trace-norm topology to a *degenerate* Gibbs semigroup for m-sectorial generators A and B, which are *not* subordinated. More precisely:

Let A, B be m-sectorial operators with vertexes at zero and such that $e^{-t\operatorname{Re} A} \in \mathscr{C}_1(\mathscr{H})$ for any $t > 0$. Then the Trotter product formula converges in the trace-norm topology to $e^{-t(A\dot{+}B)}P_0$ uniformly in $t > 0$. Generator $H = A \dot{+} B$ is the m-sectorial form-sum of A, B, and P_0 is the orthogonal projection $P_0 : \mathscr{H} \to \mathscr{H}_0 = \overline{\operatorname{dom} H}$.

2 Holomorphic Families of Semigroup Generators

We start with conditions imposed on the pair of generators A and B. Let them be densely defined m-sectorial operators in $\mathscr{H}$ with corresponding semi-angles α_A and α_B belonging to $[0, \pi/2)$. We denote by a and b the densely defined closed sectorial sesquilinear forms associated, respectively, with operators A and B.

Recall that if A is an m-sectorial operator with a vertex $\gamma = 0$ and semi-angle $\alpha_A \in [0, \pi/2)$, then we can associate with A a closed, densely defined sesquilinear sectorial form $a[u, v] : \operatorname{dom} a \times \operatorname{dom} a \to \mathbb{C}$ (with the same vertex $\gamma = 0$ and semi-angle $\alpha \in [0, \pi/2)$), such that:

$$a[u, v] = (Au, v) := (\operatorname{Re} a)[u, v] + i(\operatorname{Im} a)[u, v] . \tag{2.1}$$

Here, $u \in \operatorname{dom} A \subset \operatorname{dom} a$, $v \in \operatorname{dom} a$, and the adjoint form is $a^*[u, v] := \overline{a[v, u]}$. Then the real and imaginary parts of this form are defined correspondingly by $\operatorname{Re} a := \frac{1}{2}(a + a^*)$ and by $\operatorname{Im} a := \frac{1}{2i}(a - a^*)$. The form is symmetric if $a[u, v] = a^*[u, v]$. By virtue of the *polarisation* identity

$$a[u, v] = \frac{1}{4}\{a[u + v] - a[u - v] + i\, a[u + i\, v] - i\, a[u - i\, v]\}\,,$$

the complex sesquilinear form $a[u, v]$ is symmetric if and only if the *quadratic* form $a[u] := a[u, u]$, $u \in \operatorname{dom} a$, is real-valued. A vertex γ and a semi-angle α_A are (not uniquely) determined by the sesquilinear form a via inequalities:

$$(\operatorname{Re} a)[u] \geq \gamma\,, \quad |(\operatorname{Im} a)[u]| \leq \operatorname{tg}\alpha_A (\operatorname{Re} a)[u]\,, \quad u \in \operatorname{dom} a\,. \tag{2.2}$$

We would like to recall that $(\operatorname{Re} a)[u, v] \neq \operatorname{Re} a[u, v]$ and $(\operatorname{Im} a)[u, v] \neq \operatorname{Im} a[u, v]$ for $u, v \in \operatorname{dom} a$.

Proposition 2.1 (Representation Theorem [7, Ch.VI, §2]) *Let the sesquilinear form* $t : \operatorname{dom} t \times \operatorname{dom} t \to \mathbb{C}$, *be densely defined, closed, and sectorial on domain* $\operatorname{dom} t \subset \mathscr{H}$. *Then there exists a unique m-sectorial operator* T *with* $\operatorname{dom} T \subset \operatorname{dom} t$ *such that* $t[u, v] = (Tu, v)$ *for* $u \in \operatorname{dom} T$, $v \in \operatorname{dom} t$, *and* $\operatorname{dom} T$ *is a core of the form* t.

If in addition the form t *is symmetric and non-negative:* $t[u] \geq 0$, $u \in \operatorname{dom} t$, *then* T *is non-negative self-adjoint operator such that* $\operatorname{dom} T^{1/2} = \operatorname{dom} t$ *and* $t[u, v] = (T^{1/2}u, T^{1/2}v)$, $u, v \in \operatorname{dom} t$. *Moreover, a subset* $D \subset \operatorname{dom} t$ *is a core of* t *if and only if* D *is a core of* $T^{1/2}$.

Corollary 2.2 *The mapping:* $t \mapsto T_t(= T)$ *is a one-to-one correspondence between the set of all densely defined, closed sectorial forms and the set of all m-sectorial operators. The form* t *is bounded if and only if* T *is bounded,* $T_{t^*} = T^*$ *and* t *is symmetric if and only if operator* T *is self-adjoint.*

Since there are *no maximal* sectorial forms, the *representation theorem* yields advantages in exploiting sesquilinear forms for constructing m-sectorial operators. The first is that a densely defined sectorial operator S naturally generates a *closable* sesquilinear form $s[u, v] = (Su, v)$ for $u, v \in \operatorname{dom} S$. Let $T = T_t$ be m-sectorial operator associated with the closed form $t := \widetilde{s}$. Then T is extension of S such that $\operatorname{dom} S = \operatorname{dom} s$ is a core of t. Hence, even when the closure $\widetilde{S}$ is not m-sectorial, one can always assign to S a *minimal* m-sectorial extension T with $\operatorname{dom} T \subset \operatorname{dom} t$.

The second advantage comes from observation that contrary to the set of the closed operators $\mathscr{C}(\mathscr{H})$ the set of the closed forms $\mathscr{C}_f(\mathscr{H})$ is a *linear space*. If a and b are densely defined, closed sectorial forms, then by the *representation theorem* there exists m-sectorial operator C associated with the closed form $c = a + b$. Since in turn there are m-sectorial operators A and B associated with forms a and b, the operator $C := A \dotplus B$ is called the *form-sum* of these operators. On the other hand, if a and b are densely defined, closed sectorial forms generated by densely defined sectorial operator A and B, then the corresponding form-sum is an extension of the operator-sum: $(A + B) \subset (A \dot{+} B)$.

Recall that the form-sum $A \dot{+} B$ may exist and be well-defined even in the case when $\mathrm{dom}\, A \cap \mathrm{dom}\, B = \{0\}$. We have also seen another option, when for m-sectorial operators one has: $\mathrm{dom}\, A \cap \mathrm{dom}\, B \subset \overline{\mathrm{dom}\, a \cap \mathrm{dom}\, b} = \mathscr{H}_0 \subset \mathscr{H}$, i.e. the operator sum $A + B$ is *not* densely defined in $\mathscr{H}$. Then the m-sectorial form-sum operator $C = A \dot{+} B$ is well-defined in the Hilbert space $\mathscr{H}_0 = P_0(\mathscr{H})$, where P_0 is the orthogonal projection $P_0 : \mathscr{H} \to \mathscr{H}_0 = \overline{\mathrm{dom}\, C}$.

Now we return to the form (2.1). Since a is closed and $\gamma = 0$, the *symmetric* form $(\mathrm{Re}\, a)[u, v]$ is also closed and $(\mathrm{Re}\, a)[u, u] \geq 0$. Then by Proposition 2.1 it defines a non-negative self-adjoint operator $A_R := \mathrm{Re}\, A \geq 0$, which is called the *real* part of the operator A. Similar arguments for the closed non-negative symmetric form $(\mathrm{Re}\, a^*)[u, v]$ yield that also $A_R = \mathrm{Re}\, A^*$.

Note that for a bounded operator $A \in \mathscr{L}(\mathscr{H})$ one readily obtains the identity: $A_R = \frac{1}{2}(A + A^*)$, although in general it is *not* true for unbounded operators. On the other hand, by definitions of a, a^* and by the representation theorem we obtain that the real part of A is the form-sum: $A_R = \frac{1}{2}(A \dot{+} A^*)$, which is a non-negative self-adjoint operator with $\mathrm{dom}\, \sqrt{A_R} = \mathrm{dom}\, a$.

We also recall that with help of the real part $A_R = \mathrm{Re}\, A$ the m-sectorial operator A with vertex $\gamma = 0$ and semi-angle $\alpha \in [0, \pi/2)$ has the following representation:

$$A = A_R^{1/2}(\mathbb{1} + i\, L)A_R^{1/2} \, . \tag{2.3}$$

Here symmetric bounded operator $L \in \mathscr{L}(\mathscr{H})$ is such that $\|L\| \leq \mathrm{tg}\alpha$.

Remark 2.3 By virtue of representation (2.3) the resolvent of m-sectorial operator A is compact if and only if compact is the resolvent of its real part A_R.

Step 1 The first step in our programme of the *analytic extension method* is a complex extension of sectorial forms.

Let a and b be densely defined closed sesquilinear sectorial forms with vertex $\gamma = 0$ and semi-angles $\alpha_A, \alpha_B \in [0, \pi/2)$. We define two families of *auxiliary* closed sesquilinear forms for $z \in \mathbb{C}$ by parametrisation of imaginary parts

$$a_z := \mathrm{Re}\, a + z \mathrm{Im}\, a \, , \qquad \mathrm{dom}\, a_z = \mathrm{dom}\, a \, , \tag{2.4}$$

$$b_z := \mathrm{Re}\, b + z \mathrm{Im}\, b \, , \qquad \mathrm{dom}\, b_z = \mathrm{dom}\, b \, . \tag{2.5}$$

Lemma 2.4 *The forms $z \mapsto a_z$ and $z \mapsto b_z$ are sectorial (with $\gamma = 0$) and holomorphic functions in the strip*

$$D_{AB} := \{z \in \mathbb{C} : \ |\mathrm{Re}\, z| < \min\{\mathrm{ctg}\alpha_A, \mathrm{ctg}\alpha_B\}\}. \tag{2.6}$$

Proof Let $x, y \in \mathbb{R}$, $z = x + iy$, and $u \in \mathrm{dom}\, a$. Since the form a is sectorial, by (2.2) we obtain:

$$|x(\mathrm{Im}\, a)[u]| \leq |x|\mathrm{tg}\alpha_A(\mathrm{Re}\, a)[u]. \tag{2.7}$$

Hence, for $|x|\mathrm{tg}\alpha_A < 1$ one gets $\mathrm{Re}\, a_z = \mathrm{Re}\, a + x\mathrm{Im}\, a > 0$, i.e. $\gamma = 0$, see (2.2). Moreover, for these values of x and $u \in \mathrm{dom}\, a$, we get

$$(\mathrm{Re}\, a_z)[u] \geq (\mathrm{Re}\, a)[u] - |x(\mathrm{Im}\, a)[u]| \geq (1 - |x|\mathrm{tg}\alpha_A)(\mathrm{Re}\, a)[u]. \tag{2.8}$$

By (2.7) and (2.8), we obtain that

$$\begin{aligned} |(\mathrm{Im}\, a_z)[u]| = |y(\mathrm{Im}\, a)[u]| &\leq |y|\mathrm{tg}\alpha_A(\mathrm{Re}\, a)[u] \\ &\leq \frac{|y|\mathrm{tg}\alpha_A}{1 - |x|\mathrm{tg}\alpha_A}(\mathrm{Re}\, a_z)[u]\,, \end{aligned} \tag{2.9}$$

and by (2.2) the form a_z is sectorial for $z \in \{z \in \mathbb{C} \mid |x| < \mathrm{ctg}\alpha_A\}$.

Similarly, the form b_z is sectorial with $\gamma = 0$ for $|x| < \mathrm{ctg}\alpha_B$. Therefore, both of the forms a_z and b_z are sectorial with $\gamma = 0$ in the strip D_{AB} defined by (2.6).

Note that the families (2.4) and (2.5) are explicit functions of the complex variable $z \in \mathbb{C}$. Their derivatives with respect to this variable exist and have the form:

$$\begin{aligned} \partial_z a_z[u] &= \mathrm{Im}\, a[u] \qquad && u \in \mathrm{dom}\, a\,, \\ \partial_z b_z[u] &= \mathrm{Im}\, b[u] \qquad && u \in \mathrm{dom}\, b\,. \end{aligned}$$

For each $u \in \mathrm{dom}\, a$, or, respectively, $u \in \mathrm{dom}\, b$, they are linear holomorphic complex functions in the domain D_{AB}. □

Now we recall that a set of sesquilinear forms $\{t_z\}_{z \in D \subset \mathbb{C}}$ is called a *holomorphic family* of *type* (a) if it verifies two conditions, [7, Ch.VII, §4.2]:

(1) Each t_z is sectorial and closed with dense domain: $\mathrm{dom}\ t_z = Q$, which is independent of $z \in D$.
(2) $D \ni z \mapsto t_z[u]$ is holomorphic for each fixed $u \in Q$. By polarisation identity this implies that sesquilinear $t_z[u, v]$ is holomorphic in $z \in D$ for each fixed pair $u, v \in Q$.

Recall that the concept of *holomorphic families* of operators is based on bounded-holomorphic properties of the corresponding resolvents. To this aim, consider a family of closed operators in a Banach space: $\{A(z) \in \mathscr{C}(\mathscr{B})\}_{z \in D}$, which is defined in a neighborhood of $z_0 \in D$. If $\zeta \in \mathbb{C}$ belongs to the resolvent set $\rho(A(z_0))$, then $z \mapsto A(z)$ is called *holomorphic* at z_0 if there exists a disc $d_\varepsilon(z_0) := \{z \in \mathbb{C} : |z - z_0| < \varepsilon\}$ such that $\zeta \in \rho(A(z))$ and the resolvent $R_\zeta(A(z)) = (A(z) - \zeta\mathbb{1})^{-1}$ is bounded-holomorphic for $z \in d_\varepsilon(z_0)$, $\varepsilon > 0$.

A special class of holomorphic families of operators, that we refer to in the text, is the family $\{A(z)\}_{z \in D}$ of *type* (A).

Definition 2 A holomorphic family of type (A) is defined by two conditions:

(1) Each $A(z)$ is closed with dense domain: $\mathrm{dom}\, A(z) = \mathfrak{D}$, which is independent of $z \in D$.

(2) The mapping $z \mapsto A(z)u$ is holomorphic vector-valued function for $z \in D$ and for every $u \in \mathfrak{D}$.

Conditions (1) and (2) imply that the family $\{A(z)\}_{z\in D}$ is a holomorphic family in the bounded-holomorphic resolvent sense. The following criterion for type (A) is useful.

Proposition 2.5 ([7, Ch.VII, §2.2]) *Let A be a closable operator with domain* $\mathrm{dom}\, A = \mathfrak{D}$ *in a Banach space $\mathscr{B}$. Let $\{A^{(n)}\}_{n\geq 1}$ be operators with domains:* $\mathrm{dom}\, A^{(n)} \subset \mathfrak{D}$ *and μ, ν, η be non-negative constants such that*

$$\|A^{(n)}u\| \leq \eta^{n-1}\,(\mu\|u\| + \nu\|Au\|)\,, \quad u \in \mathfrak{D},\; n \in \mathbb{N}\,.$$

Then for $|z| < 1/\eta$ and $u \in \mathfrak{D}$ the series :

$$A(z)u = \sum_{n=0}^{\infty} z^n\, A^{(n)}u\,, \quad A := A^{(0)}\,,$$

defines an operator $A(z)$ with $\mathrm{dom}\, A(z) = \mathfrak{D}$. *If $|z| < (\nu+\eta)^{-1}$, then operators $A(z)$ are closable and the closures $\{\tilde{A}(z)\}_{\{z:|z|<(\nu+\eta)^{-1}\}}$ form a holomorphic family of type (A).*

Corollary 2.6

(i) *The forms $z \mapsto a_z$ (2.4) and $z \mapsto b_z$ (2.5) are holomorphic families of type (a) in the strip D_{AB} (2.6). Since $(a_z)^* = a_{\bar{z}}$ and $(b_z)^* = b_{\bar{z}}$ they are self-adjoint holomorphic families [7, ch. VII, Sect. 3.1]. By the representation theorem (Proposition 2.1) there exist m-sectorial operators $A(z)$ and $B(z)$ with $\gamma = 0$, which are associated with the closed sectorial forms a_z and b_z, $z \in D_{AB}$.*

(ii) *These operators create the resolvent bounded-holomorphic families, which are called holomorphic families of type (B). Operators $\{A(z)\}_{z\in D_{AB}}$ and $\{B(z)\}_{z\in D_{AB}}$ are locally uniformly m-sectorial with vertex $\gamma = 0$ and for the particular value $z = i$ yield: $A(i) = A$, $B(i) = B$. Therefore, for each $z \in D_{AB}$ the operators $A(z)$ and $B(z)$ are generators of contraction holomorphic semigroups.*

(iii) *Operator holomorphic families of type (B) inherit the property of the form self-adjoint families. This yields: $A^*(z) = A(\bar{z})$ and $B^*(z) = B(\bar{z})$.*

Remark 2.7 Let family $\{C(z)\}_{z\in D}$ be holomorphic of type (A) and have non-empty resolvent sets such that $\rho_D = \bigcap_{z\in D}\rho(C(z)) \neq \emptyset$. Then $C(z)$ has compact resolvent $R_\zeta(C(z))$ either for all $z \in D$ or for no z.

Let $\{C(z)\}_{z\in D}$ be a type (B) holomorphic family of (m-sectorial) operators. If operator $C(z_0)$ has compact resolvent for $z_0 \in D$, then resolvent of $C(z)$ is compact for any $z \in D$.

Step 2 The second step is a complex extension of the corresponding to sectorial forms the semigroup generators.

Proposition 2.8 *Let $\{C(z)\}_{z\in D}$ be a holomorphic family of type (B) with vertex $\gamma = 0$ and semi-angles $\alpha(z) < \pi/2$. Then for each $z \in D$ the operator $C(z)$ is generator of a contraction holomorphic semigroup $\{U_t(C(z))\}_{t\in S_{\theta(z)}}$ with semi-angle $\theta(z) \in [0, \pi/2)$. The function $z \mapsto U_t(C(z))$ is holomorphic in $z \in D$ for any t in the open sector S_{θ_0}, with semi-angle $\theta_0 := \inf_{z\in D}(\pi/2 - \alpha(z))$.*

Proof For each $z \in D$ the m-sectorial operator $C(z) \in \mathscr{H}(\theta(z) = \pi/2 - \alpha(z), 0)$ is generator of a contraction holomorphic semigroup, see Corollary 2.6(ii). Then by the Riesz-Dunford representation for holomorphic semigroups we have:

$$U_t(C(z)) = \frac{1}{2\pi i}\int_{\Gamma} d\zeta \, \frac{e^{-t\zeta}}{\zeta \mathbb{1} - C(z)} \,. \tag{2.10}$$

Recall that this integral is absolutely $\|\cdot\|$-convergent for $t > 0$ if the contour $\Gamma \in \bigcap_{z\in D} \rho(C(z)) := \rho_D$, running from infinity with $\arg \zeta = \alpha_{max} + \varepsilon$, and then back to infinity with $\arg \zeta = -(\alpha_{max} + \varepsilon)$, where $\alpha_{max} := \sup_{z\in D} \alpha(z)$ for some $0 < \varepsilon < \pi/2 - \alpha_{max}$. Since $\{C(z)\}_{z\in D}$ is a holomorphic family, the resolvents $\{R_\zeta(C(z))\}_{z\in D}$ form a holomorphic family for any $\zeta \in \rho_D$. Then representation (2.10) yields the estimate

$$\begin{aligned} \left\|\partial_z(\zeta\mathbb{1} - C(z))^{-1}\right\| &= \left\|\frac{1}{2\pi i}\int_{\gamma_r} dz' \, (z' - z)^{-2}(\zeta\mathbb{1} - C(z'))^{-1}\right\| \\ &\leq \frac{\tilde{M}_\varepsilon}{|\zeta| r} \,, \end{aligned} \tag{2.11}$$

for any $\zeta \in \{\zeta \in \mathbb{C} \mid -(\alpha_{max} + \varepsilon) < \arg\zeta < \alpha_{max} + \varepsilon\}$. Here γ_r is a small circle of radius r around z and

$$\tilde{M}_\varepsilon := \sup_{\substack{\zeta\in\mathbb{C}\setminus\bar{S}_{\pi/2-\theta_0+\varepsilon} \\ z'\in\gamma_r}} \left\|\zeta\,(\zeta\mathbb{1} - C(z'))^{-1}\right\| .$$

The estimate (2.11) shows that the representation (2.10) is operator-norm differentiable under the integral since the integrand is operator-valued holomorphic complex function in D.

Note that the arguments developed above are also valid for $t \in \mathbb{C}$ such that $|\arg(t\,\zeta)| < \pi/2$. Then we get the statement for any t in the open sector S_{θ_0} with semi-angle $\theta_0 = \pi/2 - \alpha_{max}$, that implies the assertion. □

Corollary 2.9 *Thus, applying Proposition 2.8 to the particular case of generators $\{A(z)\}_{z\in D_{AB}}$ and $\{B(z)\}_{z\in D_{AB}}$ in Corollary 2.6, we obtain that contraction semigroups*

$$\{U_t(A(z))\}_{t\in S_{\theta_0}} \quad \textit{and} \quad \{U_t(B(z))\}_{t\in S_{\theta_0}} \tag{2.12}$$

are holomorphic families for $z \in D^0_{AB}$, *and for any* t *in the sector* S_{θ_0}. *Here, the domain of analyticity in* z *is defined as*

$$D^0_{AB} = \{z \in \mathbb{C} : |\mathrm{Re}\, z| < \min\{\mathrm{ctg}\alpha_A, \mathrm{ctg}\alpha_B\} - \delta_0 := \Delta_R\ , \qquad (2.13)$$
$$|\mathrm{Im}\, z| < \Delta_I\}\ ,$$

for a small $\delta_0 > 0$ *and for some* $\Delta_I > 1$, *cf.* (2.6). *Then by* (2.9), *the corresponding semi-angle* $\theta_0 > 0$ *is defined by*

$$\mathrm{tg}\left(\frac{\pi}{2} - \theta_0\right) = \frac{\Delta_I \max\{\mathrm{tg}\alpha_A, \mathrm{tg}\alpha_B\}}{1 - \Delta_R \max\{\mathrm{tg}\alpha_A, \mathrm{tg}\alpha_B\}}\ . \qquad (2.14)$$

Step 3 (*Vitali theorem*, see [5, Theorem 3.14.1], [8, Theorem 17.18], or [13, §6.2]) The last step of our programme for the *analytic extension method* is the application of the Vitali theorem to holomorphic families of the Trotter approximants.

Remark 2.10 Recall that for holomorphic families of bounded operators in $\mathscr{L}(\mathscr{H})$ there is no distinction between *uniform*, *strong* or *weak* operator analyticity. Moreover, these holomorphic operator-valued functions inherit some properties known from the standard complex analysis. For example, let $\{\Phi_n(z)\}_{n\geq 1} \subset \mathscr{L}(\mathscr{H})$ be a sequence of operator-valued complex functions such that the norm

$$\|\Phi_n(z)\| < M\ ,$$

for all $n \geq 1$ and $z \in D \subset \mathbb{C}$. If $\{\Phi_n(z)\}_{n\geq 1}$ converges (in any of the three topologies) on a subset of D having a *limit point* in D, then by the *Vitali theorem* the limit $\lim_{n\to\infty} \Phi_n(z) = \Phi(z)$ exists for any $z \in D$, the convergence is uniform on any compact $K \subset D$, and the limiting function $\Phi(z)$ is holomorphic in D.

3 Holomorphic Families of the Trotter Approximants

To proceed with analysis of convergence of the Trotter product formula in the trace-norm topology we recall the *exponential* case of the Trotter-Kato product formula ([11, Theorem 3.2], or [14, Chapter 4.2]) in the operator-norm topology.

Proposition 3.1 *Let* A *and* B *be non-negative densely defined self-adjoint operators in* $\mathscr{H}$ *and the form-sum* $H = A \dot{+} B$ *be defined in the subspace* $\mathscr{H}_0 = P_0(\mathscr{H})$, *where* P_0 *is the orthogonal projection* $P_0 : \mathscr{H} \to \overline{\mathrm{dom}\, A^{1/2} \cap \mathrm{dom}\, B^{1/2}}$. *If the resolvent of* A *or of* B *is compact:* $(\mathbb{1} + A)^{-1} \vee (\mathbb{1} + B)^{-1} \in \mathscr{C}_\infty(\mathscr{H})$, *then the Trotter product formula*

$$\|\cdot\| - \lim_{n\to\infty} \left(e^{-tA/n} e^{-tB/n}\right)^n = e^{-t(A\dot{+}B)} P_0\ , \qquad (3.1)$$

converges in the operator-norm topology locally uniformly in t *away from zero.*

Now we can extend Proposition 3.1 to m-sectorial generators.

Proposition 3.2 *If A and B are m-sectorial operators (with vertex $\gamma = 0$) in a Hilbert space $\mathscr{H}$ and if $(\mathbb{1} + A)^{-1} \vee (\mathbb{1} + B)^{-1} \in \mathscr{C}_\infty(\mathscr{H})$, then the Trotter product formula converges in the operator-norm topology:*

$$\|\cdot\| - \lim_{n\to\infty} \left(e^{-tA/n} e^{-tB/n}\right)^n = e^{-t(A\dot{+}B)} P_0 \,, \tag{3.2}$$

for any $t \in S_\theta$, where $\theta = \pi/2 - \max\{\alpha_A, \alpha_B\}$. The convergence is uniform on the compact subsets of the sector S_θ. Generator $H = A \dot{+} B$ is the m-sectorial form-sum of A and B, and P_0 is the orthogonal projection $P_0 : \mathscr{H} \to \mathscr{H}_0 = \overline{\mathrm{dom}(H)}$.

Proof Following Corollary 2.6 we include operators A and B into holomorphic type (B) self-adjoint families $\{A(z)\}_{z\in D_{AB}}$ and $\{B(z)\}_{z\in D_{AB}}$ of m-sectorial operators with $\gamma = 0$, which are generators of holomorphic contraction semigroups. Then taking into account Corollary 2.9 we construct from the two families (2.12) a sequence of operator-valued uniformly bounded and holomorphic in the operator-norm topology Trotter *approximants*:

$$z \mapsto \Phi_n(t,z) := \left(e^{-tA(z)/n} e^{-tB(z)/n}\right)^n \,, \quad \|\Phi_n(t,z)\| \le 1 \,, \tag{3.3}$$

$n \in \mathbb{N}$, with domain of analyticity D^0_{AB}, which is determined by (2.13) for any t in the sector S_{θ_0} with semi-angle defined by (2.14).

Note that by condition of the proposition and by Remark 2.7 for $D = D^0_{AB}$, the $(\mathbb{1} + A(z))^{-1}$ (or $(\mathbb{1} + B(z))^{-1}$) is a holomorphic family of *compact* operators. In particular, the resolvent $\{(\mathbb{1} + A(x))^{-1}\}_{x\in\Delta_R}$ (or $\{(\mathbb{1} + B(x))^{-1}\}_{x\in\Delta_R}$) is compact.

Since generators in (3.3) are self-adjoint families of m-sectorial operators with $\gamma = 0$ [7, ch. VII, Sect. 3.1], operators $A(x)$ and $B(x)$ are self-adjoint and non-negative for each $x \in \Delta_R$. Then (2.4) and (2.5) yield that $\mathrm{dom}\, a = \mathrm{dom}\, A(x)^{1/2}$, $\mathrm{dom}\, b = \mathrm{dom}\, B(x)^{1/2}$ and that the self-adjoint non-negative form-sum $A(x) \dot{+} B(x)$ exists in domain $\mathscr{D} \subseteq \mathscr{H}_0$, which is dense in $\mathscr{H}_0 = \overline{(\mathrm{dom}\, a \cap \mathrm{dom}\, b)}$.

Now we are in the position to apply Proposition 3.1 to the sequence of approximants (3.3) for $z = x$ and $t \in S_\theta$. This yields

$$\|\cdot\| - \lim_{n\to\infty} \Phi_n(t,x) = e^{-t(A(x)\dot{+}B(x))} P_0 \,, \; x \in \Delta_R \,. \tag{3.4}$$

Therefore, operator-norm convergence (3.4) of the operator-valued uniformly bounded and holomorphic family (3.3) on the interval Δ_R, together with the Vitali theorem imply the operator-norm convergence in (3.4) for any $z \in D^0_{AB}$. Since $i \in D^0_{AB}$, we obtain (3.2) as well as $H = A(i) \dot{+} B(i)$. □

The analytic extension method of Proposition 3.2 allows to *lift up* the operator-norm convergence of the Trotter-Kato product formula for non-self-adjoint semigroups to the trace-norm convergence for non-self-adjoint Gibbs semigroups, see Theorem 4.4. The first step is the following assertion.

Proposition 3.3 *If A is an m-sectorial operator with vertex $\gamma = 0$ such that $e^{-t\,\mathrm{Re}\,A} \in \mathscr{C}_1(\mathscr{H})$ for $t > 0$, then $\{e^{-tA(z)}\}_{t\geq 0}$ is a family of holomorphic contraction Gibbs semigroups for $z \in D_A^0$, where*

$$D_A^0 = \{z \in \mathbb{C} : |\mathrm{Re}\,z| < \Delta_R(A) = \mathrm{ctg}\alpha_A - \delta_0\,,\quad |\mathrm{Im}\,z| < \Delta_I(A)\}\,, \tag{3.5}$$

for some $\delta_0 > 0$ and $\Delta_I(A) > 1$, in the sector $S_{\theta_0(A)}$ with the semi-angle $\theta_0(A)$ defined by the equation:

$$\mathrm{tg}\left(\frac{\pi}{2} - \theta_0(A)\right) = \frac{\Delta_I(A)\mathrm{tg}\alpha_A}{1 - \Delta_R(A)\mathrm{tg}\alpha_A}. \tag{3.6}$$

Moreover, the function $z \mapsto e^{-tA(z)}$ is $\|\cdot\|_1$-holomorphic in $z \in D_A^0$ for any $t \in S_{\theta_0(A)}$.

Proof Note that Proposition 4.1 yields that $e^{-tA} \in \mathscr{C}_1(\mathscr{H})$, for $t > 0$. Then by representation (4.2) the operators $\mathrm{Re}\,A$ and A have compact resolvents. By Corollary 2.6 and Remark 2.7 $\{A(z)\}_{z\in D_A^0}$ is a holomorphic family of uniformly sectorial operators and $A(z = 0) = \mathrm{Re}\,A$. Hence $A(z)$ has a compact resolvent and $A(z) \in \mathscr{H}(\theta_0(A), 0)$ for any $z \in D_A^0$, where $\theta_0(A)$ is defined by (3.6). The same is true for $\mathrm{Re}\,A(z)$.

Taking into account inequality (2.9) we obtain the estimate

$$\{1 - |\mathrm{Re}\,z|\mathrm{tg}\alpha_A\}\mathrm{Re}\,a[u] \leq \mathrm{Re}\,a_z[u]\,,\ z \in D_A^0\,,\quad u \in \mathrm{dom}\,a\,. \tag{3.7}$$

Then by the *minimax principle* for the self-adjoint operators $\mathrm{Re}\,A(z)$ and $\mathrm{Re}\,A$ and by the condition $e^{-t\mathrm{Re}\,A} \in \mathscr{C}_1(\mathscr{H})$ for $t > 0$, we conclude that $e^{-t\mathrm{Re}\,A(z)} \in \mathscr{C}_1(\mathscr{H})$ for $t > 0$ and $z \in D_A^0$. In addition, by Proposition 4.1 we have the estimate $\|e^{-tA(z)}\|_1 \leq \|e^{-t\mathrm{Re}\,A(z)}\|_1$. Finally, by (2.9), (3.5), and (3.6), the family $\{A(z)\}_{z\in D_A^0}$ consists of uniformly sectorial operators with numerical range $Nr(A(z)) \subset S_{\pi/2-\theta_0(A)}$. Thus, we infer that $\{G_t(A(z))\}_{t\in S_{\theta_0(A)}}$ is a holomorphic contraction Gibbs semigroup for any $z \in D_A^0$.

On the other hand, by Proposition 2.8 the family $\{e^{-tA(z)}\}_{z\in D_A^0}$ is $\|\cdot\|$-holomorphic (cf. Remark 2.10) for any $t \in S_{\theta_0(A)}$. Therefore, we have the representation

$$e^{-tA(z)} = \frac{1}{2\pi i}\oint_{\gamma_r} dw \frac{e^{-tA(w)}}{w - z}, \tag{3.8}$$

where the $\|\cdot\|$-convergent Cauchy integral is taken along a small circle $\gamma_r = \{w \in \mathbb{C} : |w - z| = r\} \subset D_A^0$. Taking into account that the trace-norm of the semigroup $\{G_t(A(z))\}_{t \in S_{\theta_0(A)}}$ is bounded, and using (3.7), one gets for $w \in \gamma_r$ the estimate:

$$\left\| \partial_z \frac{e^{-tA(w)}}{w - z} \right\|_1 \leq \frac{\|e^{-t \operatorname{Re} A(w)}\|_1}{r^2} \leq \|e^{-\beta t \operatorname{Re} A}\|_1 \frac{1}{r^2} ,$$

where $\beta := \inf_{w \in \gamma_r} (1 - |\operatorname{Re} w| \operatorname{tg}\alpha_A)$. Therefore, the Cauchy integral (3.8) is $\|\cdot\|_1$-differentiable. This yields that the function $z \mapsto e^{-tA(z)}$ is $\|\cdot\|_1$-holomorphic in D_A^0 for any $t \in S_{\theta_0(A)}$. □

Now we can apply the *analytic extension method* to prove the trace-norm convergence of the Trotter-Kato product formula for holomorphic Gibbs semigroups with non-self-adjoint m-sectorial generators A and B. This approach does not give error bound estimate for the convergence rate but allows to treat generators on the equal level and even for a trivial common domain: $\operatorname{dom} A \cap \operatorname{dom} B = \{0\}$.

To this aim we collect in Remark 3.4 the elements that we established for application of the *Vitali theorem* in the $\|\cdot\|_1$-topology.

Remark 3.4 Let A and B be m-sectorial operators with vertex $\gamma = 0$. Then according to definitions (2.4) and (2.5), and by virtue of Corollary 2.6, we constructed two holomorphic type (B) families of m-sectorial operators $\{A(z)\}_{z \in D_A^0}$ and $\{B(z)\}_{z \in D_B^0}$ with D_A^0 and D_B^0 defined by (3.5), such that:

(a) For $z = i$ one has: $A(i) = A$ and $B(i) = B$.
(b) For $x = \operatorname{Re} z \in [-\Delta_R, \Delta_R]$, where $\Delta_R = \min\{\operatorname{ctg}\alpha_A, \operatorname{ctg}\alpha_B\} - \delta_0$, the non-negative operators $A(x) = \operatorname{Re} A(z) \geq 0$ and $B(x) = \operatorname{Re} B(z) \geq 0$ are self-adjoint.
(c) If $e^{-t\operatorname{Re} A}$ and $e^{-t\operatorname{Re} B}$ are trace-class operators for $t > 0$, then by Proposition 3.3 the m-sectorial operators $\{A(z)\}_{z \in D_A^0}$ and $\{B(z)\}_{z \in D_B^0}$ generate two $\|\cdot\|_1$-holomorphic in D_A^0 and in D_B^0 families of holomorphic for $t \in S_{\theta_0(A)}$, respectively, for $t \in S_{\theta_0(B)}$, contraction Gibbs semigroups. Here sector $S_{\theta_0(A)}$, respectively, sector $S_{\theta_0(B)}$, are defined by condition (3.6).

4 Proof of Theorem 4.4

First, we establish a relation between generators A and $\operatorname{Re} A$ in the case of the Gibbs semigroups.

Proposition 4.1 *Let* $A \in \mathscr{H}(\pi/2 - \alpha, 0)$ *be an* m*-sectorial operator with real part* $\operatorname{Re} A \geq 0$. *If* $e^{-t \operatorname{Re} A} \in \mathscr{C}_1(\mathscr{H})$ *for* $t > 0$, *then* $e^{-tA} \in \mathscr{C}_1(\mathscr{H})$ *for* $t \in S_\theta$, $\theta = \pi/2 - \alpha$, *and*

$$\|e^{-tA}\|_1 \leq \cdots \leq \|e^{-tA/2^p}\|_{2^p}^{2^p} \leq \|e^{-tA/2^{p+1}}\|_{2^{p+1}}^{2^{p+1}} \\ \leq \cdots \leq \lim_{p\to\infty} \|e^{-tA/2^p}\|_{2^p}^{2^p} = \|e^{-t\mathrm{Re}\,A}\|_1 \,. \tag{4.1}$$

Proof First we note that by virtue of $\mathrm{Re}\,A \geq 0$ and $e^{-t\mathrm{Re}\,A} \in \mathscr{C}_1(\mathscr{H})$

$$(\mathbb{1} + \mathrm{Re}\,A)^{-1} = \int_0^\infty dt\, e^{-t(\mathbb{1}+\mathrm{Re}\,A)} \in \mathscr{C}_\infty(\mathscr{H}) \,, \tag{4.2}$$

since the right-hand side operator can be approximated in the operator-norm by finite-rank operators with any accuracy. By Remark 2.3 this implies that m-sectorial generators A and A^* also have compact resolvents. Since $\overline{\mathrm{dom}\,a} = \mathscr{H}$ for the sesquilinear form a corresponding to A, this ensures that the Trotter product formula

$$\|\cdot\| - \lim_{n\to\infty} (e^{-tA^*/2n} e^{-tA/2n})^n = e^{-t(A^* \dot{+} A)/2} \,, \quad t \in S_\theta \,, \tag{4.3}$$

converges in the operator-norm topology to the semigroup generated by the half of the form-sum of A and A^*, which coincides with non-negative self-adjoint generator $\mathrm{Re}\,A = (A \dot{+} A^*)/2 \geq 0$, see Proposition 3.2.

Second, the operator-norm convergence of the Trotter product formula (4.3) implies the convergence of all singular values of the compact operators involved in expression (4.3), i.e., for $j = 1, 2, 3, \ldots$, we have

$$s_j((e^{-tA^*/2n} e^{-tA/2n})^n) \xrightarrow[n\to\infty]{} s_j(e^{-t\mathrm{Re}\,A}). \tag{4.4}$$

Furthermore, by Gohberg and Kreĭn [4, Ch.II, Corollary 4.2], one gets the inequality

$$S_k(p) := \sum_{j=1}^{k} s_j(e^{-tA/2^p})^{2^p} \leq \sum_{j=1}^{k} s_j(e^{-tA/2^{p+1}})^{2^{p+1}} =: S_k(p+1). \tag{4.5}$$

The limit (4.4) yields the convergence of any finite sum in (4.5), i.e.,

$$S_k := \lim_{p\to\infty} S_k(p) = \sum_{j=1}^{k} s_j(e^{-t\mathrm{Re}\,A}), \tag{4.6}$$

thus, expressions (4.5) and (4.6) give

$$S_k(p) \leq S_k(p+1) \leq \cdots \leq S_k, k = 1, 2, \ldots$$

Hence, the series $\{S_k(p)\}_{k\geq 1, p\geq 0}$ is monotonously increasing in k and p. Since $e^{-t\mathrm{Re}\,A} \in \mathscr{C}_1(\mathscr{H})$ for $t > 0$, this series is bounded by $\sup_{k\geq 1} S_k = \|e^{-t\mathrm{Re}\,A}\|_1$, see (4.6). In particular,

$$\lim_{k\to\infty} S_k(0) = \|e^{-tA}\|_1, \tag{4.7}$$

and

$$\lim_{k\to\infty} S_k(p) = \|e^{-tA/2^p}\|_{2^p}^{2^p}. \tag{4.8}$$

Therefore, taking the limit $k \to \infty$ in the inequalities

$$S_k(0) \le \cdots \le S_k(p) \le S_k(p+1) \le \cdots \le S_k \le \|e^{-t\mathrm{Re}\,A}\|_1,$$

we get by virtue of the limits (4.7) and (4.8) the announced inequalities (4.1) and

$$\lim_{p\to\infty} \|e^{-tA/2^p}\|_{2^p}^{2^p} \le \|e^{-t\mathrm{Re}\,A}\|_1. \tag{4.9}$$

On the other hand, by (4.5), we also have

$$\|e^{-tA/2^p}\|_{2^p}^{2^p} \ge S_k(p),$$

or, see (4.6),

$$\lim_{p\to\infty} \|e^{-tA/2^p}\|_{2^p}^{2^p} \ge \sup_{k\ge 1} S_k = \|e^{-t\mathrm{Re}\,A}\|_1,$$

which together with *converse* inequality (4.9) implies the equality in (4.1). □

The next lemma is a generalisation of the Ginibre-Gruber inequality [3] for the trace-norm and to m-sectorial generators, cf. [2].

Lemma 4.2 *Let m-sectorial operator A be such that $e^{-t\mathrm{Re}\,A} \in \mathscr{C}_1(\mathscr{H})$ for $t > 0$, and let $V_1, V_2, \ldots, V_n$ be bounded operators on $\mathscr{H}$. For any set of positive numbers $t_1, t_2, \ldots, t_n$, we have the inequality*

$$\left\|\prod_{j=1}^{n} V_j e^{-t_j A}\right\|_1 \le \prod_{j=1}^{n} \|V_j\| \|e^{-(t_1+t_2+\ldots+t_n)\mathrm{Re}\,A/4}\|_1. \tag{4.10}$$

Proof Firstly, let $V_j \in \mathscr{C}_\infty(\mathscr{H})$ for $j = 1, 2, \ldots, n$. We set $t_m := \min\{t_j\}_{j=1}^n > 0$ and $T := \sum_{j=1}^n t_j > 0$. For any $1 \le j \le n$, we define an integer $\ell_j \in \mathbb{N}$ by

$$2^{\ell_j} t_m \le t_j \le 2^{\ell_j+1} t_m.$$

We then get $\sum_{j=1}^n 2^{\ell_j} t_m > T/2$ and

$$\prod_{j=1}^{n} V_j e^{-t_j A} = \prod_{j=1}^{n} V_j e^{-(t_j - 2^{\ell_j} t_m)A} (e^{-t_m A})^{\ell_j}. \tag{4.11}$$

By definition of the $\|\cdot\|_1$-norm and by inequalities for singular values we obtain

$$\begin{aligned}\Big\|\prod_{j=1}^{n} V_j e^{-t_j A}\Big\|_1 &= \sum_{k=1}^{\infty} s_k\Big(\prod_{j=1}^{n} V_j e^{-(t_j - 2^{\ell_j} t_m)A} (e^{-t_m A})^{2^{\ell_j}}\Big) \\ &\leq \sum_{k=1}^{\infty} \prod_{j=1}^{n} s_k\left(e^{-(t_j - 2^{\ell_j} t_m)A}\right) \left[s_k(e^{-t_m a})\right]^{2^{\ell_j}} s_k(V_j) \\ &\leq \sum_{k=1}^{\infty} s_k(e^{-t_m A})^{\sum_{j=1}^{n} 2^{\ell_j}} \prod_{j=1}^{n} \|V_j\|. \end{aligned} \tag{4.12}$$

Here, we used that $s_k(e^{-(t_j - 2^{\ell_j} t_m)A}) \leq \|e^{-(t_j - 2^{\ell_j} t_m)A}\| \leq 1$ and that $s_k(V_j) \leq \|V_j\|$. Let $N := \sum_{j=1}^{n} 2^{\ell_j}$ and $T_m := N t_m > T/2$, then inequality (4.12) yields

$$\Big\|\prod_{j=1}^{n} V_j e^{-t_j A}\Big\| \leq \left(\|e^{-T_m A/N}\|_{q=N}\right)^N \prod_{j=1}^{n} \|V_j\|. \tag{4.13}$$

In order to apply Proposition 4.1, we consider an integer $p \in \mathbb{N}$ such that $2^p \leq N < 2^{p+1}$. It then follows that $T/4 < T_m/2 < 2^p T_m/N$ and hence, we obtain

$$\begin{aligned}\left(\|e^{-T_m A/N}\|_{q=N}\right)^N &= \sum_{k=1}^{\infty} s_k^N (e^{-T_m A/N}) \\ &\leq \sum_{k=1}^{\infty} s_k^{2^p} (e^{-2^p T_m A/2^p N}) \\ &\leq \sum_{k=1}^{\infty} s_k^{2^p} (e^{-T A/2^{p+2}}), \end{aligned} \tag{4.14}$$

where we used that $s_k(e^{-T_m A/N}) = s_k(e^{-2^p T_m A/2^p N}) \leq \|e^{-T_m A/N}\| \leq 1$, and that $s_k(e^{-(t+\tau)A}) \leq \|e^{-tA}\| s_k(e^{-\tau A}) \leq s_k(e^{-\tau A})$ for any $t, \tau > 0$. Therefore, the estimates (4.13), (4.14) and inequalities (4.1) give the bound (4.10).

Secondly, let $V_j \in \mathscr{L}(\mathscr{H})$, $j = 1, 2, \ldots, n$, we set $\tilde{V}_j := V_j e^{-\varepsilon A}$ for $0 < \varepsilon < t_m$. By consequence, $\tilde{V}_j \in \mathscr{C}_1(\mathscr{H})$ and $s_k(\tilde{V}_j) \leq \|\tilde{V}_j\| \leq \|V_j\|$. If we set $\tilde{t}_j := t_j - \varepsilon$, then

$$\Big\|\prod_{j=1}^{n} V_j e^{-t_j A}\Big\|_1 \leq \prod_{j=1}^{n} \|V_j\| \|e^{-(\tilde{t}_1 + \tilde{t}_2 + \cdots + \tilde{t}_n)\mathrm{Re}\, A/4}\|_1. \tag{4.15}$$

Since the semigroup $\{e^{-t\mathrm{Re}\,A}\}_{t\geq 0}$ is $\|\cdot\|_1$-continuous for $t > 0$, we can now take in (4.15) the limit $\varepsilon \downarrow 0$, which gives the result (4.10) in the general case. □

Now we are in position to establish the following preliminary result.

Theorem 4.3 *Let $A \geq 0$ be generator of the self-adjoint Gibbs semigroup $\{G_t(A) = e^{-tA}\}_{t\geq 0}$. Then the exponential Trotter-Kato product formula converges in the trace-norm topology away from zero for Kato-functions $f(x) = g(x) = e^{-x}$ and for any self-adjoint operator $B \geq 0$ to a degenerate Gibbs semigroup:*

$$\|\cdot\|_1 - \lim_{n\to\infty} \left(e^{-tA/n}e^{-tB/n}\right)^n = e^{-tH}P_0\ , \ \ H = A \dot{+} B\ , \tag{4.16}$$

where orthogonal projection $P_0 : \mathscr{H} \to \overline{\mathrm{dom}\, H}$ and $t > 0$.

Proof Operator A is obviously m-sectorial and B is generator of a contraction semigroup. Then the form-sum operator $H = A \dot{+} B$ is self-adjoint in the subspace $\mathscr{H}_0 := \overline{\mathrm{dom}\, A^{1/2} \cap \mathrm{dom}\, B^{1/2}}$ and defines generator of a degenerate contraction semigroup $\|e^{-tH}P_0\| \leq 1$ on $\mathscr{H}$. Moreover, it is known [6] that the Trotter product formula converges strongly to $e^{-tH}P_0$ away from $t = 0$:

$$s - \lim_{n\to\infty} \left(e^{-tA/n}e^{-tB/n}\right)^n = e^{-tH}P_0\ .$$

Here P_0 is the orthogonal projection $P_0 : \mathscr{H} \to \mathscr{H}_0$.

Now we have to check that $e^{-tH} \in \mathscr{C}_1(\mathscr{H}_0)$. To this aim let $\{B_k \geq 0\}_{k\geq 1} \subset \mathscr{L}(\mathscr{H})$ be monotonously increasing sequence of positive bounded operators. Then positivity of generator A implies

$$0 \leq e^{-t(A+B_{k+1})} \leq e^{-t(A+B_k)} \leq e^{-tA}\ . \tag{4.17}$$

We suppose that the weak limit on $\mathscr{H}$:

$$w - \lim_{k\to\infty} e^{-t(A+B_k)} = e^{-tH}P_0\ , \ \ H = A \dot{+} B\ . \tag{4.18}$$

Note that the perturbed semigroup is positive and Gibbs: $e^{-t(A+B_k)} \in \mathscr{C}_{1,+}(\mathscr{H})$. Then by the lower weak semi-continuity of the *trace* and by monotonicity (4.17) one gets

$$0 \leq \mathrm{Tr}\, e^{-tH}P_0 \leq \liminf_{k\to\infty} \mathrm{Tr}\, e^{-t(A+B_k)} \leq \mathrm{Tr}\, e^{-tA}\ . \tag{4.19}$$

Consequently, the positivity of semigroups implies that $\|e^{-tH}P_0\|_1 \leq \|e^{-tA}\|_1$ on $\mathscr{H}$, or $e^{-tH} \in \mathscr{C}_1(\mathscr{H}_0)$.

Note that by the Laplace transformation of the Gibbs semigroup $\{G_t(A)\}_{t\geq 0}$ one gets that resolvent

$$(\lambda \mathbb{1} + A)^{-1} = \int_0^\infty dt\, e^{-t\lambda} e^{-tA} \in \mathscr{C}_\infty(\mathscr{H}) , \quad \lambda > 0 , \tag{4.20}$$

is compact. By Proposition 3.2 this yields the operator-norm convergence of the Trotter product formula locally uniformly away from zero and as a consequence:

$$\lim_{s\to\infty} \varepsilon(s,t) = \tag{4.21}$$

$$\lim_{s\to\infty} \sup_{(2s/(2s+1))t \le \xi \le (2s/(2s-1))t} \left\| \left(e^{-\xi A/s} e^{-\xi B/s}\right)^s - e^{-\xi H} P_0 \right\| = 0 ,$$

for $s = 2, 3, \ldots$, and $t > 0$. Hence, for immediately Gibbs semigroups we obtain by the limit (4.21) that

$$\lim_{n\to\infty} \left\| \left(e^{-tA/n} e^{-tB/n}\right)^n - e^{-tH} P_0 \right\|_1 = 0,$$

for $t \geq t_0$ and any $t_0 > 0$, that is, away from zero. □

Theorem 4.4 *Let A and B be m-sectorial operators with vertex $\gamma = 0$ in a Hilbert space $\mathscr{H}$. If $e^{-t\mathrm{Re}\, A} \vee e^{-t\mathrm{Re}\, B}$ is trace-class operator for $t > 0$, then the Trotter product formula for contraction Gibbs semigroups converges in the trace-norm topology:*

$$\| \cdot \|_1 - \lim_{n\to\infty} \left(e^{-tA/n} e^{-tB/n}\right)^n = e^{-t(A \dot{+} B)} P_0 , \tag{4.22}$$

for any $t \in S_\theta$, where $\theta = \pi/2 - \max\{\alpha_A, \alpha_B\}$. The convergence is uniform on the compact subsets of the open sector S_θ. Generator $H = A \dot{+} B$ is the m-sectorial form-sum of A and B, and P_0 is the orthogonal projection $P_0 : \mathscr{H} \to \mathscr{H}_0 = \overline{\mathrm{dom}\, H}$.

Proof Essentially we follow the line of reasoning of Proposition 3.2.

To this end we extend to the trace-norm topology the Vitali theorem about the sequence of the operator-valued functions identical to the Trotter approximants (3.3):

$$\Phi_n(t,z) = (e^{-tA(z)/n} e^{-tB(z)/n})^n , \; z \in D^0_{AB} , \quad t \in S_{\theta_0} . \tag{4.23}$$

Conditions of the Theorem, together with Corollary 2.9, Proposition 3.3 and the $\| \cdot \|_1$-continuity in z of the product $e^{-tA(z)} e^{-tB(z)}$ imply that functions $\{z \mapsto e^{-tA(z)} e^{-tB(z)}\}_{t \in S_{\theta_0}}$ are $\| \cdot \|_1$-holomorphic for $z \in D^0_{AB}$ if either of exponentials (or both) belong to the ideal $\mathscr{C}_1(\mathscr{H})$. Therefore, $\Phi_n(t,z) \in \mathscr{C}_1(\mathscr{H})$ and the function $z \mapsto \Phi_n(t,z)$ is $\| \cdot \|_1$-holomorphic in $z \in D^0_{AB}$ for any $t \in S_{\theta_0}$.

Since conditions of the Theorem are symmetric with respect to A and B, suppose that $e^{-t\,\mathrm{Re}\,A} \in \mathscr{C}_1(\mathscr{H})$. Then by Lemma 4.2 the sequence $\{\Phi_n(t, z)\}_{n\geq 1}$ is uniformly bounded for $z \in D^0_{AB}$ and $t \in S_{\theta_0}$ in the trace-norm topology:

$$\|\Phi_n(t, z)\|_1 \leq \|e^{-tB(z)/n}\|^n \|e^{-t\,\mathrm{Re}\,A(z)/4}\|_1 \leq \|e^{-\beta t\,\mathrm{Re}\,A/4}\|_1 \,. \tag{4.24}$$

Here we used that by Corollary 2.9 for $z \in D^0_{AB}$ the operator $B(z)$ is generator of contraction semigroup: $\|e^{-tB(z)}\| \leq 1$, and that by (3.7) the last inequality in (4.24) is valid for $\beta = \inf_{z\in D^0_A}(1 - |\mathrm{Re}\, z|\, \mathrm{tg}\alpha_A)$.

Note that for any real $x \in [-\Delta_R, \Delta_R]$, $\Delta_R = \min\{\mathrm{ctg}\alpha_A, \mathrm{ctg}\alpha_B\} - \delta_0$, the operators $A(x) \geq 0$ and $B(x) \geq 0$ are self-adjoint, Remark 3.4(b). Then by Theorem 4.3 the sequence $\{\Phi_n(t, x)\}_{n\geq 1}$ converges in the $\|\cdot\|_1$-norm to $e^{-tH(x)}P_0$ for $t \in S_{\theta_0}$ and for $x \in [-\Delta_R, \Delta_R]$. Here the m-sectorial operator $H(z) = A(z)\dot{+}B(z)$, with $z \in D^0_{AB}$, is the form-sum corresponding to the sum of two closed densely defined sectorial forms (2.4) and (2.5).

The interval $[-\Delta_R, \Delta_R]$ is compact in D^0_{AB}. Since the family $\{\Phi_n(t, z)\}_{n\geq 1}$ is $\|\cdot\|_1$-holomorphic in $z \in D^0_{AB}$ and $\|\cdot\|_1$-uniformly bounded in this domain by (4.24), the Vitali theorem (Remark 2.10) yields

$$\|\cdot\|_1 - \lim_{n\to\infty} (e^{-tA(z)/n} e^{-tB(z)/n})^n = e^{-tH(z)} P_0 \,, \quad t \in S_{\theta_0} \,, \tag{4.25}$$

for any $z \in D^0_{AB}$ (2.13). Since $z = i \in D^0_{AB}$, one gets the assertion (4.22) in a smaller sector: $\theta_0 < \theta$, where θ_0 is defined by (2.14).

The Trotter product formula (4.22) for $\theta_0 = \theta$ follows from (4.25) by taking in (2.14) the limits $\Delta_R \to 0$ and $\Delta_I \to 1$ (Remark 3.4 (a),(b)), that localise the point $z = i$. □

Corollary 4.5 *Let A be an m-sectorial operator with semi-angle α_A and vertex $\gamma = 0$. If $e^{-t\,\mathrm{Re}\,A}$ is a trace-class operator for $t > 0$, then the Trotter product formula for contraction Gibbs semigroups converges in the trace-norm topology:*

$$\|\cdot\|_1 - \lim_{n\to\infty} \left(e^{-tA^*/2n} e^{-tA/2n}\right)^n = e^{-t\,\mathrm{Re}\,A} \,, \tag{4.26}$$

for any $t \in S_\theta$, where $\theta = \pi/2 - \alpha_A$. The convergence is uniform on the compact subsets of the open sector S_θ.

Note that the trace-norm limit (4.26) is a *lifting up* of the operator-norm convergence in (4.3).

5 Concluding Remarks

The idea of the *analytic extension method* has been formulated by B.Simon and published by T. Kato as Addendum in [6]. There it was presented as a remark concerning an alternative way to prove the *strong* convergence of the Trotter product formula in the setting of non-subordinated generators when the common domain is not dense. In [1] it was shown that the analytic extension method allows to prove the convergence of the Trotter product formula for sectorial generators in the *operator-norm* topology.

In the present note the analytic extension method is used to prove the *trace-norm* convergence of the Trotter product formula for non-self-adjoint Gibbs semigroups. Note that similar to the paper [6] the generators A and B in the Trotter product formula (Theorem 4.4) are not subordinated. If one assumes smallness condition on, e.g., generator B with respect to A, then it is possible to obtain the error bounds estimate for the rate of the Trotter product formula convergence in the trace-norm topology [2].

References

1. V. Cachia, V.A. Zagrebnov, Operator-norm convergence of the Trotter product formula for sectorial generators. Lett. Math. Phys. **50**, 203–211 (1999)
2. V. Cachia, V.A. Zagrebnov, Trotter product formula for nonself-adjoint Gibbs semigroups. J. Lond. Math. Soc. **64**, 436–444 (2001)
3. J. Ginibre, C. Gruber, Green functions of the anisotropic Heisenberg model. Commun. Math. Phys. **11**, 198–213 (1969)
4. I.C. Gohberg, M.G. Kreĭn, *Introduction to the Theory of Linear Nonselfadjoint Operators*. Translations of Mathematical Monographs, vol. 18 (American Mathematical Society, Providence RI, 1969)
5. E. Hille, R.S. Phillips, *Functional Analysis and Semigroups*, vol. 31 (American Mathematical Society Colloquium Publications, Providence RI, 1957)
6. T. Kato, Trotter's product formula for an arbitrary pair of self-adjoint contraction semigroups, in *Topics in Functional Analysis*. Anal., Adv. Math. Suppl. Studies, ed. by I. Gohberg, M. Kac, vol. 3 (Academic, New York, 1978), pp.185–195
7. T. Kato, *Perturbation Theory for Linear Operators* (Springer, Berlin, 1980), Corrected Printing of the Second Edition
8. A.I. Markushevich, *Theory of Analytic Functions*, vol. 1 (Nauka, Moscow, 1967) (in Russian). Translated by R.A. Silverman *Theory of Functions of a Complex Variable*, vol. I, II, III, 2nd English edition, (Chelsea Publishing Co, New York, 1977)
9. H. Neidhardt, V.A. Zagrebnov, The Trotter product formula for Gibbs semigroup. Commun. Math. Phys. **131**, 333–346 (1990)
10. H. Neidhardt, V.A. Zagrebnov, On the Trotter product formula for Gibbs semigroups. Ann. der Physik **47**, 183–191 (1990)
11. H. Neidhardt, V.A. Zagrebnov, Trotter-Kato product formula and operator-norm convergence. Commun. Math. Phys. **205**, 129–159 (1999)
12. B. Simon, *Trace Ideals and Their Applications*. Mathematical Surveys and Monographs, 2nd ed, vol. 120 (American Mathematical Society, Providence RI, 2005)
13. B. Simon, *Basic Complex Analysis. A Comprehensive Course in Analysis*, Part 2A (American Mathematical Society, Providence RI, 2015)
14. V.A. Zagrebnov, *Topics in the Theory of Gibbs Semigroups*. Leuven Notes in Mathematical and Theoretical Physics, vol. 10 (Series A: Mathematical Physics) (Leuven University Press, Leuven, 2003)

www.ingramcontent.com/pod-product-compliance
Ingram Content Group UK Ltd.
Pitfield, Milton Keynes, MK11 3LW, UK
UKHW021007290726
14059UKWH00001BA/13

* 9 7 8 3 0 3 0 3 1 3 3 8 8 *